Peter J. Brockwell[†] and Alexander M. Lindner

Continuous-Parameter Time Series

De Gruyter Studies in Mathematics

Volume 98

Peter J. Brockwell[†] and Alexander M. Lindner

Continuous-Parameter Time Series

DE GRUYTER

Mathematics Subject Classification 2020
Primary: 60-01; Secondary: 60E07, 60G10, 60G12, 60G25, 60G51, 60H05, 60H10, 62F10, 62M10, 62M20

Authors
Prof. Dr. Peter J. Brockwell†

Prof. Dr. Alexander M. Lindner
Universität Ulm
Institut für Finanzmathematik
Helmholtzstr. 18
89081 Ulm
Germany
alexander.lindner@uni-ulm.de

ISBN 978-3-11-132499-9
e-ISBN (PDF) 978-3-11-132503-3
e-ISBN (EPUB) 978-3-11-132520-0
ISSN 0179-0986

Library of Congress Control Number: 2024934634

Bibliographic information published by the Deutsche Nationalbibliothek
The Deutsche Nationalbibliothek lists this publication in the Deutsche Nationalbibliografie; detailed bibliographic data are available on the Internet at http://dnb.dnb.de.

Typesetting: VTeX UAB, Lithuania
Printing and binding: CPI books GmbH, Leck

www.degruyter.com

MIX
Papier | Fördert gute Waldnutzung
FSC® C083411

To the memory of Peter Brockwell (1937–2023)

To our families

Preface

There are many excellent books devoted to discrete-parameter (in particular ARMA) time-series models, used to represent data that are observed at regularly spaced times, customarily labelled $1, 2, \ldots, N$, where N is the number of observations. There are also many excellent books devoted to continuous-time stochastic processes and stochastic differential equations.

There are relatively few, however, devoted to continuous-time stationary processes analogous to the corresponding discrete-time models for time series data. However, the use of such continuous-parameter models to represent data that are observable at any time in a specified subinterval of the real line $\mathbb{R}$, e. g., $[0, 100]$, has become increasingly widespread in recent years. This is due partly to the successful application of such models to the analysis of financial data, partly because of the prevalence of high-frequency data in many areas of science and economics and partly because of their ability to represent time series data which are observed at irregularly spaced times. Moreover, because it can be sampled at any frequency, a single continuous-parameter model can be used to represent sequences obtained by sampling a single data set at different frequencies, rather than fitting a separate discrete-time model for each sampling frequency.

Our goal in this book is to provide a self-contained account of continuous-parameter time series, starting with second-order models, the analysis of which is based heavily on Hilbert space, integration with respect to orthogonal increment processes and Fourier transforms. Mean-square linear prediction in this setting amounts to orthogonal projection in the Hilbert space of square-integrable random variables on the assumed underlying probability space.

The study of continuous-parameter weakly stationary processes goes back many years, at least as far as the work of Fowler [46], and was greatly stimulated by the works of Doob [35, 36, 37]. Like discrete-parameter weakly stationary sequences, they are characterized by their means and autocovariance functions or, if they are mean-square continuous (MSC), by their means and spectral distribution functions. The spectral representation of an MSC stationary process $(X_t)_{t\in\mathbb{R}}$ (Chapter 5) expresses X_t as an integral with respect to an orthogonal-increment process (Chapter 4). Moreover, if an MSC stationary processes has absolutely continuous spectral distribution function, then it can be represented (see Chapter 5) as a moving average of the form $X_t = \int_{-\infty}^{\infty} f(t-u)\, d\xi_u$, where ξ is again an orthogonal-increment process. If in addition X is regular (see Chapter 6), then it has a one-sided moving average representation $X_t = \int_{-\infty}^{t} f(t-u)\, d\xi_u$. We develop these results in Chapters 1–7, after first introducing the essentials of Hilbert spaces, orthogonal-increment processes and Fourier transforms, which are needed in the development. Chapter 6 presents Doob's approach to the Wold–Karhunen representation and minimum-mean-squared error linear prediction of regular stationary MSC processes and in Chapter 7 we give its detailed application to the class of processes with rational spectral density (otherwise known as second-order continuous-time ARMA or

https://doi.org/10.1515/9783111325033-201

second-order CARMA) processes. In Chapter 7 we also give R-programs for computing the autocorrelation, autocovariance and kernel functions of any specified CARMA(p, q) model with $q < p < 6$. All R-programs presented in this book can be downloaded from

https://doi.org/10.1515/9783111325033.

In order to develop continuous-parameter stationary time series models with properties that cannot be determined using only first and second-order moments, such as sample path properties and marginal distributions (possibly without finite second moments), it is natural to replace the orthogonal-increment process ξ in the moving average representations of the previous paragraph by a process with stationary, *independent* increments with specified distributions. This leads to the class of *strictly* stationary Lévy-driven moving average processes in which ξ is replaced by a Lévy process L. Before we can investigate the properties of such processes, it is necessary to develop the basics of stochastic calculus in order to define exactly what is meant by the integral with respect to a Lévy process.

In order to make the book self-contained, we develop from first principles those areas of stochastic calculus which are fundamental to our goal.

We begin, in Chapters 8 and 9, by introducing infinitely divisible distributions and exploring their properties, in particular their characterization by the Lévy–Khintchine formula and the distributions of random variables obtained by applying linear transformations and limit operations to random variables with infinitely divisible distributions. Infinitely divisible distributions are intimately related to Lévy processes, which are continuous-time analogues of random walks and, as such, constitute an important class of drivers for stochastic differential equations and integrators for stochastic integrals, as we shall see in later chapters. In fact, it is shown in Chapter 10 that there is a one-to-one correspondence between infinitely divisible distributions and Lévy processes (in law), in the sense that the marginal distributions of a Lévy process are infinitely divisible, and that, conversely, to every infinitely divisible distribution there exists a Lévy process, unique in law, whose marginal distribution at time one coincides with the given infinitely divisible distribution.

While in Chapters 8 and 9, we concentrate on infinitely divisible distributions in their own right, Lévy processes are introduced in Chapter 10 and their properties explored. In particular, the aforementioned one-to-one correspondence between Lévy processes in law and infinitely divisible distributions is established. It is also shown that every Lévy process in law has a modification that is a Lévy process, thereby establishing the existence of many such processes.

Distributional properties of Lévy processes are studied in Chapter 11. In particular, we investigate their Markov property, their behavior under time reversal, their moment properties and a strong law of large numbers.

Sample path properties of Lévy processes are treated in Chapters 12 and 13. In Chapter 12 we regard the sample paths of a d-dimensional Lévy process on a subinterval J of

$\mathbb{R}$ as random elements of the space $D(J;\mathbb{R}^d)$ of $\mathbb{R}^d$-valued càdlàg (right-continuous with finite left limits) functions on J with sigma-algebra generated by the coordinate projections. By modifying a càdlàg process (i. e., a stochastic process whose sample paths are only almost surely càdlàg) on a null set, we can always find a version which is strictly càdlàg and is indistinguishable from the original one, and by a slight abuse of language, we shall sometimes also view càdlàg processes as random elements in $D(J;\mathbb{R}^d)$. We characterize both the Poisson process and Brownian motion in terms of their sample path behavior and consider in detail the jump structure of a general Lévy process and its connection with the Lévy measure. In Chapter 13, we derive the fundamental Lévy-Itô path decomposition and, as a consequence, derive Corollary 13.16, which later plays a key role in defining the integral with respect to a Lévy process. Poisson random measures and subordinators are also introduced.

We know from Theorem 10.26 that there is a one-to-one correspondence between Lévy processes in law and infinitely divisible distributions, and from Corollary 10.42 we know that every Lévy process in law has a modification that is a Lévy process. This means that the class of Lévy processes consists of many more processes than simply linear combinations of independent Brownian motion and compound Poisson processes. In Chapter 14 we define and derive properties of some important further examples of Lévy processes in this large class, including stable Lévy processes, the gamma process, the variance gamma process and the inverse Gaussian Lévy process.

In Chapters 15 and 16, we discuss integration of deterministic functions with respect to Lévy processes. Our definition of the integral in Chapter 15 is based on the fact that every Lévy process can be written as the sum of an orthogonal-increment process and an independent process of finite variation. We start with the definition of the Lebesgue–Stieltjes integral for Lévy processes of finite variation. We then move to the integral $\int_{\mathbb{R}} f(t)\,\mathrm{d}L_t$ when $f \in L^1 \cap L^2$ and L is a one-dimensional Lévy process and then extend this to the case of matrix-valued f and $\mathbb{R}^d$-valued Lévy processes. In Chapter 16 we derive the characteristic function of the integral and also discuss its distributional properties. There is a broader definition of the integral, due to Rajput and Rosiński [73], which includes the one developed in Chapters 15 and 16 (see Remark 16.5). However, since the functions of interest to us, which are integrable in the sense of Rajput and Rosiński, are also integrable in the sense of Definition 15.10, we chose to use the approach taken in these chapters, as it uses only the concepts of integration with respect to orthogonal-increment and finite-variation processes.

In Chapter 17, the multivariate Ornstein–Uhlenbeck process satisfying the equation $\mathrm{d}X_t = AX_t\,\mathrm{d}t + \mathrm{d}L_t$, where L is a two-sided $\mathbb{R}^d$-valued Lévy process, is studied in detail. Sufficient conditions for the existence of weakly and of strictly stationary solutions are derived, both in the one-dimensional and multi-dimensional cases. We also study causal solutions and Markov stability of multivariate Ornstein–Uhlenbeck processes.

In Chapter 18, we define a CARMA process driven by a Lévy process L as a strictly stationary solution of a *formal* stochastic differential equation analogous to the difference equations defining a discrete-time ARMA sequence. By rewriting the equation in state-space form, such a solution can be interpreted rigorously as a linear combination of the components of a *state vector*, which is a strictly stationary multivariate Ornstein–Uhlenbeck process. Under the conditions $\mathbb{E}(\log^+ |L_1|) < \infty$ and no autoregressive zeros with zero real parts, we use the results of previous chapters, particularly those of Chapter 17, to establish existence and uniqueness of the process. The moving average representation of the process as well as second-order properties are derived, and the sequence obtained when the process is sampled at regular intervals is studied in detail.

Chapter 19 deals with quasi maximum likelihood (QML) estimation for Lévy-driven CARMA processes. The Kalman recursions are used to derive the Gaussian likelihood of observations made at fixed but possibly irregularly spaced times, maximization of which leads to the so-called QML estimates of the CARMA parameters. The remainder of the chapter is devoted to QML estimation for uniformly spaced data, using an alternative procedure, based on QML estimation of the parameters of the *sampled sequence*. This approach depends heavily on the relation between the parameters of the CARMA process and the ARMA sequence obtained by sampling it at regular intervals. The sampling problem was dealt with in Theorem 18.33. In this chapter, we need a corresponding result for *embedding* an ARMA sequence into a CARMA process with matching autocorrelation structure at discrete lags. This is dealt with by Theorem 19.5. Illustrative examples of QML estimation are given with R programs, which can be downloaded and copied and pasted into R in order for the reader to reproduce the results or apply them to new data sets. Programs for the simulation of CARMA processes are also given. R programs to generate a variety of Lévy increments are included in Appendix A at the end of the book, using the yuima-package of Brouste et al. [31].

Peter Brockwell and I started writing this book in 2013. From the beginning, we advanced steadily, but never felt in a hurry. In early 2023, the book was nearly complete. All parts were written and only some final polishing was required. Despite Peter's declining health, he continued to work with me on the book, and we signed a contract with De Gruyter in early July. Unfortunately, Peter died just a few days later. Peter was one of the greatest and kindest people I ever met. He was a friend, mentor and coauthor to me and it was always a true pleasure to be in his company and to work with him. I miss him very much and regret that Peter could never see the published version of the book.

I would like to thank David Berger, Merve Kutlu, Verena Schwarz and Robert Stelzer for reading parts of previous versions of the manuscript and their valuable comments on it. I am greatly indebted to Anthony Brockwell for careful reading of the final version and his help with the book. I would like to thank Steven Elliot and René Schilling for their continued interest in this book and their support in it being published within the De Gruyter Studies in Mathematics series, Nadja Schedensack from De Gruyter for her help with the preparation of the manuscript, Vilma Vaičeliūnienė from VTeX UAB for her help with the typesetting and two anonymous referees for their valuable and encouraging comments on the book proposal. Thanks to TU Braunschweig, Colorado State University,

Melbourne University and Ulm University for the hospitality during mutual visits and providing excellent working environments. Finally, I would like to thank Peter for the excellent collaboration and friendship, Peter's family for support and patience, and my own family for their long-standing support, encouragement, patience and love.

Ulm, June 2024 Alexander Lindner

Contents

List of symbols and abbreviations

In the following, we list some of the notation used throughout the book, most of which is standard. Some symbols have multiple meanings (such as D meaning the unit ball in $\mathbb{R}^d$, or the space of càdlàg functions, or differentiation), but usually the meaning should be clear from the context.

Abbreviations

acf	autocorrelation function
acvf	autocovariance function
a. e.	almost everywhere
AR(p)	autoregressive process of order p
ARMA(p, q)	autoregressive moving average process of orders p and q
a. s.	almost surely
BLP	best linear predictor
c. f.	characteristic function
CAR(p)	continuous time autoregressive process of order p
CARMA(p, q)	continuous time autoregressive moving average process of orders p and q
CMA	continuous time moving average process
i. d.	infinitely divisible
i. i. d.	independent and identically distributed
MS	mean-square
MSC	mean-square continuous
MSD	mean-square differentiable
OIP	orthogonal-increment process
OU	Ornstein–Uhlenbeck (process)
QML, QMLE	quasi maximimum likelihood, quasi maximum likelihood estimation

Numbers and sets

$\mathbb{C}$	complex numbers
$\mathbb{N}$	$1, 2, 3, \ldots$
$\mathbb{N}_0$	$0, 1, 2, 3, \ldots$
$\mathbb{Q}$	rational numbers
$\mathbb{Q}_{\text{dyad}}$	dyadic rational numbers, $x = p/2^n$ with $p \in \mathbb{Z}, n \in \mathbb{N}$
$\mathbb{R}$	real numbers
$\overline{\mathbb{R}}$	$[-\infty, \infty]$
$\mathbb{Z}$	integers
e	Euler's number
i	imaginary unit $\sqrt{-1}$
$x \wedge y$	minimum of x and y
$x \vee y$	maximum of x and y
$\bar{z}$	complex conjugate of $z \in \mathbb{C}$ or of matrix $z \in \mathbb{C}^{m \times d}$
$\mathfrak{I}(z)$	imaginary part of $z \in \mathbb{C}$
$\mathfrak{R}(z)$	real part of $z \in \mathbb{C}$
$\mathcal{B}_d$	Borel sets in $\mathbb{R}^d$

https://doi.org/10.1515/9783111325033-202

$\mathcal{B}(X)$	Borel sets in a topological space X
$\overline{B}$	topological closure of set B
∂C	topological boundary of set C
D	closed unit ball in $\mathbb{R}^d$
$\mathcal{F}_{D(J;\mathbb{R}^d)}$	σ-algebra on $D(J;\mathbb{R}^d)$ generated by coordinate projections
$\mathrm{ID}(\mathbb{R}^d)$	infinitely divisible distibutions on $\mathbb{R}^d$
S^{d-1}	$(d-1)$-dimensional unit sphere
$\subset$ and $\subseteq$	both denote inclusion, not necessarily proper inclusion
$\#A$	number of elements of a set A

Vectors, matrices and norms

$A \in \mathbb{K}^{m\times d}$	$(m \times d)$-matrix with entries in $\mathbb{K}$, where $\mathbb{K} \in \{\mathbb{R}, \mathbb{C}\}$
$v \in \mathbb{K}^d$	d-dimensional column vector in $\mathbb{K}^d$, where $\mathbb{K} = \mathbb{R}$ or $\mathbb{C}$
v^T, A^T	transpose of vector or matrix
$0_{m\times d}$	zero-matrix in $\mathbb{C}^{m\times d}$
$\mathbf{e}_p$	p^{th} unit vector in $\mathbb{C}^p$
I_p	identity matrix in $\mathbb{R}^{p\times p}$ or $\mathbb{C}^{p\times p}$
$\|A\|$	matrix norm of matrix A, sometimes natural matrix norm of A with respect to Euclidian norm, i. e. spectral norm, but often irrelevant which matrix norm is used
$\|\cdot\|$	some norm
$\lvert v\rvert$	Euclidian length of $v \in \mathbb{R}^d$
$\langle v, w\rangle$	inner product between v and w, often Euclidian inner product when $v, w \in \mathbb{R}^d$
$\langle f, g\rangle_{L^2(F)}$	$\int f(t)\overline{g(t)}\,\mathrm{d}F(t)$
$\|f\|_{L^2(F)}$	$\sqrt{\langle f, f\rangle_{L^2(F)}}$
$\langle f, g\rangle_2$	$\int_{\mathbb{R}} f(t)\overline{g(t)}\,\mathrm{d}t$
$\|f\|_1$	$\int_{\mathbb{R}} \lvert f(t)\rvert\,\mathrm{d}t$
$\|f\|_2$	$\sqrt{\int_{\mathbb{R}} \lvert f(t)\rvert^2\,\mathrm{d}t}$
$\|f\|_\infty$	$\sup_{t\in J} \lvert f(t)\rvert$ for $f : J \to \mathbb{R}^d$
$v \perp w$	v and w are orthogonal to each other
$M^\perp$	orthogonal complement to subset M of Hilbert space
$\overline{\mathrm{sp}}\{x_s, s \in S\}$	closed span of vectors x_s, $s \in S$, in a Hilbert space
$\hat{x}$	orthogonal projection of x onto some specified subspace

Functions

$\arctan(x)$	arcus tangens of x
$\mathbf{1}_B$	indicator function of set B
$\cos(x)$	cosine of x
DX_t	mean-square or pathwise derivative of X at t
$\Delta f(t)$	jump size of f at time t
$\exp(x)$, $\exp(A)$	exponential of $x \in \mathbb{C}$, matrix exponential of $A \in \mathbb{C}^{d\times d}$
$f_{\mid B}$	restriction of function f to set B
$f * g$	convolution of functions f and g
$f(t-)$	left-limit of function f at time t

$f(t+)$	right-limit of function f at time t
$\mathcal{F}f = \hat{f}$	Fourier transform of L^1- or L^2-function f
$\mathcal{F}^{-1}f$	inverse Fourier transform of L^1- or L^2-function f
$\mathcal{F}_2 f$	L^2-Fourier transform of $f \in L^2$
$\Gamma(x)$	gamma function at argument x
log	principal branch of complex logarithm
$\log^+(x)$	$\log(x \vee 1)$ for $x \in \mathbb{R}$
Log	distinguished logarithm
$\widehat{\mu}$	characteristic function of distribution μ
ϕ_X, ϕ_μ, ϕ_L	characteristic function of random vector X or distribution μ or Lévy process L
$\phi^{1/n}$	distinguished n^{th} root or principal branch of nth complex root of $\phi : \mathbb{R}^d \to \mathbb{C}$
ϕ^t	distinguished t^{th} power or principal branch of tth complex power of $\phi : \mathbb{R}^d \to \mathbb{C}$
Φ_X, Φ_μ, Φ_L	Laplace exponent of random variable $X \geq 0$ or distribution μ supported on $[0, \infty)$ or subordinator L
ψ_X, ψ_μ, ψ_L	characteristic exponent of random vector X or distribution μ or Lévy process L
sign	sign-function, taking values $-1, 0, 1$ according to sign of argument
$\sin(x)$	sine of x
$\tan(x)$	tangens of x
$\mathrm{TV}_{[a,b]}(g)$	total variation of function g over interval $[a, b]$
Y_t'	mean-square or pathwise derivative of Y at t, i. e. DY_t

Function spaces

$C^1(I; \mathbb{R})$	continuously differentiable functions from I to $\mathbb{R}$
$C_b(\mathbb{R}^d; \mathbb{K})$	bounded continuous functions from $\mathbb{R}^d$ to $\mathbb{K} \in \{\mathbb{R}, \mathbb{C}\}$
$C_c(\mathbb{R}^d; \mathbb{K})$	continuous functions from $\mathbb{R}^d$ to $\mathbb{K} \in \{\mathbb{R}, \mathbb{C}\}$ with compact support
$C_\sharp(\mathbb{R}^d; \mathbb{R})$	functions in $C_b(\mathbb{R}^d; \mathbb{R})$ vanishing in neighborhood of 0
$D(J; \mathbb{R}^d)$	càdlàg functions from interval J to $\mathbb{R}^d$
$\mathcal{D}$	$\mathbb{C}$-valued simple functions
$\mathcal{D}^{m,d}_{\text{real}}$	$\mathbb{R}^{m \times d}$-valued simple functions
ℓ^2	square-summable $\mathbb{C}$-valued sequences on $\mathbb{N}$
$L^1(\mathbb{R}; \mathbb{K}^{m \times d})$	Borel-measurable, Lebesgue-integrable $\mathbb{K}^{m \times d}$-valued functions on $\mathbb{R}$ with $\mathbb{K} \in \{\mathbb{R}, \mathbb{C}\}$
L^1	$L^1(\mathbb{R}; \mathbb{C})$
$L^1(F)$	$\mathbb{C}$-valued, w. r. t. distribution function F integrable functions on $\mathbb{R}$
$L^2(\mathbb{R}; \mathbb{K}^{m \times d})$	Borel-measurable, w. r. t. Lebesgue-measure square-integrable $\mathbb{K}^{m \times d}$-valued functions on $\mathbb{R}$ with $\mathbb{K} \in \{\mathbb{R}, \mathbb{C}\}$
$L^2 = L^2(\mathrm{d}\mu)$	$L^2(\mathbb{R}; \mathbb{C})$
$L^2((-\pi, \pi]; \mathbb{C})$	Borel-measurable, w. r. t. Lebesgue-measure square-integrable $\mathbb{C}$-valued functions on $(-\pi, \pi]$
$L^2(\mathrm{d}\lambda)$	$L^2((-\pi, \pi]; \mathbb{C})$
$L^2(F)$	$\mathbb{C}$-valued functions on $\mathbb{R}$ that are square-integrable w. r. t. distribution function F
$L^2(\Omega, \mathcal{F}, P)$	With respect to P square-integrable $\mathbb{C}$- (or $\mathbb{R}$)-valued functions on probability space $(\Omega, \mathcal{F}, P)$
$L^\infty(J; \mathbb{R}^d)$	bounded $\mathbb{R}^d$-valued functions on interval J
$\mathcal{S}$	$L^2(\Omega, \mathcal{F}, P)$

Integrals

$I(f)$	OIP integral of function f w. r. t. some OIP
$\int f(t)\,\mathrm{d}g(t)$	Lebesgue–Stieltjes integral of f w. r. t. function g
L. S. $\int_{\mathbb{R}} f(t)\,\mathrm{d}g(t)$	Lebesgue–Stieltjes integral of f w. r. t. function g
L. S. $\int_{\mathbb{R}} f(t)\,\mathrm{d}L_t$	Lebesgue–Stieltjes integral of f w. r. t. Lévy process L
$\int f(t)\,\mathrm{d}L_t$	integral of f with respect to Lévy process L
$\int f(t)\,\mathrm{d}\mu(t)$	integral of function f w. r. t. measure μ
$\int f(t)\,\mathrm{d}Z_t$	OIP integral of f w. r. t. OIP Z
O. I. P. $\int_{\mathbb{R}} f(t)\,\mathrm{d}Z_t$	OIP integral of f w. r. t. OIP Z
$\int_B$	integral over set B
$\int_a^b$	integral over set $[a,b]$ if $a \le b$, and $-\int_b^a$ when $a > b$
$\int_{-\infty}^{\infty}$	improper integral, $\lim_{n,m\to\infty}\int_{-n}^{m}$
$\mathcal{R}_2 \int g(t)Y_t\,\mathrm{d}t$	mean-square Riemann integral

Probabilistic notation

$\overset{d}{=}$	equality in law
$\to$	convergence
$\overset{d}{\to}$	convergence in distribution
$\overset{P}{\to}$	convergence in probability
$\overset{\mathcal{S}}{\to}$	L^2-limit in space $\mathcal{S} = L^2(\Omega, \mathcal{F}, P)$
$\overset{w}{\to}$	weak convergence
(ζ, γ)	standard characteristic pair of i. d. distribution or LP on $\mathbb{R}$
$(\zeta, \gamma)_c$	characteristic pair of i. d. distribution or Lévy process on $\mathbb{R}$ w. r. t. truncation function c
(A, ν, γ)	standard characteristic triplet of i. d. distribution or Lévy process
$(A, \nu, \gamma)_c$	characteristic triplet of i. d. distribution or Lévy process w. r. t. (extended) truncation function c
$\mathrm{Cov}(X)$	covariance matrix of $\mathbb{C}^d$-valued random vector X
$\mathrm{Cov}(X, Y)$	covariance between X and Y
$\mathrm{Corr}(X, Y)$	correlation between X and Y
δ_j	Dirac measure at j
ΔL_t	jump size of process L at time t
dlim	limit in distribution
$\mathbb{E}(X) = \mathbb{E}X$	expectation of X
$\mathbb{E}(X\|\mathcal{G})$	conditional expectation of X given σ-algebra $\mathcal{G}$
$\mathbb{E}(X\|Z_s; s \in \Lambda)$	conditional expectation of X given random vectors $\{Z_s; s \in \Lambda\}$
$h * L$	convolution of function h and stochastic process L
L_{t-}	left-limit of process L at time t
$\mathcal{L}(X)$	distribution (law) of random vector or stochastic process X
λ^1	one-dimensional Lebesgue-measure on $(\mathbb{R}, \mathcal{B}_1)$
Lap_X, Lap_μ	Laplace transform of random variable X or distribution μ
$\lim^{L^2(F)}_{n\to\infty}$	$L^2(F)$-limit in $L^2(F)$-space
$\lim^{\mathcal{S}}_{n\to\infty}$	$L^2(\Omega, \mathcal{F}, P)$-limit in space $\mathcal{S} = L^2(\Omega, \mathcal{F}, P)$
$\mu \otimes \nu$	product measure of measures μ and ν
$\mu * \nu$	convolution of finite measures μ and ν

μ^{*n}	n-fold convolution of finite measure μ with itself
$\mu^{*1/n}$	distinguished n^{th} convolution root of i. d. distribution μ
μ^{*t}	distinguished t^{th} (convolution) power of i. d. distribution μ
$N(B,\omega) = N^L(B,\omega)$	number of points $(s,\Delta L_s)$ falling into B for Lévy process L
$\nu_{\mid\mathcal{A}}$	measure ν restricted to sub-sigma-algebra $\mathcal{A}$
$\nu_{\mid B}$	measure ν restricted to trace-sigma algebra $B\cap\mathcal{F}$ of set B if ν lives on $\mathcal{F}$
P_M	orthogonal projection onto closed subspace M
$\mathcal{P}_M$	componentwise orthogonal projection onto M
P_X	distribution (law) of random vector or stochastic process X
plim	limit in probability
π_t	projection onto the tth coordinate
$\operatorname{supp}(\mu)$	support of probability distribution μ
$\operatorname{Var}(X)$	variance of random variable X
$\mathrm{WN}(0,\sigma^2)$	white noise sequence with mean zero and variance σ^2
$\hat{X}$	best linear predictor of X
Y_n^{Δ}	stochastic process $Y=(Y_t)_{t\in\mathbb{R}}$ sampled at $n\Delta$
$\tilde{Z}_\mu$	Fourier transform of orthogonal increment process Z at μ

Probability distributions and Lévy processes

$N(\mu,\sigma^2)$	normal distribution with mean μ and variance $\sigma^2\geq 0$
$B(\mu,\sigma^2)$	Brownian motion with drift μ and variance σ^2
$\mathrm{IG}(\delta,\gamma)$	inverse Gaussian with parameters δ and γ
$\mathrm{NIG}(\alpha,\beta,\mu,\delta)$	normal inverse Gaussian with parameters α, β, μ, δ
$\mathrm{PTS}(\alpha,\sigma,\lambda)$	positive tempered stable with parameters α, σ, λ
$S_\alpha(\sigma,\beta,\tau)$	one-dimensional α-stable with skewness parameter β, scale parameter σ and location parameter τ
$\mathrm{VG}(\sigma,\kappa,\theta)$	variance gamma with parameters σ, κ, θ

1 Time series

Abstract: In this chapter, we introduce some of the basic ideas underlying time series and stochastic processes. Of particular importance are the concepts of strictly stationary and covariance stationary processes and, for the latter, the corresponding mean and autocovariance functions. We need to distinguish carefully between empirically observed *time series data* and the *mathematical models* used to represent the data. If there is no danger of confusion, we shall refer to both simply as time series, but when necessary we shall refer to the former as time series data and to the latter as *time series models* or stochastic processes. Time series recorded at uniformly-spaced discrete times have been extensively studied by many authors. The class of so-called autoregressive moving average (ARMA) processes and their many extensions (integrated ARMA, fractionally integrated ARMA, and seasonal ARMA processes, to name a few) are widely used as mathematical models for the representation and prediction of such time series. In this book the emphasis is on continuous-parameter time series, i. e., those for which observations are, in principle, available at any time in a continuous time interval. In practice, of course, numerical observations of continuous-parameter time series can be recorded at discrete points in time only. These may be irregularly-spaced, closely-spaced, or chosen at the discretion of the observer. In this first chapter, we give some illustrative examples, basic definitions and notation, and recall some of the properties and applications of time series models, before looking more deeply into their properties in later chapters.

1.1 Examples of time series

A time series is a set of observations x_t, each one being recorded at a specified time t. A discrete-parameter time series is one in which the set of times at which observations are available is discrete, as is the case, for example, when observations are provided at fixed time intervals. Continuous-parameter time series are obtained when observations are available continuously over some time interval, e. g., $[0, T_0]$, such as in a seismogram, which records graphically the motion of the ground over time at a particular location, or an electroencephalogram, which similarly records electrical impulses in the brain.

Example 1.1 (Current through a resistor). If a deterministic sinusoidal voltage $V_t = a\cos(\nu t + \theta)$, $t \in [0, 10]$ is applied to a resistor of resistance r, then the current through the resistor is the continuous parameter time series

$$x_t = r^{-1} a\cos(\nu t + \theta), \quad t \in [0, 10],$$

shown in Figure 1.1 when $r = a = 1$, $\nu = 3$ and $\theta = \sqrt{3}/2$. The sequence obtained by sampling the current at times $0, 1, \ldots, 10$ is the discrete-parameter time series,

https://doi.org/10.1515/9783111325033-001

$$x_j = r^{-1} a \cos(\nu j + \theta), \quad j \in \{0, 1, \ldots, 10\}.$$

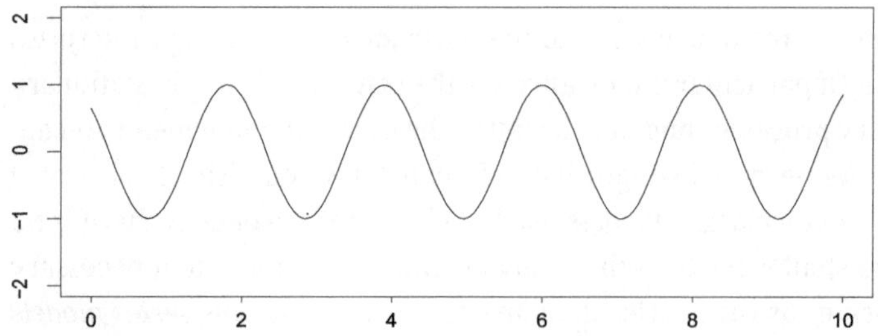

Figure 1.1: x_t, $0 \leq t \leq 10$, as defined in Example 1.1 with $r = a = 1$, $\nu = 3$ and $\theta = \sqrt{3}/2$.

This is a particularly simple deterministic continuous-parameter time series, but we shall see later that such series play an important role in the study of more complex stationary time series.

Example 1.2 (Australian All-ordinaries Stock-price Index). Neglecting periods when the Australian stock market is closed, this is a continuous-parameter time series (an average of 100 Australian stock prices), which can be sampled at any time during the trading day (10:00 until 16:00, Australian eastern standard time). Figure 1.2 shows values of the index sampled at one-minute intervals from 10:40 on October 24th, 2005, until 16:00 on October 28th, rescaled so that the starting value is 1.0.

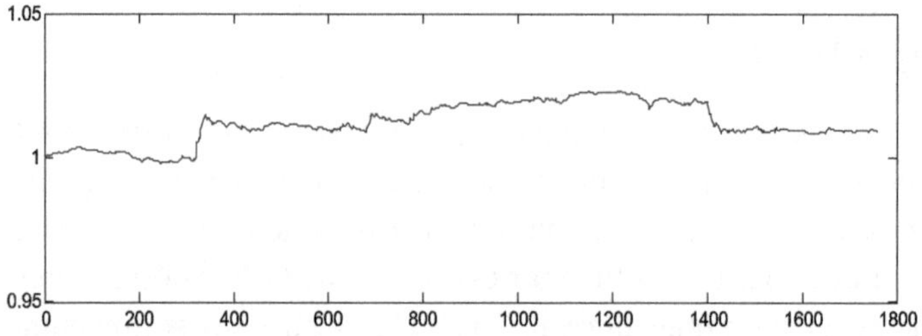

Figure 1.2: Australian All-ordinaries Index sampled at one-minute intervals for five trading days starting from October 24th, 2005, and rescaled so that the starting value is 1.0.

Example 1.3 (Geometric Brownian motion as model for stock prices). The price P_t of a stock at time t, $t \geq 0$, is sometimes modelled as

$$P_t = P_0 \exp\left[\left(\mu - \frac{\sigma^2}{2}\right)t + \sigma B_t\right], \tag{1.1}$$

where $\mu \in \mathbb{R}$ and $\sigma > 0$ are parameters that must be estimated from data, and B is standard Brownian motion, i. e., a random process with continuous sample paths, $B_0 = 0$

and increments that are independent normally distributed random variables with mean zero and variance Δ on disjoint intervals of length Δ. We shall discuss this model in more detail later in the book. For now, we simply illustrate a realization of the random process $P_t, 0 \le t \le 10$ in Figure 1.3.

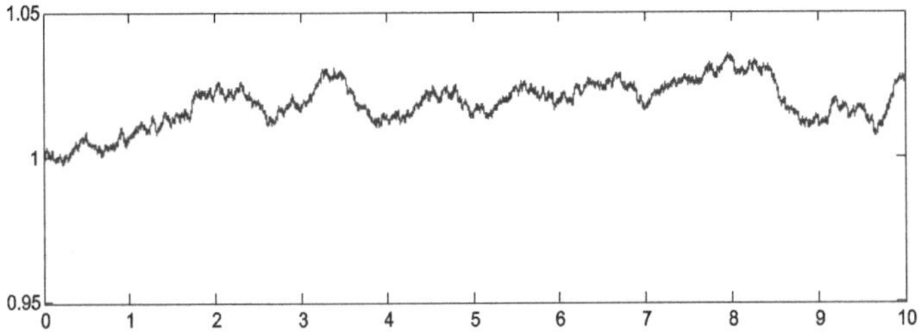

Figure 1.3: A realization of P_t, $0 \le t \le 10$, as defined in Example 1.3 with $P_0 = 1$, $\mu = 0$ and $\sigma = 0.01$.

It follows at once from the definition (1.1) of P_t that the so-called *log returns* over the time-intervals $(n\Delta, (n+1)\Delta)$, $n = 0, 1, 2, \ldots$, i. e.,

$$\log P_{(n+1)\Delta} - \log P_{n\Delta} = \left(\mu - \frac{\sigma^2}{2}\right)\Delta + \sigma(B_{(n+1)\Delta} - B_{n\Delta}),$$

constitute a sequence of independent and identically distributed normal random variables with mean $(\mu - \frac{\sigma^2}{2})\Delta$ and variance $\sigma^2\Delta$. The log returns are important from a financial point of view since

$$\log P_{(n+1)\Delta} - \log P_{n\Delta} = \log(1 + x),$$

where x is the fractional change in price, $x = [P_{(n+1)\Delta} - P_{n\Delta}]/P_{n\Delta}$, and, for small x, $\log(1 + x) \approx x$.

The model (1.1) for P_t, due to Samuelson [77], is the one for which the celebrated option-pricing formula of Black and Scholes [18] and Merton [69] was derived. It superseded the much earlier model proposed by Bachelier [4], in which P_t was represented as Brownian motion with drift, a process that has the drawback of possibly taking negative values.

Although in practice numerical values of stock prices are recorded only at discrete times, the observed log returns over fixed intervals of length Δ, such as one trading day, can be checked empirically to test the validity of a model such as (1.1). Figure 1.4, for example, shows the daily log returns of the pound sterling priced in US dollars from July 8th, 2013, for 40 trading days. (The data is from https://www.federalreserve.gov/)

It has been found that for small Δ (less than one day) the normal distribution of the log returns and their independence (as implied by the geometric Brownian motion

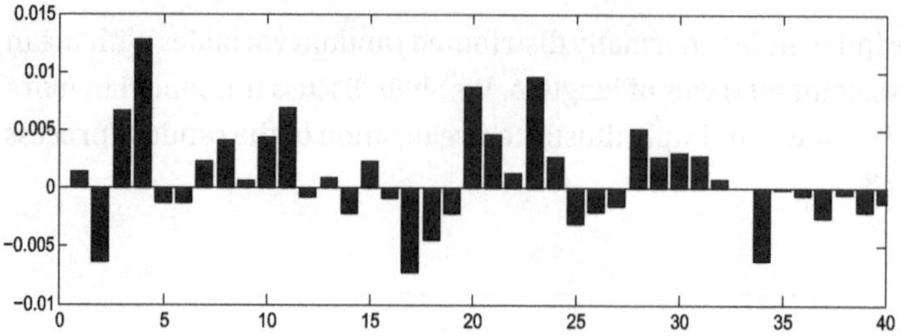

Figure 1.4: Daily log returns on the price of the pound sterling in US dollars for 40 trading days beginning July 8th, 2013.

model (1.1)) are both questionable. This has led to the development of more refined models for P_t, which will be discussed later in the book.

The fundamental aim of time series analysis is to make inferences of various kinds from empirically observed time series data. The objectives include the testing of hypotheses of interest, such as the presence or otherwise of a trend, the forecasting of future values of the time series and the determination of an underlying mathematical model that accounts for the most relevant features of the observed data. The latter can then be used in a variety of possible applications, one of which is nicely illustrated by the determination of an appropriate price for stock options based on the model (1.1) used by Black, Scholes and Merton for the purpose. Again, a mathematical model which accounts for the essential features of the data can be used to simulate future values so that a variety of future eventualities, e. g., the overflow or emptiness of a storage facility, can be assessed probabilistically based on the model.

To account for the apparent randomness observed in most observed time series data, the natural mathematical model for the representation of time series is a stochastic process. In the following section, we discuss stochastic processes and some of their important properties.

1.2 Stochastic processes

The first step in the analysis of time series data is the selection of a suitable model (or class of models) to represent the observed data. To allow for the possibly unpredictable nature of future observations, it is natural to suppose that each observed value x_t of the time series at time t is a realized value of a certain random variable X_t. These considerations suggest modeling the observed data as a realization (or part of a realization) of a stochastic process $(X_t)_{t\in T}$. For time series recorded at uniformly spaced discrete times $\{n\Delta, n = 0, 1, \ldots, N\}$ a natural index set T is $\{n\Delta, n \in \mathbb{Z}\}$ or $\{n\Delta, n = 0, 1, 2, \ldots\}$, while for realizations available on a continuous interval $[0, T_0]$ a natural index set T is the set of real numbers $\mathbb{R}$ or a subinterval of $\mathbb{R}$. We have already encountered geometric Brownian motion as a possible model for stock prices in Section 1.1. In this section, we shall define in more general terms what is meant by a stochastic process and its realizations. In later sections, we shall restrict attention to special classes of processes that are particularly useful for modeling many of the time series encountered in practice.

Definition 1.4 (Stochastic process). A *stochastic process* is a family of random variables $(X_t)_{t\in T}$ defined on a probability space $(\Omega, \mathcal{F}, P)$. (More generally, X may take values in $\mathbb{R}^n$, $\mathbb{C}$ or $\mathbb{C}^n$ for any positive integer n.)

Recalling the definition of a random variable on $(\Omega, \mathcal{F}, P)$, we note that for each fixed $t \in T$, X_t is an $\mathcal{F}$-measurable function $X_t(\cdot)$ on Ω, while, for each fixed $\omega \in \Omega$, $X_.(\omega)$ is a function on T.

Definition 1.5 (Realizations of a stochastic process). The functions $X_.(\omega)$ on T, for $\omega \in \Omega$, are known as the *realizations* or *sample paths* of the process $(X_t)_{t\in T}$.

Example 1.6 (Sinusoid with random phase and amplitude). Let A and Θ be independent random variables with $A \geq 0$ and Θ distributed uniformly on the interval $[0, 2\pi)$. A stochastic process $(X_t)_{t\in\mathbb{R}}$ can then be defined in terms of A and Θ for any fixed $\nu \geq 0$ and $r > 0$ by

$$X_t = r^{-1} A \cos(\nu t + \Theta),$$

or more expicitly,

$$X_t(\omega) = r^{-1} A(\omega) \cos(\nu t + \Theta(\omega)), \tag{1.2}$$

where ω is an element of the probability space Ω on which A and Θ are defined.

The realizations of the process defined by (1.2) are the functions of t obtained by fixing ω, i. e., functions of the form,

$$x_t = r^{-1} a \cos(\nu t + \theta),$$

where a and θ range over the possible values of $A(\omega)$ and $\Theta(\omega)$. The time series plotted in Figure 1.1 is one such realization when $r = 1$ and $\nu = 3$.

Example 1.7 (Discrete-time white noise). The sequence of random variables $(Z_n)_{n\in\mathbb{Z}}$ is called *white noise* with variance σ^2, if $\mathbb{E}(Z_n) = 0$ and $\mathbb{E}(Z_n^2) = \sigma^2 < \infty$ for each n and $\mathbb{E}(Z_m Z_n) = 0$ if $m \neq n$. The sequence is called *strict* white noise if in addition the sequence is independent. White noise is a fundamental building block in the construction of discrete-parameter time series. For example, the *discrete-time autoregression of order* 1 (known as AR(1)) with coefficient ϕ such that $|\phi| < 1$ and driven by the white noise sequence $(Z_n)_{n\in\mathbb{Z}}$ is the weakly stationary (see Definition 1.18) solution of the discrete-time AR(1) equations,

$$X_n = \phi X_{n-1} + Z_n, \quad n \in \mathbb{Z},$$

given by (Exercise 1.2)

$$X_n = \sum_{j=0}^{\infty} \phi^j Z_{n-j}, \quad n \in \mathbb{Z}, \tag{1.3}$$

where the sum converges absolutely with probability one. A detailed account of the properties and applications of these and related processes can be found, e. g., in Brockwell and Davis [22].

Example 1.8 (Homogeneous orthogonal-increment process). In continuous time, a role analogous to that of discrete-time white noise is played by *homogeneous orthogonal-increment processes*, i. e., processes $(Z_t)_{t\in\mathbb{R}}$, whose *increments* over disjoint intervals of fixed length Δ are uncorrelated with mean zero and variance $\sigma^2\Delta$ for some $\sigma > 0$. Moreover, if the increments are independent, the process is a homogeneous independent-increment process. The continuous-time analog of the AR(1) sequence in Example 1.7 is a process of the form,

$$X_t = \int\limits_{(-\infty,t]} e^{\rho(t-u)} \, dZ_u, \quad t \in \mathbb{R},$$

with $\rho < 0$. Precise definitions and properties of such integrals and corresponding integrals with respect to independent-increment processes will be given in later chapters.

Example 1.9 (Standard Brownian motion). The most celebrated process with homogeneous orthogonal (and independent) increments is *standard Brownian motion*, $(B_t)_{t\in\mathbb{R}}$, for which $B_0 = 0$ and the increments on disjoint time-intervals of length Δ are independent normally distributed random variables with variance Δ. A realization of $B_t, 0 \leq t \leq 10$, is shown in Figure 1.5. (The values plotted are actually sampled values recorded at times which are integer multiples of 10^{-4}. The true realizations are almost surely continuous and nowhere differentiable. Some of these properties will be discussed later in the book when we deal with Brownian motion in more detail.)

If standard Brownian motion is sampled at integer multiples of any fixed interval Δ, the sampled sequence $\{S_n = B_{n\Delta}, n = 0, 1, 2, \ldots\}$ is a *random walk* since $S_0 = 0$ and $S_n = S_{n-1} + Z_n$, $n \in \mathbb{N}$, where $(Z_n)_{n\in\mathbb{N}}$ is an independent and identically distributed (i. i. d.) sequence of normal random variables with mean zero and variance Δ. In this sense, standard Brownian motion can be regarded as the continuous-time analog of a discrete-time random walk with normally distributed increments.

Brownian motion with drift $\mu \in \mathbb{R}$ *and variance* $\sigma^2 > 0$ is defined as

$$B_t(\mu, \sigma^2) := \mu t + \sigma B_t, \quad t \in \mathbb{R}. \tag{1.4}$$

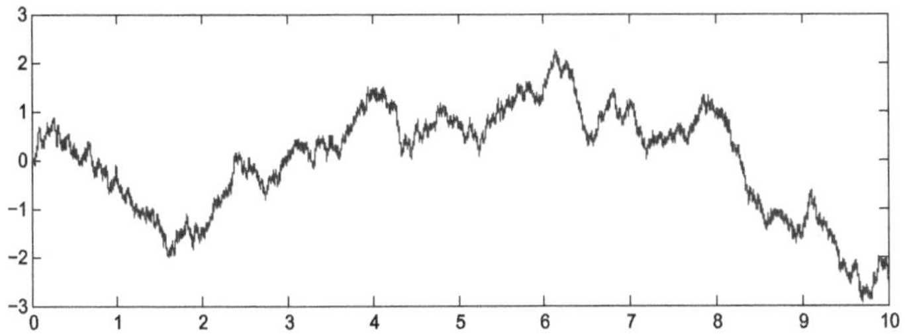

Figure 1.5: A realization of standard Brownian motion, B_t, $0 \le t \le 10$.

Example 1.10 (The Poisson process). Another continuous-time process of fundamental importance is the *Poisson process* $(N_t)_{t\in\mathbb{R}}$ with *intensity* or *jump rate* or *parameter* $\lambda > 0$. Like standard Brownian motion, it has independent increments on disjoint subintervals of $\mathbb{R}$ and is characterized by the following properties.

1. $N_0 = 0$ almost surely.
2. $N_v - N_u$, $-\infty < u < v < \infty$, has the Poisson distribution with mean $\lambda(v-u)$.
3. The increments $N_v - N_u$ on disjoint intervals $(u, v]$ are independent.
4. The sample paths of N are almost surely right-continuous step functions with jump size 1.

N_t is the number of jumps in $(0, t]$ for $t > 0$ and, for $t < 0$, $-N_t$ is the number of jumps in $(t, 0]$. A realization of $\{N_t, 0 \le t \le 10\}$ with $\lambda = 5$ is shown in Figure 1.6. The sample paths of N are right-continuous and constant, except for upward jumps of size one occurring at the jump times. The expected number of jumps in any time interval of length Δ is $\lambda\Delta$.

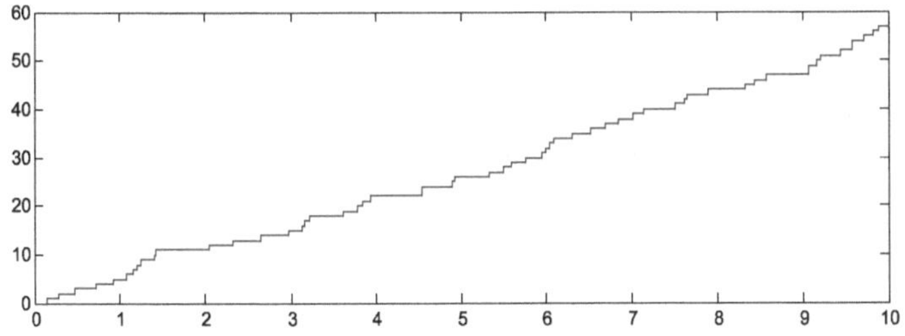

Figure 1.6: A realization of a Poisson process, N_t, $0 \le t \le 10$, with jump rate 5 per unit time.

Example 1.11 (Compound Poisson process). Figure 1.7 illustrates a realization on the time interval $[0, 10]$ of another random process whose sample paths change only by jumps. The particular process illustrated is a *compound Poisson process* with a jump rate of 5 per time unit and jump sizes that are independent normal random variables, independent of the jump times, with mean zero and variance 1. This means, more precisely, that the number of jumps in fixed disjoint intervals of length Δ are independent

Poisson random variables with mean 5Δ and that the jumps are normally distributed with mean zero and variance 1. This is a special case of a pure-jump Lévy process. The most general Lévy process is a sum of a pure-jump process and Brownian motion. Such processes play a key role in generating models for stock prices that overcome some of the limitations of the geometric Brownian motion model introduced in Example 1.3.

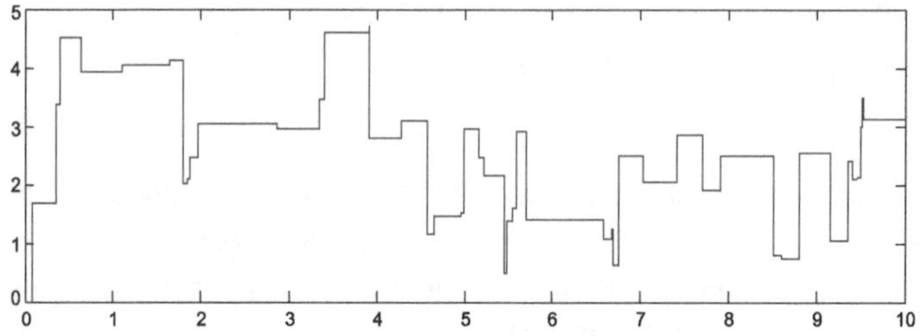

Figure 1.7: A realization of a compound Poisson process, X_t, $0 \le t \le 10$, with jump rate 5 per unit time and jump-size distribution normal with mean 0 and variance 1.

In Example 1.6, we were able to define $X_t(\omega)$ quite explicitly for each t and ω. Very frequently however we may wish (or be forced) to specify instead the collection of joint distributions of all finite-dimensional vectors $(X_{t_1}, \dots, X_{t_n})^T$, $\mathbf{t} = (t_1, \dots, t_n)^T \in T^n$, $n \in \mathbb{N}$. In such a case, we need to be sure that a stochastic process (see Definition 1.4) with the specified distributions really does exist. Kolmogorov's theorem, stated below, guarantees this existence under minimal conditions for the specified distribution functions. We shall use the theorem in Section 1.4 to establish the existence of standard Brownian motion as defined in Example 1.9. Our statement of Kolmogorov's theorem is simplified slightly by the assumption that T is a subset of $\mathbb{R}$ and hence a linearly ordered set. If T were not so ordered, an additional "permutation" condition would be required. A statement and proof of the theorem for arbitrary T can be found in numerous books on probability theory, for example in Billingsley [16] or Klenke [57].

Definition 1.12 (Distribution functions of stochastic process). Let $(X_t)_{t \in T}$ be a real-valued stochastic process and let $\mathcal{T}$ be the set of all vectors $\{\mathbf{t} = (t_1, \dots, t_n)^T \in T^n : t_1 < \cdots < t_n, n \in \mathbb{N}\}$. Then the finite-dimensional *distribution functions* of $(X_t)_{t \in T}$ are the functions $\{F_{\mathbf{t}}, \mathbf{t} \in \mathcal{T}\}$, defined for $\mathbf{t} = (t_1, \dots, t_n)^T \in \mathcal{T}$ by

$$F_{\mathbf{t}}(\mathbf{x}) := P(X_{t_1} \le x_1, \dots, X_{t_n} \le x_n), \quad \mathbf{x} = (x_1, \dots, x_n)^T \in \mathbb{R}^n.$$

Theorem 1.13 (Kolmogorov's consistency theorem). *The probability distribution functions $\{F_{\mathbf{t}}, \mathbf{t} \in \mathcal{T}\}$, with $\mathcal{T}$ as in Definition 1.12, are the distribution functions of some stochastic process $(X_t)_{t \in T}$ if and only if for every $n \in \mathbb{N}$, $\mathbf{t} = (t_1, \dots, t_n)^T \in \mathcal{T}$ and $1 \le i \le n$,*

$$\lim_{x_i \to \infty} F_{\mathbf{t}}(\mathbf{x}) = F_{\mathbf{t}(i)}(\mathbf{x}(i)), \tag{1.5}$$

where $\mathbf{t}(i)$ *and* $\mathbf{x}(i)$ *are the* $(n-1)$*-component vectors obtained by deleting the* i^{th} *components of* $\mathbf{t}$ *and* $\mathbf{x}$*, respectively.*

Remark 1.14. Theorem 1.13 holds also for $\mathbb{R}^k$-valued stochastic processes, with each x_i replaced by a vector $\mathbf{x}_i \in \mathbb{R}^k$ and $x_i \to \infty$ replaced by $\mathbf{x}_i \to \infty$ componentwise.

If $\phi_{\mathbf{t}}(\cdot)$ is the *characteristic function* corresponding to $F_{\mathbf{t}}(\cdot)$, i. e., if

$$\phi_{\mathbf{t}}(\mathbf{u}) = \mathbb{E}\, e^{i\sum_{k=1}^{n} u_k X_{t_k}}, \quad \mathbf{u} = (u_1, \ldots, u_n)^T \in \mathbb{R}^n,$$

then (1.5) can be restated in the equivalent form,

$$\phi_{\mathbf{t}}((u_1, \ldots, u_{i-1}, 0, u_{i+1}, \ldots, u_n)^T) = \phi_{\mathbf{t}(i)}((u_1, \ldots, u_{i-1}, u_{i+1}, \ldots, u_n)^T), \tag{1.6}$$

for all $n \in \mathbb{N}$, $\mathbf{t} = (t_1, \ldots, t_n)^T \in \mathcal{T}$, $u_1, \ldots, u_n \in \mathbb{R}$ and $i \in \{1, \ldots, n\}$.

Condition (1.5) is simply the "consistency" requirement that each function $F_{\mathbf{t}}(\cdot)$ should have marginal distributions which coincide with the specified lower-dimensional distribution functions. Notice, however, that the joint distribution functions of a stochastic process are not sufficient to determine properties, such as continuity and differentiability, of the sample paths. A second theorem of Kolmogorov, stated below, is a useful tool for establishing the existence of a *version* of the process with continuous sample paths under the simple moment conditions specified in the theorem.

Definition 1.15. If X and Y are stochastic processes on $(\Omega, \mathcal{F}, P)$, we say that X is a *version* of Y if

$$P(\{\omega \in \Omega : X_t(\omega) = Y_t(\omega)\}) = 1 \quad \text{for all } t.$$

Theorem 1.16 (Kolmogorov's continuity theorem). *If* $X = (X_t)_{t\in\mathbb{R}}$ *is a stochastic process on* $(\Omega, \mathcal{F}, P)$ *and for all* $T > 0$ *there exist constants* $\alpha, \beta, K > 0$ *such that*

$$\mathbb{E}(|X_t - X_s|^\alpha) \le K|t-s|^{1+\beta} \quad \text{for all } s \text{ and } t \text{ such that } |s|, |t| \le T,$$

then there exists a continuous version of X*.*

A proof can be found in Klenke [57, Thm. 21.6] or Kallenberg [55, Thm. 3.23]. Actually, it can be shown that under the condition stated above, the continuous version is locally Hölder continuous with exponent c for any $c \in (0, \beta/\alpha)$ and Theorem 1.16 is often called the *Kolmogorov–Chentsov theorem.*

1.3 Weak and strict stationarity

When dealing with a finite number of random variables with finite second moments, a useful measure of the dependence between them is given by their covariance matrix.

For a time series $(X_t)_{t\in\mathbb{R}}$, we need to extend the concept of covariance matrix to deal with infinite collections of random variables. This can be achieved by introducing the *autocovariance function* (acvf) of X.

Definition 1.17 (The autocovariance function). If $X := (X_t)_{t\in\mathbb{R}}$ is a real-valued or complex-valued process such that $\mathbb{E}(|X_t|^2) < \infty$ for each $t \in \mathbb{R}$, then the *autocovariance function* $\kappa_X(\cdot,\cdot)$ of X is defined as

$$\kappa_X(r,s) := \operatorname{Cov}(X_r, X_s) = \mathbb{E}[(X_r - \mathbb{E}X_r)(\overline{X_s} - \overline{\mathbb{E}X_s})], \quad r,s \in \mathbb{R},$$

the bar denoting complex conjugation.

Definition 1.18 (Weak stationarity). The real-valued or complex-valued process $(X_t)_{t\in\mathbb{R}}$ is said to be *weakly stationary* if

(i) $\mathbb{E}(|X_t|^2) < \infty$ for all $t \in \mathbb{R}$,
(ii) $\mathbb{E}(X_t)$ is independent of t and
(iii) $\kappa_X(t+h,t)$ is independent of t for all $h \in \mathbb{R}$.

Remark 1.19. Weak stationarity as just defined is frequently referred to in the literature as covariance stationarity, stationarity in the wide sense or second-order stationarity. In this book, the term *stationary*, without further qualification, will always mean weak stationarity. Strict stationarity will be defined later in this section.

Definition 1.20 (Autocovariance function of a stationary process). If the process $(X_t)_{t\in\mathbb{R}}$ is stationary, then $\kappa_X(r,s) = \kappa_X(r-s,0)$ for all $r,s \in \mathbb{R}$. It is therefore convenient to redefine the autocovariance function of a stationary process as the function of just one variable,

$$\gamma_X(h) := \kappa_X(h,0) = \operatorname{Cov}(X_{t+h}, X_t) \quad \text{for all } t,h \in \mathbb{R}.$$

The function $\gamma_X(\cdot)$ is then also referred to as the *autocovariance function (acvf) of the stationary process* X and $\gamma_X(h)$ as its value at "lag" h. The *autocorrelation function (acf) of* X is defined analogously as the function whose value at lag h is

$$\rho_X(h) := \frac{\gamma_X(h)}{\gamma_X(0)} = \operatorname{Corr}(X_{t+h}, X_t) \quad \text{for all } t,h \in \mathbb{R} \text{ provided } \gamma_X(0) > 0.$$

Remark 1.21. We have defined stationarity and the autocovariance function only for processes with index set $T = \mathbb{R}$. Stationarity and the autocovariance function for discrete-parameter time series with $T = \mathbb{Z}$ is defined in precisely the same way, except that $\mathbb{R}$ in Definitions 1.18 and 1.20 is replaced throughout by $\mathbb{Z}$. If we wish to model data $\{x_t, t \in T \subset \mathbb{R}\}$ as a realization of a stationary process, we can always consider it to be part of a realization of a stationary process with index set $\mathbb{R}$.

Example 1.22. The process

$$X_t = A\cos(\nu t + \Theta), \quad t \in \mathbb{R},$$

defined in Example 1.6 has the property

$$\mathbb{E}X_t = \mathbb{E}(A)\mathbb{E}\big(\cos(\nu t + \Theta)\big) = 0,$$

provided $\mathbb{E}(A) < \infty$ (by the assumed independence of A and Θ and the uniform distribution of Θ on $[0, 2\pi)$). If $\mathbb{E}(A^2) < \infty$, its acvf is

$$\kappa_X(t+h,t) = \frac{1}{2}\mathbb{E}(A^2)\mathbb{E}\big(\cos(\nu(2t+h) + 2\Theta) + \cos(\nu h)\big) = \frac{1}{2}\mathbb{E}(A^2)\cos(\nu h).$$

These calculations show that the conditions of Definition 1.18 are satisfied and hence that the process X is stationary with acvf $\gamma_X(h) = \frac{1}{2}\mathbb{E}(A^2)\cos(\nu h)$ and acf $\rho_X(h) = \cos(\nu h)$.

Example 1.23 (Differenced Poisson process). If $d > 0$ and N is a Poisson process with jump rate λ, as defined in Example 1.10, we define the *Poisson process differenced at lag d* as

$$Y_t = N_t - N_{t-d}, \quad t \in \mathbb{R}.$$

It is clear that $\mathbb{E}(Y_t) = \lambda d$ and a simple calculation using the independence of the increments of N shows that Y is stationary with the triangular autocovariance function $\gamma_Y(h) = \lambda \max(d - |h|, 0)$, $h \in \mathbb{R}$.

Example 1.24 (The stationary Poisson-driven Ornstein–Uhlenbeck process). If N is a Poisson process with jump rate λ and $a > 0$, then the *stationary Poisson-driven Ornstein–Uhlenbeck process with parameters λ and a* can be defined as the pathwise improper Riemann–Stieltjes integral,

$$Y_t = \int_{-\infty}^{t} \mathrm{e}^{-a(t-u)}\, \mathrm{d}N_u. \tag{1.7}$$

This is an example of a stochastic integral with respect to the random process N. In general, the interpretation of such integrals requires much more care than is required here and will be dealt with later in the book. However, due to the simple path structure of the Poisson process, we can write the integral (1.7) more explicitly as

$$Y_t = \sum_{i=1}^{\infty} \mathrm{e}^{-a(t-T_i)},$$

where $T_i, i = 1, 2, \ldots$ are the jump-times of the Poisson process (in descending order) prior to t. It is known that $t - T_1, T_1 - T_2, T_2 - T_3, \ldots$ form an independent and identically distributed (i. i. d.) sequence having the exponential distribution with mean λ^{-1} (i. e. with Lebesgue density of the form $\lambda \mathrm{e}^{-\lambda x}\mathbf{1}_{(0,\infty)}(x)$, $x \in \mathbb{R}$); a full proof of this fact is given later in Corollary 12.20. Hence, we can rewrite Y_t in the form,

$$Y_t = e^{-a(t-T_1)} + e^{-a(t-T_1)}e^{-a(T_1-T_2)} + e^{-a(t-T_1)}e^{-a(T_1-T_2)}e^{-a(T_2-T_3)}$$
$$+ \cdots$$
$$= \sum_{j=1}^{\infty} X_1X_2 \cdots X_j,$$

where $\{X_i, i = 1, 2, \ldots\}$ is an i. i. d. sequence with the distribution of e^{-aE}, where E has the exponential distribution with mean λ^{-1}. Since

$$\mathbb{E}(X_i) = \int_0^{\infty} \lambda e^{-\lambda x} e^{-ax}\, dx = \frac{\lambda}{\lambda + a} \quad \text{and} \quad \mathbb{E}(X_i^2) = \frac{\lambda}{\lambda + 2a},$$

it is not difficult to show that the series for Y_t converges almost surely and in mean-square (Exercise 1.3), so that Y_t has finite mean and variance. Denote these by μ and σ^2 respectively. To determine μ and σ^2, we observe that Y_t satisfies the equation,

$$Y_t = X_1 + X_1 Z,$$

where $Z = \sum_{j=2}^{\infty} X_2 \cdots X_j$ has the same distribution as Y_t (in particular $\mathbb{E}(Z) = \mu$ and $\mathrm{Var}(Z) = \sigma^2$) and is independent of X_1. Equating expectations on both sides of the last equation for Y_t immediately gives

$$\mathbb{E}(Y_t) = \mu = \frac{\mathbb{E}X_1}{1 - \mathbb{E}X_1} = \frac{\lambda}{a}.$$

Equating second moments in the same way gives

$$\mathrm{Var}(Y_t) = \sigma^2 = \frac{\lambda}{2a}.$$

For $h \geq 0$, it is clear from (1.7) that

$$Y_{t+h} = e^{-ah} Y_t + W,$$

where $W := \int_{(t,t+h]} e^{-a(t+h-u)}\, dN_u$ is independent of Y_t (this is intuitively clear and is proved rigorously in Corollary 16.8). Hence

$$\mathrm{Cov}(Y_{t+h}, Y_t) = e^{-ah}\,\mathrm{Var}(Y_t) = \frac{\lambda}{2a} e^{-ah}, \quad h \geq 0,$$

and, by the same argument, whether $h \geq 0$ or $h < 0$,

$$\mathrm{Cov}(Y_{t+h}, Y_t) = \frac{\lambda}{2a} e^{-a|h|}.$$

These properties of the process Y establish that it is stationary with mean λ/a, autocovariance function $\gamma_Y(h) = \lambda e^{-a|h|}/(2a)$, and sample paths which are countable sums of decreasing exponential functions. We shall see later (cf. Exercise 17.1) that the cumulant generating function of the marginal distribution of Y_t, $\kappa(\theta) := \log \mathbb{E}(\exp(i\theta Y_t))$, is given by

$$\kappa(\theta) = \frac{\lambda}{a} \sum_{j=1}^{\infty} \frac{1}{j} \frac{(i\theta)^j}{j!},$$

from which we see that the jth cumulant of Y_t is $\kappa_j = \lambda/(aj)$. Setting $j = 1$ and $j = 2$ gives the same expressions for the mean and variance of Y_t as calculated previously. The process (1.7) is a particular case of the stationary Lévy-driven Ornstein–Uhlenbeck processes employed by Barndorff–Nielsen and Shephard [7] to represent stochastic volatility in the modeling of log asset prices.

We shall show later that the process (1.7) is also *strictly stationary* in the sense of the following definition (which applies also to discrete-time processes indexed by $\mathbb{Z}$ if $\mathbb{R}$ is replaced by $\mathbb{Z}$ throughout).

Definition 1.25 (Strict stationarity). The time series $(X_t)_{t\in\mathbb{R}}$ is said to be *strictly stationary* if the joint distributions of $(X_{t_1}, \ldots, X_{t_k})^T$ and $(X_{t_1+h}, \ldots, X_{t_k+h})^T$ are the same for all positive integers k and for all $t_1, \ldots, t_k, h \in \mathbb{R}$.

Strict stationarity means intuitively that the graphs over two equal-length time intervals of a realization of the time series should exhibit similar statistical characteristics. For example, the proportion of ordinates not exceeding a given level x should be roughly the same for both intervals.

Example 1.26 (Differenced Brownian motion). If $d > 0$, we define *Brownian motion differenced at lag d* as

$$X_t = B_t(\mu, \sigma^2) - B_{t-d}(\mu, \sigma^2) = \mu d + \sigma(B_t - B_{t-d}), \quad t \in \mathbb{R},$$

where B is standard Brownian motion and $B_t(\mu, \sigma^2)$ was defined in (1.4). It is clear that $\mathbb{E}(X_t) = \mu d$ and a simple calculation using the independence of the increments of B shows that X is stationary with the triangular autocovariance function $\gamma_X(h) = \sigma^2 \max(d - |h|, 0)$, $h \in \mathbb{R}$. Moreover, X is also strictly stationary since the (covariance) stationarity of X implies that the mean and covariance matrix of the vector $(X_{t_1}, \ldots, X_{t_k})^T$ are the same as the mean and covariance matrix of the vector $(X_{t_1+h}, \ldots, X_{t_k+h})^T$ for every $h \in \mathbb{R}$. Since their joint distributions are Gaussian, this implies that they are the same and hence that X is strictly stationary. We see that if $\mu = \lambda$ and $\sigma = \sqrt{\lambda}$ the mean and autocovariance function of differenced Brownian motion are the same as those of the differenced Poisson process with jump rate λ.

Remark 1.27 (Stationarity and strict stationarity). If X is strictly stationary, it immediately follows, on taking $k = 1$ in Definition 1.25, that X_t has the same distribution for each $t \in \mathbb{R}$. If $\mathbb{E}(|X_t|^2) < \infty$ this implies in particular that $\mathbb{E}(X_t)$ and $\mathrm{Var}(X_t)$ are both constant.

Moreover, taking $k = 2$ in Definition 1.25, we find that X_{t+h} and X_t have the same joint distribution and hence the same covariance for all $t \in \mathbb{R}$. Thus a strictly stationary process with finite second moments is also (covariance) stationary.

The converse of the previous statement is not true. For example, if W is Brownian motion, $W = (B_t(\lambda, \sqrt{\lambda}))_{t\in\mathbb{R}}$, and N is an independent Poisson process with jump rate λ, then the process

$$Z_t := \begin{cases} N_t, & \text{if } t < 0, \\ W_t, & \text{if } t \geq 0, \end{cases}$$

has uncorrelated increments with the same mean and variance as those of both N and W. The process Z differenced at lag d is therefore stationary with mean λd and acvf $\gamma_Z(h) = \lambda \max(d - |h|, 0)$, $h \in \mathbb{R}$. However, it is clearly not strictly stationary, since the marginal distributions of the differenced process for positive and negative values of t are not the same.

There is one important case, however, in which stationarity does imply strict stationarity.

Definition 1.28 (Gaussian time series). The real-valued process $(X_t)_{t\in T}$ is a *Gaussian time series* if the distributions of all of the random vectors $(X_{t_1}, \dots, X_{t_k})^T$, $k \in \mathbb{N}$, $t_1, \dots, t_k \in T$, are multivariate normal.

If $X = (X_t)_{t\in\mathbb{R}}$ is a stationary Gaussian time series (as in Example 1.26), then X is strictly stationary, since for all $k \in \mathbb{N}$ and for all $h, t_1, \dots, t_k \in \mathbb{R}$, the random vectors $(X_{t_1}, \dots, X_{t_k})^T$ and $(X_{t_1+h}, \dots, X_{t_k+h})^T$ have the same mean and autocovariance matrix and hence the same multivariate normal distribution.

1.4 Existence of Brownian motion and the Poisson process

In this section, we make use of Theorem 1.13 to establish the existence of the two important processes, Brownian motion (Example 1.9), and the Poisson process (Example 1.10).

Definition 1.29 (Standard Brownian motion). *Standard Brownian motion* is a process $(B_t)_{t\in\mathbb{R}}$ satisfying the conditions,

(a) $B_0 = 0$ almost surely,

(b) $B_{t_2} - B_{t_1}, B_{t_3} - B_{t_2}, \dots, B_{t_n} - B_{t_{n-1}}$, are independent for every $n \in \{2, 3, \dots\}$ and every $(t_1, t_2, \dots, t_n)^T \in \mathbb{R}^n$ such that $t_1 < t_2 < \cdots < t_n$,

(c) $B_t - B_s$ has the normal distribution with mean zero and variance $t - s$ for all $s, t \in \mathbb{R}$ such that $t \geq s$, and

(d) B has continuous sample paths almost surely.

To establish the existence of such a process, we observe that conditions (a), (b) and (c) are satisfied for B_t restricted to $t \in [0,\infty)$ if and only if, for every $\mathbf{t} = (t_1, t_2, \dots, t_n)^T \in [0,\infty)^n$ such that $t_1 < t_2 < \cdots < t_n$, the characteristic function of $(B_{t_1}, \dots, B_{t_n})^T$ is

$$\begin{aligned}
\phi_{\mathbf{t}}(\mathbf{u}) &= \mathbb{E}\exp[\mathrm{i}u_1 B_{t_1} + \cdots + \mathrm{i}u_n B_{t_n}] \\
&= \mathbb{E}\exp[\mathrm{i}u_1\Delta_1 + \mathrm{i}u_2(\Delta_1 + \Delta_2) + \cdots + \mathrm{i}u_n(\Delta_1 + \cdots + \Delta_n)] \\
&\quad (\text{where } \Delta_j = B_{t_j} - B_{t_{j-1}}, j \ge 1, \text{ and } t_0 := 0) \\
&= \mathbb{E}\exp[\mathrm{i}\Delta_1(u_1 + \cdots + u_n) + \mathrm{i}\Delta_2(u_2 + \cdots + u_n) + \cdots + \mathrm{i}\Delta_n u_n] \\
&= \exp\left[-\frac{1}{2}\sum_{j=1}^{n}(u_j + \cdots + u_n)^2(t_j - t_{j-1})\right].
\end{aligned}$$

It is trivial to check that the characteristic functions $\phi_{\mathbf{t}}(\cdot)$ satisfy the consistency condition (1.6) and so by Kolmogorov's theorem there exists a process $(B_t)_{t\ge0}$ with characteristic functions $\phi_{\mathbf{t}}(\cdot)$, or equivalently with the properties (a), (b) and (c). Now take an independent copy $(B'_t)_{t\ge0}$ of $(B_t)_{t\ge0}$ and extend B_t to $t \in \mathbb{R}$ by defining $B_t := -B'_{-t}$ for $t < 0$. Then it is easily checked that also $(B_t)_{t\in\mathbb{R}}$ satisfies the properties (a), (b) and (c). A process satisfying (a), (b) and (c), but not necessarily (d), is known as *(standard) Brownian motion in law.*

To establish the existence of a version of B with continuous sample paths, i. e., satisfying the condition (d), we simply apply Kolmogorov's continuity theorem (Theorem 1.16) with $\alpha = 4$, $\beta = 1$ and $K = 3$, using the easily established fact that $\mathbb{E}(|B_t - B_s|^4) = 3|t - s|^2$.

The existence of Brownian motion $(B_t(\mu,\sigma^2))_{t\in\mathbb{R}}$ as defined by equation (1.4), i. e.,

$$B_t(\mu,\sigma^2) = \mu t + \sigma B_t, \quad t \in \mathbb{R},$$

follows at once from that of standard Brownian motion.

Definition 1.30 (Poisson process). A *Poisson process* with *jump rate* or *parameter* or *intensity* $\lambda(> 0)$ is a process $(N_t)_{t\in\mathbb{R}}$ satisfying the conditions,

(a) $N_0 = 0$ almost surely,
(b) $N_{t_2} - N_{t_1}, N_{t_3} - N_{t_2}, \dots, N_{t_n} - N_{t_{n-1}}$, are independent for every $n \in \{2, 3, \dots\}$ and every $(t_1, t_2, \dots, t_n)^T \in \mathbb{R}^n$ such that $t_1 < t_2 < \cdots < t_n$,
(c) $N_t - N_s$ has the Poisson distribution with mean $\lambda(t - s)$ for all $s, t \in \mathbb{R}$ such that $t > s$, and
(d) the sample paths of N are almost surely right-continuous step functions with jump size 1.

A process that satisfies only conditions (a)–(c), but not necessarily (d), is called a *Poisson process in law* (with jump rate λ).

Some of the conditions in Definition 1.30 can be relaxed. For example, if (a)–(c) hold, then (d) can be replaced by the fact that the sample paths of N are only right-continuous with finite left limits, and it can be shown that then the sample paths must actually be right-continuous step functions with jump size 1. This and other characterizations of the Poisson process will be proved in Theorem 12.18.

The proof for the existence of a Poisson process in law can be done by the same argument with which we showed the existence of standard Brownian motion in law (Exercise 1.4). We shall however follow a different approach which gives an explicit construction of a Poisson process satisfying conditions (a)–(d). This has the advantage of establishing the structure of the paths and jumps of a Poisson process automatically, and providing a useful representation for further investigations.

In the following, the *indicator function* of a subset A of a given set C (where C is usually clear from the context) will be denoted by $\mathbf{1}_A$, i. e.,

$$\mathbf{1}_A : C \to \mathbb{R}, \quad \mathbf{1}_A(x) = \begin{cases} 1, & \text{if} \quad x \in A, \\ 0, & \text{if} \quad x \notin A. \end{cases}$$

Recall that an *exponentially distributed* random variable *with parameter* $\lambda > 0$ is a random variable T with probability density $x \mapsto \lambda e^{-\lambda x} \mathbf{1}_{(0,\infty)}(x)$, i. e. with distribution function $P(T \leq x) = 0$ for $x \leq 0$ and $P(T \leq x) = 1 - e^{-\lambda x}$ for $x > 0$. Its expectation is λ^{-1}. Also recall that the sum of n independent exponentially distributed random variables with parameter λ is *gamma distributed with parameters n and λ*, i. e. has probability density

$$x \mapsto \frac{\lambda^n}{(n-1)!} x^{n-1} e^{-\lambda x} \mathbf{1}_{(0,\infty)}(x), \tag{1.8}$$

see, e. g., Feller [44, Section I.3].

Theorem 1.31 (Construction/existence of Poisson processes). *Let $(T_n)_{n \in \mathbb{Z}}$ be an i. i. d. sequence of exponentially distributed random variables with parameter $\lambda > 0$ on a probability space $(\Omega, \mathcal{F}, P)$. For $n \in \mathbb{Z}$ define*

$$\Gamma_n := \begin{cases} \sum_{k=1}^{n} T_k, & n > 0, \\ -\sum_{k=n}^{0} T_k, & n \leq 0, \end{cases}$$

and

$$\Omega_0 := \Big\{ \omega \in \Omega : \Gamma_n(\omega) < \Gamma_{n+1}(\omega) \ \forall\, n \in \mathbb{Z} \text{ and } \lim_{n \to \infty} \Gamma_n(\omega) = -\lim_{n \to -\infty} \Gamma_n(\omega) = \infty \Big\}.$$

Then $\Omega_0 \in \mathcal{F}$ and $P(\Omega_0) = 1$. For $t \in \mathbb{R}$ define N_t by $N_t(\omega) := 0$ if $\omega \in \Omega \setminus \Omega_0$ and for $\omega \in \Omega_0$ by

$$N_t(\omega) := \sup\{n \in \mathbb{Z} : \Gamma_n(\omega) \le t\},$$

i. e., N_t is the unique integer n such that $\Gamma_n \le t < \Gamma_{n+1}$ (on Ω_0). Then $N = (N_t)_{t\in\mathbb{R}}$ is a Poisson process with jump rate λ.

Proof. That $\Omega_0 \in \mathcal{F}$ and $P(\Omega_0) = 1$ is clear, as is the fact that each N_t is indeed a random variable so that $N = (N_t)_{t\in\mathbb{R}}$ is a stochastic process. It is also clear that $N_0 = 0$ and that the sample paths of N are right-continuous $\mathbb{Z}$-valued step functions with jump size 1, taking the value n on $[\Gamma_n, \Gamma_{n+1})$. It remains to show conditions (b) and (c) of Definition 1.30. For that, we shall first show that N restricted to $[0,\infty)$ satisfies these conditions, then that N restricted to $(-\infty, 0]$ does, and finally that N defined on $\mathbb{R}$ does the same.

(i) Let us first treat the case $(N_t)_{t\ge 0}$. Define

$$W_n := \begin{cases} \Gamma_n, & n \ge 1, \\ 0, & n = 0. \end{cases}$$

Then, $W_n = \sum_{k=1}^n T_k$ for all $n \in \mathbb{N}_0$ and $W_{n+1} = W_n + T_{n+1}$. Then, W_n is gamma-distributed for $n \ge 1$ and independent of the exponentially distributed T_{n+1}. Suppose that $t \ge 0, s \ge 0$ and $n \in \mathbb{N}$. Then, from (1.8),

$$\begin{aligned} &P(N_t = n, W_{n+1} > t + s) \\ &\quad = P(W_n \le t, W_n + T_{n+1} > t + s) \\ &\quad = \int_0^t \frac{\lambda^n}{(n-1)!} x^{n-1} \mathrm{e}^{-\lambda x} \Big(\int_{t+s-x}^{\infty} \lambda \mathrm{e}^{-\lambda y}\, \mathrm{d}y \Big) \mathrm{d}x \\ &\quad = \frac{\lambda^n}{(n-1)!} \int_0^t x^{n-1} \mathrm{e}^{-\lambda x} \mathrm{e}^{-\lambda(t+s-x)}\, \mathrm{d}x \\ &\quad = \mathrm{e}^{-\lambda(t+s)} \frac{(\lambda t)^n}{n!}, \end{aligned} \tag{1.9}$$

which also holds for $n = 0$ since

$$P(N_t = 0, W_1 > t + s) = P(T_1 > t + s) = \mathrm{e}^{-\lambda(t+s)}. \tag{1.10}$$

Letting $s = 0$ in (1.9) and (1.10), we obtain

$$P(N_t = n) = P(W_n \le t, W_n + T_{n+1} > t) = \mathrm{e}^{-\lambda t} \frac{(\lambda t)^n}{n!}, \quad t \ge 0, n \in \mathbb{N}_0, \tag{1.11}$$

so that N_t has the Poisson distribution with mean λt for $t > 0$.

Now suppose that $n \in \mathbb{N}_0$, $m \in \mathbb{N}$, and $t, s_1, \dots, s_m \ge 0$. Then, from (1.9)–(1.11), since the T_i are i. i. d. exponentially distributed and W_n and W_{n+1} are independent of $T_{n+2}, \dots, T_{n+m}$,

$$\begin{aligned}
&P(N_t = n, W_{n+1} > t + s_1, T_{n+2} > s_2, \dots, T_{n+m} > s_m)\\
&\quad = P(W_n \le t \le t + s_1 < W_{n+1}, T_{n+2} > s_2, \dots, T_{n+m} > s_m)\\
&\quad = P(W_n \le t \le t + s_1 < W_{n+1}) P(T_{n+2} > s_2, \dots, T_{n+m} > s_m)\\
&\quad = \mathrm{e}^{-\lambda(t+s_1)} \frac{(\lambda t)^n}{n!} \prod_{k=2}^{m} P(T_{n+k} > s_k)\\
&\quad = P(N_t = n) \prod_{k=1}^{m} P(T_k > s_k),
\end{aligned}$$

so that

$$\begin{aligned}
&P(W_{n+1} - t > s_1, T_{n+2} > s_2, \dots, T_{n+m} > s_m | N_t = n)\\
&\quad = P(T_1 > s_1, \dots, T_m > s_m).
\end{aligned}$$

In other words, the conditional distribution of $(W_{n+1} - t, T_{n+2} \dots, T_{n+m})^T$, given $N_t = n$, is the same as the distribution of $(T_1, \dots, T_m)^T$, expressed symbolically as

$$(W_{n+1} - t, T_{n+2}, \dots, T_{n+m})^T | N_t = n \sim (T_1, \dots, T_m)^T. \tag{1.12}$$

Now let $k \in \mathbb{N}$, $0 \le t_0 < t_1 < t_2 < \dots < t_k$ and $n_0, \dots, n_k \in \mathbb{N}_0$. Then,

$$\begin{aligned}
&P(N_{t_0} = n_0, N_{t_1} - N_{t_0} = n_1, \dots, N_{t_k} - N_{t_{k-1}} = n_k)\\
&\quad = P(N_{t_1} - N_{t_0} = n_1, \dots, N_{t_k} - N_{t_{k-1}} = n_k | N_{t_0} = n_0) P(N_{t_0} = n_0)\\
&\quad = P(N_{t_1} = n_0 + n_1, N_{t_2} = n_0 + n_1 + n_2, \dots, N_{t_k} = n_0 + \dots + n_k | N_{t_0} = n_0)\\
&\quad\quad \times P(N_{t_0} = n_0).
\end{aligned} \tag{1.13}$$

But the event $\{N_{t_1} = n_0 + n_1, \dots, N_{t_k} = n_0 + \dots + n_k\}$ can be described as

$$\{W_{n_0+n_1} \le t_1 < W_{n_0+n_1+1}, \dots, W_{n_0+\dots+n_k} \le t_k < W_{n_0+\dots+n_k+1}\},$$

which when $n_1 \ge 1$ is equal to

$$\begin{aligned}
&\Bigg\{ W_{n_0+1} - t_0 + \sum_{m=n_0+2}^{n_0+n_1} T_m \le t_1 - t_0 < W_{n_0+1} - t_0 + \sum_{m=n_0+2}^{n_0+n_1+1} T_m, \dots,\\
&\quad W_{n_0+1} - t_0 + \sum_{m=n_0+2}^{n_0+\dots+n_k} T_m \le t_k - t_0 < W_{n_0+1} - t_0 + \sum_{m=n_0+2}^{n_0+\dots+n_k+1} T_m \Bigg\}.
\end{aligned}$$

Combining this with equations (1.12) and (1.13) leads for $n_1 \ge 1$ to

$$\begin{aligned}
&P(N_{t_0} = n_0, N_{t_1} - N_{t_0} = n_1, \dots, N_{t_k} - N_{t_{k-1}} = n_k)\\
&\quad = P(W_{n_1} \le t_1 - t_0 < W_{n_1+1}, \dots, W_{n_1+\dots+n_k} \le t_k - t_0 < W_{n_1+\dots+n_k+1})
\end{aligned}$$

$$
\begin{aligned}
&\quad \times P(N_{t_0} = n_0) \\
&= P(N_{t_1-t_0} = n_1, N_{t_2-t_0} = n_1 + n_2, \ldots, N_{t_k-t_0} = n_1 + \cdots + n_k) \\
&\quad \times P(N_{t_0} = n_0) \\
&= P(N_{t_1-t_0} = n_1, N_{t_2-t_0} - N_{t_1-t_0} = n_2, \ldots, N_{t_k-t_0} - N_{t_{k-1}-t_0} = n_k) \\
&\quad \times P(N_{t_0} = n_0). \qquad (1.14)
\end{aligned}
$$

Iterating (1.14) gives for $n_0 \in \mathbb{N}_0, n_1, \ldots, n_k \in \mathbb{N}$

$$
\begin{aligned}
&P(N_{t_0} = n_0, N_{t_1} - N_{t_0} = n_1, \ldots, N_{t_k} - N_{t_{k-1}} = n_k) \\
&\quad = P(N_{t_2-t_1} = n_2, \ldots, N_{t_k-t_1} - N_{t_{k-1}-t_1} = n_k)\, P(N_{t_1-t_0} = n_1) \\
&\qquad \times P(N_{t_0} = n_0) \\
&\quad \ldots \\
&\quad = P(N_{t_k-t_{k-1}} = n_k) \cdots P(N_{t_1-t_0} = n_1)\, P(N_{t_0} = n_0). \qquad (1.15)
\end{aligned}
$$

Specialising (1.15) to $k = 1$ gives

$$P(N_{t_0} = n_0, N_{t_1} - N_{t_0} = n_1) = P(N_{t_1-t_0} = n_1)\, P(N_{t_0} = n_0)$$

for all $n \in \mathbb{N}_0$ and $n_1 \in \mathbb{N}$, and summing this over $n_0 \in \mathbb{N}_0$ yields

$$P(N_{t_1} - N_{t_0} = n_1) = P(N_{t_1-t_0} = n_1) \quad \forall\, n_1 \in \mathbb{N}, \qquad (1.16)$$

which hence is true also for $n_1 = 0$. In combination with equation (1.11), this establishes property (c) of Definition 1.30 on $[0, \infty)$. Inserting (1.16) into (1.15) then yields

$$
\begin{aligned}
&P(N_{t_0} = n_0, N_{t_1} - N_{t_0} = n_1, \ldots, N_{t_k} - N_{t_{k-1}} = n_k) \\
&\quad = P(N_{t_k} - N_{t_{k-1}} = n_k) \cdots P(N_{t_1} - N_{t_0} = n_1)\, P(N_{t_0} = n_0) \qquad (1.17)
\end{aligned}
$$

for all $n_0 \in \mathbb{N}_0$ and $n_1 \ldots, n_k \in \mathbb{N}$. Now let $B_0, \ldots, B_k \in \mathcal{B}_1$ with $0 \notin B_1, \ldots, 0 \notin B_k$. Summing (1.17) over $n_i \in B_i$ for all $i \in \{0, \ldots, k\}$ then yields

$$
\begin{aligned}
&P(N_{t_0} \in B_0, N_{t_1} - N_{t_0} \in B_1, \ldots, N_{t_k} - N_{t_{k-1}} \in B_k) \\
&\quad = P(N_{t_k} - N_{t_{k-1}} \in B_k) \cdots P(N_{t_1} - N_{t_0} \in B_1)\, P(N_{t_0} \in B_0).
\end{aligned}
$$

Since the system of all Borel sets in $\mathbb{R}$ that do not contain the origin form a generator of $\mathcal{B}_1$ that is additionally closed under taking finite intersections (i. e. a π-system), this establishes the independence of $N_{t_0}, N_{t_1-t_0}, \ldots, N_{t_k-t_{k-1}}$ for $0 \le t_0 < t_1 \ldots < t_k$, which is property (b) of Definition 1.30.

(ii) Now consider the process $(N_t)_{t \le 0}$. For $s \ge 0$ define

$$X_s := -\lim_{u \uparrow -s} N_u \quad \text{and} \quad Y_s = -N_{-s}.$$

Then $X_s = Y_s = 0$ on $\Omega \setminus \Omega_0$, and on Ω_0 we have

$$X_s = \begin{cases} 0, & 0 \leq s < -\Gamma_0, \\ n+1, & -\Gamma_{-n} \leq s < -\Gamma_{-n-1}, n \in \mathbb{N}_0, \end{cases}$$

and

$$Y_s = \begin{cases} 0, & 0 \leq s \leq -\Gamma_0, \\ n+1, & -\Gamma_{-n} < s \leq -\Gamma_{-n-1}, n \in \mathbb{N}_0. \end{cases}$$

Since $(T_n)_{n\in\mathbb{Z}}$ is an i. i. d. sequence, the sequences $(\Gamma_{n+1})_{n\in\mathbb{N}_0}$ and $(-\Gamma_{-n})_{n\in\mathbb{N}_0}$ have the same distribution, hence by construction, $(N_s)_{s\geq 0}$ and $(X_s)_{s\geq 0}$ have the same finite-dimensional distributions. Further, since $P(\Gamma_n = a) = 0$ for all $a \in \mathbb{R}$ and $n \in \mathbb{Z}$, i. e., the distribution of Γ_n has no atoms, we see that $X_s = Y_s$ almost surely for each $s \geq 0$, hence all three processes $(Y_s)_{s\geq 0}$, $(X_s)_{s\geq 0}$ and $(N_s)_{s\geq 0}$ have the same finite dimensional distributions. Now let $t_1 < t_2 < \cdots < t_n \leq 0$. Then

$$N_{t_2} - N_{t_1} = Y_{-t_1} - Y_{-t_2} \stackrel{d}{=} N_{-t_1} - N_{-t_2}$$

and the latter is Poisson distributed with mean $\lambda(t_2 - t_1)$, as we know already from part (i). Hence condition (c) of Definition 1.30 is satisfied on $(-\infty, 0]$, and condition (b) follows similarly from (i), since

$$\begin{aligned} (N_{t_2} - N_{t_1}, \ldots, N_{t_n} - N_{t_{n-1}}) &= (Y_{-t_1} - Y_{-t_2}, \ldots, Y_{-t_{n-1}} - Y_{-t_n}) \\ &\stackrel{d}{=} (N_{-t_1} - N_{-t_2}, \ldots, N_{-t_{n-1}} - N_{-t_n}). \end{aligned}$$

(iii) Finally, consider $(N_t)_{t\in\mathbb{R}}$ as defined on the whole real line, and suppose that $t_1 < t_2 < \cdots < t_n$. Then the independence of $N_{t_2} - N_{t_1}, \ldots, N_{t_n} - N_{t_{n-1}}$ is clear from (i) and (ii) if $t_1 \geq 0$ or $t_n \leq 0$, while if $t_1 < 0 < t_n$ choose $i \in \{1, \ldots, n-1\}$ such that $t_i \leq 0 < t_{i+1}$. Then $N_{t_2} - N_{t_1}, \ldots, N_{t_i} - N_{t_{i-1}}, N_0 - N_{t_i}, N_{t_{i+1}} - N_0, N_{t_{i+2}} - N_{t_{i+1}}, \ldots, N_{t_n} - N_{t_{n-1}}$ are independent by (i), (ii) and the fact that $(N_s)_{s\geq 0}$ and $(N_s)_{s\leq 0}$ are independent by construction. Hence $N_{t_2} - N_{t_1}, \ldots, N_{t_n} - N_{t_{n-1}}$ are independent also in this case. Finally, if $s \leq 0 \leq t$, then $N_t = N_t - N_0$ and $-N_s = N_0 - N_s$ are independent and Poisson distributed with mean λt and $-\lambda s$, respectively, hence $N_t - N_s$ is again Poisson distributed with mean $\lambda(t + (-s))$, since the convolution of two Poisson distributions is again Poisson distributed. This finishes the proof. □

Remark 1.32. (a) When considering more general Lévy processes in later chapters, we shall see that every Poisson process in law with jump rate $\lambda > 0$ is unique in distribution (i. e., its finite-dimensional distributions are uniquely determined). See Proposition 10.16. Hence it follows that the construction used in Theorem 1.31 gives (up to equality in distribution) all Poisson processes in law. Similarly, Brownian motion in law is unique in distribution. It is even true that *every* Poisson process can be represented as in Theorem 1.31 as will be shown by Theorem 12.18 and Remark 12.19.

(b) We have defined Brownian motion and the Poisson process indexed by $\mathbb{R}$. These are sometimes also called *two-sided* Brownian motion and the *two-sided* Poisson process respectively. The corresponding processes $(B_t)_{t\geq 0}$ and $(N_t)_{t\geq 0}$ arise by restricting the two-sided Brownian motion or Poisson process to $[0,\infty)$ in which case they are called *one-sided* Brownian motion or Poisson process, or simply Brownian motion or Poisson process, when the index set is clear.

(c) In the proof of Theorem 1.31 (and similarly in the proof of the existence of Brownian motion in law), we first showed in (i) that the given process on $[0,\infty)$ constitutes a one-sided Poisson process, and then extended it to a two-sided Poisson process. This method applies more generally for the construction of two-sided Lévy processes (in law), as discussed later in Theorem 10.17 and Remark 10.18.

1.5 Exercises

Exercise 1.1. If, in the geometric Brownian motion model (1.1) with $P_0 = 1$, $\mu = 0$, $\sigma = 0.01$, we replace B_t by $B_t + N_t$ where N is a Poisson process independent of B with jump rate $\lambda = 0.5$, find the mean and variance of the log returns, $\log P_{n+1} - \log P_n$. How would you expect this replacement to affect the sample paths of P as illustrated in Figure 1.3?

Exercise 1.2. Show that (1.3) is the unique weakly stationary solution of the autoregressive equations $X_n = \phi X_{n-1} + Z_n$ in Example 1.7 with $|\phi| < 1$, by first showing that $X_n = \sum_{j=0}^{k-1} \phi^j Z_{n-j} + \phi^k X_{n-k}$, $k = 1, 2, \ldots$. Show also that the sum in (1.3) converges absolutely with probability 1.

Exercise 1.3. Show that the series $Y_t = \sum_{j=1}^{\infty} X_1 X_2 \cdots X_j$ in Example 1.24 converges with probability 1 and in mean-square.

Exercise 1.4. Use the argument in Section 1.4, showing the existence of Brownian motion in law, to show that of the Poisson process in law, i. e., a process satisfying conditions (a), (b) and (c) in Definition 1.30, but not necessarily (d).

2 Hilbert spaces

Abstract: Hilbert spaces, especially the space $\mathcal{S} := L^2(\Omega, \mathcal{F}, P)$ of (equivalence classes of almost surely equal) square-integrable random variables on the probability space $(\Omega, \mathcal{F}, P)$, play an important role in the study of time series with finite variance and in many other areas of probability and statistics. In this chapter, we cover those aspects of Hilbert spaces and inner-product spaces which will be used in later chapters. This is by no means a comprehensive introduction to Hilbert spaces, but will provide all of the notation and tools essential to the reader of this book. Most of the material of Sections 2.1–2.5 and 2.7 is taken more or less directly from the time series book [22] by Brockwell and Davis, where detailed proofs can be found. For the reader who wishes to go more deeply into the theory of Hilbert spaces, we recommend the book of Simmons [82].

2.1 Inner-product spaces and their properties

Definition 2.1 (Inner-product space). A complex vector space $\mathcal{H}$ is an *inner-product space* if, for each pair of elements x and y in $\mathcal{H}$, there is a complex number $\langle x,y \rangle$, called the *inner product of x and y*, such that

(a) $\langle x,y\rangle = \overline{\langle y,x\rangle}$, the bar denoting complex conjugation,
(b) $\langle x+y,z\rangle = \langle x,z\rangle + \langle y,z\rangle$ for all $x,y,z \in \mathcal{H}$,
(c) $\langle ax,y\rangle = a\langle x,y\rangle$ for all $x,y \in \mathcal{H}$ and $a \in \mathbb{C}$,
(d) $\langle x,x\rangle \geq 0$ for all $x \in \mathcal{H}$,
(e) $\langle x,x\rangle = 0$ if and only if $x = 0$.

We also speak of a *complex inner-product space.*

Remark 2.2. A *real* vector space $\mathcal{H}$ is an inner-product space if for each $x,y \in \mathcal{H}$ there exists a *real* number $\langle x,y\rangle$ satisfying conditions (a)–(e), where in (c) only $a \in \mathbb{R}$ is allowed. Of course, condition (a) reduces in this case to $\langle x,y\rangle = \langle y,x\rangle$. We also speak of a *real inner-product space.*

The inner product is a natural generalization of the inner or scalar product of two vectors in n-dimensional Euclidean space. Since this is the prototypical inner-product (and Hilbert) space, it helps intuition to keep it in mind in all that follows.

Example 2.3 (Euclidean space). The set of all column vectors

$$\mathbf{x} = (x_1, \ldots, x_k)^T \in \mathbb{R}^k,$$

is a real inner-product space if we define

https://doi.org/10.1515/9783111325033-002

$$\langle \mathbf{x}, \mathbf{y} \rangle := \sum_{i=1}^{k} x_i y_i.$$

In the same way, it is easy to see that the set of all complex k-dimensional column vectors

$$\mathbf{z} = (z_1, \dots, z_k)^T \in \mathbb{C}^k,$$

is a complex inner-product space if we define

$$\langle \mathbf{w}, \mathbf{z} \rangle := \sum_{i=1}^{k} w_i \overline{z}_i.$$

Example 2.4 (The space $\mathcal{S} := L^2(\Omega, \mathcal{F}, P)$). The set C of all complex-valued random variables X on a probability space $(\Omega, \mathcal{F}, P)$ such that

$$\int |X|^2 \mathrm{d}P = \mathbb{E}(|X|^2) < \infty,$$

satisfies the conditions (a)–(d) of Definition 2.1 if we define, for any two such random variables X and Y,

$$\langle X, Y \rangle := \int X\overline{Y}\, \mathrm{d}P = \mathbb{E}(X\overline{Y}). \tag{2.1}$$

However condition (e) is not satisfied since the condition $\langle X, X \rangle = 0$ does not imply that $X(\omega) = 0$ for all $\omega \in \Omega$, but only that X is zero with probability one, i. e. that $P(X = 0) = 1$. This difficulty is circumvented by saying that the random variables X and Y are *equivalent* if $P(X = Y) = 1$. This equivalence relation partitions the set C into classes of random variables such that any two in the same class differ only on a set of probability zero. The space $\mathcal{S} = L^2(\Omega, \mathcal{F}, P)$ is the collection of these equivalence classes with the inner product of the classes $\mathcal{C}_1$ and $\mathcal{C}_2$ defined by (2.1) with X and Y any random variables in $\mathcal{C}_1$ and $\mathcal{C}_2$, respectively. Since each class is uniquely determined by specifying any one of the random variables in it, we shall continue to use the notation X, Y for elements of $\mathcal{S}$ and refer to them as random variables. It is important to remember however that X stands for the collection of all random variables which are equivalent to X, or equivalently, that X is identified only to within almost sure equality.

Definition 2.5 (Norm). The *norm* of an element x of an inner-product space is defined to be

$$\|x\| = \sqrt{\langle x, x \rangle}. \tag{2.2}$$

In the Euclidean space $\mathbb{R}^k$, the norm of the vector x is simply its length, $\|x\| = |x| = \sqrt{\sum_{i=1}^{k} x_i^2}$.

Proposition 2.6 (The Cauchy–Schwarz inequality). *If $\mathcal{H}$ is an inner-product space then*

$$|\langle x,y\rangle| \leq \|x\|\|y\| \quad \text{for all } x,y \in \mathcal{H},$$

and

$$|\langle x,y\rangle| = \|x\|\|y\| \quad \text{if and only if} \quad y=0, \text{ or } y \neq 0 \text{ and } x = \frac{\langle x,y\rangle}{\langle y,y\rangle} y.$$

Proof. See, e. g., Brockwell and Davis [22, equations (2.1.4) and (2.1.5)]. □

Corollary 2.7 (The triangle inequality). *If $\mathcal{H}$ is an inner-product space then*

$$\|x+y\| \leq \|x\| + \|y\| \quad \text{for all } x,y \in \mathcal{H}.$$

Proof.

$$\|x+y\|^2 = \langle x,x\rangle + \langle x,y\rangle + \langle y,x\rangle + \langle y,y\rangle \leq \|x\|^2 + 2\|x\|\|y\| + \|y\|^2$$

by the Cauchy–Schwarz inequality. □

Remark 2.8 (Orthogonality). Elements x and y of an inner-product space $\mathcal{H}$ are said to be *orthogonal* or *perpendicular* if

$$\langle x,y\rangle = 0.$$

Proposition 2.9 (Properties of the norm). *If $\mathcal{H}$ is a complex (respectively real) inner-product space with norm as specified in Definition 2.5, then*
(a) $\|x+y\| \leq \|x\| + \|y\|$ *for all $x,y \in \mathcal{H}$,*
(b) $\|ax\| = |a|\|x\|$ *for all $x \in \mathcal{H}$ and all $a \in \mathbb{C}$ ($a \in \mathbb{R}$),*
(c) $\|x\| \geq 0$ *for all $x \in \mathcal{H}$,*
(d) $\|x\| = 0$ *if and only if $x = 0$.*

(These properties justify the use of the terminology "norm of x" *for $\|x\|$.)*

Proof. The first property is a restatement of the triangle inequality and the others follow at once from Definition 2.5. □

Proposition 2.10 (The parallelogram law). *If $\mathcal{H}$ is an inner-product space, then*

$$\|x+y\|^2 + \|x-y\|^2 = 2\|x\|^2 + 2\|y\|^2 \quad \text{for all } x,y \in \mathcal{H}. \tag{2.3}$$

Proof. See Exercise 2.1. Note that (2.3) is not a consequence of properties (a), (b), (c) and (d) of the norm. It depends on the particular form (2.2) of the norm as defined for elements of an inner-product space. □

Definition 2.11 (Convergence in norm). A sequence $(x_n)_{n\in\mathbb{N}}$ of elements of an inner-product space $\mathcal{H}$ is said to *converge in norm* to $x \in \mathcal{H}$ if $\|x_n - x\| \to 0$ as $n \to \infty$. We write $x_n \to x$ as $n \to \infty$.

Proposition 2.12 (Continuity of inner product). *If $(x_n)_{n\in\mathbb{N}}$ and $(y_n)_{n\in\mathbb{N}}$ are sequences of elements of the inner-product space $\mathcal{H}$ such that $\|x_n - x\| \to 0$ and $\|y_n - y\| \to 0$ as $n \to \infty$, where $x, y \in \mathcal{H}$, then $\|x_n\| \to \|x\|$ and $\langle x_n, y_n\rangle \to \langle x, y\rangle$ as $n \to \infty$.*

Proof. See, e. g., Brockwell and Davis [22, Proposition 2.1.2]. □

2.2 Hilbert spaces

An inner-product space with the additional property of *completeness* is called a Hilbert space. To define completeness, we first need the concept of a Cauchy sequence.

Definition 2.13 (Cauchy sequence). A sequence $(x_n)_{n\in\mathbb{N}}$ of elements of an inner-product space is said to be a *Cauchy sequence* if

$$\|x_n - x_m\| \to 0 \quad \text{as } m, n \to \infty,$$

i. e., if for every $\epsilon > 0$ there exists a positive integer $N(\epsilon)$ such that

$$\|x_n - x_m\| < \epsilon \quad \text{for all } m, n > N(\epsilon).$$

Definition 2.14 (Hilbert space). A *(complex) Hilbert space* $\mathcal{H}$ is an inner-product space which is *complete*, i. e., an inner-product space in which every Cauchy sequence $(x_n)_{n\in\mathbb{N}}$ converges in norm to some element $x \in \mathcal{H}$. A *real Hilbert space* is a real inner-product space that is complete.

Example 2.15 (Euclidian space). Both the real inner-product space $\mathbb{R}^k$ and the complex inner-product space $\mathbb{C}^k$ are complete and hence Hilbert spaces. This follows from the fact that for a Cauchy sequence in $\mathbb{C}^k$, all components are Cauchy sequences in $\mathbb{C}$ which converge in $\mathbb{C}$ by completeness of $\mathbb{C}$.

Example 2.16 (The space $\mathcal{S} = L^2(\Omega, \mathcal{F}, P)$). The collection of complex-valued square-integrable random variables on the probability space $(\Omega, \mathcal{F}, P)$, introduced in Example 2.4 and abbreviated henceforth to $\mathcal{S}$, is an inner-product space with

$$\langle X, Y\rangle = \mathbb{E}(X\overline{Y}), \quad X, Y \in \mathcal{S},$$

and squared norm

$$\|X\|^2 = \mathbb{E}(|X|^2), \quad X \in \mathcal{S}.$$

Norm convergence in this space of the sequence of random variables $(X_n)_{n\in\mathbb{N}}$ to the limit X as $n \to \infty$ therefore means that

$$\mathbb{E}(|X_n - X|^2) = \|X_n - X\|^2 \to 0 \quad \text{as } n \to \infty,$$

and is therefore often referred to as *mean-square convergence* or *convergence in mean-square*. We shall denote the mean-square limit X of the sequence $(X_n)_{n\in\mathbb{N}}$ by

$$X = \lim_{n\to\infty}{}^{\mathcal{S}} X_n$$

and also write

$$X_n \xrightarrow{\mathcal{S}} X.$$

The completeness of $\mathcal{S}$ is stated in the next theorem.

Theorem 2.17. *The inner-product space $\mathcal{S}$ is complete and is therefore a Hilbert space.*

Proof. See, e. g., Brockwell and Davis [22, Theorem 2.10.1]. □

Remark 2.18. It will occasionally be necessary to distinguish between the spaces $\mathcal{S}$ of real-valued and complex-valued random variables. The definition of the inner product is the same for both, but, of course, in the real case, complex conjugation has no effect. In both cases, $\mathcal{S}$ is a Hilbert space.

Remark 2.19. If $(\Omega, \mathcal{F}, \mu)$ is any measure space, then (see, Rudin [75, Example 4.5]) the space of complex-valued square-integrable functions on $(\Omega, \mathcal{F}, \mu)$, i. e., the measurable functions f satisfying

$$\int |f|^2 \, d\mu < \infty,$$

with identification of functions f_1 and f_2 such that $\mu(\{f_1 \neq f_2\}) = 0$, constitute a Hilbert space $L^2(\Omega, \mathcal{F}, \mu)$ if the inner product is defined by

$$\langle f, g\rangle := \int f\overline{g} \, d\mu.$$

We shall frequently encounter the special case when Ω is the real line $\mathbb{R}$, $\mathcal{F}$ consists of the Borel sets and μ is Lebesgue measure. The corresponding Hilbert space is then abbreviated as $L^2(\mathbb{R};\mathbb{C})$ or simply L^2 and the real Hilbert space of real-valued functions in L^2 by $L^2(\mathbb{R};\mathbb{R})$.

Proposition 2.20 (Norm convergence and Cauchy criterion). *If $(x_n)_{n\in\mathbb{N}}$ is a sequence of elements of a Hilbert space $\mathcal{H}$, then $(x_n)_{n\in\mathbb{N}}$ converges in norm if and only if $\|x_n - x_m\| \to 0$ as $m, n \to \infty$.*

Proof. The sufficiency of the Cauchy criterion is simply a restatement of the completeness of $\mathcal{H}$. The necessity is an elementary consequence of the triangle inequality. Thus if $\|x_n - x\| \to 0$, then

$$\|x_n - x_m\| \le \|x_n - x\| + \|x - x_m\| \to 0 \quad \text{as } m, n \to \infty,$$

proving the necessity. □

The Cauchy criterion is used primarily to check the norm convergence of a sequence whose limit is not specified.

In the following, for a stochastic process $(X_t)_{t\in\mathbb{R}}$ indexed by the real numbers, a random variable X and given $t_0 \in [-\infty, \infty]$, we write $X_t \xrightarrow{\mathcal{S}} X$ as $t \to t_0$ if $\mathbb{E}|X_t - X|^2 \to 0$ as $t \to t_0$, i. e. if $\mathbb{E}|X_{t_n} - X|^2 \to 0$ as $n \to \infty$ for all sequences $\{t_n\}$ converging to t_0, which in turn is equivalent to $X_{t_n} \xrightarrow{\mathcal{S}} X$ for all sequences $\{t_n\}$ converging to t_0.

The completeness of $\mathcal{S}$ leads to another useful criterion for convergence given in the following proposition.

Proposition 2.21. *If $(X_t)_{t\in\mathbb{R}}$ is a second-order process (i. e., $\mathbb{E}(|X_t|^2 < \infty$ for all t) and $t_0 \in [-\infty, \infty]$, then there exists $X \in \mathcal{S}$ such that $X_t \xrightarrow{\mathcal{S}} X$ as $t \to t_0$ if and only if there exists $K \in \mathbb{C}$ such that for all sequences $\{t_n\}$ and $\{t_n'\}$ converging to t_0,*

$$\mathbb{E}[X_{t_n}\overline{X_{t_m'}}] \to K, \quad \text{as } n, m \to \infty.$$

In that case, $K \in \mathbb{R}$.

Proof. If $X_t \xrightarrow{\mathcal{S}} X$ as $t \to t_0$, then, by continuity of the inner product, $\mathbb{E}[X_{t_n}\overline{X_{t_m'}}] \to \mathbb{E}(|X|^2) \in \mathbb{R}$ as $n, m \to \infty$.

If $\mathbb{E}[X_{t_n}\overline{X_{t_m'}}] \to K \in \mathbb{C}$ as $n, m \to \infty$ for all sequences, then, in particular, $\mathbb{E}[X_{t_n}\overline{X_{t_m}}] \to K$, so that $K \in \mathbb{R}$ (take $m = n$) and

$$\mathbb{E}[|X_{t_n} - X_{t_m}|^2] \to K - 2K + K = 0$$

and so by completeness the sequence $\{X_{t_n}\}$ converges in $\mathcal{S}$ to some $X \in \mathcal{S}$. For any other sequence $\{t_n'\}$ which converges to t_0, $\{X_{t_n'}\}$ must converge to the same limit, since

$$\mathbb{E}[|X_{t_n} - X_{t_n'}|^2] \to K - 2K + K = 0$$

and

$$\mathbb{E}[|X_{t_n'} - X|^2] \le 2(\mathbb{E}[|X_{t_n'} - X_{t_n}|^2] + \mathbb{E}[|X_{t_n} - X|^2]) \to 0, \quad \text{as } n \to \infty,$$

completing the proof. □

2.3 The Projection Theorem

We first illustrate the application of orthogonal projections in Hilbert space with two simple examples and then give a general result relating best linear approximation and orthogonal projection.

Example 2.22 (Linear approximation in $\mathbb{R}^3$). Suppose we are given three vectors in $\mathbb{R}^3$, namely $\mathbf{x}_1$, $\mathbf{x}_2$ and $\mathbf{y}$, and we wish to find the linear combination $\hat{\mathbf{y}} = a_1\mathbf{x}_1 + a_2\mathbf{x}_2$ which is closest to $\mathbf{y}$ in the sense that $\hat{\mathbf{y}}$ minimizes the Euclidean distance, $\|\mathbf{y} - \mathbf{z}\|$ over all $\mathbf{z}$ in the class of real linear combinations of the vectors $\mathbf{x}_1$ and $\mathbf{x}_2$, i. e., in the closed linear span, $\overline{\text{sp}}(\mathbf{x}_1, \mathbf{x}_2)$, of $\mathbf{x}_1$ and $\mathbf{x}_2$.

In this example, it is clear geometrically that the required vector $\hat{\mathbf{y}}$ must be such that the *error vector*, $\mathbf{y} - \hat{\mathbf{y}}$, is perpendicular to the subspace $\overline{\text{sp}}(\mathbf{x}_1, \mathbf{x}_2)$ of $\mathbb{R}^3$, consisting of all linear combinations of $\mathbf{x}_1$ and $\mathbf{x}_2$. In terms of the inner product defined in Example 2.3, this is equivalent to the statement that

$$\langle \mathbf{y} - a_1\mathbf{x}_1 - a_2\mathbf{x}_2, \mathbf{x}_i \rangle = 0, \quad i = 1, 2, \tag{2.4}$$

or equivalently,

$$\begin{aligned} a_1\langle \mathbf{x}_1, \mathbf{x}_1 \rangle + a_2\langle \mathbf{x}_2, \mathbf{x}_1 \rangle &= \langle \mathbf{y}, \mathbf{x}_1 \rangle, \\ a_1\langle \mathbf{x}_1, \mathbf{x}_2 \rangle + a_2\langle \mathbf{x}_2, \mathbf{x}_2 \rangle &= \langle \mathbf{y}, \mathbf{x}_2 \rangle. \end{aligned}$$

This is a pair of linear equations for the unknown coefficients a_1 and a_2 which we know has at least one solution since there *is* a vector in $\overline{\text{sp}}(\mathbf{x}_1, \mathbf{x}_2)$ that is closest in Euclidean distance to $\mathbf{y}$. (If $\mathbf{x}_1$ and $\mathbf{x}_2$ are linearly dependent there are infinitely many solutions for (a_1, a_2), but $\hat{\mathbf{y}} = a_1\mathbf{x}_1 + a_2\mathbf{x}_2$ is the same for all of them.) Notice that the solution to the problem, namely $\hat{\mathbf{y}}$, is completely characterized by the two conditions

$$\hat{\mathbf{y}} \in \overline{\text{sp}}(\mathbf{x}_1, \mathbf{x}_2)$$

and

$$\langle \mathbf{y} - \hat{\mathbf{y}}, \mathbf{z} \rangle = 0 \quad \text{for all } \mathbf{z} \in \overline{\text{sp}}(\mathbf{x}_1, \mathbf{x}_2).$$

The latter condition means that the *error vector* is orthogonal to every vector in the closed linear span of the vectors in terms of which the approximation is to be made. It can be restated as

$$\mathbf{y} - \hat{\mathbf{y}} \perp \overline{\text{sp}}(\mathbf{x}_1, \mathbf{x}_2).$$

Example 2.23 (Linear approximation in $\mathcal{S} = L^2(\Omega, \mathcal{F}, P)$). Now suppose that Y, X_1 and X_2 are real-valued random variables in $\mathcal{S}$. If only X_1 and X_2 are observed we may wish to estimate the value of Y by using the linear combination (with real coefficients), $\hat{Y} = a_1X_1 + a_2X_2$, which minimizes the mean-square error

$$\mathbb{E}(|Y - Z|^2) = \|Y - Z\|^2$$

over all Z in the class of real linear combinations of X_1 and X_2, i. e., in the closed linear span, $\overline{\mathrm{sp}}(X_1, X_2)$, of X_1 and X_2.

A direct approach to this problem is to minimize

$$S := \mathbb{E}(|Y - a_1X_1 - a_2X_2|^2)$$

directly with respect to a_1 and a_2 by setting the appropriate derivatives equal to zero. This gives the pair of linear equations,

$$\begin{aligned} a_1\mathbb{E}(X_1^2) + a_2\mathbb{E}(X_2X_1) &= E(YX_1), \\ a_1\mathbb{E}(X_1X_2) + a_2\mathbb{E}(X_2^2) &= E(YX_2), \end{aligned}$$

for the coefficients a_1 and a_2 of the minimum mean-square error approximation to Y.

However, it is possible to use the same geometric approach to this problem as used in Example 2.22. Our aim is to find an element $\hat{Y}$ in the set

$$\mathcal{M} := \{Z \in L^2(\Omega, \mathcal{F}, P) : Z = a_1X_1 + a_2X_2 \text{ for some } a_1, a_2 \in \mathbb{R}\},$$

whose *squared distance* from Y, i. e., $\|Y - Z\|^2$, is as small as possible. By analogy with Example 2.22 we might expect $\hat{Y}$ to have the property that $Y - \hat{Y}$ is orthogonal to all elements of $\mathcal{M}$. The validity of this analogy, and the extent to which it may be applied in more general situations is established in Theorem 2.27 (the Projection Theorem) below. Applying it in the present situation gives the equations for a_1 and a_2,

$$\langle Y - a_1X_1 - a_2X_2, Z\rangle = 0 \quad \text{for all } Z \in \mathcal{M}, \tag{2.5}$$

or equivalently

$$\langle Y - a_1X_1 - a_2X_2, X_i\rangle = 0, \quad i = 1, 2,$$

which are precisely the equations derived above by direct minimization of the error sum of squares, S.

For the statement of the Projection Theorem, we still need some notation.

Definition 2.24 (Closed subspace). A linear subspace $\mathcal{M}$ of a Hilbert space $\mathcal{H}$ is said to be a *closed subspace* of $\mathcal{H}$ if $\mathcal{M}$ contains all of its limit points (i. e., if $x_n \in \mathcal{M}, x \in \mathcal{M}$ and $\|x_n - x\| \to 0$, then $x \in \mathcal{M}$).

Definition 2.25 (Orthogonal complement). The *orthogonal complement* of a subset $\mathcal{M}$ of $\mathcal{H}$ is defined as the set $\mathcal{M}^\perp$ of all elements of $\mathcal{H}$ that are orthogonal to every element of $\mathcal{M}$. Thus

$$x \in \mathcal{M}^\perp \quad \text{if and only if} \quad \langle x, y\rangle = 0 \text{ (written } x \perp y) \quad \text{for all } y \in \mathcal{M}.$$

Proposition 2.26. *If $\mathcal{M}$ is any subset of a Hilbert space $\mathcal{H}$, then $\mathcal{M}^\perp$ is a closed subspace of $\mathcal{H}$.*

Proof. Exercise 2.4. □

Theorem 2.27 (The Projection Theorem). *Let $\mathcal{M}$ be a closed subspace of the Hilbert space $\mathcal{H}$ and $x \in \mathcal{H}$.*
(a) *Then there is a unique element $\hat{x} \in \mathcal{M}$ such that*

$$\|x - \hat{x}\| = \inf_{y\in\mathcal{M}} \|x - y\|.$$

(b) *The element $\hat{x}$ from (a) is characterized by the two conditions*

$$\hat{x} \in \mathcal{M} \quad \textit{and} \quad (x - \hat{x}) \in \mathcal{M}^\perp.$$

The element $\hat{x}$ is called the (orthogonal) projection of x onto $\mathcal{M}$.

Proof. See, e. g., Brockwell and Davis [22, Theorem 2.3.1]. □

Corollary 2.28 (The projection mapping of $\mathcal{H}$ onto a closed subspace). *If $\mathcal{M}$ is a closed subspace of the Hilbert space $\mathcal{H}$ and I is the identity mapping on $\mathcal{H}$, then there is a unique mapping $P_\mathcal{M}$ from $\mathcal{H}$ to $\mathcal{M}$ such that $I - P_\mathcal{M}$ maps $\mathcal{H}$ to $\mathcal{M}^\perp$. The mapping $P_\mathcal{M}$* is onto *$\mathcal{M}$ (i. e. $P_\mathcal{M} : \mathcal{H} \to \mathcal{M}$ is surjective) and $I - P_\mathcal{M}$* is onto *$\mathcal{M}^\perp$. The mapping $P_\mathcal{M}$* is called *the* (orthogonal) projection (mapping) of $\mathcal{H}$ onto $\mathcal{M}$.

Proof. By Theorem 2.27, for each $x \in \mathcal{H}$, there is a unique $\hat{x} \in \mathcal{M}$ such that $x - \hat{x} \in \mathcal{M}^\perp$. The mapping,

$$P_\mathcal{M}x := \hat{x}, \quad x \in \mathcal{H},$$

is therefore the unique mapping with the specified properties. That $P_\mathcal{M}$ is onto $\mathcal{M}$ follows from the fact that $P_\mathcal{M}x = x$ for all $x \in \mathcal{M}$, and that $I - P_\mathcal{M}$ maps $\mathcal{H}$ onto $\mathcal{M}^\perp$ from the fact that $P_\mathcal{M}x = 0$ for $x \in \mathcal{M}^\perp$. □

Proposition 2.29 (Properties of projection mappings). *Let $\mathcal{H}$ be a Hilbert space and let $P_\mathcal{M}$ denote the projection mapping onto a closed subspace $\mathcal{M}$. Then*
(a) $P_\mathcal{M}(\alpha x + \beta y) = \alpha P_\mathcal{M}x + \beta P_\mathcal{M}y$, *for all $x, y \in \mathcal{H}$, $\alpha, \beta \in \mathbb{C}$,*
(b) $\|x\|^2 = \|P_\mathcal{M}x\|^2 + \|(I - P_\mathcal{M})x\|^2$,

(c) *each* $x \in \mathcal{H}$ *has a unique representation as a sum of an element of* $\mathcal{M}$ *and an element of* $\mathcal{M}^{\perp}$, *namely*

$$x = P_{\mathcal{M}}x + (I - P_{\mathcal{M}})x,$$

(d) $P_{\mathcal{M}}x_n \to P_{\mathcal{M}}x$ *if* $\|x_n - x\| \to 0$ *as* $n \to \infty$,
(e) $x \in \mathcal{M}$ *if and only if* $P_{\mathcal{M}}x = x$,
(f) $x \in \mathcal{M}^{\perp}$ *if and only if* $P_{\mathcal{M}}x = 0$,
(g) $(\mathcal{M}^{\perp})^{\perp} = \mathcal{M}$ *and* $P_{\mathcal{M}^{\perp}} = I - P_{\mathcal{M}}$,
and
(h) $\mathcal{M}_1 \subseteq \mathcal{M}_2$ *if and only if* $P_{\mathcal{M}_1}P_{\mathcal{M}_2}x = P_{\mathcal{M}_1}x$ *for all* $x \in \mathcal{H}$.

Proof. For (a)–(f) and (h), see, e. g., Brockwell and Davis [22, Proposition 2.3.2].

(g) Since $(I - P_{\mathcal{M}})x \in \mathcal{M}^{\perp}$ and $I - (I - P_{\mathcal{M}})x = P_{\mathcal{M}}x \in \mathcal{M}$ is orthogonal to $\mathcal{M}^{\perp}$ for all $x \in \mathcal{H}$, we see that $P_{\mathcal{M}^{\perp}} = I - P_{\mathcal{M}}$. Hence $P_{(\mathcal{M}^{\perp})^{\perp}} = I - P_{\mathcal{M}^{\perp}} = P_{\mathcal{M}}$, which clearly implies $(\mathcal{M}^{\perp})^{\perp} = \mathcal{M}$. □

Given a Hilbert space $\mathcal{H}$, a closed subspace $\mathcal{M}$ and an element $x \in \mathcal{H}$, Theorem 2.27 shows that the element of $\mathcal{M}$ closest to x is the unique element $\hat{x} \in \mathcal{M}$ such that

$$\langle x - \hat{x}, y \rangle = 0 \quad \text{for all } y \in \mathcal{M}. \tag{2.6}$$

The equations (2.4) and (2.5), which arose in Examples 2.22 and 2.23, are special cases of (2.6). In later chapters, we shall make extensive use of the equations (2.6), interpreting $\hat{x}$ as the best predictor or best approximator of x in the subspace $\mathcal{M}$. Equations (2.6) are called the *prediction equations*.

Example 2.30 (Linear prediction of stationary sequence). (This example is taken directly from [22, Example 2.3.3].) Let $(X_n)_{n\in\mathbb{Z}}$ be a stationary sequence on $(\Omega, \mathcal{F}, P)$, with mean zero and autocovariance function $\mathrm{Cov}(X_{n+h}, X_n) = \gamma(h)$, $n, h \in \mathbb{Z}$. Consider the problem of finding the linear combination $\hat{X}_{n+1} = \sum_{j=1}^{n} \phi_{nj}X_{n+1-j}$ which best approximates X_{n+1} in the sense that $\mathbb{E}|X_{n+1} - \sum_{j=1}^{n} \phi_{nj}X_{n+1-j}|^2$ is minimum. This problem is easily solved with the aid of the Projection Theorem by taking $\mathcal{H} = \mathcal{S} = L^2(\Omega, \mathcal{F}, P)$ and $\mathcal{M} = \{\sum_{j=1}^{n} a_jX_{n+1-j} : a_1, \dots, a_n \in \mathbb{C}\}$. Since minimization of $\mathbb{E}|X_{n+1} - \hat{X}_{n+1}|^2$ is identical to the minimization of the squared norm $\|X_{n+1} - \hat{X}_{n+1}\|^2$, we see at once that $\hat{X}_{n+1} = P_{\mathcal{M}}X_{n+1}$. The prediction equations (2.6) are

$$\left\langle X_{n+1} - \sum_{j=1}^{n} \phi_{nj}X_{n+1-j}, Y \right\rangle = 0, \quad \text{for all } Y \in \mathcal{M},$$

which, by the linearity of the inner product, are equivalent to the n equations,

$$\left\langle X_{n+1} - \sum_{j=1}^{n} \phi_{nj} X_{n+1-j}, X_k \right\rangle = 0, \quad k = n, n-1, \ldots, 1.$$

Recalling the definition $\langle X, Y\rangle = \mathbb{E}(X\overline{Y})$ of the inner product in $\mathcal{S}$, we see that the prediction equations can be written in the form,

$$\Gamma_n \boldsymbol{\phi}_n = \boldsymbol{\gamma}_n \tag{2.7}$$

where $\boldsymbol{\phi}_n = (\phi_{n1}, \ldots, \phi_{nn})^T$, $\boldsymbol{\gamma}_n = (\gamma(1), \ldots, \gamma(n))^T$ and $\Gamma_n = [\gamma(i-j)]_{i,j=1,\ldots,n}$. By the Projection Theorem there is at least one solution $\boldsymbol{\phi}_n$ of (2.7). If Γ_n is singular, then (2.7) has infinitely many solutions, but the Projection Theorem guarantees that every solution yields the same (uniquely defined) predictor $\hat{X}_{n+1}$.

2.4 Orthonormal sets

Definition 2.31 (Closed span). The *closed span* $\overline{\mathrm{sp}}\{x_s, s \in S\}$ of any subset $\{x_s, s \in S\}$ of a Hilbert space $\mathcal{H}$ is defined to be the smallest closed subspace of $\mathcal{H}$, which contains x_s for all $s \in S$.

Remark 2.32. The closed span of a finite set $\{x_1, \ldots, x_n\}$ is the set of all linear combinations, $y = \alpha_1 x_1 + \cdots + \alpha_n x_n$, $\alpha_1, \ldots, \alpha_n \in \mathbb{C}$ (or $\mathbb{R}$ if $\mathcal{H}$ is real), see Exercise 2.6. If for example $\mathbf{x}_1, \mathbf{x}_2 \in \mathbb{R}^3 \setminus \{0\}$ and $\mathbf{x}_1$ is not a scalar multiple of $\mathbf{x}_2$ then $\overline{\mathrm{sp}}\{\mathbf{x}_1, \mathbf{x}_2\}$ is the plane containing $\mathbf{x}_1$ and $\mathbf{x}_2$.

Remark 2.33. If $\mathcal{M} = \overline{\mathrm{sp}}\{x_1, \ldots, x_n\}$, then for any given $x \in \mathcal{H}$, $P_{\mathcal{M}}x$ is the unique element of the form

$$P_{\mathcal{M}}x = \alpha_1 x_1 + \cdots + \alpha_n x_n,$$

such that

$$\langle x - P_{\mathcal{M}}x, y\rangle = 0 \quad \text{for all } y \in \mathcal{M},$$

or equivalently such that

$$\langle x - P_{\mathcal{M}}x, x_j\rangle = 0, \quad j = 1, \ldots, n. \tag{2.8}$$

The equations (2.8) can be rewritten as a set of linear equations for $\alpha_1, \ldots, \alpha_n$, as

$$\sum_{i=1}^{n} \alpha_i \langle x_i, x_j\rangle = \langle x, x_j\rangle, \quad j = 1, \ldots, n. \tag{2.9}$$

By the Projection Theorem, the system (2.9) has at least one solution for $\alpha_1, \dots, \alpha_n$. The uniqueness of $P_{\mathcal{M}}x$ implies that all solutions of (2.9) must yield the same element $\alpha_1 x_1 + \dots + \alpha_n x_n$.

Definition 2.34 (Orthonormal set). A set $\{e_s, s \in S\}$ of elements of an inner-product space is said to be *orthonormal* if for every $r, s \in S$,

$$\langle e_r, e_s \rangle = \begin{cases} 1, & \text{if } r = s, \\ 0, & \text{if } r \neq s. \end{cases}$$

Example 2.35. (a) The set of vectors $\{(1,0,0)^T, (0,1,0)^T, (0,0,1)^T\}$ is an orthonormal set in $\mathbb{R}^3$.

(b) Every white noise sequence $(Z_n)_{n\in\mathbb{Z}} \sim WN(0,1)$ with variance 1 is an orthonormal set in $\mathcal{S}$.

From the Cauchy criterion, we obtain the following criterion for convergence of sequences of the form $\sum_{n=0}^{\infty} a_n e_n$ for an orthonormal set $\{e_n; n \in \mathbb{N}_0\}$.

Proposition 2.36. *Let $\{e_n; n \in \mathbb{N}_0\}$ be an orthonormal set in a Hilbert space $\mathcal{H}$ and $a_0, a_1, \dots \in \mathbb{C}$. Then the sequence $\sum_{j=0}^{n} a_j e_j$ converges in norm as $n \to \infty$ if and only if $\sum_{j=0}^{\infty} |a_j|^2 < \infty$, in which case the limit is denoted by $\sum_{j=0}^{\infty} a_j e_j$.*

Proof. Denote $x_n := \sum_{j=0}^{n} a_j e_j$. Then $\|x_n - x_m\| = \sum_{j=m+1}^{n} |a_j|^2$ for $m \le n$, hence the claim follows from Proposition 2.20. □

Theorem 2.37. *If $\{e_1, \dots, e_k\}$ is an orthonormal subset of the Hilbert space $\mathcal{H}$ and $\mathcal{M} = \overline{\mathrm{sp}}\{e_1, \dots, e_k\}$, then*

$$P_{\mathcal{M}}x = \sum_{i=1}^{k} \langle x, e_i \rangle e_i \quad \text{for all } x \in \mathcal{H},$$

$$\|P_{\mathcal{M}}x\|^2 = \sum_{i=1}^{k} |\langle x, e_i \rangle|^2 \quad \text{for all } x \in \mathcal{H}, \tag{2.10}$$

$$\left\| x - \sum_{i=1}^{k} \langle x, e_i \rangle e_i \right\| \le \left\| x - \sum_{i=1}^{k} c_i e_i \right\| \quad \text{for all } x \in \mathcal{H} \tag{2.11}$$

and for all $c_1, \dots, c_k \in \mathbb{C}$ (or $\mathbb{R}$ if $\mathcal{H}$ is real). Equality holds in (2.11) if and only if $c_i = \langle x, e_i \rangle$, $i = 1, \dots, k$.

The numbers $\langle x, e_i \rangle$ are sometimes called the Fourier coefficients of x relative to the set $\{e_1, \dots, e_k\}$.

Proof. See, e. g., Brockwell and Davis [22, Theorem 2.4.1]. □

Corollary 2.38 (Bessel's inequality). *If x is any element of a Hilbert space $\mathcal{H}$ and $\{e_1, \dots, e_k\}$ is an orthonormal subset of $\mathcal{H}$, then*

$$\sum_{i=1}^{k} |\langle x, e_i \rangle|^2 \le \|x\|^2.$$

Proof. This follows at once from (2.10) and Proposition 2.29(b). □

Definition 2.39 (Complete orthonormal set). If $\{e_s, s \in S\}$ is an orthonormal subset of the Hilbert space $\mathcal{H}$ and if $\mathcal{H} = \overline{\text{sp}}\{e_s, s \in S\}$, then we say that $\{e_s, s \in S\}$ is a *complete orthonormal set* or an *orthonormal basis for* $\mathcal{H}$.

Definition 2.40 (Separability). The Hilbert space $\mathcal{H}$ is *separable* if $\mathcal{H} = \overline{\text{sp}}\{e_s, s \in S\}$ with $\{e_s, s \in S\}$ a finite or countably infinite orthonormal set.

Theorem 2.41 (Parseval's identity and other properties). *If $\mathcal{H}$ is a separable Hilbert space $\mathcal{H} = \overline{\text{sp}}\{e_i, i = 1, 2, \ldots\}$, where $\{e_i, i = 1, 2, \ldots\}$ is a countably infinite orthonormal set, then*

(a) *the set of all finite linear combinations of $\{e_1, e_2, \ldots\}$ is dense in $\mathcal{H}$, i. e., for each $x \in \mathcal{H}$ and $\epsilon > 0$, there exists a positive integer k and constants $c_1, \ldots, c_k$ such that*

$$\left\| x - \sum_{i=1}^{k} c_i e_i \right\| < \epsilon,$$

(b) $x = \sum_{i=1}^{\infty} \langle x, e_i \rangle e_i$ *for each $x \in \mathcal{H}$, i. e., $\|x - \sum_{i=1}^{n} \langle x, e_i \rangle e_i\| \to 0$ as $n \to \infty$,*
(c) $\|x\|^2 = \sum_{i=1}^{\infty} |\langle x, e_i \rangle|^2$ *for each $x \in \mathcal{H}$,*
(d) $\langle x, y \rangle = \sum_{i=1}^{\infty} \langle x, e_i \rangle \langle e_i, y \rangle$ *for each $x, y \in \mathcal{H}$, and*
(e) $x = 0$ *if and only if $\langle x, e_i \rangle = 0$ for all $i \in \{1, 2, \ldots\}$.*

The result (d) *is known as* Parseval's identity.

Proof. See, e. g., Brockwell and Davis [22, Theorem 2.4.2]. □

Remark 2.42. Separable Hilbert spaces are frequently encountered as the closed spans of countable subsets of possibly non-separable Hilbert spaces.

2.5 Mean-square convergence, conditional expectation and best linear prediction in $\mathcal{S}$

We are concerned in this section with the complex Hilbert space $\mathcal{S} = L^2(\Omega, \mathcal{F}, P)$ with inner product $\langle X, Y \rangle = \mathbb{E}(X\overline{Y})$. In the special case that the random values are real-valued, the same results apply, but of course complex conjugation can be ignored. As indicated in Example 2.16, convergence of a sequence $\{X_n\}$ in $\mathcal{S}$ (often referred to as mean-square convergence) means that there exists $X \in \mathcal{S}$ such that

$$\|X_n - X\|^2 = \mathbb{E}(|X_n - X|^2) \to 0 \quad \text{as } n \to \infty,$$

written more concisely as $X_n \overset{\mathcal{S}}{\to} X$ or $X = \lim^{\mathcal{S}}_{n\to\infty} X_n$. By simply restating properties already established for norm convergence, we obtain the following proposition.

Proposition 2.43 (Properties of mean-square convergence).
(a) *X_n converges in mean-square as $n \to \infty$ if and only if $\mathbb{E}(|X_n - X_m|^2) \to 0$ as $m, n \to \infty$.*
(b) *If $X_n \overset{\mathcal{S}}{\to} X$ and $Y_n \overset{\mathcal{S}}{\to} Y$, then as $n \to \infty$,*
  (i) $\mathbb{E}X_n = \langle X_n, 1\rangle \to \langle X, 1\rangle = \mathbb{E}X$,
  (ii) $\mathbb{E}(|X_n|^2) = \langle X_n, X_n\rangle \to \langle X, X\rangle = \mathbb{E}(|X|^2)$,
  and
  (iii) $\mathbb{E}(X_n\overline{Y_n}) = \langle X_n, Y_n\rangle \to \langle X, Y\rangle = \mathbb{E}(X\overline{Y})$.

Definition 2.44 (Best mean-square predictor of Y). If $\mathcal{M}$ is a closed subspace of $\mathcal{S}$ and $Y \in \mathcal{S}$, then the *best mean-square predictor of Y in $\mathcal{M}$* is the element $\hat{Y} \in \mathcal{M}$ such that

$$\|Y - \hat{Y}\|^2 = \inf_{Z\in\mathcal{M}} \|Y - Z\|^2 = \inf_{Z\in\mathcal{M}} \mathbb{E}(|Y - Z|^2).$$

The Projection Theorem immediately identifies the unique best mean-square predictor of Y in $\mathcal{M}$ as $P_{\mathcal{M}}Y$. By imposing a little more structure on the closed subspace $\mathcal{M}$, we are led from Definition 2.44 to the notions of conditional expectation and best linear predictor.

Let us recall the definition of the conditional expectation, as it can be found, e. g., in Klenke [57, Definitions 8.11 and 8.13].

Definition 2.45 (Conditional expectations $\mathbb{E}(X|\mathcal{G})$ and $\mathbb{E}(X|Z_s; s \in \Lambda)$). Let $(\Omega, \mathcal{F}, P)$ be a probability space and $\mathcal{G} \subset \mathcal{F}$ be a sub-σ-algebra of $\mathcal{F}$. If X is an integrable random variable on $(\Omega, \mathcal{F}, P)$, then the *conditional expectation of X given $\mathcal{G}$* is defined to be any $\mathcal{G}$-measurable integrable random variable Y on Ω such that

$$\int_A X \, \mathrm{d}P = \int_A Y \, \mathrm{d}P \quad \forall\, A \in \mathcal{G}. \tag{2.12}$$

The random variable Y is denoted by $\mathbb{E}(X|\mathcal{G})$. If $\{Z_s; s \in \Lambda\}$ is a collection of random vectors (or more generally, random elements) on $(\Omega, \mathcal{F}, P)$ and X is an integrable random variable, then the *conditional expectation of X given $\{Z_s; s \in \Lambda\}$* is defined as $\mathbb{E}(X|Z_s; s \in \Lambda) := \mathbb{E}(X|\sigma(Z_s; s \in \Lambda))$, i. e., as the conditional expectation of X given the σ-algebra generated by $\{Z_s; s \in \Lambda\}$.

It is known that the conditional expectation always exists and is unique (up to almost sure equality), see, e. g., Klenke [57, Theorem 8.12]. For a sub-σ-algebra $\mathcal{G} \subset \mathcal{F}$, consider $\mathcal{M} = L^2(\Omega, \mathcal{G}, P)$, the space of all (equivalence classes of) $\mathbb{C}$-valued $\mathcal{G}$-measurable square-integral random variables on $(\Omega, \mathcal{F}, P)$. Then $\mathcal{M}$ is a linear subspace of $\mathcal{S} = L^2(\Omega, \mathcal{F}, P)$ and it is easily seen to be closed. For $X \in \mathcal{S}$ consider the best mean-square predictor $P_{\mathcal{M}}X$ of X in $\mathcal{M}$. Then $P_{\mathcal{M}}X \in \mathcal{M}$ and $\mathbb{E}((X - P_{\mathcal{M}}X)\overline{H}) = 0$ for all $H \in \mathcal{M}$, in particular, choosing $H = \mathbf{1}_A$ for $A \in \mathcal{G}$ we obtain $\mathbb{E}((X - P_{\mathcal{M}}X)\mathbf{1}_A) = 0$, i. e., $\int_A X \, \mathrm{d}P = \int_A P_{\mathcal{M}}X \, \mathrm{d}P$.

Hence $P_{\mathcal{M}}X$ is $\mathcal{G}$-measurable and satisfies equation (2.12). From the uniqueness, we conclude that the conditional expectation $\mathbb{E}(X|\mathcal{G})$ of $X \in \mathcal{S}$ given $\mathcal{G}$ is nothing else than the projection of X onto $\mathcal{M}$. In particular, $\mathbb{E}(X|\mathcal{G}) \in \mathcal{S}$ when $X \in \mathcal{S}$. The conditional expectation $\mathbb{E}(X|Z_s; s \in \Lambda)$ is often simply called the *best predictor* of X given $\{Z_s; s \in \Lambda\}$.

Remark 2.46 (Conditional expectation and best linear prediction). We have seen that the conditional expectation $\mathbb{E}(X|Z_1, \dots, Z_n)$ of $X \in \mathcal{S}$ is the minimum mean-square error predictor of X in $L^2(\Omega, \sigma(Z_1, \dots, Z_n), P)$, i. e., by the factorization lemma (e. g., Klenke [57, Corollary 1.97]) it is the best (in the mean-square sense) Borel function of $Z_1, \dots, Z_n$ for predicting X. However, the determination of conditional expectations is usually very difficult. On the other hand, if $Z_1, \dots, Z_n \in \mathcal{S}$, it is relatively easy to compute instead the projection of X on the closed *linear* span $\overline{\text{sp}}\{Z_0, Z_1, \dots, Z_n\} \subseteq L^2(\Omega, \sigma(Z_1, \dots, Z_n), P)$, where $Z_0 := 1$, since we can write

$$P_{\overline{\text{sp}}\{Z_0, Z_1, \dots, Z_n\}}(X) = \sum_{i=0}^{n} \alpha_i Z_i, \tag{2.13}$$

where $\alpha_0, \dots, \alpha_n \in \mathbb{C}$ satisfy

$$\left\langle \sum_{i=0}^{n} \alpha_i Z_i, Z_j \right\rangle = \langle X, Z_j \rangle, \quad j = 0, \dots, n,$$

or equivalently,

$$\sum_{i=0}^{n} \alpha_i \mathbb{E}(Z_i \overline{Z_j}) = \mathbb{E}(X \overline{Z_j}), \quad j = 0, \dots, n. \tag{2.14}$$

The Projection Theorem guarantees that a solution $(\alpha_0, \dots, \alpha_n)$ exists. Any solution, when substituted into (2.13) gives the required projection, known as the *best linear predictor* of X in terms of $1, Z_1, \dots, Z_n$. As a projection of X onto a *subspace* of $L^2(\Omega, \sigma(Z_1, \dots, Z_n), P)$, it can never have smaller mean-square error than $\mathbb{E}(X|Z_1, \dots, Z_n)$. Nevertheless, it is of great importance for the following reasons:

(a) it is easier to calculate than $\mathbb{E}(X|Z_1, \dots, Z_n)$,

(b) it depends only on the first and second moments, $\mathbb{E}X$, $\mathbb{E}Z_,$ $\mathbb{E}(Z_i \overline{Z_j})$ and $\mathbb{E}(X \overline{Z_j})$ of the joint distribution of $(X, Z_1, \dots, Z_n)$,

(c) if $(X, Z_1, \dots, Z_n)'$ has a multivariate normal distribution, then (see Exercise 2.11),

$$P_{\overline{\text{sp}}\{Z_0, Z_1, \dots, Z_n\}}(X) = \mathbb{E}(X|Z_1, \dots, Z_n).$$

Definition 2.47 (Best linear predictor of X in terms of $\{Z_s; s \in \Lambda\}$). Let $\Lambda \neq \emptyset$. If $X \in \mathcal{S}$ and $Z_s \in \mathcal{S}$ for all $s \in \Lambda$, then the *best linear predictor (BLP) of X in terms of* $\{Z_s;\ s \in \Lambda\}$ is defined to be the element of $\overline{\text{sp}}\{Z_s,\ s \in \Lambda\}$ with smallest mean-square distance from X. By the Projection Theorem this is just $P_{\overline{\text{sp}}\{Z_s,\ s \in \Lambda\}}X$.

Example 2.48. Suppose that $Y = X^3Z^2$ where X and Z are independent standard normal random variables. The best predictor of Y in terms of X is $\mathbb{E}(Y|X) = X^3$. On the other hand, the best linear predictor of Y in terms of $\{1, X\}$ is

$$P_{\overline{\text{sp}}\{1,X\}}Y = aX + b,$$

where, by the prediction equations (2.14),

$$\langle aX + b, X\rangle = \langle Y, X\rangle = \mathbb{E}(YX) = \mathbb{E}(X^4Z^2) = 3$$

and

$$\langle aX + b, 1\rangle = \langle Y, 1\rangle = \mathbb{E}Y = 0.$$

Hence $a\mathbb{E}X^2 + b\mathbb{E}X = 3$ and $a\mathbb{E}X + b = 0$, so that $a = 3$ and $b = 0$, hence $P_{\overline{\text{sp}}\{1,X\}}Y = 3X$. The mean-square errors of the two predictors are $\|Y - \mathbb{E}(Y|X)\|^2 = \mathbb{E}(X^6(Z^2 - 1)^2) = 30$ and $\|Y - P_{\overline{\text{sp}}\{1,X\}}Y\|^2 = \mathbb{E}(X^3Z^2 - 3X)^2 = 36$, showing the superiority of the best predictor over the best linear predictor in this case.

2.6 Best linear estimation of random vectors in $\mathcal{S}$

The random variables $X \in \mathcal{S} = L^2(\Omega, \mathcal{F}, P)$ constitute a Hilbert space with inner product defined by

$$\langle X, Y\rangle := \mathbb{E}(X\overline{Y}).$$

In this section, we are concerned with random vectors whose components all belong to $\mathcal{S}$, or equivalently with random vectors $\mathbf{X} = (X_1, \dots, X_n)^T$ such that $\sum_{k=1}^n \mathbb{E}(|X_k|^2) < \infty$.

Definition 2.49. If $\mathbf{X}$ is a random vector $\mathbf{X} = (X_1, \dots, X_n)^T$ such that each component $X_i \in \mathcal{S}$, and $\mathcal{M}$ is any closed subspace of $\mathcal{S}$, we shall say, by an abuse of notation whose meaning should nevertheless be clear and unambiguous, that

$$\mathbf{X} \in \mathcal{M} \quad \text{if } X_i \in \mathcal{M}, i = 1, \dots, n,$$

and

$$\mathbf{X} \perp \mathcal{M} \quad \text{if, for all } Y \in \mathcal{M}, \langle X_i, Y\rangle = 0, i = 1, \dots, n.$$

In other words, when speaking of inclusion and orthogonality of random vectors (indicated in bold-face) relative to closed subspaces of $\mathcal{S}$, we shall always mean *component-wise* inclusion and orthogonality respectively.

Definition 2.50. Keeping in mind Definition 2.49, we define the *componentwise (orthogonal) projection* of $\mathbf{X}$ onto $\mathcal{M}$ by

$$\mathcal{P}_{\mathcal{M}}\mathbf{X} := (P_{\mathcal{M}}X_1, \dots, P_{\mathcal{M}}X_n)^T. \tag{2.15}$$

It follows from Theorem 2.27 that $\mathcal{P}_{\mathcal{M}}\mathbf{X}$ is characterized by the properties,

$$\mathcal{P}_{\mathcal{M}}\mathbf{X} \in \mathcal{M} \tag{2.16}$$

and

$$\mathbf{X} - \mathcal{P}_{\mathcal{M}}\mathbf{X} \perp \mathcal{M}, \tag{2.17}$$

and that $\mathcal{P}_{\mathcal{M}}\mathbf{X}$ is the unique random vector in $\mathcal{M}$ satisfying

$$\|\mathbf{X} - \mathcal{P}_{\mathcal{M}}\mathbf{X}\|^2 = \min_{\mathbf{U}\in\mathcal{M}} \|\mathbf{X} - \mathbf{U}\|^2,$$

where, for any $\mathbf{V} = (V_1, \dots, V_n)^T \in \mathcal{S}$ and any integer n, we define

$$\|\mathbf{V}\|^2 = \sum_{k=1}^{n} \mathbb{E}|V_k|^2.$$

We shall refer to $\mathcal{P}_{\mathcal{M}}\mathbf{X}$ as the *best linear estimate* of $\mathbf{X}$ in $\mathcal{M}$.

Example 2.51. If $\mathbf{Y} = (Y_1, \dots, Y_m)^T$ and $\mathbf{X} = (X_1, \dots, X_n)^T$ are random vectors all of whose components are in $\mathcal{S}$ and $\mathcal{M}_Y$ is the closed subspace $\overline{\text{sp}}\{Y_1, \dots, Y_m\}$ of $\mathcal{S}$, then clearly $\mathbf{X} \in \mathcal{M}_Y$ if and only if there is a matrix $C \in \mathbb{C}^{n\times m}$ such that $\mathbf{X} = C\mathbf{Y}$. Moreover $\mathbf{X} \perp \mathcal{M}_Y$ if and only if the matrix $\mathbb{E}(\mathbf{X}\overline{\mathbf{Y}}^T)$ is zero. By (2.16) and (2.17), the componentwise projection of $\mathbf{X}$ onto $\mathcal{M}_Y$ is given by,

$$\mathcal{P}_{\mathcal{M}_Y}\mathbf{X} = \Phi\mathbf{Y},$$

where Φ is any $n \times m$ matrix satisfying

$$\Phi[\mathbb{E}(Y_i\overline{Y_j})]_{i,j=1,\dots,m} = [\mathbb{E}(X_i\overline{Y_j})]_{i=1,\dots,n;j=1,\dots,m}. \tag{2.18}$$

If the matrix $[\mathbb{E}(Y_i\overline{Y_j})]_{i,j=1,\dots,m}$ is non-singular the matrix Φ is uniquely defined by (2.18). If not, there will be multiple solutions for Φ, but $\Phi\mathbf{Y}$ will be the same for all of them.

Example 2.52. Suppose now that $\mathbf{Y}_s = (Y_{s1}, \dots, Y_{sm})^T$, $s \in \mathbb{R}$, is an m-variate second-order process with $\mathbb{E}(\mathbf{Y}_s) = \mathbf{0}$ and $\mathbf{X} = (X_1, \dots, X_n)^T$ is a random vector in $\mathcal{S}$. Let $\mathcal{S}_t^Y$ denote the closed linear span in $\mathcal{S}$ of the components of $(\mathbf{Y}_s)_{s\leq t}$, i. e., the set of all linear combinations, $\sum_{j=1}^k C_j\mathbf{Y}_{t_j}$, with $k \in \mathbb{N}$, $t_j \leq t$, and C_j a $1 \times m$ matrix with complex-valued components, together with the mean-square limits of all such linear combinations.

Taking $\mathcal{M} = \mathcal{S}_t^Y$ in (2.15), we shall write $\hat{\mathbf{X}}_t$ for $\mathcal{P}_{\mathcal{M}}\mathbf{X}$. Then $\hat{\mathbf{X}}_t$ is the random n-component vector in $\mathcal{S}_t^Y$, each of whose components have minimum mean-square error as estimates of the corresponding components of $\mathbf{X}$. By (2.17), $\hat{\mathbf{X}}_t$ is the unique n-component random vector in $\mathcal{S}_t^Y$ such that

$$\mathbb{E}(\hat{\mathbf{X}}_t\overline{\mathbf{Y}_s}^T) = \mathbb{E}(\mathbf{X}\overline{\mathbf{Y}_s}^T), \quad \text{for all } s \le t.$$

We shall later consider the determination of $\hat{\mathbf{X}}_t$ for various classes of processes $\mathbf{Y}_s$, $s \in \mathbb{R}$.

2.7 Hilbert space isomorphisms

Definition 2.53 (Isomorphism). An *isomorphism* of the Hilbert space $\mathcal{H}_1$ onto the Hilbert space $\mathcal{H}_2$ is a one-to-one mapping T of $\mathcal{H}_1$ onto $\mathcal{H}_2$ such that for all $f_1, f_2 \in \mathcal{H}_1$,
(a) $T(af_1 + bf_2) = aTf_1 + bTf_2$ for all scalars a and b
and
(b) $\langle Tf_1, Tf_2\rangle = \langle f_1, f_2\rangle$.

We say that $\mathcal{H}_1$ and $\mathcal{H}_2$ are *isomorphic* if there is an isomorphism T of $\mathcal{H}_1$ onto $\mathcal{H}_2$. The inverse mapping T^{-1} is then an isomorphism of $\mathcal{H}_2$ onto $\mathcal{H}_1$.

Remark 2.54. Conditions (a) and (b) are equivalent (Exercise 2.15) to the conditions (a) and
(c) $\|Tf\| = \|f\|$ for all $f \in \mathcal{H}$.

For this reason, the isomorphism is often referred to as a *linear isometry*.

Example 2.55 (The space l^2). Let l^2 denote the complex Hilbert space of sequences $\{z_n, n = 1, 2, \ldots\}$, $z_n \in \mathbb{C}$, $\sum_{n=1}^{\infty} |z_n|^2 < \infty$, with inner product

$$\langle \{y_n\}, \{z_n\}\rangle = \sum_{i=1}^{\infty} y_i \overline{z_i}.$$

(For the proof that l^2 is a separable Hilbert space see Exercise 2.13.) If now $\mathcal{H}$ is any Hilbert space with an orthonormal basis $\{e_n, n = 1, 2, \ldots\}$, then the mapping $T : \mathcal{H} \to l^2$ defined by

$$Th = \{\langle h, e_n\rangle\}$$

is an isomorphism of $\mathcal{H}$ onto l^2 (see Exercise 2.14). Thus every separable infinite-dimensional Hilbert space is isomorphic to l^2.

Properties of isomorphisms. Suppose T is an isomorphism of $\mathcal{H}_1$ onto $\mathcal{H}_2$. We then have the following properties, all of which follow at once from the definitions:

(i) If $\{e_n\}$ is a (complete) orthonormal set in $\mathcal{H}_1$ then $\{Te_n\}$ is a (complete) orthonormal set in $\mathcal{H}_2$.
(ii) $\|Tx\| = \|x\|$ for all $x \in \mathcal{H}_1$.
(iii) $\|x_n - x\| \to 0$ if and only if $\|Tx_n - Tx\| \to 0$.
(iv) $\{x_n\}$ is a Cauchy sequence if and only if $\{Tx_n\}$ is a Cauchy sequence.
(v) $TP_{\overline{\text{sp}}\{x_\lambda,\lambda\in\Lambda\}}(x) = P_{\overline{\text{sp}}\{Tx_\lambda,\lambda\in\Lambda\}}(Tx)$.

The last property is the basis for the spectral theory of prediction of a stationary sequence $(X_n)_{n\in\mathbb{Z}}$, in which we use the fact that the mapping

$$TX_n := \exp(in\,\cdot)$$

determines an isomorphism of a certain Hilbert space of random variables onto a Hilbert space $L^2((-\pi,\pi],\mathcal{B}((-\pi,\pi]),\mu)$ of functions on $(-\pi,\pi]$ with μ a finite measure. The problem of computing projections in the former space can then be transformed by means of (v) into the problem of computing projections in the latter. For details, see, e. g., Brockwell and Davis [22, Section 5.6].

2.8 Fourier transforms

For the analysis of continuous-parameter time series $(Y_t)_{t\in\mathbb{R}}$, the *Fourier transform* $\mathcal{F}f : \mathbb{R} \to \mathbb{C}$ of the function $f : \mathbb{R} \to \mathbb{C}$ plays a key role. Formally, it is defined as

$$\mathcal{F}f(\omega) = \frac{1}{\sqrt{2\pi}} \int_{\mathbb{R}} e^{-i\omega t} f(t)\,dt, \quad \omega \in \mathbb{R}, \tag{2.19}$$

however, care must be taken regarding the sense in which the integral is defined and the class of functions for which it exists. The *inverse Fourier transform* of the function f is defined to be

$$\mathcal{F}^{-1}f(t) = \frac{1}{\sqrt{2\pi}} \int_{\mathbb{R}} e^{i\omega t} f(\omega)\,d\omega, \quad t \in \mathbb{R}, \tag{2.20}$$

where the same care must be taken as in the interpretation of (2.19). Excellent accounts of Fourier transforms can be found in the books of Rudin [75], Folland [45] and Brémaud [19]. Different but equivalent definitions are frequently used, one of which eliminates the factors $\frac{1}{\sqrt{2\pi}}$ and replaces $i\omega t$ by $2\pi i\omega t$ in both (2.19) and (2.20).

In this section, we shall use the abbreviations $L^1 = L^1(\mathbb{R};\mathbb{C})$ and $L^2 = L^2(\mathbb{R};\mathbb{C})$ to denote the integrable and square-integrable complex-valued functions on $(\mathbb{R},\mathcal{B}_1,\mu)$, where μ denotes Lebesgue measure on the Borel-σ-algebra $\mathcal{B}_1$ of $\mathbb{R}$. The inner product of the functions f and g in L^2 is denoted by

$$\langle f, g\rangle_2 := \int_{\mathbb{R}} f(t)\overline{g}(t)\,\mathrm{d}t,$$

and the L^2-*norm* of f by $\|f\|_2 = \sqrt{\langle f, f\rangle_2}$. The L^1-*norm* of $f \in L^1$ is defined as $\|f\|_1 := \int_{\mathbb{R}} |f(t)|\,\mathrm{d}t$.

2.8.1 The Fourier transform of an L^1 function

If $f \in L^1$, then $f(\cdot)\exp(-\mathrm{i}\omega\,\cdot) \in L^1$ and the definition (2.19) specifies $\mathcal{F}f(\omega)$, as a Lebesgue integral for each $\omega \in \mathbb{R}$. Moreover, if $f \in L^1$, then $\mathcal{F}f$ is bounded and continuous, since $|\mathcal{F}f| \le \int_{\mathbb{R}} |f(t)|\,\mathrm{d}t$ and $\int_{\mathbb{R}}(\mathrm{e}^{-\mathrm{i}t(\omega+\delta)} - \mathrm{e}^{-\mathrm{i}t\omega})f(t)\,\mathrm{d}t \to 0$ as $\delta \to 0$ by dominated convergence. $\mathcal{F}f$ also has the property $\mathcal{F}f(\omega) \to 0$ as $\omega \to \pm\infty$ (see the Riemann–Lebesgue lemma, Theorem 2.60, below).

Example 2.56 (The centered Gaussian density). If, for $\sigma > 0$,

$$\phi_\sigma(t) = \frac{1}{\sigma\sqrt{2\pi}}\exp\left(-\frac{t^2}{2\sigma^2}\right), \tag{2.21}$$

then substitution in (2.19) gives

$$\psi_\sigma(\omega) := \mathcal{F}\phi_\sigma(\omega) = \frac{1}{\sqrt{2\pi}}\exp\left(-\frac{1}{2}\sigma^2\omega^2\right). \tag{2.22}$$

Since $\psi_\sigma \in L^1$, $\mathcal{F}^{-1}\psi_\sigma$ is well defined by (2.20) as a Lebesgue integral and a simple calculation shows that

$$\mathcal{F}^{-1}\psi_\sigma = \phi_\sigma.$$

Essentially the same calculation shows also that

$$\mathcal{F}\psi_\sigma = \phi_\sigma. \tag{2.23}$$

Example 2.57 (The two-sided exponential function). If we define, for $\lambda > 0$,

$$f(t) := e^{-\lambda|t|},$$

then it is easily found that

$$\mathcal{F}f(\omega) = \frac{1}{\sqrt{2\pi}}\frac{2\lambda}{\lambda^2 + \omega^2}.$$

Example 2.58 (The rectangular function). If

$$f(t) = \mathbf{1}_{[-T/2,T/2]}(t), \quad 0 < T < \infty, \tag{2.24}$$

then a simple calculation shows that

$$\mathcal{F}f(\omega) = \frac{T}{\sqrt{2\pi}} \operatorname{sinc}\left(\frac{\omega T}{2}\right), \tag{2.25}$$

where

$$\operatorname{sinc}(x) := \begin{cases} \frac{\sin(x)}{x}, & \text{if } x \neq 0, \\ 1, & \text{if } x = 0. \end{cases}$$

Notice that in this example (as distinct from the previous two) $\mathcal{F}f \notin L^1$ (Exercise 2.16).

The following theorem specifies a number of fundamental properties of the Fourier transform when applied to L^1 functions. One of the crucial properties is the mapping of convolutions into products (with an additional factor $\sqrt{2\pi}$ as in property (e) below). The *convolution* of two L^1 functions f and g is defined as

$$f * g(t) := \int_{\mathbb{R}} f(t-u)g(u)\,\mathrm{d}u = \int_{\mathbb{R}} g(t-u)f(u)\,\mathrm{d}u. \tag{2.26}$$

Fubini's theorem shows that if $f \in L^1$ and $g \in L^1$, then $f * g \in L^1$.

Theorem 2.59. *If $f \in L^1$, $\alpha \in \mathbb{R}$ and $\lambda > 0$, then*
(a) $g(t) = e^{i\alpha t}f(t) \Rightarrow \mathcal{F}g(\omega) = \mathcal{F}f(\omega - \alpha)$,
(b) $g(t) = f(t-\alpha) \Rightarrow \mathcal{F}g(\omega) = e^{-i\alpha\omega}\mathcal{F}f(\omega)$,
(c) $g(t) = \overline{f(-t)} \Rightarrow \mathcal{F}g = \overline{\mathcal{F}f}$,
(d) $g(t) = f(t/\lambda) \Rightarrow \mathcal{F}g(\omega) = \lambda\mathcal{F}f(\lambda\omega)$,
(e) $g \in L^1, h = f * g \Rightarrow \mathcal{F}h = \sqrt{2\pi}(\mathcal{F}f)(\mathcal{F}g)$,
(f) $g \in L^1 \Rightarrow \int_{\mathbb{R}} f(x)(\mathcal{F}g)(x)\mathrm{d}x = \int_{\mathbb{R}} (\mathcal{F}f)(x)g(x)\mathrm{d}x$, *and*
(g) $g(t) = -itf(t)$ *and* $g \in L^1 \Rightarrow \mathcal{F}f$ *is differentiable with derivative* $(\mathcal{F}f)' = \mathcal{F}g$.

Proof. Properties (a) through (d) follow by direct substitution into the definition (2.19). Properties (e) and (f) follow by application of Fubini's theorem and property (g) by application of the dominated convergence theorem. Details can be found in Rudin [75, Theorem 9.2]. □

Theorem 2.60 (Riemann–Lebesgue lemma). *If $f \in L^1$, then*

$$\lim_{|\omega|\to\infty} |\mathcal{F}f(\omega)| = 0.$$

Proof. Example 2.58 and Theorem 2.59(b) show that if g is the indicator function of any finite interval, then $|\mathcal{F}g(\omega)| \leq K/|\omega|$, $\omega \in \mathbb{R}$, for some $K > 0$. The same is clearly true if g is any finite linear combination, with possibly complex-valued coefficients, of indicator functions of finite intervals. Since the class $\mathcal{C}$ of such functions is dense in L^1 (e.g., by

Rudin [75, Theorem 3.14]) it follows that for each $f \in L^1$ there is a sequence $\{g_n\}$ of functions in C and a sequence of constants $K_n \in (0, \infty)$ such that,

$$\lim_{n\to\infty} \int_{\mathbb{R}} |f(t) - g_n(t)|\, dt = 0$$

and

$$|\mathcal{F}g_n(\omega)| \le \frac{K_n}{|\omega|}.$$

From the inequality

$$|\mathcal{F}f(\omega) - \mathcal{F}g_n(\omega)| \le \int_{\mathbb{R}} |f(t) - g_n(t)| dt,$$

it follows that

$$\begin{aligned} |\mathcal{F}f(\omega)| &\le |\mathcal{F}g_n(\omega)| + \int_{\mathbb{R}} |f(t) - g_n(t)| dt \\ &\le \frac{K_n}{|\omega|} + \int_{\mathbb{R}} |f(t) - g_n(t)| dt. \end{aligned}$$

Hence

$$\limsup_{|\omega|\to\infty} |\mathcal{F}f(\omega)| \le \int_{\mathbb{R}} |f(t) - g_n(t)| dt$$

and, since n can be arbitrarily large, the left-hand side must be zero. □

In applications of the Fourier transform it is frequently important to recover the function f from its Fourier transform $\mathcal{F}f$. This is complicated for $f \in L^1$ by the fact that $\mathcal{F}f$ is not necessarily in L^1 (as in Example 2.58). If this is the case, then the inverse Fourier transform of $\mathcal{F}f$ as specified by (2.20) is not defined as a Lebesgue integral. This problem can be circumvented by imposing conditions on f as in the inversion theorem below. Before proving the theorem, we need the following lemma.

Lemma 2.61. *If $f \in L^1$ is piecewise continuous on $\mathbb{R}$ and ϕ_σ is the Gaussian density defined in* (2.21), *then*

$$\lim_{\sigma\downarrow 0} f * \phi_\sigma(x) = \frac{1}{2}[f(x+) + f(x-)], \quad x \in \mathbb{R},$$

where $f(x+)$ and $f(x-)$ denote the right- and left-limits of f at x, respectively.

Proof. We can write

$$f * \phi_\sigma(x) - \frac{1}{2}[f(x+) + f(x-)] = \int_{-\infty}^{0} [f(x-y) - f(x+)]\phi_\sigma(y)\,\mathrm{d}y$$
$$+ \int_{0}^{\infty} [f(x-y) - f(x-)]\phi_\sigma(y)\,\mathrm{d}y. \tag{2.27}$$

We shall show the convergence of the second integral to 0 as $\sigma \downarrow 0$. The argument for the first integral is the same.

Given $\epsilon > 0$, choose $c > 0$ such that $|f(x-y) - f(x-)| < \epsilon/2$ for $0 < y < c$. Then $|\int_0^c [f(x-y) - f(x-)]\phi_\sigma(y)\,\mathrm{d}y| < \epsilon/2$ and

$$\left| \int_{c}^{\infty} [f(x-y) - f(x-)]\phi_\sigma(y)\,\mathrm{d}y \right| \le \phi_\sigma(c)(\|f\|_1 + |f(x-)|) < \epsilon/2$$

for σ sufficiently small. Hence the second integral in (2.27) converges to zero as $\sigma \downarrow 0$ and similarly so does the first. □

Remark 2.62. If $f : \mathbb{R} \to \mathbb{C}$ is bounded and Borel measurable, and $g \in L^1$, we can still define the *convolution* of f and g by (2.26). Now if f is additionally continuous at x, then an argument similar to the proof of Lemma 2.61 (Exercise 2.17) shows that

$$\lim_{\sigma \downarrow 0} f * \phi_\sigma(x) = f(x).$$

Theorem 2.63 (Fourier inversion theorem).

(a) *If $f \in L^1$ is piecewise continuous and if f is defined at its points of discontinuity by $f(t) = \frac{1}{2}[f(t+) + f(t-)]$, then*

$$f(t) = \lim_{\sigma \downarrow 0} \int_{\mathbb{R}} \psi_\sigma(\omega) e^{it\omega} \mathcal{F}f(\omega)\,\mathrm{d}\omega, \tag{2.28}$$

where ψ_σ was defined in (2.22).

(b) *If $f \in L^1$ is piecewise continuous and $\mathcal{F}f \in L^1$, then f is continuous and*

$$f(t) = \frac{1}{\sqrt{2\pi}} \int_{\mathbb{R}} e^{it\omega} \mathcal{F}f(\omega)\,\mathrm{d}\omega. \tag{2.29}$$

Proof. (a) Applying Theorem 2.59(f), equation (2.23) and Theorem 2.59(a) to the right-hand side of (2.28) and using Lemma 2.61 gives

$$\lim_{\sigma \downarrow 0} \int_{\mathbb{R}} \psi_\sigma(\omega) e^{it\omega} \mathcal{F}f(\omega)\,\mathrm{d}\omega = \lim_{\sigma \downarrow 0} \int_{\mathbb{R}} \phi_\sigma(\omega - t) f(\omega)\,\mathrm{d}\omega = f(t).$$

(b) If $\mathcal{F}f \in L^1$, we can apply the dominated convergence theorem to evaluate the limit in (2.28) as (2.29). The function f is therefore continuous, since it follows from (2.29) that $\tilde{f}(t) := f(-t)$, $t \in \mathbb{R}$, is the Fourier transform of the L^1 function $\mathcal{F}f$ and is therefore continuous. □

Remark 2.64. Without the assumption that f is piecewise continuous, a deeper proof (see, e. g., Rudin [75, Theorem 9.11]) shows that if $f \in L^1$ and $\mathcal{F}f \in L^1$, then $f(t) = \frac{1}{\sqrt{2\pi}} \int_{\mathbb{R}} e^{\mathrm{i}t\omega} \mathcal{F}f(\omega) d\omega$ almost everywhere.

Example 2.65. The inversion formula (2.29) can be applied to recover the original functions in both Example 2.56 and Example 2.57 since they are continuous and their Fourier transforms are in L^1. It cannot be applied in Example 2.58 since $\mathcal{F}f \notin L^1$, however we can apply the inversion formula (2.28) in this case to the Fourier transform (2.25) to obtain the original function (2.24) modified so that the values at the discontinuities are 1/2 instead of 1.

2.8.2 The Fourier transform of an L^2 function

If $f \in L^2$, it is not immediately clear how to define its Fourier transform, since the existence of the Lebesgue integral (2.19) requires the condition $f \in L^1$. However, starting with the Fourier transforms of functions $f \in L^1 \cap L^2$, we shall construct an isomorphism $\mathcal{F}_2$ of the Hilbert space L^2 onto itself, which uniquely extends the mapping $\mathcal{F} : L^1 \cap L^2 \to L^2$. The extended mapping $\mathcal{F}_2$ on L^2 is frequently referred to as the *Plancherel transform* or the *L^2-Fourier transform* (in contrast to the *L^1-Fourier transform*, as defined in Section 2.8.1). It has a more symmetric relationship with its inverse than does the Fourier transform on L^1.

The establishment of the isomorphism and explicit representations of the mapping and its inverse are contained in the proof and statement of the following theorem.

Theorem 2.66. *To each function $f \in L^2$ there is a function $\hat{f} \in L^2$ such that the following properties hold.*

(a) *If $f \in L^1 \cap L^2$, then $\hat{f} = \mathcal{F}f$, as defined in Section 2.8.1.*
(b) *For every $f \in L^1 \cap L^2$, $\|\hat{f}\|_2 = \|f\|_2$.*
(c) *The mapping $f \mapsto \mathcal{F}f$ of $L^1 \cap L^2$ into L^2 has a unique extension to a Hilbert space isomorphism of L^2 onto L^2, which we shall denote by $\mathcal{F}_2$. We have $\hat{f} = \mathcal{F}_2 f$ for all $f \in L^2$.*
(d) *If we define, for $f \in L^2$ and $A > 0$,*

$$\eta_A(\omega) := \frac{1}{\sqrt{2\pi}} \int_{-A}^{A} e^{-\mathrm{i}t\omega} f(t)\,\mathrm{d}t \quad \textit{and} \quad \xi_A(t) := \frac{1}{\sqrt{2\pi}} \int_{-A}^{A} e^{\mathrm{i}t\omega} \mathcal{F}_2 f(\omega)\,\mathrm{d}\omega,$$

then

$$\|\eta_A - \mathcal{F}_2 f\|_2 \to 0 \quad and \quad \|\xi_A - f\|_2 \to 0 \quad as\ A \to \infty.$$

The proof of the theorem makes use of three lemmas, which we present first, since they are of interest in their own right.

Definition 2.67 (Translate of the function f on $\mathbb{R}$). For any function f on $\mathbb{R}$ and any $x \in \mathbb{R}$, we define the *x-translate* f_x *of* f to be the function

$$f_x(t) := f(t + x), \quad t \in \mathbb{R}.$$

Lemma 2.68. *If $f \in L^2$ the mapping $T : x \mapsto f_x$ from $\mathbb{R}$ into L^2 is uniformly continuous.*

Proof. By Theorem 3.14 of Rudin [75], for any $\epsilon > 0$, there exists $A > 0$ and a continuous function g on $\mathbb{R}$ with support in $[-A, A]$ such that $\|f - g\|_2 < \epsilon/3$. By the uniform continuity of g, there exists $\delta(\epsilon) \in (0, A]$ such that

$$|g(t + x) - g(t + y)| < \epsilon/(6\sqrt{A}) \quad \text{whenever } |x - y| < \delta(\epsilon).$$

By the translation invariance of Lebesgue measure

$$\int_{\mathbb{R}} |g(t + x) - g(t + y)|^2 \,\mathrm{d}t = \int_{\mathbb{R}} |g(t + x - y) - g(t)|^2 \,\mathrm{d}t < \frac{\epsilon^2}{36A} 2(A + \delta(\epsilon)) \le \frac{\epsilon^2}{9},$$

i. e.

$$\|g_x - g_y\|_2 < \epsilon/3 \quad \text{whenever } |x - y| < \delta(\varepsilon).$$

Finally, by the triangle inequality,

$$\begin{aligned} \|f_x - f_y\|_2 &\le \|f_x - g_x\|_2 + \|g_x - g_y\|_2 + \|g_y - f_y\|_2 \\ &= 2\|f - g\|_2 + \|g_x - g_y\| \\ &< \epsilon \quad \text{if } |x - y| < \delta(\epsilon), \end{aligned}$$

again using the translation invariance of Lebesgue measure. □

Lemma 2.69. *If $f \in L^1$ and ϕ_σ and ψ_σ are defined as in Example 2.56, then $f * \phi_\sigma$ is continuous and*

$$f * \phi_\sigma(t) = \int_{\mathbb{R}} \psi_\sigma(\omega) \mathcal{F}f(\omega) e^{it\omega} \mathrm{d}\omega.$$

Proof. Since $f \in L^1$ and $\phi_\sigma \in L^1$ the convolution $f * \phi_\sigma \in L^1$ and its Fourier transform, by Theorem 2.59(e) and (2.22), is $\sqrt{2\pi}\psi_\sigma \mathcal{F}f$. Moreover, $\psi_\sigma \mathcal{F}f \in L^1$, since $\mathcal{F}f$ is bounded and continuous and $\psi_\sigma \in L^1$. The convolution $f * \phi_\sigma$ is continuous since

$$f * \phi_\sigma(t+h) - f * \phi_\sigma(t) = \int_{\mathbb{R}} f(y)[\phi_\sigma(t+h-y) - \phi_\sigma(t-y)]\mathrm{d}y,$$

which converges to zero as $h \to 0$ by dominated convergence. Applying Theorem 2.63(b) to $f * \phi_\sigma$ gives the required result. □

Lemma 2.70. *If $f \in L^2$, then the function $t \mapsto \int_{\mathbb{R}} |f(t-u)|\phi_\sigma(u)\,\mathrm{d}u$, $t \in \mathbb{R}$, is in L^2, hence we can still define $f * \phi_\sigma$ by (2.26), and then $f * \phi_\sigma \in L^2$. Further,*

$$\lim_{\sigma \downarrow 0} \|f * \phi_\sigma - f\|_2 = 0.$$

Proof. Observing that $\phi_\sigma \geq 0$ and has Lebesgue integral 1, an application of Hölder's inequality shows

$$\left(\int_{\mathbb{R}} |f(t-u)|\phi_\sigma(u)\,\mathrm{d}u\right)^2 \leq \int_{\mathbb{R}} |f(t-u)|^2 \phi_\sigma(u)\,\mathrm{d}u = \int_{\mathbb{R}} |f(u)|^2 \phi_\sigma(t-u)\,\mathrm{d}u.$$

Hence, by Fubini's theorem,

$$\int_{\mathbb{R}} \left(\int_{\mathbb{R}} |f(t-u)|\phi_\sigma(u)\,\mathrm{d}u\right)^2 \mathrm{d}t \leq \int_{\mathbb{R}} \int_{\mathbb{R}} \phi_\sigma(t-u)\,\mathrm{d}t\,|f(u)|^2\,\mathrm{d}u = \|f\|_2^2 < \infty.$$

This implies that $f * \phi_\sigma$ can be defined and is in L^2. Using a similar argument, we have

$$\begin{aligned} |f * \phi_\sigma(t) - f(t)|^2 &\leq \left|\int_{\mathbb{R}} |f(t-y) - f(t)|\phi_\sigma(y)\,\mathrm{d}y\right|^2 \\ &\leq \int_{\mathbb{R}} |f(t-y) - f(t)|^2 \phi_\sigma(y)\,\mathrm{d}y \end{aligned}$$

and hence

$$\|f * \phi_\sigma - f\|_2^2 \leq \int_{\mathbb{R}} \|f_{-y} - f\|^2 \phi_\sigma(y)\mathrm{d}y = g * \phi_\sigma(0),$$

where $g(y) := \|f_{-y} - f\|^2$ is bounded (since $f \in L^2$) and continuous by Lemma 2.68. Applying Remark 2.62 completes the proof. □

We are now in a position to give the proof of Theorem 2.66.

Proof of Theorem 2.66. For $f \in L^1 \cap L^2$, we define the functions

$$\hat{f}(\cdot) := \mathcal{F}f(\cdot), \quad \tilde{f}(\cdot) := \overline{f(-\cdot)} \quad \text{and} \quad g := f * \tilde{f}.$$

Then $g(t)$ can be written as

$$g(t) = \int_{\mathbb{R}} f(t-y)\overline{f(-y)}\,dy = \int_{\mathbb{R}} f(t+y)\overline{f(y)}\,dy = \langle f_t, f\rangle_2,$$

where f_t is the t-translate of f as specified in Definition 2.67. Since the mapping $t \mapsto f_t$ is continuous by Lemma 2.68, continuity of the inner product in L^2 implies that g is continuous. By the Cauchy–Schwarz inequality,

$$|g(t)| \le \|f_t\|_2 \|f\|_2 = \|f\|_2^2,$$

showing that g is bounded. Since g is the convolution of the two L^1 functions f and $\tilde{f}$, it follows also that $g \in L^1$. We can therefore write, by Lemma 2.69,

$$g * \phi_\sigma(0) = \int_{\mathbb{R}} \psi_\sigma(\omega)\mathcal{F}g(\omega)\,d\omega. \tag{2.30}$$

Since g is continuous and bounded, we can apply Remark 2.62 to conclude that

$$\lim_{\sigma \downarrow 0} g * \phi_\sigma(0) = g(0) = \|f\|_2^2. \tag{2.31}$$

To evaluate the corresponding limit of the right-hand side of (2.30) observe, from Theorem 2.59(c) and (e), that $\mathcal{F}g = \sqrt{2\pi}|\hat{f}|^2 \ge 0$ and that $\psi_\sigma(\omega) \uparrow 1/\sqrt{2\pi}$ as $\sigma \downarrow 0$. Hence, by monotone convergence,

$$\lim_{\sigma \downarrow 0} \int_{\mathbb{R}} \psi_\sigma(\omega)\mathcal{F}g(\omega)d\omega = \int_{\mathbb{R}} |\hat{f}(\omega)|^2\,d\omega = \|\hat{f}\|_2^2. \tag{2.32}$$

It follows at once from (2.30)–(2.32) that $\hat{f} \in L^2$ and $\|\hat{f}\| = \|f\|$.

To show that the mapping $f \mapsto \hat{f}$ extends to an isomorphism of L^2 onto L^2, we first observe that $L^1 \cap L^2$ is dense in L^2 (see Rudin [75, Theorem 3.14]). Let us show that the set $\mathcal{M} = \{\hat{f} : f \in L^1 \cap L^2\}$ is dense in L^2. To do this, we use the fact that the functions $t \mapsto e^{ita}\psi_\sigma(t)$ are in $L^1 \cap L^2$ for all real a and all $\sigma > 0$. Their Fourier transforms,

$$\phi_\sigma(a-t) = \frac{1}{\sqrt{2\pi}} \int_{\mathbb{R}} e^{ita}\psi_\sigma(t)e^{-it\omega}\,d\omega,$$

are therefore in $\mathcal{M}$. If $w \in \mathcal{M}^\perp$ (the orthogonal complement of $\mathcal{M}$ in L^2), then

$$\phi_\sigma * \overline{w}(a) = \int_{\mathbb{R}} \phi_\sigma(a-t)\overline{w(t)}\,dt = \langle \phi_\sigma(a-\cdot), w(\cdot)\rangle = 0 \quad \text{for all } \sigma > 0.$$

This implies, by Lemma 2.70, that $w = 0$ and hence that the orthogonal complement of the closure $\overline{\mathcal{M}}$ of $\mathcal{M}$ in L^2 is the zero element of L^2 (since $\overline{\mathcal{M}}^\perp \subseteq \mathcal{M}^\perp$). Hence for any $f \in L^2, f = P_{\overline{\mathcal{M}}} f \in \overline{\mathcal{M}}$, showing that $\overline{\mathcal{M}} = L^2$.

We have shown now that $\mathcal{F}$ is a linear isometry from the dense subset $L^1 \cap L^2$ of L^2 onto the dense subset $\mathcal{M}$. A simple Cauchy sequence argument shows the existence of a unique extension of $\mathcal{F}$ on $L^1 \cap L^2$ to a linear isometry $\mathcal{F}_2$ of L^2 onto itself, which is therefore a Hilbert space isomorphism (Exercise 2.18). The preservation of the inner product under this isomorphism immediately gives *Parseval's identity*,

$$\int_{\mathbb{R}} f(t)\overline{g(t)}\mathrm{d}t = \int_{\mathbb{R}} \mathcal{F}_2 f(\omega)\overline{\mathcal{F}_2 g(\omega)}\mathrm{d}\omega, \quad f, g \in L^2. \tag{2.33}$$

Finally, to prove (d), let $\mathbf{1}_{[-A,A]}$ denote the indicator function of the interval $[-A, A]$, $0 < A < \infty$. Then, if $f \in L^2, f\mathbf{1}_{[-A,A]} \in L^1 \cap L^2$ and, by (b),

$$\eta_A := \mathcal{F}(f\mathbf{1}_{[-A,A]}) = \mathcal{F}_2(f\mathbf{1}_{[-A,A]}).$$

Since $\|f\mathbf{1}_{[-A,A]} - f\|_2 \to 0$ as $A \to \infty$ by dominated convergence, it follows that

$$\|\eta_A - \mathcal{F}_2 f\|_2 = \|\mathcal{F}_2(f\mathbf{1}_{[-A,A]} - f)\|_2 \to 0, \quad \text{as } A \to \infty.$$

The other half of (d) is proved in the same way, observing that also $\mathcal{F}_2^{-1}$ is a linear isometry. □

Remark 2.71. The Fourier transform of a function $f \in L^2$ will always mean $\mathcal{F}_2 f$. If $f \in L^1 \cap L^2$ then $\mathcal{F}_2 f$ is (Lebesgue) almost everywhere equal to the integral (2.19) and is in L^2 but not necessarily in L^1. However, by the isomorphism, $\mathcal{F}_2^{-1}\mathcal{F}_2 f = f$. If $g \in L^2$ but not in L^1 and if a function in $L^2 \cap L^1$ is known to have Fourier transform g, then this function is necessarily $\mathcal{F}_2^{-1} g$. This observation is sometimes useful for identifying the inverse Fourier transform of a function in L^2 that is not also in L^1.

2.9 Exercises

Exercise 2.1. Prove the parallelogram law (2.3).

Exercise 2.2. If $\{X_n, n = 0, \pm 1, ...\}$ is a stationary sequence with mean zero and autocovariance function γ, and $(a_n)_{n\in\mathbb{N}}$ a sequence of complex numbers, show that $Y_n = \sum_{k=1}^n a_k X_k$ converges in mean-square if $\sum_{i=1}^\infty \sum_{j=1}^\infty |a_i a_j \gamma(i-j)|$ is finite.

Exercise 2.3. Show that if $\{X_n, n = 0, \pm 1, \ldots\}$ is stationary and $|\theta| < 1$, then $\sum_{j=1}^m \theta^j X_{t+1-j}$ converges in mean-square for each t as $m \to \infty$.

Exercise 2.4. Prove Proposition 2.26.

Exercise 2.5. If $\mathcal{M}$ is a subset of a Hilbert space $\mathcal{H}$, show that $\mathcal{M} \subseteq (\mathcal{M}^\perp)^\perp$.

Exercise 2.6. If $\mathcal{H}$ is a complex Hilbert space and $x_i \in \mathcal{H}$, $i = 1,\dots,n$, show that $\overline{\text{sp}}\{x_1,\dots,x_n\} = \{\sum_{j=1}^{n} a_j x_j : a_j \in \mathbb{C}, j = 1,\dots,n\}$. Similarly, if $\mathcal{H}$ is a real Hilbert space and $x_i \in \mathcal{H}$, $i = 1,\dots,n$, show that $\overline{\text{sp}}\{x_1,\dots,x_n\} = \{\sum_{j=1}^{n} a_j x_j : a_j \in \mathbb{R}, j = 1,\dots,n\}$.

Exercise 2.7. If $X_n = Z_n - \theta Z_{n-1}$, where $|\theta| < 1$ and $\{Z_n, n = 0, \pm 1, \dots\}$ is a sequence of uncorrelated random variables, each with mean 0 and variance $\sigma^2 \in (0,\infty)$, show, by checking the prediction equations, that the best mean-square predictor of X_{n+1} in $\overline{\text{sp}}\{X_j, -\infty < j \le n\}$ is

$$\hat{X}_{n+1} = -\sum_{j=1}^{\infty} \theta^j X_{n+1-j}.$$

What is the mean-square error of $\hat{X}_{n+1}$?

Exercise 2.8. If X_n is defined as in Exercise 2.7 with $\theta = 1$, find the best mean-square predictor of X_{n+1} in $\overline{\text{sp}}\{X_j, 1 \le j \le n\}$ and the corresponding mean-square error.

Exercise 2.9. If $X_n = \phi_1 X_{n-1} + \phi_2 X_{n-2} + \cdots + \phi_p X_{n-p} + Z_n$, $n = 0, \pm 1, \dots$, where $\{Z_n\}$ is a sequence of uncorrelated random variables, each with mean zero and variance $\sigma^2 \in (0,\infty)$ and such that Z_n is uncorrelated with $\{X_j, j < n\}$ for each n, use the prediction equations to show that the best mean-square predictor of X_{n+1} in $\overline{\text{sp}}\{X_j, -\infty < j \le n\}$ is

$$\hat{X}_{n+1} = \phi_1 X_n + \phi_2 X_{n-1} + \cdots + \phi_p X_{n+1-p}.$$

Exercise 2.10. (a) Consider the Hilbert space $\mathcal{H} = \overline{\text{sp}}\{x_1, x_2, \dots\}$ and let $x \in \mathcal{H}$. Show that $\mathcal{H}$ is separable and that

$$P_{\overline{\text{sp}}\{x_1,\dots,x_n\}} x \to x \quad \text{as } n \to \infty.$$

(b) If $\{X_n, n = 0, \pm 1, \dots\}$ is a stationary sequence, show that

$$P_{\overline{\text{sp}}\{X_j, -\infty<j\le n\}} X_{n+1} = \lim_{r\to\infty}{}^{\mathcal{S}}\, P_{\overline{\text{sp}}\{X_j, n-r<j\le n\}} X_{n+1}.$$

Exercise 2.11. If $(X, Z_1, \dots, Z_n)^T$ has a multivariate normal distribution, show that

$$P_{\overline{\text{sp}}\{1,Z_1,\dots,Z_n\}} X = \mathbb{E}(X|Z_1,\dots,Z_n).$$

Exercise 2.12. (a) Let $X, X_1, \dots, X_n \in \mathcal{S}$ with mean zero. Show that

$$P_{\overline{\text{sp}}\{1,X_1,\dots,X_n\}}(X) = P_{\overline{\text{sp}}\{X_1,\dots,X_n\}}(X).$$

(b) Let $X, X_1, \dots, X_n \in \mathcal{S}$ and define $Y = X - \mathbb{E}(X)$, $Y_i = X_i - \mathbb{E}(Y_i)$. Show that

$$P_{\overline{\text{sp}}\{1,X_1,\dots,X_n\}}(X) = \mathbb{E}(X) + P_{\overline{\text{sp}}\{Y_1,\dots,Y_n\}}(Y).$$

Exercise 2.13. Show that the space l^2 (Example 2.55) is a separable Hilbert space.

Exercise 2.14. If $\mathcal{H}$ is any Hilbert space with orthonormal basis $\{e_n, n = 1, 2, \dots\}$, show that the mapping defined by $Th := \{\langle h, e_n \rangle, n = 1, 2, \dots\}$, $h \in \mathcal{H}$, is an isomorphism of $\mathcal{H}$ onto l^2.

Exercise 2.15. Prove Remark 2.54.

Exercise 2.16. Show that the Fourier transform of the rectangular function defined in Example 2.58 is not in L^1.

Exercise 2.17. Prove the statement in Remark 2.62.

Exercise 2.18. Show that the linear isometry $\mathcal{F}$ defined in the proof of Theorem 2.66 from the dense subset $L^1 \cap L^2$ of L^2 onto the dense subset $\mathcal{M}$ of L^2 has a unique extension to a linear isometry $\mathcal{F}_2$ of L^2 onto itself and deduce Parseval's identity,

$$\int_{\mathbb{R}} f(t)\overline{g(t)}\mathrm{d}t = \int_{\mathbb{R}} \mathcal{F}_2 f(\omega)\overline{\mathcal{F}_2 g(\omega)}\mathrm{d}\omega, \quad f, g \in L^2.$$

Exercise 2.19. Use Theorem 2.66(d) to show that the properties (a)–(d) of $\mathcal{F}f$, $f \in L^1$, stated in Theorem 2.59, continue to hold for $\mathcal{F}_2$, $f \in L^2$, remembering, however, that if $f \in L^2$ and $f \notin L^1$, then $\mathcal{F}_2 f$ is defined only in L^2 and not pointwise. Derive similar statements for (e), when $f \in L^2, g \in L^1$ and for (f), when $f, g \in L^2$.

3 Second-order stochastic processes

Abstract: In this chapter, we shall examine the properties of complex-valued stochastic processes $(Y_t)_{t\in\mathbb{R}}$ defined on a probability space $(\Omega, \mathcal{F}, P)$, which satisfy the conditions $\mathbb{E}(|Y_t|^2) < \infty$, $t \in \mathbb{R}$. As we saw in Chapter 2, random variables with finite second moments (or more precisely the equivalence classes of almost surely equal random variables they represent) constitute a Hilbert space $\mathcal{S} := L^2(\Omega, \mathcal{F}, P)$ with inner product $\langle X, Y\rangle := \mathbb{E}(X\overline{Y})$ (see Theorem 2.17). The process $(Y_t)_{t\in\mathbb{R}}$ is referred to as a *second-order (stochastic) process* and can be regarded, as t varies, as describing a path in $\mathcal{S}$. In this chapter, we shall be concerned with the properties of the process that depend only on the first- and second-order moment functions $\mathbb{E}(Y_t)$, $t \in \mathbb{R}$ and $\mathbb{E}(Y_r\overline{Y_s})$, $r, s \in \mathbb{R}$. These include continuity, differentiability and integrability in mean-square.

3.1 The mean and autocovariance functions

A stochastic process $(Y_t)_{t\in\mathbb{R}}$ is a *second-order (stochastic) process* if $Y_t \in \mathcal{S}$ (i. e., $\mathbb{E}(|Y_t|^2) < \infty$) for all $t \in \mathbb{R}$. For a second-order stochastic process $Y = (Y_t)_{t\in\mathbb{R}}$, we define its *mean function* as

$$\mu_Y(t) := \langle Y_t, 1\rangle = \mathbb{E}Y_t, \quad t \in \mathbb{R},$$

its *second-moment function* as

$$M_Y(s,t) := \langle Y_s, Y_t\rangle = \mathbb{E}(Y_s\overline{Y_t}), \quad s,t \in \mathbb{R},$$

and its *autocovariance function* as

$$\kappa_Y(s,t) := \big\langle Y_s - \mu(s), Y_t - \mu(t)\big\rangle = M_Y(s,t) - \mu_Y(s)\overline{\mu_Y(t)}, \quad s,t \in \mathbb{R},$$

omitting the subscript Y whenever the process concerned is clear from the context.

Clearly, $\kappa_Y(t,t) = \mathbb{E}|Y_t - \mu_Y(t)|^2 \geq 0$ and $|\kappa_Y(r,s)| \leq \sqrt{\kappa_Y(r,r)\kappa_Y(s,s)}$ by the Cauchy–Schwarz inequality. These properties are therefore necessary for a function κ on $\mathbb{R} \times \mathbb{R}$ to be an autocovariance function. However, they are not sufficient. They are in fact consequences of the necessary and sufficient condition given by the following theorem.

Theorem 3.1 (Characterization of autocovariance functions). *A complex-valued function κ on $\mathbb{R} \times \mathbb{R}$ is the autocovariance function of some (possibly complex-valued) second-order process Y with index set $\mathbb{R}$ if and only if κ is* non-negative definite, *i. e., if and only if for every positive integer n, every $c_1, \ldots, c_n \in \mathbb{C}$, and every $t_1, \ldots, t_n \in \mathbb{R}$,*

https://doi.org/10.1515/9783111325033-003

$$\sum_{j=1}^{n}\sum_{k=1}^{n} c_j\overline{c_k}\kappa(t_j,t_k) \geq 0. \tag{3.1}$$

The underlying second-order stochastic process can be chosen to have mean 0.

Proof. The necessity of (3.1) is clear, since if κ is the autocovariance function of the second-order process Y, then the left-hand side of (3.1) is $\mathbb{E}|\sum_{j=1}^{n} c_j(Y_{t_j} - \mu_Y(t_j))|^2$, which is non-negative.

The sufficiency of (3.1) is less clear. If κ satisfies (3.1) then, taking $n = 1$ and $c_1 = 1$ in (3.1), we see that $\kappa(t_1,t_1) \geq 0$. Moreover, taking $n = 2$, $c_1 = 1+\mathrm{i}$ and $c_2 = 1-\mathrm{i}$, we see that $\kappa(t_1,t_2)-\kappa(t_2,t_1)$ has zero real part and taking $c_1 = c_2 = 1+\mathrm{i}$, we see that $\kappa(t_1,t_2)+\kappa(t_2,t_1)$ is real. Hence the condition (3.1) implies that κ is Hermitian, i. e., that $\kappa(t_1,t_2) = \overline{\kappa(t_2,t_1)}$.

Now let κ_1 and κ_2 be the real and imaginary parts of the Hermitian function κ and let $\Lambda^{(n)}$ be the $2n \times 2n$ matrix,

$$\Lambda^{(n)} = \Lambda^{(n)}(t_1,\ldots,t_n) = \frac{1}{2}\begin{bmatrix} K_1^{(n)} & K_2^{(n)} \\ -K_2^{(n)} & K_1^{(n)} \end{bmatrix},$$

where $K_r^{(n)} = [\kappa_r(t_j,t_k)]_{j,k=1}^{n}$, $r = 1,2$. The Hermitian property of κ implies that $K_1^{(n)T} = K_1^{(n)}$ and $K_2^{(n)T} = -K_2^{(n)}$, and hence that $\Lambda^{(n)}$ is a real symmetric matrix. Furthermore, the matrix $\Lambda^{(n)}$ is non-negative in the sense that for any $\mathbf{a}$ and $\mathbf{b}$ in $\mathbb{R}^n$,

$$[\mathbf{a}^T \ \ \mathbf{b}^T]\,\Lambda^{(n)}\begin{bmatrix}\mathbf{a}\\ \mathbf{b}\end{bmatrix} = \frac{1}{2}(\mathbf{a}^T K_1^{(n)}\mathbf{a} + \mathbf{a}^T K_2^{(n)}\mathbf{b} - \mathbf{b}^T K_2^{(n)}\mathbf{a} + \mathbf{b}^T K_1^{(n)}\mathbf{b}) \geq 0$$

since the right-hand side can be expressed (Exercise 3.1(a)) as $\frac{1}{2}(\mathbf{a}+\mathrm{i}\mathbf{b})^T(K_1^{(n)}+\mathrm{i}K_2^{(n)})(\mathbf{a}-\mathrm{i}\mathbf{b})$ and since κ is non-negative definite in the sense of (3.1). Kolmogorov's consistency theorem, Theorem 1.13, together with Remark 1.14, now guarantees the existence of a bivariate Gaussian process $((Y_t,Z_t)^T)_{t\in\mathbb{R}}$ with

$$(Y_{t_1},\ldots,Y_{t_n},Z_{t_1},\ldots,Z_{t_n})^T \sim N(\mathbf{0},\Lambda^{(n)}(t_1,\ldots,t_n))$$

(see Exercise 3.1(b) for details). Setting $X_t = Y_t - \mathrm{i}Z_t$, we have

$$\mathbb{E}X_t = \mathbb{E}Y_t - \mathrm{i}\,\mathbb{E}Z_t = 0, \quad \text{for all } t,$$

and, from the definition of $\Lambda^{(2)}(t+h,t)$,

$$\begin{aligned}
&\mathbb{E}(X_{t+h}\overline{X}_t)\\
&\quad = \mathbb{E}(Y_{t+h}Y_t) - \mathrm{i}\,\mathbb{E}(Z_{t+h}Y_t) + \mathrm{i}\,\mathbb{E}(Y_{t+h}Z_t) + \mathbb{E}(Z_{t+h}Z_t)\\
&\quad = \frac{1}{2}(\kappa_1(t+h,t) - \mathrm{i}(-\kappa_2(t+h,t)) + \mathrm{i}\kappa_2(t+h,t) + \kappa_1(t+h,t))\\
&\quad = \kappa(t+h,t).
\end{aligned}$$

Thus κ is the autocovariance function of the second-order process X with mean zero. □

Corollary 3.2 (Characterization of acvf of stat. processes). *A complex-valued function γ on $\mathbb{R}$ is the acvf of some stationary process Y (see Definitions* 1.18 *and* 1.20*) if and only if γ is* non-negative definite *in the sense that for every positive integer n, every $c_1, \dots, c_n \in \mathbb{C}$, and every $t_1, \dots, t_n \in \mathbb{R}$,*

$$\sum_{j=1}^{n} \sum_{k=1}^{n} c_j \overline{c_k} \gamma(t_j - t_k) \geq 0. \tag{3.2}$$

Proof. γ is the acvf of a stationary process if and only if $\kappa(s,t) := \gamma(s-t)$, $s, t \in \mathbb{R}$, satisfies (3.1). □

Remark 3.3. If κ is real-valued, then (3.1) is equivalent to the conditions
(i) $\kappa(t_j, t_k) = \kappa(t_k, t_j)$, for all $t_j, t_k \in \mathbb{R}$, and
(ii) for every positive integer n, every $c_1, \dots, c_n \in \mathbb{R}$, and every $t_1, \dots, t_n \in \mathbb{R}$

$$\sum_{j=1}^{n} \sum_{k=1}^{n} c_j c_k \, \kappa(t_j, t_k) \geq 0,$$

see Exercise 3.2.

3.2 Mean-square continuity

The second-order process $Y = (Y_t)_{t \in \mathbb{R}}$ is said to be *mean-square continuous* (MSC) at t if $\|Y_{t+h} - Y_t\| \to 0$ as $h \to 0$ and *mean-square continuous* (MSC) if it is mean-square continuous at all $t \in \mathbb{R}$. (Recall that for any $X \in \mathcal{S}$, $\|X\| := \sqrt{\langle X, X \rangle} = \sqrt{\mathbb{E}(|X|^2)}$.)

Theorem 3.4 (Characterization of MSC). *For a zero-mean second-order process $Y = (Y_t)_{t \in \mathbb{R}}$ with second-moment function M, the following statements are equivalent:*
(i) *Y is mean-square continuous,*
(ii) *M is continuous on $\mathbb{R} \times \mathbb{R}$, and*
(iii) *M is continuous at all diagonal points $(t,t) \in \mathbb{R} \times \mathbb{R}$.*

Proof. If Y is MSC, $(t,u) \in \mathbb{R} \times \mathbb{R}$ and $(r,s) \in \mathbb{R} \times \mathbb{R}$, then

$$|M(t,u) - M(r,s)| = |\langle Y_t, Y_u \rangle - \langle Y_r, Y_s \rangle| \to 0 \quad \text{as } r \to t \text{ and } s \to u,$$

by continuity of the inner-product. Thus M is continuous on $\mathbb{R} \times \mathbb{R}$, showing that (i) implies (ii). Obviously (ii) implies (iii). To see that (iii) implies (i), we write

$$\begin{aligned} \|Y_t - Y_s\|^2 &= \langle Y_t, Y_t \rangle - \langle Y_t, Y_s \rangle - \langle Y_s, Y_t \rangle + \langle Y_s, Y_s \rangle \\ &= M(t,t) - M(t,s) - M(s,t) + M(s,s), \end{aligned}$$

which converges to zero as $s \to t$ if M is continuous at (t, t). Hence, if M is continuous at all diagonal points in $\mathbb{R} \times \mathbb{R}$, then Y is MSC, completing the proof. □

Corollary 3.5. *If Y is a second-order process with continuous mean function μ and autocovariance function κ that is continuous at all points $(t, t) \in \mathbb{R} \times \mathbb{R}$, then Y is mean-square continuous.*

Proof.

$$\begin{aligned}
&\|Y_t - Y_s\|^2 \\
&\quad = \|(Y_t - \mu(t)) - (Y_s - \mu(s)) + \mu(t) - \mu(s)\|^2 \\
&\quad \leq 2\|(Y_t - \mu(t)) - (Y_s - \mu(s))\|^2 + 2|\mu(t) - \mu(s)|^2 \to 0 \quad \text{as } s \to t,
\end{aligned}$$

by the theorem and the continuity of μ. □

Proposition 3.6 (Characterization of MSC for stationary processes). *If γ is the autocovariance function of some stationary process Y (i. e., if γ is non-negative definite as in Corollary 3.2), then the following are equivalent:*
(i) *Y is mean-square continuous,*
(ii) *γ is uniformly continuous on $\mathbb{R}$,*
(iii) *γ is continuous at 0.*

Proof. That (ii) implies (iii) is obvious, and the implication (iii) $\Longrightarrow$ (i) follows from

$$\begin{aligned}
\|Y_{t+h} - Y_t\|^2 &= \langle Y_{t+h} - \mu_Y - Y_t + \mu_Y, Y_{t+h} - \mu_Y - Y_t + \mu_Y\rangle \\
&= 2\gamma(0) - \gamma(h) - \overline{\gamma(h)} \to 0 \quad \text{as } h \to 0.
\end{aligned}$$

That (i) implies (ii) follows from the mean-square continuity at 0 and

$$\begin{aligned}
|\gamma(t+h) - \gamma(t)| &= |\langle Y_{t+h} - \mu_Y, Y_0 - \mu_Y\rangle - \langle Y_t - \mu_Y, Y_0 - \mu_Y\rangle| \\
&= |\langle Y_{t+h} - Y_t, Y_0 - \mu_Y\rangle| \\
&\leq \sqrt{\|Y_{t+h} - Y_t\|^2 \|Y_0 - \mu_Y\|^2} \\
&= \gamma(0)\|Y_h - Y_0\| \to 0 \quad \text{uniformly in } t \text{ as } h \to 0.
\end{aligned}$$

□

Most of the results for second-order processes derived later in this book refer to MSC processes. An example of a stationary process that is not MSC is given in the following example.

Example 3.7 (A stationary process that is not MSC). Suppose Y_t, $t \in \mathbb{R}$, is a family of uncorrelated random variables with mean zero and variance 1. Then Y has autocovariance function $\gamma(h) = 1$ if $h = 0$ and 0 otherwise. By the preceding corollary, Y is therefore not MSC.

Example 3.8 (The Poisson process). If $(Y_t)_{t\in\mathbb{R}}$ is a Poisson process with jump rate λ, then its mean function is $\mu(t) = \lambda t$, $t \in \mathbb{R}$ and its covariance function is

$$\kappa(r,s) = \begin{cases} \lambda(|r| \wedge |s|), & \text{if } rs \geq 0, \\ 0, & \text{if } rs < 0. \end{cases}$$

By Corollary 3.5, Y is MSC.

3.3 Mean-square differentiability

The second-order process $Y = (Y_t)_{t\in\mathbb{R}}$ is said to be *mean-square differentiable* (*MSD*) at $t \in \mathbb{R}$ if $\frac{1}{h}[Y_{t+h} - Y_t]$ converges in mean-square (i. e., in $\mathcal{S}$) as $h \to 0$. In the case of existence, the limit is called the *mean-square derivative of Y at t* and denoted by

$$Y_t' = \lim_{h\to 0}{}^{\mathcal{S}} \frac{1}{h}[Y_{t+h} - Y_t].$$

Theorem 3.9 (Characterization of MSD). *A second-order process Y is mean-square differentiable at t if and only if*

$$\frac{1}{hh'}[\langle Y_{t+h}, Y_{t+h'}\rangle - \langle Y_t, Y_{t+h'}\rangle - \langle Y_{t+h}, Y_t\rangle + \langle Y_t, Y_t\rangle] \tag{3.3}$$

has a finite limit as $h, h' \to 0$.

Proof. By Proposition 2.21 convergence of (3.3) as $h, h' \to 0$ is necessary and sufficient for the existence of the mean-square limit as $h \to 0$ of $\frac{1}{h}[Y_{t+h} - Y_t]$. □

Lemma 3.10. *If Y is a second-order process with covariance function κ and if $M(s,t) := \langle Y_s, Y_t\rangle$, $s, t \in \mathbb{R}$, has continuous second partial derivatives on $(a,b) \times (a,b)$, then for all $(s,t) \in (a,b) \times (a,b)$,*

$$\frac{\partial^2}{\partial s \partial t} M(s,t) = \frac{\partial^2}{\partial t \partial s} M(s,t)$$

and both are equal to

$$\lim_{h,h'\to 0} \frac{1}{hh'}[\langle Y_{s+h}, Y_{t+h'}\rangle - \langle Y_s, Y_{t+h'}\rangle - \langle Y_{s+h}, Y_t\rangle + \langle Y_s, Y_t\rangle]. \tag{3.4}$$

Proof. The first statement is a direct consequence of Schwarz's theorem for commutativity of partial differentiation. The second follows from the fundamental theorem of calculus using

$$M(s+h, t+h') - M(s, t+h') - M(s+h, t) + M(s, t)$$
$$= \int_s^{s+h} \frac{\partial M}{\partial s}(u, t+h')\,\mathrm{d}u - \int_s^{s+h} \frac{\partial M}{\partial s}(u, t)\,\mathrm{d}u$$
$$= \int_s^{s+h}\int_t^{t+h'} \frac{\partial^2 M}{\partial t \partial s}(u, v)\,\mathrm{d}v\,\mathrm{d}u. \qquad \square$$

Corollary 3.11 (Sufficient conditions for MSD). *If $M(s,t)$ has continuous second partial derivatives on $(a,b) \times (a,b)$, then Y is mean-square differentiable on (a,b) and the mean-square derivative $(Y_t')_{t\in(a,b)}$ has covariance function*

$$\frac{\partial^2}{\partial s \partial t} M(s,t) = \frac{\partial^2}{\partial t \partial s} M(s,t), \quad (s,t) \in (a,b) \times (a,b).$$

Proof. By Lemma 3.10, (3.3) has a finite limit, $\frac{\partial^2}{\partial s \partial t} M(s,t)|_{s=t}$, as $h, h' \to 0$. By continuity of the inner-product,

$$\langle Y_s', Y_t'\rangle = \left\langle \operatorname{lim}^{\mathcal{S}}_{h\to 0} \frac{1}{h}(Y_{s+h} - Y_s), \operatorname{lim}^{\mathcal{S}}_{h'\to 0} \frac{1}{h'}(Y_{t+h'} - Y_t)\right\rangle$$

is equal to (3.4), i. e., to the partial derivatives as claimed. $\square$

Remark 3.12. Since $M(s,t) = \kappa(s,t) + \mu(s)\overline{\mu(t)}$, a sufficient condition for M to have continuous second partial derivatives is for κ to have continuous second partial derivatives and for μ to be twice continuously differentiable.

Corollary 3.13 (Characterization of MSD for stationary processes). *Let $Y = (Y_t)_{t\in\mathbb{R}}$ be a stationary process with autocovariance function γ. Then the following are equivalent:*

(i) *Y is mean-square differentiable at some $t \in \mathbb{R}$.*
(ii) *Y is mean-square differentiable at all $t \in \mathbb{R}$.*
(iii) *The limit $\lim_{h,h'\to 0} \frac{1}{hh'}[\gamma(h-h') - \gamma(-h') - \gamma(h) + \gamma(0)]$ exists and is finite.*

If one and hence all of the above conditions are satisfied, then γ is twice differentiable on $\mathbb{R}$ and the mean-square derivative Y_t' and Y_t satisfy

$$\mathbb{E}(Y_t') = 0, \quad \operatorname{Cov}(Y_{t+h}', Y_t) = \gamma'(h), \quad \operatorname{Cov}(Y_{t+h}', Y_t') = -\gamma''(h)$$

for all $t, h \in \mathbb{R}$. In particular, $(Y_t')_{t\in\mathbb{R}}$ is stationary with autocovariance function $-\gamma''$.

Proof. Since $M(s,t) = \gamma(s-t) + \mu\bar{\mu}$, where $\mu = \mathbb{E}(Y_t)$, the necessary and sufficient condition for mean-square differentiability at fixed $t \in \mathbb{R}$, which is given by Theorem 3.9, is the finiteness of the limit in (iii). Since this condition is independent of t it ensures that Y is MSD at all t.

Now assume that one and hence all of the conditions (i)–(iii) are satisfied and let $h, t \in \mathbb{R}$. By continuity of the inner product,

$$\mathbb{E}(Y_t') = \lim_{\delta \to 0} \frac{1}{\delta} \langle Y_{t+\delta} - Y_t, 1 \rangle = 0$$

and

$$\begin{aligned} \mathrm{Cov}(Y_{t+h}', Y_t) = \langle Y_{t+h}', Y_t \rangle &= \left\langle \lim_{\delta \to 0}^{\mathcal{S}} \frac{1}{\delta}(Y_{t+h+\delta} - Y_{t+h}), Y_t \right\rangle \\ &= \lim_{\delta \to 0} \frac{\gamma(h+\delta) - \gamma(h)}{\delta}, \end{aligned}$$

showing that γ is differentiable at h and that $\mathrm{Cov}(Y_{t+h}', Y_t) = \gamma'(h)$. Using again the continuity of the inner product, we obtain

$$\begin{aligned} \mathrm{Cov}(Y_{t+h}', Y_t') &= \left\langle \lim_{\delta \to 0}^{\mathcal{S}} \frac{1}{\delta}(Y_{t+h+\delta} - Y_{t+h}), \lim_{u \to 0}^{\mathcal{S}} \frac{1}{u}(Y_{t+u} - Y_t) \right\rangle \\ &= \lim_{\delta \to 0} \lim_{u \to 0} \frac{1}{\delta u} [\gamma(h+\delta-u) - \gamma(h+\delta) - \gamma(h-u) + \gamma(h)] \\ &= \lim_{\delta \to 0} \frac{-\gamma'(h+\delta) + \gamma'(h)}{\delta}, \end{aligned}$$

showing that γ is twice differentiable at h and we obtain the desired form of $\mathrm{Cov}(Y_{t+h}', Y_t)$. □

Remark 3.14. Let Y be a stationary process with autocovariance function γ and mean μ. By Corollary 3.13, a necessary condition for Y to be MSD is that γ is twice differentiable. On the other hand, a sufficient condition for Y to be MSD is that γ is twice continuously differentiable in a neighbourhood of 0. This follows from Corollaries 3.11 and 3.13, since $M(s,t) = \gamma(s-t) + \mu\bar{\mu}$ is then twice continuously differentiable in a neighborhood of $(0,0)^T$, hence Y_t is MSD in a neighborhood of 0 and hence everywhere.

Example 3.15 (The Poisson process). For the Poisson process of Example 3.8,

$$\langle Y_r, Y_s \rangle = \begin{cases} \lambda(|r| \wedge |s|) + \lambda^2 rs, & \text{if } rs \geq 0, \\ \lambda^2 rs, & \text{if } rs < 0. \end{cases}$$

For $t \geq 0$ and $0 < h < h'$, a straightforward calculation shows that the expression (3.3) reduces (Exercise 3.4) to $\frac{1}{hh'}(\lambda h + \lambda^2 hh')$, which clearly has no finite limit as $h, h' \to 0$. By Theorem 3.9 we conclude that the Poisson process is not MSD at any $t \geq 0$ and an analogous argument shows that the same is true for all $t < 0$.

Example 3.16 (Stationary processes). The autocovariance function of the differenced Poisson process is $\gamma(h) = \lambda \max(d - |h|, 0)$ (see Example 1.23), which is not differentiable at 0. Hence the differenced Poisson process is nowhere mean-square differentiable.

The same is true for the differenced Brownian motion. Similarly, the autocovariance function of the stationary Poisson-driven Ornstein–Uhlenbeck process of Example 1.24 is nowhere mean-square differentiable since its autocovariance function is given by $\gamma(h) = \lambda e^{-a|h|}/(2a)$, which is not differentiable at 0. The autocovariance function of the stationary process $X_t = A\cos(\nu t + \Theta)$ of Example 1.22 is $\frac{1}{2}\mathbb{E}(A^2)\cos(\nu h)$, which is twice continuously differentiable, hence $(X_t)_{t\in\mathbb{R}}$ is mean-square differentiable everywhere.

3.4 Mean-square Riemann integrability

The mean-square Riemann integral is defined, for certain $\mathcal{S}$-valued functions on $\mathbb{R}$, as follows:

Definition 3.17. (a) Suppose that $(Y_t)_{t\in\mathbb{R}}$ is a zero-mean second-order process with autocovariance function κ, g is a complex-valued function on $\mathbb{R}$ and $-\infty < a < b < \infty$. Let $\{\pi_n := (t_0 = a < t_1 < \cdots < t_n = b)\}$ be a sequence of partitions of $[a, b]$ such that $\Delta(\pi_n) := \max_{1\le j\le n}(t_j - t_{j-1}) \to 0$ as $n \to \infty$. If

$$I_n := \sum_{j=0}^{n-1} g(t_j) Y_{t_j}(t_{j+1} - t_j)$$

converges in $\mathcal{S}$ as $n \to \infty$ to some random variable I, which is the same for all sequences of partitions π_n such that $\Delta(\pi_n) \to 0$ as $n \to \infty$, then we define the *mean-square Riemann integral* of $g(t)Y_t$ on $[a, b]$ as

$$\mathcal{R}_2 \int_a^b g(t) Y_t \,\mathrm{d}t := \lim_{n\to\infty}{}^{\mathcal{S}} I_n.$$

(b) If Y is a second-order process with $\mathbb{E}Y_t = \mu(t)$, then we define

$$\mathcal{R}_2 \int_a^b g(t) Y_t \,\mathrm{d}t := \mathcal{R}_2 \int_a^b g(t)(Y_t - \mu(t)) \,\mathrm{d}t + \int_a^b g(t)\mu(t)\mathrm{d}t,$$

provided the first integral on the right exists and the function $g(\cdot)\mu(\cdot)$ is Riemann integrable on $[a, b]$. (If Y is MSC and g is continuous, the second integral exists since the mean function of an MSC process is necessarily continuous.)

Theorem 3.18 (Existence of mean-square Riemann-integral). *If Y is a zero-mean second-order process with covariance function κ, g a complex-valued function on $\mathbb{R}$ and $-\infty < a < b < \infty$, then $\mathcal{R}_2 \int_a^b g(t) Y_t \,\mathrm{d}t$ exists if the Riemann integral, $\int_a^b \int_a^b g(r)\kappa(r, s)\overline{g(s)} \,\mathrm{d}r \,\mathrm{d}s$, exists. The expected value of the integral is then zero and its variance is equal to $\int_a^b \int_a^b g(r)\kappa(r, s)\overline{g(s)} \,\mathrm{d}r \,\mathrm{d}s$.*

Proof. Let $\{\pi_n = (t_0 = a < t_1 < \cdots < t_n = b)\}$ and $\{\pi'_m = (u_0 = a < u_1 < \cdots < u_m = b)\}$ be sequences of partitions of $[a, b]$ such that $\Delta(\pi_n) \to 0$ and $\Delta(\pi'_m) \to 0$ as $n, m \to \infty$. If I_n and I'_m are the corresponding approximating sums as in Definition 3.17, then

$$\mathbb{E}[I_n \overline{I'_m}] = \sum_{j=0}^{n-1} \sum_{k=0}^{m-1} g(t_j)\kappa(t_j, u_k)\overline{g(u_k)}(t_{j+1} - t_j)(u_{k+1} - u_k).$$

It follows from this expression that $\mathbb{E}[I_n \overline{I'_m}]$ converges as $n, m \to \infty$ to the Riemann integral, $\int_a^b \int_a^b g(t)\kappa(t, u)\overline{g(u)}\,\mathrm{d}t\mathrm{d}u$, and so, by the same argument used in the proof of Proposition 2.21,

$$\lim_{n\to\infty}{}^{\mathcal{S}} I_n = \lim_{m\to\infty}{}^{\mathcal{S}} I'_m = \mathcal{R}_2 \int_a^b g(t)Y_t\,\mathrm{d}t,$$

establishing the existence of the integral. Continuity of the inner product then ensures that its mean and variance are as asserted in the theorem. □

Corollary 3.19 (Existence of mean-square R.-integral for stat. processes). *If Y is a mean-square continuous zero-mean stationary process with autocovariance function γ and $-\infty < a < b < \infty$, then $\mathcal{R}_2 \int_a^b Y_t\,\mathrm{d}t$ exists and its variance is given by $\int_0^{b-a}(b - a - y)(\gamma(y) + \overline{\gamma(y)})\,\mathrm{d}y$.*

Proof. Since Y is MSC, the function γ is continuous. Hence $\kappa(r, s) = \gamma(r - s)$ is Riemann integrable (and Lebesgue-integrable) over $[a, b]^2$ and a routine calculation gives $\int_a^b \int_a^b \gamma(r - s)\,\mathrm{d}r\,\mathrm{d}s = \int_0^{b-a}(b - a - y)(\gamma(y) + \overline{\gamma(y)})\,\mathrm{d}y$. □

Remark 3.20. The same argument used in the proof of Theorem 3.18 shows that if Y is a zero-mean second-order process and the Riemann integral $\int_a^b \int_a^b \kappa(r, s)\,\mathrm{d}r\,\mathrm{d}s$ exists for *every* $[a, b]$ such that $-\infty < a < b < \infty$, then if $[a, b]$ and $[c, d]$ are two such intervals,

$$\operatorname{Cov}\left(\mathcal{R}_2 \int_a^b Y_u\,\mathrm{d}u,\ \mathcal{R}_2 \int_c^d Y_v\,\mathrm{d}v\right) = \int_a^b \int_c^d \kappa(u, v)\,\mathrm{d}u\,\mathrm{d}v. \tag{3.5}$$

Example 3.21 (The Poisson process). If Y is the Poisson process of Example 3.8 and $0 < a < b$, then the integral $\mathcal{R}_2 \int_a^b (Y_t - \mu(t))\,\mathrm{d}t$ exists by Theorem 3.18 since $\kappa(r, s) = \lambda(r \wedge s)$ is Riemann integrable on $[a, b] \times [a, b]$. Since $\mu(t) = \lambda t$ is integrable on $[a, b]$,

$$\mathcal{R}_2 \int_a^b Y_t\,\mathrm{d}t = \mathcal{R}_2 \int_a^b (Y_t - \mu(t))\,\mathrm{d}t + \frac{\lambda}{2}(b^2 - a^2),$$

where, again by Theorem 3.18, the first term on the right has mean zero and variance,

$$\int_a^b \int_a^b \lambda(r \wedge s)\, \mathrm{d}r\, \mathrm{d}s = \frac{\lambda}{3}(b-a)^2(b+2a).$$

3.5 Mean-square and pathwise properties

It is natural to ask what the mean-square properties of a process $(Y_t)_{t\in\mathbb{R}}$ tell us, if anything, about the corresponding pathwise properties of the realizations $Y_t(\omega)$, $t \in \mathbb{R}$, for each $\omega \in \Omega$, except possibly on a set of probability zero.

(i) Mean-square continuity of Y on a non-empty interval (a, b) does not imply almost sure continuity of Y on (a, b), as is evident from the sample paths of the Poisson process, which we know from Example 3.8 to be mean-square continuous at all $t \in \mathbb{R}$.

(ii) If the sample paths, $Y_t(\omega)$, $t \in \mathbb{R}$, are Riemann integrable on $[a, b]$ for all ω outside a set of probability zero, then

$$\mathcal{R}_2 \int_a^b Y_t\, \mathrm{d}t = \int_a^b Y_t\, \mathrm{d}t, \quad \text{almost surely,} \tag{3.6}$$

since both sides are the limits in probability of the same Riemann sums and are therefore equal with probability one. In particular (3.6) holds if Y is a Poisson process or Brownian motion.

(iii) Mean-square differentiability of Y on a non-empty interval (a, b) does not imply almost sure differentiability of Y on (a, b), as illustrated by the process $Y_t := \int_0^t N_u\, \mathrm{d}u$, $t \geq 0$, where N is a Poisson process with jump rate 1. (Mean-square differentiability of Y at each $t > 0$ follows from Corollary 3.11 and Remark 3.12, since a straightforward calculation using (3.5) shows that

$$\mathrm{Cov}\left(\mathcal{R}_2 \int_0^s N_u\, \mathrm{d}u,\ \mathcal{R}_2 \int_0^t N_v\, \mathrm{d}v \right) = \begin{cases} \frac{s^2}{6}(3t-s), & \text{if } 0 < s \leq t, \\ \frac{t^2}{6}(3s-t), & \text{if } s > t > 0, \end{cases}$$

and the right-hand side has continuous second partial derivatives with respect to s and t on $(0, \infty) \times (0, \infty)$. Further, the mean function of $\mathcal{R}_2 \int_0^t N_u\, \mathrm{d}u$, $t \geq 0$, is $\int_0^t u\, \mathrm{d}u = t^2/2$, which is twice continuously differentiable.) The pathwise derivative of Y on $(0, \infty)$ however exists only piecewise with probability one.

3.6 Exercises

Exercise 3.1. (a) In the proof of Theorem 3.1, show that

$$\mathbf{a}^T K_1^{(n)} \mathbf{a} + \mathbf{a}^T K_2^{(n)} \mathbf{b} - \mathbf{b}^T K_2^{(n)} \mathbf{a} + \mathbf{b}^T K_1^{(n)} \mathbf{b} = (\mathbf{a} + \mathrm{i}\mathbf{b})^T (K_1^{(n)} + \mathrm{i}K_2^{(n)})(\mathbf{a} - \mathrm{i}\mathbf{b}).$$

(b) For $\mathbf{t} = (t_1, \dots, t_n)^T \in \mathbb{R}^n$ consider an $\mathbb{R}^{2n}$-valued Gaussian distribution with mean $\mathbf{0}$ and covariance matrix $\Lambda^{(n)}(\mathbf{t})$, as defined in the proof of Theorem 3.1 (observe that $\Lambda^{(n)}(\mathbf{t})$ is non-negative definite, hence such a distribution exists). For $\mathbf{x} = (x_1, \dots, x_{i-1}, 0, x_{i+1}, \dots, x_n)^T \in \mathbb{R}^n$ and $\mathbf{y} = (y_1, \dots, y_{i-1}, 0, y_{i+1}, \dots, y_n)^T \in \mathbb{R}^n$, and $\mathbf{x}(i)$, $\mathbf{y}(i)$ and $\mathbf{t}(i)$, as defined in Theorem 1.13, show that

$$[\mathbf{x}^T, \mathbf{y}^T]\,\Lambda^{(n)}(\mathbf{t}) \begin{bmatrix} \mathbf{x} \\ \mathbf{y} \end{bmatrix} = [\mathbf{x}(i)^T, \mathbf{y}(i)^T]\,\Lambda^{(n-1)}(\mathbf{t}(i)) \begin{bmatrix} \mathbf{x}(i) \\ \mathbf{y}(i) \end{bmatrix}.$$

Using Kolmogorov's consistency theorem in the form (1.6) (generalized to the corresponding 2-dimensional setting, as outlined in Remark 1.14), deduce the existence of a bivariate Gaussian process $((Y_t, Z_t)^T)_{t\in\mathbb{R}}$ such that $(Y_{t_1}, \dots, Y_{t_n}, Z_{t_1}, \dots, Z_{t_n})^T$ is $N(\mathbf{0}, \Lambda^{(n)}(t_1, \dots, t_n))$-distributed.

Exercise 3.2. Prove Remark 3.3.

Exercise 3.3. Let $Y = (Y_t)_{t\in\mathbb{R}}$ be a second-order stochastic process that is mean-square differentiable at $t \in \mathbb{R}$.

(a) Show that it is also mean-square continuous at t.

(b) Show that the mean function μ_Y of Y is differentiable at t with $\mu_Y'(t) = \mathbb{E}(Y_t')$, where Y_t' denotes the mean-square derivative at t.

Exercise 3.4. Show that the Poisson process is not mean-square differentiable at any $t \in \mathbb{R}$. (See Example 3.15.)

Exercise 3.5. Verify the calculation in Section 3.5(iii).

The following result shows that the fundamental theorem of calculus extends to mean-square Riemann integrals and mean-square differentiability.

Exercise 3.6. Let $Y = (Y_t)_{t\in\mathbb{R}}$ be a zero-mean second-order process that is mean-square Riemann integrable over every compact interval. For fixed $a \in \mathbb{R}$, define the process $Z = (Z_t)_{t\in\mathbb{R}}$ by $Z_t := \mathcal{R}_2 \int_a^t Y_s \, ds$ when $t \geq a$ and $Z_t := -\mathcal{R}_2 \int_t^a Y_s \, ds$ when $t < a$. Assume that Y is mean-square continuous at some fixed $t_0 \in \mathbb{R}$. Show that Z is a second-order mean-zero process that is mean-square differentiable at t_0 with mean-square derivative Y_{t_0}.
Hint: Using Theorem 3.18, calculate the variance of $\frac{1}{h}(Z_{t_0+h} - Z_{t_0}) - Y_{t_0}$ as $h \to 0$.

4 Orthogonal increment processes

Abstract: Orthogonal increment processes, $(Z_t)_{t\in\mathbb{R}}$, play a key role in the spectral representation of stationary time series with continuous time parameter as well as in the construction of time-domain integral representations of such processes, where they play a role analogous to that of white noise (more precisely, to the sequence of cumulative sums of white noise) as building blocks from which to construct models with more complex covariance structures.

4.1 Preliminaries

In the previous chapter, we studied the second-order properties of processes $(X_t)_{t\in\mathbb{R}}$ such that $\mathbb{E}|X_t|^2 < \infty$ for all $t \in \mathbb{R}$. The weakly stationary complex-valued processes $(X_t)_{t\in\mathbb{R}}$ constitute a subclass for which $\mathbb{E}(X_t)$ is independent of t, and $\mathbb{E}[X_{t+h}\overline{X_t}]$ is independent of t for all $h \in \mathbb{R}$. ($\overline{X_t}$ denotes the complex conjugate of X_t.)

As before, we are concerned with processes defined on a probability space $(\Omega, \mathcal{F}, P)$ and we denote by $\mathcal{S}$ the Hilbert space of complex-valued square-integrable random variables, $L^2(\Omega, \mathcal{F}, P)$, with the inner product of any two elements X and Y defined as $\langle X, Y\rangle := \mathbb{E}(X\overline{Y})$. (More precisely $\mathcal{S}$ consists of equivalence classes, each consisting of a collection of almost surely equal random variables, but for our purposes each class can be represented by a single random variable belonging to the class.) The squared norm corresponding to this inner product is $\|X\|^2 := \langle X, X\rangle = \mathbb{E}|X|^2$ and the random variables X and Y are said to be orthogonal if $\langle X, Y\rangle = 0$. A sequence $(X_n)_{n\in\mathbb{N}}$ of random variables in $\mathcal{S}$ is said to converge in $\mathcal{S}$ to the random variable X as $n \to \infty$ if $\|X_n - X\| \to 0$ as $n \to \infty$, in which case we say that $X = \lim^{\mathcal{S}}_{n\to\infty} X_n$.

4.2 Definition and elementary properties

Definition 4.1. A *right-continuous orthogonal increment process* (*OIP*) on $\mathbb{R}$ is a complex-valued stochastic process $(Z_u)_{u\in\mathbb{R}}$ defined on a probability space $(\Omega, \mathcal{F}, P)$ such that

$$\|Z_u\|^2 = \mathbb{E}|Z_u|^2 < \infty, \quad u \in \mathbb{R}, \tag{4.1}$$

$$Z_0 = 0 \quad \text{almost surely}, \tag{4.2}$$

$$\mathbb{E}(Z_u) = 0, \quad u \in \mathbb{R}, \tag{4.3}$$

$$\|Z_{u+\delta} - Z_u\| \to 0 \quad \text{as } \delta \downarrow 0, u \in \mathbb{R}, \tag{4.4}$$

and

$$\langle Z_{u_4} - Z_{u_3}, Z_{u_2} - Z_{u_1}\rangle = 0, \quad \text{if } u_1 \le u_2 \le u_3 \le u_4, \tag{4.5}$$

where the inner product is defined by $\langle X, Y\rangle := \mathbb{E}(X\overline{Y})$.

https://doi.org/10.1515/9783111325033-004

The property (4.1) means that for each t, $Z_t \in \mathcal{S}$. Property (4.3) can be restated as $\langle Z_u, 1\rangle = 0$ and (4.5) (in conjunction with (4.3)) as

$$\operatorname{Cov}(Z_{u_4} - Z_{u_3}, Z_{u_2} - Z_{u_1}) = 0, \quad \text{if } u_1 \le u_2 \le u_3 \le u_4,$$

i. e., increments over non-overlapping intervals are uncorrelated. The term *orthogonal increment process* (*OIP*) will henceforth always mean *right-continuous* orthogonal increment process, i. e., it will be assumed that (4.4) holds.

Proposition 4.2 (Distribution function associated with OIP). *If $(Z_u)_{u\in\mathbb{R}}$ is an OIP, then there is a unique non-decreasing right-continuous function F on $\mathbb{R}$ such that*

$$F(0) = 0 \tag{4.6}$$

and

$$F(v) - F(u) = \|Z_v - Z_u\|^2, \quad -\infty < u \le v < \infty. \tag{4.7}$$

Proof. For F to satisfy (4.6) and (4.7), it is clear, on setting $u = 0$ and observing (4.2), that $F(v) = \|Z_v\|^2$ when $v \ge 0$ and, on setting $v = 0$, that $F(u) = -\|Z_u\|^2$ when $u < 0$. In order for F to be a right-continuous function satisfying (4.6) and (4.7), it must therefore be the function,

$$F(v) = \begin{cases} \|Z_v\|^2, & \text{if } v \ge 0, \\ -\|Z_v\|^2, & \text{if } v < 0. \end{cases} \tag{4.8}$$

To verify the monotonicity and right-continuity of F as defined by (4.8), we easily find from the orthogonality of the increments of Z that,

$$F(v) = F(u) + \|Z_v - Z_u\|^2, \quad -\infty < u \le v < \infty,$$

showing that F satisfies (4.7) and is indeed non-decreasing. In particular, for $-\infty < v \le v + \delta < \infty$,

$$F(v+\delta) = F(v) + \|Z_{v+\delta} - Z_v\|^2 \to F(v) \quad \text{as } \delta \downarrow 0,$$

by the assumed right-continuity of Z. Thus the right-continuity of F follows from that of Z. □

Remark 4.3. The distribution function F defined in (4.8) will be referred to as the *distribution function associated with the OIP Z*. The equations (4.5) and (4.7) are often combined into the shorthand version,

$$\mathbb{E}(dZ_u\, d\overline{Z}_v) = \delta_{u,v}\, dF(u),$$

with $\delta_{u,v}$ denoting the Kronecker symbol.

Example 4.4 (Brownian motion). Brownian motion $(B_t)_{t\in\mathbb{R}}$ with zero mean and variance σ^2 is an orthogonal increment process on $\mathbb{R}$ with the properties
1. $B_0 = 0$ almost surely,
2. B_t is normally distributed with mean 0 and variance $\sigma^2|t|$.

The associated distribution function is

$$F(t) = \sigma^2 t, \quad t \in \mathbb{R}.$$

Example 4.5 (Centered Poisson process). A Poisson process $(N_t)_{t\in\mathbb{R}}$ with intensity $\lambda > 0$ has the properties,
1. $N_0 = 0$ almost surely,
2. $N_v - N_u$, $-\infty < u < v < \infty$, has the Poisson distribution with mean $\lambda(v-u)$,
3. the increments $N_v - N_u$ on disjoint intervals $(u, v]$ are independent.

As a consequence of these properties, the *centered Poisson process*, $(N_t - \lambda t)_{t\in\mathbb{R}}$ is an OIP with associated distribution function,

$$F(t) = \lambda t, \quad t \in \mathbb{R}.$$

If $\lambda = \sigma^2$, the associated distribution function is the same as that of the Brownian motion in the previous example.

Remark 4.6. In spite of the similarity between their associated distribution functions, the sample paths of Brownian motion and the centered Poisson process are quite different. With a probability of one, Brownian sample paths are continuous, while the centered Poisson paths are linear, except for jumps of size one, of which the number in any interval of length ℓ has the Poisson distribution with mean $\lambda\ell$.

In Examples 4.4 and 4.5, the associated distribution functions where of the form $F(t) = ct$ for some $c \geq 0$. Can other distribution functions also arise? This is indeed the case. Any right-continuous and non-decreasing function is the distribution function associated with some OIP, and we can even choose the OIP as a Gaussian process and have independent increments.

Theorem 4.7 (Existence of OIPs). *Let F be a real-valued non-decreasing function on $\mathbb{R}$ satisfying $F(0) = 0$. Then there exists (on some probability space) a Gaussian process $(Z_t)_{t\in\mathbb{R}}$ such that*
(i) *$Z_0 = 0$ almost surely,*
(ii) *Z has independent increments, i. e. $Z_{t_2} - Z_{t_1}, Z_{t_3} - Z_{t_2}, \ldots, Z_{t_n} - Z_{t_{n-1}}$, are independent for every $(t_1, \ldots, t_n)^T \in \mathbb{R}^n$ such that $t_1 < \cdots < t_n$, and*
(iii) *$Z_t - Z_s$ has the normal distribution with mean zero and variance $F(t) - F(s)$ for all $s, t \in \mathbb{R}$ such that $t \geq s$.*

If the function F is additionally right-continuous, then so is Z, and hence Z is an orthogonal increment process with associated distribution function F.

Proof. In complete analogy to the proof of the existence of Brownian motion in Section 1.4, we observe that the properties (i), (ii) and (iii) are satisfied for a Gaussian process (Z_t), restricted to $t \in [0, \infty)$ if and only if for every $\mathbf{t} = (t_1, t_2, \ldots, t_n)^T \in [0, \infty)^n$ such that $t_1 < t_2 < \cdots < t_n$, the characteristic function of $(Z_{t_1}, \ldots, Z_{t_n})^T$ is

$$\phi_{\mathbf{t}}(\mathbf{u}) = \exp\left[-\frac{1}{2}\sum_{j=1}^{n}(u_j + \cdots + u_n)^2(F(t_j) - F(t_{j-1}))\right]$$

(where $t_0 := 0$ and $\mathbf{u} = (u_1, \ldots, u_n)^T \in \mathbb{R}^n$). Again, it is trivial to check that the characteristic functions $\phi_{\mathbf{t}}(\cdot)$ satisfy the consistency condition (1.6) and so, according to Kolmogorov's consistency theorem (Theorem 1.13), there exists a process $(Z_t)_{t\geq 0}$ with characteristic functions $\phi_{\mathbf{t}}(\cdot)$, or equivalently with the properties (i), (ii) and (iii). Now define the function $G : [0, \infty) \to \mathbb{R}$ by $G(t) = -F(-t)$. Then G is non-decreasing with $G(0) = 0$ and by the previous argument there exists a process $(W_t)_{t\geq 0}$ satisfying conditions (i), (ii) and
(iii′) $W_t - W_s \overset{d}{=} N(0, G(t) - G(s))$ for $0 \leq s \leq t$.

By constructing W on a different probability space and then taking the product space, we can arrange it such that $(Z_t)_{t\geq 0}$ and $(W_t)_{t\geq 0}$ become independent. Now extend Z_t to the negative real line by setting $Z_t := -W_{-t}$ for $t < 0$. Then, for $s \leq t \leq 0$,

$$Z_t - Z_s = W_{-s} - W_{-t} \overset{d}{=} N(0, G(-s) - G(-t)) = N(0, F(t) - F(s)),$$

so that $(Z_t)_{t\leq 0}$ satisfies conditions (i), (ii) and (iii) and it is then easily checked that also $(Z_t)_{t\in\mathbb{R}}$ satisfies conditions (i), (ii) and (iii) on the whole real line. That $(Z_t)_{t\in\mathbb{R}}$ is then a Gaussian process, i. e., that all its finite dimensional distributions are Gaussian, is clear since Z has independent and normally distributed increments.

Now assume that F is additionally right-continuous. That Z as constructed satisfies (4.1), (4.2), (4.3) and (4.5) is clear from (i) – (iii). Further, $\|Z_t - Z_s\|^2 = F(t) - F(s)$ for $s \leq t$ by (iii), from which (4.4) follows. Hence, Z is an OIP with associated distribution function F. □

4.3 Integration with respect to an OIP

We next show how to construct stochastic integrals of the form, $I(f) = \int_{\mathbb{R}} f(u)\, \mathrm{d}Z_u$, where $(Z_u)_{u\in\mathbb{R}}$ is an OIP defined on the probability space $(\Omega, \mathcal{F}, P)$ and f a suitable function. Denote the distribution function associated with Z by F. Since F is non-decreasing and right-continuous, there exists a unique measure μ_F on $(\mathbb{R}, \mathcal{B}_1 = \mathcal{B}(\mathbb{R}))$ such that

$$F(b) - F(a) = \mu_F((a,b]), \quad \text{for all } -\infty < a \le b < \infty \tag{4.9}$$

(see, e. g., Billingsley [16, Theorem 12.4]). The (Lebesgue–Stieltjes) integral of a deterministic Borel-measurable function $f : \mathbb{R} \to \mathbb{C}$ with respect to F is then the measure-theoretic integral with respect to μ_F, i. e.,

$$\int_{\mathbb{R}} f(t)\,\mathrm{d}F(t) := \int_{\mathbb{R}} f(t)\,\mathrm{d}\mu_F(t),$$

provided the right-hand side exists. We say that f is *(square-) integrable with respect to* F if it is (square-) integrable with respect to μ_F, and we denote by

$$L^2(F) := L^2((\mathbb{R}, \mathcal{B}_1, \mu_F); \mathbb{C})$$

the space of all complex-valued functions on $\mathbb{R}$ that are square-integrable with respect to F, i. e., with respect to μ_F. By Remark 2.19, the space $L^2(F)$ is a Hilbert space with inner product

$$\langle f, g\rangle_{L^2(F)} = \int_{\mathbb{R}} f(u)\overline{g}(u)\,\mathrm{d}F(u) = \int_{\mathbb{R}} f(u)\overline{g}(u)\,\mathrm{d}\mu_F(u)$$

and squared norm $\|f\|^2_{L^2(F)} := \langle f, f\rangle_{L^2(F)}$ (if functions that are μ_F-almost everywhere equal are identified). Convergence in this space of a sequence of functions f_n to f as $n \to \infty$ means that $\|f_n - f\|^2_{L^2(F)} \to 0$ as $n \to \infty$ and will also be written as $f = \lim^{L^2(F)}_{n\to\infty} f_n$.

Our goal is to define the stochastic integral

$$I(f) = \int_{\mathbb{R}} f(u)\,\mathrm{d}Z_u$$

when Z is an OIP with distribution function F and $f \in L^2(F)$. The integral $I(f)$ will then be a random variable in $\mathcal{S}$.

We define the integral first for a class of simple functions with compact support, then extend the definition by continuity to the closure in $L^2(F)$ of this class and finally show that this closure is in fact $L^2(F)$ itself.

Step 1. Let $\mathcal{D}$ denote the class of simple functions $f \in L^2(F)$ with support $[-T, T]$, for some $T \in (0, \infty)$, which can be written as

$$f(u) = \sum_{i=0}^{n} f_i \mathbf{1}_{(u_i, u_{i+1}]}(u), \quad -T = u_0 < u_1 < \cdots < u_{n+1} = T, \tag{4.10}$$

for some $n \in \mathbb{N}$ and some partition, $\pi(T) = (-T = u_0 < u_1 < \cdots < u_{n+1} = T)$ of $[-T, T]$. For such functions it is natural to define $I(f)$ as

$$I(f) = \sum_{i=0}^{n} f_i [Z_{u_{i+1}} - Z_{u_i}]. \tag{4.11}$$

The definition is consistent since there is a unique partition $\pi_0(T) = (-T = t_0 < \cdots < t_{m+1} = T)$ of $[-T, T]$ such that

$$f(u) = \sum_{i=0}^{m} g_i \mathbf{1}_{(t_i, t_{i+1}]} \quad \text{and} \quad g_i \neq g_{i+1}, \quad i = 0, \ldots, m$$

(unless f is the zero-function). All other representations of f having the form (4.10) are obtained by reexpressing one or more of the indicator functions $\mathbf{1}_{(t_i, t_{i+1}]}$ as a sum of indicator functions of adjoining intervals. However, this makes no difference to the value of $I(f)$ and hence the definition (4.11) is the same for all representations (4.10) of f.

Proposition 4.8. *The mapping defined by* (4.11) *of $\mathcal{D}$ into $\mathcal{S}$ is linear and preserves inner products.*

Proof. The linearity of the mapping is clear. To establish the preservation of inner products, we observe that any f and g in $\mathcal{D}$ can be represented using a common value of T and a common partition of $[-T, T]$. Consequently, we can write

$$f(u) = \sum_{i=0}^{n} f_i \mathbf{1}_{(u_i, u_{i+1}]}(u) \quad \text{and}$$
$$g(u) = \sum_{i=0}^{n} g_i \mathbf{1}_{(u_i, u_{i+1}]}(u).$$

The inner product of $I(f)$ and $I(g)$ in $\mathcal{S}$ is thus

$$\begin{aligned}\langle I(f), I(g) \rangle &= \left\langle \sum_{i=0}^{n} f_i [Z_{u_{i+1}} - Z_{u_i}], \sum_{i=0}^{n} g_i [Z_{u_{i+1}} - Z_{u_i}] \right\rangle \\ &= \sum_{i=0}^{n} f_i \overline{g}_i [F(u_{i+1}) - F(u_i)],\end{aligned}$$

by the orthogonality of the increments of Z and Proposition 4.2. But the last expression can be written as

$$\int_{\mathbb{R}} f(u) \overline{g}(u) \mathrm{d}F(u) = \langle f, g \rangle_{L^2(F)},$$

the inner product in $L^2(F)$ of f and g. Hence the mapping I on $\mathcal{D}$ preserves the inner product. □

Step 2. Let $\overline{\mathcal{D}}$ be the closure in $L^2(F)$ of the set $\mathcal{D}$. If $f \in \overline{\mathcal{D}}$ then there is a sequence $(f_n)_{n \in \mathbb{N}}$ of elements of $\mathcal{D}$ such that $f = \lim_{n \to \infty}^{L^2(F)} f_n$. We therefore define $I(f)$ for $f \in \overline{\mathcal{D}}$ as the limit in $\mathcal{S}$,

$$I(f) = \lim_{n\to\infty}{}^{\mathcal{S}}\, I(f_n), \tag{4.12}$$

after first checking (a) that the limit exists and (b) that the limit is the same for all sequences $(f_n)_{n\in\mathbb{N}}$ such that $\|f_n - f\|_{L^2(F)} \to 0$.

To check (a) observe that for f_m and f_n in $\mathcal{D}$, Proposition 4.8 implies that

$$\|I(f_n) - I(f_m)\| = \|I(f_n - f_m)\| = \|f_n - f_m\|_{L^2(F)}$$

so if $\|f_n - f\|_{L^2(F)} \to 0$, then $(I(f_n))_{n\in\mathbb{N}}$ is a Cauchy sequence and therefore convergent in $\mathcal{S}$.

To check (b), suppose that $\|f_n - f\|_{L^2(F)} \to 0$ and $\|g_n - f\|_{L^2(F)} \to 0$, where $f_n \in \mathcal{D}$ and $g_n \in \mathcal{D}$. Then the sequence $f_1, g_1, f_2, g_2, \dots$ must be convergent in $L^2(F)$ and therefore the sequence $I(f_1), I(g_1), I(f_2), I(g_2), \dots,$ must converge in $\mathcal{S}$. But this is not possible unless the subsequences $(I(f_n))_{n\in\mathbb{N}}$ and $(I(g_n))_{n\in\mathbb{N}}$ have the same $\mathcal{S}$ limit.

To check that the extended mapping I defined by (4.12) on $\overline{\mathcal{D}}$ is linear and preserves the inner product, suppose that $f^{(i)} \in \overline{\mathcal{D}}$, $i = 1, 2$, and $(f_n^{(i)})_{n\in\mathbb{N}}$, $i = 1, 2$, are sequences of functions in $\mathcal{D}$ such that $\|f_n^{(i)} - f^{(i)}\|_{L^2(F)} \to 0$, $i = 1, 2$. Then, by the linearity of I on $\mathcal{D}$, for any complex-valued scalars a_1 and a_2,

$$\begin{aligned} I(a_1 f^{(1)} + a_2 f^{(2)}) &= \lim_{n\to\infty}{}^{\mathcal{S}}\, I(a_1 f_n^{(1)} + a_2 f_n^{(2)}) \\ &= \lim_{n\to\infty}{}^{\mathcal{S}}\, (a_1 I(f_n^{(1)}) + a_2 I(f_n^{(2)})) \\ &= a_1 I(f^{(1)}) + a_2 I(f^{(2)}), \end{aligned}$$

and by the continuity of the inner product,

$$\begin{aligned} \langle I(f^{(1)}), I(f^{(2)}) \rangle &= \left\langle \lim_{n\to\infty}{}^{\mathcal{S}}\, I(f_n^{(1)}), \lim_{n\to\infty}{}^{\mathcal{S}}\, I(f_n^{(2)}) \right\rangle \\ &= \lim_{n\to\infty} \langle f_n^{(1)}, f_n^{(2)} \rangle_{L^2(F)} \\ &= \langle f^{(1)}, f^{(2)} \rangle_{L^2(F)}. \end{aligned}$$

We have thus established that I as defined by (4.12) on $\overline{\mathcal{D}}$ is linear and preserves the inner product.

Step 3. It remains only to show that $\overline{\mathcal{D}} = L^2(F)$. But $\mathcal{D}$ is clearly dense in $C_c(\mathbb{R}; \mathbb{C})$, the set of complex-valued continuous functions on $\mathbb{R}$ with compact support, and $C_c(\mathbb{R})$ is dense in $L^2(F)$ (see Rudin [75, Theorem 3.14]). This implies that the closure of $\mathcal{D}$ in $L^2(F)$ is indeed $L^2(F)$. The mapping I defined on $\overline{\mathcal{D}}$ by (4.11) and (4.12) is thus a linear inner-product preserving mapping of $L^2(F)$ into $\mathcal{S}$. This leads to the following definition.

Definition 4.9 (The stochastic integral). If Z is an OIP on $\mathbb{R}$ with associated distribution function F and if $f \in L^2(F)$, then the *stochastic integral*,

$$I(f) = \int_{\mathbb{R}} f(u)\,\mathrm{d}Z_u,$$

of f with respect to the OIP Z is defined to be the image of f under the linear, inner-product preserving mapping of $L^2(F)(=\overline{\mathcal{D}})$ into $\mathcal{S}$, defined by (4.11) and (4.12). This implies, in particular, that the mapping I is a *linear isometry*, i. e. a linear mapping such that

$$\mathbb{E}(|I(f)|^2) = \|I(f)\|^2 = \|f\|^2_{L^2(F)}, \quad \text{for all } f \in L^2(F),$$

sometimes called the *Itô isometry*, of $L^2(F)$ onto the closed subspace $\{I(f), f \in L^2(F)\}$ of $\mathcal{S}$.

Proposition 4.10 (Properties of the stochastic integral). *If Z is an OIP with associated distribution function F, if f, g, f_n, g_n are all in $L^2(F)$, and if $f = \lim^{L^2(F)}_{n\to\infty} f_n$ and $g = \lim^{L^2(F)}_{n\to\infty} g_n$, then the following properties hold.*

(i)

$$\mathbb{E}(I(f)) = 0. \tag{4.13}$$

(ii)

$$I(a_1 f + a_2 g) = a_1 I(f) + a_2 I(g) \quad \textit{for all } a_1, a_2 \in \mathbb{C}.$$

(iii)

$$I(f) = \lim_{n\to\infty}{}^{\mathcal{S}}\, I(f_n). \tag{4.14}$$

(iv)

$$\mathbb{E}(I(f)\overline{I(g)}) = \int_{\mathbb{R}} f(u)\overline{g(u)}\,\mathrm{d}F(u). \tag{4.15}$$

(v)

$$\mathbb{E}(I(f_n)\overline{I(g_n)}) \to \mathbb{E}(I(f)\overline{I(g)}) = \int_{\mathbb{R}} f(u)\overline{g(u)}\,\mathrm{d}F(u) \quad \textit{as } n \to \infty.$$

Proof. (i) From (4.11) it is clear that $E(I(f)) = \langle I(f), 1\rangle = 0$ for all $f \in \mathcal{D}$. It is therefore also true for $f \in L^2(F)$ by the denseness of $\mathcal{D}$ and the continuity of the inner product.

(ii) This is a restatement of the linearity of I.

(iii) This is a consequence of $\|f_n - f\|_{L^2(F)} = \|I(f_n) - I(f)\|$.

(iv) This is a restatement of the preservation of the inner product by I.

(v) This follows by continuity of the inner product. □

Example 4.11 (Continuous-time moving averages). If $(Z_u)_{u\in\mathbb{R}}$ is an OIP with associated distribution function, $F(t) = ct$, $t \in \mathbb{R}$, for some $c > 0$, then the OIP has weakly stationary increments in the sense that the mean and variance of $Z_{t+h} - Z_t$ are independent of t (in fact, the mean is zero and the variance is $c|h|$). If $g \in L^2(F)$, then, by (4.13) and (4.15), the process,

$$X_t := \int_{\mathbb{R}} g(t-u)\, \mathrm{d}Z(u), \quad t \in \mathbb{R}, \tag{4.16}$$

is a weakly stationary process with zero mean and autocovariance function,

$$\gamma_X(h) = \mathbb{E}(X_{t+h}\overline{X}_t) = c\int_{\mathbb{R}} g(t+h)\overline{g(t)}\, \mathrm{d}t, \quad h \in \mathbb{R}.$$

Since $g(\cdot + h)$ converges in L^2 to g as $h \to 0$ by Lemma 2.68, the continuity of the inner product implies that γ_X is continuous at 0, hence X is MSC by Proposition 3.6 (alternatively, mean-square continuity can be deduced directly from (4.16) and the continuity of the stochastic integral). Processes of the form (4.16) are known as *continuous-time moving average* (or *CMA*) *processes*. By appropriate choice of the kernel function g, they can be made to exhibit a very large range of autocorrelation functions for the representation of dependent data.

Example 4.12 (The causal stationary Gaussian CAR(1) process). If in Example 4.11 we take Z to be the Brownian motion B of Example 4.4 and g to be the function

$$g(t) = \mathrm{e}^{-at}\mathbf{1}_{[0,\infty)}(t), \quad a > 0,$$

we obtain the *causal stationary Gaussian* CAR(1) *process* (also known as the *causal stationary Gaussian Ornstein–Uhlenbeck process*),

$$X_t = \int_{\mathbb{R}} \mathrm{e}^{-a(t-u)}\mathbf{1}_{(-\infty,t]}(u)\, \mathrm{d}B_u, \tag{4.17}$$

with mean zero and autocovariance function,

$$\mathbb{E}(X_{t+h}X_t) = \frac{\sigma^2}{2a}\mathrm{e}^{-a|h|}.$$

This process X is said to be *causal* (or more precisely *a causal function of B*), since, for each t, X_t depends only on the increments of B_u, $u \le t$.

4.4 OIP processes with densities

In this section, we show how the stochastic integral induces new OIPs from a given one. For an OIP $Z = (Z_t)_{t\in\mathbb{R}}$, a Borel set $A \subset \mathbb{R}$ and a function $f : \mathbb{R} \to \mathbb{C}$ such that $\mathbf{1}_A f \in L^2(F)$ we define

$$\int_A f(u)\, \mathrm{d}Z_u := \int_{\mathbb{R}} f(t)\mathbf{1}_A(u)\, \mathrm{d}Z_u,$$

and call it the *stochastic integral of f with respect to Z over the set A*.

Theorem 4.13 (Density of OIP). *Let $Z = (Z_t)_{t\in\mathbb{R}}$ be an OIP with associated distribution function F. Let g be a complex-valued function on $\mathbb{R}$ that is locally in $L^2(F)$, i. e., such that $g\mathbf{1}_{[-T,T]} \in L^2(F)$ for all $T > 0$. Define the stochastic process $M = (M_t)_{t\in\mathbb{R}}$ by*

$$M_t = \begin{cases} \int_{(0,t]} g(u)\,\mathrm{d}Z_u, & \text{if } t \geq 0, \\ -\int_{(t,0]} g(u)\,\mathrm{d}Z_u, & \text{if } t < 0. \end{cases}$$

(a) *Then M is an orthogonal increment process with associated distribution function G given by*

$$G(t) = \begin{cases} \int_{(0,t]} |g(u)|^2\,\mathrm{d}F(u), & \text{if } t \geq 0, \\ -\int_{(t,0]} |g(u)|^2\,\mathrm{d}F(u), & \text{if } t < 0. \end{cases}$$

We say that M has density g with respect to Z.

(b) *If $\psi \in L^2(G)$ then $\psi g \in L^2(F)$ and*

$$\int_{\mathbb{R}} \psi(t)\,\mathrm{d}M_t = \int_{\mathbb{R}} \psi(t)g(t)\,\mathrm{d}Z_t. \tag{4.18}$$

(c) *The density of M with respect to Z is F-almost everywhere unique, i. e., if h is another function that is locally in $L^2(F)$ and such that M also has density h with respect to Z, then $h = g$ μ_F-almost everywhere, where μ_F is the measure associated with F via* (4.9).

Proof. (a) From the definition of the stochastic integral, we see immediately that $\mathbb{E}|M_t|^2 < \infty$, that $\mathbb{E}M_t = 0$ and that $M_0 = 0$ (almost surely). Further, from the definition we see that $M_t - M_s = \int_{\mathbb{R}} \mathbf{1}_{(s,t]}(u)g(u)\,\mathrm{d}Z_u$ for $s \leq t$. If $t_1 \leq t_2 \leq t_3 \leq t_4$ then (4.15) shows that

$$\langle M_{t_4} - M_{t_3}, M_{t_2} - M_{t_1}\rangle = \int_{\mathbb{R}} \mathbf{1}_{(t_3,t_4]}(u)g(u)\,\overline{\mathbf{1}_{(t_1,t_2]}(u)g(u)}\,\mathrm{d}F(u) = 0,$$

so that M has orthogonal increments, and similarly that

$$\|M_{t_2} - M_{t_1}\|^2 = \int_{\mathbb{R}} \mathbf{1}_{(t_1,t_2]}(u)|g(u)|^2\,\mathrm{d}F(u). \tag{4.19}$$

Letting $t_2 \downarrow t_1$ the right-hand side of (4.19) converges to 0 by the dominated convergence theorem, so that M is right-continuous. Hence, M is an OIP. Setting $t_1 = 0$ or $t_2 = 0$ in (4.19), we also derive the desired form of the associated distribution function G of M.

(b) That $\psi g \in L^2(F)$ for $\psi \in L^2(G)$ is clear, since $\int_{\mathbb{R}} |\psi(u)|^2\,\mathrm{d}G(u)$ is equal to $\int_{\mathbb{R}} |\psi(u)|^2|g(u)|^2\,\mathrm{d}F(u)$. If $\psi = \mathbf{1}_{(s,t]}$ for some $s < t$, then

$$\int_{\mathbb{R}} \mathbf{1}_{(s,t]}(u)\,\mathrm{d}M_u = M_t - M_s = \int_{\mathbb{R}} \mathbf{1}_{(s,t]}(u)g(u)\,\mathrm{d}Z_u,$$

so that (4.18) is true for indicator functions of intervals. By linearity, it is then also true for any $\psi \in \mathcal{D}$. For general $\psi \in L^2(G)$, choose a sequence $(\psi_n)_{n\in\mathbb{N}}$ of functions in $\mathcal{D}$ that converges in $L^2(G)$ to ψ (which is again possible by Rudin [75, Theorem 3.14]). Then $\psi_n g$ converges in $L^2(F)$ to ψg, and the continuity of the stochastic integral (equation (4.14)) then implies the validity of (4.18) also for general $\psi \in L^2(G)$.

(c) For $t \geq 0$, we have $M_t = \int_{(0,t]} g(u)\,\mathrm{d}Z_u = \int_{(0,t]} h(u)\,\mathrm{d}Z_u$, hence $\int_{(0,t]}(g(u) - h(u))\,\mathrm{d}Z_u = 0$. Taking the variance, we obtain $0 = \int_{(0,t]} |g(u) - h(u)|^2\,\mathrm{d}F(u)$ from (4.15). A similar argument works for $t < 0$. □

Next, we establish a converse to Theorem 4.13. Suppose that we have an OIP $M = (M_t)_{t\in\mathbb{R}}$ with associated distribution function G which is absolutely continuous with respect to some other right-continuous non-decreasing function F satisfying $F(0) = 0$, i. e., such that there exists a non-negative function h which is locally integrable with respect to F such that $G(t) - G(s) = \int_{(s,t]} h(u)\,\mathrm{d}F(u)$ for all $s \leq t$. Can we construct an OIP $Z = (Z_t)_{t\in\mathbb{R}}$ with associated distribution function F such that M has a density with respect to Z? By the previous theorem, an obvious choice to construct such a process would be to define Z_t as $\int_{\mathbb{R}} \frac{1}{g(t)}\,\mathrm{d}M_t$, where g is a Borel measurable function such that $|g|^2 = h$. However, if $g(t) = 0$ we have to be more careful with this construction. This can be addressed by enlarging the probability space.

Corollary 4.14. *Let $M = (M_t)_{t\in\mathbb{R}}$ be an OIP with associated distribution function G. Let F be a real-valued, non-decreasing right-continuous function on $\mathbb{R}$ with $F(0) = 0$ and assume that G is absolutely continuous with respect to F, i. e., has a density $h \geq 0$ with respect to F such that $G(t) - G(s) = \int_{(s,t]} h(u)\,\mathrm{d}F(u)$ for all $s \leq t$. Let $g : \mathbb{R} \to \mathbb{C}$ be an arbitrary Borel-measurable function such that $|g|^2 = h$. Then there exist an OIP $Z = (Z_t)_{t\in\mathbb{R}}$, possibly defined on an enlarged probability space, such that Z has associated distribution function F and such that M has density g with respect to Z.*

Proof. By enlarging the probability space and using Theorem 4.7, we can construct an OIP $W^3 = (W^3_t)_{t\in\mathbb{R}}$ with associated distribution function F such that W^3 is uncorrelated with M. Denote $B := \{t \in \mathbb{R} : g(t) \neq 0\}$. Observe that

$$\int\limits_{(-T,T]} \frac{1}{|g(u)|^2} \mathbf{1}_B(u)\,\mathrm{d}G(u) = \int\limits_{(-T,T]} \frac{1}{|g(u)|^2} \mathbf{1}_B(u)\,|g(u)|^2\,\mathrm{d}F(u)$$

is finite for all $T > 0$, so that $\frac{1}{g}\mathbf{1}_B$ is locally in $L^2(G)$. Define the process $W^1 = (W^1_t)_{t\in\mathbb{R}}$ by

$$W^1_t := \begin{cases} \int_{(0,t]} \frac{1}{g(u)} \mathbf{1}_B(u)\,\mathrm{d}M_u, & \text{if } t \geq 0, \\ -\int_{(t,0]} \frac{1}{g(u)} \mathbf{1}_B(u)\,\mathrm{d}M_u, & \text{if } t < 0. \end{cases}$$

By Theorem 4.13, W^1 is an OIP with associated distribution function F_{W^1} satisfying

$$F_{W^1}(t) - F_{W^1}(s) = \int_{(s,t]} \frac{1}{|g(u)|^2} \mathbf{1}_B(u)\,\mathrm{d}G(u)$$
$$= \int_{(s,t]} \frac{1}{|g(u)|^2} \mathbf{1}_B(u)\,|g(u)|^2\,\mathrm{d}F(u) = \int_{(s,t]} \mathbf{1}_B(u)\,\mathrm{d}F(u)$$

for all $s \le t$. Next, define $W^2 = (W_t^2)_{t\in\mathbb{R}}$ by

$$W_t^2 := \begin{cases} \int_{(0,t]} \mathbf{1}_{\mathbb{R}\setminus B}(u)\,\mathrm{d}W_u^3, & \text{if } t \ge 0, \\ -\int_{(t,0]} \mathbf{1}_{\mathbb{R}\setminus B}(u)\,\mathrm{d}W_u^3, & \text{if } t < 0. \end{cases}$$

By Theorem 4.13, W^2 is an OIP with associated distribution function F_{W^2} satisfying

$$F_{W^2}(t) - F_{W^2}(s) = \int_{(s,t]} \mathbf{1}_{\mathbb{R}\setminus B}(u)\,\mathrm{d}F(u), \quad s \le t.$$

By construction of the stochastic integral, we see that W^1 is uncorrelated with W^3 and hence uncorrelated with W^2. Define $Z = (Z_t)_{t\in\mathbb{R}}$ by $Z_t := W_t^1 + W_t^2$. Then Z is an OIP with associated distribution function $F_Z = F_{W^1} + F_{W^2} = F$, see Exercise 4.2 (a). Since $g = g\mathbf{1}_B$ is locally in $L^2(F) = L^2(F_{W^1}) \cap L^2(F_{W^2})$, we conclude from Exercise 4.2 (b) and (4.18) for $t \ge 0$ that

$$\int_{(0,t]} g(u)\,\mathrm{d}Z_u = \int_{(0,t]} g(u)\mathbf{1}_B(u)\,\mathrm{d}W_u^1 + \int_{(0,t]} g(u)\mathbf{1}_B(u)\,\mathrm{d}W_u^2$$
$$= \int_{(0,t]} g(u)\mathbf{1}_B(u)\,\frac{1}{g(u)}\,\mathrm{d}M_u + \int_{(0,t]} g(u)\mathbf{1}_B(u)\,\mathrm{d}W_u^2.$$

But

$$\int_{\mathbb{R}} \left|g(u)\mathbf{1}_B(u)\right|^2 \mathrm{d}F_{W^2}(u) = \int_{\mathbb{R}} \left|g(u)\mathbf{1}_B(u)\right|^2 \mathbf{1}_{\mathbb{R}\setminus B}(u)\,\mathrm{d}F(u) = 0,$$

so that the integral with respect to W^2 above is 0. Similarly,

$$\int_{\mathbb{R}} \left|\mathbf{1}_{\mathbb{R}\setminus B}(u)\right|^2 \mathrm{d}G(u) = \int_{\mathbb{R}} \mathbf{1}_{\mathbb{R}\setminus B}(u)\left|g(u)\right|^2 \mathrm{d}F(u) = 0,$$

so that $\int_{(0,t]} \mathbf{1}_{\mathbb{R}\setminus B}(u)\,\mathrm{d}M_u = 0$, resulting in

$$\int_{(0,t]} g(u)\,\mathrm{d}Z_u = \int_{(0,t]} \mathbf{1}_B(u)\,\mathrm{d}M_u + \int_{(0,t]} \mathbf{1}_{\mathbb{R}\setminus B}(u)\,\mathrm{d}M_u = M_t - M_0 = M_t$$

for $t \ge 0$. A similar calculation holds for $t < 0$, so that M has density g with respect to Z. □

Remark 4.15. If in the situation of Corollary 4.14 the density h is everywhere strictly positive, then an inspection of the proof shows that Z can be taken as the process with density $1/g$ with respect to M and there is no need to enlarge the probability space. In general, however, it may be necessary to enlarge the probability space. As a simple example, consider $\Omega = \{0, 1\}$, $\mathcal{F}$ the power set on Ω and P an arbitrary probability measure on $(\Omega, \mathcal{F})$. Then the process $M = (M_t)_{t\in\mathbb{R}}$, defined on $(\Omega, \mathcal{F}, P)$ by $M_t = 0$, is an OIP with associated distribution function $G(t) = 0$. This distribution function is absolutely continuous with respect to $F(t) = t$, but it is clear that on $(\Omega, \mathcal{F}, P)$ there cannot exist an OIP with associated distribution function F.

4.5 The Fourier transform of an OIP process

We next introduce Doob's definition [37, p. 434] of the Fourier transform of an orthogonal-increment process. It will play an important role later in Theorem 5.13, linking the spectral and continuous moving average representations of a mean-square continuous weakly stationary process.

Definition 4.16 (The Fourier transform of an OIP). Given an OIP Z on $\mathbb{R}$ with associated distribution function $F(t) = v^2 t$ for some $v^2 \geq 0$, we define its *Fourier transform* to be the process $\widetilde{Z} = (\widetilde{Z}_\mu)_{\mu\in\mathbb{R}}$ defined by

$$\widetilde{Z}_\mu := \frac{1}{\sqrt{2\pi}} \int_{\mathbb{R}} \frac{1}{-\mathrm{i}t}(\mathrm{e}^{-\mathrm{i}\mu t} - 1)\, \mathrm{d}Z_t, \quad \mu \in \mathbb{R}$$

(the integrand being interpreted as its limit at $t = 0$, i. e., as μ).

The process $\widetilde{Z}$ is hence determined by $\widetilde{Z}_0 = 0$ and

$$\widetilde{Z}_{\mu_2} - \widetilde{Z}_{\mu_1} = \frac{1}{\sqrt{2\pi}} \int_{\mathbb{R}} \frac{1}{-\mathrm{i}t}(\mathrm{e}^{-\mathrm{i}\mu_2 t} - \mathrm{e}^{-\mathrm{i}\mu_1 t})\, \mathrm{d}Z_t, \quad \mu_1 \leq \mu_2 \in \mathbb{R}. \tag{4.20}$$

We shall show in the following lemma that $\widetilde{Z}$ is an OIP with associated distribution function F and that Z is the *inverse Fourier transform* of $\widetilde{Z}$, defined by

$$\frac{1}{\sqrt{2\pi}} \int_{\mathbb{R}} \frac{1}{\mathrm{i}\mu}(\mathrm{e}^{\mathrm{i}\mu t} - 1)\, \mathrm{d}\widetilde{Z}_\mu, \quad t \in \mathbb{R}. \tag{4.21}$$

The motivation for the definition comes from the quest for a process $\widetilde{Z}$ such that the (possibly non-existent) derivatives $\widetilde{Z}'$ and Z' satisfy the formal equations,

$$\widetilde{Z}'_\mu = \frac{1}{\sqrt{2\pi}} \int_{\mathbb{R}} \mathrm{e}^{-\mathrm{i}\mu t} Z'_t\, \mathrm{d}t \quad \text{and} \quad Z'_t = \frac{1}{\sqrt{2\pi}} \int_{\mathbb{R}} \mathrm{e}^{\mathrm{i}\mu t} \widetilde{Z}'_\mu\, \mathrm{d}\mu.$$

Formal integration of the latter equations leads to the preceding meaningful definitions of the Fourier transform (4.20) and its inverse (4.21).

Lemma 4.17 (OIP property of (inverse) Fourier transform). *Given an OIP Z on $\mathbb{R}$ with distribution function $F(t) = v^2 t$ for some $v^2 \geq 0$, both the Fourier transform $\widetilde{Z}$ of Z as well as the inverse Fourier transform of Z are OIPs with distribution function F. Moreover, Z is the inverse Fourier transform of $\widetilde{Z}$.*

Proof. Let $\lambda_1 \leq \lambda_2$ and $\mu_1 \leq \mu_2$. An easy calculation shows that the L^2 (and also L^1) Fourier transform of the indicator function $\mathbf{1}_{(\lambda_1,\lambda_2]}$ is given by $\mathcal{F}_2\mathbf{1}_{(\lambda_1,\lambda_2]}(t) = \frac{1}{-\mathrm{i}t\sqrt{2\pi}}(\mathrm{e}^{-\mathrm{i}\lambda_2 t} - \mathrm{e}^{-\mathrm{i}\lambda_1 t})$. Hence, (4.20) and (4.15) imply

$$\mathbb{E}[(\widetilde{Z}(\lambda_2) - \widetilde{Z}(\lambda_1))\overline{(\widetilde{Z}(\mu_2) - \widetilde{Z}(\mu_1))}] = v^2 \int_{\mathbb{R}} \mathcal{F}_2\mathbf{1}_{(\lambda_1,\lambda_2]}(t)\,\overline{\mathcal{F}_2\mathbf{1}_{(\mu_1,\mu_2]}(t)}\,\mathrm{d}t.$$

By the isomorphism induced by the L^2 Fourier transform (Theorem 2.66), i. e., by Parseval's identity (2.33),

$$\mathbb{E}[(\widetilde{Z}(\lambda_2) - \widetilde{Z}(\lambda_1))\,\overline{(\widetilde{Z}(\mu_2) - \widetilde{Z}(\mu_1))}] = v^2 \int_{\mathbb{R}} \mathbf{1}_{(\lambda_1,\lambda_2]}(t)\overline{\mathbf{1}_{(\mu_1,\mu_2]}(t)}\,\mathrm{d}t,$$

from which it follows at once that $\widetilde{Z}$ is an orthogonal-increment process with associated distribution function $t \mapsto v^2 t$, i. e., with distribution function F.

The proof that the inverse Fourier transform of Z is an OIP with associated distribution function F is similar, using the inverse L^2 Fourier transform of $\mathbf{1}_{(\lambda_1,\lambda_2]}$ rather than the L^2 Fourier transform.

Now let $\widetilde{Z}$ be the Fourier transform of Z. The inverse Fourier transform of $\widetilde{Z}$ takes the value 0 at $t = 0$, hence the fact that Z is the inverse Fourier transform of $\widetilde{Z}$ is equivalent to

$$Z_{t_2} - Z_{t_1} = \frac{1}{\sqrt{2\pi}} \int_{\mathbb{R}} \frac{1}{\mathrm{i}\mu}(\mathrm{e}^{\mathrm{i}\mu t_2} - \mathrm{e}^{\mathrm{i}\mu t_1})\,\mathrm{d}\widetilde{Z}_\mu, \quad t_1 \leq t_2 \in \mathbb{R}. \tag{4.22}$$

Equation (4.22) is a special case (with $f = \mathbf{1}_{(t_1,t_2]}$) of the equation,

$$\int_{\mathbb{R}} f(t)\,\mathrm{d}Z_t = \int_{\mathbb{R}} \mathcal{F}_2^{-1} f(t)\,\mathrm{d}\widetilde{Z}_t, \quad f \in L^2, \tag{4.23}$$

which we shall now establish. (The function $\mathcal{F}_2^{-1} f$ is the inverse Fourier transform of the L^2 function f.) We know that (4.23) holds if $\mathcal{F}_2^{-1} f = \mathbf{1}_{(\mu_1,\mu_2]}$, since it is then a restatement of equation (4.20). The class of functions $\mathcal{F}_2^{-1} f$ for which (4.23) holds, is a closed linear subspace of L^2 and since it contains every function of the form $\mathbf{1}_{(\mu_1,\mu_2]}$ it is in fact equal to L^2. The equation (4.23) therefore holds for all $\mathcal{F}_2^{-1} f \in L^2$ or equivalently for all $f \in L^2$. □

Remark 4.18. Equation (4.23) continues to hold if on the right-hand side $\widetilde{Z}$ is replaced by the *inverse* Fourier transform of Z and $\mathcal{F}_2^{-1}f$ is replaced by $\mathcal{F}_2 f$, the Fourier transform of f (Exercise 4.3).

4.6 Exercises

Exercise 4.1. (a) Use Proposition 4.10 to show that X_t, as defined by (4.17) is normally distributed with $\mathbb{E}(X_t) = 0$ and that $\mathbb{E}(X_{t+h}X_t) = \sigma^2 \mathrm{e}^{-a|h|}/(2a)$, $t, h \in \mathbb{R}$, from which it follows that the process X is stationary.

(b) Using the fact that an $\mathbb{R}^d$-valued random vector is Gaussian if and only if all its linear combinations are Gaussian, show that X is a Gaussian process.

(c) By evaluating $\mathbb{E}[(X_{t+h} - \mathrm{e}^{-ah}X_t)X_s]$, $s \le t < t + h$, show that

$$X_{t+h} = \mathrm{e}^{-ah}X_t + Z_{t,h}, \quad t \in \mathbb{R}, h > 0,$$

where $Z_{t,h}$ is uncorrelated with X_s for every $s \le t$. Using (b), deduce that $Z_{t,h}$ is independent of $\{X_s, s \le t\}$.

(d) Deduce that X is a Markov process and determine the conditional mean and conditional variance of X_{t+h} given $X_s, s \le t$.

(e) Show that the process X is strictly stationary.

Exercise 4.2. Let $Z = (Z_t)_{t\in\mathbb{R}}$ and $W = (W_t)_{t\in\mathbb{R}}$ be two uncorrelated OIPs with associated distribution functions F and G, respectively.

(a) Show that $Z + W$ is an OIP with associated distribution function $F + G$.

(b) Let $f \in L^2(F) \cap L^2(G)$. Show that $f \in L^2(F + G)$ and that

$$\int_{\mathbb{R}} f(t)\,\mathrm{d}(Z_t + W_t) = \int_{\mathbb{R}} f(t)\,\mathrm{d}Z_t + \int_{\mathbb{R}} f(t)\,\mathrm{d}W_t.$$

Exercise 4.3. Verify that the inverse Fourier transform of an OIP with distribution function $F(t) = \nu^2 t$ is an OIP with distribution function F, as mentioned in the proof of Lemma 4.17. Also prove Remark 4.18.

5 Spectral theory of MSC stationary processes

Abstract: The spectral representation of a mean-square continuous, weakly stationary random process is the stochastic analog of the Fourier representation of a deterministic function. There is also a corresponding spectral representation of the autocovariance function of the process, known as Bochner's theorem, which is the continuous-time analog of Herglotz's theorem. These representations play an essential role in the study of weakly stationary processes, especially in connection with linear filters and the analysis of multivariate weakly stationary processes. Since the theory to be developed in this chapter is essentially second-order theory, we shall continue to use the term *stationary* to mean *weakly stationary* throughout the chapter.

5.1 Spectral representation of the autocovariance function

We showed in Corollary 3.2 that a complex-valued function γ on $\mathbb{R}$ is the acvf of some stationary process Y if and only if γ is non-negative definite, i. e., if and only if γ satisfies (3.2). The process Y was also shown in Proposition 3.6 to be mean-square continuous (MSC) if and only if γ is continuous at 0.

In this section, we shall derive the following frequency-domain characterization of autocovariances of MSC stationary processes.

Theorem 5.1 (Bochner's theorem). *A complex-valued function γ defined on $\mathbb{R}$ is continuous at* 0 *and non-negative definite (i. e., is the autocovariance function of some MSC stationary process X) if and only if*

$$\gamma(h) = \int_{\mathbb{R}} e^{ih\lambda}\, dF(\lambda), \quad h \in \mathbb{R}, \tag{5.1}$$

where F is a uniquely defined right-continuous, non-decreasing, bounded function on $\mathbb{R}$ with $F(0) = 0$. The function F is called the spectral distribution function *of γ or of X. If there is a non-negative $f \in L^1$ such that $F(\lambda) - F(-\infty) = \int_{(-\infty,\lambda]} f(u)\, du$, $\lambda \in \mathbb{R}$, then f is called a* spectral density *of γ or of X.*

Observe that the spectral distribution function of an MSC stationary process is necessarily bounded, while the distribution function associated to an OIP can be bounded or unbounded.

Proof. If γ has the representation (5.1), then it is clear by dominated convergence that γ is continuous at 0. Moreover, if $u_1, \ldots, u_n \in \mathbb{R}$ and $z_1, \ldots, z_n \in \mathbb{C}$, then

https://doi.org/10.1515/9783111325033-005

$$\sum_{j,k=1}^{n} z_j \gamma(u_j - u_k)\bar{z}_k = \int_{\mathbb{R}} \sum_{j,k=1}^{n} z_j \bar{z}_k \exp[\mathrm{i}\lambda(u_j - u_k)]\, \mathrm{d}F(\lambda)$$
$$= \int_{\mathbb{R}} \left| \sum_{j=1}^{n} z_j \exp[\mathrm{i}\lambda u_j] \right|^2 \mathrm{d}F(\lambda) \geq 0,$$

so that γ is also non-negative definite and therefore, by Corollary 3.2 and Proposition 3.6, is the autocovariance function of a mean-square continuous stationary process.

Conversely, assume that γ is non-negative definite and continuous at 0. Then γ is an acvf by Corollary 3.2, in particular bounded by $\gamma(0)$. Hence, if $\gamma(0) = 0$, then $\gamma(h) = 0$ for all h and (5.1) is trivially satisfied with $F(\lambda) = 0$, and uniqueness of F follows from the fact that a finite measure is uniquely determined by its characteristic function. Hence, in the following, assume that $\gamma(0) \neq 0$. By Proposition 3.6, γ is uniformly continuous on $\mathbb{R}$ and using approximating Riemann sums and (3.2), we conclude that the integral

$$f_T(\lambda) := \frac{1}{2\pi T} \int_0^T \int_0^T \gamma(u - v)\, \mathrm{e}^{-\mathrm{i}\lambda(u-v)}\, \mathrm{d}u\, \mathrm{d}v$$
$$= \frac{1}{2\pi T} \lim_{n\to\infty} \sum_{j,k=1}^{n} \gamma(jT/n - kT/n)\, \mathrm{e}^{-\mathrm{i}\lambda jT/n}\, \overline{\mathrm{e}^{-\mathrm{i}\lambda kT/n}}\, (T/n)^2$$

is non-negative for $T > 0$ and $\lambda \in \mathbb{R}$. By making a change of variable, we can rewrite $\sqrt{2\pi} f_T(\lambda)$ as

$$\sqrt{2\pi} f_T(\lambda) = \frac{1}{\sqrt{2\pi}} \int_{\mathbb{R}} g\left(\frac{x}{T}\right) \gamma(x) \mathrm{e}^{-\mathrm{i}\lambda x} \mathrm{d}x, \tag{5.2}$$

where

$$g(x) := (1 - |x|)\mathbf{1}_{[-1,1]}(x).$$

We next show that the non-negative function f_T is integrable on $\mathbb{R}$ and is hence the density of a finite measure on the Borel subsets of $\mathbb{R}$. Consider the integral, with $c > 0$,

$$\int_{\mathbb{R}} \mathrm{e}^{-c|\lambda|} f_T(\lambda)\, \mathrm{d}\lambda = \frac{1}{2\pi} \int_{\mathbb{R}} g\left(\frac{x}{T}\right) \gamma(x) \int_{\mathbb{R}} \mathrm{e}^{-c|\lambda| - \mathrm{i}\lambda x}\, \mathrm{d}\lambda\, \mathrm{d}x$$
$$= \int_{\mathbb{R}} g\left(\frac{x}{T}\right) \gamma(x) \frac{c}{\pi(c^2 + x^2)}\, \mathrm{d}x$$
$$\leq \gamma(0) \int_{\mathbb{R}} \frac{c}{\pi(c^2 + x^2)}\, \mathrm{d}x = \gamma(0),$$

where the interchange of the order of integration on the first line is justified by Fubini's theorem (observe that $g(\cdot/T)\gamma$ is bounded with compact support). Letting $c \downarrow 0$ and using the monotone convergence theorem, we find that $\int_{\mathbb{R}} f_T(\lambda)d\lambda \le \gamma(0) < \infty$.

Since f_T is integrable, application of the Fourier inversion formula, Theorem 2.63(b), to (5.2) gives

$$g\left(\frac{x}{T}\right)\gamma(x) = \int_{\mathbb{R}} f_T(\lambda)e^{ix\lambda}\,d\lambda, \tag{5.3}$$

and in particular, setting $x = 0$,

$$\gamma(0) = \int_{\mathbb{R}} f_T(\lambda)\,d\lambda.$$

Since f_T is non-negative, this implies that $f_T/\gamma(0)$ is the probability density of some random variable X_T which, by (5.3) has characteristic function,

$$\phi_T(x) = \mathbb{E}(e^{ixX_T}) = g\left(\frac{x}{T}\right)\frac{\gamma(x)}{\gamma(0)}, \quad x \in \mathbb{R}.$$

As $T \to \infty$, $\phi_T(x)$ converges to the limit $\gamma(x)/\gamma(0)$, which is continuous at 0 and is hence, by Lévy's continuity theorem, the characteristic function of some random variable with (non-decreasing right-continuous) probability distribution function G, i. e.,

$$\gamma(x) = \int_{\mathbb{R}} e^{i\lambda x}\gamma(0)\,dG(x).$$

Defining the function F by

$$F(\lambda) = \gamma(0)(G(\lambda) - G(0)), \quad \lambda \in \mathbb{R}, \tag{5.4}$$

establishes (5.1). The uniqueness of F follows from the fact that the probability distribution function G is uniquely determined by its characteristic function $\gamma/\gamma(0)$. □

Example 5.2 (Harmonic process). The complex-valued process defined by

$$X_t = \sum_{j=1}^{n} A(\lambda_j)\,e^{it\lambda_j}, \quad t \in \mathbb{R},$$

where $-\infty < \lambda_1 < \lambda_2 < \cdots < \lambda_n < \infty$ and the coefficients $A(\lambda_j)$ are uncorrelated zero-mean random variables with

$$\mathbb{E}|A(\lambda_j)|^2 = \sigma_j^2, \quad j = 1, \ldots, n,$$

is stationary with mean zero and autocovariance function,

$$\gamma_X(h) = \mathbb{E}(X_{t+h}\overline{X_t}) = \sum_{j=1}^{n} \sigma_j^2 \mathrm{e}^{\mathrm{i}h\lambda_j} = \int_{\mathbb{R}} \mathrm{e}^{\mathrm{i}h\lambda}\, \mathrm{d}F(\lambda),$$

where

$$F(\lambda) = \sum_{j:\lambda_j \le \lambda} \sigma_j^2 - \sum_{j:\lambda_j \le 0} \sigma_j^2.$$

Example 5.3 (Continuous-time moving average). The continuous-time moving average $X_t = \int_{\mathbb{R}} g(t-u)\,\mathrm{d}Z(u)$, where Z is an OIP with associated distribution function $G(t) = ct$, $t \in \mathbb{R}$, for some $c > 0$, and $g \in L^2(G)$ (see Example 4.11) is MSC, has zero mean and autocovariance function

$$\gamma_X(h) = \mathbb{E}(X_{t+h}\overline{X_t}) = c\int_{\mathbb{R}} g(t+h)\overline{g(t)}\mathrm{d}t = c\int_{\mathbb{R}} g_h(t)\overline{g(t)}\mathrm{d}t, \tag{5.5}$$

where $g_h(\cdot) := g(\cdot + h)$. The L^2 Fourier transform $\mathcal{F}_2 g_h$ is related to $\mathcal{F}_2 g$ by $\mathcal{F}_2 g(\mu) = \mathrm{e}^{\mathrm{i}h\mu}\mathcal{F}_2 g(\mu)$, see Exercise 2.19. Hence, applying Parseval's identity (2.33) to (5.5), gives

$$\gamma_X(h) = c\int_{\mathbb{R}} \mathrm{e}^{\mathrm{i}\mu h}\left|\mathcal{F}_2 g(\mu)\right|^2 \mathrm{d}\mu,$$

from which we conclude immediately that the spectral distribution function F of X is absolutely continuous with spectral density

$$f(\mu) = c\left|\mathcal{F}_2 g(\mu)\right|^2, \quad \mu \in \mathbb{R}.$$

For the causal CAR(1) process of Example 4.12, we have $\gamma_X(h) = \frac{c}{2a}\mathrm{e}^{-a|h|}$, $g(t) = \mathrm{e}^{-at}\mathbf{1}_{[0,\infty)}(t)$, and $\mathcal{F}_2 g(\mu) = \frac{1}{\sqrt{2\pi}}(a + \mathrm{i}\mu)^{-1}$. Hence the spectral density is

$$f_{OU}(\mu) = \frac{c}{2\pi(a^2 + \mu^2)}, \quad \mu \in \mathbb{R}.$$

We shall show below in Theorem 5.13 that absolute continuity of the spectral distribution function F of the MSC weakly stationary process X is in fact necessary and sufficient for X to have a continuous-time moving average representation.

Remark 5.4. Bochner's theorem essentially decomposes the autocovariance function of an MSC stationary process into weighted sinusoidal components of various frequencies. This is most clearly seen in Example 5.2, where the spectral distribution function assigns masses only to the frequencies of the sinusoidal components appearing in the definition of the process X. In fact there is a simple relation between the mass σ_j^2 assigned to frequency λ_j and the random coefficient $A(\lambda_j)$ of $\mathrm{e}^{\mathrm{i}\lambda_j t}$ in the definition of X, namely $\sigma_j^2 = \mathbb{E}|A(\lambda_j)|^2$. The following section deals with the way in which an *arbitrary* zero-mean MSC stationary process can be represented essentially as a weighted linear combination of sinusoids with random coefficients, and the connection of this representation with the spectral distribution function appearing in Bochner's theorem.

5.2 The spectral representation of a stationary process

Orthogonal increment processes with bounded associated distribution function give rise to mean-zero MSC stationary processes in a canonical way.

Proposition 5.5. *Let $Z = (Z_\lambda)_{\lambda\in\mathbb{R}}$ be an OIP with associated distribution function F and assume that F is bounded. Define*

$$X_t := \int_{\mathbb{R}} \mathrm{e}^{\mathrm{i}t\lambda}\,\mathrm{d}Z_\lambda, \quad t \in \mathbb{R}.$$

Then $X = (X_t)_{t\in\mathbb{R}}$ is a zero-mean MSC stationary process with spectral distribution function F.

Proof. First, observe that $\mathrm{e}^{\mathrm{i}t\cdot} \in L^2(F)$ since F is bounded, so that the stochastic integral is indeed defined. That X has mean zero is clear from (4.13), as is $\mathbb{E}|X_t|^2 < \infty$ by construction. From (4.15) we further have for $h, t \in \mathbb{R}$

$$\mathbb{E}(X_{t+h}\overline{X_t}) = \int_{\mathbb{R}} \mathrm{e}^{\mathrm{i}(t+h)\lambda}\,\mathrm{e}^{-\mathrm{i}t\lambda}\,\mathrm{d}F(\lambda) = \int_{\mathbb{R}} \mathrm{e}^{\mathrm{i}h\lambda}\,\mathrm{d}F(\lambda).$$

This shows that X is stationary. By Bochner's theorem, X is also MSC and its spectral distribution function is F. □

In this section, we shall show the converse of Proposition 5.5, i. e., if $(X_t)_{t\in\mathbb{R}}$ is a zero-mean MSC stationary process and if F is the spectral distribution function in the representation (5.1) of its autocovariance function, then X itself has the *spectral representation*,

$$X_t = \int_{\mathbb{R}} \mathrm{e}^{\mathrm{i}t\lambda}\,\mathrm{d}Z_\lambda \quad \text{with probability one, for each } t \in \mathbb{R}, \tag{5.6}$$

where Z is a uniquely defined (right-continuous) orthogonal increment process with associated distribution function F. If $\mathbb{E}(X_t) = \mu \neq 0$, then (5.6) implies of course that $X - \mu$ has such a representation, so we shall assume for now that $\mathbb{E}(X_t) = 0$.

Before giving the proof, we need some preliminary notation and results. The identification of the OIP Z will be achieved by defining a certain isomorphism between the closed linear spans $\mathcal{H}_X := \overline{\mathrm{sp}}\{X_t, t \in \mathbb{R}\} \subset \mathcal{S} = L^2(\Omega, \mathcal{F}, P)$ and $\mathcal{K} := \overline{\mathrm{sp}}\{\mathrm{e}^{\mathrm{i}t\cdot}, t \in \mathbb{R}\}$, the latter of which will be seen to be the same as $L^2(F)$. This isomorphism will provide a link between random variables in the *time domain* and functions on $\mathbb{R}$ in the *frequency domain*. Its existence and construction are established in the proof of the following theorem.

Theorem 5.6. *If F is the spectral distribution function of the zero-mean MSC stationary process X, then there is a unique isomorphism T of $\overline{\mathcal{H}_X} := \overline{\text{sp}}\{X_t, t \in \mathbb{R}\}$ onto $L^2(F)$ such that*

$$TX_t = e^{it\cdot}, \quad t \in \mathbb{R}.$$

Proof. Let $\mathcal{H}_X := \text{sp}\{X_t, t \in \mathbb{R}\}$ and $\mathcal{K} := \text{sp}\{e^{it\cdot}, t \in \mathbb{R}\}$ denote the (not necessarily closed) subspaces $\mathcal{H}_X \subset \mathcal{S}$ and $\mathcal{K} \subset L^2(F)$ consisting of finite linear combinations of $X_t, t \in \mathbb{R}$, and $e^{it\cdot}, t \in \mathbb{R}$, respectively. We first show that the mapping

$$T\left(\sum_{j=1}^{n} a_j X_{t_j}\right) := \sum_{j=1}^{n} a_j e^{it_j\cdot}, \tag{5.7}$$

defines an isomorphism between $\mathcal{H}_X$ and $\mathcal{K}$. To check that T is well-defined by (5.7), suppose that $\| \sum_{j=1}^{n} a_j X_{t_j} - \sum_{k=1}^{m} b_k X_{s_k} \| = 0$. Then, by definition of the $L^2(F)$ norm and Bochner's theorem,

$$\begin{aligned}
&\left\| \sum_{j=1}^{n} a_j e^{it_j\cdot} - \sum_{k=1}^{m} b_k e^{is_k\cdot} \right\|_{L^2(F)}^2 \\
&= \int_{\mathbb{R}} \left(\sum_{j=1}^{n} a_j\, e^{it_j\lambda} - \sum_{k=1}^{m} b_k\, e^{is_k\lambda} \right) \overline{\left(\sum_{u=1}^{n} a_u\, e^{it_u\lambda} - \sum_{v=1}^{m} b_v\, e^{is_v\lambda} \right)}\, dF(\lambda) \\
&= \sum_{j,u=1}^{n} a_j \overline{a_u} \gamma(t_j - t_u) - \sum_{j=1}^{n}\sum_{v=1}^{m} a_j \overline{b_v} \gamma(t_j - s_v) - \sum_{k=1}^{m}\sum_{u=1}^{n} b_k \overline{a_u} \gamma(s_k - t_u) \\
&\quad + \sum_{k,v=1}^{m} b_k \overline{b_v} \gamma(s_k - s_v) \\
&= \mathbb{E}\left[\left(\sum_{j=1}^{n} a_j X_{t_j} - \sum_{k=1}^{m} b_k X_{s_k} \right) \overline{\left(\sum_{u=1}^{n} a_u X_{t_u} - \sum_{v=1}^{m} b_v X_{s_v} \right)} \right] = 0,
\end{aligned}$$

showing that (5.7) defines T consistently on $\mathcal{H}_X$. The linearity of T then follows at once. Moreover,

$$\begin{aligned}
&\left\langle T\left(\sum_{j=1}^{n} a_j X_{t_j}\right), T\left(\sum_{k=1}^{m} b_k X_{s_k}\right) \right\rangle_{L^2(F)} \\
&= \sum_{j=1}^{n}\sum_{k=1}^{m} a_j \overline{b_k} \langle e^{it_j\cdot}, e^{is_k\cdot} \rangle_{L^2(F)}
\end{aligned}$$

$$= \sum_{j=1}^{n} \sum_{k=1}^{m} a_j \bar{b}_k \int_{\mathbb{R}} e^{i(t_j - s_k)\lambda} dF(\lambda)$$
$$= \sum_{j=1}^{n} \sum_{k=1}^{m} a_j \bar{b}_k \langle X_{t_j}, X_{s_k} \rangle$$
$$= \left\langle \sum_{j=1}^{n} a_j X_j, \sum_{k=1}^{m} b_k X_{s_k} \right\rangle,$$

showing that T does in fact define an isomorphism between $\mathcal{H}_X$ and $\mathcal{K}$.

We show next that the mapping T can be extended uniquely to an isomorphism from $\overline{\mathcal{H}_X}$ onto $\overline{\mathcal{K}}$. If $Y \in \overline{\mathcal{H}_X}$ then there is a sequence $Y_n \in \mathcal{H}_X$ such that $\|Y_n - Y\| \to 0$. This implies that Y_n is a Cauchy sequence and hence, since T is norm-preserving, the sequence $\{TY_n\}$ is Cauchy in $L^2(F)$. The sequence $\{TY_n\}$ therefore converges in norm to an element of $\overline{\mathcal{K}}$. If T is to be norm preserving on $\overline{\mathcal{H}_X}$, we must define

$$TY := \lim_{n\to\infty}^{L^2(F)} TY_n. \tag{5.8}$$

This is a consistent definition of T on $\overline{\mathcal{H}}$ since if $\|\tilde{Y}_n - Y\| \to 0$ then the sequence $TY_1, T\tilde{Y}_1, TY_2, T\tilde{Y}_2, \ldots$ is convergent, implying that the sequences $\{TY_n\}$ and $\{T\tilde{Y}_n\}$ have the same limit, namely TY. The fact that T extended to $\overline{\mathcal{H}}$ is linear and preserves inner products follows by exactly the same argument as for the corresponding properties of the stochastic integral extended to $\overline{\mathcal{D}}$ in Step 2 of the construction of the integral in Section 4.3.

It remains only to show that $\overline{\mathcal{K}} = L^2(F)$. But clearly any function in $C_c(\mathbb{R}; \mathbb{C})$, the set of continuous functions on $\mathbb{R}$ with compact support, can be approximated arbitrarily well by functions in $\mathcal{K}$, and $C_c(\mathbb{R}; \mathbb{C})$ is dense in $L^2(F)$ by Theorem 3.14 of Rudin [75]. This implies that the closure of $\mathcal{K}$ in $L^2(F)$ is indeed $L^2(F)$. The mapping T defined on $\overline{\mathcal{H}_X}$ by (5.7) and (5.8) is thus a linear inner-product preserving mapping of the closed linear span $\overline{\mathcal{H}_X} = \overline{\mathrm{sp}}\{X_t, t \in \mathbb{R}\}$ onto $L^2(F)$. This completes the proof. □

Theorem 5.6 is particularly useful in the theory of linear prediction. It is also the key to the identification of the orthogonal increment process Z appearing in the representation (5.6), as shown in the following proposition.

Proposition 5.7. *If T is defined as in Theorem 5.6, then the process W defined by*

$$W_\lambda := T^{-1}(\mathbf{1}_{(-\infty,\lambda]}(\cdot)), \quad \lambda \in \mathbb{R},$$

is right-continuous with orthogonal increments, which satisfy

$$\mathbb{E}(|W_\mu - W_\lambda|^2) = F(\mu) - F(\lambda), \quad \lambda \le \mu,$$

where F is the spectral distribution function of X.

Proof. For each $\lambda \in \mathbb{R}$, W_λ is a well-defined element of $\overline{\text{sp}}\{X_t, t \in \mathbb{R}\}$ by Theorem 5.6. Hence $\langle W_\lambda, 1\rangle = 0$ and $\langle W_\lambda, W_\lambda\rangle < \infty$ for each $\lambda \in \mathbb{R}$ and, if $-\infty < \lambda_1 \le \lambda_2 \le \lambda_3 \le \lambda_4 < \infty$, then

$$\begin{aligned}\langle W_{\lambda_4} - W_{\lambda_3}, W_{\lambda_2} - W_{\lambda_1}\rangle &= \langle TW_{\lambda_4} - TW_{\lambda_3}, TW_{\lambda_2} - TW_{\lambda_1}\rangle_{L^2(F)} \\ &= \langle \mathbf{1}_{(\lambda_3,\lambda_4]}(\cdot), \mathbf{1}_{(\lambda_1,\lambda_2]}(\cdot)\rangle_{L^2(F)} \\ &= \int_{\mathbb{R}} \mathbf{1}_{(\lambda_3,\lambda_4]}(t)\mathbf{1}_{(\lambda_1,\lambda_2]}(t)\, \mathrm{d}F(t) = 0,\end{aligned}$$

completing the proof that W has orthogonal increments. An almost identical calculation shows that if $\lambda \le \mu$, then $\mathbb{E}(|W_\mu - W_\lambda|^2) = F(\mu) - F(\lambda)$ and hence, by the right-continuity of F, that W is also right-continuous. □

It is now a simple matter to establish the spectral representation (5.6).

Theorem 5.8 (The Spectral Representation). *If $(X_t)_{t\in\mathbb{R}}$ is a zero-mean MSC stationary process with spectral distribution function F, then there exists a (right-continuous) orthogonal increment process Z such that*
(i) $\mathbb{E}(|Z_\lambda|^2) = |F(\lambda)|$, $\lambda \in \mathbb{R}$,
and
(ii) $X_t = \int_{\mathbb{R}} \mathrm{e}^{\mathrm{i}\lambda t}\, \mathrm{d}Z_\lambda$ *with probability one*, $t \in \mathbb{R}$.

Proof. Define $Z_\lambda := W_\lambda - W_0$, where W is the process defined in Proposition 5.7. Then Z has the same increments as W, but satisfies the additional condition $Z_0 = 0$. Hence Z is an orthogonal increment process. Let I be the isomorphism,

$$I(f) = \int_{\mathbb{R}} f(\lambda)\, \mathrm{d}Z_\lambda,$$

of $\overline{\mathcal{D}} = L^2(F)$ onto $I(\overline{\mathcal{D}}) \subset \mathcal{S}$ specified in Definition 4.9. For $f \in \mathcal{D}$, i. e., of the form $f = \sum_{i=0}^n f_i \mathbf{1}_{(u_i,u_{i+1}]}$, as in (4.10), we see at once from Proposition 5.7 that

$$T^{-1}(f) = \sum_{i=0}^{n} f_i(Z_{u_{i+1}} - Z_{u_i}) = I(f).$$

This relationship remains valid for all $f \in \overline{\mathcal{D}} = L^2(F)$, since both I and T^{-1} are isomorphisms. Therefore, $I = T^{-1}$ (i. e., $TI(f) = f$ for all $f \in L^2(F)$) and hence, by Theorem 5.6,

$$X_t = I(\mathrm{e}^{\mathrm{i}t\cdot}) = \int_{\mathbb{R}} \mathrm{e}^{\mathrm{i}t\lambda}\, \mathrm{d}Z_\lambda,$$

giving the required representation of X_t. The first assertion of the theorem follows at once from Proposition 5.7. □

Corollary 5.9 (Uniqueness of Spectral Representation). *If X is a zero-mean MSC stationary process and Y is a (right-continuous) OIP with bounded associated distribution function such that*

$$X_t = \int_{\mathbb{R}} e^{it\lambda} dY_\lambda \quad \textit{with probability one, for each } t \in \mathbb{R},$$

then

$$P(Y_\lambda = Z_\lambda) = 1 \quad \textit{for each } \lambda \in \mathbb{R},$$

where Z was defined in Theorem 5.8.

Proof. By assumption,

$$X_t = \int_{\mathbb{R}} e^{it\lambda} dY_\lambda = \int_{\mathbb{R}} e^{it\lambda} dZ_\lambda \quad \text{with probability one.} \tag{5.9}$$

Proposition 5.5 then implies that the distribution function associated with Y is F, the spectral distribution function of X, and the same is true for Z. If we define, for $f \in L^2(F)$,

$$I_Y(f) = \int_{\mathbb{R}} f(\lambda)\, dY_\lambda$$

and

$$I_Z(f) = \int_{\mathbb{R}} f(\lambda)\, dZ_\lambda$$

then we have from (5.9)

$$P(I_Y(e^{it\cdot}) = I_Z(e^{it\cdot})) = 1, \quad t \in \mathbb{R}.$$

Since I_Y and I_Z are equal on $\operatorname{sp}\{e^{it\cdot}, t \in \mathbb{R}\}$, which is dense in $L^2(F)$ (see the last paragraph of the proof of Theorem 5.6) it follows that $I_Y(f) = I_Z(f)$ for all $f \in L^2(F)$. Choosing $f(\cdot) = \mathbf{1}_{(\lambda_1,\lambda_2]}(\cdot)$, $\lambda_1 \le \lambda_2$, we see that Y and Z have the same increments with probability one. Since $Y_0 = Z_0 = 0$, we conclude that

$$P(Y_\lambda = Z_\lambda) = 1, \quad \text{for each } \lambda \in \mathbb{R},$$

completing the proof. □

Theorem 5.8 and Corollary 5.9 justify the name *spectral process* (of a zero-mean MSC stationary process X) for the unique OIP Z satisfying $X_t = \int_{\mathbb{R}} e^{it\lambda}\, dZ_\lambda$.

Remark 5.10. In the course of the proof of Theorem 5.8, we have shown that $Y \in \overline{\mathrm{sp}}\{X_t, t \in \mathbb{R}\}$ if and only if there exists a function $f \in L^2(F)$ such that $Y = I_Z(f) = \int_{\mathbb{R}} f(\lambda)\,\mathrm{d}Z_\lambda$. This means that I_Z *is an isomorphism of* $L^2(F)$ *onto* $\overline{\mathrm{sp}}\{X_t, t \in \mathbb{R}\} \subset \mathcal{S}$ *with the property that* $I_Z(\mathrm{e}^{\mathrm{i}t\cdot}) = X_t$. Also, observe that $I_Z = T^{-1}$, where T is the isomorphism of Theorem 5.6.

Remark 5.11. The argument used in the proof of Theorem 5.8 is an existence proof that does not reveal in an explicit manner how Z is constructed. In the next section we give a formula for obtaining Z from X.

Let $Z = (Z_\lambda)_{\lambda\in\mathbb{R}}$ be an OIP with associated distribution function F. Then for given $\lambda_0 \in \mathbb{R}$, we can define the *mean-square left-limit* $Z_{\lambda_0^-}$ as

$$Z_{\lambda_0^-} := \lim_{n\to\infty}{}^{\mathcal{S}}\, Z_{\lambda_n}, \tag{5.10}$$

where $\{\lambda_n\}$ is any sequence such that $\lambda_n \uparrow \lambda_0$. To check that (5.10) makes sense, we note first that $\{Z_{\lambda_n}\}$ is a Cauchy sequence, since $\|Z_{\lambda_n} - Z_{\lambda_m}\|^2 = |F(\lambda_n) - F(\lambda_m)| \to 0$ as $m, n \to \infty$ and hence the limit in (5.10) exists. Moreover, if $\{\nu_n\}$ is any sequence such that $\nu_n \uparrow \lambda_0$ as $n \to \infty$, then $\|Z_{\lambda_n} - Z_{\nu_n}\|^2 = |F(\lambda_n) - F(\nu_n)| \to 0$ as $n \to \infty$ so that the limit in (5.10) is the same for all non-decreasing sequences with limit λ_0. We use the notion $Z_{\lambda_0^-}$ rather than Z_{λ_0-} to distinguish it from a pathwise left-limit (if existent almost surely). Observe, however, that if pathwise left-limit and mean-square left-limit both exist, then they must be equal, since both are limits in probability.

Proposition 5.12. *Let X be a zero-mean MSC stationary process with spectral distribution function F, and let $Z = (Z_\lambda)_{\lambda\in\mathbb{R}}$ be the spectral process in the spectral representation of X as in* (5.6)*. Suppose that F has a discontinuity at $\lambda_0 \in \mathbb{R}$. Then, for each $t \in \mathbb{R}$, with probability one,*

$$X_t = \int_{\mathbb{R}\setminus\{\lambda_0\}} \mathrm{e}^{\mathrm{i}t\lambda}\,\mathrm{d}Z_\lambda + (Z_{\lambda_0} - Z_{\lambda_0^-})\mathrm{e}^{\mathrm{i}t\lambda_0},$$

where the two terms on the right are uncorrelated and

$$\mathbb{E}|Z_{\lambda_0} - Z_{\lambda_0^-}|^2 = \mathrm{Var}(Z_{\lambda_0} - Z_{\lambda_0^-}) = F(\lambda_0) - F(\lambda_0-).$$

Proof. For $\delta > 0$ define $\lambda_{\pm\delta} := \lambda_0 \pm \delta$. Then, by the spectral representation of X,

$$X_t = \int_{\mathbb{R}\setminus(\lambda_{-\delta},\lambda_\delta]} \mathrm{e}^{\mathrm{i}t\nu}\mathrm{d}Z_\nu + \int_{(\lambda_{-\delta},\lambda_\delta]} \mathrm{e}^{\mathrm{i}t\nu}\mathrm{d}Z_\nu, \quad t \in \mathbb{R}. \tag{5.11}$$

Notice that the two terms are uncorrelated by (4.15) (and (4.13)) since the regions of integration are disjoint. As $\delta \to 0$, the first term converges in mean-square to $\int_{\mathbb{R}\setminus\{\lambda_0\}} \mathrm{e}^{\mathrm{i}t\nu}\mathrm{d}Z_\nu$

since $e^{it\cdot}\mathbf{1}_{\mathbb{R}\setminus(\lambda_{-\delta},\lambda_\delta]}$ converges in $L^2(F)$ to $e^{it\cdot}\mathbf{1}_{\mathbb{R}\setminus\{\lambda_0\}}$. To see how the last term of (5.11) behaves as $\delta \to 0$, we use that $\| \int_{(\lambda_{-\delta},\lambda_\delta]} e^{it\nu}dZ_\nu - e^{it\lambda_0}(Z_{\lambda_0} - Z_{\lambda_0^-})\|$ can be bounded by

$$\Bigg\| \int\limits_{(\lambda_{-\delta},\lambda_\delta]} e^{it\nu}dZ_\nu - e^{it\lambda_0}(Z_{\lambda_\delta} - Z_{\lambda_{-\delta}})\Bigg\| + \|Z_{\lambda_\delta} - Z_{\lambda_{-\delta}} - (Z_{\lambda_0} - Z_{\lambda_0^-})\|. \tag{5.12}$$

As $\delta \to 0$, the second term of (5.12) goes to zero by the right-continuity of Z and the definition of $Z_{\lambda_0^-}$. The first term of (5.12) can be written as

$$\Bigg\|\int\limits_{\mathbb{R}}(e^{it\nu} - e^{it\lambda_0})\mathbf{1}_{(\lambda_{-\delta},\lambda_\delta]}(\nu)\, dZ_\nu\Bigg\| = \|(e^{it\cdot} - e^{it\lambda_0})\mathbf{1}_{(\lambda_{-\delta},\lambda_\delta]}(\cdot)\|_{L^2(F)}$$

and hence converges to 0 as $\delta \to 0$ by dominated convergence. Hence, we deduce from (5.12) that

$$\int\limits_{(\lambda_{-\delta},\lambda_\delta]} e^{it\nu}dZ_\nu \overset{S}{\to} e^{it\lambda_0}(Z_{\lambda_0} - Z_{\lambda_0^-}) \quad \text{as } \delta \to 0.$$

The continuity of the inner product and the orthogonality of the two integrals in (5.11) guarantee that their mean-square limits are also orthogonal. Moreover,

$$\mathbb{E}|Z_{\lambda_0} - Z_{\lambda_0^-}|^2 = \lim_{\lambda\uparrow\lambda_0} \|Z_{\lambda_0} - Z_\lambda\| = F(\lambda_0) - F(\lambda_0-),$$

completing the proof of the proposition. □

If the spectral distribution function has k points of discontinuity at $\lambda_1, \dots, \lambda_k$, then a simple extension of the argument used in the proof of Proposition 5.12 gives the representation

$$X_t = \int\limits_{\mathbb{R}\setminus\{\lambda_1,\dots,\lambda_k\}} e^{it\lambda}dZ_\lambda + \sum_{j=1}^{k}(Z_{\lambda_j} - Z_{\lambda_j^-})e^{it\lambda_j},$$

where the $(k+1)$ terms on the right are uncorrelated. The process

$$Y_t = \sum_{j=1}^{k}(Z_{\lambda_j} - Z_{\lambda_j^-})e^{it\lambda_j}$$

corresponding to the discontinuous part of F is said to be *deterministic*, since observations of Y up to time t completely determine the future values of Y. The process in Example 5.2 is such a process.

The next theorem uses the Fourier transform of an OIP (Definition 4.16) to show that the MSC weakly stationary process Y has a continuous-time moving average representation with respect to an OIP having associated distribution function $F(t) = t$ if and only the spectral distribution function of Y is absolutely continuous.

Theorem 5.13 (Characterization of CMA). *The zero-mean MSC weakly stationary process Y has a moving average representation of the form,*

$$Y_t = \int_{\mathbb{R}} c(t-u)\,\mathrm{d}\xi_u, \quad t \in \mathbb{R},$$

where $c \in L^2$ and ξ is an OIP with associated distribution function $F(t) = t$ (on a possibly enlarged probability space), if and only if the spectral distribution function G of Y is absolutely continuous with respect to Lebesgue-measure, i. e., if and only if there exists a non-negative $h \in L^1$ such that $G(t) - G(s) = \int_{(s,t]} h(u)\,\mathrm{d}u$ for all $s \leq t$.

Proof. If Y has a representation $Y_t = \int_{\mathbb{R}} c(t-u)\,\mathrm{d}\xi_u$ as described, then by Examples 4.11 and 5.3, Y is weakly stationary with spectral density $|\mathcal{F}_2 g|^2$, so that its spectral distribution function is absolutely continuous.

Conversely, let G be the spectral distribution function of the MSC stationary process $Y = (Y_t)_{t\in\mathbb{R}}$, such that G is absolutely continuous with respect to Lebesgue measure. Choose a function $g \in L^2(\mathbb{R})$ such that $G(t) - G(s) = \int_{(s,t]} |g(u)|^2\,\mathrm{d}u$ for all $s \leq t$. From the spectral representation, there exists an OIP M with associated distribution function G such that

$$Y_t = \int_{\mathbb{R}} \mathrm{e}^{\mathrm{i}t\mu}\,\mathrm{d}M_\mu, \quad t \in \mathbb{R}.$$

By Corollary 4.14, there exists an OIP $Z = (Z_t)_{t\in\mathbb{R}}$ that is possibly defined on an enlarged probability space, such that Z has associated distribution function $F(t) = t$ and such that M has density g with respect to Z. Using equation (4.18), the spectral representation of Y in terms of M can then be rewritten as

$$Y_t = \int_{\mathbb{R}} \mathrm{e}^{\mathrm{i}t\mu} g(\mu)\,\mathrm{d}Z_\mu, \quad t \in \mathbb{R}. \tag{5.13}$$

Denote by ξ the inverse Fourier transform of Z, as defined in equation (4.21). By Theorem 2.59 and Exercise 2.19, the Fourier transform of $\mathrm{e}^{\mathrm{i}t\cdot} g(\cdot)$ is $c(t-\cdot)$ where $c = \mathcal{F}_2 g$ is the Fourier transform of g. Hence, applying Lemma 4.17 and Remark 4.18 to (5.13), we obtain $Y_t = \int_{\mathbb{R}} c(t-s)\,\mathrm{d}\xi_s$ for $t \in \mathbb{R}$ as claimed. □

5.3 Inversion formulae

For a zero-mean MSC stationary process X with autocovariance function γ, Theorems 5.1 and 5.8 and Corollary 5.9 established the existence and uniqueness of the spectral distribution function F and the spectral process Z (with $F(0) = 0$ and $Z_0 = 0$ with probability one) and the relation between them, namely

$$\mathbb{E}(|Z_\lambda|^2) = |F(\lambda)|, \quad \lambda \in \mathbb{R}.$$

In this section, we show how the increments of $F(\lambda)$ and Z_λ can be recovered from γ and X respectively. Recall the notion of the mean-square Riemann integral $\mathcal{R}_2 \int$ from Section 3.4.

Theorem 5.14 (Inversion formula). *Let X be a zero-mean MSC stationary process with autocovariance function γ, spectral distribution function F and spectral process Z. If ν and ω are continuity points of the distribution function F such that $\nu < \omega$ then, as $A \to \infty$,*

$$\frac{1}{2\pi} \int_{-A}^{A} \left(\int_{\nu}^{\omega} \mathrm{e}^{-ity} \mathrm{d}y \right) \gamma(t) \, \mathrm{d}t \to F(\omega) - F(\nu) \tag{5.14}$$

and

$$\frac{1}{2\pi} \mathcal{R}_2 \int_{-A}^{A} \left(\int_{\nu}^{\omega} \mathrm{e}^{-ity} \mathrm{d}y \right) X_t \, \mathrm{d}t \xrightarrow{\mathcal{S}} Z_\omega - Z_\nu. \tag{5.15}$$

Proof. The first result is immediate from (5.1), (5.4) and Lévy's inversion formula for characteristic functions (see Billingsley [16, Theorem 26.2]). For the second assertion, let first g be an arbitrary continuous function on $[-A, A]$. Then the Riemann-integral $\int_{-A}^{A} \int_{-A}^{A} g(r)\gamma(r-s)\overline{g(s)} \, \mathrm{d}r \, \mathrm{d}s$ exists (since also γ is continuous), hence the mean-square Riemann integral $\mathcal{R}_2 \int_{-A}^{A} g(t) X_t \, \mathrm{d}t$ exists by Theorem 3.18. Denote by T the isomorphism defined in Theorem 5.6, and let $\{\pi_n := (t_0 = -A < t_1 < \cdots < t_n = A)\}$ be a sequence of partitions of $[-A, A]$ such that $\max_{1 \le j \le n}(t_j - t_{j-1}) \to 0$ as $n \to \infty$. Then

$$\begin{aligned}
\mathcal{R}_2 \int_{-A}^{A} g(t) X_t \, \mathrm{d}t &= \lim_{n\to\infty}{}^{\mathcal{S}} \sum_{j=0}^{n-1} g(t_j) X_{t_j} (t_{j+1} - t_j) \\
&= \lim_{n\to\infty}{}^{\mathcal{S}} T^{-1}\left(\sum_{j=0}^{n-1} g(t_j) \mathrm{e}^{it_j \cdot} (t_{j+1} - t_j) \right) \\
&= T^{-1}\left(\lim_{n\to\infty}{}^{L^2(F)} \sum_{j=0}^{n-1} g(t_j) \mathrm{e}^{it_j \cdot} (t_{j+1} - t_j) \right) \\
&= T^{-1}\left(\int_{-A}^{A} g(t) \mathrm{e}^{it\cdot} \, \mathrm{d}t \right).
\end{aligned} \tag{5.16}$$

Now let $v < \omega$. Since $t \mapsto \int_v^\omega e^{-ity}\,dy$ is continuous, we can rewrite the left-hand side of (5.15) as $T^{-1}(\phi_A)$, where

$$\phi_A(\cdot) = \frac{1}{2\pi}\int_{-A}^{A} e^{it\cdot}\left(\int_v^\omega e^{-ity}\,dy\right) dt.$$

Using Fubini's theorem, a simple calculation shows that

$$\phi_A(u) = \frac{1}{\pi}\int_{A(u-\omega)}^{A(u-v)} \frac{\sin y}{y}\,dy, \quad u \in \mathbb{R}.$$

Since $\lim_{T\to\infty}\int_0^T (\sin y)/y\,dy = \lim_{T\to\infty}\int_{-T}^0 (\sin y)/y\,dy = \pi/2$ (see e. g. Billingley [16, equation (18.11)]), we see that the ϕ_A are uniformly bounded and that $\phi_A(u)$ converges as $A \to \infty$ to 1 for $v < u < \omega$ and to 0 for $u \notin [-v, \omega]$. Hence, provided that v and ω are not discontinuity points of F, ϕ_A converges in $L^2(F)$ to $\mathbf{1}_{(v,\omega]}$ as $A \to \infty$ by dominated convergence, and so by the continuity properties of the isomorphism,

$$T^{-1}(\phi_A) \overset{S}{\to} T^{-1}(\mathbf{1}_{(v,\omega]}) = \int_{\mathbb{R}} \mathbf{1}_{(v,\omega]}(\lambda)\,dZ_\lambda = Z_\omega - Z_v, \quad A \to \infty,$$

as required. □

If a mean-zero MSC stationary process has a spectral density f, then $\gamma(h) = \int_{\mathbb{R}} e^{ih\lambda} f(\lambda)\,d\lambda$ for all $h \in \mathbb{R}$, so that $\lim_{|t|\to\infty}\gamma(t) = 0$ by the Riemann–Lebesgue lemma (Theorem 2.60). Hence $\lim_{|t|\to\infty}\gamma(t) = 0$ is a necessary condition for a zero-mean MSC stationary process to have a spectral density. Using (5.14), we can also give sufficient conditions for the existence of a spectral density.

Theorem 5.15 (Sufficient condition for existence of spectral density). *Let X be a zero-mean MSC stationary process with autocovariance function γ.*

(a) *Assume that $\gamma \in L^2$. Then the Fourier transform $\mathcal{F}_2\gamma$ of γ is non-negative, in $L^1 \cap L^2$ and X has a spectral density, given by $\frac{1}{\sqrt{2\pi}}\mathcal{F}_2\gamma$.*

(b) *Assume that $\gamma \in L^1$. Then X has a bounded and continuous spectral density f, given by $\frac{1}{\sqrt{2\pi}}$ times the L^1-Fourier transform $\mathcal{F}\gamma$ of f, i. e.,*

$$f(\omega) = \frac{1}{2\pi}\int_{\mathbb{R}} e^{-i\omega h}\gamma(h)\,dh, \quad \omega \in \mathbb{R}. \tag{5.17}$$

Proof. Denote the spectral distribution function of X by F.

(a) Let $v < \omega$. By Theorem 2.66(d), $\frac{1}{\sqrt{2\pi}}\int_{-A}^{A} e^{-it\cdot}\gamma(t)\,dt$ converges in L^2 to $\mathcal{F}_2\gamma$ as $A \to \infty$, hence $\frac{1}{2\pi}\mathbf{1}_{(v,\omega]}\int_{-A}^{A} e^{-it\cdot}\gamma(t)\,dt$ converges in L^2 to $\frac{1}{\sqrt{2\pi}}\mathbf{1}_{(v,\omega]}\,\mathcal{F}_2\gamma$ as $A \to \infty$, and since

$[\nu, \omega]$ is bounded, this convergence is also in L^1. This and equation (5.14) then imply, if ν and ω are continuity points of F, that

$$\frac{1}{\sqrt{2\pi}} \int_{\nu}^{\omega} \mathcal{F}_2\gamma(u)\,du = \lim_{A\to\infty} \frac{1}{2\pi} \int_{\nu}^{\omega} \int_{-A}^{A} e^{-itu}\gamma(t)\,dt\,du = F(\omega) - F(\nu). \tag{5.18}$$

Since all but countably many points are continuity points of F, since F is increasing and since the left-hand side of (5.18) is continuous in both ν and ω, this implies, that also the right-hand side must be continuous in both ν and ω, so that F is continuous, hence (5.18) is true for all $\nu, \omega \in \mathbb{R}$ with $\nu < \omega$. Since F is increasing, this implies that $\mathcal{F}_2\gamma \geq 0$ Lebesgue-almost everywhere (e. g., by the Fundamental Theorem of Calculus for Lebesgue integrals, see Rudin [75, Theorem 7.11], or by Remark 6.3(c) below). Consequently,

$$\frac{1}{\sqrt{2\pi}} \int_{\mathbb{R}} \mathcal{F}_2\gamma(u)\,du = \lim_{T\to\infty} \frac{1}{\sqrt{2\pi}} \int_{-T}^{T} \mathcal{F}_2\gamma(u)\,du = \lim_{T\to\infty} (F(T) - F(-T)) < \infty.$$

This shows that $\mathcal{F}_2\gamma \in L^1$. Equation (5.18) now implies that F is absolutely continuous with density $\frac{1}{\sqrt{2\pi}}\mathcal{F}_2\gamma$, giving the claim.

(b) Now assume that $\gamma \in L^1$. Since γ is bounded as an ACVF, it is also in L^2 (see Exercise 5.2). Hence by (a), X has spectral density $f := \frac{1}{\sqrt{2\pi}}\mathcal{F}_2\gamma$. But since $\gamma \in L^1 \cap L^2$, Theorem 2.66(a) implies that f is given by (5.17), and hence bounded and continuous as the Fourier transform of an L^1-function. □

The following proposition provides a useful criterion for checking whether or not an absolutely integrable function γ on $\mathbb{R}$ with absolutely integrable Fourier transform is the autocovariance function of an MSC stationary process.

Corollary 5.16 (Characterization for being an integrable ACVF). *Let γ be a continuous complex-valued function on $\mathbb{R}$ such that $\gamma \in L^1$ or $\gamma \in L^2$ and denote the L^1- or L^2-Fourier transform of γ by $\mathcal{F}\gamma$. Then γ is the acvf of a zero-mean MSC stationary process if and only if $\mathcal{F}\gamma$ is Lebesgue-almost everywhere non-negative and in L^1. In that case, $\frac{1}{\sqrt{2\pi}}\mathcal{F}\gamma$ is the spectral density of γ.*

Proof. That γ being an acvf implies $\mathcal{F}\gamma \geq 0$, $\mathcal{F}\gamma \in L^1$ and that $\frac{1}{\sqrt{2\pi}}\mathcal{F}\gamma$ is the spectral density of γ follows from Theorem 5.15. Conversely, assume that $\mathcal{F}\gamma \geq 0$ and $\mathcal{F}\gamma \in L^1$. Then, if $\gamma \in L^1$, Theorem 2.63(b) implies $\gamma(h) = \frac{1}{\sqrt{2\pi}} \int_{\mathbb{R}} e^{ih\omega}\mathcal{F}\gamma(\omega)\,d\omega$ for $h \in \mathbb{R}$, so that γ is an acvf of mean-zero MSC stationary process by Bochner's theorem. If $\gamma \in L^2$, then $\gamma(\cdot)$ is the L^2-limit of $\frac{1}{\sqrt{2\pi}} \int_{-A}^{A} e^{i\omega\cdot}\mathcal{F}\gamma(\omega)\,d\omega$ by Theorem 2.66(d). However, since $\mathcal{F}\gamma$ was assumed to be integrable, this L^2-limit is equal to $\frac{1}{\sqrt{2\pi}} \int_{\mathbb{R}} e^{ih\omega}\mathcal{F}\gamma(\omega)\,d\omega$, $h \in \mathbb{R}$. Hence γ is again the acvf of a mean-zero MSC stationary process by Bochner's theorem. □

5.4 Linear transformations of MSC stationary processes

Let $X = (X_t)_{t\in\mathbb{R}}$ be a zero-mean MSC stationary process with spectral distribution function F. By the spectral representation (Theorem 5.8 and Corollary 5.9), there exists a unique orthogonal increment process $Z = (Z_\lambda)_{\lambda\in\mathbb{R}}$ with associated distribution function F such that X has spectral representation

$$X_t = \int_{\mathbb{R}} \mathrm{e}^{\mathrm{i}t\lambda}\,\mathrm{d}Z_\lambda, \quad t \in \mathbb{R}.$$

Now let $\phi \in L^2(F)$. Then the process $Y = (Y_t)_{t\in\mathbb{R}}$, defined by

$$Y_t := \int_{\mathbb{R}} \mathrm{e}^{\mathrm{i}t\lambda}\phi(\lambda)\,\mathrm{d}Z_\lambda, \quad t \in \mathbb{R},$$

is said to be *obtained from X by application of the linear filter with transfer function ϕ.*

Denote by M the OIP with density ϕ with respect to Z (see Theorem 4.13). Then M has associated distribution function G satisfying $G(t)-G(s) = \int_{(s,t]} |\phi(u)|^2\,\mathrm{d}F(u)$ for $s \le t$, and

$$Y_t = \int_{\mathbb{R}} \mathrm{e}^{\mathrm{i}t\lambda}\,\mathrm{d}M_\lambda, \quad t \in \mathbb{R},$$

by equation (4.18), which is the spectral representation of Y. In particular, by Proposition 5.5, Y is a zero-mean MSC stationary process with spectral distribution function G. Two particularly important examples that arise in connection with mean-square linear prediction are the following.

Example 5.17 (Differentiation). Theorem 3.9 provided necessary and sufficient conditions for the existence of the mean-square (MS) derivative of a not necessarily stationary second-order process and Corollary 3.13 gave a characterization for stationary processes. Remark 3.14 showed that the autocovariance function being twice continuously differentiable in a neighbourhood of 0 is sufficient for the existence of the MS derivative of a stationary process at all $t \in \mathbb{R}$. In this example, we consider MS differentiability of a zero-mean MSC stationary process from a spectral point of view.

Existence of the MS derivative DX_t at t of a mean-zero MSC stationary process X (with spectral distribution function F and spectral process Z) means that $\lim^{\mathcal{S}}_{\Delta\to 0} \Delta^{-1}(X_{t+\Delta} - X_t)$ exists. By the isomorphism between $\overline{\mathrm{sp}}\{X_t,\ t \in \mathbb{R}\}$ and $L^2(F)$ (Remark 5.10), this is equivalent to existence of the limit in $L^2(F)$ as $\Delta \to 0$ of $\Delta^{-1}(\mathrm{e}^{\mathrm{i}(t+\Delta)\cdot} - \mathrm{e}^{\mathrm{i}t\cdot})$. If the limit exists, it must equal (to within a set of F-measure zero) the pointwise limit $\mathrm{i}\mu\mathrm{e}^{\mathrm{i}t\mu}$, $\mu \in \mathbb{R}$, implying that $\int_{\mathbb{R}} \mu^2\,\mathrm{d}F(\mu) < \infty$.

Conversely if $\int_{\mathbb{R}} \mu^2\,\mathrm{d}F(\mu) < \infty$ then, by dominated convergence, $\Delta^{-1}(\mathrm{e}^{\mathrm{i}(t+\Delta)\mu} - \mathrm{e}^{\mathrm{i}t\mu})$, $\mu \in \mathbb{R}$, converges in $L^2(F)$ as $\Delta \to 0$ to $\mathrm{i}\mu\mathrm{e}^{\mathrm{i}t\mu}$, $\mu \in \mathbb{R}$, and so by Proposition 4.10(iii), $\Delta^{-1}(X_{t+\Delta} - X_t)$ converges in $\mathcal{S}$ to $\int_{\mathbb{R}} \mathrm{i}\mu\mathrm{e}^{\mathrm{i}t\mu}\,\mathrm{d}Z_\mu$.

In other words, DX_t exists if and only if $\int_{\mathbb{R}} \mu^2 \, dF(\mu) < \infty$ and is obtained from X_t by application of the linear filter with transfer function, $\phi(\mu) = i\mu$, i. e.,

$$DX_t = \int_{\mathbb{R}} i\mu e^{it\mu} \, dZ_\mu.$$

In particular, DX_t is not only mean-zero stationary (as was already known from Corollary 3.13), but also MSC.

The same argument applied to DX_t shows immediately that the second MS derivative of X_t exists if and only if $\int_{\mathbb{R}} \mu^4 \, dF(\mu) < \infty$ and then

$$D^2 X_t = \int_{\mathbb{R}} (i\mu)^2 e^{it\mu} \, dZ_\mu.$$

More generally, the jth MS derivative of X_t exists if and only if

$$\int_{\mathbb{R}} \mu^{2j} \, dF(\mu) < \infty \tag{5.19}$$

and then

$$D^j X_t = \int_{\mathbb{R}} (i\mu)^j e^{it\mu} dZ_\mu, \quad j \in \mathbb{N}.$$

Remark 5.18. An important special case arises when X_t has the representation

$$X_t = \int_{\mathbb{R}} e^{it\mu} g(\mu) \, dZ^*_\mu,$$

where Z^* is an OIP with associated distribution function $G(\mu) = \mu$ and $g \in L^2$. The necessary and sufficient condition (5.19) becomes $\int_{\mathbb{R}} |\mu|^{2j} |g(\mu)|^2 \, d\mu < \infty$ and, provided it is satisfied,

$$D^j X_t = \int_{\mathbb{R}} (i\mu)^j e^{it\mu} g(\mu) \, dZ^*_\mu, \quad j \in \mathbb{N}.$$

As a next example, we consider the convolution of functions with stationary processes under suitable conditions.

Definition 5.19. Let $X_t = \int_{\mathbb{R}} e^{it\mu} \, dZ_\mu$, $t \in \mathbb{R}$, be a zero-mean MSC stationary process with absolutely continuous spectral distribution function F and bounded spectral density. Let $h \in L^2$ be locally bounded and have at most finitely many discontinuity points. The *convolution of h and X* is defined as

$$\int_{\mathbb{R}} h(u)X_{t-u}\,\mathrm{d}u := \lim_{R\to\infty}{}^{\mathcal{S}}\,\mathcal{R}_2\int_{-R}^{R} h(u)X_{t-u}\,\mathrm{d}u, \qquad t\in\mathbb{R},$$

where the existence of the mean-square Riemann integral on the right is established below with the aid of Theorem 3.18 and its convergence in mean-square as $R\to\infty$ with the aid of the isomorphism T constructed in Theorem 5.6.

We first observe that for fixed $t\in\mathbb{R}$, the process $(X_{t-u})_{u\in\mathbb{R}}$ is a zero-mean MSC stationary process with acvf $h\mapsto\gamma_X(-h)$, where γ_X is the acvf of X. Since h is locally bounded and has only finitely many discontinuity points, the Riemann integral $\int_{-R}^{R}\int_{-R}^{R} h(r)\gamma_X(-(r-s))\overline{h(s)}\,\mathrm{d}r\,\mathrm{d}s$ exists, so that the mean-square Riemann integral $\mathcal{R}_2\int_{-R}^{R} h(u)X_{t-u}\,\mathrm{d}u$ exists by Theorem 3.18.

Denote by T the isomorphism defined in Theorem 5.6. Then

$$\mathcal{R}_2\int_{-R}^{R} h(u)X_{t-u}\,\mathrm{d}u = T^{-1}\left(\int_{-R}^{R} h(u)\,\mathrm{e}^{\mathrm{i}(t-u)\cdot}\,\mathrm{d}u\right)$$

by exactly the same argument that led to (5.16). Since $h\in L^2$, Theorem 2.66(d) implies that $\int_{-R}^{R} h(u)\,\mathrm{e}^{\mathrm{i}(t-u)\cdot}\,\mathrm{d}u$ converges in L^2 as $R\to\infty$ to $\sqrt{2\pi}\,\mathrm{e}^{\mathrm{i}t\cdot}\mathcal{F}_2h$, where $\mathcal{F}_2h$ is the Fourier transform of h. Since the spectral density of X is bounded, we also have

$$\int_{-R}^{R} h(u)\,\mathrm{e}^{\mathrm{i}(t-u)\cdot}\,\mathrm{d}u \overset{L^2(F)}{\longrightarrow} \sqrt{2\pi}\,\mathrm{e}^{\mathrm{i}t\cdot}\mathcal{F}_2h, \qquad \text{as } R\to\infty.$$

Since T is an isomorphism, this implies the existence of the mean-square limit $\lim^{\mathcal{S}}_{R\to\infty}\mathcal{R}_2\int_{-R}^{R} h(u)X_{t-u}\,\mathrm{d}u$ with value $T^{-1}(\sqrt{2\pi}\,\mathrm{e}^{\mathrm{i}t\cdot}\mathcal{F}_2h)$. But by Remark 5.10, T^{-1} is equal to the stochastic integral operator I_Z, so that

$$\int_{\mathbb{R}} h(u)X_{t-u}\,\mathrm{d}u = \int_{\mathbb{R}} \sqrt{2\pi}\,\mathcal{F}_2h\,(\mu)\,\mathrm{e}^{\mathrm{i}t\mu}\,\mathrm{d}Z_\mu.$$

This shows that convolution as specified in Definition 5.19 is a well-defined linear transformation with transfer function $\sqrt{2\pi}\,\mathcal{F}_2h$.

5.5 Exercises

Exercise 5.1. Establish whether or not each of the following functions on $\mathbb{R}$ is nonnegative definite. Determine also whether or not each of them is the acvf of an MSC stationary process, and for those which are, determine the spectral distribution function.

(i) $g(h) = 1$, if $h = 0$, and 0 otherwise,
(ii) $g(h) = 1$, if $-1 < h < 1$, and 0 otherwise,
(iii) $g(h) = 1 - |h|$, if $-1 < h < 1$, and 0 otherwise,
(iv) $g(h) = \frac{\sigma^2}{2a} e^{-a|h|}$, $\sigma, a > 0$,
(v) $g(h) = \cos(h\lambda_0)$, $\lambda_0 \in \mathbb{R}$,
(vi) $g(h) = \cos(h\lambda_0)\frac{\sin(\delta h)}{\delta h}$, $\lambda_0 \in \mathbb{R}$, $\delta > 0$.

Exercise 5.2. (a) Let γ be the acvf of a stationary process such that $\gamma \in L^1$. Show that $\gamma \in L^2$.
(b) Using Bochner's theorem (and inspiration from Theorem 5.15), find an acvf of a mean-zero MSC stationary process that is in L^2, but not in L^1.

Exercise 5.3. If $(X_t)_{t\in\mathbb{R}}$ and $(Y_t)_{t\in\mathbb{R}}$ are uncorrelated MSC stationary processes with autocovariance functions γ_X and γ_Y and spectral distribution functions F_X and F_Y, respectively, show that the process $W = X + Y$ is stationary and MSC with autocovariance function $\gamma_X + \gamma_Y$ and spectral distribution function $F_X + F_Y$.

Exercise 5.4. Let W be the process defined by

$$W_t := A\cos(\pi t/3) + B\sin(\pi t/3) + X_t, \quad t \in \mathbb{R},$$

where X is the process defined in Example 4.12, A and B are uncorrelated with mean zero and variance v^2 and X_t is uncorrelated with A and B for each $t \in \mathbb{R}$. Find the autocovariance function and spectral distribution function of W.

Exercise 5.5. Complete the proof of the assertion made in the last paragraph of Proposition 5.7 that if $\lambda \le \mu$, then

$$\|W_\mu - W_\lambda\|^2 = F(\mu) - F(\lambda),$$

thereby showing that the spectral process Z of the zero-mean MSC process Y is a right-continuous orthogonal increment process with associated distribution function F, the spectral distribution function of the autocovariance function of Y.

6 Mean-square linear prediction of weakly stationary processes

Abstract: In this chapter, we consider linear prediction of a zero-mean weakly stationary mean-square continuous (MSC) process $(Y_t)_{t\in\mathbb{R}}$. The goal is to determine for $h \ge 0$ the element of $\mathcal{N}_t$, the closed linear span of $(Y_u)_{u\le t}$, with minimum mean-square distance in $\mathcal{S} = L^2(\Omega, \mathcal{F}, P)$ from Y_{t+h}. We shall refer to this as the *h-step best linear predictor* (*BLP*) of Y_{t+h}. By Theorem 2.27 this is the orthogonal projection, $P_{\mathcal{N}_t} Y_{t+h}$. In the course of determining the required projection and its mean-square error, we shall need to make assumptions about the process Y and to study its structure and second-order properties in some detail. We follow the approach of Doob [37], who introduces a discrete-time sequence X related to Y. This leads to a representation of the process Y (Theorem 6.26), which can be expressed in terms of the coefficients in the Wold representation (Theorem 6.5) of the sequence X. This representation is the key to the determination of $P_{\mathcal{N}_t} Y_{t+h}$. (If $\mathbb{E}Y_t = \mu \ne 0$ then the best linear predictor of Y_{t+h} in terms of $\{1, Y_u, u \le t\}$ is found by adding μ to the h-step BLP of Y^*_{t+h}, where Y^* is the zero-mean process $Y - \mu$.)

6.1 The discrete-time case

In this section, we briefly summarize the corresponding results for the determination of the *best linear predictor* (*BLP*) of X_n in terms of $X_j, j \le n-h, h \in \{0, 1, 2, \ldots\}$, when $(X_n)_{n\in\mathbb{Z}}$ is a zero-mean complex-valued, weakly stationary sequence. For proofs of the results see Doob [37], and for the general notions regarding spectral theory of discrete time series, see Brockwell and Davis [22, Chapter 4]. We shall denote by $\gamma(h)$, $h \in \mathbb{Z}$, and $F(\lambda)$, $-\pi < \lambda \le \pi$, the autocovariance and spectral distribution functions of X respectively (where F is normalized such that $F(0) = 0$, rather than $F(-\pi) = 0$). The spectral distribution function is then necessarily bounded. Thus we can write

$$\mathbb{E}(X_{n+h}\overline{X}_n) = \gamma(h) = \int_{(-\pi,\pi]} e^{i\lambda h}\, dF(\lambda), \quad h \in \mathbb{Z},$$

and

$$X_n = \int_{(-\pi,\pi]} e^{i\lambda n}\, dZ_\lambda,$$

where $(Z_\lambda)_{-\pi<\lambda\le\pi}$ is an orthogonal-increment process (OIP) with $\mathbb{E}|Z_{\lambda_2} - Z_{\lambda_1}|^2 = F(\lambda_2) - F(\lambda_1)$ for $-\pi < \lambda_1 \le \lambda_2 \le \pi$ (the definition of an OIP on $(-\pi, \pi]$ is completely analogous to that of OIPs on $\mathbb{R}$, as defined in Definition 4.1). When F is absolutely continuous with respect to Lebesgue-measure, its density f is called the *spectral density* of X.

https://doi.org/10.1515/9783111325033-006

Remark 6.1. We shall continue to measure frequencies in radians per unit time. Doob [37] measures them in cycles per unit time, however, it is a simple matter to translate results from one formulation to the other.

As in the continuous-time case, the mapping

$$I(g) = \int_{(-\pi,\pi]} g(\lambda)\,\mathrm{d}Z_\lambda, \quad g \in L^2(F),$$

is an isomorphism of $L^2(F)$ (the measurable functions on $(-\pi,\pi]$, which are square-integrable with respect to F) onto the closed linear span in $\mathcal{S}$ of the random variables $\{X_n, n \in \mathbb{Z}\}$. Under this isomorphism

$$I(\mathrm{e}^{\mathrm{i}n\cdot}) = X_n, \quad n \in \mathbb{Z}.$$

It is convenient now to introduce the following subspaces of $\mathcal{S}$ and the corresponding subspaces of $L^2(F)$ under the isomorphism I^{-1}.

$$\left.\begin{aligned} &\mathcal{M} := \overline{\operatorname{sp}}\{X_k, k \in \mathbb{Z}\}, && \widetilde{\mathcal{M}} := \overline{\operatorname{sp}}\{\mathrm{e}^{\mathrm{i}k\cdot}, k \in \mathbb{Z}\} = L^2(F),\\ &\mathcal{M}_n := \overline{\operatorname{sp}}\{X_k, k \in \mathbb{Z} \cap (-\infty, n]\}, && \widetilde{\mathcal{M}}_n := \overline{\operatorname{sp}}\{\mathrm{e}^{\mathrm{i}k\cdot}, k \in \mathbb{Z} \cap (-\infty, n]\},\\ &\mathcal{M}_{-\infty} := \bigcap_{n\in\mathbb{Z}} \mathcal{M}_n, && \widetilde{\mathcal{M}}_{-\infty} := \bigcap_{n\in\mathbb{Z}} \widetilde{\mathcal{M}}_n. \end{aligned}\right\} \tag{6.1}$$

The isomorphism allows us to compute projections (in particular predictors) of random variables in $\mathcal{M}$ by computing corresponding projections of functions in $\widetilde{\mathcal{M}}$ and applying the mapping I. For example, the best linear predictor $P_{\mathcal{M}_{n-h}} X_n$, $h > 0$, is

$$P_{\mathcal{M}_{n-h}} X_n = I(P_{\widetilde{\mathcal{M}}_{n-h}} \mathrm{e}^{\mathrm{i}n\cdot}),$$

and its *mean-square error* is given by

$$\sigma_h^2 := \mathbb{E}[|X_n - P_{\mathcal{M}_{n-h}} X_n|^2] = \int_{(-\pi,\pi]} |\mathrm{e}^{\mathrm{i}n\lambda} - (P_{\widetilde{\mathcal{M}}_{n-h}} \mathrm{e}^{\mathrm{i}n\cdot})(\lambda)|^2\,\mathrm{d}F(\lambda). \tag{6.2}$$

Since $(X_n)_{n\in\mathbb{Z}}$ is stationary, an easy calculation (see Exercise 6.1(a)) shows that σ_h^2, as defined in (6.2), indeed does not depend on n.

We turn now to a fundamental representation of the most general weakly stationary sequence X which not only provides insight into the structure of the sequence, but also provides a very nice representation of the h-step BLP of X_n and its mean-square error. Before doing so, we recall that there are weakly stationary sequences for which

$$\sigma_1^2 = \mathbb{E}(|X_n - P_{\mathcal{M}_{n-1}} X_n|^2),$$

is strictly positive. For example, if $(X_n)_{n\in\mathbb{Z}} \sim \mathrm{WN}(0,\sigma^2)$, i. e., if $(X_n)_{n\in\mathbb{Z}}$ is a sequence of uncorrelated random variables, each with mean zero and variance $\sigma^2 > 0$, then $\sigma_1^2 = \sigma^2 > 0$. On the other hand, if A and B are zero-mean uncorrelated random variables, each with variance σ^2, then the harmonic process, $X_n = A\cos(n\pi/2) + B\sin(n\pi/2)$, $n \in \mathbb{Z}$, has one-step mean-square error $\sigma_1^2 = 0$ since $X_{n+1} = X_{n-3}$ for all $n \in \mathbb{Z}$ (see Exercise 6.2).

Definition 6.2 (Deterministic and regular sequences). A zero-mean weakly stationary sequence $(X_n)_{n\in\mathbb{Z}}$ for which the one-step BLP, $P_{\mathcal{M}_{n-1}}X_n$, of X_n, has mean-square error $\sigma_1^2 = 0$ is called *deterministic*. If $\sigma_1^2 > 0$, the sequence is said to be *regular*.

Before we continue, we summarize some results regarding the Lebesgue decomposition of locally finite measures and differentiability of the corresponding distribution functions in the following remark.

Remark 6.3. (a) Let ρ be a measure on $(\mathbb{R}, \mathcal{B}_1)$. Then ρ is *absolutely continuous* with respect to Lebesgue measure, if every Lebesgue null-set in $\mathcal{B}_1$ is also a ρ-null set. If ρ is additionally σ-finite (e. g., if it is locally finite), then by the Radon-Nikodym theorem this is equivalent to the existence of a Lebesgue density of ρ, i. e., to the existence of a $\mathcal{B}_1 \to \mathcal{B}_1$-measurable function $h : \mathbb{R} \to [0,\infty)$ such that $\rho(A) = \int_A h(s)\,ds$ for all $A \in \mathcal{B}_1$ (this density is Lebesgue-almost everywhere unique). In contrast, a measure ρ on $(\mathbb{R}, \mathcal{B}_1)$ is called *singular* with respect to Lebesgue measure if there is some set $B \in \mathcal{B}_1$ with Lebesgue measure 0 such that $\rho(\mathbb{R}\setminus B) = 0$. We then say that ρ is *carried by the set B*. Now *Lebesgue's Decomposition Theorem* (e. g., Cohn [33, Theorem 4.3.2]) states that to every σ-finite measure ρ on $(\mathbb{R}, \mathcal{B}_1)$ there exist unique measures ρ_{ac} and ρ_{sing} on $(\mathbb{R}, \mathcal{B}_1)$ such that $\rho = \rho_{\mathrm{ac}} + \rho_{\mathrm{sing}}$, such that ρ_{ac} is absolutely continuous with respect to Lebesgue measure and such that ρ_{sing} is singular with respect to Lebesgue measure. They are called the *absolutely continuous part* and *singular part of* ρ, respectively.

(b) Let $H : \mathbb{R} \to \mathbb{R}$ be a right-continuous non-decreasing function. Then there is a unique locally finite measure ρ on $(\mathbb{R}, \mathcal{B}_1)$ with distribution function ρ. Then the distribution functions belonging to ρ_{ac} and to ρ_{sing} are called the *absolutely continuous part of* H and the *singular part of* H, respectively. They are unique up to addition of constants and will be denoted by H_{ac} and H_{sing}, respectively. The Lebesgue decomposition can then be restated as $H = H_{\mathrm{ac}} + H_{\mathrm{sing}} + \text{const.}$, where the constant can be taken to be 0 if suitable versions of H_{ac} and H_{sing} are chosen. By a *(Lebegue) density* of H_{ac} we shall always mean a density of ρ_{ac}, and we will say that H_{sing} is *carried* by a Lebesgue null set $S \in \mathcal{B}_1$ if the same is true for ρ_{sing}.

(c) Let $H : \mathbb{R} \to \mathbb{R}$ be a right-continuous non-decreasing function. A theorem of Lebesgue (e. g., Cohn [33, Theorem 6.3.3]) then shows that H is Lebesgue-almost everywhere differentiable. It can be shown that the exceptional Lebesgue-null set on which H is not differentiable is a Borel set, and that the function

$$h : \mathbb{R} \to \mathbb{R}, \quad h(t) := \begin{cases} H'(t), & H \text{ differentiable at } t, \\ 0, & \text{otherwise,} \end{cases}$$

is $\mathcal{B}_1 \to \mathcal{B}_1$-measurable (cf., Exercise 6.4). Another result of Lebesgue (combine Theorem 6.2.3 with Lemma 6.3.1 in Cohn [33]) then shows that the function h defined above is the Lebesgue density of the absolutely continuous part H_{ac} of H. One often simply writes H' for the function h.

(d) Parts (a)–(c) hold similarly for measures and distribution functions defined on $(-\pi, \pi]$.

Having clarified in particular the meaning of the derivative of a distribution function as the density of the absolutely continuous part, we can now state the following theorem. It provides a useful inequality for the mean-square error σ_h^2 of the h-step BLP of X, subject to the conditions stated in the theorem.

Theorem 6.4. *If* $(Z_n)_{n\in\mathbb{Z}} \sim \mathrm{WN}(0,1)$, $c_0 \neq 0$, $\sum_{n=0}^{\infty} |c_n|^2 < \infty$ *and*

$$X_n := \sum_{j=0}^{\infty} c_j Z_{n-j}, \quad n \in \mathbb{Z}, \tag{6.3}$$

then X *is weakly stationary with spectral density* f_X *given by* $f_X(\lambda) = (2\pi)^{-1} |\sum_{j=0}^{\infty} c_j\, e^{-ij\lambda}|^2$. *The process* X *defined by* (6.3) *is regular, with h-step mean-square prediction error* σ_h^2 *satisfying*

$$\sigma_h^2 \geq \sum_{j=0}^{h-1} |c_j|^2. \tag{6.4}$$

Furthermore, the inequality continues to hold if, instead of supposing that X *is defined by* (6.3), *we assume only that* X *has spectral distribution function* F *(not necessarily absolutely continuous) such that*

$$F'(\lambda) = \frac{1}{2\pi} \left| \sum_{j=0}^{\infty} c_j\, e^{-ij\lambda} \right|^2 \quad \textit{Lebesgue a. e., where } c_0 \neq 0 \textit{ and } \sum_{j=0}^{\infty} |c_j|^2 < \infty.$$

Under either of the two preceding assumptions on X, *there cannot be equality in* (6.4) *if* $\sum_{j=0}^{\infty} c_j z^j = 0$ *for some* $z \in \mathbb{C}$ *with* $|z| < 1$.

Proof. See Doob [37, Theorem XII.4.1]. □

The next theorem provides a fundamental characterization of the structure of regular weakly stationary sequences and shows in particular that regular sequences may contain a non-zero deterministic component.

Theorem 6.5 (The Wold decomposition). *If $X = (X_n)_{n\in\mathbb{Z}}$ is a zero-mean weakly stationary sequence with one-step mean-square prediction error σ_1^2, where $\sigma_1 > 0$, and the subspaces $\mathcal{M}_n$ and $\mathcal{M}_{-\infty}$ are defined as in* (6.1), *then X_n can be expressed as*

$$X_n = \sum_{j=0}^{\infty} \psi_j Z_{n-j} + V_n, \quad n \in \mathbb{Z}, \tag{6.5}$$

where
(i) $\psi_0 = \sigma_1$ *and* $\sum_{j=0}^{\infty} |\psi_j|^2 < \infty$,
(ii) $(Z_n)_{n\in\mathbb{Z}} \sim \mathrm{WN}(0,1)$,
(iii) $Z_n \in \mathcal{M}_n$ *for each* $n \in \mathbb{Z}$,
(iv) $\mathbb{E}(Z_m \overline{V}_n) = 0$ *for all* $m, n \in \mathbb{Z}$,
(v) $V_n \in \mathcal{M}_{-\infty}$ *for all* $n \in \mathbb{Z}$, *and*
(vi) $(V_n)_{n\in\mathbb{Z}}$ *is deterministic.*

(The properties (v) *and* (vi) *are not the same since $\mathcal{M}_{-\infty}$ is defined in terms of the sequence X, not in terms of V.)*

The sequences $(\psi_j)_{j\in\mathbb{N}_0}$, $(Z_n)_{n\in\mathbb{Z}}$ and $V = (V_n)_{n\in\mathbb{Z}}$ are uniquely determined by (6.5) *and the conditions* (i)–(vi).

Defining $U = (U_n)_{n\in\mathbb{Z}}$ by $U_n := \sum_{j=0}^{\infty} \psi_j Z_{n-j}$, $n \in \mathbb{Z}$, the spectral distribution functions, F_U and F_V, of the sequences U and V are respectively the absolutely continuous and singular components of the spectral distribution function F of X (i. e., $F_U = F_{\mathrm{ac}}$ and $F_V = F_{\mathrm{sing}}$, apart from possibly a constant). In particular,

$$F'(\lambda) = F_U'(\lambda) = (2\pi)^{-1} \left| \sum_{j=0}^{\infty} \psi_j \, \mathrm{e}^{-\mathrm{i}j\lambda} \right|^2 \quad \textit{Lebesgue a. e.}$$

and

$$F(\lambda_2) - F(\lambda_1) = (2\pi)^{-1} \int_{-\lambda_1}^{\lambda_2} \left| \sum_{j=0}^{\infty} \psi_j \, \mathrm{e}^{-\mathrm{i}j\theta} \right|^2 \mathrm{d}\theta + F_V(\lambda_2) - F_V(\lambda_1) \tag{6.6}$$

for $-\pi < \lambda_1 \le \lambda_2 \le \pi$.

Proof. See Doob [37, Theorem XII.4.2]. □

Corollary 6.6. *From the Wold decomposition* (6.5) *it follows at once that*

$$P_{\mathcal{M}_{n-h}} X_n = \sum_{j=h}^{\infty} \psi_j Z_{n-j} + V_n$$

and

$$\sigma_h^2 = \sum_{j=0}^{h-1} |\psi_j|^2. \tag{6.7}$$

Theorem 6.4, together with (6.6) and (6.7), shows that the coefficients ψ_j in the Wold decomposition have the additional property that

$$\psi(z) := \sum_{j=0}^{\infty} \psi_j z^j \neq 0 \quad \text{for all } z \in \mathbb{C} \text{ such that } |z| < 1. \tag{6.8}$$

The next theorem relates the regularity of a weakly stationary sequence X and its Wold coefficients to the spectral distribution function F of X.

Theorem 6.7. *The zero-mean weakly stationary sequence X is regular if and only if its spectral distribution function F satisfies the conditions $F'(\lambda) > 0$ Lebesgue almost everywhere and*

$$\int_{-\pi}^{\pi} \log F'(\lambda)\, d\lambda > -\infty.$$

If X is regular, then $\psi_0, \psi_1, \psi_2, \ldots$, as defined in Theorem 6.5, are uniquely determined by the conditions,

$$\psi_0 = \sqrt{2\pi} \exp\left\{ \frac{1}{4\pi} \int_{-\pi}^{\pi} \log F'(\lambda)\, d\lambda \right\}, \tag{6.9}$$

$$\sum_{j=0}^{\infty} |\psi_j|^2 < \infty, \tag{6.10}$$

$$\sum_{j=0}^{\infty} \psi_j z^j \neq 0 \quad \textit{for all } z \in \mathbb{C} \textit{ such that } |z| < 1, \quad \textit{and} \tag{6.11}$$

$$F'(\lambda) = (2\pi)^{-1} \left| \sum_{j=0}^{\infty} \psi_j\, e^{-ij\lambda} \right|^2 \quad \textit{Lebesgue almost everywhere.} \tag{6.12}$$

The constants ψ_j in the representation (6.5) can be found from the spectral distribution function F via the relation

$$\sum_{j=0}^{\infty} \overline{\psi}_j z^j = \exp\left\{ a_0 + 2 \sum_{j=1}^{\infty} a_j z^j \right\}, \quad |z| < 1, \tag{6.13}$$

where

$$a_j = \frac{1}{4\pi} \int_{-\pi}^{\pi} \log(2\pi F'(\lambda))\, e^{-ij\lambda}\, d\lambda. \tag{6.14}$$

Proof. See Doob [37, Theorem XII.4.3]. □

Remark 6.8. Equation (6.9) is known as *Kolmogorov's formula* for $\psi_0 = \sigma_1$, the square root of the one-step mean-square prediction error of X.

Example 6.9 (Causal ARMA(p,q) process). Let $p, q \in \mathbb{N}_0$, $\phi_0, \dots, \phi_p, \theta_0, \dots, \theta_q \in \mathbb{C}$ with $\phi_0 = \theta_0 = 1$ and $\phi_p, \theta_q \neq 0$. Consider the ARMA(p,q) equation

$$X_n - \phi_1 X_{n-1} - \cdots - \phi_p X_{n-p} = W_0 + \theta_1 W_{n-1} + \cdots + \theta_q W_{n-q}, \tag{6.15}$$

where $(W_n)_{n\in\mathbb{Z}} \sim \mathrm{WN}(0, \nu^2)$ with $\nu^2 > 0$. Define the characteristic polynomials ϕ and θ by

$$\phi(z) = 1 - \phi_1 z - \cdots - \phi_p z^p \quad \text{and} \quad \theta(z) = 1 + \theta_1 z + \cdots + \theta_q z^q.$$

Denoting the zeros of ϕ and θ, counted with multiplicity, by $\xi_1^{-1}, \dots, \xi_p^{-1}$ and $\eta_1^{-1}, \dots, \eta_q^{-1}$, respectively, we can write

$$\phi(z) = \prod_{k=1}^{p} (1 - \xi_k z) \quad \text{and} \quad \theta(z) = \prod_{k=1}^{q} (1 - \eta_k z).$$

Now assume that ϕ has no zeros on the unit circle, i. e., that $|\xi_k| \neq 1$ for all $k = 1, \dots, p$. Then it is well known (e. g., Brockwell and Davis [22, Theorem 3.1.3]) that the ARMA(p,q) equation (6.15) has a unique stationary solution $X = (X_t)_{t\in\mathbb{Z}}$, given by $X_t = \sum_{j\in\mathbb{Z}} \beta_j W_{t-j}$, where the coefficients β_j are the Laurent coefficients of $\theta(z)/\phi(z)$ expanded in a neighbourhood of the unit circle around zero, i. e. $\sum_{j\in\mathbb{Z}} \beta_j z^j = \theta(z)/\phi(z)$ for all $z \in \mathbb{C}$ with $\delta_1 < |z| < \delta_2$ for some $\delta_1 < 1 < \delta_2$. This solution is called an *ARMA(p,q)-process*. It is further well known (e. g. [22, Theorem 4.4.2]) that the spectral distribution function F_X of this X is absolutely continuous, where the spectral density is given by

$$f_X(\lambda) = F_X'(\lambda) = \frac{\nu^2}{2\pi} \frac{|\theta(\mathrm{e}^{-\mathrm{i}\lambda})|^2}{|\phi(\mathrm{e}^{-\mathrm{i}\lambda})|^2} = \frac{\nu^2}{2\pi} \frac{\prod_{k=1}^{q} |1 - \eta_k \mathrm{e}^{-\mathrm{i}\lambda}|^2}{\prod_{k=1}^{p} |1 - \xi_k \mathrm{e}^{-\mathrm{i}\lambda}|^2}. \tag{6.16}$$

From this it is easy to see that $\int_{-\pi}^{\pi} \log f_X(\lambda)\, \mathrm{d}\lambda > -\infty$, so that X is regular by Theorem 6.7. Now assume *additionally* that all zeros of ϕ lie *outside the unit circle*, i. e., that $|\xi_k| < 1$ for all $k \in \{1, \dots, p\}$ (so that X will be *causal*), and that all zeros of θ lie *outside or on the unit circle*, i. e., that $|\eta_k| \leq 1$ for all $k \in \{1, \dots, q\}$. Then $\theta(z)/\phi(z)$ is holomorphic on a ball of the form $\{z \in \mathbb{C} : |z| < \delta_2\}$ where $\delta_2 > 1$ and the coefficients β_j obtained above are the Taylor coefficients of $\theta(z)/\phi(z)$ when expanded around zero, so we have

$$X_t = \sum_{j=0}^{\infty} \beta_j W_{t-j}, \quad \text{where } \beta(z) := \sum_{j=0}^{\infty} \beta_j z^j = \frac{\theta(z)}{\phi(z)}, |z| < \delta_2.$$

Now define

$$\psi_j := \nu\beta_j, \quad j \in \mathbb{N}_0 \quad \text{and} \quad Z_t := \nu^{-1}W_t, \quad t \in \mathbb{Z}.$$

Then $(Z_t) \sim \mathrm{WN}(0,1)$, $\sum_{j=0}^{\infty} |\psi_j|^2 < \infty$, $\sum_{j=0}^{\infty} \psi_j z^j = \nu\theta(z)/\phi(z) \neq 0$ for all $|z| < 1$ and

$$\frac{1}{2\pi}\left|\sum_{j=0}^{\infty} \psi_j \mathrm{e}^{-\mathrm{i}j\lambda}\right|^2 = \frac{\nu^2}{2\pi}|\beta(\mathrm{e}^{-\mathrm{i}\lambda})|^2 = f_X(\lambda),$$

so that Equations (6.10)–(6.12) are satisfied. For the verification of (6.9), a simple calculation (cf., Exercise 6.5) shows that $\int_{-\pi}^{\pi} \log|1 - \alpha\mathrm{e}^{-\mathrm{i}\lambda}|^2 \,\mathrm{d}\lambda = 0$ whenever $|\alpha| \leq 1$, hence $\int_{-\pi}^{\pi} \log f_X(\lambda)\,\mathrm{d}\lambda$ can be calculated as

$$2\pi \log \frac{\nu^2}{2\pi} + \sum_{k=1}^{p} \int_{-\pi}^{\pi} \log|1 - \eta_k \mathrm{e}^{-\mathrm{i}\lambda}|^2 \,\mathrm{d}\lambda - \sum_{k=1}^{q} \int_{-\pi}^{\pi} \log|1 - \xi_k \mathrm{e}^{-\mathrm{i}\lambda}|^2 \,\mathrm{d}\lambda$$

which is equal to $2\pi \log \frac{\nu^2}{2\pi}$. Hence the right-hand side of (6.9) is equal to ν. On the other hand, $\beta_0 = \beta(0) = \theta(0)/\phi(0) = 1$ so that $\psi_0 = \nu$, showing that (6.9) is satisfied. Theorem 6.7 now shows that the constants ψ_j thus obtained are indeed the coefficients in the Wold-decomposition. It can be shown that the previously defined white noise $(Z_t)_{t\in\mathbb{Z}}$ is indeed the white noise appearing in the Wold decomposition, so that $X_t = \sum_{j=0}^{\infty} \psi_j Z_{t-j}$ is the Wold decomposition. This is easily seen when all zeros of θ do not only lie outside or on the unit circle, but lie indeed *outside* (i. e., $|\eta_k| < 1$ for all k), since then the ARMA process is *invertible* and hence $W_t = \sum_{j=0}^{\infty} \pi_j X_{t-j} \in \mathcal{M}_n$, where the coefficients π_j are determined by $\sum_{j=0}^{\infty} \pi_j z^j = \phi(z)/\theta(z)$ for $|z| < 1$ (e. g., [22, Theorem 3.1.2]). If some of the zeros of θ lie *on the unit circle* (but still none inside), then the ARMA process is not invertible, but it is still true that $W_t \in \mathcal{M}_t$ and hence $Z_t \in \mathcal{M}_t$, a proof of which follows by an argument as outlined in Problem 3.8 of Brockwell and Davis [22]. We omit the details, since later we will only need the coefficients ψ_j in the Wold decomposition, but not the actual noise.

Returning to the general setting, the linear manifolds $\widetilde{\mathcal{M}}_n$ and $\widetilde{\mathcal{M}}_{-\infty}$ defined in (6.1) clearly only depend on the spectral distribution function F, independently of the probability space on which the random sequence X is defined. The following theorem characterizes them directly in terms of F.

Theorem 6.10. *Let F be a bounded non-decreasing right-continuous function on $(-\pi,\pi]$ (i. e., a bounded distribution function). Denote by F_{sing} the singular part of F and let $S \subset (-\pi,\pi]$ be a Borel set of Lebesgue measure zero which carries F_{sing}, i. e., such that $\int_{(-\pi,\pi]\setminus S} \mathrm{d}F_{\mathrm{sing}} = 0$. Then the following statements hold.*

(a) *If $F'(\lambda) = 0$ only on a set of Lebesgue measure zero and*

$$\int_{(-\pi,\pi]} \log F'(\lambda)\,\mathrm{d}\lambda > -\infty,$$

there is one and only one sequence $(\psi_j)_{j\in\mathbb{N}_0}$ *satisfying the conditions* (6.9)–(6.12) *and it is determined by* (6.13) *and* (6.14).

(b) *The subspace* $\widetilde{\mathcal{M}}_{-\infty}$ *consists of all measurable functions* α *on* $(-\pi,\pi]$ *which vanish except on a set of Lebesgue measure zero and belong to* $L^2(F_{\text{sing}})$. *The subspace* $\widetilde{\mathcal{M}}_n$ *consists of all functions of the form*

$$h(\lambda) := \frac{e^{in\lambda}}{\sum_{j=0}^{\infty}\psi_j e^{-ij\lambda}}\left(\sum_{j=0}^{\infty}\gamma_j e^{-ij\lambda}\right)\mathbf{1}_{(-\pi,\pi]\setminus S}(\lambda) + \alpha(\lambda)\mathbf{1}_S(\lambda), \quad \lambda \in (-\pi,\pi],$$

where $\alpha \in \widetilde{\mathcal{M}}_{-\infty}$ *and* $(\gamma_j)_{j\in\mathbb{N}_0}$ *is a square-summable sequence of complex numbers.*

Proof. See Doob [37, Theorem XII.4.4]. □

6.2 The continuous-time case

We shall assume henceforth in this section, without further comment, that $Y = (Y_t)_{t\in\mathbb{R}}$ is a zero-mean complex-valued weakly stationary process with autocovariance function $\gamma(h)$, $h \in \mathbb{R}$, where γ is continuous at 0 (implying that Y is mean-square continuous).

We shall denote by $G(\mu)$, $-\infty < \mu < \infty$, the spectral distribution function of Y. By the spectral representation (Theorems 5.1 and 5.8) we can write

$$\mathbb{E}(Y_{t+h}\overline{Y}_t) = \gamma(h) = \int_{\mathbb{R}} e^{i\mu h}\,dG(\mu), \quad t,h \in \mathbb{R},$$

and

$$Y_t = \int_{\mathbb{R}} e^{it\mu}\,dZ_\mu, \quad t \in \mathbb{R},$$

where $(Z_\mu)_{\mu\in\mathbb{R}}$ is an orthogonal-increment process with associated distribution function G. The mapping

$$I(g) = \int_{\mathbb{R}} g(\mu)\,dZ_\mu, \quad g \in L^2(G), \tag{6.17}$$

is an isomorphism of $L^2(G)$ (the measurable functions on $\mathbb{R}$, which are square-integrable with respect to G) onto the closed linear span in $\mathcal{S}$ of the random variables $\{Y_t, t \in \mathbb{R}\}$. Under this isomorphism

$$I(e^{it\cdot}) = Y_t, \quad t \in \mathbb{R}.$$

The mapping I is the inverse of the mapping T introduced in Theorem 5.6 (see Remark 5.10).

As in Section 6.1, it will be convenient to introduce the following subspaces of $\mathcal{S}$ and the corresponding subspaces of $L^2(G)$ under the isomorphism T.

$$\left.\begin{aligned} &\mathcal{N} := \overline{\text{sp}}\{Y_t, -\infty < t < \infty\}, && \widetilde{\mathcal{N}} := \overline{\text{sp}}\{e^{it\cdot}, -\infty < t < \infty\} = L^2(G), \\ &\mathcal{N}_t := \overline{\text{sp}}\{Y_s, -\infty < s \le t\}, && \widetilde{\mathcal{N}}_t := \overline{\text{sp}}\{e^{is\cdot}, -\infty < s \le t\}, \\ &\mathcal{N}_{-\infty} := \bigcap_{t\in\mathbb{R}} \mathcal{N}_t, && \widetilde{\mathcal{N}}_{-\infty} := \bigcap_{t\in\mathbb{R}} \widetilde{\mathcal{N}}_t. \end{aligned}\right\} \tag{6.18}$$

The isomorphism allows us to compute projections (in particular predictors) of random variables in $\mathcal{N}$ by computing corresponding projections of functions in $\widetilde{\mathcal{N}}$ and applying the mapping I. For example the best linear predictor $P_{\mathcal{N}_t} Y_{t+h}$ for $h \ge 0$ is

$$P_{\mathcal{N}_t} Y_{t+h} = I(P_{\widetilde{\mathcal{N}}_t} e^{i(t+h)\cdot}), \tag{6.19}$$

and its mean-square error is given by

$$\sigma_h^2 := \mathbb{E}[|Y_{t+h} - P_{\mathcal{N}_t} Y_{t+h}|^2] = \int_{\mathbb{R}} \left|e^{i(t+h)\mu} - (P_{\widetilde{\mathcal{N}}_t} e^{i(t+h)\cdot})(\mu)\right|^2 dG(\mu). \tag{6.20}$$

As in the discrete case, it is easily seen that σ_h^2 does not depend on t, see Exercise 6.1(b). From this we quickly obtain the following result.

Lemma 6.11. *The following are equivalent:*
(i) $\sigma_h^2 = 0$ *for some* $h > 0$.
(ii) $\sigma_h^2 = 0$ *for all* $h > 0$.
(iii) $\mathcal{N}_t = \mathcal{N}_s$ *for all* $s, t \in \mathbb{R}$.
(iv) $\mathcal{N}_0 = \mathcal{N}$.

Proof. The implications (iii) $\Longrightarrow$ (ii) $\Longrightarrow$ (i) are clear. To see that (i) implies (iii), observe that $\mathcal{N}_t \subset \mathcal{N}_s$ for $t \le s$. Now let $\sigma_{h_0}^2 = 0$ for some $h_0 > 0$. For $h \in (0, h_0]$ and $t \in \mathbb{R}$, we have $\mathcal{N}_{t+h-h_0} \subset \mathcal{N}_t$ and hence

$$\sigma_h^2 = \mathbb{E}[|Y_{t+h} - P_{\mathcal{N}_t} Y_{t+h}|^2] \le \mathbb{E}[|Y_{t+h} - P_{\mathcal{N}_{t+h-h_0}} Y_{t+h}|^2] = \sigma_{h_0}^2 = 0.$$

This implies $Y_{t+h} \in \mathcal{N}_t$ for all $h \in (0, h_0]$, from which we conclude $\mathcal{N}_{t+h_0} \subset \mathcal{N}_t$, hence $\mathcal{N}_{t+h} = \mathcal{N}_t$ for all $t \in \mathbb{R}$ and all $h \in (0, h_0]$. Replacing t by $t + h_0$, we deduce that also $\mathcal{N}_{t+h} = \mathcal{N}_{t+h_0} = \mathcal{N}_t$ for all $h \in (h_0, 2h_0]$, and continuing this way, we see that $\mathcal{N}_t = \mathcal{N}_{t+h}$ for all $t \in \mathbb{R}$ and $h > 0$.

That (iii) implies (iv) is clear. Conversely, if $\mathcal{N}_0 = \mathcal{N}$, then $Y_h \in \mathcal{N} = \mathcal{N}_0$ for all $h > 0$, hence $P_{\mathcal{N}_0} Y_h = Y_h$ so that $\sigma_h^2 = 0$ for all $h > 0$, showing that (iv) implies (ii). □

Lemma 6.11 gives rise to the following definition.

Definition 6.12 (Deterministic and regular processes). The mean-square continuous complex-valued zero-mean stationary stochastic process $Y = (Y_t)_{t\in\mathbb{R}}$ is called *regular* if the h-step mean-square prediction error σ_h^2 in (6.20) is strictly greater than 0 for some and hence for all $h > 0$. Otherwise it is called *deterministic*.

The reader should take care with the notion of a deterministic process. Frequently, by a *deterministic stochastic process* one simply means a stochastic process $(Y_t)_{t\in\mathbb{R}}$ for which Y_t is constant almost surely for each t. This is clearly a different notion than the one in Definition 6.12, which refers to the fact that the future value Y_{t+h} is completely determined by the values of the process up to time t. For example, if Y_0 is a standard normally distributed random variable and $Y_t := Y_0$ for all $t \in \mathbb{R}$, then $(Y_t)_{t\in\mathbb{R}}$ is deterministic in the sense of Definition 6.12, but clearly Y_t is not constant almost surely.

Our aim in this section is to express $P_{\mathcal{N}_t} Y_{t+h}$ and σ_h^2 in terms of Y and its defining properties. In the course of doing so, we shall also clarify, as in Section 6.1, the structure and properties of Y itself.

We shall follow Doob [37] in solving the continuous-time problem by relating it to a corresponding discrete-time problem for which the essentials of the solution have been given in Section 6.1. In order to distinguish clearly between the continuous-time and discrete-time problems, we shall use G exclusively for the spectral distribution function of the continuous-time process Y and F for the spectral distribution function of a corresponding discrete-time sequence X, which is defined below. We shall also use $\lambda \in (-\pi, \pi]$ to denote the discrete-time frequency variable (as in Section 6.1) and $\mu \in \mathbb{R}$ to denote the corresponding continuous-time frequency variable. Both λ and μ are frequencies measured in radians per unit time, unlike the frequencies in [37], which are measured in cycles per unit time.

Given the continuous-time process $Y = (Y_t)_{t\in\mathbb{R}}$ with spectral distribution function G on $\mathbb{R}$, we introduce a mean-zero weakly stationary discrete time sequence $X = (X_n)_{n\in\mathbb{Z}}$ with spectral distribution function F on $(-\pi, \pi]$, where

$$F(\lambda) = G(\mu(\lambda)), \quad \lambda \in (-\pi, \pi], \tag{6.21}$$

and

$$\mu(\lambda) := \tan\frac{\lambda}{2}, \quad \text{i.e.} \quad \lambda(\mu) = 2\arctan(\mu). \tag{6.22}$$

Observe that (the measure associated with) F cannot have an atom at π. We shall refer to X as the *Doob sequence* associated with the process Y. The relation between μ and λ can be rewritten as

$$\mu = -\mathrm{i}\frac{\mathrm{e}^{\mathrm{i}\lambda} - 1}{\mathrm{e}^{\mathrm{i}\lambda} + 1},$$

from which it follows that

$$e^{i\lambda} = \frac{1 + i\mu}{1 - i\mu}. \tag{6.23}$$

Remark 6.13. Denoting by $f(\cdot)$ an element of the subspace $\widetilde{\mathcal{M}}$ of $L^2(F)$ defined in (6.1), the mapping $f(\cdot) \mapsto f(\lambda(\cdot))$ of $\widetilde{\mathcal{M}}$ onto the subspace

$$\widetilde{\mathcal{N}}^* := \overline{\text{sp}}\left\{\left(\frac{1 + i\cdot}{1 - i\cdot}\right)^n, n \in \mathbb{Z}\right\}$$

of $L^2(G)$ is clearly a Hilbert space isomorphism (see Exercise 6.6). Defining

$$\widetilde{\mathcal{N}}_0^* := \overline{\text{sp}}\left\{\left(\frac{1 + i\cdot}{1 - i\cdot}\right)^n, n \in \mathbb{Z} \cap (-\infty, 0]\right\},$$

we shall show in the following lemma that $\widetilde{\mathcal{N}}_0^* = \widetilde{\mathcal{N}}_0$, where $\widetilde{\mathcal{N}}_0$ was defined in (6.18).

Lemma 6.14. $\widetilde{\mathcal{N}}_0^* = \widetilde{\mathcal{N}}_0$, *where* $\widetilde{\mathcal{N}}_0^*$ *and* $\widetilde{\mathcal{N}}_0$ *were defined in Remark* 6.13 *and equation* (6.18), *respectively.*

Proof. We observe first that

$$\frac{1 - i\mu}{1 + i\mu} = -1 + \frac{2}{1 + i\mu} = -1 + 2\int_{-\infty}^{0} e^{t(1+i\mu)}\, dt,$$

so that $((1 + i\cdot)/(1 - i\cdot))^{-n}$ for $n \in \mathbb{N}$ is the pointwise limit as $m \to \infty$ of the functions $g_{n,m}(\cdot) = (-1 + 2\sum_{k=-m^2}^{-1} m^{-1}e^{k/m}e^{i(k/m)\cdot})^n$, which are uniformly bounded for fixed n. Hence $(\frac{1+i\cdot}{1-i\cdot})^{-n}$ is the $L^2(F)$-limit of the $g_{n,m}$, and since $g_{n,m} \in \widetilde{\mathcal{N}}_0$, so is $(\frac{1+i\cdot}{1-i\cdot})^{-n}$ for each $n \in \mathbb{N}$. This clearly implies $\widetilde{\mathcal{N}}_0^* \subset \widetilde{\mathcal{N}}_0$.

To show the reverse inclusion, consider the function

$$f(z) := \exp\left(t\frac{z - 1}{z + 1}\right), \quad z \in \mathbb{C} \setminus \{-1\},$$

which, for $t \in \mathbb{R}$, is a holomorphic function of $1/z$ for all $z \in \mathbb{C}$ such that $|z| > 1$. Hence, there is a sequence $(c_k)_{k\in\mathbb{N}_0}$ of complex numbers such that

$$f(z) = \sum_{k=0}^{\infty} c_k z^{-k} = \sum_{k=0}^{\infty} c_k |z|^{-k}(e^{i\lambda})^{-k}$$

for $z = |z|e^{i\lambda}$ with $|z| > 1$ and $\lambda \in (-\pi, \pi]$. Fixing $|z| > 1$ and viewing the above as a function of $\lambda \in (-\pi, \pi]$, this function is in $\widetilde{\mathcal{M}}_0$ (the series converges uniformly in λ, hence also in $L^2(F)$). Applying the isomorphism of Remark 6.13, this function is clearly mapped into $\widetilde{\mathcal{N}}_0^*$. In other words, for fixed $t \in \mathbb{R}$ and $|z| > 1$, by equation (6.23), the function

$$g_{t,|z|} : \mathbb{R} \to \mathbb{C}, \quad \mu \mapsto g_{t,|z|}(\mu) := \exp\Big(t\Big(|z|\frac{1+\mathrm{i}\mu}{1-\mathrm{i}\mu} - 1\Big)\Big(|z|\frac{1+\mathrm{i}\mu}{1-\mathrm{i}\mu} + 1\Big)^{-1}\Big),$$

is in $\widetilde{\mathcal{N}}_0^*$. It is easily seen that $g_{t,|z|}(\mu)$ converges to $\mathrm{e}^{\mathrm{i}t\mu}$ as $|z| \downarrow 1$. Since $\Re(\frac{w-1}{w+1}) > 0$ for $w \in \mathbb{C}$ with $|w| > 1$ we have $|g_{t,|z|}(\mu)| \le 1$ for $t < 0$ and $|z| > 1$. Hence, by dominated convergence, when $t < 0$, $g_{t,|z|}$ does not only converge pointwise to $\mathrm{e}^{\mathrm{i}t\cdot}$ as $|z| \downarrow 1$, but also in $L^2(G)$. Hence, $\mathrm{e}^{\mathrm{i}t\cdot} \in \widetilde{\mathcal{N}}_0^*$ for $t \le 0$, showing that $\widetilde{\mathcal{N}}_0 \subset \widetilde{\mathcal{N}}_0^*$, which, with the reverse inclusion, completes the proof of the lemma. □

Remark 6.15. A slight modification of the proof of Lemma 6.14 (considering the function f defined there as a holomorphic function in $\{z \in \mathbb{C} : |z| < 1\}$, letting $|z| \uparrow 1$ and observing that $\Re(\frac{w-1}{w+1}) < 0$ for $|w| < 1$) shows also the equality of the following closed subspaces of $L^2(G)$, namely,

$$\overline{\mathrm{sp}}\Big\{\Big(\frac{1+\mathrm{i}\cdot}{1-\mathrm{i}\cdot}\Big)^n, n \in \mathbb{Z} \cap [0,\infty)\Big\} = \overline{\mathrm{sp}}\{\mathrm{e}^{\mathrm{i}t\cdot}, 0 \le t < \infty\}.$$

Together with Lemma 6.14 this implies that the mapping $f(\cdot) \mapsto f(\lambda(\cdot))$ defined in Remark 6.13 is a Hilbert space isomorphism of $\widetilde{\mathcal{M}} = L^2(F)$ onto $\widetilde{\mathcal{N}} = L^2(G)$ (of course, the latter assertion can be proved also directly using equation (6.21) and the transformation rule for integration). Thus we have shown that under this mapping $\widetilde{\mathcal{M}}$ and $\widetilde{\mathcal{N}}$ are isomorphic as are $\widetilde{\mathcal{M}}_0$ and $\widetilde{\mathcal{N}}_0$. These observations lead at once to the following criterion for regularity of the continuous-time process Y.

Theorem 6.16 (Paley–Wiener conditions). *The process Y is regular if and only if the derivative G' of its spectral distribution function G vanishes only on a set of Lebesgue measure zero and*

$$\int_{\mathbb{R}} \frac{\log G'(\mu)}{1+\mu^2}\,\mathrm{d}\mu > -\infty. \tag{6.24}$$

In other words, if G is the spectral distribution function of Y, then Y is regular if and only if the density g of the absolutely continuous part G_{ac} of G vanishes only on a set of Lebesgue measure zero and if $\int_{\mathbb{R}} \frac{g(\mu)}{1+\mu^2}\,\mathrm{d}\mu > -\infty$.

Proof. From Lemma 6.11, the continuous-time process Y is deterministic if and only if $\mathcal{N}_0 = \mathcal{N}$. Using the isomorphism I introduced in (6.17), this is equivalent to $\widetilde{\mathcal{N}}_0 = \widetilde{\mathcal{N}}$. Similarly, by an argument completely analogous to that of Lemma 6.11 (see Exercise 6.3), the corresponding discrete-time sequence $X = (X_n)_{n\in\mathbb{Z}}$ is deterministic if and only if $\mathcal{M}_0 = \mathcal{M}$, which using the isomorphism introduced in Section 6.1 is equivalent to $\widetilde{\mathcal{M}}_0 = \widetilde{\mathcal{M}}$. However, because of the isomorphism $f(\cdot) \mapsto f(\lambda(\cdot))$, by Remark 6.15 this condition is equivalent to the condition $\widetilde{\mathcal{N}}_0 = \widetilde{\mathcal{N}}$. In other words, the continuous-time process Y is deterministic (or regular) if and only if the corresponding sequence X is deterministic (or regular).

The necessary and sufficient conditions for regularity of X are given in Theorem 6.7, namely $F'(\lambda) > 0$ Lebesgue almost everywhere and $\int_{(-\pi,\pi]} \log F'(\lambda)\mathrm{d}\lambda > -\infty$. Expressing these conditions in terms of G, we see that they are equivalent respectively to the conditions $G'(\mu) > 0$ Lebesgue almost everywhere and

$$\int_{\mathbb{R}} \frac{\log[G'(\mu)(1+\mu^2)/2]}{(1+\mu^2)/2}\,\mathrm{d}\mu > -\infty.$$

But the latter condition is equivalent to (6.24) since $\int_{\mathbb{R}} \frac{2\log[(1+\mu^2)/2]}{(1+\mu^2)}\,\mathrm{d}\mu < \infty$ (cf. Exercise 6.7). The equivalent formulation follows from Remark 6.3. □

Remark 6.17. We have already introduced the spaces $L^2(F)$ and $L^2(G)$ of measurable square-integrable functions on $(-\pi,\pi]$ with respect to F and on $\mathbb{R}$ with respect to G, respectively. We shall also use the spaces

$$L^2(\mathrm{d}\lambda) := L^2((-\pi,\pi];\mathbb{C}) \quad \text{and} \quad L^2(\mathrm{d}\mu) := L^2 = L^2(\mathbb{R};\mathbb{C})$$

of measurable square-integrable functions on $(-\pi,\pi]$ and $\mathbb{R}$, respectively, where $\mathrm{d}\lambda$ denotes Lebesgue measure on $(-\pi,\pi]$ and $\mathrm{d}\mu$ denotes Lebesgue measure on $\mathbb{R}$. In the analysis of the weakly stationary sequence X, a key role is played by the orthonormal sequence $\{\frac{1}{\sqrt{2\pi}}\mathrm{e}^{\mathrm{i}j\lambda}, j \in \mathbb{Z}\}$ of functions on $(-\pi,\pi]$ in $L^2(\mathrm{d}\lambda)$. An analogous role is played in the analysis of the corresponding weakly stationary process Y by the sequence $\{\frac{1}{\sqrt{\pi}}\frac{(1-\mathrm{i}\mu)^j}{(1+\mathrm{i}\mu)^{j+1}}, j \in \mathbb{Z}\}$ of functions on $\mathbb{R}$. The following lemma establishes the orthonormality of this sequence in $L^2(\mathrm{d}\mu)$.

Lemma 6.18. *The sequence of functions,*

$$\left\{f_j(\mu) := \frac{1}{\sqrt{\pi}}\frac{(1-\mathrm{i}\mu)^j}{(1+\mathrm{i}\mu)^{j+1}}, j \in \mathbb{Z}\right\},$$

is an orthonormal basis in $L^2(\mathrm{d}\mu)$, i. e., an orthonormal system in $L^2(\mathrm{d}\mu)$ that is complete.

Proof. Making the transformation of variables defined by (6.22), and noting that $\lambda'(\mu) = 2(1+\mu^2)^{-1} = 2(1+\mathrm{i}\mu)^{-1}(1-\mathrm{i}\mu)^{-1}$, we obtain

$$\frac{1}{2\pi}\int_{(-\pi,\pi]} \mathrm{e}^{(k-j)\mathrm{i}\lambda}\,\mathrm{d}\lambda = \int_{\mathbb{R}} f_j(\mu)\overline{f_k(\mu)}\,\mathrm{d}\mu.$$

The orthonormality of $\{\frac{1}{\sqrt{2\pi}}\mathrm{e}^{\mathrm{i}j\lambda}, j \in \mathbb{Z}\}$ thus yields that of $\{f_j(\mu), j \in \mathbb{Z}\}$. To see that the system is complete, let $g \in L^2 = L^2(\mathrm{d}\mu)$ such that $\int_{\mathbb{R}} g(\mu)\overline{f_k(\mu)}\,\mathrm{d}\mu = 0$ for all $k \in \mathbb{Z}$. Define $h(\lambda) := g(\mu(\lambda))(1+\mathrm{i}\mu(\lambda))/\sqrt{2}$ for $\lambda \in (-\pi,\pi)$. Since $\mu'(\lambda) = (1+\mu(\lambda)^2)/2$, we obtain

$$\int_{-\pi}^{\pi} |h(\lambda)|^2 \, \mathrm{d}\lambda = \int_{-\pi}^{\pi} |g(\mu(\lambda))|^2 \mu'(\lambda) \, \mathrm{d}\lambda = \int_{-\infty}^{\infty} |g(\mu)|^2 \, \mathrm{d}\mu < \infty,$$

so that $h \in L^2((-\pi, \pi]; \mathbb{C})$ and further

$$\begin{aligned} 0 &= \frac{1}{\sqrt{\pi}} \int_{\mathbb{R}} g(\mu) \frac{(1+\mathrm{i}\mu)^k}{(1-\mathrm{i}\mu)^{k+1}} \, \mathrm{d}\mu \\ &= \frac{1}{\sqrt{\pi}} \int_{-\pi}^{\pi} g(\mu(\lambda)) \mathrm{e}^{\mathrm{i}\lambda k} \frac{\mu'(\lambda)}{1-\mathrm{i}\mu(\lambda)} \, \mathrm{d}\lambda \\ &= \frac{1}{\sqrt{2\pi}} \int_{-\pi}^{\pi} h(\lambda) \mathrm{e}^{\mathrm{i}\lambda k} \, \mathrm{d}\lambda. \end{aligned}$$

Since this is true for all $k \in \mathbb{Z}$, we conclude that $h = 0$ Lebesgue almost everywhere (see, e. g., Brockwell and Davis [22, Thm. 2.11.2]), so that $g = 0$ Lebesgue almost everywhere. Hence the orthogonal complement in L^2 of $\overline{\mathrm{sp}}\{f_j : j \in \mathbb{Z}\}$ contains only the zero-function, implying $L^2 = \overline{\mathrm{sp}}\{f_j : j \in \mathbb{Z}\}$. Hence $\{f_j, j \in \mathbb{Z}\}$ is complete, hence an orthonormal basis. □

Corollary 6.19. *Let $(a_j)_{j \in \mathbb{N}_0}$ be a sequence of complex numbers such that $\sum_{j=0}^{\infty} |a_j|^2 < \infty$ and define the function $a = \sum_{j=0}^{\infty} a_j \mathrm{e}^{-\mathrm{i}j\cdot} \in L^2(\mathrm{d}\lambda)$ with convergence in the space $L^2(\mathrm{d}\lambda)$.*

(a) *Then the function $\mu \mapsto \frac{a(\lambda(\mu))}{\sqrt{\pi}(1+\mathrm{i}\mu)}$ is in $L^2 = L^2(\mathrm{d}\mu)$ and has the orthonormal expansion in $L^2(\mathrm{d}\mu)$,*

$$\frac{a(\lambda(\mu))}{\sqrt{\pi}(1+\mathrm{i}\mu)} = \sum_{j=0}^{\infty} a_j f_j(\mu), \quad \mu \in \mathbb{R}.$$

(b) *Denote by $A_0, \dots, A_j$ the coefficients in the partial fraction expansion,*

$$f_j(\mu) = \frac{1}{\sqrt{\pi}} \frac{(1-\mathrm{i}\mu)^j}{(1+\mathrm{i}\mu)^{j+1}} = \sum_{r=0}^{j} \frac{A_r}{(1+\mathrm{i}\mu)^{r+1}},$$

and define $\phi_j \in L^1 \cap L^2$ for $j \in \mathbb{N}_0$ by

$$\phi_j(t) = \sqrt{2\pi}\, \mathrm{e}^{-t} \sum_{r=0}^{j} \frac{t^r}{r!} A_r \mathbf{1}_{(0,\infty)}(t). \tag{6.25}$$

Then ϕ_j is the inverse Fourier transform of $f_j \in L^2$, the sequence $\{\phi_j, j \in \mathbb{N}_0\}$ is an orthonormal sequence in L^2 and $\frac{a(\lambda(\cdot))}{\sqrt{\pi}(1+\mathrm{i}\cdot)}$ is the Fourier transform of $\sum_{j=0}^{\infty} a_j \phi_j$.

(c) *The sequence $\{\phi_j, j \in \mathbb{N}_0\}$ is an orthonormal basis of the subspace $\{h \in L^2 : h(t) = 0 \ \forall\, t < 0\}$.*

Proof. (a) For $n \in \mathbb{N}$, define $a^{(n)} := \sum_{j=0}^{n} a_j e^{-ij\cdot} \in L^2(d\lambda)$. Then

$$\frac{a^{(n)}(\lambda(\mu))}{\sqrt{\pi}(1+i\mu)} = \sum_{j=0}^{n} a_j f_j(\mu) \tag{6.26}$$

by (6.23). Since $\{f_j : j \in \mathbb{Z}\}$ is an orthonormal sequence, by Proposition 2.36 the right-hand side of (6.26) converges in $L^2(d\mu)$ as $n \to \infty$, hence so must the left-hand side. But since $a^{(n)}$ converges in $L^2(d\lambda)$ to a, a subsequence $a^{(n_k)}$ converges Lebesgue almost everywhere to a, hence $a^{(n_k)}(\lambda(\mu))/(\sqrt{\pi}(1+i\mu))$ converges Lebesgue almost everywhere to $a(\lambda(\mu))/(\sqrt{\pi}(1+i\mu))$. This shows that $a(\lambda(\cdot))/(\sqrt{\pi}(1+i\cdot)) \in L^2(d\mu)$ and its orthonormal expansion in $L^2(d\mu)$.

(b) A straightforward calculation shows that

$$\frac{1}{\sqrt{2\pi}} \int_{\mathbb{R}} e^{-i\mu t} \phi_j(t)\, dt = f_j(\mu).$$

Hence, f_j is the Fourier transform of $\phi_j \in L^1 \cap L^2$, so that ϕ_j is the inverse Fourier transform of $f_j \in L^2$. We further obtain

$$\sum_{j=0}^{n} a_j f_j(\mu) = \frac{1}{\sqrt{2\pi}} \int_{\mathbb{R}} e^{-it\mu} \sum_{j=0}^{n} a_j \phi_j(t)\, dt. \tag{6.27}$$

Since ϕ_j is the inverse Fourier transform of $f_j \in L^2$ and since the functions f_j are orthonormal, Theorem 2.66(c) implies that the same is true of the functions $\phi_j \in L^2$. Consequently, $\sum_{j=0}^{n} a_j \phi_j$ converges in L^2 as $n \to \infty$ by Proposition 2.36. The right-hand side of (6.27) is the L^2-Fourier transform of $\sum_{j=0}^{n} a_j \phi_j$, which by continuity converges to the L^2-Fourier transform of $\sum_{j=0}^{\infty} a_j \phi_j$. On the other hand, according to (a), the left-hand side of (6.27) converges in L^2 to $\frac{a(\lambda(\cdot))}{\sqrt{\pi}(1+i\cdot)}$, completing the proof of (b).

(c) For $-j \in \mathbb{N}$, defined ϕ_j as the inverse (L^2-)Fourier transform of f_j, extending the relation proved in (b) to all $j \in \mathbb{Z}$. Since $\{f_j, j \in \mathbb{Z}\}$ is an orthonormal basis and the L^2-Fourier transform is an isometry from L^2 onto L^2, also $\{\phi_j, j \in \mathbb{Z}\}$ is an orthonormal basis of $L^2(\mathbb{R})$. From (b), we know that each ϕ_j with $j \in \mathbb{N}_0$ is supported in $[0, \infty)$. Now let $-j \in \mathbb{N}$. Since

$$f_j(\mu) = \frac{1}{\sqrt{\pi}} \frac{(1-i\mu)^j}{(1+i\mu)^{j+1}} = \frac{1}{\sqrt{\pi}} \left(\frac{1+i\mu}{1-i\mu}\right)^{|j|-1} \frac{1}{1-i\mu} = f_{|j|-1}(-\mu),$$

we obtain

$$f_j(\mu) = \frac{1}{\sqrt{2\pi}} \int_{\mathbb{R}} e^{-i(-\mu)t} \phi_{|j|-1}(t)\, dt = \frac{1}{\sqrt{2\pi}} \int_{\mathbb{R}} e^{-i\mu t} \phi_{|j|-1}(-t)\, dt,$$

so that f_j is the Fourier transform of $\phi_{|j|-1}(-\cdot)$. This implies $\phi_j(t) = \phi_{|j|-1}(-t)$ so that ϕ_j is supported in $(-\infty, 0]$ for $-j \in \mathbb{N}$. Since $\{\phi_j : j \in \mathbb{Z}\}$ is complete, we conclude $\overline{\text{sp}}\{\phi_j : j \in \mathbb{N}_0\} = \{h \in L^2(\mathbb{R}) : h(t) = 0 \; \forall \; t < 0\}$, as required. □

Lemma 6.20. *If $(a_j)_{j\in\mathbb{N}_0}$ is a square-summable complex-valued sequence, then*

$$\sum_{j=0}^{\infty} a_j z^j \neq 0 \quad \textit{for all } |z| < 1, \tag{6.28}$$

if and only if

$$\int_{\mathbb{R}} \mathrm{e}^{-\mathrm{i}wt} a^*(t)\,\mathrm{d}t \neq 0 \quad \textit{for all } w \in \mathbb{C} : \Im(w) < 0, \tag{6.29}$$

where

$$a^*(t) := \sum_{j=0}^{\infty} a_j \phi_j(t)$$

and ϕ_j, the inverse Fourier transform of f_j, was specified in Corollary 6.19. *Observe that $a^* \in L^2$ and that $a^*(t) = 0$ for $t < 0$, so that the integral in* (6.29) *is well defined.*

Proof. By making the transformation

$$z = \frac{1 - \mathrm{i}w}{1 + \mathrm{i}w},$$

the condition (6.28) can be rewritten as

$$\sum_{j=0}^{\infty} a_j \left(\frac{1 - \mathrm{i}w}{1 + \mathrm{i}w}\right)^j \neq 0 \quad \text{for } \Im(w) < 0,$$

or equivalently,

$$\sum_{j=0}^{\infty} a_j \frac{(1 - \mathrm{i}w)^j}{(1 + \mathrm{i}w)^{j+1}} \neq 0 \quad \text{for } \Im(w) < 0. \tag{6.30}$$

But

$$\frac{1}{\sqrt{\pi}} \frac{(1 - \mathrm{i}w)^j}{(1 + \mathrm{i}w)^{j+1}} = f_j(w) = \frac{1}{\sqrt{2\pi}} \int_{\mathbb{R}} \mathrm{e}^{-\mathrm{i}wt} \phi_j(t)\,\mathrm{d}t, \quad w \in \mathbb{R}, \tag{6.31}$$

and (6.31) clearly continues to hold for $w \in \mathbb{C}$ such that $\Im(w) < 0$ by the same calculation that led to (6.31), since the integrand decreases exponentially as $t \to \infty$ and is zero for $t < 0$. Substitution from (6.31) into (6.30) gives the required equivalence of (6.28) and (6.29). □

Definition 6.21. If G is the spectral distribution function of a regular weakly stationary process, F is the corresponding discrete time spectral distribution function as defined by (6.21), $\psi_0, \psi_1, \psi_2, \dots$ are the Wold coefficients corresponding to F as defined in Theorem 6.5 (the discrete time sequence is also regular, as seen in the proof of Theorem 6.16), then we define

$$\psi_c(t) := \sum_{j=0}^{\infty} \psi_j \phi_j(t), \quad t \in \mathbb{R},$$

and

$$\widetilde{\psi_c}(\mu) := \sum_{j=0}^{\infty} \psi_j f_j(\mu), \quad \mu \in \mathbb{R},$$

the convergence being in L^2 (cf. Proposition 2.36), so that $\widetilde{\psi_c}$ is the Fourier transform of ψ_c by Corollary 6.19.

The following theorem decomposes G into its absolutely continuous and singular components, expressing the absolutely continuous component in terms of the function $\widetilde{\psi_c}$.

Theorem 6.22. *Let G be the spectral distribution function of a regular weakly stationary process, and denote by G_{ac} and G_{sing} the absolutely continuous and singular parts of G (cf. Remark* 6.3*). Let ψ_c and $\widetilde{\psi_c}$ be as in Definition* 6.21.

(a) *Then*

$$\psi_c(t) = 0 \quad \forall\, t \le 0 \text{ and } \int_{\mathbb{R}} |\psi_c(t)|^2 \,\mathrm{d}t < \infty, \tag{6.32}$$

$$\int_{\mathbb{R}} \psi_c(t) \mathrm{e}^{-iwt} \,\mathrm{d}t \neq 0 \quad \forall\, w \in \mathbb{C} : \Im(w) < 0, \tag{6.33}$$

$$\log\left[\frac{1}{\sqrt{\pi}} \int_{\mathbb{R}} \psi_c(t)\, \mathrm{e}^{-t} \,\mathrm{d}t\right] = \frac{1}{4\pi} \int_{\mathbb{R}} \frac{\log[G'(\mu)(1+\mu^2)/2]}{(1+\mu^2)/2} \,\mathrm{d}\mu \tag{6.34}$$

(where the integral on the left of the last equation is real and strictly positive).

(b) *The density of the absolutely continuous part G_{ac} is given by $|\widetilde{\psi_c}(\cdot)|^2$, i. e.,*

$$G(b) - G(a) = \int_a^b |\widetilde{\psi_c}(\theta)|^2 \,\mathrm{d}\theta + G_{sing}(b) - G_{sing}(a), \quad a \le b \in \mathbb{R}. \tag{6.35}$$

In particular, $\widetilde{\psi_c} \neq 0$ Lebesgue almost everywhere.

(c) *Let $S \in \mathcal{B}_1$ be a Lebesgue null set that carries the singular part G_{sing}, i. e., such that $\int_{\mathbb{R}\setminus S} \mathrm{d}G_{sing} = 0$. Then the subspace $\widetilde{\mathcal{N}}_{-\infty}$ consists of all measurable functions f on $\mathbb{R}$*

that vanish except on a set of Lebesgue measure zero and belong to $L^2(G_{\text{sing}})$. *For each* $t \in \mathbb{R}$, *the subspace* $\widetilde{\mathcal{N}}_t$ *consists of all functions of the form*

$$h(\mu) := \frac{e^{it\mu}}{\widetilde{\psi_c}(\mu)}(\mathcal{F}_2\gamma)(\mu)\,\mathbf{1}_{\mathbb{R}\setminus S}(\mu) + \beta(\mu)\,\mathbf{1}_S(\mu), \quad \mu \in \mathbb{R},$$

where $\beta \in \widetilde{\mathcal{N}}_{-\infty}$ *and* $\gamma \in L^2(\mathbb{R})$ *with* $\gamma(t) = 0$ *for all* $t < 0$.

Proof. (a) Equation (6.32) follows from (6.25), the fact that $\sum_{j=0}^{\infty} |\psi_j|^2 < \infty$ and the orthonormality of $\{\phi_j\}$ in L^2. Equation (6.33) follows from Lemma 6.20 and the property of the Wold coefficients ψ_j that $\sum_{j=0}^{\infty} \psi_j z^j \neq 0$ for all $z \in \mathbb{C}$ such that $|z| < 1$ (cf. equation (6.11)).

Setting $w = -i$ in (6.31) gives $\int_{\mathbb{R}} e^{-t}\phi_j(t)dt = 1/\sqrt{2}$ if $j = 0$ and 0 otherwise. Hence

$$\frac{1}{\sqrt{\pi}}\int_{\mathbb{R}} e^{-t}\psi_c(t)\,dt = \frac{1}{\sqrt{\pi}}\int_{\mathbb{R}} e^{-t}\sum_{j=0}^{\infty}\psi_j\phi_j(t)\,dt = \frac{\psi_0}{\sqrt{2\pi}},$$

which is strictly positive by the definition of ψ_0 in Theorem 6.5. From (6.9), (6.21) and (6.22), we obtain

$$\log\frac{\psi_0}{\sqrt{2\pi}} = \frac{1}{4\pi}\int_{-\pi}^{\pi}\log F'(\lambda)\,d\lambda = \frac{1}{4\pi}\int_{\mathbb{R}}\frac{\log[G'(\mu)(1+\mu^2)/2]}{(1+\mu^2)/2}\,d\mu,$$

establishing the proof of relation (6.34).

(b) To show that $|\widetilde{\psi_c}|^2$ is the density of G_{ac}, by Remark 6.3, it is enough to show that $G'(\mu) = |\widetilde{\psi_c}(\mu)|^2$ Lebesgue almost everywhere in μ. To see this, recall from (6.12) that $F'(\lambda) = \frac{1}{2\pi}|\sum_{j=0}^{\infty}\psi_j e^{-ij\lambda}|^2$ Lebesgue almost everywhere, so that

$$\begin{aligned}
G'(\mu) &= F'(\lambda(\mu))\lambda'(\mu)\\
&= \frac{1}{2\pi}\left|\sum_{j=0}^{\infty}\psi_j\left(\frac{1-i\mu}{1+i\mu}\right)^j\right|^2\frac{2}{1+\mu^2}\\
&= \frac{1}{\pi}\left|\sum_{j=0}^{\infty}\psi_j\frac{(1-i\mu)^j}{(1+i\mu)^{j+1}}\right|^2\\
&= \left|\sum_{j=0}^{\infty}\psi_j f_j(\mu)\right|^2 = |\widetilde{\psi_c}(\mu)|^2
\end{aligned}$$

as required. Since the underlying process is regular, we have $G' \neq 0$ Lebesgue almost everywhere by Theorem 6.16, hence also $\widetilde{\psi_c} \neq 0$ Lebesgue almost everywhere.

(c) Denote by ν the measure associated with the distribution function F on $(-\pi, \pi]$ and by ρ the measure associated with the distribution function of G on $\mathbb{R}$. Also denote

$S' := \lambda(S) = \{2\arctan(s) : s \in S\}$. Then $S' \in \mathcal{B}_1 \cap (-\pi,\pi]$ is a Lebesgue null set (since λ is a C^1-diffemorphism), and since $\nu((a,b]) = \rho((\mu(a),\mu(b)])$ for all $-\pi < a \le b \le \pi$, it is easily seen that the absolutely continuous part ν_{ac} of ν and the singular part ν_{sing} are the image measures under the mapping λ of ρ_{ac} and ρ_{sing}, respectively. In particular, ν_{sing} (equivalently F_{sing}, the associated distribution function) is carried by S'.

Recall that $\widetilde{\mathcal{M}}_0$ and $\widetilde{\mathcal{N}}_0$ are isomorphic under the mapping $f(\cdot) \mapsto f(\lambda(\cdot))$ defined in Remark 6.13 (cf. Remark 6.15). Hence $h \in \widetilde{\mathcal{N}}_0$ if and only if there is some function $H \in \widetilde{\mathcal{M}}_0$ such that $h = H(\lambda(\cdot))$. By the characterization obtained in Theorem 6.10, $H \in \widetilde{\mathcal{M}}_0$ is equivalent to the fact that there exist $\alpha \in \widetilde{\mathcal{M}}_{-\infty}$ and a square-summable sequence $(a_n)_{n\in\mathbb{N}_0}$ such that

$$H = \left(\sum_{j=0}^{\infty} \psi_j \mathrm{e}^{-\mathrm{i}j\cdot}\right)^{-1} \left(\sum_{j=0}^{\infty} a_j \mathrm{e}^{-\mathrm{i}j\cdot}\right) \mathbf{1}_{(-\pi,\pi]\setminus S'} + \alpha \mathbf{1}_{S'}.$$

Hence, by Corollary 6.19(a), $h \in \widetilde{\mathcal{N}}_0$ if and only if there are $\alpha \in \widetilde{\mathcal{M}}_{-\infty}$ and a square-summable sequence $(a_j)_{j\in\mathbb{N}_0}$ such that h can be expressed as

$$h = \left(\sum_{j=0}^{\infty} \psi_j f_j\right)^{-1} \left(\sum_{j=0}^{\infty} a_j f_j\right) \mathbf{1}_{\mathbb{R}\setminus S} + \alpha(\lambda(\cdot))\mathbf{1}_S,$$

equivalently (by Corollary 6.19(b) and Definition 6.21), that

$$h = \widetilde{\psi_{\mathrm{c}}}^{-1} \mathcal{F}_2\left(\sum_{j=0}^{\infty} a_j \phi_j\right) \mathbf{1}_{\mathbb{R}\setminus S} + \alpha(\lambda(\cdot))\mathbf{1}_S.$$

Since $\{\phi_j, j \in \mathbb{N}_0\}$ is an orthonormal basis of $L^2_+ := \{f \in L^2 : f(t) = 0 \ \forall\ t < 0\}$ by Corollary 6.19(c), the set of all $\sum_{j=0}^{\infty} a_j\phi_j$ with square-summable sequences $(a_j)_{j\in\mathbb{N}_0}$ is equal to L^2_+ (cf. Theorem 2.41). Denote by $\mathcal{H}$ the set of measurable functions on $\mathbb{R}$ that vanish on a set of Lebesgue measure zero and belong to $L^2(G_{\mathrm{sing}})$. Then it is easily seen that a measurable function β is in $\mathcal{H}$ if and only if it can be written in the form $\beta = \alpha(\lambda(\cdot))$ with α vanishing on a set of Lebesgue measure zero and belonging to $L^2(F_{\mathrm{sing}})$, i. e., $\alpha \in \widetilde{\mathcal{M}}_{-\infty}$ by Theorem 6.10. Summing up, we have shown that a measurable function h belongs to $\widetilde{\mathcal{N}}_0$ if and only if it can be written in the form

$$h = \frac{\mathcal{F}_2\gamma}{\widetilde{\psi_{\mathrm{c}}}} \mathbf{1}_{\mathbb{R}\setminus S} + \beta \mathbf{1}_S, \quad \text{with } \gamma \in L^2_+ \text{ and } \beta \in \mathcal{H}.$$

Since for general $t \in \mathbb{R}$ the elements of $\widetilde{\mathcal{N}}_t$ are those of $\widetilde{\mathcal{N}}_0$ multiplied by $\mathrm{e}^{\mathrm{i}t\cdot}$, it follows that $h \in \widetilde{\mathcal{N}}_t$ if and only if it can be represented in the form

$$h = \mathrm{e}^{\mathrm{i}t\cdot} \frac{\mathcal{F}_2\gamma}{\widetilde{\psi_{\mathrm{c}}}} \mathbf{1}_{\mathbb{R}\setminus S} + \beta \mathbf{1}_S \quad \text{with } \gamma \in L^2_+ \text{ and } \beta \in \mathcal{H}.$$

The claim will hence follow if we can show that $\widetilde{\mathcal{N}}_{-\infty} = \mathcal{H}$. That $\mathcal{H} \subset \widetilde{\mathcal{N}}_{-\infty}$ is clear from the representation of $\widetilde{\mathcal{N}}_t$. Conversely, let $h \in \widetilde{\mathcal{N}}_{-\infty}$. Then for each $t \in \mathbb{R}$ there exists some $\gamma_t \in L^2_+$ such that $\widetilde{\psi_c}\, h = e^{it\cdot}\mathcal{F}_2(\gamma_t)$ (Lebesgue almost everywhere). Using Theorem 2.59(b) first and then taking the inverse Fourier transform, this is equivalent to the existence of some $\gamma_t \in L^2_+$ such that $\mathcal{F}_2^{-1}(\widetilde{\psi_c}\, h) = \gamma_t(\cdot + t)$ (Lebesgue almost everywhere). But $\gamma_t(\cdot + t)$ is supported in $[-t, \infty)$. Letting $t \to -\infty$, we see that $\mathcal{F}_2^{-1}(\widetilde{\psi_c}\, h)$ must be zero Lebesgue almost everywhere, hence so must be h. Recalling the representation of $\widetilde{\mathcal{N}}_t$, this shows $\widetilde{\mathcal{N}}_{-\infty} = \mathcal{H}$, thus finishing the proof. □

Remark 6.23. The representation (6.35) of G is the continuous-time analogue of the representation (6.6) of F.

With the aid of Theorem 6.22 and equation (6.19), it is now easy to determine the frequency-domain form of the h-step predictor.

Corollary 6.24 (Frequency-domain form of BLP). *Let $Y = (Y_t)_{t\in\mathbb{R}}$ be a zero-mean stationary MSC regular process with spectral distribution function G and spectral process Z. Denote by G_{sing} the singular part of G and let $S \in \mathcal{B}_1$ be a Lebesgue null set that carries G_{sing}. Then the h-step best linear predictor of Y_{t+h}, $h \geq 0$, is given by*

$$P_{\mathcal{N}_t} Y_{t+h} = \int_{\mathbb{R}\setminus S} \frac{e^{it\mu}}{\widetilde{\psi_c}(\mu)} \mathcal{F}_2(\psi_c(\cdot + h)\mathbf{1}_{[0,\infty)}(\cdot))(\mu)\, dZ_\mu + \int_S e^{i(t+h)\mu}\, dZ_\mu, \tag{6.36}$$

and the prediction mean-square error is

$$\sigma_h^2 = \int_0^h |\psi_c(s)|^2\, ds. \tag{6.37}$$

Proof. By Theorem 6.22(c), the h-step best linear predictor $P_{\widetilde{\mathcal{N}}_t} e^{i(t+h)\cdot}$ of $e^{i(t+h)\cdot}$ in $L^2(G)$ has the form

$$(P_{\widetilde{\mathcal{N}}_t} e^{i(t+h)\cdot})(\mu) = \frac{e^{it\mu}}{\widetilde{\psi_c}(\mu)}(\mathcal{F}_2\gamma)(\mu)\mathbf{1}_{\mathbb{R}\setminus S}(\mu) + \beta(\mu)\mathbf{1}_S(\mu)$$

for some $\beta \in \widetilde{\mathcal{N}}_{-\infty}$ and $\gamma \in L^2(\mathbb{R})$ with $\gamma(t) = 0$ for all $t < 0$. For the determination of β and γ, using Theorem 6.22(b), we compute the mean-square distance to $e^{i(t+h)\cdot}$ in $L^2(G)$ as

$$\begin{aligned}
&\int_{\mathbb{R}} \left| e^{i(t+h)\mu} - \frac{e^{it\mu}}{\widetilde{\psi_c}(\mu)}(\mathcal{F}_2\gamma)(\mu)\mathbf{1}_{\mathbb{R}\setminus S}(\mu) + \beta(\mu)\mathbf{1}_S(\mu) \right|^2 dG(\mu) \\
&= \int_{\mathbb{R}\setminus S} \left| e^{i(t+h)\mu} - \frac{e^{it\mu}}{\widetilde{\psi_c}(\mu)}(\mathcal{F}_2\gamma)(\mu) \right|^2 |\widetilde{\psi_c}(\mu)|^2\, d\mu
\end{aligned}$$

$$
\begin{aligned}
&+ \int_S \left|e^{i(t+h)\mu} - \beta(\mu)\right|^2 dG_{\text{sing}}(\mu) \\
&= \int_{\mathbb{R}} \left|e^{ih\mu}\widetilde{\psi_c}(\mu) - (\mathcal{F}_2\gamma)(\mu)\right|^2 d\mu + \int_S \left|e^{i(t+h)\mu} - \beta(\mu)\right|^2 dG_{\text{sing}}(\mu) \\
&= \int_{\mathbb{R}} \left|\mathcal{F}_2(\psi_c(\cdot + h) - \gamma)(\mu)\right|^2 d\mu + \int_S \left|e^{i(t+h)\mu} - \beta(\mu)\right|^2 dG_{\text{sing}}(\mu) \\
&= \int_{\mathbb{R}} \left|\psi_c(t+h) - \gamma(t)\right|^2 dt + \int_S \left|e^{i(t+h)\mu} - \beta(\mu)\right|^2 dG_{\text{sing}}(\mu),
\end{aligned}
$$

where we used Parseval's identity (2.33) in the last line. Since $e^{i(t+h)\cdot}\mathbf{1}_S \in \widetilde{N}_{-\infty}$ and since $\psi_c(t) = \gamma(t) = 0$ for $t < 0$, the latter expression is clearly minimized when $\beta = e^{i(t+h)\cdot}\mathbf{1}_S$ and $\gamma(t) = \psi_c(t+h)$ for $t > 0$, in which case the minimal value is

$$
\int_{\mathbb{R}} \left|\psi_c(t+h) - \psi_c(t+h)\mathbf{1}_{[0,\infty)}(t)\right|^2 dt = \int_{-h}^{0} \left|\psi_c(t+h)\right|^2 dt = \int_0^h \left|\psi_c(s)\right|^2 ds.
$$

Hence

$$
(P_{\widetilde{N}_t} e^{i(t+h)\cdot})(\mu) = \frac{e^{it\mu}}{\widetilde{\psi_c}(\mu)} \mathcal{F}_2(\psi_c(\cdot + h)\mathbf{1}_{[0,\infty)}(\cdot))(\mu)\mathbf{1}_{\mathbb{R}\setminus S}(\mu) + e^{i(t+h)\mu}\mathbf{1}_S(\mu),
$$

and an application of (6.19) yields (6.36) and (6.37). □

Example 6.25 (The 2nd-order CAR(1) process). The 2nd-order CAR(1) process of Example 4.12 has autocovariance function, $\gamma(h) = \frac{\nu^2}{2a}e^{-a|h|}$ where $a > 0$ and $\nu > 0$. By Example 5.3, its spectral distribution function is absolutely continuous with spectral density

$$
G'(\mu) = \frac{\nu^2}{2\pi(a^2 + \mu^2)}.
$$

Since G satisfies the conditions of Theorem 6.16 the process is regular. To determine the functions ψ_c and $\widetilde{\psi_c}$ appearing in Theorem 6.22, we first observe that the corresponding discrete-time distribution function F is absolutely continuous with density,

$$
\begin{aligned}
F'(\lambda) &= G'(\mu(\lambda))\mu'(\lambda) \\
&= \frac{\nu^2}{4\pi \cos^2(\lambda/2)(a^2 + \tan^2(\lambda/2))} \\
&= \frac{\nu^2}{2\pi(a^2 + 1 + (a^2 - 1)\cos\lambda)},
\end{aligned}
$$

and by (6.16) we can identify this as the spectral density of a causal discrete-time AR(1) sequence,

$$X_n = \varphi X_{n-1} + W_n, \quad n \in \mathbb{Z},$$

where $\varphi = -(a-1)/(a+1)$, and $\{W_n\} \sim \mathrm{WN}(0, \nu^2(1-\varphi^2)/(2a))$. By Example 6.9, the Wold representation (6.5) of X_n is therefore given by

$$X_n = \sum_{j=0}^{\infty} \psi_j Z_{n-j}, \quad \{Z_n\} \sim \mathrm{WN}(0,1),$$

where $\psi_j = \nu\sqrt{(1-\varphi^2)/(2a)}\varphi^j$. By Definition 6.21,

$$\widetilde{\psi_c}(\mu) = \nu\sqrt{(1-\varphi^2)/(2a)} \sum_{j=0}^{\infty} \varphi^j f_j(\mu),$$

where $f_j(\mu)$ was defined in Lemma 6.18. Straightforward algebra then gives

$$\widetilde{\psi_c}(\mu) = \frac{\nu}{\sqrt{2\pi}} \frac{1}{a + \mathrm{i}\mu}, \tag{6.38}$$

and

$$G'(\mu) = \left|\widetilde{\psi_c}(\mu)\right|^2 = \frac{\nu^2}{2\pi(a^2+\mu^2)},$$

as we already knew. Further, by Definition 6.21, ψ_c is the inverse Fourier transform of the L^2 function $\widetilde{\psi_c}$, hence

$$\psi_c(t) = \nu\, \mathrm{e}^{-at} \mathbf{1}_{(0,\infty)}(t).$$

From this, we conclude

$$\mathcal{F}_2(\psi_c(\cdot + h)\mathbf{1}_{[0,\infty)}(\cdot))(\mu) = \frac{\nu}{\sqrt{2\pi}} \frac{\mathrm{e}^{-ah}}{a + \mathrm{i}\mu},$$

so that (6.36) and (6.37) result in

$$P_{\mathcal{N}_t} Y_{t+h} = \int_{\mathbb{R}} \mathrm{e}^{\mathrm{i}t\mu} \mathrm{e}^{-ah}\, \mathrm{d}Z_\mu = \mathrm{e}^{-ah} Y_t$$

and

$$\sigma_h^2 = \int_0^h \nu^2 \mathrm{e}^{-2as}\, \mathrm{d}s = \gamma(0)(1 - \mathrm{e}^{-2ah}).$$

We now come to the main result of this section, the continuous-time analog of the Wold decomposition, sometimes known as the Wold-Karhunen decomposition. It specifies the general structure of a *regular* weakly stationary process Y. To the absolutely continuous component of G corresponds a *one-sided* moving average component of Y (as opposed to the generally two-sided representation of Theorem 5.13 in which regularity was not assumed) and to the singular component of G corresponds an added uncorrelated deterministic component (as is the case for regular weakly stationary sequences).

Recall from Definition 6.21 that

$$\widetilde{\psi_c}(\mu) = \sum_{j=0}^{\infty} \psi_j f_j(\mu),$$

where $\psi_0, \psi_1, \psi_2, \ldots,$ are the Wold coefficients of the Doob sequence X derived from Y and the functions f_j are

$$f_j(\mu) = \frac{1}{\sqrt{\pi}} \frac{(1 - \mathrm{i}\mu)^j}{(1 + \mathrm{i}\mu)^{j+1}} = \sum_{r=0}^{j} \frac{A_r}{(1 + \mathrm{i}\mu)^{r+1}},$$

where the right-hand side is just the partial fraction expansion of $f_j(\mu)$. The function ψ_c in the following representation is the inverse Fourier transform of $\widetilde{\psi_c}$ and given by $\sum_{j=0}^{\infty} \psi_j \phi_j(t)$ as in Definition 6.21, where the ϕ_j where defined in equation (6.25).

Theorem 6.26 (Wold–Karhunen representation). *The zero-mean stationary MSC regular process $Y = (Y_t)_{t\in\mathbb{R}}$ can be expressed as*

$$Y_t = \int_{(-\infty,t]} \varphi(t - s)\, \mathrm{d}\xi_s + V_t =: U_t + V_t, \quad t \in \mathbb{R}, \tag{6.39}$$

where
(i) $\varphi \in L^2(\mathbb{R}; \mathbb{C})$ *such that* $\varphi(t) = 0$ *for all* $t < 0$,
(ii) ξ *is an OIP with associated distribution function* $H(t) = t$,
(iii) *the increments of* ξ *satisfy* $\xi_{t_2} - \xi_{t_1} \in \mathcal{N}_t$ *whenever* $t_1 \le t_2 \le t$,
(iv) *each* ξ_t *is uncorrelated with* V_s *for* $s, t \in \mathbb{R}$,
(v) $V_t \in \mathcal{N}_{-\infty}$ *for all* $t \in \mathbb{R}$, *and*
(vi) $(V_t)_{t\in\mathbb{R}}$ *is a deterministic zero-mean stationary MSC process.*

In each such representation, $(U_t)_{t\in\mathbb{R}}$ *and* $(V_t)_{t\in\mathbb{R}}$ *are uniquely determined by* (6.39) *and properties* (i)–(vi), *and if* $Y_t = \int_{-\infty}^{t} \varphi^*(t - s)\, \mathrm{d}\xi_s^* + V_t$ *is another representation satisfying properties* (i)–(vi), *then there exists* $\theta \in \mathbb{C}$ *with* $|\theta| = 1$ *such that* $\varphi^* = \theta\varphi$ *and* $\xi_t^* = \theta^{-1}\xi_t$. *In particular, the orthogonal increment process is uniquely determined by* φ.

A possible choice for φ *is* ψ_c *as defined in Definition* 6.21, *so that*

$$Y_t = \int_{\mathbb{R}} \psi_c(t - s)\, \mathrm{d}\xi_s + V_t = \int_{(-\infty,t]} \psi_c(t - s)\, \mathrm{d}\xi_s + V_t, \tag{6.40}$$

in which case the orthogonal increment process ξ is given by

$$\xi_s = \int_{\mathbb{R}\setminus S} \frac{e^{is\mu} - 1}{\sqrt{2\pi}\, i\mu\widetilde{\psi}_c(\mu)}\, dZ_\mu, \quad s \in \mathbb{R}, \tag{6.41}$$

where Z is the spectral process of Y and $S \in \mathcal{B}_1$ a Lebesgue-null set, which carries the singular part of the spectral distribution function G of Y.

Observe that property (iii) is not the same as saying that $\xi_s \in \mathcal{N}_t$ for each $s \leq t$. While for $t \geq 0$ and $s \leq t$ we can write $\xi_s = \xi_s - \xi_0 \in \mathcal{N}_t$ by (iii), a similar argument cannot be used for $t < 0$.

Proof. We start by showing the existence of such a representation with $\varphi = \psi_c$ and ξ given by (6.41), and then show uniqueness.

Step 1: Existence.

Denote by G the spectral distribution function of Y and let $S \in \mathcal{B}_1$ be a Lebesgue-null set which carries the singular part of G, and denote by G_{ac} and G_{sing} the absolutely continuous and singular part of G, respectively. By Theorem 6.22, G_{ac} has density $|\widetilde{\psi}_c(\cdot)|^2$, so

$$G_{ac}(b) - G_{ac}(a) = \int_{(a,b]\setminus S} dG(\mu) = \int_{(a,b]\setminus S} |\widetilde{\psi}_c(\mu)|^2\, d\mu \quad \text{and}$$

$$G_{sing}(b) - G_{sing}(b) = \int_{(a,b]\cap S} dG(\mu)$$

for $a \leq b$. Denote by Z the spectral process of Y, and by Z^{ac} and Z^{sing} the OIPs with density $\mathbf{1}_{\mathbb{R}\setminus S}$ and $\mathbf{1}_S$ with respect to Z, respectively (cf., Theorem 4.13). For each $t \in \mathbb{R}$, define

$$U_t := \int_{\mathbb{R}\setminus S} e^{it\mu}\, dZ_\mu = \int_{\mathbb{R}} e^{it\mu}\, dZ^{ac}_\mu \quad \text{and}$$

$$V_t := \int_S e^{it\mu}\, dZ_\mu = \int_{\mathbb{R}} e^{it\mu}\, dZ^{sing}_\mu.$$

Then $Y_t = U_t + V_t$ and by Proposition 5.5 and Theorem 4.13, $U = (U_t)_{t\in\mathbb{R}}$ and $V = (V_t)_{t\in\mathbb{R}}$ are zero-mean stationary MSC processes with spectral distribution functions G_{ac} and G_{sing}, and spectral processes Z^{ac} and Z^{sing}, respectively. Since the process Y is regular, the functions ψ_c and $\widetilde{\psi}_c$ of Definition 6.21 have the properties specified in Theorem 6.22, in particular, ψ_c satisfies (i) and $\widetilde{\psi}_c \neq 0$ Lebesgue almost everywhere.

By Corollary 4.14 and Remark 4.15, there exists an OIP $\widetilde{\xi} = (\widetilde{\xi}_t)_{t\in\mathbb{R}}$ defined on the same probability space with associated distribution function $H(t) = t$, such that Z^{ac} has density $\widetilde{\psi}_c$ with respect to $\widetilde{\xi}$. Denote by ξ the inverse Fourier transform of $\widetilde{\xi}$, as defined in (4.21). By Lemma 4.17, ξ is an OIP with associated distribution function $H(t) = t$. By Theorem 4.13 and equation (4.23),

$$U_t = \int_{\mathbb{R}} e^{it\mu}\,\widetilde{\psi_c}(\mu)\,d\widetilde{\xi}_\mu = \int_{\mathbb{R}} \psi_c(t-s)\,d\xi_s = \int_{(-\infty,t]} \psi_c(t-s)\,d\xi_s.$$

Thus we have the representation (6.39) (and (6.40)) and properties (i) and (ii). Further, since V has spectral distribution function G^{sing}, it must be deterministic by the Paley-Wiener conditions (Theorem 6.16), which is property (vi).

Since $\widetilde{\xi}$ has density $1/\widetilde{\psi_c}$ with respect to Z^{ac}, by Theorem 4.13, we can write

$$\widetilde{\xi}_{t_2} - \widetilde{\xi}_{t_1} = \int_{(t_1,t_2]} \frac{1}{\widetilde{\psi_c}(\mu)}\,dZ_\mu^{\text{ac}} = \int_{(t_1,t_2]\setminus S} \frac{1}{\widetilde{\psi_c}(\mu)}\,dZ_\mu$$

for $t_1 \le t_2 \in \mathbb{R}$. On the other hand, since $V_s = \int_S e^{is\mu}\,dZ_\mu$ for $s \in \mathbb{R}$, equation (4.15) shows $\mathbb{E}((\widetilde{\xi}_{t_2} - \widetilde{\xi}_{t_1})\overline{V_s}) = 0$, so that the increments of $\widetilde{\xi}$ are orthogonal to V. Since

$$\xi_t = \frac{1}{\sqrt{2\pi}} \int_{\mathbb{R}} \frac{1}{i\mu}(e^{i\mu t} - 1)\,d\widetilde{\xi}_\mu, \quad t \in \mathbb{R},$$

it follows from the definition of the stochastic integral (by considering first simple functions as in (4.10) and then extending it to general functions) that each ξ_t is orthogonal to each V_s, giving (iv). Rewriting the above representation of ξ_t with the aid of Theorem 4.13 as

$$\xi_t = \frac{1}{\sqrt{2\pi}} \int_{\mathbb{R}} \frac{1}{i\mu}(e^{i\mu t} - 1)\frac{1}{\widetilde{\psi_c}(\mu)}\,dZ_\mu^{\text{ac}} = \int_{\mathbb{R}\setminus S} \frac{e^{it\mu} - 1}{\sqrt{2\pi}\,i\mu\widetilde{\psi_c}(\mu)}\,dZ_\mu$$

gives (6.41).

To establish property (iii), let $t_1 \le t_2 \le t$ and define

$$\Psi(\mu) := \frac{e^{it\mu}}{\widetilde{\psi_c}(\mu)} \int_{t-t_2}^{t-t_1} e^{-i\mu u}\,du\,\mathbf{1}_{\mathbb{R}\setminus S}(\mu) = \frac{e^{it_2\mu} - e^{it_1\mu}}{\widetilde{\psi_c}(\mu)i\mu}\mathbf{1}_{\mathbb{R}\setminus S}(\mu)$$

for $\mu \in \mathbb{R}$. By Theorem 6.22, $\Psi \in \widetilde{\mathcal{N}}_t$, and since

$$\xi_{t_2} - \xi_{t_1} = \int_{\mathbb{R}\setminus S} \frac{e^{it_2\mu} - e^{it_1\mu}}{\sqrt{2\pi}\,i\mu\widetilde{\psi_c}(\mu)}\,dZ_\mu = \int_{\mathbb{R}} \frac{\Psi(\mu)}{\sqrt{2\pi}}\,dZ_\mu,$$

the isomorphism (6.17) between $\widetilde{\mathcal{N}}_t$ and $\mathcal{N}_t$ shows that $\xi_{t_2} - \xi_{t_1}$ belongs to $\mathcal{N}_t$, thus establishing (iii). To see (v), observe that $e^{it\cdot}\mathbf{1}_S \in \widetilde{\mathcal{N}}_{-\infty}$ for all $t \in \mathbb{R}$ by Theorem 6.22. Hence

$V_t = \int_S e^{it\mu}\, dZ_\mu$ belongs to $\mathcal{N}_{-\infty}$, again by the isomorphism (6.17). This finishes the proof of the existence of the representation with the specific choice given by (6.40) and (6.41).

Step 2: Uniqueness.

Now let $\varphi^* \in L^2$, ξ^* an OIP with distribution function $H(t) = t$ and $V^* = (V_t^*)_{t\in\mathbb{R}}$ such that properties (i) to (vi) are satisfied and such that

$$Y_t = \int_{(-\infty,t]} \varphi^*(t-s)\, d\xi_s^* + V_t^* =: U_t^* + V_t^* .$$

Let U_t, V_t, ψ_c and ξ_t be defined as in Step 1. We have to show that $V_t^* = V_t$ and that there is $\theta \in \mathbb{C}$ with $|\theta| = 1$ such that $\varphi^* = \theta\psi_c$ and $\xi_t^* = \theta^{-1}\xi_t$.

Let $s \le t_1 \le t_2$. Then $\xi_{t_2}^* - \xi_{t_1}^* = \int_{(t_1,t_2]} d\xi_\mu^*$ and $U_s^* = \int_{(-\infty,s]} \varphi^*(s-\mu)\, d\xi_\mu^*$ are orthogonal, since the domains of integration are disjoint. By condition (iv), $\xi_{t_2}^* - \xi_{t_1}^*$ is also orthogonal to V_s, hence also to Y_s. Since this is true for all $s \le t_1$, we conclude that $\xi_{t_2}^* - \xi_{t_1}^*$ is orthogonal to $\mathcal{N}_s$ for each $s \le t_1$, hence also orthogonal to $\mathcal{N}_{-\infty}$. However, this means that all increments of ξ^* are orthogonal to $\mathcal{N}_{-\infty}$. From the construction of the stochastic integral (considering first simple functions and then the general case) we conclude that $U_t^* = \int_{(-\infty,t]} \varphi^*(t-s)\, d\xi_s^*$ is orthogonal to $\mathcal{N}_{-\infty}$. Since $V_t^* \in \mathcal{N}_{-\infty}$, we conclude that V_t^* is the orthogonal projection of $Y_t = U_t^* + V_t^*$ onto $\mathcal{N}_{-\infty}$, so that V_t^* and hence $U_t^* = Y_t - V_t^*$ are uniquely determined by the process Y and properties (i)–(vi). This means $V_t^* = V_t$ and $U_t^* = U_t$.

Our next goal is to show that ξ^* has a density g with respect to ξ. To achieve this, denote

$$\mathcal{U}_t := \overline{\text{sp}}\{U_s : s \le t\}, \quad t \in \mathbb{R}.$$

Since each U_s is orthogonal to $\mathcal{N}_{-\infty}$, as seen above, we see that $\mathcal{U}_t$ is orthogonal to $\mathcal{N}_{-\infty}$ (meaning that any element from $\mathcal{U}_t$ is orthogonal to any other element from $\mathcal{N}_{-\infty}$). Since $Y_t = U_t + V_t$, we conclude that $\mathcal{N}_t \subset \mathcal{U}_t \oplus \mathcal{N}_{-\infty}$ (the $\oplus$ meaning the vector space direct sum of the orthogonal subspaces $\mathcal{U}_t$ and $\mathcal{N}_{-\infty}$).

Now let $s \le t$. Then $\xi_t^* - \xi_s^* \in \mathcal{N}_t$ by property (iii), and it is orthogonal to $\mathcal{N}_{-\infty}$ as seen above. Hence $\xi_t^* - \xi_s^* \in \mathcal{U}_t$. On the other hand, $U_u = \int_{(-\infty,u]} \varphi^*(u-\mu)\, d\xi_\mu^*$ is orthogonal to $\xi_t^* - \xi_s^*$ for every $u \le s$, hence $\xi_t^* - \xi_s^*$ is orthogonal to $\mathcal{U}_s$. Choose a sequence $W_n \in \text{sp}\{U_u : u \le t\}$, $n \in \mathbb{N}$, that converges in $\mathcal{S}$ to $\xi_t^* - \xi_s^*$. Since $U_u = \int_{(-\infty,u]} \psi_c(u-\mu)\, d\xi_\mu$ for $u \le t$, there is some function $g_n \in L^2$ with support contained in $(-\infty, t]$ such that $W_n = \int_{(-\infty,t]} g_n(\mu)\, d\xi_\mu$. Since $(W_n)_{n\in\mathbb{N}}$ is a Cauchy sequence in $\mathcal{S}$ and since the stochastic integral defines an isomorphism, also $(g_n)_{n\in\mathbb{N}}$ is a Cauchy sequence in L^2, hence converges to some $g_{s,t} \in L^2$ with support in $(-\infty, t]$. Then

$$\xi_t^* - \xi_s^* = \int_{(-\infty,t]} g_{s,t}(\mu)\, d\xi_\mu .$$

Now decompose the integral as $\int_{(-\infty,s]} g_{s,t}(\mu)\,d\xi_\mu + \int_{(s,t]} g_{s,t}(\mu)\,d\xi_\mu$. The same arguments that we used before also apply to ξ, so that increments of the form $\xi_v - \xi_s$ for $v \geq s$ are orthogonal to $\mathcal{U}_s$ and increments of the form $\xi_u - \xi_w$ for $w \leq u \leq s$ are in $\mathcal{U}_s$. Hence $\int_{(-\infty,s]} g_{s,t}(\mu)\,d\xi_\mu \in \mathcal{U}_s$ and $\int_{(s,t]} g_{s,t}(\mu)\,d\xi_\mu$ is orthogonal to $\mathcal{U}_s$. Since $\xi_t^* - \xi_s^*$ is also orthogonal to $\mathcal{U}_s$, we conclude that $\int_{(-\infty,s]} g_{s,t}(\mu)\,d\xi_\mu = 0$. Since $0 = \mathbb{E}|\int_{(-\infty,s]} g_{s,t}(\mu)\,d\xi_\mu|^2 = \int_{(-\infty,s]} |g_{s,t}(\mu)|^2\,d\mu$ by equation (4.15), we conclude that $g_{s,t}(\mu) = 0$ for $\mu \leq s$ (Lebesgue-almost everywhere). Now if $f_{s,t} \in L^2$ is another function such that $\xi_t^* - \xi_s^* = \int_{\mathbb{R}} f_{s,t}(\mu)\,d\xi_\mu$, then $\int_{\mathbb{R}} (f_{s,t}(\mu) - g_{s,t}(\mu))\,d\xi_\mu = 0$, and taking the variance we see again from equation (4.15) that $f_{s,t} = g_{s,t}$ Lebesgue-almost everywhere. Thus we have shown that for every $s \leq t$, there exists a (Lebesgue-almost everywhere) unique $g_{s,t} \in L^2$ such that $g_{s,t}$ is supported in $[s,t]$ and such that

$$\xi_t^* - \xi_s^* = \int_{\mathbb{R}} g_{s,t}(\mu)\,d\xi_\mu = \int_{(s,t]} g_{s,t}(\mu)\,d\xi_\mu.$$

Now let $s_2 \leq s_1 \leq t_1 \leq t_2$. Then

$$\begin{aligned}
&\int_{(s_2,t_2]} g_{s_2,t_2}(\mu)\,d\xi_\mu \\
&= \xi_{t_2}^* - \xi_{s_2}^* \\
&= (\xi_{s_1}^* - \xi_{s_2}^*) + (\xi_{t_1}^* - \xi_{s_1}^*) + (\xi_{t_2}^* - \xi_{t_1}^*) \\
&= \int_{(s_2,s_1]} g_{s_2,s_1}(\mu)\,d\xi_\mu + \int_{(s_1,t_1]} g_{s_1,t_1}(\mu)\,d\xi_\mu + \int_{(t_1,t_2]} g_{t_1,t_2}(\mu)\,d\xi_\mu.
\end{aligned}$$

The uniqueness of the functions $g_{s,t}$ then implies that $g_{s_2,t_2}(\mu) = g_{s_1,t_1}(\mu)$ for (Lebesgue-almost) all $\mu \in (s_1, t_1]$. Hence we can extend $g_{s,t}$ to the left and to the right and obtain the existence of some function g which is locally in L^2 such that $g(\mu) = g_{s,t}(\mu)$ Lebesgue-almost everywhere for all $\mu \in (s,t]$, when $s \leq t$ are fixed. Hence $\xi_t^* - \xi_s^* = \int_{(s,t]} g(\mu)\,d\xi_\mu$, showing that g is the density of ξ^* with respect to ξ.

From representations (6.39) and (6.40) and Theorem 4.13 we now obtain

$$U_t = \int_{(-\infty,t]} \psi_c(t-\mu)\,d\xi_\mu = \int_{(-\infty,t]} \varphi^*(t-\mu)\,d\xi_\mu^* = \int_{(-\infty,t]} \varphi^*(t-\mu)\,g(\mu)\,d\xi_\mu$$

for each fixed $t \in \mathbb{R}$. Taking the variance leads, for each fixed $t \in \mathbb{R}$, to $\psi_c(t-\mu) = \varphi^*(t-\mu)\,g(\mu)$ for Lebesgue-almost all $\mu \in \mathbb{R}$. Now denote $N := \{(t,\mu)^T \in \mathbb{R}^2 : \psi_c(t-\mu) \neq \varphi^*(t-\mu)g(\mu)\}$. An application of Tonelli's theorem to the indicator function of N then shows that N has two-dimensional Lebesgue measure 0, and a further application of Tonelli's theorem then shows that for Lebesgue-almost all $\mu \in \mathbb{R}$, the set $\{t \in \mathbb{R} : \psi_c(t-\mu) \neq \varphi^*(t-\mu)g(\mu)\}$ has one-dimensional Lebesgue measure 0. Hence for Lebesgue-almost all $\mu \in \mathbb{R}$ we have $\psi_c(v) = \varphi^*(v)g(\mu)$ Lebesgue-almost everywhere in $v \in \mathbb{R}$. Since ψ_c is not the zero-function (as a consequence of the regularity of Y), this is only possible if g is constant (Lebesgue-almost everywhere). Hence there is $k \in \mathbb{C}$ such that $\psi_c = k\varphi^*$. Since both $|\widetilde{\psi_c}|^2 = |\mathcal{F}_2\psi_c|^2$ and $|\mathcal{F}_2\varphi^*|^2$ are the spectral density of U by Example 5.3 and representations (6.40) and (6.39), we conclude that $|k| = 1$. Choosing $\theta = 1/k$ then shows $\varphi^* = \theta\psi_c$. That $\xi^* = \theta^{-1}\xi$ then follows from the fact (as shown above) that ξ^* has density g with respect to ξ and we have identified g to be constant and equal to k. □

A representation of a regular process Y as in Theorem 6.26 is called a *Wold–Karhunen representation* of Y. It is unique in V, and up to multiplication by a constant factor of modulus 1 unique in φ and ξ. The representation as in (6.40) with $\varphi = \psi_c$ will be called the *canonical Wold–Karhunen representation*. It can be characterized as the unique Wold–Karhunen representation for which $\int_{\mathbb{R}} \varphi(t)\,\mathrm{e}^{-t}\,\mathrm{d}t$ is real and strictly positive (see Theorem 6.22).

In Corollary 6.24, we already obtained a frequency-domain form of the h-step best linear predictor $P_{\mathcal{N}_t} Y_{t+h}$, $h \geq 0$. With the aid of the Wold–Karhunen representation, we now find another representation of the h-step best linear predictor.

Corollary 6.27 (Wold–Karhunen form of BLP). *In the situation of Theorem* 6.26, *with representation* (6.40), *the h-step best linear predictor of Y_{t+h}, $h \geq 0$, is*

$$P_{\mathcal{N}_t} Y_{t+h} = \int_{(-\infty,t]} \psi_c(t+h-s)\,\mathrm{d}\xi_s + V_{t+h}, \tag{6.42}$$

and the prediction mean-square error is given by (6.37).

Proof. The quantity specified on the right-hand side of (6.42) is clearly in $\mathcal{N}_t$ by properties (iii) and (v), and the prediction error

$$Y_{t+h} - \int_{(-\infty,t]} \psi_c(t+h-s)\,\mathrm{d}\xi_s - V_{t+h} = \int_{(t,t+h]} \psi_c(t+h-s)\,\mathrm{d}\xi_s$$

is orthogonal to $\mathcal{N}_t$, since it involves the increments of ξ on (t,∞) only and these are orthogonal to each Y_s, $s \leq t$, as seen in the proof of uniqueness in Theorem 6.26. This shows (6.42). The mean-square error of the predictor is

$$\mathbb{E}|Y_{t+h} - P_{\mathcal{N}_t} Y_{t+h}|^2 = \int_t^{t+h} |\psi_c(t+h-s)|^2\,\mathrm{d}s = \int_0^h |\psi_c(s)|^2\,\mathrm{d}s,$$

as claimed. □

Corollary 6.28. *If Y is regular with absolutely continuous spectral distribution function G, then V_t is zero in* (6.40), *so Y_t has the one-sided moving average representation,*

$$Y_t = \int_{(-\infty,t]} \psi_c(t-s)\,d\xi_s, \quad t \in \mathbb{R},$$

where ξ is given by (6.41) *with $S := \emptyset$.*

Proof. If G is absolutely continuous, the singular part of G is carried by $S := \emptyset$. Since $V_t = \int_\emptyset e^{it\mu}\,dZ_\mu$, as seen in the proof of Theorem 6.26, we conclude $V_t = 0$. □

Example 6.29 (The 2nd-order CAR(1) process). We found in Example 6.25 that for the process Y with autocovariance function, $\gamma(h) = \frac{v^2}{2a} e^{-a|h|}$ where $a > 0$ and $v > 0$, the spectral distribution function is absolutely continuous with density $G'(\mu) = \frac{v^2}{2\pi(a^2+\mu^2)}$, the process is regular and

$$\psi_c(t) = v\,e^{-at}\mathbf{1}_{(0,\infty)}(t).$$

Substituting in (6.42) and (6.37) gives, for $h \geq 0$,

$$P_{\mathcal{N}_t} Y_{t+h} = \int_{(-\infty,t]} v\,e^{-a(t+h-s)}\,d\xi_s = e^{-ah} Y_t$$

and

$$\sigma_h^2 = \int_0^h v^2 e^{-2as}\,ds = \gamma(0)(1 - e^{-2ah}),$$

in accordance with the findings of Example 6.25. In this example, it is trivial to express the predictor directly in terms of the process Y rather than in terms of the unobserved OIP ξ. In the following chapter, we shall derive an analogous expression for a process having a spectral density that is a more general rational function of μ than in this example.

Remark 6.30 (Processes given as one-sided moving average process). Let $k \in L^2$ be non-identical zero with $k(t) = 0$ for all $t < 0$ and ξ be some given OIP with associated distribution function $H(t) = t$. Define

$$Y_t := \int_{(-\infty,t]} k(t-s)\,d\xi_s, \quad t \in \mathbb{R}. \tag{6.43}$$

We know from Example 5.3 that $Y = (Y_t)_{t\in\mathbb{R}}$ is zero-mean stationary MSC with absolutely continuous spectral distribution function and spectral density given by $|\mathcal{F}_2 k|^2$. Since $k(t) = 0$ for $t < 0$ but k is not identically zero, it is possible to show, using the theory of Hardy spaces, that Y satisfies the Paley–Wiener condition $\int_{\mathbb{R}} \log|(\mathcal{F}_2 k)(\mu)|^2/(1+\mu^2)\,d\mu > -\infty$, see, e. g., Kawata [56, Theorem 6.6.1] or Duren [39, Chapter11]. Hence, Y is regular by Theorem 6.16. It is now tempting to assume that (6.43) is actually the Wold–Karhunen decomposition of Y. This however is not true in general. As a simple example, suppose that not only $k(t) = 0$ for all $t < 0$, but even that $k(t) = 0$ for all $t \leq 1$ (but k not the zero-function). Then $Y_t = \int_{(-\infty,t-1]} k(t-s)\,\mathrm{d}\xi_s$ and it follows that each Y_t is in $\overline{\mathrm{sp}}\{\xi_s - \xi_u : u \leq s \leq t-1\}$, in particular is orthogonal to $\xi_t - \xi_{t-1}$. Hence $\mathcal{N}_t$ is orthogonal to $\xi_t - \xi_{t-1} \neq 0$, implying $\xi_t - \xi_{t-1} \notin \mathcal{N}_t$. We see that condition (iii) of Theorem 6.26 is violated, so that (6.43) cannot be the Wold–Karhunen representation of Y. More complicated examples of processes given by one-sided moving average processes, which are not the Wold–Karhunen representation, can be constructed using Blaschke products and the theory of Hardy spaces over the upper half-plane (or, equivalently, the lower-half plane), see Duren [39, Chapter 11].

6.3 Exercises

Exercise 6.1. (a) Let $(X_n)_{n\in\mathbb{Z}}$ be a mean-zero stationary sequence and define $\mathcal{M}_n$ as in (6.1). For $n \in \mathbb{Z}$ and $h \in \mathbb{N}$, let $c_{j,k}$, $k \in \mathbb{N}$, $j = 0, \dots, m_k$, be complex coefficients such that the best linear predictor of X_n given $\mathcal{M}_{n-h}$ is given by

$$P_{\mathcal{M}_{n-h}} X_n = \operatorname{lim}^{\mathcal{S}}_{k\to\infty} \sum_{j=0}^{m_k} c_{j,k} X_{n-h-j}.$$

Show that

$$P_{\mathcal{M}_{n-h+1}} X_{n+1} = \operatorname{lim}^{\mathcal{S}}_{k\to\infty} \sum_{j=0}^{m_k} c_{j,k} X_{n+1-h-j}$$

and deduce that

$$\mathbb{E}|X_n - P_{\mathcal{M}_{n-h}} X_n|^2 = \mathbb{E}|X_{n+1} - P_{\mathcal{M}_{n+1-h}} X_{n+1}|^2,$$

so that the mean-square error σ_h^2, as defined in (6.2), does not depend on n.

(b) Let $(Y_t)_{t\in\mathbb{R}}$ be a mean-zero stationary process and define $\mathcal{N}_t$ as in (6.18). Mimicking the proof of (a) above, show that

$$\mathbb{E}|Y_{t+h} - P_{\mathcal{N}_t} Y_{t+h}|^2 = \mathbb{E}|Y_{s+h} - P_{\mathcal{N}_s} Y_{s+h}|^2$$

for $s, t \in \mathbb{R}$ and $h \geq 0$, so that the mean-square error σ_h^2 as defined in (6.20) is independent of t.

Exercise 6.2. Show that if A and B are zero-mean uncorrelated random variables, each with variance σ^2, then the harmonic process, $X_n = A\cos(n\pi/2) + B\sin(n\pi/2)$, $n \in \mathbb{Z}$, has one-step mean-square error $\sigma_1^2 = 0$ by showing that $X_{n+1} = X_{n-3}$ for all $n \in \mathbb{Z}$.

Exercise 6.3. Let $X = (X_n)_{n\in\mathbb{Z}}$ be a zero-mean complex-valued stationary sequence. Mimicking the proof of Lemma 6.11, and using the notations of Section 6.1, show that the following are equivalent:
(i) X is deterministic.
(ii) $\sigma_h^2 = 0$ for all $h \in \mathbb{N}$.
(iii) $\mathcal{M}_t = \mathcal{M}_s$ for all $s, t \in \mathbb{Z}$.
(iv) $\mathcal{M}_0 = \mathcal{M}$.

Exercise 6.4. Let $H : \mathbb{R} \to \mathbb{R}$ be a right-continuous non-decreasing function. For $t \in \mathbb{R}$, define

$$D^+H(t) := \limsup_{h\downarrow 0} \frac{H(t+h)-H(t)}{h}, \quad D_+H(t) := \liminf_{h\downarrow 0} \frac{H(t+h)-H(t)}{h}$$

and

$$D^-H(t) := \limsup_{h\uparrow 0} \frac{H(t+h)-H(t)}{h}, \quad D\_H(t) := \liminf_{h\uparrow 0} \frac{H(t+h)-H(t)}{h}.$$

(a) Show that D^+H, D_+H, D^-H and D_H are $\mathcal{B}_1 \to \mathcal{B}(\overline{\mathbb{R}})$-measurable functions.
(b) Show that H is differentiable at t if and only if $D^+H(t), D_+H(t), D^-H(t)$ and $D\_H(t)$ are all equal and finite. Deduce that the set C of all points in which H is differentiable is a Borel set and that the function $\mathbb{R} \to \mathbb{R}$, $t \mapsto \mathbf{1}_C(t)H'(t)$ is $\mathcal{B}_1 \to \mathcal{B}_1$-measurable.

Exercise 6.5. (a) Let $\eta \in \mathbb{C}$ with $|\eta| < 1$. Show that $\int_{-\pi}^{\pi} \log(1-\eta e^{-i\lambda})\, d\lambda = 0$, where $\log(\cdot)$ denotes the principal branch of the complex logarithm.
Hint: Use the logarithmic expansion $\log(1+z) = -\sum_{k=1}^{\infty}(-z)^k/k$ for $|z| < 1$.
(b) Let $\eta \in \mathbb{C}$ with $|\eta| \leq 1$. Show that $\log|1-\eta e^{-i\cdot}|^2$ is Lebesgue-integrable over $[-\pi, \pi]$ and that $\int_{-\pi}^{\pi} \log|1-\eta e^{-i\lambda}|^2\, d\lambda = 0$.
Hint: For $|\eta| = 1$ apply Lebesgue's dominated convergence theorem with $\eta_n = (1 - n^{-1})\eta$.

Exercise 6.6. Verify the statement of Remark 6.13, i. e., that the mapping $f(\cdot) \mapsto f(\lambda(\cdot))$ is a Hilbert space isomorphism of $\widetilde{\mathcal{M}}$ (as a subspace of $L^2(F)$) onto $\widetilde{N}^*$ (as a subspace of $L^2(G)$), with the notations used in Remark 6.13.

Exercise 6.7. Show that

$$\int_{\mathbb{R}} \frac{2\log[(1+\mu^2)/2]}{(1+\mu^2)}\, d\mu < \infty.$$

Exercise 6.8. Suppose that $X_t = \int_{(-\infty,t]} e^{a(t-u)}\, dB_u$, $t \in \mathbb{R}$, where $a < 0$ and B is standard Brownian motion.

(a) Show that the inverse Fourier transform of $g(u) := \mathrm{e}^{\alpha(t-u)}\mathbf{1}_{(-\infty,t]}(u)$ is

$$(\mathcal{F}_2^{-1}(g))(\mu) = \frac{\mathrm{e}^{\mathrm{i}\mu t}}{\sqrt{2\pi}(\mathrm{i}\mu - \alpha)}, \quad \mu \in \mathbb{R}. \tag{6.44}$$

(b) Use (4.23) to deduce that

$$X_t = \frac{1}{\sqrt{2\pi}} \int_{\mathbb{R}} \mathrm{e}^{\mathrm{i}\mu t} \frac{1}{\mathrm{i}\mu - \alpha}\, \mathrm{d}\widetilde{B}_\mu, \quad t \in \mathbb{R},$$

where $\widetilde{B}$ is the Fourier transform of B as specified in Definition 4.16.

(c) Deduce that the spectral process of X is the OIP whose density with respect to $\widetilde{B}$ is given by $(\sqrt{2\pi}(\mathrm{i}\mu - \alpha))^{-1}$, $\mu \in \mathbb{R}$.

Exercise 6.9. Let $\alpha < 0$. Use Remark 2.71 and Exercise 6.8 to show that the Fourier transform of the L^2-function on the right of (6.44) is $g(u) := \mathrm{e}^{\alpha(t-u)}\mathbf{1}_{(-\infty,t]}(u)$, $u \in \mathbb{R}$. Then use (4.23) to deduce that if Y has the spectral representation,

$$Y_t = \frac{1}{\sqrt{2\pi}} \int_{\mathbb{R}} \mathrm{e}^{\mathrm{i}\mu t} \frac{1}{\mathrm{i}\mu - \alpha}\, \mathrm{d}Z_\mu, \quad t \in \mathbb{R},$$

where Z is an OIP with associated distribution function $H(t) = t$, then Y has the moving average representation,

$$Y_t = \int_{(-\infty,t]} \mathrm{e}^{\alpha(t-u)} \mathrm{d}Z_u^*, \quad t \in \mathbb{R},$$

where Z^* is the inverse Fourier transform of Z.

Exercise 6.10. If $X_t = \int_{(t,\infty)} \mathrm{e}^{\alpha(t-u)}\, \mathrm{d}B_u$, $t \in \mathbb{R}$, where $\alpha > 0$ and B is standard Brownian motion, determine the spectral representation of X.

Exercise 6.11. Verify the expression (6.38) for $\widetilde{\psi_c}$.

7 Second-order CARMA processes

Abstract: In this chapter, we consider an important class of continuous-time analogs of the discrete-time ARMA(p, q) sequences driven by uncorrelated discrete-time white noise. The continuous-time analogs are known as second-order CARMA processes or processes with rational spectral density driven by orthogonal-increment processes, and are defined in terms of their spectral density. Like the ARMA sequences, they are important because of their ability to approximate a wide range of spectral densities and autocovariance functions. In this chapter, we shall determine their second-order properties, their Wold–Karhunen expansions and their linear minimum mean-square error predictors. R-programs are provided in Section 7.4 for the calculation and plotting of the autocovariance and kernel functions. Later, in Chapter 18, we shall use stochastic differential equations to introduce the class of Lévy-driven CARMA processes, which may have infinite second moments and consequently no spectral density. The approach of Chapter 18 also permits us to analyze pathwise, as opposed to second-order, properties of Lévy-driven CARMA processes. In order to achieve this, the study of integration with respect to Lévy processes and solution of Lévy-driven stochastic differential equations (presented in the following chapters) will be required.

7.1 Definition and properties

Definition 7.1. A *second-order* CARMA(p, q) *process* is a weakly stationary mean-square continuous process $Y = (Y_t)_{t\in\mathbb{R}}$ with constant mean, say $\mathbb{E}Y_t = m$, absolutely continuous spectral distribution function and spectral density,

$$f(\mu) = \frac{\nu^2}{2\pi}\left|\frac{b(i\mu)}{a(i\mu)}\right|^2, \quad \mu \in \mathbb{R}, \tag{7.1}$$

where $\nu > 0$ and $a(z)$ and $b(z)$ are polynomials,

$$a(z) = \prod_{j=1}^{p}(z - \alpha_j) = z^p + a_1 z^{p-1} + \cdots + a_{p-1}z + a_p$$

and

$$b(z) = \prod_{j=1}^{q}(z - \beta_j) = b_0 + b_1 z + \cdots + b_{q-1}z^{q-1} + z^q, \quad 0 \le q < p,$$

with complex coefficients and no common factors. (When $q = 0$, the polynomial $b(z)$ is understood as $b(z) = 1$.)

https://doi.org/10.1515/9783111325033-007

Remark 7.2. (a) We shall assume throughout this chapter that $m = 0$, the adjustments to allow for non-zero m being quite straightforward.

(b) We shall assume that the real parts of the zeros of $a(z)$ are all strictly negative and that the real parts of the zeros of $b(z)$ are all non-positive. This involves no loss of generality, since the signs of the real parts of any zeros in the open right half-plane can be reversed without changing f. Moreover, $a(z) \neq 0$ if $\Re(z) = 0$ for integrability of f.

Remark 7.3. (a) It is a trivial matter to check that $f(\mu) \neq 0$ except on a finite set and that $\int_{\mathbb{R}} \frac{|\log f(\mu)|}{1+\mu^2}\,\mathrm{d}\mu < \infty$. Hence a second-order CARMA process satisfies the Paley–Wiener conditions, hence is regular by Theorem 6.16.

(b) By Example 5.17, a second-order CARMA process is mean-square differentiable if and only if $\int_{\mathbb{R}} \mu^2 f(\mu)\,\mathrm{d}\mu < \infty$, i. e., if and only if $q < p - 1$.

Example 7.4 (A CAR(2) process). If $a(z) = (z + 0.4 - \mathrm{i})(z + 0.4 + \mathrm{i})$, $b(z) = 1$ and $\nu = 1$, then Y is a CARMA(2, 0) (or CAR(2)) process, and from (7.1) has the spectral density,

$$f(\mu) = \frac{1}{2\pi(\mu^4 - 1.68\mu^2 + (1.16)^2)}, \qquad \mu \in \mathbb{R},$$

where the frequency μ is measured in radians per unit time. The spectral density f is shown for $0 \le \mu \le 12$ in Figure 7.1. Its most striking feature is the peak at $\mu = \sqrt{3.36}/2 = 0.9165$, corresponding to oscillatory behavior with a period $4\pi/\sqrt{3.36} = 6.855$. The spectral density at negative frequency values is $f(-\mu) = f(\mu)$, $\mu > 0$. Since $q < p - 1$ it follows that Y is mean-square differentiable.

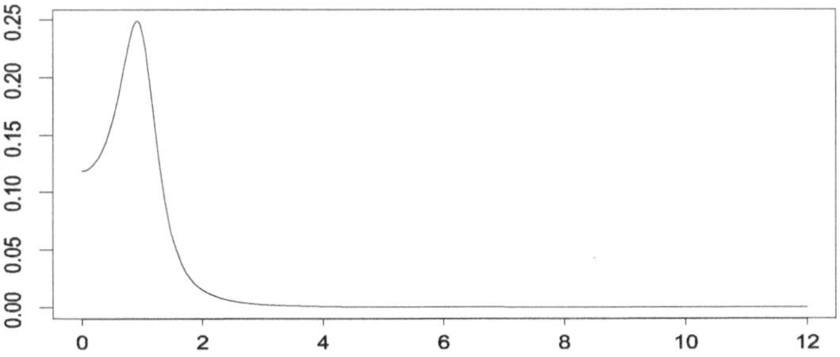

Figure 7.1: The spectral density of the CAR(2) process of Example 7.4 for frequencies 0 through 12 radians per unit time.

Example 7.5 (A CARMA(2, 1) process). If $a(z) = (z + 0.5)(z + 1)$, $b(z) = z + 0.25$ and $\nu = 1$, then Y is a CARMA(2, 1) process, and from (7.1) has the spectral density,

$$f(\mu) = \frac{\mu^2 + 0.0625}{2\pi(\mu^4 + 1.25\mu^2 + 0.25)}, \qquad \mu \in \mathbb{R}.$$

The spectral density f is shown for $0 \leq \mu \leq 12$ in Figure 7.2. The spectral density at negative frequency values is $f(-\mu) = f(\mu)$, $\mu > 0$. There is a slightly less pronounced peak than in Example 7.4 at $\mu = (3\sqrt{5} - 1)^{1/2}/4 = 0.5973$. Since $q = p - 1$, it follows that Y is not mean-square differentiable.

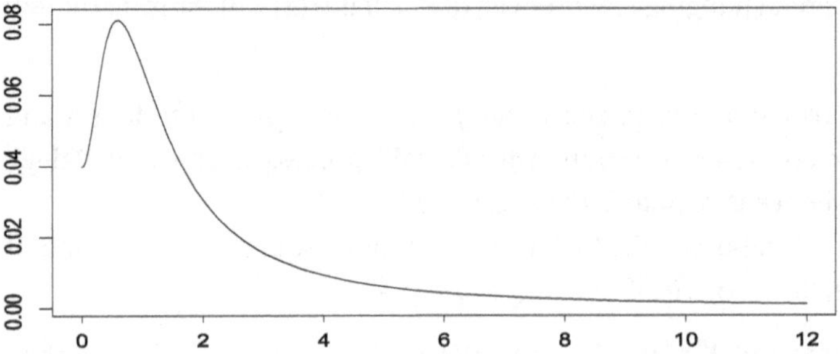

Figure 7.2: The spectral density of the CARMA(2, 1) process of Example 7.5 for frequencies 0 through 12 radians per unit time.

Proposition 7.6 (acvf of CARMA). *Consider a second-order* CARMA(p, q) *process* Y *as in Definition 7.1 (with the assumptions of Remark 7.2) and denote by* $\gamma(h) = \mathbb{E}(Y_{t+h}\overline{Y_t})$, $h \in \mathbb{R}$, *its autocovariance function.*

(a) *Then*

$$\gamma(h) = \nu^2 \sum_{\alpha \in \mathbb{C}: a(\alpha)=0} \operatorname{Res}_{z=\alpha}\left(\frac{b(z)b^*(z)}{a(z)a^*(z)} e^{zh}\right), \quad h \geq 0, \tag{7.2}$$

where

$$a^*(z) := \prod_{j=1}^{p}(-z - \overline{\alpha}_j), \quad b^*(z) := \prod_{j=1}^{q}(-z - \overline{\beta}_j),$$

and $\operatorname{Res}_{z=\alpha} g(z)$ *denotes the residue of a meromorphic function* g *at* α. *For* $h < 0$, *the autocovariance function is obtained from*

$$\gamma(h) = \overline{\gamma(-h)}.$$

If the multiplicity, $m(\alpha)$, *of each of the zeros of* $a(z)$ *is one, then* $\gamma(h)$ *takes the especially simple form,*

$$\gamma(h) = \nu^2 \sum_{\alpha \in \mathbb{C}: a(\alpha)=0} \frac{b(\alpha)b^*(\alpha)}{a'(\alpha)a^*(\alpha)} e^{\alpha h}, \quad h \geq 0. \tag{7.3}$$

If, in addition, the coefficients a_j *and* b_j *of the polynomials* $a(z)$ *and* $b(z)$ *are all real, then*

$$\gamma(h) = v^2 \sum_{\alpha \in \mathbb{C}: a(\alpha)=0} \frac{b(\alpha)b(-\alpha)}{a'(\alpha)a(-\alpha)} \mathrm{e}^{\alpha|h|}, \quad h \in \mathbb{R}. \tag{7.4}$$

(b) *If the coefficients of the polynomials $a(z)$ and $b(z)$ are all real (regardless if the zeros of $a(z)$ are simple or not), then γ is an even function, i. e., $\gamma(-h) = \gamma(h)$ for all $h \in \mathbb{R}$.*

Proof. From the spectral density (7.1), we immediately find that

$$\gamma(h) = \frac{v^2}{2\pi} \int_{\mathbb{R}} \mathrm{e}^{\mathrm{i}h\mu} \left| \frac{b(\mathrm{i}\mu)}{a(\mathrm{i}\mu)} \right|^2 d\mu = \frac{v^2}{2\pi} \int_{\mathbb{R}} \mathrm{e}^{\mathrm{i}h\mu} \frac{b(\mathrm{i}\mu)b^*(\mathrm{i}\mu)}{a(\mathrm{i}\mu)a^*(\mathrm{i}\mu)} d\mu. \tag{7.5}$$

The latter is equal to the complex path integral $\frac{v^2}{2\pi \mathrm{i}} \int_{-\mathrm{i}\infty}^{\mathrm{i}\infty} \frac{b(z)b^*(z)}{a(z)a^*(z)} \mathrm{e}^{hz}\, \mathrm{d}z$. Denoting by $\rho_R : [\pi/2, 3\pi/2] \to \mathbb{C}, t \mapsto R\,\mathrm{e}^{\mathrm{i}t}$, the half-circle in the left half-plane with radius R, a standard computation shows that $\int_{\rho_R} \frac{b(z)b^*(z)}{a(z)a^*(z)} \mathrm{e}^{hz}\, \mathrm{d}z$ converges to 0 as $R \to \infty$ when $h \geq 0$. Hence (7.2) follows from an application of the residue theorem. That $\gamma(h) = \overline{\gamma(-h)}$ for $h \in \mathbb{R}$ is clear from $\mathrm{Cov}(Y_{t+h}, Y_t) = \overline{\mathrm{Cov}(Y_t, Y_{t+h})}$. Equation (7.3) is an immediate consequence from (7.2) by evaluating the residues at simple poles.

If all coefficients of $a(z)$ and $b(z)$ are real, then (7.5) can be rewritten as $\gamma(h) = \frac{v^2}{2\pi} \int_{\mathbb{R}} \mathrm{e}^{\mathrm{i}h\mu} \frac{b(\mathrm{i}\mu)b(-\mathrm{i}\mu)}{a(\mathrm{i}\mu)a(-\mathrm{i}\mu)} d\mu$, from which it is easily seen that $\gamma(h) = \gamma(-h)$. If all zeros of $a(z)$ have multiplicity one, then the same calulations that led to (7.3) give (7.4) for $h \geq 0$, hence also for $h < 0$ since γ is then an even function. ☐

Example 7.7 (CAR(2)). If $a(z) = (z+0.4-\mathrm{i})(z+0.4+\mathrm{i})$, $b(z) = 1$ and $v = 1$, as in Example 7.4, then from (7.4) the CAR(2) process Y has the autocovariance function,

$$\gamma(h) = \frac{\sqrt{1.16}}{1.856} \cos(|h| - \arctan(0.4))\, \mathrm{e}^{-0.4|h|}, \quad h \in \mathbb{R},$$

a damped sinusoid with damping factor $\mathrm{e}^{-.4|h|}$ and period 2π, shown for $0 \leq h \leq 25$ in Figure 7.3. An easy direct calculation (see Exercise 7.1) shows that γ is twice continuously differentiable on $\mathbb{R}$, in accordance with the sufficient condition of Remark 3.14 for mean-square differentiability of Y.

Example 7.8 (CARMA(2, 1)). If $a(z) = (z + 0.5)(z + 1)$, $b(z) = z + 0.25$ and $v = 1$, as in Example 7.5, then Y is a CARMA(2, 1) process, and from (7.4) has the autocovariance function,

$$\gamma(h) = \frac{5}{8}\mathrm{e}^{-|h|} - \frac{1}{4}\mathrm{e}^{-|h|/2}, \quad h \in \mathbb{R},$$

a linear combination of a rapidly decaying and a more slowly decaying exponential function, shown for $0 \leq h \leq 25$ in Figure 7.4. It is easily seen that γ is not differentiable at 0, so that the necessary condition of Corollary 3.13 for mean-square differentiability of Y is violated.

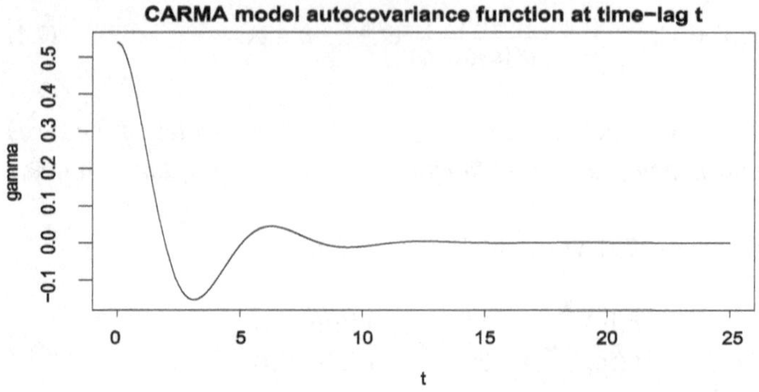

Figure 7.3: The autocovariance function of the CAR(2) process of Example 7.4 for time-lags 0 through 25.

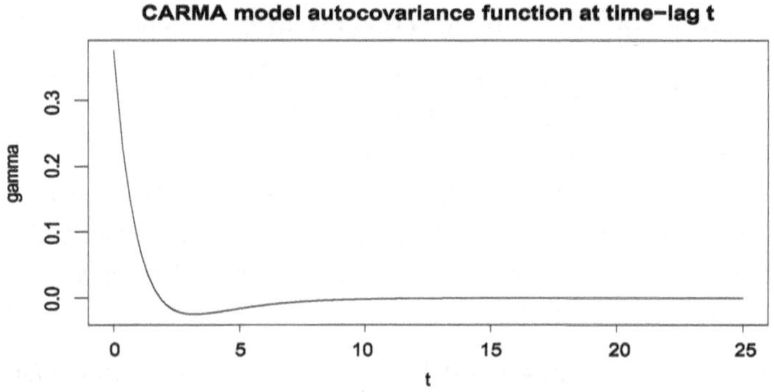

Figure 7.4: The autocovariance function of the CARMA(2, 1) process of Example 7.5 for time-lags 0 through 25.

7.2 The Wold–Karhunen representation

In order to determine the Wold–Karhunen representation of the regular second-order CARMA process Y with spectral density (7.1), we first determine the Wold coefficients of the Doob sequence $X = \{X_n, n \in \mathbb{Z}\}$ determined by the frequency transformation specified in (6.21) and (6.22).

Lemma 7.9. *If Y is the second-order* CARMA(p, q) *process with spectral density* (7.1) *and $X = \{X_n, n \in \mathbb{Z}\}$ is the sequence determined by the frequency transformation specified in* (6.21) *and* (6.22)*, then X is an ARMA sequence with Wold representation,*

$$X_n = \sum_{j=0}^{\infty} \psi_j Z_{n-j}, \quad \{Z_n\} \sim \mathrm{WN}(0, 1),$$

where

$$\sum_{j=0}^{\infty} \psi_j z^j = \sigma(1+z)^{p-q-1} \frac{\prod_{j=1}^{q}(1+\eta_j z)}{\prod_{j=1}^{p}(1+\xi_j z)}, \qquad |z| \leq 1, \tag{7.6}$$

with

$$\xi_j := \frac{\alpha_j + 1}{\alpha_j - 1}, \quad \eta_j := \frac{\beta_j + 1}{\beta_j - 1} \quad \text{and} \quad \sigma^2 := 2\nu^2 \frac{|b(1)|^2}{|a(1)|^2}. \tag{7.7}$$

(Note that $a(1)$, $b(1)$, $\alpha_j - 1$ and $\beta_j - 1$ cannot be zero because of our assumption that the zeros of $a(\cdot)$ all have strictly negative real parts and the zeros of $b(\cdot)$ all have non-positive real parts.)

Proof. Making the transformation and using (7.1), a simple calculation shows that the spectral density f_X of X is given by

$$\begin{aligned} f_X(\lambda) &= f(\mu(\lambda))\mu'(\lambda) \\ &= \frac{\nu^2}{\pi} \frac{|b(1)|^2}{|a(1)|^2} \frac{\prod_{j=1}^{q} |1 + \frac{\beta_j+1}{\beta_j-1} \mathrm{e}^{-\mathrm{i}\lambda}|^2}{\prod_{j=1}^{p} |1 + \frac{\alpha_j+1}{\alpha_j-1} \mathrm{e}^{-\mathrm{i}\lambda}|^2} \left(|1+\mathrm{e}^{-\mathrm{i}\lambda}|^2\right)^{p-q-1} \\ &= \frac{\sigma^2}{2\pi} \frac{\prod_{j=1}^{q} |1 + \eta_j \mathrm{e}^{-\mathrm{i}\lambda}|^2}{\prod_{j=1}^{p} |1 + \xi_j \mathrm{e}^{-\mathrm{i}\lambda}|^2} \left(|1+\mathrm{e}^{-\mathrm{i}\lambda}|^2\right)^{p-q-1}, \quad \lambda \in (-\pi, \pi), \end{aligned}$$

where in the last line we used (7.7). As remarked in Example 6.9 (see equation (6.16)), this is the spectral density of the ARMA$(p, p-1)$-process

$$X_n - \phi_1 X_{n-1} - \cdots - \phi_p X_{n-p} = W_n + \theta_1 Z_{n-1} + \cdots + \theta_{p-1} W_{n-p+1}, \quad n \in \mathbb{Z},$$

where

$$\begin{aligned} 1 - \phi_1 z - \cdots - \phi_p z^p &= \prod_{j=1}^{p} (1 + \xi_j z), \\ 1 + \theta_1 z + \cdots + \theta_{p-1} z^{p-1} &= (1+z)^{p-q-1} \prod_{j=1}^{q} (1 + \eta_j z), \quad z \in \mathbb{C}, \end{aligned}$$

and $(W_n) \sim \mathrm{WN}(0, \sigma^2)$. Since $\Re(\alpha_j) < 0$ and $\Re(\beta_j) \leq 0$ by our standing assumption in Remark 7.2, (7.7) gives $|\xi_j| < 1$ and $|\eta_j| \leq 1$. As shown in Example 6.9, the Wold-decomposition of X is given by $X_n = \sum_{j=0}^{\infty} \psi_j Z_{n-j}$, where $(Z_n) \sim \mathrm{WN}(0,1)$ and $(\psi_j)_{j \in \mathbb{N}_0}$ satisfies (7.6). □

For the determination of the Wold–Karhunen representation, we still need the functions ψ_c and $\widetilde{\psi_c}$.

Lemma 7.10. *For the second-order* CARMA(p, q) *process as specified in Definition* 7.1, *the functions ψ_c and $\widetilde{\psi_c}$ of Definition* 6.21 *are given by*

$$\psi_{\mathrm{c}}(t) = \nu^* \sum_{\alpha\in\mathbb{C}:a(\alpha)=0} \mathrm{Res}_{z=\alpha}(b(z)\,\mathrm{e}^{zt}/a(z))\mathbf{1}_{[0,\infty)}(t), \quad t \in \mathbb{R}, \tag{7.8}$$

and

$$\widetilde{\psi_{\mathrm{c}}}(\mu) = \frac{\nu^*}{\sqrt{2\pi}} \frac{b(\mathrm{i}\mu)}{a(\mathrm{i}\mu)}, \quad \mu \in \mathbb{R}, \tag{7.9}$$

where

$$\nu^* := \nu \frac{a(1)|b(1)|}{b(1)|a(1)|}.$$

Proof. From Definition 6.21, we have

$$\widetilde{\psi_{\mathrm{c}}}(\mu) = \sum_{j=1}^{\infty} \psi_j f_j(\mu) = \frac{1}{\sqrt{\pi}(1+\mathrm{i}\mu)} \sum_{j=0}^{\infty} \psi_j \frac{(1-\mathrm{i}\mu)^j}{(1+\mathrm{i}\mu)^j}.$$

A straightforward calculation using (7.6) then leads to (7.9). The function ψ_{c} is now the inverse Fourier transform of $\widetilde{\psi_{\mathrm{c}}}$. Hence, by Theorem 2.66(d), ψ_{c} is the L^2-limit of $(2\pi)^{-1}\int_{-T}^{T} \mathrm{e}^{\mathrm{i}\mu\cdot}\nu^* b(\mathrm{i}\mu)/a(\mathrm{i}\mu)\,\mathrm{d}\mu$, and we conclude that there is a sequence $(T_n)_{n\in\mathbb{N}}$ converging to $+\infty$ such that

$$\psi_{\mathrm{c}}(t) = \frac{\nu^*}{2\pi} \int_{-T_n}^{T_n} \mathrm{e}^{\mathrm{i}\mu t} \frac{b(\mathrm{i}\mu)}{a(\mathrm{i}\mu)}\,\mathrm{d}\mu = \frac{\nu^*}{2\pi\mathrm{i}} \int_{-T_n\mathrm{i}}^{T_n\mathrm{i}} \mathrm{e}^{tz} \frac{b(z)}{a(z)}\,\mathrm{d}z$$

Lebesgue-almost everywhere in $t \in \mathbb{R}$. Denote by $\rho_{T_n} : [\pi/2, 3\pi/2] \to \mathbb{C}$, $s \mapsto T_n\mathrm{e}^{\mathrm{i}s}$ the half-circle in the left half-plane with radius T_n. As in the proof of Proposition 7.6, we see that $\int_{\rho_{T_n}} \mathrm{e}^{tz}b(z)/a(z)\,\mathrm{d}z$ converges for $t > 0$ to 0 as $n \to \infty$, hence (7.8) follows for $t > 0$ from the residue theorem. For $t < 0$, (7.8) follows similarly by using a corresponding semi-circle in the right half-plane and the fact that $b(z)\mathrm{e}^{zt}/a(z)$ is holomorphic in a neighbourhood of the right half-plane, in particular has no poles there (since all of the zeros of $a(\cdot)$ lie in the left half-plane). Finally, the value at $t = 0$ is irrelevant, since ψ_{c} is defined only Lebesgue-almost everywhere. □

The Wold–Karhunen representation is unique only up to a constant factor of modulus 1. If we are not aiming for the canonical Wold–Karhunen representation, we can choose this factor to our liking.

Theorem 7.11 (Wold–Karhunen representation of a CARMA process). *If Y is the second-order* CARMA(p, q) *process with spectral density* (7.1), *then a Wold–Karhunen representation of Y is given by*

$$Y_t = \int_{\mathbb{R}} k(t-s)\,\mathrm{d}\xi_s = \int_{(-\infty,t]} k(t-s)\,\mathrm{d}\xi_s, \tag{7.10}$$

where

$$k(t) = \nu \sum_{\alpha \in \mathbb{C}: a(\alpha)=0} \operatorname{Res}_{z=\alpha}\left[e^{zt} \frac{b(z)}{a(z)} \right] \mathbf{1}_{[0,\infty)}(t)$$

and the sum of residues on the right is over the zeros α of $a(z)$, all of which are in the open left half of the complex plane. ξ is the orthogonal increment process given by

$$\xi_t = \frac{1}{\nu} \int_{\mathbb{R}} \frac{a(i\mu)(e^{it\mu} - 1)}{i\mu\, b(i\mu)}\, dZ_\mu, \quad t \in \mathbb{R},$$

with associated distribution function $H(t) = t$, where Z is the spectral process of Y. The Fourier transform of k is given by

$$\mathcal{F}_2 k(\mu) = \frac{\nu}{\sqrt{2\pi}} \frac{b(i\mu)}{a(i\mu)}, \quad \mu \in \mathbb{R}.$$

Proof. By Theorem 6.26 and Corollary 6.28, with V the zero process, we can write Y as

$$Y_t = \int_{\mathbb{R}} \psi_c(t-s)\, d\xi'_s = \int_{(-\infty,t]} \psi_c(t-s)\, d\xi'_s,$$

where $\{\xi'_t, t \in \mathbb{R}\}$, is the orthogonal increment driving process given by (6.41), which has associated distribution function $H(t) = t$. Since $k(t) = \nu/\nu^* \psi_c(t)$ with $|\nu/\nu^*| = 1$, the claim then follows from Lemma 7.10 with $\xi_t = \nu^*/\nu\, \xi'_t$. □

Definition 7.12 (Kernel of the CARMA process). With k and ξ as in Theorem 7.11, we define the *kernel* of the second-order CARMA(p, q) process Y as $g = k/\nu$ and the *driving orthogonal increment process (driving OIP)* as $\eta = \nu\xi$. In these terms, the representation (7.10) of Y can be written in the equivalent form,

$$Y_t = \int_{(-\infty,t]} g(t-s)\, d\eta_s, \tag{7.11}$$

where η is an OIP with associated distribution function $H(t) = \nu^2 t$. The advantage of this formulation is that the kernel g depends only on the CARMA coefficients and not on the driving process η. Specifically, we have

$$g(t) = \sum_{\alpha \in \mathbb{C}: a(\alpha)=0} \operatorname{Res}_{z=\alpha}\left[e^{zt} \frac{b(z)}{a(z)} \right] \mathbf{1}_{[0,\infty)}(t) \tag{7.12}$$

and

$$\eta_t = \int_{\mathbb{R}} \frac{a(i\mu)(e^{i\mu} - 1)}{i\mu b(i\mu)}\, dZ_\mu, \quad t \in \mathbb{R}.$$

If the p zeros, $\alpha_1, \ldots, \alpha_p$, of $a(z)$ all have multiplicity 1, then g simplifies to

$$g(t) = \sum_{j=1}^{p} \frac{b(\alpha_j)}{a'(\alpha_j)} e^{\alpha_j t} \mathbf{1}_{[0,\infty)}(t), \tag{7.13}$$

where a' is the derivative of a.

We shall meet the kernel g again in Chapter 18 when we consider Lévy-driven CARMA processes, which do not necessarily have finite second moments and consequently cannot in general be defined in terms of a spectral density function.

Example 7.13 (CAR(2)). For the CAR(2) process of Example 7.4 with $a(z) = (z+0.4-i)(z+0.4+i)$, $b(z) = 1$, we find from (7.13) that the kernel g in the representation (7.11) is given by

$$g(t) = \sum_{\alpha \in \mathbb{C}: a(\alpha)=0} \frac{b(\alpha)}{a'(\alpha)} e^{\alpha t} \mathbf{1}_{[0,\infty)}(t) = e^{-0.4t} \sin(t) \mathbf{1}_{[0,\infty)}(t),$$

which, like the corresponding autocovariance function, is a damped sinusoid with damping factor $e^{-0.4t}$ and period 2π, shown for $0 \le t \le 25$ in Figure 7.5.

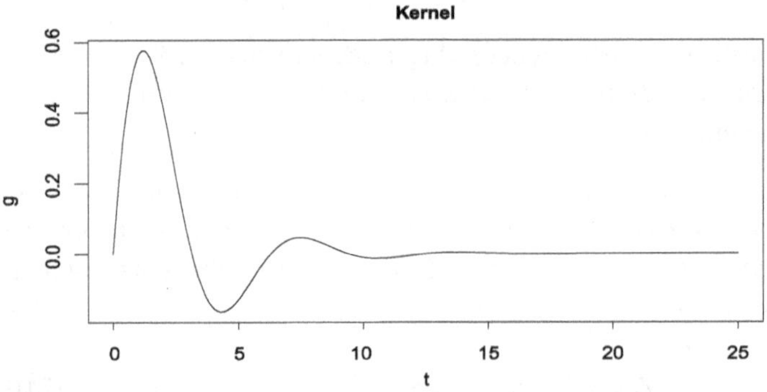

Figure 7.5: The CARMA kernel function $g(t)$, $t \le 25$, in the representation (7.11) of the CAR(2) process of Example 7.4.

Example 7.14 (CARMA(2, 1)). For the CARMA(2, 1) process of Example 7.5 with $a(z) = (z+0.5)(z+1)$, $b(z) = z+0.25$, we again find from (7.13) that the kernel g in the representation (7.11) is given by

$$g(t) = \sum_{\alpha \in \mathbb{C}: a(\alpha)=0} \frac{b(\alpha)}{a'(\alpha)} e^{\alpha t} \mathbf{1}_{[0,\infty)}(t) = \frac{1}{2}(3e^{-t} - e^{-t/2}) \mathbf{1}_{[0,\infty)}(t),$$

which, like the corresponding autocovariance function, is a linear combination of a rapidly decaying and a more slowly decaying exponential function, shown for $0 \le t \le 25$ in Figure 7.6.

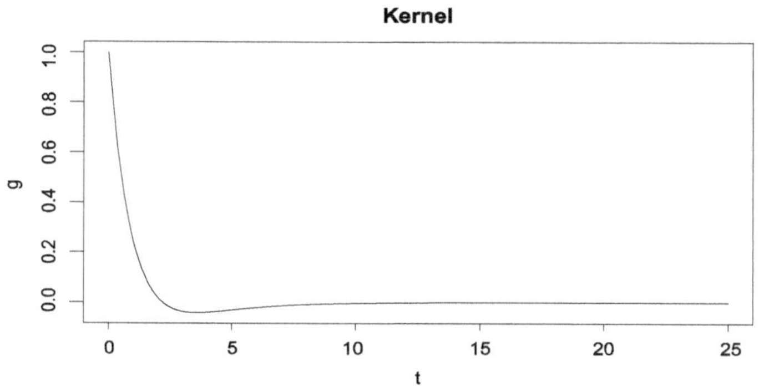

Figure 7.6: The CARMA kernel function $g(t)$, $t \leq 25$, in the representation (7.11) of the CARMA(2, 1) process of Example 7.5.

We now obtain another representation of the kernel function using matrix exponentials, which will play a key role in Chapter 18 when we consider Lévy-driven CARMA processes with possibly infinite second moments. Recall the definition of the exponential of a complex matrix $A \in \mathbb{C}^{p\times p}$, namely the complex $p \times p$-matrix,

$$\exp(A) = \mathrm{e}^A := \sum_{k=0}^{\infty} \frac{1}{k!} A^k, \tag{7.14}$$

where the sum converges absolutely in each component (here, $A^0 = I_p$, the *identity matrix* in $\mathbb{C}^{p\times p}$). It is often simply called the *matrix exponential* of A. In the following, we collect some standard facts for the matrix exponential, which will be used throughout the book (in particular, in this chapter and Chapters 17–19). It is well known from standard books on matrix analysis (e. g., Bernstein [13, Prop. 11.1.5]), that

$$\mathrm{e}^A \mathrm{e}^B = \mathrm{e}^B \mathrm{e}^A = \mathrm{e}^{A+B} \quad \text{if } AB = BA$$

for $A, B \in \mathbb{C}^{p\times p}$, in particular

$$\mathrm{e}^{Ax} \mathrm{e}^{Ay} = \mathrm{e}^{Ay} \mathrm{e}^{Ax} = \mathrm{e}^{A(x+y)} \quad \forall\, x, y \in \mathbb{C},$$

and e^{Ax} is invertible with inverse e^{-Ax}. Furthermore, the function $x \mapsto \mathrm{e}^{Ax}$, $x \in \mathbb{R}$, is infinitely often differentiable with derivative

$$\frac{\mathrm{d}}{\mathrm{d}x} \mathrm{e}^{Ax} = A\mathrm{e}^{Ax} = \mathrm{e}^{Ax} A. \tag{7.15}$$

In particular, by the fundamental theorem of calculus,

$$\int_s^t A\mathrm{e}^{Au}\, \mathrm{d}u = \mathrm{e}^{At} - \mathrm{e}^{As} \quad \forall\, s, t \in \mathbb{R}. \tag{7.16}$$

Later, we shall need some facts on the growth behaviour of the matrix exponential. For their preparation, we recall some basic facts from matrix analysis (as can be found, e. g., in Bernstein [13]). Let $A \in \mathbb{C}^{p\times p}$ and denote by $\lambda_1,\dots,\lambda_r \in \mathbb{C}$ (where $r \leq p$) its distinct eigenvalues. Then A can be transformed to Jordan canonical form as

$$J = SAS^{-1} = \begin{bmatrix} J_1 & 0 & \cdots & 0 \\ 0 & J_2 & \cdots & 0 \\ \vdots & \ddots & \ddots & \vdots \\ 0 & 0 & \cdots & J_{m_1+\cdots+m_r} \end{bmatrix}, \tag{7.17}$$

where $S \in \mathbb{C}^{p\times p}$ is invertible and where $J_{m_1+\cdots+m_{i-1}+j}$, $i \in \{1,\dots,r\}, j \in \{1,\dots,m_i\}$, is a Jordan block

$$J_{m_1+\cdots+m_{i-1}+j} = \begin{bmatrix} \lambda_i & 1 & 0 & \cdots & 0 \\ 0 & \lambda_i & 1 & \cdots & 0 \\ \vdots & \ddots & \ddots & \ddots & \vdots \\ 0 & \cdots & 0 & \lambda_i & 1 \\ 0 & \cdots & 0 & 0 & \lambda_i \end{bmatrix} \in \mathbb{C}^{d_{i,j}\times d_{i,j}} \tag{7.18}$$

of dimension $d_{i,j}$ corresponding to the eigenvalue λ_i of A (we have m_i Jordan blocks corresponding to λ_i, and $\sum_{i=1}^{r}\sum_{j=1}^{m_i} d_{i,j} = p$). The matrix J is the Jordan canonical form of A (unique up to permutations of the Jordan blocks) and $A = S^{-1}JS$ is the *Jordan canonical decomposition of* A. The matrix exponential of a Jordan block $J_k \in \mathbb{C}^{d_{i,j}\times d_{i,j}}$, corresponding to the eigenvalue λ_i is given by

$$e^{J_k t} = e^{\lambda_i t} \begin{bmatrix} 1 & t & \frac{t^2}{2} & \cdots & \frac{t^{d_{i,j}-2}}{(d_{i,j}-2)!} & \frac{t^{d_{i,j}-1}}{(d_{i,j}-1)!} \\ 0 & 1 & t & \ddots & \frac{t^{d_{i,j}-3}}{(d_{i,j}-3)!} & \frac{t^{d_{i,j}-2}}{(d_{i,j}-2)!} \\ 0 & 0 & 1 & \ddots & \frac{t^{d_{i,j}-4}}{(d_{i,j}-4)!} & \frac{t^{d_{i,j}-3}}{(d_{i,j}-3)!} \\ \vdots & \vdots & \ddots & \ddots & \ddots & \vdots \\ 0 & 0 & 0 & \ddots & 1 & t \\ 0 & 0 & 0 & \cdots & 0 & 1 \end{bmatrix} \in \mathbb{C}^{d_{i,j}\times d_{i,j}}, \quad t \in \mathbb{R}. \tag{7.19}$$

With the aid of this exponential, we can compute the matrix exponential of A via (7.17) for $t \in \mathbb{R}$ as

$$e^{At} = \exp(S^{-1}(Jt)S) = S^{-1}e^{Jt}S = S^{-1} \begin{bmatrix} e^{J_1 t} & 0 & \cdots & 0 \\ 0 & e^{J_2 t} & \cdots & 0 \\ \vdots & \ddots & \ddots & \vdots \\ 0 & 0 & \cdots & e^{J_{m_1+\cdots+m_r} t} \end{bmatrix} S.$$

From the form of the matrix exponential of J_k, we conclude that for every $\varepsilon > 0$ and $t_0 \in \mathbb{R}$ there is some $K_{\varepsilon,t_0} < \infty$ such that

$$\|e^{At}\| \leq K_{\varepsilon,t_0}\, e^{(\varepsilon+\max\{\Re(\lambda_1),\dots,\Re(\lambda_r)\})t} \quad \forall\, t \geq t_0 \tag{7.20}$$

and

$$\|e^{At}\| \leq K_{\varepsilon,t_0}\, e^{(\min\{\Re(\lambda_1),\dots,\Re(\lambda_r)\}-\varepsilon)t} \quad \forall\, t \leq t_0. \tag{7.21}$$

Another known representation of the matrix exponential $\exp(A)$ is by contour integrals. To derive it, observe that for $\lambda \in \mathbb{C}$ with $\lambda > \|A\|$ (the natural matrix norm of A with respect to the Euclidian norm defined by $\|A\| := \sup_{w \in \mathbb{C}^d : |w| \leq 1} |Aw|$, i. e., the spectral norm of A) we have

$$\sum_{k=0}^{\infty} \lambda^{-k-1} A^k = (\lambda I_p - A)^{-1} \tag{7.22}$$

by the Neumann series expansion. Now let $\kappa_R : [0, 2\pi] \to \mathbb{C},\ s \mapsto Re^{is}$ be the circle around 0 with radius R, and suppose that $R > \|A\|$. Then all eigenvalues of A lie inside κ_R, and from Cauchy's integral formula (e. g., Conway [34, Corollary IV.5.9]) we obtain with $g(z) := e^{zt}$, $z \in \mathbb{C}$, that $e^{At} = g(A)$ can be expressed as

$$\sum_{k=0}^{\infty} \frac{1}{k!} \frac{d^k g}{dx^k}(0)\, A^k = \sum_{k=0}^{\infty} \frac{1}{2\pi i} \int_{\kappa_R} \frac{g(z)}{z^{k+1}}\, dz\, A^k = \frac{1}{2\pi i} \int_{\kappa_R} g(z) \sum_{k=0}^{\infty} z^{-k-1} A^k\, dz,$$

where the interchange of summation and integration is justified by uniform convergence of the series. Inserting (7.22) gives the Cauchy integral representation of the matrix exponential

$$e^{At} = \frac{1}{2\pi i} \int_{\rho} e^{zt} (zI_p - A)^{-1}\, dz,$$

where ρ is any simple closed contour enclosing the eigenvalues of A in the anticlockwise sense (it is first proved for $\rho = \kappa_R$ with $R > \|A\|$ and then extended to general ρ, since the integral is independent of the particular path, again by Cauchy's integral formula). The above formula also holds (with the same proof) when $g(z) = e^{zt}$ is replaced by a general entire function, resulting in

$$g(A) = \frac{1}{2\pi i} \int_{\rho} g(z)(zI_p - A)^{-1}\, dz. \tag{7.23}$$

We are particularly interested in the matrix exponential of companion matrices. It will be applied to $a(z)$ and $b(z)$ later, but we use $c(z)$ and $w(z)$ since we formulate the results without restrictions on the zeros (i. e., without the assumptions of Remark 7.2).

Proposition 7.15 (Matrix exponential of companion matrix). *Let* $c_1, \ldots, c_p \in \mathbb{C}$, $w_0, \ldots, w_{p-1} \in \mathbb{C}$ *and denote* $c_0 := 1$. *Define the polynomials*

$$c(z) = z^p + c_1 z^{p-1} + \cdots + c_{p-1} z + c_p = \sum_{k=0}^{p} c_k z^{p-k}, \quad z \in \mathbb{C},$$

and

$$w(z) = w_0 + w_1 z + \cdots + w_{p-1} z^{p-1}, \quad z \in \mathbb{C}.$$

Further, define the vectors $\mathbf{w}, \mathbf{e}_p \in \mathbb{C}^p$ *and the matrix* $C \in \mathbb{C}^{p\times p}$ *by*

$$C = \begin{bmatrix} 0 & 1 & 0 & \cdots & 0 \\ 0 & 0 & 1 & \cdots & 0 \\ \vdots & \vdots & \vdots & \ddots & \vdots \\ 0 & 0 & 0 & \cdots & 1 \\ -c_p & -c_{p-1} & -c_{p-2} & \cdots & -c_1 \end{bmatrix}, \quad \mathbf{e}_p = \begin{bmatrix} 0 \\ 0 \\ \vdots \\ 0 \\ 1 \end{bmatrix}, \quad \mathbf{w} = \begin{bmatrix} w_0 \\ w_1 \\ \vdots \\ w_{p-2} \\ w_{p-1} \end{bmatrix},$$

where for $p = 1$ *the matrix* C *is to be understood as* $C = (-c_1)$.

(a) *Then* $c(\cdot)$ *is the characteristic polynomial of the matrix* C, *i. e.,* $c(z) = \det(zI_p - C)$ *for* $z \in \mathbb{C}$.

(b) *For each* $k \in \{1, \ldots, p\}$ *define the polynomial*

$$r_k(z) := \sum_{j=0}^{p-k} c_{p-k-j} z^j, \quad z \in \mathbb{C}. \tag{7.24}$$

Then the (j,k)*-entry of* e^{Ct} *for* $t \in \mathbb{R}$ *is given by*

$$(e^{Ct})_{jk} = \frac{1}{2\pi i} \int_\rho \frac{r_k(z) z^{j-1} e^{tz}}{c(z)}\, dz = \sum_{\lambda\in\mathbb{C}: c(\lambda)=0} \operatorname{Res}_{z=\lambda}[r_k(z) z^{j-1} e^{tz}/c(z)], \tag{7.25}$$

where ρ *is any simple closed contour enclosing the zeros of the polynomial* $c(\cdot)$ *in the anticlockwise sense. In particular, the* (j,p)*-element of* e^{Ct} *is given by*

$$(e^{Ct})_{jp} = \frac{1}{2\pi i} \int_\rho \frac{z^{j-1} e^{zt}}{c(z)}\, dz = \sum_{\lambda\in\mathbb{C}: c(\lambda)=0} \operatorname{Res}_{z=\lambda}[z^{j-1} e^{tz}/c(z)] \tag{7.26}$$

and

$$\mathbf{w}^T e^{Ct} \mathbf{e}_p = \frac{1}{2\pi i} \int_\rho e^{tz} \frac{w(z)}{c(z)}\, dz = \sum_{\lambda\in\mathbb{C}: c(\lambda)=0} \operatorname{Res}_{z=\lambda}[e^{tz} w(z)/c(z)]. \tag{7.27}$$

The matrix C above is called the *companion matrix of the polynomial* $c(\cdot)$. The representation (7.25) for the elements of the matrix exponential is due to Eller [41, Theorem 2.1], where a more general formula for holomorphic functions of companion matrices is obtained.

Proof. (a) is a standard fact on companion matrices and can be found in many text books on matrix analysis. It is easily proved by induction on p, cf., Exercise 7.2.

For the proof (b), denote by $\mathbf{e}_j$ the jth unit vector in $\mathbb{C}^p$. From the form of C, we have $\mathbf{e}_j^T C = \mathbf{e}_{j+1}^T$ for $j \in \{1, \ldots, p-1\}$ and hence

$$\mathbf{e}_1^T C^{j-1} = \mathbf{e}_j^T, \quad \forall\, j \in \{1, \ldots, p\}. \tag{7.28}$$

We claim that

$$r_k(C)\,\mathbf{e}_p = \mathbf{e}_k, \quad \forall\, k \in \{1, \ldots, p\}. \tag{7.29}$$

This is clear for $k = p$, since $r_p(z) = c_0 = 1$. Now assume that (7.29) is proved for $k \in \{2, \ldots, p\}$. Since $r_{k-1}(z) = c_{p-k+1} + z\, r_k(z)$, we then obtain

$$r_{k-1}(C)\mathbf{e}_p = c_{p-k+1}\mathbf{e}_p + C\, r_k(C)\mathbf{e}_p = c_{p-k+1}\mathbf{e}_p + C\mathbf{e}_k = \mathbf{e}_{k-1},$$

where we used the induction hypothesis in the second equality and the specific form of C in the last equality. Hence (7.29) is proved.

Now let $j, k \in \{1, \ldots, p\}$. Using (7.28) and (7.29), the (j, k)-element of e^{Ct} is given by

$$(\mathrm{e}^{Ct})_{jk} = \mathbf{e}_j^T\, \mathrm{e}^{Ct}\, \mathbf{e}_k = \mathbf{e}_1^T C^{j-1}\, \mathrm{e}^{Ct}\, r_k(C)\, \mathbf{e}_p,$$

which by (7.23) can be written as

$$\begin{aligned}(\mathrm{e}^{Ct})_{jk} &= \mathbf{e}_1^T \frac{1}{2\pi \mathrm{i}} \int_{\rho} z^{j-1} r_k(z) \mathrm{e}^{tz} (zI_p - C)^{-1}\, \mathrm{d}z\, \mathbf{e}_p \\ &= \frac{1}{2\pi \mathrm{i}} \int_{\rho} z^{j-1} r_k(z) \mathrm{e}^{tz} \mathbf{e}_1^T (zI_p - C)^{-1} \mathbf{e}_p\, \mathrm{d}z,\end{aligned} \tag{7.30}$$

where ρ is a simple closed contour enclosing the eigenvalues of C, i. e., the zeros of $c(\cdot)$, in the anticlockwise sense. To simplify this further, observe that by Cramér's rule, $(zI_p - C)^{-1}$ is given by $(\det(zI_p - C))^{-1} \operatorname{adj}(zI_p - C)$, where $\operatorname{adj}(zI_p - C)$ is the adjugate matrix of $zI_p - C$. But $\det(zI_p - C) = c(z)$ by (a) and the $(1, p)$-element of $\operatorname{adj}(zI_p - C)$ is obtained as $(-1)^{1+p}$ times the determinant of the matrix, which is obtained by deleting the first column and the pth row of $zI_p - C$, hence is equal to $(-1)^{1+p}(-1)^{p-1} = 1$. We conclude that $\mathbf{e}_1^T (zI_p - C)^{-1} \mathbf{e}_p = 1/c(z)$. Inserting this in (7.30) gives the first equality in (7.25), and the second inequality follows from the residue theorem. Equation (7.26) is an immediate consequence of (7.25) since $r_p(z) = 1$, and (7.27) then follows from (7.26). □

Returning to the setting of second-order CARMA processes, again with our standing assumption of Remark 7.2, we immediately obtain the following.

Corollary 7.16 (Alternative representation of kernel). *The kernel g of* (7.12) *in the representation* (7.11) *can be written as*

$$g(t) = \mathbf{b}^T e^{At} \mathbf{e}_p \mathbf{1}_{[0,\infty)}(t), \tag{7.31}$$

where $A \in \mathbb{C}^{p\times p}$ and $\mathbf{b}, \mathbf{e}_p \in \mathbb{C}^p$ are defined by

$$A = \begin{bmatrix} 0 & 1 & 0 & \cdots & 0 \\ 0 & 0 & 1 & \cdots & 0 \\ \vdots & \vdots & \vdots & \ddots & \vdots \\ 0 & 0 & 0 & \cdots & 1 \\ -a_p & -a_{p-1} & -a_{p-2} & \cdots & -a_1 \end{bmatrix}, \quad \mathbf{e}_p = \begin{bmatrix} 0 \\ 0 \\ \vdots \\ 0 \\ 1 \end{bmatrix}, \quad \mathbf{b} = \begin{bmatrix} b_0 \\ b_1 \\ \vdots \\ b_{p-2} \\ b_{p-1} \end{bmatrix}. \tag{7.32}$$

Here, $a_1, \dots, a_p, b_0, \dots, b_{p-1}$ are the coefficients of the polynomials $a(z)$ and $b(z)$ in Definition 7.1 with $b_q := 1$ and $b_j := 0$ for $j > q$. For $p = 1$ the matrix A is to be understood as $A = (-a_1)$.

In Section 7.4, we will provide R-programs to compute the kernel and autocovariance functions of CARMA processes, which make use of the representation (7.31). The latter representation will turn out very useful also in the following section on prediction of CARMA processes.

7.3 Prediction

Since (7.11) is the Wold–Karhunen representation of Y (apart from a constant factor), it is clear from Corollary 6.27 that the h-step best linear predictor of Y_{t+h}, $h > 0$, given $\mathcal{N}_t = \overline{\text{sp}}\,\{Y_s : s \le t\}$, is given by

$$P_{\mathcal{N}_t} Y_{t+h} = \int\limits_{(-\infty,t]} g(t+h-u)\,\mathrm{d}\eta_u, \tag{7.33}$$

and that the prediction mean-square error is

$$\sigma_h^2 := \mathbb{E}(|Y_{t+h} - P_{\mathcal{N}_t} Y_{t+h}|^2) = \nu^2 \int_0^h |g(t)|^2\,\mathrm{d}t. \tag{7.34}$$

This is the *Wold–Karhunen form* of the predictor. With the aid of the spectral process Z of Y, by Corollary 6.24, we can rewrite (7.33) in the *frequency-domain form*,

$$P_{\mathcal{N}_t} Y_{t+h} = \int_{\mathbb{R}} \frac{a(\mathrm{i}\mu)}{b(\mathrm{i}\mu)} \int_{(-\infty,t]} g(t+h-s)\,\mathrm{e}^{\mathrm{i}\mu s}\,\mathrm{d}s\,\mathrm{d}Z_\mu. \tag{7.35}$$

Here we used that $g \in L^1(\mathbb{R};\mathbb{C})$ by (7.31) and (7.20), that

$$\mathrm{e}^{\mathrm{i}t\mu}\mathcal{F}_2(g(\cdot+h)\mathbf{1}_{[0,\infty)}(\cdot))(\mu) = \frac{1}{\sqrt{2\pi}} \int_{(-\infty,t]} \mathrm{e}^{\mathrm{i}\mu s} g(h+t-s)\,\mathrm{d}s$$

and that $\mathcal{F}_2 g(\mu) = b(\mathrm{i}\mu)/(\sqrt{2\pi}a(\mathrm{i}\mu))$ by Theorem 7.11.

In the following, we shall derive a more explicit (in terms of $\{Y_s, s \leq t\}$) expression for $P_{\mathcal{N}_t} Y_{t+h}$ than (7.33) and (7.35). For this purpose, we will utilize the representation (7.31) of the kernel.

Proposition 7.17 (State-space representation). *Let $Y = (Y_t)_{t\in\mathbb{R}}$ be a second-order CARMA(p,q) process as in Definition 7.1, with polynomials $a(z)$ and $b(z)$ as defined in (7.1) (under the assumptions of Remark 7.2). Define $\mathbf{b}$, $\mathbf{e}_p$ and A as in (7.32). Then all entries of e^{At} decrease exponentially as $t \to \infty$, and there exists an orthogonal increment process $\zeta = (\zeta_t)_{t\in\mathbb{R}}$ with associated distribution function $H(t) = t$ such that*

$$Y_t = \nu \mathbf{b}^T \mathbf{X}_t, \quad t \in \mathbb{R}, \tag{7.36}$$

where

$$\mathbf{X}_t := \int_{(-\infty,t]} \mathrm{e}^{A(t-u)}\mathbf{e}_p\,\mathrm{d}\zeta_u, \quad t \in \mathbb{R}, \tag{7.37}$$

with the integral being understood componentwise.

The representation (7.36), (7.37) of Y is called a *state-space representation* of Y, with the unobserved vector process X being called a *state vector*.

Proof. Since the eigenvalues of A are the zeros of $a(\cdot)$ by Proposition 7.15(a), which all have strictly negative real parts, the components of e^{At} decrease exponentially as $t \to \infty$ by (7.20), so that the components of $\mathrm{e}^{At}\mathbf{1}_{[0,\infty)}(t)$ are in $L^1(\mathbb{R};\mathbb{C}) \cap L^2(\mathbb{R};\mathbb{C})$, so that the integral in (7.37) will be defined. Let $\zeta := \xi$ be the OIP of the Wold–Karhunen representation (7.10) obtained in Theorem 7.11, with associated distribution function $H(t) = t$. Then $Y_t = \nu \int_{(-\infty,t]} g(t-s)\,\mathrm{d}\zeta_s$ by (7.11). Using (7.31), we conclude

$$Y_t = \nu \int_{(-\infty,t]} \mathbf{b}^T \mathrm{e}^{A(t-s)}\mathbf{e}_p\,\mathrm{d}\zeta_s = \nu\mathbf{b}^T \int_{(-\infty,t]} \mathrm{e}^{A(t-s)}\mathbf{e}_p\,\mathrm{d}\zeta_s$$

which is the representation (7.36), (7.37). □

Remark 7.18. In the existence proof of Proposition 7.17 we used as ζ the OIP appearing in the Wold–Karhunen representation of Y. For the following considerations, it is however not necessary that ζ is the OIP from the Wold–Karhunen representation, but only that the process Y satisfies (7.36), (7.37) with *some* OIP ζ with associated distribution function $H(t) = t$. Later, in Chapter 18, we will use a state-space representation as in (7.36), (7.37) to define Lévy driven CARMA processes (with possibly infinite moments). Observe that if ζ is an OIP with associated distribution function $H(t) = t$, if polynomials $a(z) = z^p + a_1 z^{p-1} + \cdots + a_p$ and $b(z) = b_0 + b_1 z + \cdots + b_{q-1} z^{q-1} + z^q$ are given such that all zeros of a have strictly negative parts and those of b non-positive real parts, and if $\mathbf{b}$, $\mathbf{e}_p$ and A are defined as in (7.32), then it is clear from Example 5.3 that the process Y defined by (7.36) and (7.37) is indeed a second-order CARMA process with spectral density given by (7.1).

A nice feature of the state vector $\mathbf{X}$ is that its first component already determines the others:

Theorem 7.19 (Relation between components of state vector). *Let $Y = (Y_t)_{t\in\mathbb{R}}$ be as in Proposition 7.17 with state-space vector $\mathbf{X}$ and state-space representation (7.36), (7.37). For $j \in \{1,\dots,p\}$ denote the jth component of $\mathbf{X}$ by X^{j-1}. Then each of the $X^0,\dots,X^{p-2}$ are mean-zero weakly stationary MSC processes, and, if $p \geq 2$, $X^0,\dots,X^{p-2}$ are mean-square differentiable with mean square derivative $DX^j = X^{j+1}$, $j \in \{0,\dots,p-2\}$. In particular, X^0 is $(p-1)$-times mean square differentiable with $D^j X^0 = X^j$, $j \in \{0,\dots,p-1\}$. The process $\mathbf{X}$ is hence given by*

$$\mathbf{X}_t = [X_t^0,\dots,X_t^{p-1}]^T = [D^0, D^1,\dots,D^{p-1}]^T X_t^0. \tag{7.38}$$

Proof. Denote by $\mathbf{e}_j$ the jth unit vector in $\mathbb{C}^p$, $j \in \{1,\dots,p\}$. Then

$$X_t^{j-1} = \mathbf{e}_j^T \mathbf{X}_t = \int_{(-\infty,t]} \mathbf{e}_j^T \mathrm{e}^{A(t-u)} \mathbf{e}_p \, \mathrm{d}\zeta_u, \quad t \in \mathbb{R},$$

and it follows from Example 4.11 that X^{j-1} is a mean-zero weakly stationary MSC process. Now fix $t \in \mathbb{R}$ and define

$$\varphi^{j-1}(u) := \mathbf{e}_j^T \mathrm{e}^{A(t-u)} \mathbf{e}_p \mathbf{1}_{(-\infty,t]}(u), \quad u \in \mathbb{R},$$

so that $X_t^{j-1} = \int_{\mathbb{R}} \varphi^{j-1}(u)\,\mathrm{d}\zeta_u$. Denote by $\widetilde{\zeta}$ the Fourier transform of the OIP ζ and by $\mathcal{F}_2^{-1}\varphi^{j-1}$ the inverse Fourier transform of φ^{j-1}. Then

$$X_t^{j-1} = \int_{\mathbb{R}} (\mathcal{F}_2^{-1}\varphi^{j-1})(\mu)\,\mathrm{d}\widetilde{\zeta}_\mu$$

by (4.23). For the calculation of $\mathcal{F}_2^{-1}\varphi^{j-1}$, let ρ be a simple closed contour in the open left half-plane enclosing all zeros of $a(\cdot)$ in the anticlockwise sense. By (7.26) and an application of Fubini's theorem (which is possible since ρ lies strictly in the open left half-plane and by the growth of the integrands) we then have

$$(\mathcal{F}_2^{-1}\varphi^{j-1})(\mu) = \frac{1}{\sqrt{2\pi}} \int_{-\infty}^{t} \frac{\mathrm{e}^{\mathrm{i}u\mu}}{2\pi\mathrm{i}} \int_{\rho} \frac{z^{j-1}}{a(z)} \mathrm{e}^{(t-u)z} \,\mathrm{d}z \,\mathrm{d}u$$
$$= \frac{\mathrm{e}^{\mathrm{i}t\mu}}{\sqrt{2\pi}} \frac{1}{2\pi\mathrm{i}} \int_{\rho} \frac{z^{j-1}}{a(z)(\mathrm{i}\mu - z)} \,\mathrm{d}z.$$

By the residue theorem,

$$\frac{1}{2\pi\mathrm{i}} \int_{\rho} \frac{z^{j-1}}{a(z)(\mathrm{i}\mu - z)} \,\mathrm{d}z = \sum_{\lambda \in \mathbb{C}: a(\lambda)=0} \mathrm{Res}_{z=\lambda} \left[\frac{z^{j-1}}{a(z)(\mathrm{i}\mu - z)} \right]$$

(observe that all zeros of $a(\cdot)$ lie in the open left half-plane). On the other hand, if $\kappa_R : [0, 2\pi] \to \mathbb{C}$ is the circle given by $\kappa_R(s) = R\mathrm{e}^{\mathrm{i}s}$, then for large enough R, we have

$$\frac{1}{2\pi\mathrm{i}} \int_{\kappa_R} \frac{z^{j-1}}{a(z)(\mathrm{i}\mu - z)} \,\mathrm{d}z = \sum_{\lambda \in \mathbb{C}: a(\lambda)(\mathrm{i}\mu-\lambda)=0} \mathrm{Res}_{z=\lambda} \left[\frac{z^{j-1}}{a(z)(\mathrm{i}\mu - z)} \right],$$

i. e., we have the additional residue at $\lambda = \mathrm{i}\mu$. Since $\int_{\kappa_R} z^{j-1}/(a(z)(\mathrm{i}\mu - z)) \,\mathrm{d}z$ tends to 0 as $R \to \infty$ (a fact that is often expressed by saying that the residue of $z^{j-1}/(a(z)(\mathrm{i}\mu - z))$ at infinity is zero), we conclude that

$$\sum_{\lambda \in \mathbb{C}: a(\lambda)=0} \mathrm{Res}_{z=\lambda} \left[\frac{z^{j-1}}{a(z)(\mathrm{i}\mu - z)} \right] = -\mathrm{Res}_{z=\mathrm{i}\mu} \left[\frac{z^{j-1}}{a(z)(\mathrm{i}\mu - z)} \right] = \frac{(\mathrm{i}\mu)^{j-1}}{a(\mathrm{i}\mu)}.$$

Altogether, we have shown that

$$(\mathcal{F}_2^{-1}\varphi^{j-1})(\mu) = \frac{\mathrm{e}^{\mathrm{i}t\mu}}{\sqrt{2\pi}} \frac{(\mathrm{i}\mu)^{j-1}}{a(\mathrm{i}\mu)}$$

and hence we obtain the frequency domain representation

$$X_t^{j-1} = \frac{1}{\sqrt{2\pi}} \int_{\mathbb{R}} \mathrm{e}^{\mathrm{i}t\mu} \frac{(\mathrm{i}\mu)^{j-1}}{a(\mathrm{i}\mu)} \,\mathrm{d}\tilde{\zeta}_\mu \quad \forall\, t \in \mathbb{R}\; \forall\, j \in \{1, \dots, p\}. \tag{7.39}$$

Denote by Z^0 the spectral process of X^0. From (7.39), we see that Z^0 has density $1/(\sqrt{2\pi}\, a(\mathrm{i}\cdot))$ with respect to $\tilde{\zeta}$, and that X^{j-1} is obtained from X^0 by application of the linear filter with transfer function $(\mathrm{i}\cdot)^{j-1}$ (cf., Section 5.4). We also have $\int_{\mathbb{R}} |\mu|^{2(p-1)}/|a(\mathrm{i}\mu)|^2 \,\mathrm{d}\mu < \infty$, so that X^0 is $(p-1)$-times mean-square differentiable with $D^j X^0 = X^j$ by Example 5.17, giving the claim. □

For the prediction formula, we still have to determine X^0 from Y. This is done in the following proposition, under the additional assumption that not only do the zeros of $b(\cdot)$ have non-positive real part, but actually have strictly negative real part.

Proposition 7.20 (First component of state vector). *In addition to the assumptions and notations of Proposition* 7.17 *and Theorem* 7.19, *assume that all zeros of b have strictly negative real part. Then the first component X^0 of $\mathbf{X}$ is given by*

$$X_t^0 = \begin{cases} \nu^{-1} \int_{[0,\infty)} \kappa(u)\, Y_{t-u}\, \mathrm{d}u, & \text{if } q > 0, \\ \nu^{-1} Y_t, & \text{if } q = 0, \end{cases} \tag{7.40}$$

where

$$\kappa(t) := \Big[\sum_{\beta \in \mathbb{C}: b(\beta)=0} \operatorname{Res}_{z=\beta}\big(\mathrm{e}^{zt}/b(z)\big) \Big] \mathbf{1}_{[0,\infty)}(t),$$

and the sum of the residues of $\mathrm{e}^{zt}/b(z)$ is at the distinct zeros, β, of $b(z)$. The term $\int_{[0,\infty)} \kappa(u)\, Y_{t-u}\, \mathrm{d}u$ is the convolution of the function κ and the process Y as defined in Definition 5.19, *with the integral being a limit of mean-square Riemann integrals. Introducing the matrix*

$$B = \begin{bmatrix} 0 & 1 & 0 & \cdots & 0 \\ 0 & 0 & 1 & \cdots & 0 \\ \vdots & \vdots & \vdots & \ddots & \vdots \\ 0 & 0 & 0 & \cdots & 1 \\ -b_0 & -b_1 & -b_2 & \cdots & -b_{q-1} \end{bmatrix} \in \mathbb{C}^{q\times q}$$

when $q \geq 1$ (to be understood as $B = (-b_0)$ when $q = 1$), the function $\kappa(t)$ can also be written as the $(1, q)$-element of e^{Bt} times $\mathbf{1}_{[0,\infty)}(t)$. Finally, if the zeros, $\beta_1, \dots, \beta_q$, of $b(z)$ all have multiplicity 1, *then*

$$\kappa(t) = \sum_{k=1}^{q} \frac{\mathrm{e}^{\beta_k t}}{b'(\beta_k)} \mathbf{1}_{[0,\infty)}(t).$$

Proof. With $\tilde{\zeta}$ as defined in the proof of Theorem 7.19, we have from (7.36) and (7.39) that

$$Y_t = \nu \mathbf{b}^T \mathbf{X}_t = \frac{\nu}{\sqrt{2\pi}} \int_{\mathbb{R}} \mathrm{e}^{\mathrm{i}t\mu} \frac{b(\mathrm{i}\mu)}{a(\mathrm{i}\mu)}\, \mathrm{d}\tilde{\zeta}_\mu.$$

Since the spectral process Z^0 of X^0 has density $1/(\sqrt{2\pi}\, a(\mathrm{i}\cdot))$ with respect to $\tilde{\zeta}$ as observed in the proof of Theorem 7.19, we conclude that Y is obtained from X^0 by application of the linear filter with transfer function $\nu b(\mathrm{i}\cdot)$ if $q > 0$ and $Y = \nu X^0$ if $q = 0$. Hence X^0 is obtained from Y by application of the linear filter with transfer function $(\nu b(\mathrm{i}\cdot))^{-1}$ when $q > 0$. By the discussion following Definition 5.19, this corresponds to convolution of Y with $\frac{1}{\sqrt{2\pi}}$ times the inverse Fourier transform of $\nu^{-1}/b(\mathrm{i}\cdot)$, which can be evaluated by contour integration as the function $\nu^{-1}\kappa$ with κ as specified in the proposition as the sum of residues. The alternative expression as the $(1, q)$-element of the matrix exponential of Bt follows from (7.26), from which we also conclude that $\kappa \in L^2(\mathbb{R}; \mathbb{C})$ is locally bounded on $\mathbb{R}$ and continuous on $\mathbb{R} \setminus \{0\}$, so that the conditions required in the definition of the convolution in Definition 5.19 are indeed met. The remaining assertions are clear. □

Having expressed the state vector in terms of the past of Y, it is now a simple matter to derive the h-step best linear predictor of a second-order CARMA process. This is then the *state-space form* of the predictor.

Theorem 7.21 (Prediction in terms of state vector). *Under the assumptions of Proposition 7.20, if* $\mathbf{X}$ *is defined as in* (7.37), *then for* $h > 0$ *the h-step best linear predictor of the second-order* CARMA(p, q)*-process* Y *given* $\overline{\text{sp}}(Y_s : s \leq t)$ *is given by* $P_{\mathcal{N}_t} Y_{t+h} = \nu \mathbf{b}^T e^{Ah} \mathbf{X}_t$, *or more explicitly,*

$$P_{\mathcal{N}_t} Y_{t+h} = \begin{cases} \mathbf{b}^T e^{Ah} [D^0, D^1, \dots, D^{p-1}]^T \int_{[0,\infty)} \kappa(u) Y_{t-u}\, du, & \text{if } q > 0, \\ \mathbf{b}^T e^{Ah} [D^0, D^1, \dots, D^{p-1}]^T Y_t, & \text{if } q = 0, \end{cases} \tag{7.41}$$

where D *denotes mean-square differentiation with respect to* t, κ *was specified in Proposition* 7.20 *and the integral denotes convolution of* κ *and* Y. *The prediction mean-square error can be expressed as*

$$\sigma_h^2 = \mathbb{E}(|Y_{t+h} - P_{\mathcal{N}_t} Y_{t+h}|^2) = \nu^2 \int_0^h \mathbf{b}^T e^{Au} \mathbf{e}_p \mathbf{e}_p^T e^{\overline{A}^T u} \overline{\mathbf{b}}\, du. \tag{7.42}$$

Proof. From (7.37), it follows at once that

$$\mathbf{X}_{t+h} = e^{Ah} \mathbf{X}_t + \int_{(t,t+h]} e^{A(h+t-u)} \mathbf{e}_p\, d\zeta_u.$$

The components of the first term belong to $\overline{\text{sp}}\{Y_s, s \leq t\}$ by (7.40) and (7.38) and the components of the second term are orthogonal to $\overline{\text{sp}}\{Y_s, s \leq t\}$, since the increments of ζ on (t, ∞) are orthogonal to Y_s for each $s \leq t$ (the latter by (7.36) and (7.37) and the fact that ζ is an OIP). Hence

$$\mathcal{P}_{\mathcal{N}_t} \mathbf{X}_{t+h} = e^{Ah} \mathbf{X}_t$$

(recall Definition 2.50 for the orthogonal projection of random vectors). The result (7.41) then follows at once from this, (7.40) and (7.38). The prediction error is

$$Y_{t+h} - P_{\mathcal{N}_t} Y_{t+h} = \nu \mathbf{b}^T (\mathbf{X}_{t+h} - e^{Ah} \mathbf{X}_t) = \nu \mathbf{b}^T \int_{(t,t+h]} e^{A(t+h-u)} \mathbf{e}_p\, d\zeta_u,$$

so by (4.15) the mean-square prediction error is

$$\sigma_h^2 = \mathbb{E}(|Y_{t+h} - P_{\mathcal{N}_t} Y_{t+h}|^2) = \nu^2 \int_0^h \mathbf{b}^T e^{Au} \mathbf{e}_p \mathbf{e}_p^T e^{\bar{A}^T u} \bar{\mathbf{b}}\, du,$$

an alternative expression for the value of σ_h^2 previously specified in (7.34). □

Remark 7.22. If the orthogonal-increment process ζ in (7.37) is standard Brownian motion (or more generally, a Lévy process with mean zero and finite second moment), then the mean-square integral $\int_{[0,\infty)} \kappa(u) Y_{t-u}\, du$ in (7.41) can be interpreted pathwise by (3.6), the exponential decay of κ, the growth behavior of ζ (which grows at most linearly by the strong law of large numbers, cf., Theorem 11.21 below) and hence the growth behaviour of Y (cf., equation (17.10) below). Similarly, the mean-square derivatives in (7.41) can be interpreted pathwise, as we shall see later in Remark 18.5.

Remark 7.23. By (7.25), the row vector $\mathbf{b}^T e^{Ah}$ can be expressed as

$$\mathbf{b}^T e^{Ah} = \sum_{\alpha \in \mathbb{C}: a(\alpha)=0} \operatorname{Res}_{z=\alpha}\left[[r_1(z), r_2(z), \ldots, r_p(z)] \frac{b(z) e^{zh}}{a(z)} \right],$$

where $r_k(z) = \sum_{j=0}^{p-k} a_{p-k-j} z^j$ as in (7.24). If the zeros $\alpha_1, \ldots, \alpha_p$ of $a(\cdot)$ all have multiplicity 1 then this simplifies to

$$\mathbf{b}^T e^{Ah} = \sum_{j=1}^{p} [r_1(\alpha_j), r_2(\alpha_j), \ldots, r_p(\alpha_j)] \frac{b(\alpha_j)}{a'(\alpha_j)} e^{\alpha_j h}. \tag{7.43}$$

These expressions can be used in conjunction with (7.41) to give alternative expressions for the best linear predictor.

Example 7.24 (CAR(1)). For the process considered in Example 6.25, $p = 1, q = 0, b(z) = 1$, $a(z) = z + a$, $\gamma(h) = \frac{\nu^2}{2a} e^{-a|h|}$, $A = -a$ and $\mathbf{X}_t = \nu^{-1} Y_t$. Direct substitution into (7.41) and (7.42) gives

$$P_{\mathcal{N}_t} Y_{t+h} = e^{-ah} Y_t, \quad h > 0,$$

and

$$\sigma_h^2 = \nu^2 \int_0^h e^{-2au}\, du = \gamma(0)(1 - e^{-2ah}),$$

in accordance with the results of Example 6.25.

Example 7.25 (CARMA(2, 1)). Consider prediction for a CARMA(2, 1) process with $b(z) = z - \beta_1$ and $a(z) = (z - \alpha_1)(z - \alpha_2)$, where α_1, α_2 and β_1 are all strictly negative real numbers, $\alpha_1 \neq \alpha_2$ and $\beta_1 \neq \alpha_i, i = 1, 2$. With r_1 and r_2 defined as in Remark 7.23, we see from (7.43) that

$$\mathbf{b}^T \mathrm{e}^{Ah} = \sum_{j=1}^{2} [r_1(\alpha_j), r_2(\alpha_j)] \frac{b(\alpha_j)}{a'(\alpha_j)} \mathrm{e}^{\alpha_j h},$$

where $r_1(\alpha_1) = -\alpha_2$, $r_1(\alpha_2) = -\alpha_1$ and $r_2(\alpha_j) = 1$. By Proposition 7.20, $\kappa(t) = \mathrm{e}^{\beta_1 t}\mathbf{1}_{[0,\infty)}(t)$ so that

$$\int_{[0,\infty)} \kappa(u) Y_{t-u}\,\mathrm{d}u = \mathrm{e}^{\beta_1 t} \int_{(-\infty,t]} \mathrm{e}^{-\beta_1 u} Y_u\,\mathrm{d}u$$

and

$$[D^0, D^1]^T \int_{[0,\infty)} \kappa(u) Y_{t-u}\,\mathrm{d}u = \begin{bmatrix} \mathrm{e}^{\beta_1 t} \int_{(-\infty,t]} \mathrm{e}^{-\beta_1 u} Y_u\,\mathrm{d}u \\ Y_t + \beta_1 \mathrm{e}^{\beta_1 t} \int_{(-\infty,t]} \mathrm{e}^{-\beta_1 u} Y_u\,\mathrm{d}u \end{bmatrix}$$

(that the mean-square derivative of the first line is indeed given by the second line is easily checked by applying the definition of the mean-square derivative and Exercise 3.6). Substituting in (7.41) and using the expression (7.43) for $\mathbf{b}^T \mathrm{e}^{Ah}$ we find that the required predictor is

$$P_{\mathcal{N}_t} Y_{t+h} = \frac{(\alpha_2 - \beta_1)(\alpha_1 - \beta_1)}{\alpha_2 - \alpha_1} (\mathrm{e}^{\alpha_1 h} - \mathrm{e}^{\alpha_2 h}) \int_{(-\infty,t]} \mathrm{e}^{\beta_1 (t-u)} Y_u\,\mathrm{d}u$$
$$+ \left(\frac{\alpha_1 - \beta_1}{\alpha_1 - \alpha_2} \mathrm{e}^{\alpha_1 h} + \frac{\alpha_2 - \beta_1}{\alpha_2 - \alpha_1} \mathrm{e}^{\alpha_2 h} \right) Y_t.$$

Similarly, substituting into (7.42), we find that the mean-squared error is

$$\sigma_h^2 = \nu^2 \int_0^h \left(\frac{\alpha_1 - \beta_1}{\alpha_1 - \alpha_2} \mathrm{e}^{\alpha_1 u} + \frac{\alpha_2 - \beta_1}{\alpha_2 - \alpha_1} \mathrm{e}^{\alpha_2 u} \right)^2 \mathrm{d}u.$$

Just as equation (7.31) provided an alternative representation for the kernel function, with the aid of the state-space representation we can also derive an alternative representation for the autocovariance function, obtained in Proposition 7.6.

Proposition 7.26 (Alternative representation of autocovariance function). *Consider a second-order* CARMA(p, q) *process Y as in Definition* 7.1 *(with the assumptions of Remark* 7.2*) and denote by $\gamma(h) = \mathbb{E}(Y_{t+h}\overline{Y_t})$, $h \in \mathbb{R}$, its autocovariance function. Let $\mathbf{A}$, $\mathbf{e}_p$ and $\mathbf{b}$ be defined as in* (7.32). *Then*

$$\gamma(h) = \nu^2 \mathbf{b}^T \mathrm{e}^{Ah} \Sigma \overline{\mathbf{b}} \quad \forall\, h \geq 0, \tag{7.44}$$

where

$$\Sigma := \int_{[0,\infty)} \mathrm{e}^{Au}\, \mathbf{e}_p \mathbf{e}_p^T\, \mathrm{e}^{\overline{A}^T u}\,\mathrm{d}u.$$

If additionally all of the coefficients $a_1, \dots, a_p$ of the polynomial $a(z)$ are real, then Σ is symmetric and $\mathbb{R}^{p\times p}$-valued, and if v_j denotes the jth column of Σ ($j \in \{1, \dots, p\}$), then v_1 is the unique solution of the equation

$$a(-A)v_1 = \mathbf{e}_p, \tag{7.45}$$

with the matrix

$$a(-A) = (-A)^p + a_1(-A)^{p-1} + \cdots + a_{p-1}(-A) + a_p I_p$$

being invertible, and the other columns of Σ are given by

$$v_j = -Av_{j-1} = (-A)^{j-1}v_1, \quad j = 2, \dots, p. \tag{7.46}$$

With $a_0 := 0$, the matrix $a(-A)$ can then also be expressed as

$$a(-A) = 2 \sum_{j\in\{0,\dots,p\}, j \text{ even}} a_{p-j}A^j = -2 \sum_{j\in\{0,\dots,p\}, j \text{ odd}} a_{p-j}A^j. \tag{7.47}$$

Proof. Let $\mathbf{X}$ be a state-vector such that (7.36) and (7.37) hold. Denoting the components of $\mathbf{X}$ by $X^0, \dots, X^{p-1}$, equation (7.36) gives

$$\gamma(h) = v^2 \sum_{j,k=0}^{p-1} b_j \,\mathbb{E}(X^j_{t+h}\overline{X^k_t})\,\overline{b_k} \tag{7.48}$$

for $h \in \mathbb{R}$. Using (7.37) and the Itô isometry (4.15), $\mathbb{E}(X^j_{t+h}\overline{X^k_t})$ can be expressed for $h \geq 0$ as (with $\mathbf{e}_i$ denoting the ith unit vector in $\mathbb{C}^p$)

$$\begin{aligned}
&\mathbb{E}\Big(\int_{(-\infty,t+h]} \mathbf{e}^T_{j+1} \mathrm{e}^{A(t+h-u)}\,\mathbf{e}_p\,\mathrm{d}\zeta_u\,\overline{\int_{(-\infty,t]} \mathbf{e}^T_p \mathrm{e}^{A^T(t-u)}\,\mathbf{e}_{k+1}\,\mathrm{d}\zeta_u}\Big)\\
&= \int_{(-\infty,t]} \mathbf{e}^T_{j+1}\,\mathrm{e}^{A(t+h-u)}\,\mathbf{e}_p\mathbf{e}^T_p\,\mathrm{e}^{\overline{A}^T(t-u)}\,\mathbf{e}_{k+1}\,du\\
&= \mathbf{e}^T_{j+1}\mathrm{e}^{Ah} \int_{[0,\infty)} \mathrm{e}^{Au}\,\mathbf{e}_p\mathbf{e}^T_p\,\mathrm{e}^{\overline{A}^T}\,du\,\mathbf{e}_{k+1}.
\end{aligned}$$

Inserting this in (7.48) and using the definition of Σ gives (7.44).

Now assume that all of the coefficients $a_1, \dots, a_p$ are real. Then e^{Au} is $\mathbb{R}^{p\times p}$-valued, hence so is Σ, and Σ is clearly symmetric. Further,

$$
\begin{aligned}
A\Sigma + \Sigma A^T &= \lim_{s\to\infty} \int_0^s (A\mathrm{e}^{At}\,\mathbf{e}_p\mathbf{e}_p^T \mathrm{e}^{A^T t} + \mathrm{e}^{At}\,\mathbf{e}_p\mathbf{e}_p^T\,\mathrm{e}^{A^T t}A^T)\,\mathrm{d}t \\
&= \lim_{s\to\infty} \int_0^s \frac{\mathrm{d}}{\mathrm{d}t}(\mathrm{e}^{At}\,\mathbf{e}_p\mathbf{e}_p^T\,\mathrm{e}^{A^T t})\,\mathrm{d}t \\
&= \lim_{s\to\infty} (\mathrm{e}^{As}\,\mathbf{e}_p\mathbf{e}_p^T\,\mathrm{e}^{A^T s} - \mathbf{e}_p\mathbf{e}_p^T) = -\mathbf{e}_p\mathbf{e}_p^T.
\end{aligned}
$$

The latter equation is a so called *Lyapunov equation.* To solve it, we need the Kronecker product and the vec-operation (vectorization). If $C \in \mathbb{R}^{m\times n}$ with columns $c_1, \dots, c_n$, then the *vectorization* $\operatorname{vec}(C)$ of C is defined as the column vector $[c_1^T, \dots, c_n^T]^T \in \mathbb{R}^{mn}$, i. e., is obtained by stacking the columns of C on top of one another. Furthermore, if $F \in \mathbb{R}^{m_1\times n_1} = (f_{ij})_{i=1,\dots,m_1;j=1,\dots,n_1}$ and $G \in \mathbb{R}^{m_2,n_2}$ are matrices, then the *Kronecker product* $F \otimes G$ of F with G is defined to be the $m_1m_2 \times n_1n_2$-matrix

$$
F \otimes G = \begin{bmatrix} f_{11}G & \cdots & f_{1,n_1}G \\ \vdots & \ddots & \vdots \\ f_{m_1,1}G & \cdots & f_{m_1,n_1}G \end{bmatrix}.
$$

The Kronecker product and the vectorization have many interesting properties, in particular

$$
\operatorname{vec}(FHG) = (G^T \otimes F)\operatorname{vec}(H) \quad \forall\, F \in \mathbb{R}^{m_1\times n_1}, H \in \mathbb{R}^{n_1\times n_2},\ G \in \mathbb{R}^{n_2\times m_2},
$$

see, e. g., Bernstein [13, Section 7.1]. Rewriting the derived Lyapunov equation as $A\Sigma I_p + I_p\Sigma A^T = -\mathbf{e}_p\mathbf{e}_p^T$ this leads to the equation

$$
(I_p \otimes A + A \otimes I_p)\operatorname{vec}(\Sigma) = -\operatorname{vec}(\mathbf{e}_p\mathbf{e}_p^T). \tag{7.49}
$$

But $\operatorname{vec}(\Sigma) = [v_1^T, \dots, v_p^T]^T$ and $\operatorname{vec}(\mathbf{e}_p\mathbf{e}_p^T)$ is the $(p^2)^{\text{th}}$-unit vector in $\mathbb{R}^{p^2}$. From the first $p-1$ row blocks in (7.49) we hence obtain $Av_1+v_2 = 0, \dots, Av_{p-1}+v_p = 0$, giving (7.46). Evaluation of the last row block leads to $Av_p - a_pv_1 - \dots - a_1v_p = -\mathbf{e}_p$ and inserting $v_j = (-A)^{j-1}v_1$ yields (7.45). To see that $a(-A)$ is invertible, assume the contrary. Then $\det a(-A) = 0$. Now let $S \in \mathbb{C}^{p\times p}$ be an invertible matrix such that $J = SAS^{-1}$ is the Jordan canonical form of A as in (7.17). Then $\det(a(-J)) = 0$. But from the specific form (7.18) of the Jordan blocks, this implies that there is an eigenvalue λ of A such that $\sum_{j=0}^p a_{p-j}(-\lambda)^j = 0$. Since the eigenvalues of A are the zeros of $a(z)$ by Proposition 7.15(a), all of which were assumed to have strictly negative real part, we obtain a contradiction. This shows that $a(-A)$ is invertible and consequently the solution v_1 of (7.45) is unique. Finally, equation (7.47) follows from

$$\sum_{j\in\{0,\dots,p\},j\text{ even}} a_{p-j}A^j + \sum_{j\in\{0,\dots,p\},j\text{ odd}} a_{p-j}A^j = a(A) = 0,$$

where we used the Cayley–Hamilton theorem. □

The representation (7.44) together with (7.45) and (7.46) is particularly suited for the numerical calculation of the autocovariance function of a CARMA process, and will be used in the R-program 3 in the next section. The proof of Proposition 7.26 was inspired by Betser et al. [14], where also alternative ways for the numerical solution of Lyapunov equations involving companion matrices are studied. Another representation of the matrix Σ of Proposition 7.26 can be found in Arató [3, p. 128].

7.4 R-programs to compute the kernel and autocovariance function

In this section, we present three R-programs for the specification of the parameters and the calculation of the kernel and the autocovariance function of second-order CARMA processes. They only work for CARMA models with *real-valued* coefficients, and such that the assumptions of Remark 7.2 are satisfied. Before using the R programs in this book, be sure to load the R-package expm. For the simulation programs in Section 19.5, you must also load the package yuima of Brouste et al. [31] to take advantage of its ability to generate sample increments for a wide variety of Lévy processes. All R programs presented in this book can be downloaded from

https://doi.org/10.1515/9783111325033.

Program 1 requires entry of the CARMA model zeros or alternatively the coefficients, the required maximum lag tau and the interval h between successive values at which the kernel and autocovariance function are calculated in Programs 2 and 3. It calculates the coefficients and eigenvalues of the matrix A. It also calculates the vectors $\mathbf{b}$ and $\mathbf{e}_p$. The parameters T and delta used in Program 1 do not play a role yet, but will become important only later when doing the simulations of Section 19.5.

```
# PROGRAM 1: CARMA parameter set-up
# To enter the CARMA model specify p in the range 1--5 and q in the range
# 0--(p-1).
# Set the autoregressive zeros lambda[1],...,lambda[p] (the eigenvalues of A)
# and the moving average zeros mu[1], ..., mu[q] to the required values.
# There must be no common zeros and all must have negative real parts.
# Complex auroregressive or moving average zeros must be accompanied
# by their conjugates.
# T=total time interval, h=discretization step,
# tau=kernel truncation time, delta=sampling interval
# The matrix A and its eigenvalues (if p>1) are printed
# followed by the vectors b and ep.
# CARMA coefficients can also be entered directly as in the example
```

```
# in option OR below.
# EITHER SPECIFY CARMA ZEROS:

p=3;q=2;lambda<-rep(0,p);mu<-rep(0,q)
T=3000; delta=1; tau=100; h=0.01
a<-rep(0,p); b<-rep(0,p);b[q+1]<-1
lambda[1]= -.1; lambda[2] = -.2+1i; lambda[3] = -.2-1i; lambda[4]=0; lambda[5]=0
mu[1]= -.3; mu[2]= -.5; mu[3]= 0; mu[4]= 0

for(i in 1:q){b[q]<-b[q]-mu[i]}
if(q>0){for(i in 1:(q-1))
         {for (j in (i+1):q) {b[q-1]<-b[q-1]+mu[i]*mu[j]} }}
if(q>1){for(i in 1:(q-2))
         {for (j in (i+1):(q-1))
           {for (k in (j+1):q){b[q-2]<-b[q-2]-mu[i]*mu[j]*mu[k]} }}}
if(q>2){for(i in 1:(q-3))
         {for (j in (i+1):(q-2))
           {for (k in (j+1):(q-1)){for (l in (k+1):q)
             {b[q-3]<-b[q-3]+mu[i]*mu[j]*mu[k]*mu[l]}} }}}
for(i in 1:p){a[1]<-a[1]-lambda[i]}
if(p>1){for (i in 1:(p-1))
         {for (j in (i+1):p){a[2]<-a[2]+lambda[i]*lambda[j]}}}
if(p>2){for (i in 1:(p-2))
         {for (j in (i+1):(p-1))
           {for(k in (j+1):p){{a[3]<-a[3]-lambda[i]*lambda[j]*lambda[k]}}}}}
if(p>3){for (i in 1:(p-3))
         {for (j in (i+1):(p-2))
           {for (k in(j+1):(p-1))
             {for (l in (k+1):p)
               {a[4]<-a[4]+lambda[i]*lambda[j]*lambda[k]*lambda[l]}}}}}
if(p>4){for (i in 1:(p-4))
         {for (j in (i+1):(p-3))
           {for (k in(j+1):(p-2))
             {for (l in (k+1):(p-1))
               {for (m in (l+1):p)
                 {{a[5]<-a[5]-lambda[i]*lambda[j]*lambda[k]*lambda[l]*lambda[m]
                   }}}}}}}

# OR SPECIFY CARMA COEFFICIENTS:
# p, q, a=(a[1],a[2],..,a[p]), b=(b[1],b[2],...,b[q+1]=1,b[q+2]=0,...,b[p]=0) can
# be entered directly if the autoregressive and moving average zeros satisfy the
# required conditions set out above, e.g.
# p= 3; q=2; a<-c(0.5, 1.08, 0.104); b<-c(0.15, 0.8, 1)
# T=3000; delta=1; tau=100; h=0.01
# THEN:

A=matrix(rep(0,len=p^2),nrow=p);
if (p==2){A[1,]=c(0,1);A[2,]=-c(a[2],a[1])}
if (p==3){A[1,]=c(0,1,0);A[2,]=c(0,0,1);A[3,]=-c(a[3],a[2],a[1])}
if (p==4){A[1,]=c(0,1,0,0);A[2,]=c(0,0,1,0);A[3,]=c(0,0,0,1);
          A[4,]=-c(a[4],a[3],a[2],a[1])}
if(p==5){A[1,]=c(0,1,0,0,0);A[2,]=c(0,0,1,0,0);A[3,]=c(0,0,0,1,0);
         A[4,]=c(0,0,0,0,1); A[5,]=-c(a[5],a[4],a[3],a[2],a[1])};
A<-Re(A);b<- Re(b)
ep<-rep(0,p);ep[p]<-1
if(p==1){"A=-a[1]"; A[1,1] <- -a[1]; Re(A)}
if(p>1){eigen(A)$values};if(p>1){A}; b; ep
```

Program 2 computes the kernel g at times n*h, n = 1, . . . , tau/h with the aid of Corollary 7.16, where the maximum lag tau was set (in Program 1) to 100 and the spacing h between calculated values to 0.01.

```
# PROGRAM 2: CARMA kernel and truncation time tau
# Evaluates and plots the CARMA kernel at multiples of h up to time tau.
# Rerun Programs 1 and 2 after increasing tau if necessary until the ratio,
# (kernel at time tau)/(max absolute value up to tau), is small,
# say less than 10-5. tau is the time at which the kernel, g, will be truncated.
# BE SURE TO LOAD THE R-PACKAGE 'expm' FROM NOW ON
u<-c(0:(tau/h+1))
g<-matrix(1:(tau/h+1),nrow=1,byrow=TRUE)
if(p==1){ for (i in 1:(tau/h+1)){g[i]= exp(-a[1]*u[i]*h)}}
if(p>1){for (i in 1:(tau/h+1)){g[i]<- t(b) %*% expm(A*u[i]*h) %*% ep}}
t<-c(0:(tau/h))*h
plot(t,g,type="l",main="Kernel")
"The␣following␣is␣the␣ratio␣g(tau)/max(abs(g(t)),t<tau)."
g[tau/h]/max(abs(g))
"To␣reduce␣it␣rerun␣Programs␣1␣and␣2␣with␣larger␣tau."
"Once␣the␣ratio␣is␣OK␣go␣to␣Program␣3."
```

Program 3 computes the autocorrelation function kappa and autocovariance function gamma at lags 0, h, 2h, . . . , tau/h * h = tau using the method of Proposition 7.26 when $\nu^2 = 1$. The variance of Y_t is then gamma[1]. For $\nu^2 \neq 1$ the corresponding autocovariance function is obtained by multiplying gamma of Program 3 by ν^2.

```
# PROGRAM 3: CARMA model autocorrelation function (ACF)
# Sigma (below) is the covariance matrix of the stationary state vector when
# Var(L(1))=1. kappa is the model ACF of the simulated Y. It is independent of
# Var(L(1)) and well-defined even when Var(L(1)) is infinite. It is compared with
# the sample autocorrelation function of the simulated Y in Program 6 below.
Sigma= matrix(rep(0,len=p^2),nrow=p);
M = matrix(rep(0,len=p^2),nrow=p);
N = diag(p);
if (p==1){Sigma[1,1]<-1/(2*a[1])}
if (p>1){for (i in 1:p){M <- M + a[p+1-i] * N;  N <- (-A) %*% N};
         M <- M+N;
         # M such computed corresponds to the matrix a(-A)
         Sigma[,1] <- solve(M,ep);
         for (j in 2:p){Sigma[,j] <- (-A) %*% Sigma[,j-1]}
         # Sigma should be symmetric; it will now be symmetrised to account
         # for numerical imprecision
         for (i in 1:(p-1))
           {for (j in (i+1):p)
             {Sigma[i,j] <- 0.5 * (Sigma[i,j] + Sigma[j,i]);
              Sigma[j,i] <- Sigma[i,j]}};
         Sigma <- Re(Sigma)};
u<-c(0:(tau/h))
gamma=matrix(1:((tau/h)+1),nrow=1,byrow=TRUE)
for(i in 1:((tau/h)+1))
    {gamma[i]<- t(b) %*% expm(A*u[i]*h) %*% Sigma %*% b}
kappa<-gamma/gamma[1]
gamma=gamma[1]*kappa
t<-c(0:(tau/h))*h
plot(t,kappa,main="CARMA␣model␣autocorrelation␣function␣at␣time-lag␣t",type="l")
Sys.sleep(5)
```

The programs are executed by copying and pasting the appropriate parts into R. The graphs presented in Figures 7.3–7.6 were calculated with these programs, however with $\tau = 25$ rather than $\tau = 100$.

7.5 Exercises

Exercise 7.1. Consider the autocovariance function

$$\gamma(h) = \frac{\sqrt{1.16}}{1.856}\cos(|h| - \arctan(0.4))\,\mathrm{e}^{-0.4|h|}, \qquad h \in \mathbb{R},$$

of the CAR(2) process considered in Examples 7.4 and 7.7. Show by direct calculation that γ is twice continuously differentiable on $\mathbb{R}$, but not three times differentiable at 0.

Exercise 7.2. Let $c(z)$ and $C \in \mathbb{C}^{p\times p}$ be given as in Proposition 7.15. Show by induction on p that $\det(zI_p - C) = c(z)$ by calculating the determinant along its first column.

Exercise 7.3. For the CARMA(2, 1) process of Example 7.25 determine the kernel g and the increments of the orthogonal increment process η in the representation (7.11) of Y. Express the predictor $P_{\mathcal{N}_t}Y_{t+h}$ in terms of the spectral process Z of Y. Verify that the mean-square error σ_h^2 of $P_{\mathcal{N}_t}Y_{t+h}$, as given by (7.34), is the same as the value obtained in Example 7.25.

Exercise 7.4. If g is the CARMA kernel in the representation (7.11), show that $g(0) = 1$ if $q = p - 1$ and that $g(0) = 0$ if $q < p - 1$. In particular, the kernel is continuous when $q < p - 1$.
Hint: Rewrite the sum of residues in (7.12) with $t = 0$ as a contour integral over the circle $R\mathrm{e}^{\mathrm{i}t}$, $t \in [0, 2\pi]$ and let $R \to \infty$.

Exercise 7.5. Show that the first component X^0 of $\mathbf{X}$, as defined by (7.37), is a second-order CAR(p) process.

8 Infinitely divisible distributions

Abstract: In this chapter, we introduce infinitely divisible distributions and explore some of their properties. Infinitely divisible distributions are intimately related to Lévy processes, which are continuous-time analogues of random walks and as such constitute an important class of drivers for stochastic differential equations and integrators for stochastic integrals. In fact, it can be shown that there is a one-to-one correspondence between infinitely divisible distributions and Lévy processes (in law), in the sense that the marginal distributions of a Lévy process are infinitely divisible, and that, conversely, to every infinitely divisible distribution there exists a Lévy process, unique in law, whose marginal distribution at time one coincides with the given infinitely divisible distribution. In this chapter, we concentrate on infinitely divisible distributions in their own right and derive some of their properties. We characterize them as limit distributions of compound Poisson distributions and introduce the distinguished logarithm and the distinguished n^{th} root, which will turn out useful when proving the Lévy–Khintchine formula in the next chapter. Our exposition of infinitely divisible distributions and Lévy processes in this and the next chapters is mostly inspired by Sato [79], but also by Itô [51], Kyprianou [61] and Applebaum [2].

8.1 Infinite divisibility in Lévy processes

In the following, by a *distribution on* $\mathbb{R}^d$ we will always mean a probability measure on $(\mathbb{R}^d, \mathcal{B}_d)$, where $\mathcal{B}_d$ denotes the Borel-σ-algebra of subsets of $\mathbb{R}^d$. The *distribution* or *law* of a random vector X will be denoted by $\mathcal{L}(X)$ or P_X. The *convolution* of two distributions μ_1 and μ_2 on $\mathbb{R}^d$ will be denoted by $\mu_1 * \mu_2$.

Definition 8.1. A distribution μ on $\mathbb{R}^d$ is called *infinitely divisible*, abbreviated as *i. d.*, if for every $n \in \mathbb{N}$ there exists a distribution μ_n on $\mathbb{R}^d$, such that

$$\mu_n^{*n} := \underbrace{\mu_n * \cdots * \mu_n}_{n \text{ times}} = \mu.$$

In other words, the distribution $\mathcal{L}(X)$ of a d-dimensional random vector X is i. d., if for every $n \in \mathbb{N}$ there exist independent and identically distributed vectors $X_{n,1}, \ldots, X_{n,n} : \Omega_n \to \mathbb{R}^d$ on a probability space $(\Omega_n, \mathcal{F}_n, P_n)$, such that

$$\mathcal{L}(X_{n,1} + \cdots + X_{n,n}) = \mathcal{L}(X).$$

By a slight abuse of notation, we shall occasionally also say that the random vector X is infinitely divisible. The set of all infinitely divisible distributions on $\mathbb{R}^d$ will be denoted by $\mathrm{ID}(\mathbb{R}^d)$.

https://doi.org/10.1515/9783111325033-008

The importance of infinitely divisible distributions lies in their deep connection with Lévy processes (in law), which will be studied in detail in Chapter 10. For the moment, we confine ourselves to showing that the marginal distributions of Lévy processes are infinitely divisible. A stochastic process $L = (L_t)_{t \geq 0}$ on a probability space $(\Omega, \mathcal{F}, P)$ with $L_t(\omega) \in \mathbb{R}^d$ for each $\omega \in \Omega$ is called a *(one-sided) Lévy process* (in $\mathbb{R}^d$) or a *(one-sided) $\mathbb{R}^d$-valued Lévy process* if it has the following properties.

(a) The process L has *independent increments*, i. e.,

$$L_{t_1} - L_{t_0}, L_{t_2} - L_{t_1}, \ldots, L_{t_n} - L_{t_{n-1}}$$

are independent for all $n \in \mathbb{N}$ and $0 \leq t_0 < t_1 < \cdots < t_n$.

(b) The process L has *stationary increments*, i. e.,

$$L_{t+h} - L_t \stackrel{d}{=} L_h - L_0 \quad \forall\, t, h \geq 0.$$

(c) $L_0 = 0$ almost surely.

(d) The process has almost surely càdlàg paths on $[0, \infty)$, i. e., the sample paths are almost surely right-continuous on $[0, \infty)$ and have finite left-limits on $(0, \infty)$.

Brownian motion as in Definition 1.29 and the Poisson process as in Definition 1.30 are one-sided Lévy processes in $\mathbb{R}$ when restricted to $[0, \infty)$. For general one-sided Lévy processes, we have the connection with infinitely divisible distributions shown in the following proposition.

Proposition 8.2. *Let $L = (L_t)_{t \geq 0}$ be a one-sided Lévy process in $\mathbb{R}^d$. Then the distribution $\mathcal{L}(L_t)$ is infinitely divisible for all $t \geq 0$.*

Proof. For $t \neq 0$ and $n \in \mathbb{N}$ let

$$X_{n,1} := L_{t/n} - L_0, \quad X_{n,2} := L_{2t/n} - L_{t/n}, \quad \ldots, \quad X_{n,n} := L_{nt/n} - L_{(n-1)t/n}.$$

Then $X_{n,1}, \ldots, X_{n,n}$ are independent and identically distributed and $L_t = X_{n,1} + \cdots + X_{n,n}$, so that $\mathcal{L}(L_t)$ is infinitely divisible for $t \neq 0$. That $\mathcal{L}(L_0) = \delta_0$ is infinitely divisible is clear. □

Since one-dimensional Brownian motion and the Poisson process are Lévy processes, it follows immediately that both the (one-dimensional) normal distribution $N(a, \sigma^2)$ as well as the Poisson distribution with parameter (i. e., mean) $\lambda \in (0, \infty)$ are infinitely divisible.

8.2 First examples and elementary properties of infinitely divisible distributions

In this section, we give some elementary examples of infinitely divisible distributions and some properties of infinitely divisible distributions in general. We have already seen that the Poisson distribution is infinitely divisible and that the one-dimensional Gaussian distribution is infinitely divisible. The latter generalizes easily to higher dimensions.

Example 8.3. The Gaussian distribution $N(\gamma, A)$, with $\gamma \in \mathbb{R}^d$ and A a non-negative definite symmetric matrix in $\mathbb{R}^{d\times d}$, is infinitely divisible. This follows from the simple observation that the n-fold convolution of $N(n^{-1}\gamma, n^{-1}A)$ with itself is $N(\gamma, A)$.

Infinitely divisibility of random vectors is often more easily checked using characteristic functions. For this we need some notation.

Notation 8.4. The Euclidean inner product of $x, z \in \mathbb{R}^d$ will be denoted by $\langle x, z\rangle = x^T z$. We write $|x| = \sqrt{\langle x, x\rangle}$. The *Fourier transform* or *characteristic function* (abbreviated as *c.f.*) of a finite measure σ on $(\mathbb{R}^d, \mathcal{B}_d)$ is defined as

$$\widehat{\sigma} : \mathbb{R}^d \to \mathbb{C}, \quad z \mapsto \widehat{\sigma}(z) := \int_{\mathbb{R}^d} \mathrm{e}^{\mathrm{i}\langle z,x\rangle}\, \sigma(\mathrm{d}x) = \int_{\mathbb{R}^d} \mathrm{e}^{\mathrm{i}z^T x}\, \sigma(\mathrm{d}x).$$

In particular, the *characteristic function* of a random vector $X : \Omega \to \mathbb{R}^d$ with distribution $\mathcal{L}(X)$, is defined as

$$\phi(z) := \widehat{\mathcal{L}(X)}(z) = \mathbb{E}\,\mathrm{e}^{\mathrm{i}\langle z,X\rangle} = \mathbb{E}\,\mathrm{e}^{\mathrm{i}z^T X}, \quad z \in \mathbb{R}^d.$$

Since the characteristic function of a convolution is the product of the corresponding characteristic functions, we immediately have the following result.

Lemma 8.5. *A distribution μ on $\mathbb{R}^d$ is infinitely divisible if and only if for every $n \in \mathbb{N}$ there exists a distribution μ_n on $\mathbb{R}^d$ such that*

$$\widehat{\mu}(z) = (\widehat{\mu_n}(z))^n \quad \forall\, z \in \mathbb{R}^d.$$

Our next example is the compound Poisson distribution, defined as follows.

Definition 8.6 (Compound Poisson distribution). Let M be a Poisson-distributed random variable with mean $a > 0$, and $(Y_i)_{i\in\mathbb{N}}$ an i. i. d. sequence of $\mathbb{R}^d$-valued random variables, independent of M, all defined on the same probability space $(\Omega, \mathcal{F}, P)$, such that the distribution σ of Y_1 satisfies $\sigma(\{0\}) = 0$. Then the distribution of the $\mathbb{R}^d$-valued random variable $X = \sum_{i=1}^M Y_i$ is called a *compound Poisson distribution (with parameter $a > 0$ and jump size distribution σ)*. We shall refer to $\alpha = a\sigma(\mathbb{R}^d)$ as the *jump rate* of the distribution.

Obviously, every compound Poisson distribution in $\mathbb{R}$ with jump size distribution $\sigma = \delta_1$, the Dirac measure at 1, is a Poisson distribution. Compound Poisson distributions are infinitely divisible and they can be characterized as follows in terms of their characteristic function.

Theorem 8.7 (C. f. and i. d. of compound Poisson distribution). *Let $\mu = \mathcal{L}(X)$ be a distribution on $\mathbb{R}^d$ with characteristic function $\phi_X(z) = \widehat{\mu}(z) = \mathbb{E}e^{i\langle z,X\rangle}$, $z \in \mathbb{R}^d$. Suppose that $a > 0$ and σ is a distribution on $\mathbb{R}^d$ with $\sigma(\{0\}) = 0$. Then μ is a compound Poisson distribution with parameter a and jump size distribution σ if and only if*

$$\widehat{\mu}(z) = \exp(a(\widehat{\sigma}(z) - 1)) = \exp\Big(\int_{\mathbb{R}^d} (e^{i\langle z,x\rangle} - 1)(a\sigma)(dx)\Big), \quad z \in \mathbb{R}^d. \tag{8.1}$$

Moreover, every compound Poisson distribution is infinitely divisible.

Proof. Let $\mu = \mathcal{L}(X)$ be a compound Poisson distribution with parameter a and jump size distribution σ. Let M and Y_i be as in Definition 8.6 and define $S_n := \sum_{i=1}^n Y_i$ for $n \in \mathbb{N}_0$, so that $X = S_M$. Then for $z \in \mathbb{R}^d$,

$$\begin{aligned}
\mathbb{E}\, e^{i\langle z,X\rangle} &= \mathbb{E}\,\mathbb{E}(e^{i\langle z,X\rangle}\,|M) \\
&= \sum_{n=0}^{\infty} \underbrace{P(M = n)}_{M\sim\mathrm{Pois}(a)} \underbrace{\mathbb{E}(e^{i\langle z,X\rangle}\,|M = n)}_{=\mathbb{E}\, e^{i\langle z,S_n\rangle}} \\
&= \sum_{n=0}^{\infty} e^{-a}\frac{a^n}{n!}(\widehat{\sigma}(z))^n = \exp(a(\widehat{\sigma}(z) - 1)).
\end{aligned}$$

This shows that the characteristic function of a compound Poisson distribution with parameter a and jump size distribution σ is given by (8.1). Conversely, if μ is the distribution with characteristic function specified by (8.1), then it is necessarily, by the same argument, the characteristic function of the compound Poisson distribution with parameter a and jump size distribution σ.

Now let μ be the compound Poisson distribution with parameter a and jump size distribution σ. For each $n \in \mathbb{N}$, let μ_n be the compound Poisson distribution with parameter a/n and jump size distribution σ. Then $\widehat{\mu}(z) = (\widehat{\mu_n}(z))^n$ for $z \in \mathbb{R}^d$ by (8.1), showing that μ is infinitely divisible. □

We shall see later in Example 9.19, as a consequence of a more general result in Proposition 9.18, that the parameter a and the jump size distribution σ of a compound Poisson distribution μ are uniquely determined by μ.

The first property of infinitely divisible distributions that we shall establish is that the convolution of two infinitely divisible distributions is again infinitely divisible.

Lemma 8.8 (Convolution of i. d. distributions). *Suppose that $\mu_1, \mu_2 \in \mathrm{ID}(\mathbb{R}^d)$. Then $\mu_1 * \mu_2 \in \mathrm{ID}(\mathbb{R}^d)$, i. e., the convolution of two infinitely divisible distributions is again infinitely* *divisible. By induction, the convolution of a finite number of infinitely divisible distributions is also infinitely divisible.*

Proof. It follows, since $\mu_i \in \mathrm{ID}(\mathbb{R}^d)$, $i \in \{1,2\}$, that for every $n \in \mathbb{N}$ there exist distributions $\mu_{i,n}$ on $\mathbb{R}^d$ such that $\mu_i = \mu_{i,n}^{*n}$. Then $\mu_{1,n} * \mu_{2,n}$ is a distribution on $\mathbb{R}^d$ and, since convolution is commutative,

$$(\mu_{1,n} * \mu_{2,n})^{*n} = \mu_{1,n}^{*n} * \mu_{2,n}^{*n} = \mu_1 * \mu_2.$$

It follows that $\mu_1 * \mu_2 \in \mathrm{ID}(\mathbb{R}^d)$. □

Corollary 8.9. *Suppose that $\gamma \in \mathbb{R}^d$, that A is a real symmetric non-negative definite matrix in $\mathbb{R}^{d\times d}$ and that ν is a finite measure on $(\mathbb{R}^d, \mathcal{B}_d)$ such that $\nu(\{0\}) = 0$. Then $\phi : \mathbb{R}^d \to \mathbb{C}$, defined by*

$$z \mapsto \phi(z) := \exp\left(\mathrm{i}\langle \gamma, z\rangle - \frac{1}{2}\langle z, Az\rangle + \int_{\mathbb{R}^d} (\mathrm{e}^{\mathrm{i}\langle z,x\rangle} - 1)\,\nu(\mathrm{d}x) \right),$$

is the characteristic function of the infinitely divisible distribution $N(\gamma, A)$, if $\nu(\mathbb{R}^d) = 0$ and, if $\nu(\mathbb{R}^d) > 0$, the convolution of the $N(\gamma, A)$-distribution and the compound Poisson distribution with jump rate $a\sigma(\mathbb{R}^d) = \nu(\mathbb{R}^d)$ and jump size distribution $\sigma = \nu/\nu(\mathbb{R}^d)$.

Proof. If $\nu(\mathbb{R}^d) > 0$ it is clear from Theorem 8.7 that ϕ is the characteristic function of the convolution of an $N(\gamma, A)$-distribution and a compound Poisson distribution with jump rate $\nu(\mathbb{R}^d)$ and jump size distribution $(\nu(\mathbb{R}^d))^{-1}\nu$. Its infinite divisibility follows from Example 8.3, Theorem 8.7 and Lemma 8.8. The conclusion in the case when $\nu(\mathbb{R}^d) = 0$ is clear. □

Corollary 8.9 gives already a wide range of infinitely divisible distributions, and we shall see in the next section in Corollary 8.30 that every infinitely divisible distribution μ can be obtained as the limit of a sequence of compound Poisson distributions. One such sequence we shall specify in terms of μ. Before we do that, we shall determine some conditions that an infinitely divisible distribution must satisfy.

Proposition 8.10 (Unbounded support of i. d. distribution). *A distribution μ on $\mathbb{R}^d$ with bounded support is infinitely divisible if and only if it is a Dirac measure.*

Proof. That a Dirac measure is infinitely divisible is immediate from the definition or alternatively from Example 8.3 since $\delta_\gamma = N(\gamma, 0)$, where 0 is the zero-matrix in $\mathbb{R}^{d\times d}$. Conversely, let μ be an infinitely divisible distribution with bounded support, and let X be a random vector with distribution $\mathcal{L}(X) = \mu$. Then there is $C \in [0, \infty)$ such that $|X| \le C$ almost surely. For each $n \in \mathbb{N}$, let $X_{n,1}, \ldots, X_{n,n}$ be independent and identically distributed random vectors such that $\mathcal{L}(X) = \mathcal{L}(X_{n,1} + \cdots + X_{n,n})$. Write $X = (X^1, \ldots, X^d)^T$ and $X_{n,i} = (X_{n,i}^1, \ldots, X_{n,i}^d)^T$. Fix $j \in \{1, \ldots, d\}$. Then it is easy to see that $|X_{n,1}^j| \le C/n$ a. s. In particular, X^j and $X_{n,i}^j$ have finite variance and

$$\mathrm{Var}(X^j) = \mathrm{Var}(X^j_{n,1} + \cdots + X^j_{n,n}) = n\ \mathrm{Var}(X^j_{n,1}) \le n\mathbb{E}(X^j_{n,1})^2 \le nC^2/n^2.$$

The latter tends to 0 as $n \to \infty$, so that $\mathrm{Var}(X^j) = 0$, hence $\mu = \mathcal{L}(X)$ must be a Dirac measure. □

Proposition 8.11 (Zero-free c. f. of i. d. distribution). *The characteristic function of an infinitely divisible distribution has no zeros, i. e.,*

$$\widehat{\mu}(z) \neq 0 \quad \forall\, z \in \mathbb{R}^d, \forall\, \mu \in \mathrm{ID}(\mathbb{R}^d).$$

Proof. Suppose that $\mu \in \mathrm{ID}(\mathbb{R}^d)$, X is a random vector with distribution $\mathcal{L}(X) = \mu$, and for each $n \in \mathbb{N}$, μ_n is a distribution on $\mathbb{R}^d$ such that $\mu_n^{*n} = \mu$. Let X_n be a random vector such that $\mathcal{L}(X_n) = \mu_n$. Now let Y be an independent copy of X and Y_n an independent copy of X_n, and define $Z := X - Y$ and $Z_n := X_n - Y_n$. Then $(\mathcal{L}(Z_n))^{*n} = \mathcal{L}(Z)$ for every n, so that $\mathcal{L}(Z) \in \mathrm{ID}(\mathbb{R}^d)$. The characteristic function of Z is given by

$$\phi_Z(z) = \phi_X(z)\phi_{-X}(z) = \mathbb{E}\, e^{i\langle z,X\rangle}\, \overline{\mathbb{E}\, e^{i\langle z,X\rangle}} = \left|\mathbb{E}\, e^{i\langle z,X\rangle}\right|^2 = |\phi_X(z)|^2 \ge 0,$$

and similarly

$$\phi_{Z_n}(z) = |\phi_{X_n}(z)|^2 \ge 0, \quad z \in \mathbb{R}^d.$$

Since $(\mathcal{L}(Z_n))^{*n} = \mathcal{L}(Z)$, for every $n \in \mathbb{N}$, we have $\phi_Z(z) = [\phi_{Z_n}(z)]^n$ and hence, since $\phi_{Z_n}(z)$ and $\phi_Z(z)$ are non-negative,

$$\phi_{Z_n}(z) = [\phi_Z(z)]^{1/n} \quad \forall\, z \in \mathbb{R}^d, n \in \mathbb{N}.$$

Since $\phi_Z(z) \in [0,1]$, the limit

$$\psi(z) := \lim_{n\to\infty} [\phi_Z(z)]^{1/n} = \lim_{n\to\infty} \phi_{Z_n}(z)$$

exists for each $z \in \mathbb{R}^d$ and is given by

$$\lim_{n\to\infty} \phi_{Z_n}(z) = \psi(z) = \begin{cases} 1, & \phi_Z(z) \neq 0, \\ 0, & \phi_Z(z) = 0. \end{cases}$$

But since $\phi_Z(0) = 1$ and ϕ_Z is continuous, we must have $\phi_Z(z) \neq 0$ for z in a neighborhood of 0 and hence $\psi(z) = 1$ for z in a neighborhood of 0, so that ψ is continuous in a neighborhood of 0. It then follows from Lévy's continuity theorem that ψ is the characteristic function of a random vector, W say. As a characteristic function, ψ must be continuous on $\mathbb{R}^d$, so we conclude that $\psi(z) = 1$ for all $z \in \mathbb{R}^d$ and hence that $|\phi_X(z)|^2 = \phi_Z(z) \neq 0$ for all $z \in \mathbb{R}^d$. Hence $\phi_X = \widehat{\mu}$ has no zeros on $\mathbb{R}^d$. □

Example 8.12. (a) Suppose that $a < b$ and that X is uniformly distributed on $[a, b]$. Its characteristic function ϕ_X at $z \neq 0$ is then given by

$$\phi_X(z) = \frac{1}{b-a} \int_a^b e^{izx}\, dx = \frac{1}{b-a} \frac{1}{iz} e^{iza}(e^{iz(b-a)} - 1),$$

which has zeros on $\mathbb{R}$, e. g., $\phi_X(2\pi/(b-a)) = 0$. Hence $\mathcal{L}(X)$ is not infinitely divisible. Alternatively, this can be shown from Proposition 8.10.

(b) Suppose that $a, b \in \mathbb{R}^d$ and $a \neq b$. Consider the two-point distribution $\mu = p\delta_a + (1-p)\delta_b$ with $p \in (0, 1)$. Then

$$\widehat{\mu}(z) = p\, e^{i\langle z,a\rangle} + (1-p)\, e^{i\langle z,b\rangle} = p\, e^{i\langle z,a\rangle}\left(1 + \frac{1-p}{p} e^{i\langle z,b-a\rangle}\right),$$

and clearly $\widehat{\mu}$ has zeros if and only if $p = 1/2$. Hence, by Proposition 8.11, $\mu \notin \mathrm{ID}(\mathbb{R}^d)$ if $p = 1/2$. From Proposition 8.10 we see that if $p \neq 1/2$, $p\delta_a + (1-p)\delta_b \notin \mathrm{ID}(\mathbb{R}^d)$ even though $\widehat{\mu}$ has no zeros in that case. Hence, the condition of Proposition 8.11 is necessary but not sufficient for a distribution to be infinitely divisible.

8.3 The distinguished logarithm and n^{th} root

The aim of this section is to show that the nth convolution root of an infinitely divisible distribution is unique and that $\mathrm{ID}(\mathbb{R}^d)$ is closed under weak convergence. In order to do that, we first have to show the existence of the "distinguished logarithm" and the "distinguished n^{th} root". This is more complicated in $\mathbb{R}^d$ than in $\mathbb{R}$, and we will need the concept of winding numbers for the proof. We start with Lemma 8.14, which can be seen as a "lifting theorem" and has further generalizations in the notion of coverings in algebraic topology. We recall the principal branch of the complex logarithm.

Definition 8.13 (Principal branch of the logarithm). We denote the principal branch of the complex logarithm by log, and recall that it is the unique holomorphic function $\log : \mathbb{C} \setminus (-\infty, 0] \to \mathbb{C}$ such that $\exp(\log z) = z$ for all $z \in \mathbb{C} \setminus (-\infty, 0]$ and such that log maps $\mathbb{C} \setminus (-\infty, 0]$ bijectively onto $\{z \in \mathbb{C} : |\Im z| < \pi\}$. For $|z| < 1$ we have

$$\log(1+z) = \sum_{j=1}^{\infty} (-1)^{j-1} j^{-1} z^j \tag{8.2}$$

by the logarithmic series.

Lemma 8.14 (Lifting lemma for the exponential). *Let $\gamma : [a, b] \to \mathbb{C}$ be a continuous function such that $\gamma(t) \neq 0$ for all $t \in [a, b]$. Then there exists a continuous function $\alpha : [a, b] \to \mathbb{C}$ such that $\exp(\alpha(t)) = \gamma(t)$ for all $t \in [a, b]$. This function α is unique up to addition of integer multiples of $2\pi i$, so that a continuous function $\beta : [a, b] \to \mathbb{C}$ satisfies*

$\exp(\beta(t)) = \gamma(t)$ *for all* $t \in [a,b]$ *if and only if there is* $k \in \mathbb{Z}$ *such that* $\beta(t) = \alpha(t) + 2k\pi\mathrm{i}$ *for all* $t \in [a,b]$.

Proof. The uniqueness assertion is clear, since $\exp(\alpha(t) + r(t)) = \exp(\alpha(t))$ for some function $r : [a,b] \to \mathbb{C}$ if and only if $r(t) \in 2\pi\mathrm{i}\mathbb{Z}$ for all t, and since $t \mapsto \beta(t) = \alpha(t) + r(t)$ is continuous if and only if r is continuous, and this can only happen if r is constant.

To show the existence, assume without loss of generality that $a \neq b$, and let $\rho := \inf\{|\gamma(t)| : t \in [a,b]\}$. Then $\rho > 0$ since γ is continuous on $[a,b]$. Further, γ is uniformly continuous on $[a,b]$, so that there exists $\delta > 0$ such that $|\gamma(t) - \gamma(t')| < \varepsilon := \rho/2$ whenever $t, t' \in [a,b]$ with $|t - t'| < \delta$. Now choose a finite sequence $(t_k)_{k=0,\ldots,m}$ such that $a = t_0 < t_1 < \cdots < t_m = b$ and $|t_k - t_{k-1}| < \delta$ for all $k \in \{1,\ldots,m\}$. Choose $c \in \mathbb{C}$ such that $\exp(c) = \gamma(a)$ and define $\alpha(a) := c$. Now assume that the function α is already constructed on $[t_0, t_k]$ for some $k \in \{0,\ldots,m-1\}$, i. e. α is continuous on $[t_0, t_k]$ and $\exp(\alpha(t)) = \gamma(t)$ for all $t \in [t_0, t_k]$ (for $k = 0$ this is satisfied). Then we extend α to $[t_0, t_{k+1}]$ by defining

$$\alpha(t) := \alpha(t_k) + \log \frac{\gamma(t)}{\gamma(t_k)}, \qquad t \in (t_k, t_{k+1}].$$

Since

$$\left| \frac{\gamma(t)}{\gamma(t_k)} - 1 \right| = \left| \frac{\gamma(t) - \gamma(t_k)}{\gamma(t_k)} \right| \leq \frac{\varepsilon}{\rho} = \frac{1}{2} \quad \forall\, t \in [t_k, t_{k+1}],$$

$t \mapsto \log \frac{\gamma(t)}{\gamma(t_k)}$ is defined and continuous on $[t_k, t_{k+1}]$. Further, since $\log \frac{\gamma(t_k)}{\gamma(t_k)} = \log 1 = 0$, it follows that α is continuous on $[0, t_{k+1}]$. Finally, for $t \in (t_k, t_{k+1}]$, we have

$$\exp(\alpha(t)) = \exp(\alpha(t_k)) \exp\left(\log \frac{\gamma(t)}{\gamma(t_k)}\right) = \gamma(t_k) \frac{\gamma(t)}{\gamma(t_k)} = \gamma(t),$$

completing the induction step for the construction of α. □

Definition 8.15. A *complex curve* is a continuous function $\gamma : [a,b] \to \mathbb{C}$ on some compact interval $[a,b] \subset \mathbb{R}$. The curve is *closed* if $\gamma(a) = \gamma(b)$.

Corollary 8.16. *Let* $\gamma : [a,b] \to \mathbb{C}$ *be a closed complex curve such that* $\gamma(t) \neq 0$ *for all* $t \in [a,b]$. *Let* $\alpha : [a,b] \to \mathbb{C}$ *and* $\beta : [a,b] \to \mathbb{C}$ *be two complex curves such that* $\exp(\alpha(t)) = \exp(\beta(t)) = \gamma(t)$ *for all* $t \in [a,b]$ *(such curves exist by Lemma* 8.14*). Then there exists an integer* $n(\gamma, 0)$*, depending only on* γ *(and not on* α *or* β*), such that*

$$\alpha(b) - \alpha(a) = \beta(b) - \beta(a) = 2\pi\mathrm{i}\, n(\gamma, 0).$$

Proof. That $\alpha(b) - \alpha(a) = \beta(b) - \beta(a)$ is clear, since $\alpha - \beta$ is constant by Lemma 8.14, and $\alpha(b) - \alpha(a) \in 2\pi\mathrm{i}\mathbb{Z}$ since $\exp(\alpha(b)) = \gamma(b) = \gamma(a) = \exp(\alpha(a))$. □

Definition 8.17. For a closed complex curve $\gamma : [a,b] \to \mathbb{C}$ such that $\gamma(t) \neq 0$ for all $t \in [a,b]$, the unique integer $n(\gamma, 0)$ appearing in Corollary 8.16 is called the *winding number of γ around 0*, or alternatively the *index of γ around 0.*

Remark 8.18. If the closed curve γ in Definition 8.17 is additionally C^1, then it can be shown that $n(\gamma, 0) = \frac{1}{2\pi i} \int_\gamma \frac{1}{z}\,dz$, cf., Exercise 8.1. In complex analysis, the winding number is often defined by this formula.

Example 8.19. Suppose that $a < b \leq c < d$ are real numbers such that $d - c = b - a$ and $\gamma : [a,d] \to \mathbb{C} \setminus \{0\}$ is a closed complex curve satisfying

$$\gamma(t) = \begin{cases} \gamma(b), & t \in [b,c], \\ \gamma(b + c - t), & t \in [c,d]. \end{cases}$$

Then if you set out from $\gamma(a)$ at time a and walk along the curve in $\mathbb{C}$ in such a way that your location is $\gamma(t)$ at time t, you reach $\gamma(b)$ at time b, remain there until time c and then retrace your steps back to $\gamma(a)$, which you reach at time d. It follows that if $\alpha : [a,b] \to \mathbb{C}$ is a curve such that $\exp(\alpha(t)) = \gamma(t)$ for all $t \in [a,b]$, then $\tilde{\alpha} : [a,d] \to \mathbb{C}$, defined by

$$\tilde{\alpha}(t) := \begin{cases} \alpha(t), & t \in [a,b], \\ \alpha(b), & t \in [b,c], \\ \alpha(b + c - t), & t \in [c,d], \end{cases}$$

is a complex curve such that $\exp(\tilde{\alpha}(t)) = \gamma(t)$ for all $t \in [a,d]$. Since $\tilde{\alpha}(a) = \tilde{\alpha}(d)$, we conclude $n(\gamma, 0) = 0$.

Lemma 8.20. *Let $\gamma_1, \gamma_2 : [a,b] \to \mathbb{C}$ be two complex curves such that $\gamma_1(t) \neq 0$ for all $t \in [a,b]$ and $|\gamma_2(t) - \gamma_1(t)| < |\gamma_1(t)|$ for all $t \in [a,b]$. Then the following are true:*
(a) *$\gamma_2(t) \neq 0$ and $|\frac{\gamma_2(t)}{\gamma_1(t)} - 1| < 1$ for all $t \in [a,b]$, so that $\log \frac{\gamma_2(t)}{\gamma_1(t)}$ is well-defined.*
(b) *If $\alpha_1 : [a,b] \to \mathbb{C}$ is a complex curve with $\exp(\alpha_1(t)) = \gamma_1(t)$ for all $t \in [a,b]$, then $\alpha_2 : [a,b] \to \mathbb{C}$, defined by*

$$\alpha_2(t) := \alpha_1(t) + \log \frac{\gamma_2(t)}{\gamma_1(t)},$$

is a complex curve with $\exp(\alpha_2(t)) = \gamma_2(t)$ for all $t \in [a,b]$.
(c) *Suppose in addition that γ_1 and γ_2 are closed. Then $n(\gamma_1, 0) = n(\gamma_2, 0)$.*

Proof. Define $\rho(t) := \frac{\gamma_2(t)}{\gamma_1(t)}$ for $t \in [a,b]$. Then ρ is continuous, and

$$|\rho(t) - 1| = \left| \frac{\gamma_2(t) - \gamma_1(t)}{\gamma_1(t)} \right| < 1 \quad \forall\, t \in [a,b].$$

It follows that $\rho(t) \neq 0$ and hence that $\gamma_2(t) \neq 0$ for all $t \in [a,b]$, and that $\rho(t)$ is in the domain of definition of the principal branch of the logarithm. In particular, $t \mapsto \log \rho(t)$

is continuous. Let $\alpha_1 : [a, b] \to \mathbb{C}$ be a complex curve such that $\exp(\alpha_1(t)) = \gamma_1(t)$, and let $\alpha_2(t) := \alpha_1(t) + \log \rho(t)$. Then α_2 is continuous and

$$\exp(\alpha_2(t)) = \exp(\alpha_1(t)) \exp(\log \rho(t)) = \gamma_1(t)\rho(t) = \gamma_2(t).$$

If, in addition, γ_1 and γ_2 are closed, then $\rho(b) = \frac{\gamma_2(b)}{\gamma_1(b)} = \frac{\gamma_2(a)}{\gamma_1(a)} = \rho(a)$ and hence

$$\begin{aligned} 2\pi i\, n(\gamma_2, 0) &= \alpha_2(b) - \alpha_2(a) = \alpha_1(b) - \alpha_1(a) + \log \rho(b) - \log \rho(a) \\ &= \alpha_1(b) - \alpha_1(a) = 2\pi i\, n(\gamma_1, 0), \end{aligned}$$

finishing the proof. □

Theorem 8.21 (Distinguished logarithm). *Let $\varphi : \mathbb{R}^d \to \mathbb{C}$ be a continuous function such that $\varphi(0) = 1$ and $\varphi(z) \neq 0$ for all $z \in \mathbb{R}^d$. Then there exists a unique continuous function $\psi : \mathbb{R}^d \to \mathbb{C}$ such that $\psi(0) = 0$ and $\exp(\psi(z)) = \varphi(z)$ for all $z \in \mathbb{R}^d$.*

Proof. To show uniqueness, suppose that ψ_1 and ψ_2 are two functions with the desired properties. Since $\exp(\psi_1(z)) = \varphi(z) = \exp(\psi_2(z))$ for all $z \in \mathbb{R}^d$, we conclude that $\psi_1(z) - \psi_2(z) \in 2\pi i\mathbb{Z}$ for all z, and since ψ_1 and ψ_2 are continuous, this shows that $\psi_1 - \psi_2$ must be constant. Since $\psi_1(0) = \psi_2(0) = 0$, we conclude that $\psi_1 = \psi_2$.

To show the existence, define, for each $z \in \mathbb{R}^d$, the function $\beta_z(t) := tz$, $t \in [0, 1]$, which describes a straight line from 0 to z. Define $\gamma_z := \varphi \circ \beta_z : [0, 1] \to \mathbb{C} \setminus \{0\}$. Then γ_z is a complex curve not taking the value 0, and it follows from Lemma 8.14 that there is a unique complex curve $\alpha_z : [0, 1] \to \mathbb{C}$ such that $\exp(\alpha_z(t)) = \gamma_z(t)$ for all $t \in [0, 1]$ and $\alpha_z(0) = 0$. Define

$$\psi(z) := \alpha_z(1), \quad z \in \mathbb{R}^d.$$

Then $\exp(\psi(z)) = \exp(\alpha_z(1)) = \gamma_z(1) = \varphi(z)$, $\psi(0) = \alpha_0(1) = \alpha_0(0) = 0$ since γ_0 and hence α_0 are constant, and we only need to show that ψ is continuous. Fix $z_0 \in \mathbb{R}^d$. For any $z \in \mathbb{R}^d$, define the closed curve $\beta_{z_0,z} : [0, 3] \to \mathbb{R}^d$ by

$$\beta_{z_0,z}(t) := \begin{cases} tz_0 = \beta_{z_0}(t), & t \in [0, 1], \\ (t-1)z + (2-t)z_0, & t \in [1, 2], \\ (3-t)z = \beta_z(3-t), & t \in [2, 3]. \end{cases}$$

Then if you set out from $\beta_{z,z_0}(0)$ at time 0 and walk along the curve in such a way that your location at time t is $\beta_{z,z_0}(t)$, you will move first on a straight line from 0 to z_0, then on a straight line from z_0 to z, and finally return on a straight line from z to 0. Consequently

$$\gamma_{z_0,z} := \varphi \circ \beta_{z_0,z}$$

describes a closed complex curve such that $\gamma_{z_0,z}(t) \neq 0$ for all $t \in [0,3]$, and $\gamma_{z_0,z}(0) = \varphi(0) = 1$. Let $\alpha_{z_0,z} : [0,3] \to \mathbb{C}$ be the unique complex curve which, by Lemma 8.14, exists and satisfies $\alpha_{z_0,z}(0) = 0$ and $\exp(\alpha_{z_0,z}(t)) = \gamma_{z_0,z}(t)$ for all $t \in [0,3]$. Now define

$$\tilde{\alpha}_{z_0,z} : [0,1] \to \mathbb{C}, \quad \tilde{\alpha}_{z_0,z}(t) := \alpha_{z_0,z}(3-t) - 2\pi \mathrm{i}\, n(\gamma_{z_0,z}, 0).$$

Then

$$\exp(\tilde{\alpha}_{z_0,z}(t)) = \exp(\alpha_{z_0,z}(3-t)) = \gamma_{z_0,z}(3-t) = \varphi(\beta_z(t)) = \gamma_z(t) \quad \forall\, t \in [0,1]$$

and $\tilde{\alpha}_{z_0,z}(0) = \alpha_{z_0,z}(3) - 2\pi \mathrm{i}\, n(\gamma_{z_0,z}, 0) = \alpha_{z_0,z}(0) = 0$ by the definition of the winding number. By the uniqueness assertion in Lemma 8.14, we conclude that $\tilde{\alpha}_{z_0,z} = \alpha_z$ and hence that

$$\psi(z) = \tilde{\alpha}_{z_0,z}(1) = \alpha_{z_0,z}(2) - 2\pi \mathrm{i}\, n(\gamma_{z_0,z}, 0). \tag{8.3}$$

From Example 8.19, we have $n(\gamma_{z_0,z_0}, 0) = 0$. Since φ is uniformly continuous on $\{w \in \mathbb{C} : |w| \le |z_0| + 1\}$ and since $\inf\{|\varphi(w)| : |w| \le |z_0| + 1\} > 0$, for every $\varepsilon \in (0,1)$, there is $\delta(\varepsilon) > 0$ such that

$$|\gamma_{z_0,z}(t) - \gamma_{z_0,z_0}(t)| < \varepsilon |\gamma_{z_0,z_0}(t)| \quad \forall\, t \in [0,3], z \in \mathbb{C} : |z - z_0| < \delta(\varepsilon). \tag{8.4}$$

From Lemma 8.20(c), we conclude that $n(\gamma_{z_0,z}, 0) = n(\gamma_{z_0,z_0}, 0) = 0$ for all $z \in \mathbb{C} : |z - z_0| < \delta(\varepsilon)$, and since $\alpha_{z_0,z}(0) = \alpha_{z_0,z_0}(0) = 0$, the uniqueness assertion of Lemma 8.14 together with Lemma 8.20(b) shows that

$$\alpha_{z_0,z}(t) - \alpha_{z_0,z_0}(t) = \log \frac{\gamma_{z_0,z}(t)}{\gamma_{z_0,z_0}(t)} \quad \forall\, t \in [0,3], z \in \mathbb{C} : |z - z_0| < \delta(\varepsilon).$$

Equations (8.3) and (8.4) then show that

$$|\psi(z) - \psi(z_0)| \le \sup\{|\log(1+w)| : w \in \mathbb{C}, |w| \le \varepsilon\}.$$

Letting $\varepsilon \to 0$ gives continuity of ψ at z_0. □

Definition 8.22 (Distinguished logarithm). Let $\varphi : \mathbb{R}^d \to \mathbb{C}$ be a continuous function such that $\varphi(z) \neq 0$ for all $z \in \mathbb{R}^d$. Then the unique continuous function $\psi : \mathbb{R}^d \to \mathbb{C}$ satisfying $\psi(0) = 0$ and $\exp(\psi(z)) = \varphi(z)$ for all $z \in \mathbb{R}^d$, the existence of which was established in Theorem 8.21, is called the *distinguished logarithm* of φ and is denoted by

$$\operatorname{Log} \varphi(z) := \psi(z).$$

By construction, the distinguished and principal logarithms satisfy $\operatorname{Log} \varphi(z) = \log \varphi(z)$ in a neighborhood of 0, but for z further from the origin they may differ.

Corollary 8.23. *Let $\varphi, \varphi_n : \mathbb{R}^d \to \mathbb{C}$, $n \in \mathbb{N}$, be continuous functions such that $\varphi(0) = \varphi_n(0) = 1$ for all $n \in \mathbb{N}$, and such that $\varphi(z) \neq 0$ and $\varphi_n(z) \neq 0$ for all $n \in \mathbb{N}$ and $z \in \mathbb{R}^d$. Suppose also that φ_n converges locally uniformly, i. e., uniformly on every compact subset of $\mathbb{R}^d$, to φ as $n \to \infty$. Then $\operatorname{Log} \varphi_n$ converges locally uniformly to $\operatorname{Log} \varphi$.*

Proof. For $z \in \mathbb{R}^d$ define $\beta_z(t) := tz$, $t \in [0,1]$, and $\gamma_z := \varphi \circ \beta_z$. Let $\alpha_z : [0,1] \to \mathbb{C}$ be the unique complex curve such that $\alpha_z(0) = 0$ and $\exp(\alpha_z(t)) = \gamma_z(t)$ for all $t \in [0,1]$. Similarly, for each $n \in \mathbb{N}$, define $\gamma_z^{(n)} := \varphi_n \circ \beta_z$, and let $\alpha_z^{(n)} : [0,1] \to \mathbb{C}$ be the unique complex curve satisfying $\alpha_z^{(n)}(0) = 0$ and $\exp(\alpha_z^{(n)}(t)) = \gamma_z^{(n)}(t)$ for all $t \in [0,1]$ (which exists for each n by Lemma 8.14). By the construction of the distinguished logarithm in the proof of Theorem 8.21, we have

$$\operatorname{Log} \varphi(z) = \alpha_z(1) \quad \text{and} \quad \operatorname{Log} \varphi_n(z) = \alpha_z^{(n)}(1) \quad \forall\, z \in \mathbb{R}^d\ \forall\, n \in \mathbb{N}.$$

Define $K_T := [-T, T]^d \subset \mathbb{R}^d$ for $T > 0$. Since φ_n converges uniformly to φ on K_T as $n \to \infty$, and since $\inf\{|\varphi(z)| : z \in K_T\} > 0$, we have

$$\sup_{z \in K_T} \sup_{t \in [0,1]} \frac{|\gamma_z^{(n)}(t) - \gamma_z(t)|}{|\gamma_z(t)|} \to 0 \quad \text{as } n \to \infty.$$

Since $\alpha_z^{(n)}(0) = \alpha_z(0)$, it follows from Lemma 8.20(b) and the uniqueness assertion in Lemma 8.14 that

$$\operatorname{Log} \varphi_n(z) - \operatorname{Log} \varphi(z) = \alpha_z^{(n)}(1) - \alpha_z(1) = \log \frac{\gamma_z^{(n)}(1)}{\gamma_z(1)} = \log \frac{\varphi_n(z)}{\varphi(z)},$$

for $z \in K_T$ and sufficiently large n, and the latter converges to 0 as $n \to \infty$, uniformly on K_T. Since this is true for all $T > 0$, we obtain local uniform convergence of $\operatorname{Log} \varphi_n$ to $\operatorname{Log} \varphi$ as $n \to \infty$. □

Corollary 8.24 (Distinguished n^{th} root). *Let $\varphi : \mathbb{R}^d \to \mathbb{C}$ be a continuous function such that $\varphi(0) = 1$ and $\varphi(z) \neq 0$ for all $z \in \mathbb{R}^d$. Then for each $n \in \mathbb{N}$, there exists a unique continuous function $g_n : \mathbb{R}^d \to \mathbb{C}$ such that $g_n(0) = 1$ and $[g_n(z)]^n = \varphi(z)$ for all $z \in \mathbb{R}^d$. It is given by*

$$g_n(z) = \exp\left(\frac{1}{n} \operatorname{Log} \varphi(z)\right), \quad z \in \mathbb{R}^d. \tag{8.5}$$

Proof. If g_n is the function specified in (8.5), then it is clear that g_n is continuous, that

$$[g_n(z)]^n = \exp(\operatorname{Log} \varphi(z)) = \varphi(z) \quad \forall\, z \in \mathbb{R}^d, \quad \text{and that} \quad g_n(0) = \exp(0) = 1.$$

To show uniqueness, suppose that h_n is another continuous function with these properties. Since $[h_n(z)]^n = [g_n(z)]^n$, for each $z \in \mathbb{R}^d$ there exists a unique $k(z) \in \{0, 1, \ldots, n-1\}$ such that $h_n(z) = g_n(z) \exp(2\pi i k(z)/n)$. Since both h_n and g_n are continuous and since

$g_n(z) \neq 0$ for all z, also $k : \mathbb{R}^d \to \{0, 1, \dots, n-1\}$ must be continuous and hence constant. Since $h_n(0) = g_n(0) = 1$, we conclude that $k(z) = k(0) = 0$ for all $z \in \mathbb{R}^d$, so that $h_n = g_n$. □

Definition 8.25. Suppose that $\varphi : \mathbb{R}^d \to \mathbb{C}$ is a continuous function such that $\varphi(0) = 1$ and $\varphi(z) \neq 0$ for all $z \in \mathbb{R}^d$. Suppose also that $n \in \mathbb{N}$. The unique continuous function g_n appearing in Corollary 8.24 such that $g_n(0) = 1$ and $[g_n(z)]^n = \varphi(z)$ for all $z \in \mathbb{R}^d$ is called the *distinguished* n^{th} *root* of φ and is denoted by $\varphi^{1/n}$.

We can now apply our results to the characteristic function ϕ of a probability distribution. Each characteristic function ϕ is well known to be continuous and to satisfy $\phi(0) = 1$. In the following, we shall always use the term *distribution* to mean *probability distribution.*

Theorem 8.26 (Characterization of i. d. distributions). *A distribution μ on $\mathbb{R}^d$ is infinitely divisible if and only if the characteristic function $\phi = \widehat{\mu}$ of μ has no zeros on $\mathbb{R}^d$ and for each $n \in \mathbb{N}$ the distinguished n^{th} root of ϕ is the characteristic function of a probability distribution. Further, if μ is infinitely divisible, then for each $n \in \mathbb{N}$ the distribution μ_n such that $\mu_n^{*n} = \mu$ is unique, and the characteristic function of μ_n is the distinguished n^{th} root of the characteristic function of μ.*

Proof. If μ is such that $\phi(z) = \widehat{\mu}(z) \neq 0$ for all $z \in \mathbb{R}^d$, and if for each $n \in \mathbb{N}$, $z \mapsto \phi(z)^{1/n}$ defines the characteristic function of a distribution, μ_n say, then μ_n^{*n} has characteristic function $z \mapsto [\phi(z)^{1/n}]^n = \phi(z)$, so that $\mu_n^{*n} = \mu$, implying that μ is infinitely divisible.

Conversely, if μ is infinitely divisible, then $\phi(z) \neq 0$ for all $z \in \mathbb{R}^d$ by Proposition 8.11 and for each $n \in \mathbb{N}$, there exists a distribution μ_n such that $\mu_n^{*n} = \mu$. Denote by ϕ_n the characteristic function of μ_n. Then ϕ_n is continuous, $\phi_n(0) = 1$ and $[\phi_n(z)]^n = \phi(z)$ for all $z \in \mathbb{R}^d$. This shows that ϕ_n is the distinguished n^{th} root of ϕ, showing that μ_n is unique and that $\phi^{1/n}$ is a characteristic function. □

Definition 8.27 (n^{th} convolution root). Suppose that μ is a distribution on $\mathbb{R}^d$ and that $n \in \mathbb{N}$. A distribution μ_n on $\mathbb{R}^d$ is called an nth *convolution root* of μ, if $\mu_n^{*n} = \mu$. If μ is infinitely divisible, then μ_n is unique by Theorem 8.26 and is therefore called *the* n^{th} convolution root; it is denoted by $\mu^{*1/n}$.

Remark 8.28. By definition, a distribution μ is infinitely divisible if and only if it has an nth convolution root for every $n \in \mathbb{N}$, which then is unique, as we have seen. If a distribution is not infinitely divisible, then convolution roots of some order n may still exist, but need not be unique. An example of this phenomenon is given in Exercise 8.4.

We next show that $\mathrm{ID}(\mathbb{R}^d)$ is closed under weak convergence.

Theorem 8.29 (Closedness of $\mathrm{ID}(\mathbb{R}^d)$). *The class $\mathrm{ID}(\mathbb{R}^d)$ of infinitely divisible distributions on $\mathbb{R}^d$ is closed under weak convergence, i. e., if $(\mu_n)_{n\in\mathbb{N}}$ is a sequence of infinitely divisible distributions on $\mathbb{R}^d$ converging weakly to some distribution μ on $\mathbb{R}^d$ as $n \to \infty$,* *then μ is also infinitely divisible. In that case, $\mu_n^{*1/m}$ converges weakly to $\mu^{*1/m}$ as $n \to \infty$ for every $m \in \mathbb{N}$.*

Proof. Denote the characteristic function of μ_n by ϕ_n and that of μ by ϕ. Since μ_n converges weakly to μ, $\phi_n(z)$ converges locally uniformly to $\phi(z)$ as $n \to \infty$. Hence, for each $m \in \mathbb{N}$, $|\phi_n(z)|^{2/m}$ converges locally uniformly to $|\phi(z)|^{2/m}$. As shown in the proof of Proposition 8.11, the functions $|\phi_n|^2$ and $|\phi|^2$ are characteristic functions of some distributions, $|\phi_n|^2$ is infinitely divisible and $|\phi_n|^{2/m}$ is the distinguished m^{th} root of $|\phi_n|^2$, hence $z \mapsto |\phi_n(z)|^{2/m}$ is the characteristic function of a probability distribution by Theorem 8.26. Since the limit as $n \to \infty$, namely $|\phi(z)|^{2/m}$, is continuous in z with $|\phi(0)|^{2/m} = 1$, it follows from Lévy's continuity theorem that $z \mapsto |\phi(z)|^{2/m}$ is the characteristic function of some distribution. But since $[|\phi(z)|^{2/m}]^m = |\phi(z)|^2$, it follows that $z \mapsto |\phi(z)|^2$ is the characteristic function of some infinitely divisible distribution, hence $|\phi(z)|^2 \neq 0$ for all $z \in \mathbb{R}^d$ and consequently $\phi(z) \neq 0$ for all $z \in \mathbb{R}^d$.

Since ϕ_n converges locally uniformly to ϕ and since ϕ has no zeros, its distinguished logarithm is defined and, by Corollary 8.23, $\operatorname{Log}\phi_n(z)$ converges locally uniformly to $\operatorname{Log}\phi(z)$. Hence the distinguished m^{th} root $[\phi_n(z)]^{1/m}$ of ϕ_n converges locally uniformly to the distinguished m^{th} root $[\phi(z)]^{1/m}$ of ϕ. Again, by Lévy's continuity theorem, $\phi^{1/m}$ is the characteristic function of a distribution. From Theorem 8.26 we then find that μ is infinitely divisible, and we have seen that $\mu_n^{*1/m}$ converges weakly to $\mu^{*1/m}$ for $n \to \infty$. □

As an application, we can characterize all infinitely divisible distributions as weak limits of compound Poisson distributions:

Corollary 8.30 (Characterization of i. d. distributions). *A distribution μ on $\mathbb{R}^d$ is infinitely divisible if and only if it is the weak limit of a sequence of compound Poisson distributions. Moreover, if μ is infinitely divisible and $\mu \neq \delta_0$, then one possible sequence of approximating compound Poisson distributions is μ_n, $n = 1, 2, \ldots$, where μ_n has jump rate $n\mu^{*1/n}(\mathbb{R}^d \setminus \{0\})$ and jump size distribution $(\mu^{*1/n}(\mathbb{R}^d \setminus \{0\}))^{-1}\mathbf{1}_{\mathbb{R}^d\setminus\{0\}}(x)\mu^{*1/n}(\mathrm{d}x)$.*

Proof. Since every compound Poisson distribution is infinitely divisible by Corollary 8.9, so is the weak limit of compound Poisson distributions by Theorem 8.29, so that it only remains to show the converse. If $\mu = \delta_0$, let μ_n be a Poisson distribution with parameter $1/n$, which is a particular compound Poisson distribution, and it follows easily that μ_n converges weakly to μ. Now suppose that μ is infinitely divisible and $\mu \neq \delta_0$. Then $\mu^{*1/n}$ exists for each $n \in \mathbb{N}$ and $\mu^{*1/n}(\mathbb{R}^d \setminus \{0\}) > 0$. Define ϕ_n by

$$\begin{aligned}
\phi_n(z) &:= \exp(n(\widehat{\mu^{*1/n}}(z) - 1)) \\
&= \exp\Big(n \int_{\mathbb{R}^d} (\mathrm{e}^{\mathrm{i}\langle z,x\rangle} - 1)\,\mu^{*1/n}(\mathrm{d}x)\Big) \\
&= \exp\Big(\int_{\mathbb{R}^d\setminus\{0\}} (\mathrm{e}^{\mathrm{i}\langle z,x\rangle} - 1) n\mu^{*1/n}(\mathrm{d}x)\Big), \quad z \in \mathbb{R}^d.
\end{aligned}$$

By Theorem 8.7, ϕ_n is the characteristic function of a compound Poisson distribution μ_n with parameter $n\mu^{*1/n}(\mathbb{R}^d \setminus \{0\})$ and jump size distribution $(\mu^{*1/n}(\mathbb{R}^d \setminus \{0\}))^{-1}$ $\mathbf{1}_{\mathbb{R}^d\setminus\{0\}}(x)\mu^{*1/n}(dx)$. Since

$$n(\widehat{\mu^{*1/n}}(z) - 1) = n\left(\exp\left(\frac{1}{n}\operatorname{Log}\widehat{\mu}(z)\right) - 1\right)$$
$$= n\sum_{k=1}^{\infty}\frac{n^{-k}(\operatorname{Log}\widehat{\mu}(z))^k}{k!} \to \operatorname{Log}\widehat{\mu}(z) \quad \text{as } n \to \infty$$

for each $z \in \mathbb{R}^d$ it follows that $\phi_n(z)$ converges to $\exp(\operatorname{Log}\widehat{\mu}(z)) = \widehat{\mu}(z)$. Hence μ_n converges weakly to μ as $n \to \infty$, finishing the proof. □

8.4 Exercises

Exercise 8.1. Let $\gamma : [a,b] \to \mathbb{C}$ be a closed complex curve that is continuously differentiable and such that $\gamma(t) \neq 0$ for all $t \in [a,b]$. Show that $n(\gamma, 0) = \frac{1}{2\pi i}\int_\gamma \frac{1}{z}\,dz$.

Exercise 8.2. When $d = 1$, a much simpler proof of Theorem 8.21 can be given by applying Lemma 8.14 to $\phi_{|[-T,T]}$ (the restriction of ϕ to $[-T,T]$) and then letting $T \to \infty$. Carry out the details.

Exercise 8.3. Let U be an open and convex subset of $\mathbb{R}^d$ containing the origin and $\varphi : U \to \mathbb{C}$ a continuous function such that $\varphi(0) = 1$ and $\varphi(z) \neq 0$ for all $z \in U$. Show that there exists a unique continuous function $\psi : U \to \mathbb{C}$ such that $\psi(0) = 0$ and $\exp(\psi(z)) = \varphi(z)$ for all $z \in U$. Furthermore, show that for every $n \in \mathbb{N}$ there is a unique continuous function $g_n : U \to \mathbb{C}$ such that $g_n(0) = 1$ and $[g_n(z)]^n = \varphi(z)$ for all $z \in U$.
Hint: Adapt the proofs of Theorem 8.21 and Corollary 8.24.

Exercise 8.4 (See [44], p. 506). Define $c_0 := 1/2$, $c_k := (1 - \cos(k\pi))/(\pi^2 k^2)$ for $k \in \mathbb{Z} \setminus \{0\}$ and consider the probability distribution μ defined by $\mu := \sum_{k\in\mathbb{Z}} c_k \delta_k$.

(a) Show that the characteristic function of μ is a 2π-periodic function satisfying $\widehat{\mu}(z) = 1 - |z|/\pi$ for $|z| \leq \pi$.
Hint: Compute the Fourier coefficients of the function $[-\pi,\pi] \to \mathbb{R}$, $z \mapsto 1 - |z|/\pi$.
(b) Define the discrete probability measure $\nu := \sum_{k\in\mathbb{Z}\setminus\{0\}} 2c_k\delta_{k/2}$. Show that the characteristic function of ν satisfies

$$\widehat{\nu}(z) = 2\widehat{\mu}(z/2) - 1 \quad \forall\, z \in \mathbb{R}.$$

Using (a), deduce that $\widehat{\mu}(z) = |\widehat{\nu}(z)|$ for all $z \in \mathbb{R}$.
(c) Show that $\mu * \mu = \nu * \nu$, although $\mu \neq \nu$. Is $\mu * \mu$ infinitely divisible?

Exercise 8.5. Let $n \in \mathbb{N}$. Show that the nth convolution root of an infinitely divisible distribution is infinitely divisible.

9 The Lévy–Khintchine formula for infinitely divisible distributions

Abstract: In this chapter, we continue our study of infinitely divisible distributions. In particular, we prove the fundamental Lévy–Khintchine formula, which characterizes infinitely divisible distributions in terms of their characteristic functions, using the characteristic triplet, consisting of a Gaussian variance, a Lévy measure and a location parameter. We also study linear transformations, independence and weak convergence of infinitely divisible distributions in terms of the characteristic triplet.

9.1 The Lévy–Khintchine formula

In this section, we shall derive the fundamental Lévy–Khintchine formula that characterizes all infinitely divisible distributions by expressing the characteristic function of each as a unique member of an explicitly defined family. In order to establish this characterization, we must first introduce the Lévy measure.

Definition 9.1 (Lévy measure). A *Lévy measure* on $\mathbb{R}^d$ is a (positive) measure ν on $(\mathbb{R}^d, \mathcal{B}_d)$ such that $\nu(\{0\}) = 0$,

$$\int\limits_{|x|\leq 1} |x|^2\, \nu(\mathrm{d}x) < \infty \quad \text{and} \quad \nu(\{x \in \mathbb{R}^d : |x| > 1\}) < \infty. \tag{9.1}$$

If ν has a Lebesgue density f, we shall call f a *Lévy density*.

Remark 9.2. Equations (9.1) can be summarized as

$$\int\limits_{\mathbb{R}^d} (|x|^2 \wedge 1)\, \nu(\mathrm{d}x) < \infty,$$

which, in the case of the existence of a Lebesgue density f, can be written as

$$\int\limits_{\mathbb{R}^d} (|x|^2 \wedge 1) f(x)\, \mathrm{d}x < \infty.$$

It is also easy to see that the equations (9.1) are equivalent to

$$\int\limits_{U} |x|^2\, \nu(\mathrm{d}x) < \infty \quad \text{and} \quad \nu(\mathbb{R}^d \setminus U) < \infty \tag{9.2}$$

for some (and equivalently all) bounded neighbourhood(s) U of 0, i. e., a measure ν on $(\mathbb{R}^d, \mathcal{B}_d)$ is a Lévy measure if $\nu(\{0\}) = 0$, $\nu(B) < \infty$ is finite for every Borel set B

https://doi.org/10.1515/9783111325033-009

with strictly positive distance from 0, and the integral of $|x|^2$ with respect to ν over any bounded neighbourhood of 0 is finite.

Example 9.3. Every finite measure ν on $\mathbb{R}^d$ with $\nu(\{0\}) = 0$ is a Lévy measure. In $\mathbb{R}$, the measure ν with Lebesgue density $x \mapsto cx^{-1}e^{-ax}\mathbf{1}_{(0,\infty)}(x)\,dx$, where c and a are strictly positive constants, is a Lévy measure that is infinite (i. e., $\nu(\mathbb{R}) = \infty$). Similarly, $x \mapsto |x|^{-2}$ is the Lévy density of some Lévy measure in $\mathbb{R}$, but $x \mapsto |x|^{-3}$ and $x \mapsto |x|^{-1}$ are not Lévy densities.

For the Lévy–Khintchine representation, we also need to introduce a truncation function.

Definition 9.4 (Truncation function). A function $c : \mathbb{R}^d \to \mathbb{R}^d$ is called a *truncation function* (on $\mathbb{R}^d$), if it is Borel measurable, bounded and satisfies $c(0) = 0$ and

$$\limsup_{x\to 0, x\neq 0} \frac{|c(x) - x|}{|x|^2} < \infty. \tag{9.3}$$

We shall denote by $D := \{x \in \mathbb{R}^d : |x| \leq 1\}$ the closed unit ball in $\mathbb{R}^d$. The function $\mathbb{R}^d \to \mathbb{R}^d$, $x \mapsto \mathbf{1}_D(x)x$ is then obviously a truncation function, and we shall call it the *standard truncation function*.

Remark 9.5. Equation (9.3) implies that $x \mapsto |x|^{-2}|c(x) - x|$ is bounded in a neighbourhood of 0. Since every truncation function c is also bounded, it follows also that $x \mapsto |x|^{-2}|c(x) - x|$ is bounded outside neighborhoods of 0. Hence, a function $c : \mathbb{R}^d \to \mathbb{R}^d$ with $c(0) = 0$ is a truncation function if and only if it is Borel measurable, bounded and

$$\sup_{x\in\mathbb{R}^d\setminus\{0\}} \frac{|c(x) - x|}{|x|^2} < \infty. \tag{9.4}$$

Observe also that $\lim_{x\to 0, x\neq 0} c(x) = 0$ as a consequence of (9.3). Hence c is continuous at 0.

Example 9.6. It is easy to see that the functions

$$x \mapsto \mathbf{1}_D(x)x \quad \text{(standard truncation function)}, \tag{9.5}$$

$$x \mapsto \frac{x}{1 + |x|^2}, \quad \text{and} \tag{9.6}$$

$$x \mapsto [\mathbf{1}_D(x) + (2 - |x|)\mathbf{1}_{(1,2]}(|x|)]x \tag{9.7}$$

are truncation functions on $\mathbb{R}^d$. In the special case when $d = 1$,

$$x \mapsto \sin x \tag{9.8}$$

is also a truncation function. Observe that the standard truncation function in (9.5) is not continuous, while the truncation functions in (9.6)–(9.8) are continuous.

Lemma 9.7. *Let ν be a Lévy measure on $\mathbb{R}^d$ and $c : \mathbb{R}^d \to \mathbb{R}^d$ a truncation function. Then there exists a constant $C \in [0, \infty)$ such that*

$$|\mathrm{e}^{\mathrm{i}\langle z,x\rangle} - 1 - \mathrm{i}\langle z, c(x)\rangle| \leq C \min\{[|x|^2(|z|^2 + |z|)], 2 + |z|\} \quad \forall\, x, z \in \mathbb{R}^d. \tag{9.9}$$

In particular,

$$\int_{\mathbb{R}^d} |\mathrm{e}^{\mathrm{i}\langle z,x\rangle} - 1 - \mathrm{i}\langle z, c(x)\rangle|\, \nu(\mathrm{d}x) < \infty \quad \forall\, z \in \mathbb{R}^d, \tag{9.10}$$

and $z \mapsto \int_{\mathbb{R}^d} (\mathrm{e}^{\mathrm{i}\langle z,x\rangle} - 1 - \mathrm{i}\langle z, c(x)\rangle)\, \nu(\mathrm{d}x)$ is a continuous function on $\mathbb{R}^d$.

Proof. Since $\lim_{y\to 0, y\in\mathbb{R}} y^{-2}(\mathrm{e}^{\mathrm{i}y} - 1 - \mathrm{i}y) = -1/2$, there is a constant $C_1 < \infty$ such that $|\mathrm{e}^{\mathrm{i}y} - 1 - \mathrm{i}y| \leq C_1 y^2$ for all $y \in \mathbb{R}$, which implies that $|\mathrm{e}^{\mathrm{i}\langle z,x\rangle} - 1 - \mathrm{i}\langle z, x\rangle| \leq C_1 |\langle z, x\rangle|^2 \leq C_1 |x|^2 |z|^2$. The triangle inequality together with (9.4) then shows that there is $C_2 < \infty$ such that $|\mathrm{e}^{\mathrm{i}\langle z,x\rangle} - 1 - \mathrm{i}\langle z, c(x)\rangle| \leq C_2 |x|^2(|z|^2 + |z|)$ for all $x, z \in \mathbb{R}^d$. Together with the fact that c is bounded this then gives equation (9.9). Equation (9.10) is a direct consequence of (9.9) and (9.2), as is the continuity of $z \mapsto \int_{\mathbb{R}} (\mathrm{e}^{\mathrm{i}\langle z,x\rangle} - 1 - \mathrm{i}\langle z, c(x)\rangle)\nu(\mathrm{d}x)$ by virtue of the continuity lemma (the latter is a simple consequence of the dominated convergence theorem, see, e. g., Bauer [10, Lemma 16.1]). □

The next result is already the first half of the Lévy–Khintchine representation, showing that all characteristic functions having a particular form are characteristic functions of infinitely divisible distributions. It generalizes Corollary 8.9.

Proposition 9.8. *Suppose that $\gamma \in \mathbb{R}^d$ and that A is a real symmetric non-negative definite $d \times d$ matrix. Suppose also that ν is a Lévy measure on $(\mathbb{R}^d, \mathcal{B}_d)$ and that $c : \mathbb{R}^d \to \mathbb{R}^d$ is a fixed truncation function. Then $\phi : \mathbb{R}^d \to \mathbb{C}$, defined by*

$$\phi(z) := \exp\Big(\mathrm{i}\langle \gamma, z\rangle - \frac{1}{2}\langle z, Az\rangle + \int_{\mathbb{R}^d} (\mathrm{e}^{\mathrm{i}\langle z,x\rangle} - 1 - \mathrm{i}\langle z, c(x)\rangle)\, \nu(\mathrm{d}x)\Big), \tag{9.11}$$

is the characteristic function of an infinitely divisible distribution. The triplet (A, ν, γ) is uniquely determined by ϕ and c, in the sense that if $\gamma' \in \mathbb{R}^d$, $A' \in \mathbb{R}^{d\times d}$ is a symmetric non-negative definite real matrix and ν' is a Lévy measure on $(\mathbb{R}^d, \mathcal{B}_d)$ such that ϕ satisfies both (9.11) *and*

$$\phi(z) = \exp\Big(\mathrm{i}\langle \gamma', z\rangle - \frac{1}{2}\langle z, A'z\rangle + \int_{\mathbb{R}^d} (\mathrm{e}^{\mathrm{i}\langle z,x\rangle} - 1 - \mathrm{i}\langle z, c(x)\rangle)\, \nu'(\mathrm{d}x)\Big)$$

for all $z \in \mathbb{R}^d$, then $(A', \nu', \gamma') = (A, \nu, \gamma)$.

Proof. For $n \in \mathbb{N}$, define the measure ν_n by $\nu_n(B) := \nu(B \cap \{x \in \mathbb{R}^d : |x| \geq 1/n\})$, $B \in \mathcal{B}_d$, i. e., by restricting ν to $\{x \in \mathbb{R}^d : |x| \geq 1/n\}$ and setting it to be 0 otherwise. Then ν_n is a finite measure. Since c is integrable with respect to ν_n, it follows from Corollary 8.9 that $z \mapsto \phi_n(z)$, defined by

$$\begin{aligned}\phi_n(z) &:= \exp\Big(\mathrm{i}\langle z, \gamma\rangle - \frac{1}{2}\langle z, Az\rangle + \int_{\mathbb{R}^d} (\mathrm{e}^{\mathrm{i}\langle z,x\rangle} - 1 - \mathrm{i}\langle z, c(x)\rangle)\, \nu_n(\mathrm{d}x)\Big)\\ &= \exp\Big(\mathrm{i}\Big\langle z, \gamma - \int_{\mathbb{R}^d} c(x)\, \nu_n(\mathrm{d}x)\Big\rangle - \frac{1}{2}\langle z, Az\rangle + \int_{\mathbb{R}^d} (\mathrm{e}^{\mathrm{i}\langle z,x\rangle} - 1)\, \nu_n(\mathrm{d}x)\Big)\end{aligned}$$

is the characteristic function of an infinitely divisible distribution, μ_n say. Since by (9.9)

$$\Big|\int_{\mathbb{R}^d} (\mathrm{e}^{\mathrm{i}\langle z,x\rangle} - 1 - \mathrm{i}\langle z, c(x)\rangle)\, \nu(\mathrm{d}x) - \int_{\mathbb{R}^d} (\mathrm{e}^{\mathrm{i}\langle z,x\rangle} - 1 - \mathrm{i}\langle z, c(x)\rangle)\, \nu_n(\mathrm{d}x)\Big|$$

can be bounded by $C(|z|^2 + |z|) \int_{|x|<1/n} |x|^2\, \nu(\mathrm{d}x)$, which by (9.1) tends to 0 as $n \to \infty$, it follows that ϕ_n converges pointwise to ϕ as $n \to \infty$. Since ϕ is continuous by Lemma 9.7, Lévy's continuity theorem implies that it is the characteristic function of some distribution μ, which is the weak limit of μ_n. The infinite divisibility of μ then follows from Theorem 8.29.

To show the uniqueness of (A, ν, γ), define

$$\begin{aligned}\psi(z) &:= \mathrm{i}\langle \gamma, z\rangle - \frac{1}{2}\langle z, Az\rangle + \int_{\mathbb{R}^d} (\mathrm{e}^{\mathrm{i}\langle z,x\rangle} - 1 - \mathrm{i}\langle z, c(x)\rangle)\, \nu(\mathrm{d}x),\\ \psi'(z) &:= \mathrm{i}\langle \gamma', z\rangle - \frac{1}{2}\langle z, A'z\rangle + \int_{\mathbb{R}^d} (\mathrm{e}^{\mathrm{i}\langle z,x\rangle} - 1 - \mathrm{i}\langle z, c(x)\rangle)\, \nu'(\mathrm{d}x),\end{aligned}$$

and suppose that $\phi(z) = \exp(\psi(z)) = \exp(\psi'(z))$ for all $z \in \mathbb{R}^d$. Since both ψ and ψ' are continuous and satisfy $\psi(0) = \psi'(0) = 0$, they coincide with the distinguished logarithm of ϕ, and hence are uniquely determined by ϕ, so that $\psi = \psi'$. Since $|z|^{-2}(\mathrm{e}^{\mathrm{i}\langle z,x\rangle} - 1 - \mathrm{i}\langle z, c(x)\rangle) \to 0$ as $|z| \to \infty$ by (9.9), it follows from Lebesgue's dominated convergence theorem together with (9.9) and (9.1) that $|z|^{-2} \int_{\mathbb{R}^d} (\mathrm{e}^{\mathrm{i}\langle z,x\rangle} - 1 - \mathrm{i}\langle z, c(x)\rangle)\, \nu(\mathrm{d}x) \to 0$ as $|z| \to \infty$, and similarly for ν'. In particular, for fixed $z \in \mathbb{R}^d \setminus \{0\}$,

$$-\frac{1}{2}\langle z, Az\rangle = \lim_{s\to\infty} s^{-2}\psi(sz) = \lim_{s\to\infty} s^{-2}\psi'(sz) = -\frac{1}{2}\langle z, A'z\rangle,$$

so that $A = A'$. Hence

$$\begin{aligned}\beta(z) &:= \mathrm{i}\langle \gamma, z\rangle + \int_{\mathbb{R}^d} (\mathrm{e}^{\mathrm{i}\langle z,x\rangle} - 1 - \mathrm{i}\langle z, c(x)\rangle)\, \nu(\mathrm{d}x)\\ &= \mathrm{i}\langle \gamma', z\rangle + \int_{\mathbb{R}^d} (\mathrm{e}^{\mathrm{i}\langle z,x\rangle} - 1 - \mathrm{i}\langle z, c(x)\rangle)\, \nu'(\mathrm{d}x) \quad \forall z \in \mathbb{R}^d\end{aligned}$$

is uniquely determined by ϕ. For $z, w \in \mathbb{R}^d$, we then have

$$\beta(z) - \beta(z+w) = \int_{\mathbb{R}^d} (e^{i\langle z,x\rangle} - e^{i\langle z+w,x\rangle} + i\langle w, c(x)\rangle)\nu(dx) - i\langle w, \gamma\rangle.$$

Suppose that $h > 0$. (For the purpose of this proof it suffices to take $h = 1$, but the derived formulas for general h will be convenient for the proof of Lemma 9.12). Since $\int_{[-h,h]^d} \langle w, \gamma\rangle \, dw = \int_{[-h,h]^d} \langle w, c(x)\rangle \, dw = 0$, an application of Fubini's theorem (which is valid because of (9.9)) gives

$$\begin{aligned} &\int_{[-h,h]^d} (\beta(z) - \beta(z+w)) \, dw \\ &= \int_{\mathbb{R}^d} \int_{[-h,h]^d} (e^{i\langle z,x\rangle} - e^{i\langle z+w,x\rangle}) \, dw \, \nu(dx) \\ &= \int_{\mathbb{R}^d} e^{i\langle z,x\rangle} \int_{[-h,h]^d} (1 - e^{i\langle w,x\rangle}) \, dw \, \nu(dx). \end{aligned} \tag{9.12}$$

Writing $x = (x_1, \dots, x_d)^T$ and $w = (w_1, \dots, w_d)^T$, define

$$\begin{aligned} g_h(x) &:= \int_{[-h,h]^d} (1 - e^{i\langle w,x\rangle}) \, dw = (2h)^d - \prod_{j=1}^{d} \int_{[-h,h]} e^{iw_j x_j} \, dw_j \\ &= (2h)^d \left(1 - \prod_{j=1}^{d} \frac{\sin(hx_j)}{hx_j}\right), \end{aligned} \tag{9.13}$$

where $\frac{\sin(hx_j)}{hx_j}$ is understood to be 1 when $x_j = 0$. We see that $g_h(x) > 0$ for all $x \neq 0$, that g_h is bounded and that g_h is integrable with respect to ν as a consequence of (9.1) and the fact that

$$\prod_{j=1}^{d} \frac{\sin(hx_j)}{hx_j} = \prod_{j=1}^{d} \left(1 - \frac{1}{6}h^2 x_j^2 + O(|x_j|^4)\right) = 1 - \frac{1}{6}h^2|x|^2 + O(|x|^4) \tag{9.14}$$

as $|x| \to 0$. Hence $\int_{[-h,h]^d} (\beta(z) - \beta(z+w)) \, dw = \int_{\mathbb{R}^d} e^{i\langle z,x\rangle} g_h(x) \, \nu(dx)$ is the characteristic function of the finite positive measure $g_h(x)\,\nu(dx)$, and since β is uniquely determined by ϕ, it follows that the measure $g_h(x)\,\nu(dx)$ is also uniquely determined by ϕ. Since $\nu(\{0\}) = 0$ and $g_h(x) > 0$ for $x \neq 0$, it follows that ν is uniquely determined by ϕ and hence that $\nu = \nu'$. Since $A = A'$, it also follows immediately that $\gamma = \gamma'$. □

Remark 9.9. The uniqueness assertion in Proposition 9.8 holds even when the matrix A is symmetric but not non-negative definite. More precisely, suppose that $c : \mathbb{R}^d \to \mathbb{R}^d$ is a fixed truncation function, $\gamma, \gamma' \in \mathbb{R}^d$, A, A' are real symmetric $d \times d$ matrices, and ν, ν' are two Lévy measures on $\mathbb{R}^d$. Suppose also that

$$\exp\Big(\mathrm{i}\langle \gamma, z\rangle - \frac{1}{2}\langle z, Az\rangle + \int_{\mathbb{R}^d} (\mathrm{e}^{\mathrm{i}\langle z,x\rangle} - 1 - \mathrm{i}\langle z, c(x)\rangle)\, \nu(\mathrm{d}x)\Big)$$
$$= \exp\Big(\mathrm{i}\langle \gamma', z\rangle - \frac{1}{2}\langle z, A'z\rangle + \int_{\mathbb{R}^d} (\mathrm{e}^{\mathrm{i}\langle z,x\rangle} - 1 - \mathrm{i}\langle z, c(x)\rangle)\, \nu'(\mathrm{d}x)\Big)$$

for all $z \in \mathbb{R}^d$. Then $(A', \nu', \gamma') = (A, \nu, \gamma)$. The proof is exactly the same as that of Proposition 9.8.

Our next goal is to show that the characteristic function of every infinitely divisible distribution can be expressed in the form (9.11). We already know from Corollary 8.30 that every infinitely divisible distribution is a weak limit of compound Poisson distributions, and that each compound Poisson distribution has a representation as in (9.11). So our strategy will be to show that the limit of infinitely divisible distributions with representation (9.11) also has a representation of the form (9.11). In doing so, we shall need the notions of tightness and weak convergence for finite measures. They are a natural extension of those for probability measures, but for completeness we add their definitions and properties.

Definition 9.10. Let $(\rho_n)_{n\in\mathbb{N}}$ be a sequence of finite (positive) measures on $(\mathbb{R}^d, \mathcal{B}_d)$, and let ρ be a finite measure on $(\mathbb{R}^d, \mathcal{B}_d)$.

(a) The sequence $(\rho_n)_{n\in\mathbb{N}}$ converges *weakly* to ρ, written $\rho_n \xrightarrow{w} \rho$ $(n \to \infty)$, if

$$\lim_{n\to\infty} \int_{\mathbb{R}^d} f(x)\, \rho_n(\mathrm{d}x) = \int_{\mathbb{R}^d} f(x)\, \rho(\mathrm{d}x) \quad \forall f \in C_b(\mathbb{R}^d; \mathbb{R}),$$

where $C_b(\mathbb{R}^d; \mathbb{R})$ denotes the set of all real-valued bounded and continuous functions on $\mathbb{R}^d$.

(b) The sequence $(\rho_n)_{n\in\mathbb{N}}$ is *tight*, if for every $\varepsilon > 0$ there exists a compact set $K \subset \mathbb{R}^d$ such that

$$\rho_n(\mathbb{R}^d \setminus K) \le \varepsilon \quad \forall n \in \mathbb{N}.$$

(c) The sequence $(\rho_n)_{n\in\mathbb{N}}$ is *uniformly bounded* if $\sup_{n\in\mathbb{N}} \rho_n(\mathbb{R}^d) < \infty$.

By considering real and imaginary parts separately, we see that $C_b(\mathbb{R}^d; \mathbb{R})$ in the definition of weak convergence can be replaced by $C_b(\mathbb{R}^d; \mathbb{C})$, the space of all complex-valued bounded continuous functions on $\mathbb{R}^d$.

We next establish Prokhorov's theorem for finite measures, which is a simple generalization of Prokhorov's theorem for probability measures and can be found, for example, in Klenke [57, Theorem 13.29]. We assume that the reader is acquainted with Prokhorov's theorem for probability measures (see, e. g., Shiryaev [81, Theorem III.2.1]).

Theorem 9.11 (Prokhorov's theorem for finite measures). *Let $(\rho_n)_{n\in\mathbb{N}}$ be a sequence of finite measures on $(\mathbb{R}^d, \mathcal{B}_d)$. Then the following are equivalent:*

(i) *Every subsequence $(\rho_{n_k})_{k\in\mathbb{N}}$ of $(\rho_n)_{n\in\mathbb{N}}$ contains a subsequence that converges weakly to some finite measure.*

(ii) *$(\rho_n)_{n\in\mathbb{N}}$ is tight and uniformly bounded.*

Proof. Recall that Prokhorov's theorem for probability measures states that a sequence of probability measures is tight if and only if every subsequence of the sequence has a subsequence that converges weakly to a probability measure (see, e. g., Shiryaev [81, Theorem III.2.1]).

To show that (ii) implies (i), let $(\rho_n)_{n\in\mathbb{N}}$ be tight and uniformly bounded and let $(\rho_{n_k})_{k\in\mathbb{N}}$ be a subsequence. If $\liminf_{k\to\infty}\rho_{n_k}(\mathbb{R}^d) = 0$, then a subsequence of (ρ_{n_k}) obviously converges to the zero measure, and there is nothing to prove. So suppose that $\liminf_{k\to\infty}\rho_{n_k}(\mathbb{R}^d) > 0$, and by choosing a further subsequence if necessary, that $\inf_{k\in\mathbb{N}}\rho_{n_k}(\mathbb{R}^d) > 0$. Define $\mu_k := (\rho_{n_k}(\mathbb{R}^d))^{-1}\rho_{n_k}$ for $k \in \mathbb{N}$. Then $(\mu_k)_{k\in\mathbb{N}}$ is a sequence of probability measures, and since $(\rho_{n_k})_{k\in\mathbb{N}}$ is tight, so is $(\mu_k)_{k\in\mathbb{N}}$ as a consequence of $\inf_{k\in\mathbb{N}}\rho_{n_k}(\mathbb{R}^d) > 0$. By Prokhorov's theorem for probability measures, there exists a subsequence $(\mu_{k_l})_{l\in\mathbb{N}}$ and a probability measure μ such that $\mu_{k_l} \xrightarrow{w} \mu$ as $l \to \infty$. Since $0 < \inf_{k\in\mathbb{N}}\rho_{n_k}(\mathbb{R}^d) \le \sup_{n\in\mathbb{N}}\rho_n(\mathbb{R}^d) < \infty$ by uniform boundedness, there exists a further subsequence $(\rho_{n_{k_{l_m}}})_{m\in\mathbb{N}}$ such that $\rho_{n_{k_{l_m}}}(\mathbb{R}^d)$ converges to some $c \in (0,\infty)$ as $m \to \infty$. For $f \in C_b(\mathbb{R}^d;\mathbb{R})$ we then have

$$\lim_{m\to\infty}\int_{\mathbb{R}^d} f(x)\,\rho_{n_{k_{l_m}}}(dx) = \lim_{m\to\infty}\Big(\rho_{n_{k_{l_m}}}(\mathbb{R}^d)\int_{\mathbb{R}^d} f(x)\,\mu_{k_{l_m}}(dx)\Big)$$

which is equal to $c\int_{\mathbb{R}^d} f(x)\,\mu(dx)$, showing that $\rho_{n_{k_{l_m}}}$ converges weakly to $c\mu$ as $m \to \infty$.

We next show that (i) implies (ii). Let $(\rho_n)_{n\in\mathbb{N}}$ satisfy condition (i). Suppose that $(\rho_n)_{n\in\mathbb{N}}$ is not uniformly bounded. Then there exists a subsequence $(\rho_{n_k})_{k\in\mathbb{N}}$ with $\rho_{n_k}(\mathbb{R}^d) \to \infty$ as $k \to \infty$. But this clearly contradicts (i) with $f = \mathbf{1}_{\mathbb{R}^d}$. Now suppose that $(\rho_n)_{n\in\mathbb{N}}$ is not tight. Then there exists $\varepsilon > 0$ and a subsequence $(\rho_{n_k})_{k\in\mathbb{N}}$ such that $\rho_{n_k}(\mathbb{R}^d \setminus [-k,k]^d) \ge \varepsilon$ for all $k \in \mathbb{N}$. Let $(\rho_{n_{k_l}})_{l\in\mathbb{N}}$ be a subsequence such that $\rho_{n_{k_l}} \xrightarrow{w} \rho$ as $l \to \infty$ for some finite measure ρ. Let $T > 0$ be such that $\rho(\mathbb{R}^d \setminus [-T,T]^d) < \varepsilon$. Let $f : \mathbb{R}^d \to [0,1]$ be continuous with the properties $0 \le f \le 1$, $f(x) = 0$ for $x \in [-T,T]^d$ and $f(x) = 1$ for $x \in \mathbb{R}^d \setminus [-2T,2T]^d$. Then $\int_{\mathbb{R}^d} f(x)\,\rho_{n_k}(dx) \ge \rho_{n_k}(\mathbb{R}^d \setminus [-k,k]^d) \ge \varepsilon$ for $k \ge 2T$, but $\int_{\mathbb{R}^d} f(x)\,\rho(dx) \le \rho(\mathbb{R}^d \setminus [-T,T]^d) < \varepsilon$. This contradicts $\rho_{n_{k_l}} \xrightarrow{w} \rho$. □

In order to show that limits of infinitely divisible distributions with characteristic functions of the form (9.11) again have characteristic functions of the form (9.11), we first establish the following lemma. It will be improved in Section 9.3, but is sufficient for our present purpose.

Lemma 9.12. *Let $c : \mathbb{R}^d \to \mathbb{R}^d$ be a fixed truncation function, and for each $n \in \mathbb{N}$, suppose that $\gamma_n \in \mathbb{R}^d$, A_n is a real symmetric non-negative definite $d \times d$ matrix and ν_n is a Lévy measure on $(\mathbb{R}^d, \mathcal{B}_d)$. Define the finite Borel measure ρ_n on $(\mathbb{R}^d, \mathcal{B}_d)$ by $\rho_n(\mathrm{d}x) := (|x|^2 \wedge 1)\,\nu_n(\mathrm{d}x)$ and let μ_n be the infinitely divisible distribution with characteristic function*

$$z \mapsto \widehat{\mu}_n(z) = \exp\Big(\mathrm{i}\langle\gamma_n, z\rangle - \frac{1}{2}\langle z, A_n z\rangle + \int_{\mathbb{R}^d} (\mathrm{e}^{\mathrm{i}\langle z,x\rangle} - 1 - \mathrm{i}\langle z, c(x)\rangle)\,\nu_n(dx)\Big)$$

(cf., Proposition 9.8). Assume that μ_n converges weakly to some distribution μ as $n \to \infty$. Then $(\rho_n)_{n\in\mathbb{N}}$ is tight and uniformly bounded and $(A_n)_{n\in\mathbb{N}}$ is bounded in the sense that $\sup_{n\in\mathbb{N}, i,j\in\{1,\dots,d\}} |A_n(i,j)| < \infty$, where $A_n(i,j)$ denotes the (i,j)-entry of A_n.

Proof. Suppose that $h > 0$, g_h is defined as in (9.13) and

$$\beta_n(z) := \mathrm{i}\langle\gamma_n, z\rangle + \int_{\mathbb{R}^d} (\mathrm{e}^{\mathrm{i}\langle z,x\rangle} - 1 - \mathrm{i}\langle z, c(x)\rangle)\,\nu_n(\mathrm{d}x), \quad z \in \mathbb{R}^d.$$

Since $\beta_n(0) = 0$, we deduce from (9.12) that

$$\begin{aligned} - \int_{[-h,h]^d} \operatorname{Log}\widehat{\mu}_n(z)\,\mathrm{d}z &= \frac{1}{2}\int_{[-h,h]^d} \langle z, A_n z\rangle\,\mathrm{d}z + \int_{[-h,h]^d} (\beta_n(0) - \beta_n(z))\,\mathrm{d}z \\ &= \frac{1}{2}\int_{[-h,h]^d} \langle z, A_n z\rangle\,\mathrm{d}z + \int_{\mathbb{R}^d} g_h(x)\,\nu_n(\mathrm{d}x) \\ &= \frac{1}{2}\int_{[-h,h]^d} \langle z, A_n z\rangle\,\mathrm{d}z + \int_{\mathbb{R}^d} \frac{g_h(x)}{1\wedge|x|^2}\,\rho_n(\mathrm{d}x). \end{aligned} \tag{9.15}$$

Since μ_n converges weakly to μ, $\operatorname{Log}\widehat{\mu}(z)$ converges uniformly on $[-h,h]^d$ to $\operatorname{Log}\widehat{\mu}(z)$ as $n \to \infty$ by Corollary 8.23 (observe that μ is infinitely divisible by Theorem 8.29 and hence $\widehat{\mu}$ is zero-free by Proposition 8.11). Moreover, since $-\int_{[-h,h]^d} \operatorname{Log}\widehat{\mu}(z)\,\mathrm{d}z < \infty$, $g_h(x) \geq 0$ and $\langle z, A_n z\rangle \geq 0$, we deduce that

$$\sup_{n\in\mathbb{N}} \int_{[-h,h]^d} \langle z, A_n z\rangle\,\mathrm{d}z < \infty \quad \text{and} \quad \sup_{n\in\mathbb{N}} \int_{\mathbb{R}^d} \frac{g_h(x)}{1\wedge|x|^2}\rho_n(\mathrm{d}x) < \infty. \tag{9.16}$$

Since $\inf_{x\in\mathbb{R}^d} \frac{g_h(x)}{1\wedge|x|^2} > 0$ by (9.13) and (9.14), it follows that $\sup_{n\in\mathbb{N}} \rho_n(\mathbb{R}^d) < \infty$, so that $(\rho_n)_{n\in\mathbb{N}}$ is uniformly bounded. To see that $(\rho_n)_{n\in\mathbb{N}}$ is tight, observe that $\lim_{h\downarrow 0}(2h)^{-d}\int_{[-h,h]^d} \operatorname{Log}\widehat{\mu}(z)\,\mathrm{d}z = 0$ since $\operatorname{Log}\widehat{\mu}(0) = 0$ and $\operatorname{Log}\widehat{\mu}$ is continuous. Hence, by (9.15), for any $\varepsilon > 0$ there exist $n_0 \in \mathbb{N}$ and $h_0 > 0$ such that

$$(2h_0)^{-d}\int_{\mathbb{R}^d} \frac{g_{h_0}(x)}{1\wedge|x|^2}\,\rho_n(\mathrm{d}x) < \varepsilon/2 \quad \forall\, n \geq n_0.$$

From (9.13) it follows that there exists $T_1 \geq 1$ such that $(2h_0)^{-d} g_{h_0}(x) \geq 1/2$ for all $x \in \mathbb{R}^d$ such that $|x| \geq T_1$. Hence

$$\rho_n(\mathbb{R}^d \setminus [-T_1, T_1]^d) \leq \int_{\mathbb{R}^d} 2\frac{(2h_0)^{-d} g_{h_0}(x)}{1 \wedge |x|^2} \rho_n(\mathrm{d}x) \leq 2\varepsilon/2 = \varepsilon \quad \forall\, n \geq n_0.$$

Since all measures ρ_n are finite, there exists $T \geq T_1$ such that $\rho_n(\mathbb{R}^d \setminus [-T, T]^d) \leq \varepsilon$ for all $n \in \mathbb{N}$, giving tightness of $(\rho_n)_{n\in\mathbb{N}}$.

To show the boundedness of $(A_n)_{n\in\mathbb{N}}$, define

$$K := \sup_{n\in\mathbb{N}} \int_{[-1,1]^d} \langle z, A_n z\rangle \,\mathrm{d}z,$$

which is finite by (9.16), and define $\delta := \int_{x\in\mathbb{R}^d:|x|\leq 1} x_1^2 \,\mathrm{d}(x_1, \ldots, x_n) > 0$. Since each A_n is non-negative definite, for each $n \in \mathbb{N}$ there exists an orthogonal matrix $S_n \in \mathbb{R}^{d\times d}$ such that $S_n A_n S_n^T = \Lambda_n$ is a diagonal matrix with the non-negative eigenvalues of A_n on the diagonal and the largest eigenvalue λ_n as the $(1, 1)$ component. By the change of variable formula for the Lebesgue-integral (see, e. g., Cohn [33, Theorem 6.1.7]), we conclude that

$$\begin{aligned}
\lambda_n &= \frac{\lambda_n}{\delta} \int_{|x|\leq 1} x_1^2 \,\mathrm{d}x \leq \frac{1}{\delta} \int_{|x|\leq 1} x^T \Lambda_n x \,\mathrm{d}x \\
&= \frac{1}{\delta} \int_{|z|\leq 1} (S_n z)^T \Lambda_n (S_n z) \,\mathrm{d}z \\
&= \frac{1}{\delta} \int_{|z|\leq 1} z^T A_n z \,\mathrm{d}z \leq \frac{1}{\delta} \int_{[-1,1]^d} \langle z, A_n z\rangle \,\mathrm{d}z \leq \frac{K}{\delta} \quad \forall\, n \in \mathbb{N}.
\end{aligned}$$

Thus the sequence of largest eigenvalues of $(A_n)_{n\in\mathbb{N}}$ is bounded. But the largest eigenvalue of a real symmetric non-negative definite matrix is its spectral norm, and by the equivalence of norms in $\mathbb{R}^{d\times d}$ it follows that $(\sup_{i,j=1,\ldots,d} |A_n(i,j)|)_{n\in\mathbb{N}}$ must be bounded. □

We can now give the famous Lévy–Khintchine representation of infinitely divisible distributions.

Theorem 9.13 (Lévy–Khintchine representation). *Let $c : \mathbb{R}^d \to \mathbb{R}^d$ be a fixed truncation function. The following results then hold.*

(a) *For every infinitely divisible distribution μ on $\mathbb{R}^d$ there exists a unique triplet (A, ν, γ), consisting of a symmetric non-negative definite matrix $A \in \mathbb{R}^{d\times d}$, a Lévy measure ν on $\mathbb{R}^d$ and a constant $\gamma \in \mathbb{R}^d$ such that the characteristic function $\widehat{\mu}$ of μ has the representation*

$$\widehat{\mu}(z) = \exp\Big(\mathrm{i}\langle \gamma, z\rangle - \frac{1}{2}\langle z, Az\rangle + \int_{\mathbb{R}^d} (\mathrm{e}^{\mathrm{i}\langle z,x\rangle} - 1 - \mathrm{i}\langle z, c(x)\rangle)\, \nu(\mathrm{d}x)\Big) \tag{9.17}$$

for $z \in \mathbb{R}^d$.

(b) *Conversely, for every symmetric non-negative definite matrix* $A \in \mathbb{R}^{d\times d}$, *Lévy measure* ν *on* $\mathbb{R}^d$ *and constant* $\gamma \in \mathbb{R}^d$, *the right-hand side of* (9.17) *is the characteristic function of an infinitely divisible distribution.*

Proof. Part (b) and the uniqueness assertion in (a) have already been proved in Proposition 9.8, so we only have to show existence of the triplet. To do so let μ be an infinitely divisible distribution on $\mathbb{R}^d$, and assume that $\mu \neq \delta_0$ (if $\mu = \delta_0$ the existence is obvious). For each $n \in \mathbb{N}$, let μ_n be the compound Poisson distribution with jump size distribution $(\mu^{*1/n}(\mathbb{R}^d \setminus \{0\}))^{-1}\mathbf{1}_{\mathbb{R}^d\setminus\{0\}}(x)\mu^{*1/n}(\mathrm{d}x)$ and jump rate $n\mu^{*1/n}(\mathbb{R}^d \setminus \{0\})$. We know from Corollary 8.30 that μ_n converges weakly to μ as $n \to \infty$. Further, denoting the (finite) Lévy measure ν_n by $\nu_n(\mathrm{d}x) := n\mathbf{1}_{\mathbb{R}^d\setminus\{0\}}(x)\mu^{*1/n}(\mathrm{d}x)$, we obtain $\widehat{\mu}_n(z) = \exp(\int_{\mathbb{R}^d}(\mathrm{e}^{\mathrm{i}\langle z,x\rangle} - 1)\nu_n(\mathrm{d}x))$. Define the truncation function $\widetilde{c} : \mathbb{R}^d \to \mathbb{R}^d$ by

$$\widetilde{c}(x) = [\mathbf{1}_{[0,1]}(|x|) + (2 - |x|)\mathbf{1}_{(1,2]}(|x|)]x$$

as in (9.7). Then

$$\begin{aligned} \operatorname{Log}\widehat{\mu}(z) &= \lim_{n\to\infty} \operatorname{Log}\widehat{\mu}_n(z) \\ &= \lim_{n\to\infty}\Big[\int_{\mathbb{R}^d} \Big(\mathrm{e}^{\mathrm{i}\langle z,x\rangle} - 1 - \mathrm{i}\langle z, \widetilde{c}(x)\rangle + \frac{1}{2}\langle z, \widetilde{c}(x)\rangle^2 \Big)\, \nu_n(\mathrm{d}x) \\ &\quad - \frac{1}{2}\int_{\mathbb{R}^d} \langle z, \widetilde{c}(x)\rangle^2\, \nu_n(\mathrm{d}x) + \mathrm{i}\langle \beta_n, z\rangle \Big] \quad \forall\, z \in \mathbb{R}^d, \end{aligned} \tag{9.18}$$

where

$$\beta_n := \int_{\mathbb{R}^d} \widetilde{c}(x)\, \nu_n(\mathrm{d}x), \quad n \in \mathbb{N}.$$

For each $z \in \mathbb{R}^d$ define the function $u_z : \mathbb{R}^d \to \mathbb{C}$ by

$$u_z(0) := 0 \quad \text{and} \quad u_z(x) := \frac{\mathrm{e}^{\mathrm{i}\langle z,x\rangle} - 1 - \mathrm{i}\langle z, \widetilde{c}(x)\rangle + \frac{1}{2}\langle z, \widetilde{c}(x)\rangle^2}{1 \wedge |x|^2} \quad \text{for } x \neq 0.$$

Since $\lim_{y\to 0, y\in\mathbb{R}} y^{-2}(\mathrm{e}^{\mathrm{i}y} - 1 - \mathrm{i}y + y^2/2) = 0$ and since $\widetilde{c}$ is continuous and $\widetilde{c}(x)$ agrees with x for small x, the function u_z is continuous and bounded for any $z \in \mathbb{R}^d$. Define the measures ρ_n by $\rho_n(\mathrm{d}x) = (1 \wedge |x|^2)\, \nu_n(\mathrm{d}x)$. Since μ_n converges weakly to μ as $n \to \infty$, the sequence $(\rho_n)_{n\in\mathbb{N}}$ is uniformly bounded and tight by Lemma 9.12. By Prokhorov's theorem (Theorem 9.11), there exist a finite measure ρ on $\mathbb{R}^d$ and a subsequence $(\rho_{n_k})_{k\in\mathbb{N}}$ that converges weakly to ρ. Define the Lévy measure ν on $\mathbb{R}^d$ by $\nu(dx) = \mathbf{1}_{\mathbb{R}^d\setminus\{0\}}(x)\frac{1}{1\wedge|x|^2}\,\rho(dx)$ (the measure ρ may have a point mass at 0, but it does not enter the definition of ν). Then, observing that $u_z(0) = 0$,

$$\begin{aligned}
\lim_{k\to\infty} \int_{\mathbb{R}^d} &\left(e^{i\langle z,x\rangle} - 1 - i\langle z,\tilde{c}(x)\rangle + \frac{1}{2}\langle z,\tilde{c}(x)\rangle^2\right)\nu_{n_k}(dx) \\
&= \lim_{k\to\infty} \int_{\mathbb{R}^d} u_z(x)\,\rho_{n_k}(dx) = \int_{\mathbb{R}^d} u_z(x)\,\rho(dx) \\
&= \int_{\mathbb{R}^d} \left(e^{i\langle z,x\rangle} - 1 - i\langle z,\tilde{c}(x)\rangle + \frac{1}{2}\langle z,\tilde{c}(x)\rangle^2\right)\nu(dx) \\
&= \int_{\mathbb{R}^d} (e^{i\langle z,x\rangle} - 1 - i\langle z,\tilde{c}(x)\rangle)\,\nu(dx) + \frac{1}{2}\int_{\mathbb{R}^d} \langle z,\tilde{c}(x)\rangle^2\,\nu(dx).
\end{aligned}$$

Together with (9.18) this gives

$$\begin{aligned}
\operatorname{Log}\widehat{\mu}(z) - &\int_{\mathbb{R}^d} (e^{i\langle z,x\rangle} - 1 - i\langle z,\tilde{c}(x)\rangle)\,\nu(dx) \\
&= \lim_{k\to\infty}\left(i\langle \beta_{n_k}, z\rangle - \frac{1}{2}\int_{\mathbb{R}^d} \langle z,\tilde{c}(x)\rangle^2(\nu_{n_k}(dx) - \nu(dx))\right).
\end{aligned} \tag{9.19}$$

Taking real and imaginary parts in (9.19) shows the convergence of $\langle \beta_{n_k}, z\rangle$ and of $\int_{\mathbb{R}^d}\langle z,\tilde{c}(x)\rangle^2(\nu_{n_k}(dx) - \nu(dx))$ as $k \to \infty$ for every $z \in \mathbb{R}^d$. In particular, there is $\beta \in \mathbb{R}^d$ such that $\beta_{n_k} \to \beta$ as $k \to \infty$. To identify the second limit, define

$$B_n := \int_{\mathbb{R}^d\setminus\{0\}} \frac{1}{1\wedge|x|^2}\tilde{c}(x)\tilde{c}(x)^T\,(\rho_n(dx) - \rho(dx)) \in \mathbb{R}^{d\times d}.$$

Then B_n is obviously symmetric and

$$\begin{aligned}
\int_{\mathbb{R}^d} \langle z,\tilde{c}(x)\rangle^2\,(\nu_{n_k}(dx) - \nu(dx)) &= \int_{\mathbb{R}^d\setminus\{0\}} \frac{z^T\tilde{c}(x)\tilde{c}(x)^T z}{1\wedge|x|^2}(\rho_{n_k}(dx) - \rho(dx)) \\
&= z^T B_{n_k} z \quad \forall\, z \in \mathbb{R}^d,
\end{aligned}$$

so that $z^T B_{n_k} z$ converges as $k \to \infty$ for all $z \in \mathbb{R}^d$. Using the polarisation identity for quadratic forms, we have

$$2w^T B_{n_k} z = w^T B_{n_k} z + z^T B_{n_k} w = (w+z)^T B_{n_k}(w+z) - w^T B_{n_k} w - z^T B_{n_k} z$$

showing that $w^T B_{n_k} z$ converges as $k \to \infty$ for all $w, z \in \mathbb{R}^d$. Choosing appropriate unit vectors for w and z we see that B_{n_k} converges to some matrix $B \in \mathbb{R}^{d\times d}$ as $k \to \infty$, which is then necessarily symmetric. Together with (9.19) this gives

$$\operatorname{Log}\widehat{\mu}(z) = \int_{\mathbb{R}^d} \left(e^{i\langle z,x\rangle} - 1 - i\langle z, \widetilde{c}(x)\rangle\right) \nu(dx) + i\langle \beta, z\rangle - \frac{1}{2}\langle z, Bz\rangle \tag{9.20}$$

for all $z \in \mathbb{R}^d$, where ν is a Lévy measure, $\beta \in \mathbb{R}^d$ and $B \in \mathbb{R}^{d\times d}$ is symmetric. To show that this is the desired representation for the particular truncation function $\widetilde{c}$, we still need to prove that B is non-negative definite. To do so we define, for $z \in \mathbb{R}^d$ and $\varepsilon > 0$, the functions $h_z, \alpha_{z,\varepsilon} : \mathbb{R}^d \to \mathbb{R}$ by

$$h_z(x) := \frac{\langle z, \widetilde{c}(x)\rangle^2}{1 \wedge |x|^2} \mathbf{1}_{\mathbb{R}\setminus\{0\}}(x) \quad \text{and} \quad \alpha_{z,\varepsilon}(x) := h_z(x)\left(\frac{|x|}{\varepsilon} \wedge 1\right).$$

Then h_z is bounded on $\mathbb{R}^d$ and continuous on $\mathbb{R}^d\setminus\{0\}$, and $\alpha_{z,\varepsilon}$ is bounded and continuous on $\mathbb{R}^d$ with $\alpha_{z,\varepsilon}(0) = 0$. For $z \in \mathbb{R}^d$, we then have

$$\begin{aligned}
z^T B z &= \lim_{k\to\infty} z^T B_{n_k} z \\
&= \lim_{k\to\infty} \int_{\mathbb{R}^d\setminus\{0\}} h_z(x)\left(\rho_{n_k}(dx) - \rho(dx)\right) \\
&= \lim_{k\to\infty} \int_{\mathbb{R}^d\setminus\{0\}} \left(h_z(x) - \alpha_{z,\varepsilon}(x)\right)\left(\rho_{n_k}(dx) - \rho(dx)\right) \\
&\quad + \lim_{k\to\infty} \int_{\mathbb{R}^d} \alpha_{z,\varepsilon}(x)\left(\rho_{n_k}(dx) - \rho(dx)\right) \\
&= \lim_{k\to\infty} \int_{\mathbb{R}^d\setminus\{0\}} \left(h_z(x) - \alpha_{z,\varepsilon}(x)\right)\left(\rho_{n_k}(dx) - \rho(dx)\right)
\end{aligned}$$

since $\rho_{n_k} \overset{w}{\to} \rho$ as $k \to \infty$. But $h_z(x) - \alpha_{z,\varepsilon}(x) = 0$ for $|x| \geq \varepsilon$ and

$$0 \leq h_z(x) - \alpha_{z,\varepsilon}(x) \leq h_z(x) \leq \frac{|z|^2|\widetilde{c}(x)|^2}{|x|^2} = |z|^2 \quad \text{for } 0 < |x| \leq \varepsilon \leq 1.$$

Hence

$$z^T B z \geq -|z|^2 \rho(\{x \in \mathbb{R}^d : 0 < |x| \leq |\varepsilon|\}) \quad \forall\, z \in \mathbb{R}^d$$

for any $\varepsilon \in (0,1]$ and, letting $\varepsilon \to 0$, finiteness of ρ shows that $z^T B z \geq 0$, so that B is non-negative definite.

Finally, returning to our original truncation function c, define

$$\gamma := \beta + \int_{\mathbb{R}^d} (c(x) - \tilde{c}(x))\, \nu(\mathrm{d}x).$$

This is well defined since $|c(x) - \tilde{c}(x)| \le C(1 \wedge |x|^2)$ for all $x \in \mathbb{R}^d$ and for some constant C by the boundedness of c and (9.3). From (9.20), we then obtain

$$\operatorname{Log} \widehat{\mu}(z) = \left(\mathrm{i}\langle \gamma, z\rangle - \frac{1}{2}\langle z, Bz\rangle + \int_{\mathbb{R}^d} (\mathrm{e}^{\mathrm{i}\langle z,x\rangle} - 1 - \mathrm{i}\langle z, c(x)\rangle)\, \nu(\mathrm{d}x) \right)$$

for all $z \in \mathbb{R}^d$, with $\gamma \in \mathbb{R}^d$, ν a Lévy measure and B non-negative definite, completing the proof. □

Definition 9.14. Suppose that $\mu \in \mathrm{ID}(\mathbb{R}^d)$ and $c : \mathbb{R}^d \to \mathbb{R}^d$ is a fixed truncation function.

(a) Then the unique triplet (A, ν, γ), consisting of a non-negative definite symmetric matrix $A \in \mathbb{R}^{d\times d}$, a Lévy measure ν on $\mathbb{R}^d$ and a constant $\gamma \in \mathbb{R}^d$ such that

$$\widehat{\mu}(z) = \exp\left(\mathrm{i}\langle \gamma, z\rangle - \frac{1}{2}\langle z, Az\rangle + \int_{\mathbb{R}^d} (\mathrm{e}^{\mathrm{i}\langle z,x\rangle} - 1 - \mathrm{i}\langle z, c(x)\rangle)\, \nu(\mathrm{d}x) \right) \tag{9.21}$$

for all $z \in \mathbb{R}^d$, is called the *characteristic triplet of μ with respect to c*, or the *Lévy–Khintchine triplet of μ with respect to c*, or the *generating triplet of μ with respect to c*. The non-negative definite symmetric matrix A is called the *Gaussian variance of μ*, and ν is called the *Lévy measure of μ*. The representation (9.21) is called the *Lévy–Khintchine representation of μ with respect to c* or *Lévy–Khintchine formula.* In order to denote the dependence on c, we also write $(A, \nu, \gamma)_c$ for the characteristic triplet with respect to c, and sometimes we even write $(A, \nu, \gamma^c)_c$ in order to make the dependence on c even clearer (since γ is the only term that depends on c, as we show below in Proposition 9.17). We also call γ^c the *location parameter of μ (with respect to the truncation function c).*

(b) When c is the standard truncation function $x \mapsto \mathbf{1}_D(x)x$ as in Definition 9.4, we call the corresponding characteristic triplet *the (standard) characteristic, generating or Lévy–Khintchine triplet* of μ and simply write (A, ν, γ) (omitting explicit mention of the truncation function). The representation

$$\widehat{\mu}(z) = \exp\left(\mathrm{i}\langle \gamma, z\rangle - \frac{1}{2}\langle z, Az\rangle + \int_{\mathbb{R}^d} (\mathrm{e}^{\mathrm{i}\langle z,x\rangle} - 1 - \mathrm{i}\langle z, x\rangle \mathbf{1}_D(x))\, \nu(\mathrm{d}x) \right) \tag{9.22}$$

for all $z \in \mathbb{R}^d$ is then called *the (standard) Lévy–Khintchine representation of μ.*

(c) The function $\Psi(z) = \Psi_\mu(z) := \operatorname{Log}\widehat{\mu}(z)$, $z \in \mathbb{R}^d$, i. e.,

$$\Psi(z) = \mathrm{i}\langle \gamma^c, z\rangle - \frac{1}{2}\langle z, Az\rangle + \int_{\mathbb{R}^d} \left(\mathrm{e}^{\mathrm{i}\langle z,x\rangle} - 1 - \mathrm{i}\langle z, c(x)\rangle\right)\nu(\mathrm{d}x), \quad z \in \mathbb{R}^d,$$

is called the *characteristic exponent of* μ and is independent of the specific truncation function c.

(d) When X is a random vector with $\mathcal{L}(X) \in \mathrm{ID}(\mathbb{R}^d)$, we also speak of the characteristic triplet and characteristic exponent of X (meaning that of $\mathcal{L}(X)$) and write Ψ_X for $\Psi_{\mathcal{L}(X)}$.

Example 9.15. Suppose that $\mu_1 \in \mathrm{ID}(\mathbb{R}^d)$ and $\mu_2 \in \mathrm{ID}(\mathbb{R}^d)$ have characteristic triplets with respect to a fixed truncation function c given by $(A_1, \nu_1, \gamma_1)_c$ and $(A_2, \nu_2, \gamma_2)_c$ respectively. Suppose also that their characteristic exponents are Ψ_{μ_1} and Ψ_{μ_2} respectively. Then it is easy to see that $\mu_1 * \mu_2$ has characteristic exponent $\Psi_{\mu_1} + \Psi_{\mu_2}$ and characteristic triplet $(A_1 + A_2, \nu_1 + \nu_2, \gamma_1 + \gamma_2)_c$ with respect to c.

Example 9.16. We list here a number of illustrative examples.

(a) The (standard) characteristic triplet of an $N(\gamma, A)$ distribution is $(A, 0, \gamma)$.

(b) If μ is the compound Poisson distribution on $\mathbb{R}^d$ with parameter $\alpha > 0$ and jump size distribution σ then

$$\begin{aligned}\operatorname{Log}\widehat{\mu}(z) &= \alpha \int_{\mathbb{R}^d} (\mathrm{e}^{\mathrm{i}\langle z,x\rangle} - 1)\sigma(\mathrm{d}x) \\ &= \mathrm{i}\left\langle z, \alpha \int_{|x|\le 1} x\,\sigma(\mathrm{d}x)\right\rangle + \int_{\mathbb{R}^d} \left(\mathrm{e}^{\mathrm{i}\langle z,x\rangle} - 1 - \mathrm{i}\langle z, \mathbf{1}_D(x)x\rangle\right)\alpha\,\sigma(\mathrm{d}x)\end{aligned}$$

so that μ has (standard) characteristic triplet $(0, \alpha\sigma, \alpha\int_{|x|\le 1} x\,\sigma(\mathrm{d}x))$.

(c) If μ is the Poisson distribution with mean $\alpha > 0$, i. e., if $\mu(\{k\}) = \mathrm{e}^{-\alpha}\alpha^k/k!$, $k \in \mathbb{N}_0$, then it is a compound Poisson distribution with jump rate α and jump size distribution δ_1 and so, by (b), its standard characteristic triplet is $(0, \alpha\delta_1, \alpha)$.

(d) If μ is the geometric distribution (on $\mathbb{R}$) with parameter $p \in (0,1)$, i. e., if $\mu(\{k\}) = pq^k$ for $k \in \mathbb{N}_0$, where $q := 1 - p$, then

$$\widehat{\mu}(z) = \sum_{k=0}^{\infty} \mathrm{e}^{\mathrm{i}kz} pq^k = p\sum_{k=0}^{\infty} (q\mathrm{e}^{\mathrm{i}z})^k = \frac{p}{1 - q\mathrm{e}^{\mathrm{i}z}} \qquad \forall\, z \in \mathbb{R}.$$

Since the values of $\widehat{\mu}$ are restricted to $\{w \in \mathbb{C} : \Re w > 0\}$, the distinguished logarithm and the principal branch of the logarithm of $\widehat{\mu}$ coincide, and we can use the logarithmic expansion (8.2) to conclude that

$$\operatorname{Log}\widehat{\mu}(z) = \log p - \log(1 - q\mathrm{e}^{\mathrm{i}z}) = \log p + \sum_{j=1}^{\infty} j^{-1} q^j \mathrm{e}^{\mathrm{i}zj}.$$

Define $\nu := \sum_{j=1}^{\infty} j^{-1} q^j \delta_j$. Then ν is a finite Lévy measure with

$$\nu(\mathbb{R}) = \sum_{j=1}^{\infty} j^{-1} q^j = -\log(1 - q\mathrm{e}^{i0}) = -\log(1-q) = -\log p.$$

Hence

$$\operatorname{Log} \widehat{\mu}(z) = \int_{\mathbb{R}} (\mathrm{e}^{\mathrm{i}zx} - 1)\, \nu(\mathrm{d}x) = \mathrm{i}zq + \int_{\mathbb{R}} (\mathrm{e}^{\mathrm{i}zx} - 1 - \mathrm{i}zx\mathbf{1}_D(x))\, \nu(\mathrm{d}x)$$

for all $z \in \mathbb{R}$. We conclude that μ is infinitely divisible with characteristic triplet $(0, \nu, q)$ and that μ is a compound Poisson distribution with parameter $-\log p$ and jump size distribution $(-\log p)^{-1}\nu$.

(e) If μ is the exponential distribution (on $\mathbb{R}$) with parameter $\beta > 0$, i. e., if μ has Lebesgue density $x \mapsto \beta\mathrm{e}^{-\beta x}\mathbf{1}_{(0,\infty)}(x)$, then

$$\widehat{\mu}(z) = \int_0^{\infty} \beta\mathrm{e}^{-\beta x}\mathrm{e}^{\mathrm{i}zx}\, \mathrm{d}x = \beta \frac{\mathrm{e}^{(\mathrm{i}z-\beta)x}}{\mathrm{i}z - \beta}\bigg|_{x=0}^{x=\infty} = \frac{\beta}{\beta - \mathrm{i}z}, \qquad z \in \mathbb{R},$$

the values of which are restricted to $\{w \in \mathbb{C} : \Re w > 0\}$. Hence the distinguished logarithm and the principal branch of the logarithm of $\widehat{\mu}$ coincide and we obtain, for $z \in \mathbb{R}$,

$$\begin{aligned}
\operatorname{Log} \widehat{\mu}(z) &= \log \widehat{\mu}(z) = \log(\beta - \mathrm{i}0) - \log(\beta - \mathrm{i}z) \\
&= -\int_0^z \frac{\mathrm{d}}{\mathrm{d}t} \log(\beta - \mathrm{i}t)\, \mathrm{d}t = \int_0^z \frac{\mathrm{i}}{\beta - \mathrm{i}t}\, \mathrm{d}t \\
&= \int_0^z \mathrm{i} \int_0^{\infty} \mathrm{e}^{\mathrm{i}tx}\mathrm{e}^{-\beta x}\, \mathrm{d}x\, \mathrm{d}t = \int_0^{\infty} \mathrm{e}^{-\beta x} \int_0^z \mathrm{i}\,\mathrm{e}^{\mathrm{i}tx}\, \mathrm{d}t\, \mathrm{d}x \\
&= \int_0^{\infty} (\mathrm{e}^{\mathrm{i}xz} - 1)\frac{\mathrm{e}^{-\beta x}}{x}\, \mathrm{d}x,
\end{aligned}$$

where we have used the fundamental theorem of calculus and Fubini's theorem. Define the measure ν to have Lebesgue density $x \mapsto \frac{\mathrm{e}^{-\beta x}}{x}\mathbf{1}_{(0,\infty)}(x)$. This is not a finite measure, but a Lévy measure, and it follows that

$$\operatorname{Log} \widehat{\mu}(z) = \mathrm{i}\frac{1 - \mathrm{e}^{-\beta}}{\beta} z + \int_{\mathbb{R}} (\mathrm{e}^{\mathrm{i}xz} - 1 - \mathrm{i}zx\mathbf{1}_D(x))\, \nu(\mathrm{d}x),$$

so that the exponential distribution with parameter $\beta > 0$ is infinitely divisible with standard characteristic triplet $(0, \nu, \frac{1-\mathrm{e}^{-\beta}}{\beta})$. Observe that μ cannot be a compound Poisson distribution by the uniqueness of the characteristic triplet, since ν is infinite.

Many other important distributions are infinitely divisible. We refer to Sato [79] or Steutel and van Harn [83] for further examples of infinitely divisible distributions.

The characteristic triplet depends on the choice of the truncation function, but it is only the location parameter that depends on the choice. The precise dependence is established in the following proposition.

Proposition 9.17. *Suppose that $c : \mathbb{R}^d \to \mathbb{R}^d$ and $c' : \mathbb{R}^d \to \mathbb{R}^d$ are two truncation functions and $\mu \in \mathrm{ID}(\mathbb{R}^d)$. Let $(A, \nu, \gamma)_c$ be the characteristic triplet of μ with respect to c and let $(A', \nu', \gamma')_{c'}$ be the characteristic triplet with respect to c'. Then*

$$A' = A, \quad \nu' = \nu \quad \textit{and} \quad \gamma' = \gamma + \int_{\mathbb{R}^d} (c'(x) - c(x))\, \nu(\mathrm{d}x).$$

Proof. This follows easily from the fact that

$$\begin{aligned} &\mathrm{i}\langle \gamma, z\rangle + \int_{\mathbb{R}^d} (\mathrm{e}^{\mathrm{i}\langle z,x\rangle} - 1 - \mathrm{i}\langle z, c(x)\rangle)\, \nu(\mathrm{d}x) \\ &\quad = \mathrm{i}\langle \gamma, z\rangle + \int_{\mathbb{R}^d} \mathrm{i}\langle z, c'(x) - c(x)\rangle\, \nu(\mathrm{d}x) + \int_{\mathbb{R}^d} (\mathrm{e}^{\mathrm{i}\langle z,x\rangle} - 1 - \mathrm{i}\langle z, c'(x)\rangle)\, \nu(\mathrm{d}x), \end{aligned}$$

the uniqueness of the characteristic triplet for fixed truncation functions and the fact that $c - c'$ is integrable with respect to ν by boundedness and (9.4). □

It is sometimes convenient to use *extended truncation functions*, which do not satisfy all of the properties specified in Definition 9.4, but nevertheless give a Lévy–Khintchine-like representation when the Lévy measure satisfies suitable additional properties. The most relevant are the extended truncation functions $c_0(x) = 0$, $x \in \mathbb{R}^d$, and $c_m(x) = x$, $x \in \mathbb{R}^d$, to which the following proposition applies.

Proposition 9.18. *Suppose that $\mu \in \mathrm{ID}(\mathbb{R}^d)$ with standard characteristic triplet (A, ν, γ).*

(a) *Consider the extended truncation function $c_0 : \mathbb{R}^d \to \mathbb{R}^d$, $x \mapsto 0$, and suppose that the Lévy measure ν of μ is such that $\int_{|x|\le 1} |x|\, \nu(\mathrm{d}x) < \infty$. Then $x \mapsto \mathrm{e}^{\mathrm{i}\langle z,x\rangle} - 1$ is integrable with respect to ν for each $z \in \mathbb{R}^d$, and*

$$\begin{aligned} \widehat{\mu}(z) &= \exp\Big(\mathrm{i}\langle \gamma_0, z\rangle - \frac{1}{2}\langle z, Az\rangle + \int_{\mathbb{R}^d} (\mathrm{e}^{\mathrm{i}\langle z,x\rangle} - 1)\, \nu(\mathrm{d}x)\Big) \qquad (9.23) \\ &= \exp\Big(\mathrm{i}\langle \gamma_0, z\rangle - \frac{1}{2}\langle z, Az\rangle + \int_{\mathbb{R}^d} (\mathrm{e}^{\mathrm{i}\langle z,x\rangle} - 1 - \mathrm{i}\langle z, c_0(x)\rangle)\, \nu(\mathrm{d}x)\Big), \end{aligned}$$

where

$$\gamma_0 = \gamma - \int_{|x|\le 1} x\, \nu(\mathrm{d}x). \qquad (9.24)$$

Conversely, if a distribution μ' on $\mathbb{R}^d$ satisfies

$$\widehat{\mu'}(z) = \exp\left(\mathrm{i}\langle \gamma'_0, z\rangle - \frac{1}{2}\langle z, A'z\rangle + \int_{\mathbb{R}^d} (\mathrm{e}^{\mathrm{i}\langle z,x\rangle} - 1)\, \nu'(\mathrm{d}x) \right), \quad z \in \mathbb{R}^d, \tag{9.25}$$

for some $\gamma'_0 \in \mathbb{R}^d$, some symmetric and non-negative definite matrix $A' \in \mathbb{R}^{d\times d}$ and a Lévy measure ν' such that $\int_{|x|\le 1} |x|\, \nu'(\mathrm{d}x) < \infty$, then μ' is infinitely divisible, (A', ν', γ'_0) is unique and the standard characteristic triplet of μ is given by $(A', \nu', \gamma' = \gamma'_0 + \int_{|x|\le 1} x\, \nu'(\mathrm{d}x))$.

(b) *Consider the extended truncation function $c_m : \mathbb{R}^d \to \mathbb{R}^d$, $x \mapsto x$, and suppose the Lévy measure ν of μ is such that $\int_{|x|>1} |x|\, \nu(\mathrm{d}x) < \infty$. Then $x \mapsto \mathrm{e}^{\mathrm{i}\langle z,x\rangle} - 1 - \mathrm{i}\langle z, x\rangle$ is integrable with respect to ν for each $z \in \mathbb{R}^d$, and*

$$\begin{aligned}\widehat{\mu}(z) &= \exp\left(\mathrm{i}\langle \gamma_m, z\rangle - \frac{1}{2}\langle z, Az\rangle + \int_{\mathbb{R}^d} (\mathrm{e}^{\mathrm{i}\langle z,x\rangle} - 1 - \mathrm{i}\langle z, x\rangle)\, \nu(\mathrm{d}x) \right) \\ &= \exp\left(\mathrm{i}\langle \gamma_m, z\rangle - \frac{1}{2}\langle z, Az\rangle + \int_{\mathbb{R}^d} (\mathrm{e}^{\mathrm{i}\langle z,x\rangle} - 1 - \mathrm{i}\langle z, c_m(x)\rangle)\, \nu(\mathrm{d}x) \right),\end{aligned} \tag{9.26}$$

where

$$\gamma_m = \gamma + \int_{|x|>1} x\, \nu(\mathrm{d}x). \tag{9.27}$$

Conversely, if a distribution μ' on $\mathbb{R}^d$ satisfies

$$\widehat{\mu'}(z) = \exp\left(\mathrm{i}\langle \gamma'_m, z\rangle - \frac{1}{2}\langle z, A'z\rangle + \int_{\mathbb{R}^d} (\mathrm{e}^{\mathrm{i}\langle z,x\rangle} - 1 - \mathrm{i}\langle z, x\rangle)\, \nu'(\mathrm{d}x) \right),$$

for all $z \in \mathbb{R}^d$, for some $\gamma'_m \in \mathbb{R}^d$, some symmetric and non-negative definite matrix $A' \in \mathbb{R}^{d\times d}$ and a Lévy measure ν' such that $\int_{|x|>1} |x|\, \nu'(\mathrm{d}x) < \infty$, then μ' is infinitely divisible, (A', ν', γ'_m) is unique and the standard characteristic triplet of μ is given by $(A', \nu', \gamma' = \gamma'_m - \int_{|x|>1} x\, \nu'(\mathrm{d}x))$.

Proof. (a) Since $\lim_{y\in\mathbb{R}, y\to 0} y^{-1}(\mathrm{e}^{\mathrm{i}y} - 1) = \mathrm{i}$, we have $|\mathrm{e}^{\mathrm{i}\langle z,x\rangle} - 1| \le C|z|\, |x| \wedge 2$ for some constant C, so that $\mathrm{e}^{\mathrm{i}\langle z,x\rangle} - 1$ is integrable with respect to ν. Equation (9.23) with γ_0 as defined in (9.24) is then obvious from (9.22), using an argument similar to that in Proposition 9.17. Conversely, if μ' satisfies (9.25), then also

$$\widehat{\mu'}(z) = \exp\Big(\mathrm{i}\langle \gamma_0', z\rangle + \mathrm{i}\Big\langle \int_{|x|\le 1} x\,\nu'(\mathrm{d}x), z\Big\rangle - \frac{1}{2}\langle z, A'z\rangle$$
$$+ \int_{\mathbb{R}^d} (\mathrm{e}^{\mathrm{i}\langle z,x\rangle} - 1 - \mathrm{i}\langle z, x\rangle \mathbf{1}_D(x))\,\nu'(\mathrm{d}x)\Big),$$

so that μ is infinitely divisible with standard characteristic triplet (A', ν', γ') where $\gamma' = \gamma_0' + \int_{|x|\le 1} x\,\nu'(\mathrm{d}x)$. The uniqueness of (A', ν', γ_0') then follows from the uniqueness of the standard characteristic triplet. The proof of (b) is completely analogous. □

Example 9.19. Let μ be a compound Poisson distribution on $\mathbb{R}^d$. Then its parameter α and its jump size distribution σ are uniquely determined by μ. This follows from Proposition 9.18(a) together with (8.1).

Definition 9.20. Suppose that $\mu \in \mathrm{ID}(\mathbb{R}^d)$ and that its standard characteristic triplet is (A, ν, γ). If $\int_{|x|\le 1} |x|\,\nu(\mathrm{d}x) < \infty$, then the constant γ_0 of (9.24) is called the *drift* of μ and we denote the characteristic triplet with respect to the extended truncation function $c_0 = 0$ by $(A, \nu, \gamma_0)_0$. If $\int_{|x|>1} |x|\,\nu(\mathrm{d}x) < \infty$, then the constant γ_m in (9.27) is called the *center* of μ and we denote the characteristic triplet with respect to the extended truncation function $c_m(x) = x$ by $(A, \nu, \gamma_m)_m$.

It will be shown later in Remark 11.16 that $\int_{|x|>1} |x|\,\nu(\mathrm{d}x) < \infty$ is equivalent to $\int_{\mathbb{R}^d} |x|\,\mu(\mathrm{d}x) < \infty$ and that, under this condition, $\gamma_m = \int_{\mathbb{R}^d} x\,\mu(\mathrm{d}x)$, i. e., the center of μ is the mean of μ, i. e., the expectation of X when $\mathcal{L}(X) = \mu$.

Example 9.21. Returning to Example 9.16, we see from the characteristic triplets given there that the drift of an $N(\gamma, A)$ distribution is γ, the drift of a compound Poisson distribution is 0, the drift of the geometric distribution is 0, and the drift of the exponential distribution is 0. Although these examples all have drift, we shall see later that there are many infinitely divisible distributions, such as the stable distributions with index ≤ 1, for which the drift does not exist (i. e., for which $\int_{|x|\le 1} |x|\,\nu(\mathrm{d}x) = \infty$), see Lemma 14.7 below.

In one dimension, it is sometimes convenient to work with a "characteristic pair" instead of a characteristic triplet. This is achieved by considering $(1 \wedge x^2)\,\nu(\mathrm{d}x)$ instead of ν and by adding a point mass of size A at 0 to this new measure. This is made precise in the following theorem.

Theorem 9.22. *Let $c : \mathbb{R} \to \mathbb{R}$ be a truncation function on $\mathbb{R}$ which satisfies*

$$\lim_{x\to 0} \frac{c(x) - x}{x^2} = 0 \tag{9.28}$$

(a slightly stronger condition than (9.3)). Then $\lim_{x\to 0} x^{-2}(\mathrm{e}^{\mathrm{i}zx} - 1 - \mathrm{i}zc(x))$ exists for each $z \in \mathbb{R}$ and is equal to $-z^2/2$. Let μ be a distribution on $\mathbb{R}$. Then $\mu \in \mathrm{ID}(\mathbb{R})$ if and only if there exist $\gamma \in \mathbb{R}$ and a finite measure ζ on $(\mathbb{R}, \mathcal{B}_1)$ such that

$$\widehat{\mu}(z) = \exp\left\{ \mathrm{i}\gamma z + \int_{\mathbb{R}} \frac{\mathrm{e}^{\mathrm{i}zx} - 1 - \mathrm{i}zc(x)}{1 \wedge x^2} \,\zeta(\mathrm{d}x) \right\} \quad \forall\, z \in \mathbb{R}, \tag{9.29}$$

where the integrand is interpreted as $-z^2/2$ for $x = 0$. In this case, the pair (ζ, γ) is unique and related to the characteristic triplet $(A, \nu, \gamma')_c$ of μ with respect to c by

$$\gamma = \gamma' \quad \textit{and} \quad \zeta(\mathrm{d}x) = A\,\delta_0(\mathrm{d}x) + (1 \wedge x^2)\,\nu(\mathrm{d}x). \tag{9.30}$$

Proof. By a Taylor expansion, $\lim_{x\to 0} x^{-2}(\mathrm{e}^{\mathrm{i}zx} - 1 - \mathrm{i}zx) = -z^2/2$, hence (9.28) gives $\lim_{x\to 0} x^{-2}(\mathrm{e}^{\mathrm{i}zx} - 1 - \mathrm{i}zc(x)) = -z^2/2$. Now if $A \in [0, \infty)$ and ν is a Lévy measure on $\mathbb{R}$, then

$$\begin{aligned}
-Az^2/2 &+ \int_{\mathbb{R}} (\mathrm{e}^{\mathrm{i}zx} - 1 - \mathrm{i}zc(x))\,\nu(\mathrm{d}x) \\
&= -Az^2/2 + \int_{\mathbb{R}\setminus\{0\}} \frac{\mathrm{e}^{\mathrm{i}zx} - 1 - \mathrm{i}zc(x)}{1 \wedge x^2}(1 \wedge x^2)\,\nu(\mathrm{d}x) \\
&= \int_{\mathbb{R}} \frac{\mathrm{e}^{\mathrm{i}zx} - 1 - \mathrm{i}zc(x)}{1 \wedge x^2}\,\zeta(\mathrm{d}x), \quad z \in \mathbb{R},
\end{aligned}$$

where ζ is defined as $\zeta(\mathrm{d}x) := A\delta_0(\mathrm{d}x) + (1 \wedge x^2)\,\nu(\mathrm{d}x)$. The measure ζ is then obviously finite and it follows that $\mu \in \mathrm{ID}(\mathbb{R})$ has a representation as in (9.29) with γ and ζ given by (9.30). Conversely, if ζ is a finite measure, then ν defined by $\nu(\mathrm{d}x) := \mathbf{1}_{\mathbb{R}\setminus\{0\}}(x)\frac{1}{1\wedge x^2}\,\nu(\mathrm{d}x)$ is obviously a Lévy measure, and it follows that every distribution with representation (9.29) has a Lévy–Khintchine representation with characteristic triplet $(A, \nu, \gamma)_c$, where $A := \zeta(\{0\})$. Together with the uniqueness of the characteristic triplet, this completes the proof. □

Definition 9.23. Suppose that $\mu \in \mathrm{ID}(\mathbb{R})$ and $c : \mathbb{R} \to \mathbb{R}$ is a truncation function satisfying (9.28). Then the unique pair (ζ, γ), consisting of the finite measure ζ on $\mathbb{R}$ and the constant $\gamma \in \mathbb{R}$ such that (9.29) holds, is called the *characteristic pair of μ with respect to c* and denoted by $(\zeta, \gamma)_c$. When $c(x) = \mathbf{1}_D(x)x$, we also speak of the *standard characteristic pair* of μ and simply write (ζ, γ).

Remark 9.24. Some statements become easier in dimension 1 and more insight can be gained using characteristic pairs. For example, Lemma 9.12 can be restated in terms of the characteristic pair by saying that if $\mu_n \in \mathrm{ID}(\mathbb{R})$ with characteristic pair $(\zeta_n, \gamma_n)_c$ converges weakly, then $(\zeta_n)_{n\in\mathbb{N}}$ must be tight and uniformly bounded. This could have been used in the proof of Theorem 9.13 by considering the function $x \mapsto (1\wedge x^2)^{-1}(\mathrm{e}^{\mathrm{i}zx} - 1 - \mathrm{i}z\bar{c}(x))$ rather than u_z. The reader is encouraged to carry out the details of this simplified proof in Exercise 9.2.

9.2 Linear transformations and independence of infinitely divisible distributions

In this section, we characterize independence of infinitely divisible distributions in terms of the characteristic triplet and also show how the characteristic triplet of an infinitely divisible distribution changes under affine linear transformations. We start with the latter.

Theorem 9.25 (Characteristic triplet under linear transformations). *Suppose that $\mu = \mathcal{L}(X) \in \mathrm{ID}(\mathbb{R}^d)$ is the distribution of an $\mathbb{R}^d$-valued random vector X with characteristic exponent Ψ_X and characteristic triplet $(A, \nu, \gamma^{c_1})_{c_1}$ with respect to the truncation function $c_1 : \mathbb{R}^d \to \mathbb{R}^d$. Suppose also that $C \in \mathbb{R}^{m\times d}$, that $b \in \mathbb{R}^m$ and that $c_2 : \mathbb{R}^m \to \mathbb{R}^m$ is a truncation function in $\mathbb{R}^m$. If*

$$Y = b + CX,$$

then $\mathcal{L}(Y) \in \mathrm{ID}(\mathbb{R}^m)$ has characteristic exponent

$$\Psi_Y : \mathbb{R}^m \to \mathbb{C}, \quad z \mapsto \Psi_Y(z) = \Psi_X(C^T z) + \mathrm{i}\langle b, z\rangle$$

and characteristic triplet $(A_Y, \nu_Y, \gamma_Y^{c_2})_{c_2}$, where

$$A_Y = CAC^T, \tag{9.31}$$

$$\nu_Y(\mathrm{d}y) = (C(\nu))(\mathrm{d}y)\,\mathbf{1}_{\mathbb{R}^m\setminus\{0\}}(y) = (\nu \circ C^{-1})(\mathrm{d}y)\,\mathbf{1}_{\mathbb{R}^m\setminus\{0\}}(y), \tag{9.32}$$

$$\gamma_Y^{c_2} = b + C\gamma^{c_1} + \int_{\mathbb{R}^d} (c_2(Cx) - Cc_1(x))\,\nu(\mathrm{d}x). \tag{9.33}$$

Notice that ν_Y is the image measure $C(\nu)$ of ν under the linear mapping $C : \mathbb{R}^d \to \mathbb{R}^m$, restricted to $\mathbb{R}^m \setminus \{0\}$ with $\nu_Y(\{0\})$ defined to be 0.

Proof. It is clear that $\Psi_Y(z) = \Psi_X(C^T z) + \mathrm{i}\langle b, z\rangle$, since $\mathbb{E}\,\mathrm{e}^{\mathrm{i}\langle z, CX+b\rangle} = \mathbb{E}\mathrm{e}^{\mathrm{i}\langle C^T z, X\rangle}\,\mathrm{e}^{\mathrm{i}\langle b, z\rangle}$. Moreover, $A_Y \in \mathbb{R}^{m\times m}$, as defined in (9.31), is obviously symmetric and non-negative definite, and ν_Y, as defined in (9.32), is a Lévy measure on $\mathbb{R}^m$, since $\nu_Y(\{0\}) = 0$ and

$$\begin{aligned}\int_{\mathbb{R}^m} (|y|^2 \wedge 1)\,\nu_Y(\mathrm{d}y) &= \int_{\mathbb{R}^m} (|y|^2 \wedge 1)\,(\nu \circ C^{-1})(\mathrm{d}y)\\ &= \int_{\mathbb{R}^d} (|(Cx)|^2 \wedge 1)\,\nu(\mathrm{d}x)\\ &\le (\|C\|^2 \vee 1) \int_{\mathbb{R}^d} (|x|^2 \wedge 1)\,\nu(\mathrm{d}x) < \infty,\end{aligned}$$

where $\|C\|$ denotes the spectral norm of C. In addition, $x \mapsto c_2(Cx) - Cc_1(x)$ is Borel measurable, bounded on $\mathbb{R}^d$ and bounded by a constant times $|x|^2$ in a neighborhood of

0 by (9.3). It is therefore integrable with respect to ν, so that $\gamma_Y^{c_2}$ as specified by (9.33) is well defined.

To see that $\mathcal{L}(Y) \in \mathrm{ID}(\mathbb{R}^m)$ with characteristic triplet $(A_Y, \nu_Y, \gamma_Y^{c_2})_{c_2}$, let (A, ν, γ) be the standard characteristic triplet of X so that, by Proposition 9.17,

$$\gamma = \gamma^{c_1} + \int_{\mathbb{R}^d} (\mathbf{1}_D(x)x - c_1(x))\,\nu(\mathrm{d}x), \tag{9.34}$$

where $D = \{x \in \mathbb{R}^d : |x| \le 1\}$. Then

$$\begin{aligned}
&\Psi_Y(z) \\
&= \mathrm{i}\langle b, z\rangle + \Psi_X(C^T z) \\
&= \mathrm{i}\langle b, z\rangle + \mathrm{i}\langle \gamma, C^T z\rangle - \frac{1}{2}\langle C^T z, AC^T z\rangle \\
&\quad + \int_{\mathbb{R}^d} (\mathrm{e}^{\mathrm{i}\langle C^T z, x\rangle} - 1 - \mathrm{i}\langle C^T z, \mathbf{1}_D(x)x\rangle)\,\nu(\mathrm{d}x) \\
&= \mathrm{i}\langle b + C\gamma, z\rangle - \frac{1}{2}\langle z, CAC^T z\rangle \\
&\quad + \int_{\mathbb{R}^d} (\mathrm{e}^{\mathrm{i}\langle z, Cx\rangle} - 1 - \mathrm{i}\langle z, c_2(Cx)\rangle + \mathrm{i}\langle z, c_2(Cx) - \mathbf{1}_D(x)Cx\rangle)\,\nu(\mathrm{d}x) \\
&= \mathrm{i}\Big\langle z, b + C\gamma + \int_{\mathbb{R}^d} (c_2(Cx) - \mathbf{1}_D(x)Cx)\,\nu(\mathrm{d}x)\Big\rangle - \frac{1}{2}\langle z, CAC^T z\rangle \\
&\quad + \int_{\mathbb{R}^m} (\mathrm{e}^{\mathrm{i}\langle z, y\rangle} - 1 - \mathrm{i}\langle z, c_2(y)\rangle)\,(\nu \circ C^{-1})(\mathrm{d}y).
\end{aligned}$$

Together with (9.34) this implies that Y is infinitely divisible with characteristic triplet given by (9.31)–(9.33). □

If $\mathcal{L}(X) = \mu \in \mathrm{ID}(\mathbb{R}^d)$ has a drift or center, then so has $\mathcal{L}(b + CX)$, and the latter's drift or center is related to that of $\mathcal{L}(X)$ by the following corollary.

Corollary 9.26. *Suppose that X is a random vector, $\mathcal{L}(X) \in \mathrm{ID}(\mathbb{R}^d)$, $C \in \mathbb{R}^{m\times d}$, $b \in \mathbb{R}^m$, and $Y := b + CX$. Denote the Lévy measure of $\mathcal{L}(X)$ by ν and the Lévy measure of $\mathcal{L}(Y)$ by ν_Y.*

(a) *If $\int_{|x|\le 1} |x|\,\nu(\mathrm{d}x) < \infty$, then also $\int_{|x|\le 1} |x|\,\nu_Y(\mathrm{d}x) < \infty$, and the drift γ_Y^0 of $\mathcal{L}(Y)$ is related to the drift γ^0 of μ by $\gamma_Y^0 = b + C\gamma^0$.*

(b) *If $\int_{|x|>1} |x|\,\nu(\mathrm{d}x) < \infty$, then also $\int_{|x|>1} |x|\,\nu_Y(\mathrm{d}x) < \infty$, and the center γ_Y^m of $\mathcal{L}(Y)$ is related to the center γ^m of $\mathcal{L}(X)$ by $\gamma_Y^m = b + C\gamma^m$.*

Proof. If $\int_{|x|\le 1} |x|\,\nu(dx) < \infty$, then, as in the proof of Theorem 9.25, we have

$$\int_{\mathbb{R}^m} (|y| \wedge 1)\,\nu_Y(dy) \le (\|C\| \wedge 1) \int_{\mathbb{R}^d} (|x| \wedge 1)\,\nu(dx) < \infty.$$

If $\int_{|x|>1} |x|\,\nu(dx) < \infty$, then by continuity of $x \mapsto Cx$ there is some $\varepsilon \in (0,1)$ such that $\{x \in \mathbb{R}^d : |Cx| > 1\} \subset \{x \in \mathbb{R}^d : |x| \ge \varepsilon\}$ and

$$\begin{aligned}\int_{|y|>1} |y|\,\nu_Y(dy) &= \int_{|y|>1} |y|\,(\nu \circ C^{-1})(dy) = \int_{\{x\in\mathbb{R}^d:|Cx|>1\}} |Cx|\,\nu(dx) \\ &\le \|C\| \int_{\{x\in\mathbb{R}^d:|x|\ge\varepsilon\}} |x|\,\nu(dx) < \infty.\end{aligned}$$

Hence if the drift or center of $\mathcal{L}(X)$ exists, then so does the drift or center of $\mathcal{L}(Y)$, respectively. The equations $\gamma_Y^0 = b + C\gamma^0$ and $\gamma_Y^m = b + C\gamma^m$ then follow by analogy with the corresponding calculations in the proof of Theorem 9.25, or alternatively by applying (9.24) and (9.27), respectively. □

Example 9.27. Suppose that $\mu = \mathcal{L}((X_1,X_2)^T) \in \mathrm{ID}(\mathbb{R}^2)$ has standard characteristic triplet (A,ν,γ), where $A = \begin{pmatrix} a_{11} & a_{12} \\ a_{12} & a_{22}\end{pmatrix} \in \mathbb{R}^{2\times 2}$, and $\gamma = (\gamma_1,\gamma_2)^T$. Letting $C = (1,0) \in \mathbb{R}^{1\times 2}$, we see that $\mathcal{L}(X_1) = \mathcal{L}(C(X_1,X_2)^T) \in \mathrm{ID}(\mathbb{R})$ with standard characteristic triplet $(a_{11},\nu_{X_1},\beta_1)$, where $\nu_{X_1}(dy) = (\nu \circ C^{-1})(dy)\mathbf{1}_{\mathbb{R}\setminus\{0\}}(y)$ is the projection of ν onto $\mathbb{R} \setminus \{0\}$, i. e., $\nu_{X_1}(B) = \nu(B \times \mathbb{R})$ for all $B \in \mathcal{B}_1$ with $0 \notin B$, and

$$\beta_1 = \gamma_1 + \int_{\mathbb{R}^2} (\mathbf{1}_{\{|x_1|\le 1\}}(x_1,x_2) - \mathbf{1}_{\{x_1^2+x_2^2\le 1\}}(x_1,x_2))x_1\,\nu(d(x_1,x_2)).$$

Observe that in general β_1 differs from the first component γ_1 of γ, when using the standard characteristic triplets. However, when the drift $\gamma^0 = (\gamma_1^0,\gamma_2^0)^T$ or center $\gamma^m = (\gamma_1^m,\gamma_2^m)^T$ of μ exists, then γ_1^0 is the drift of $\mathcal{L}(X_1)$ and γ_1^m is the center of $\mathcal{L}(X_1)$, respectively. Similar formulas hold for the characteristic triplet of $\mathcal{L}(X_2)$.

It is often important to know the form of the characteristic triplet of a random vector X with $\mathcal{L}(X) \in \mathrm{ID}(\mathbb{R}^d)$ and independent components. The following theorem deals with this question.

Theorem 9.28 (Inf. divisibility and indep.). *Let $X = (X_1,\dots,X_d)^T$ be a random vector in $\mathbb{R}^d$ with distribution μ. Then the following are equivalent:*
(i) *$X_1,\dots,X_d$ are independent and $\mathcal{L}(X_i) \in \mathrm{ID}(\mathbb{R})$ for all $i \in \{1,\dots,d\}$.*
(ii) *$X_1,\dots,X_d$ are independent and $\mu \in \mathrm{ID}(\mathbb{R}^d)$.*
(iii) *$X_1,\dots,X_d$ are pairwise independent and $\mu \in \mathrm{ID}(\mathbb{R}^d)$.*
(iv) *$\mu \in \mathrm{ID}(\mathbb{R}^d)$ with standard characteristic triplet (A,ν,γ), where A is diagonal and ν is supported on the axes. (The fact that ν is supported on the axes means $\nu(\mathbb{R}^d \setminus$*

$\bigcup_{j=1}^d E_j) = 0$, where $E_j := \{0\}^{j-1} \times \mathbb{R} \times \{0\}^{d-j}$ is the jth axis in $\mathbb{R}^d$, or equivalently that $\nu(\{(x_1, \ldots, x_d)^T \in \mathbb{R}^d : x_i x_j \neq 0$ for some $i \neq j\}) = 0$.)

Proof. That (ii) implies (i) is clear from Theorem 9.25, and that (i) implies (ii) follows from the fact that if $\mu_i := \mathcal{L}(X_i) \in \mathrm{ID}(\mathbb{R})$ for each $i \in \{1, \ldots, d\}$ then for each $n \in \mathbb{N}$ there exist distributions $\mu_{i,n}$ with $\mu_{i,n}^{*n} = \mu_i$. It is then easy to see that $(\mu_{1,n} \otimes \cdots \otimes \mu_{d,n})^{*n} = \mu_1 \otimes \cdots \otimes \mu_d = \mu$, showing that $\mu \in \mathrm{ID}(\mathbb{R}^d)$.

To show the equivalence of (ii) and (iv) denote the standard characteristic triplet of μ by (A, ν, γ), where $\gamma = (\gamma_1, \ldots, \gamma_d)^T$ and $A = (a_{jk})_{j,k=1,\ldots,d}$. Denote by ν_j the Lévy measure of $\mu_j := \mathcal{L}(X_j), j \in \{1, \ldots, d\}$. Then μ_j has standard characteristic triplet (a_{jj}, ν_j, β_j) for some $\beta_j \in \mathbb{R}$. Define $\widetilde{\beta} := (\beta_1, \ldots, \beta_d)^T \in \mathbb{R}^d$, the diagonal matrix $\widetilde{A} := \mathrm{diag}(a_{11}, \ldots, a_{dd}) \in \mathbb{R}^{d \times d}$ and the Lévy measure $\widetilde{\nu}$ on $\mathbb{R}^d$ by

$$\widetilde{\nu}(\mathrm{d}(x_1, \ldots, x_d)) := \sum_{j=1}^d \nu_j(\mathrm{d}x_j) \prod_{k=1, k \neq j}^d \delta_0(\mathrm{d}x_k).$$

(It is easy to see that $\widetilde{\nu}$ is indeed a Lévy measure on $\mathbb{R}^d$ and that $\widetilde{A}$ is non-negative definite.) For $z = (z_1, \ldots, z_d)^T \in \mathbb{R}^d$ we then have

$$\begin{aligned}
&\sum_{j=1}^d \Psi_{\mu_j}(z_j) \\
&= \sum_{j=1}^d \left[\mathrm{i} z_j \beta_j - \frac{1}{2} z_j a_{jj} z_j + \int_{\mathbb{R}} (\mathrm{e}^{\mathrm{i} z_j x_j} - 1 - \mathrm{i} z_j x_j \mathbf{1}_{[-1,1]}(x_j)) \, \nu_j(\mathrm{d}x_j) \right] \\
&= \mathrm{i} \langle \widetilde{\beta}, z \rangle - \frac{1}{2} \langle z, \widetilde{A} z \rangle \\
&\quad + \sum_{j=1}^d \int_{\mathbb{R}^d} (\mathrm{e}^{\mathrm{i}\langle z, x \rangle} - 1 - \mathrm{i} \langle z, x \rangle \mathbf{1}_{\{|x| \le 1\}}(x)) \nu_j(\mathrm{d}x_j) \prod_{k=1, k \neq j}^d \delta_0(\mathrm{d}x_k) \\
&= \mathrm{i} \langle \widetilde{\beta}, z \rangle - \frac{1}{2} \langle z, \widetilde{A} z \rangle + \int_{\mathbb{R}^d} (\mathrm{e}^{\mathrm{i}\langle z, x \rangle} - 1 - \mathrm{i} \langle z, x \rangle \mathbf{1}_{\{|x| \le 1\}}(x)) \widetilde{\nu}(\mathrm{d}x).
\end{aligned}$$

Now $X_1, \ldots, X_d$ are independent if and only if $\phi_\mu(z_1, \ldots, z_d) = \prod_{j=1}^d \phi_{\mu_j}(z_j)$, or equivalently $\Psi_\mu(z_1, \ldots, z_d) = \sum_{j=1}^d \Psi_{\mu_j}(z_j)$, for all $(z_1, \ldots, z_d)^T \in \mathbb{R}^d$. By the previous calculation and the uniqueness of the characteristic triplet, we see that $X_1, \ldots, X_d$ are independent if and only if $(A, \nu, \gamma) = (\widetilde{A}, \widetilde{\nu}, \widetilde{\beta})$. But $\widetilde{A}$ is diagonal and $\widetilde{\nu}$ is concentrated on the axes, showing that (ii) implies (iv). Conversely, if (iv) holds, then A is diagonal and hence equal to $\widetilde{A}$. Since ν is concentrated on the axes, we can write $\nu(\mathrm{d}(x_1, \ldots, x_d)) = \sum_{j=1}^d \nu_j'(\mathrm{d}x_j) \prod_{k=1,\ldots,d; k \neq j} \delta_0(\mathrm{d}x_k)$ for some one-dimensional Lévy measures $\nu_1', \ldots, \nu_d'$. But Theorem 9.25 then shows that $\nu_j' = \nu_j$ and

$$\beta_j = \gamma_j + \int_{\mathbb{R}^d} (\mathbf{1}_{\{|x_j|\le 1\}}(x_j) - \mathbf{1}_{\{|x|\le 1\}}(x))x_j\, \nu(dx) = \gamma_j$$

since ν is concentrated on the axes. Hence we have $A = \widetilde{A}$, $\nu = \widetilde{\nu}$ and $\gamma = \widetilde{\beta}$, showing that $X_1, \dots, X_d$ are independent.

Finally, it is clear that (ii) implies (iii). For the converse, suppose that $d \ge 2$, that $\mu \in \mathrm{ID}(\mathbb{R}^d)$ and that $X_1, \dots, X_d$ are pairwise independent. Write $A = (a_{ij})_{i,j=1,\dots,d}$. Fix k and m such that $1 \le k < m \le d$. By Theorem 9.25, $\mathcal{L}((X_k, X_m)^T) \in \mathrm{ID}(\mathbb{R}^2)$ with standard characteristic triplet

$$\left(\begin{pmatrix} a_{kk} & a_{km} \\ a_{km} & a_{mm} \end{pmatrix}, \nu_{km}, \beta_{km}\right)$$

for some $\beta_{km} \in \mathbb{R}^2$ and

$$\nu_{km}(d(x_k, x_m)) = (\nu \circ C_{km}^{-1})(d(x_k, x_m))\, \mathbf{1}_{\{x_k^2 + x_m^2 \neq 0\}}(x_k, x_m),$$

where $C_{km}(x_1, \dots, x_m) = (x_k, x_m)^T$. From the independence of X_k and X_m and the part already proved, we conclude that $a_{km} = 0$ and that ν_{km} is concentrated on the axes, i. e., $\nu_{km}(\{(x_k, x_m)^T \in \mathbb{R}^2 : x_k x_m \neq 0\}) = 0$. From the definition of ν_{km} we conclude that $\nu(\{(x_1, \dots, x_d)^T \in \mathbb{R}^d : x_k x_m \neq 0\}) = 0$ and, letting k and m vary over all $k < m$, we conclude that $\nu(\{(x_1, \dots, x_d)^T \in \mathbb{R}^d : x_i x_j \neq 0 \text{ for some } i < j\}) = 0$. However, the latter means that ν is concentrated on the axes, and since $a_{km} = 0$ for all $k < m$, we also see that A must be a diagonal matrix. Hence (iii) implies (iv) and hence also (ii). □

It is easy to extend Theorem 9.28 to the case where $X = (X_1^T, \dots, X_d^T)^T$ is a random vector in $\mathbb{R}^{d_1} \times \cdots \times \mathbb{R}^{d_m}$. The precise formulation is left to the reader (see Exercise 9.3).

9.3 Convergence of infinitely divisible distributions

In this section, we characterize weak convergence of infinitely divisible distributions in terms of the convergence of their characteristic triplets. For this we need the following function space and lemma.

Notation 9.29. Denote by $C_\sharp(\mathbb{R}^d; \mathbb{R})$ the set of all bounded continuous functions $f : \mathbb{R}^d \to \mathbb{R}$ such that $f(x) = 0$ for all x in a neighborhood of 0.

Lemma 9.30. *Let $(\nu_n)_{n\in\mathbb{N}}$ be a sequence of Lévy measures and ν a Lévy measure on $\mathbb{R}^d$. Let ρ_n and ρ be finite measures on $\mathbb{R}^d$ such that $\mathbf{1}_{\mathbb{R}^d\setminus\{0\}}(x)\rho_n(dx) = \mathbf{1}_{\mathbb{R}^d\setminus\{0\}}(x)(1 \wedge |x|^2)\, \nu_n(dx)$ and $\mathbf{1}_{\mathbb{R}^d\setminus\{0\}}(x)\rho(dx) = \mathbf{1}_{\mathbb{R}^d\setminus\{0\}}(x)(1 \wedge |x|^2)\, \nu(dx)$. This means that the measures ρ_n and ρ on sets excluding 0 have density $(1 \wedge |x|^2)$ with respect to ν_n or ν, but they may have an atom at 0. Suppose that ρ_n converges weakly to ρ as $n \to \infty$. Then*

$$\lim_{n\to\infty} \int_{\mathbb{R}^d} f(x)\, \nu_n(\mathrm{d}x) = \int_{\mathbb{R}^d} f(x)\, \nu(\mathrm{d}x) \quad \forall f \in C_\sharp(\mathbb{R}^d;\mathbb{R}).$$

Proof. Let $f \in C_\sharp(\mathbb{R}^d;\mathbb{R})$. Then $x \mapsto \frac{f(x)}{1\wedge|x|^2}$ is in $C_b(\mathbb{R}^d;\mathbb{R})$ and we obtain

$$\int_{\mathbb{R}^d} f(x)\, \nu_n(\mathrm{d}x) = \int_{\mathbb{R}^d} \frac{f(x)}{1\wedge|x|^2}\, \rho_n(\mathrm{d}x) \to \int_{\mathbb{R}^d} \frac{f(x)}{1\wedge|x|^2}\, \rho(\mathrm{d}x) = \int_{\mathbb{R}^d} f(x)\, \nu(\mathrm{d}x)$$

as $n \to \infty$. □

Now we can characterize the convergence of infinitely divisible distributions in terms of convergence of their characteristic triplets:

Theorem 9.31 (Convergence of infinitely divisible distributions on $\mathbb{R}^d$). *Let $c : \mathbb{R}^d \to \mathbb{R}^d$ be a fixed truncation function that is continuous on $\mathbb{R}^d$ and satisfies*

$$\lim_{x\to 0} \frac{c(x)-x}{|x|^2} = 0 \tag{9.35}$$

(which is (9.28) in the special case $d = 1$, and is slightly stronger than (9.3)). Let $(\mu_n)_{n\in\mathbb{N}}$ be a sequence in $\mathrm{ID}(\mathbb{R}^d)$ with characteristic triplets $(A_n, \nu_n, \gamma_n^c)_c$ with respect to c, and let μ in $\mathrm{ID}(\mathbb{R}^d)$ have characteristic triplet $(A, \nu, \gamma^c)_c$ with respect to c. Then μ_n converges weakly to μ as $n \to \infty$ if and only if the following conditions are all satisfied.

$$A_n + \int_{\mathbb{R}^d} c(x)c(x)^T\, \nu_n(\mathrm{d}x) \to A + \int_{\mathbb{R}^d} c(x)c(x)^T\, \nu(\mathrm{d}x), \quad \text{as } n \to \infty, \tag{9.36}$$

$$\int_{\mathbb{R}^d} f(x)\, \nu_n(\mathrm{d}x) \to \int_{\mathbb{R}^d} f(x)\, \nu(\mathrm{d}x), \quad \text{as } n \to \infty, \forall f \in C_\sharp(\mathbb{R}^d;\mathbb{R}), \tag{9.37}$$

and

$$\gamma_n^c \to \gamma^c, \quad \text{as } n \to \infty. \tag{9.38}$$

Proof. The proof is largely similar to the proof of Theorem 9.13.

For fixed $z \in \mathbb{R}^d$, define the function $u_z : \mathbb{R}^d \to \mathbb{R}$ by

$$u_z(0) := 0 \quad \text{and} \quad u_z(x) := \frac{\mathrm{e}^{\mathrm{i}\langle z,x\rangle} - 1 - \mathrm{i}\langle z, c(x)\rangle + \frac{1}{2}\langle z, c(x)\rangle^2}{1\wedge|x|^2} \quad \text{for } x \neq 0.$$

Since $\lim_{y\to 0, y\in\mathbb{R}} y^{-2}(\mathrm{e}^{\mathrm{i}y} - 1 - \mathrm{i}y + y^2/2) = 0$, the function $x \mapsto (1\wedge|x|^2)^{-1}(\mathrm{e}^{\mathrm{i}\langle z,x} - 1 - \mathrm{i}\langle z,x\rangle + \frac{1}{2}\langle z,x\rangle^2)$ is continuous at 0, where it is defined to be 0. From (9.35), we see that $\lim_{x\to 0} |x|^{-2}\langle z, c(x) - x\rangle = 0$ and that

$$\lim_{x\to 0} \frac{\langle z, c(x)\rangle^2 - \langle z,x\rangle^2}{|x|^2} = \lim_{x\to 0} z^T \frac{(c(x)-x)c(x)^T + x(c(x)^T - x^T)}{|x|^2} z = 0,$$

so that u_z is continuous at 0. By continuity and boundedness of c, u_z is then obviously bounded and continuous on $\mathbb{R}^d$. Define

$$M_n := A_n + \int_{\mathbb{R}^d} c(x)\,c(x)^T\,\nu_n(\mathrm{d}x), \quad M := A + \int_{\mathbb{R}^d} c(x)c(x)^T\,\nu(\mathrm{d}x) \in \mathbb{R}^{d\times d}$$

and the finite measures ρ_n and ρ by

$$\rho_n(\mathrm{d}x) := (1 \wedge |x|^2)\,\nu_n(\mathrm{d}x) \quad \text{and} \quad \rho(\mathrm{d}x) := (1 \wedge |x|^2)\,\nu(\mathrm{d}x)$$

(observe that $\rho_n(\{0\}) = \rho(\{0\}) = 0$). Then

$$\begin{aligned}
\operatorname{Log}\widehat{\mu_n}(z) &= \mathrm{i}\langle \gamma_n^c, z\rangle - \frac{1}{2}z^T A_n z - \frac{1}{2}\int_{\mathbb{R}^d} z^T c(x)c(x)^T z\,\nu_n(\mathrm{d}x) \\
&\quad + \int_{\mathbb{R}^d}\left(\mathrm{e}^{\mathrm{i}\langle z,x\rangle} - 1 - \mathrm{i}\langle z, c(x)\rangle + \frac{1}{2}\langle z, c(x)\rangle^2\right)\nu_n(\mathrm{d}x) \\
&= \mathrm{i}\langle \gamma_n^c, z\rangle - \frac{1}{2}z^T M_n z + \int_{\mathbb{R}^d} u_z(x)\,\rho_n(\mathrm{d}x)
\end{aligned} \tag{9.39}$$

and similarly

$$\operatorname{Log}\widehat{\mu}(z) = \mathrm{i}\langle \gamma^c, z\rangle - \frac{1}{2}z^T M z + \int_{\mathbb{R}^d} u_z(x)\,\rho(\mathrm{d}x). \tag{9.40}$$

(a) Suppose that $\mu_n \xrightarrow{w} \mu$ as $n \to \infty$. By Lemma 9.12, $(\rho_n)_{n\in\mathbb{N}}$ is tight and uniformly bounded and hence, by Prokhorov's theorem (Theorem 9.11), there exists a finite measure $\overline{\rho}$ on $\mathbb{R}^d$ and a subsequence $(\rho_{n_k})_{k\in\mathbb{N}}$ which converges weakly to $\overline{\rho}$. Define the measure $\overline{\nu}$ on $\mathbb{R}^d$ by $\overline{\nu}(\mathrm{d}x) := \mathbf{1}_{\mathbb{R}^d\setminus\{0\}}(x)(1 \wedge |x|^2)^{-1}\,\overline{\rho}(\mathrm{d}x)$. Then $\overline{\nu}$ is a Lévy measure. Since ρ_{n_k} converges weakly to $\overline{\rho}$, we have $\int_{\mathbb{R}^d} f(x)\,\nu_{n_k}(\mathrm{d}x) \to \int_{\mathbb{R}^d} f(x)\,\overline{\nu}(\mathrm{d}x)$ as $k \to \infty$ for all $f \in C_\sharp(\mathbb{R}^d;\mathbb{R})$ by Lemma 9.30. Further, since u_z is bounded and continuous, $\int_{\mathbb{R}^d} u_z(x)\,\rho_{n_k}(\mathrm{d}x)$ converges to $\int_{\mathbb{R}^d} u_z(x)\overline{\rho}(x)$ as $k \to \infty$. Since also $\operatorname{Log}\widehat{\mu_{n_k}}(z)$ converges, equation (9.39) shows that $\mathrm{i}\langle \gamma_{n_k}^c, z\rangle - \frac{1}{2}z^T M_{n_k} z$ converges as $k \to \infty$. Taking real and imaginary parts, it follows that both $\langle \gamma_{n_k}^c, z\rangle$ and $z^T M_{n_k} z$ converge as $k \to \infty$ for all $z \in \mathbb{R}^d$. Hence $\gamma_{n_k}^c$ converges to some $\overline{\gamma}^c$ and, using the polarization identity

$$2w^T M_{n_k} z = w^T M_{n_k} z + z^T M_{n_k} w = (w+z)^T M_{n_k}(w+z) - w^T M_{n_k} w - z^T M_{n_k} z,$$

for $w, z \in \mathbb{R}^d$ and taking for w and z the unit vectors in $\mathbb{R}^d$, it follows that M_{n_k} converges to some matrix $\overline{M}$ as $k \to \infty$, which is then necessarily non-negative definite. Summing up, we have

$$M_{n_k} \to \overline{M}, \quad \gamma^c_{n_k} \to \overline{\gamma}^c \quad \text{and} \quad \int_{\mathbb{R}^d} f(x)\, \nu_{n_k}(\mathrm{d}x) \to \int_{\mathbb{R}^d} f(x)\, \overline{\nu}(\mathrm{d}x)$$

for all $f \in C_\sharp(\mathbb{R}^d;\mathbb{R})$ as $k \to \infty$. Let $\overline{A} := \overline{M} - \int_{\mathbb{R}^d} c(x)c(x)^T\, \overline{\nu}(\mathrm{d}x)$. Then $\overline{A}$ is symmetric. We also have

$$\begin{aligned}
\operatorname{Log}\widehat{\mu}(z) &= \lim_{k\to\infty} \operatorname{Log}\widehat{\mu_{n_k}}(z) \\
&= \lim_{k\to\infty}\left(\mathrm{i}\langle \gamma^c_{n_k}, z\rangle - \frac{1}{2} z^T M_{n_k} z + \int_{\mathbb{R}^d} u_z(x)\, \rho_{n_k}(\mathrm{d}x)\right) \\
&= \mathrm{i}\langle \overline{\gamma}^c, z\rangle - \frac{1}{2} z^T \overline{M} z + \int_{\mathbb{R}^d} u_z(x)\, \overline{\rho}(\mathrm{d}x) \\
&= \mathrm{i}\langle \overline{\gamma}^c, z\rangle - \frac{1}{2} z^T \overline{A} z + \int_{\mathbb{R}^d} \left(\mathrm{e}^{\mathrm{i}\langle z,x\rangle} - 1 - \mathrm{i}\langle z, c(x)\rangle\right) \overline{\nu}(\mathrm{d}x),
\end{aligned}$$

where the last line follows in the same way as (9.39). We still need to show that $\overline{A}$ is non-negative definite. However, since μ is infinitely divisible with characteristic triplet $(A, \nu, \gamma^c)_c$, we know from Remark 9.9 that

$$\overline{\gamma}^c = \gamma^c, \quad \overline{A} = A \quad \text{and} \quad \overline{\nu} = \nu,$$

so that $\overline{A}$ is non-negative definite, and $\overline{M} = M$. We have shown that there exists a subsequence $(n_k)_{k\in\mathbb{N}}$ such that (9.36)–(9.38) hold along this subsequence. Now suppose that one of (9.36)–(9.38), say (9.37) for example, does not hold for the full sequence. Then there exists some $f_0 \in C_\sharp(\mathbb{R}^d;\mathbb{R})$, some $\varepsilon > 0$ and a subsequence $(n_l)_{l\in\mathbb{N}}$ such that $|\int_{\mathbb{R}^d} f_0(x)\, \nu_{n_l}(\mathrm{d}x) - \int_{\mathbb{R}^d} f_0(x)\, \nu(\mathrm{d}x)| \geq \varepsilon$ for all $l \in \mathbb{N}$. Applying the argument just given to the sequence $(\nu_{n_l})_{l\in\mathbb{N}}$, there exists a further subsequence $(n_{l_k})_{k\in\mathbb{N}}$ such that $\int_{\mathbb{R}^d} f(x)\, \nu_{n_{l_k}} \to \int_{\mathbb{R}^d} f(x)\, \nu(\mathrm{d}x)$ for all $f \in C_\sharp(\mathbb{R}^d;\mathbb{R})$, which is clearly a contradiction. So (9.37) must hold for the full sequence and, by the same argument, so must (9.36) and (9.38).

(b) Now suppose that (9.36)–(9.38) hold. Then

$$\operatorname{Log}\widehat{\mu_n}(z) - \int_{\mathbb{R}^d} u_z(x)\, \rho_n(\mathrm{d}x) \to \operatorname{Log}\widehat{\mu}(z) - \int_{\mathbb{R}^d} u_z(x)\, \rho(\mathrm{d}x)$$

by (9.36), (9.38), (9.39) and (9.40), hence $\mu_n \xrightarrow{w} \mu$ as $n \to \infty$ if we can show that $\int_{\mathbb{R}^d} u_z(x)\, \rho_n(\mathrm{d}x) \to \int_{\mathbb{R}^d} u_z(x)\, \rho(\mathrm{d}x)$ as $n \to \infty$. To see this, observe first that by the triangle inequality, for $x \neq 0$,

$$\left|\frac{|c(x)|}{|x|} - 1\right| = \frac{||c(x)| - |x||}{|x|} \leq \frac{|c(x) - x|}{|x|^2}|x|$$

which converges to 0 as $x \to 0$ by (9.35). Hence there is $\varepsilon \in (0,1]$ such that $|c(x)| \geq |x|/2$ for all $|x| \leq \varepsilon$. From (9.36) we see that $z^T M_n z$ converges to $z^T M z$ as $n \to \infty$ for each $z \in \mathbb{R}^d$. However, since $z^T A_n z \geq 0$, it follows that $z^T \int_{\mathbb{R}^d} c(x)c(x)^T \nu_n(\mathrm{d}x)\, z$ must be bounded in n for each $z \in \mathbb{R}^d$. Taking z to be the jth unit vector in $\mathbb{R}^d$ and writing $c(x) = (c_1(x), \dots, c_d(x))^T$, it follows that $\int_{\mathbb{R}^d} (c_j(x))^2\, \nu_n(\mathrm{d}x)$ is bounded in n for each $j \in \{1, \dots, d\}$. Summing over $j \in \{1, \dots, d\}$ and using the inequality $|c(x)| \geq |x|/2$ for $0 \leq |x| \leq \varepsilon \leq 1$, we find that $\int_{|x| \leq \varepsilon} |x|^2\, \nu_n(\mathrm{d}x)$ is bounded in n, implying that the same is true of $\rho_n(\{x \in \mathbb{R}^d : |x| \leq \varepsilon\})$. Suppose that $0 < \delta \leq \varepsilon$ and define, for $x \in \mathbb{R}^d$,

$$h_\delta(x) := \begin{cases} 0, & |x| \leq \delta/2, \\ \frac{|x| - \delta/2}{\delta/2}, & |x| \in (\delta/2, \delta), \\ 1, & |x| \geq \delta. \end{cases}$$

Then h_δ is continuous, $0 \leq h_\delta \leq 1$, and the mapping $x \mapsto h_\delta(x) u_z(x)(1 \wedge |x|^2)$ is in $C_\sharp(\mathbb{R}^d; \mathbb{R})$. Now, for each $\delta \in (0, \varepsilon]$ and $z \in \mathbb{R}^d$, we have by (9.37)

$$\begin{aligned} &\limsup_{n\to\infty} \left| \int_{\mathbb{R}^d} u_z(x)\, \rho_n(\mathrm{d}x) - \int_{\mathbb{R}^d} u_z(x)\, \rho(\mathrm{d}x) \right| \\ &\leq \limsup_{n\to\infty} \left| \int_{\mathbb{R}^d} u_z(x) h_\delta(x)\, \rho_n(\mathrm{d}x) - \int_{\mathbb{R}^d} u_z(x) h_\delta(x)\, \rho(\mathrm{d}x) \right| \\ &\quad + \limsup_{n\to\infty} \left| \int_{|x|\leq\delta} u_z(x)(1 - h_\delta(x))\, \rho_n(\mathrm{d}x) - \int_{|x|\leq\delta} u_z(x)(1 - h_\delta(x))\, \rho(\mathrm{d}x) \right| \\ &\leq \sup_{|x|\leq\delta} |u_z(x)| \Big(\sup_{n\in\mathbb{N}} \rho_n(\{x \in \mathbb{R}^d : |x| \leq \delta\}) + \rho(\{x \in \mathbb{R}^d : |x| \leq \delta\}) \Big). \end{aligned}$$

But $\sup_{n\in\mathbb{N}} \rho_n(\{x \in \mathbb{R}^d : |x| \leq \delta\}) + \rho(\{x \in \mathbb{R}^d : |x| \leq \delta\})$ is bounded in $\delta \in (0, \varepsilon]$, and $\sup_{|x|\leq\delta} |u_z(x)|$ tends to $u_z(0) = 0$ as $\delta \to 0$ by continuity of u_z. Hence $\int_{\mathbb{R}^d} u_z(x)\, \rho_n(\mathrm{d}x) \to \int_{\mathbb{R}^d} u_z(x)\, \rho(\mathrm{d}x)$ and consequently $\mu_n \xrightarrow{w} \mu$ as $n \to \infty$. □

Of course, Theorem 9.31 also holds in dimension 1, but in the one-dimensional case a much neater convergence condition with a much easier proof can be obtained in terms of the characteristic pairs as defined in Definition 9.23. The condition is derived in the proof of the next theorem.

Theorem 9.32 (Convergence of infinitely divisible distributions on $\mathbb{R}$). *Let $c : \mathbb{R} \to \mathbb{R}$ be a fixed truncation function that is continuous and satisfies* (9.28). *Let $(\mu_n)_{n\in\mathbb{N}}$ be a sequence in* $\mathrm{ID}(\mathbb{R})$ *with characteristic pairs $(\zeta_n, \gamma_n^c)_c$ with respect to c, and let μ be in* $\mathrm{ID}(\mathbb{R})$ *with characteristic pair (ζ, γ^c) with respect to c. Then μ_n converges weakly to μ as $n \to \infty$ if and only if ζ_n converges weakly to ζ and γ_n^c converges to γ^c as $n \to \infty$.*

Proof. For each $z \in \mathbb{R}$, define the function $v_z : \mathbb{R} \to \mathbb{R}$ by

$$v_z(0) := -z^2/2 \quad \text{and} \quad v_z(x) := \frac{e^{izx} - 1 - izc(x)}{1 \wedge x^2} \quad \text{for } x \neq 0.$$

As seen in the proof of Theorem 9.22, v_z is continuous on $\mathbb{R}$ and obviously bounded. Further, by (9.29),

$$\operatorname{Log} \widehat{\mu_n}(z) = i\gamma_n^c z + \int_{\mathbb{R}} v_z(x)\, \zeta_n(dx) \quad \forall\, z \in \mathbb{R},$$

and similarly for $\log \widehat{\mu}(z)$. Now if $\gamma_n^c \to \gamma^c$ and $\zeta_n \xrightarrow{w} \zeta$ as $n \to \infty$, then obviously $\operatorname{Log} \widehat{\mu_n}(z) \to \operatorname{Log} \widehat{\mu}(z)$, since v_z is continuous and bounded, so that $\mu_n \xrightarrow{w} \mu$ as $n \to \infty$.

Conversely, suppose that $\mu_n \xrightarrow{w} \mu$ as $n \to \infty$. Suppose for the moment that ζ_n does not converge weakly to ζ. Then there exist $\varepsilon > 0$, a function $f_0 \in C_b(\mathbb{R}; \mathbb{R})$ and a subsequence (n_k) such that $|\int_{\mathbb{R}} f_0(x)\, \zeta_{n_k}(dx) - \int_{\mathbb{R}} f_0(x)\, \zeta(dx)| \geq \varepsilon$ for all $k \in \mathbb{N}$. Since μ_{n_k} converges weakly to μ, it follows from Lemma 9.12 and the definition of the characteristic pair that $(\zeta_{n_k})_{k \in \mathbb{N}}$ is tight and uniformly bounded. By Prokhorov's theorem (Theorem 9.11), there exist a finite measure $\overline{\zeta}$ on $\mathbb{R}$ and a subsequence $(n_{k_l})_{l \in \mathbb{N}}$ such that $\zeta_{n_{k_l}}$ converges weakly to $\overline{\zeta}$ as $l \to \infty$, and we obviously must have $\overline{\zeta} \neq \zeta$. Since v_z is continuous, $i\gamma_{n_{k_l}}^c z = \operatorname{Log} \widehat{\mu_{n_{k_l}}}(z) - \int_{\mathbb{R}} v_z(x) \zeta_{n_{k_l}}(dx)$ converges for all $z \in \mathbb{R}$, hence $\gamma_{n_{k_l}}^c$ must converge to some $\overline{\gamma}^c$. From the part already proved we conclude that $\mu_{n_{k_l}}$ converges to an infinitely divisible distribution $\overline{\mu}$ with characteristic pair $(\overline{\zeta}, \overline{\gamma}^c)$. However, since $\overline{\zeta} \neq \zeta$, we must have $\overline{\mu} \neq \mu$ by the uniqueness of the characteristic pair, which obviously is impossible. Hence, as $n \to \infty$, $\zeta_n \xrightarrow{w} \zeta$, implying that $\gamma_n^c \to \gamma^c$ as well. □

Remark 9.33. Neither Theorem 9.31 nor Theorem 9.32 remain true when the assumption of the continuity of the truncation function c is dropped, or when the truncation function c satisfies only (9.3) rather than (9.35). Counterexamples are given in Sato [79, p. 428]. See also Exercise 9.5.

9.4 Exercises

Exercise 9.1. A measure ρ on $\mathbb{R}^d$ is *symmetric* if $\rho(B) = \rho(-B)$ for every $B \in \mathcal{B}_d$. Suppose that $\mu \in \mathrm{ID}(\mathbb{R}^d)$ with standard characteristic triplet (A, ν, γ). Show that μ is symmetric if and only if ν is symmetric and $\gamma = 0$.

Exercise 9.2. Give an alternative proof of Theorem 9.13 in dimension 1 by using characteristic pairs as suggested in Remark 9.24.

Exercise 9.3. Let $X = (X_1^T, \dots, X_m^T)^T$ be a random vector in $\mathbb{R}^{d_1} \times \cdots \times \mathbb{R}^{d_m}$ with distribution μ and define $d := d_1 + \cdots + d_m$. Show that the following are equivalent:

(i) $X_1, \dots, X_m$ are independent and $\mathcal{L}(X_i) \in \mathrm{ID}(\mathbb{R}^{d_i})$ for all $i \in \{1, \dots, m\}$.
(ii) $X_1, \dots, X_m$ are independent and $\mu \in \mathrm{ID}(\mathbb{R}^d)$.
(iii) $X_1, \dots, X_m$ are pairwise independent and $\mu \in \mathrm{ID}(\mathbb{R}^d)$.
(iv) $\mu \in \mathrm{ID}(\mathbb{R}^d)$ with standard characteristic triplet (A, ν, γ), where A is the block matrix $(A_{jk})_{j,k=1,\dots,m}$ with $A_{jk} \in \mathbb{R}^{d_j \times d_k}$ and $A_{jk} = 0, j \neq k$, and $\nu(\{(x_1^T, \dots, x_m^T)^T \in \mathbb{R}^d : |x_i| \cdot |x_j| \neq 0 \text{ for some } i \neq j\}) = 0$.

Exercise 9.4. Suppose that $\mu := \mathcal{L}(X) \in \mathrm{ID}(\mathbb{R})$ has standard characteristic triplet (a, ν, γ) and that $b \in \mathbb{R}$. Show that $\mathcal{L}(Y) := \mathcal{L}((X, bX)^T) \in \mathrm{ID}(\mathbb{R}^2)$ and calculate its characteristic triplet. If the drift or center of X exists, calculate the drift or center respectively of Y.

Exercise 9.5. For each $n \in \mathbb{N}$ let $\mu_n \in \mathrm{ID}(\mathbb{R})$ with standard characteristic triplet $(0, \delta_{1+1/n}, 0)$ and let $\mu \in \mathrm{ID}(\mathbb{R})$ with standard characteristic triplet $(0, \delta_1, 1)$. Show that μ_n converges weakly to μ as $n \to \infty$, although the location parameters of μ_n do not converge to the location parameter of μ. Hence the assumption of continuity of the truncation function cannot be dropped in Theorems 9.31 and 9.32.

Exercise 9.6. Let $(\mu_n)_{n\in\mathbb{N}}$ be a sequence in $\mathrm{ID}(\mathbb{R}^d)$ with characteristic triplets (A_n, ν_n, γ_n) that converges weakly to some $\mu \in \mathrm{ID}(\mathbb{R}^d)$ as $n \to \infty$ with characteristic triplet (A, ν, γ). Show that $\limsup_{n\to\infty} z^T A_n z \le z^T A z$ for all $z \in \mathbb{R}^d$.
Hint: Apply (9.36) with $c_\varepsilon(x) = [\mathbf{1}_{|x|\le\varepsilon}(x) + (2\varepsilon - |x|)\mathbf{1}_{\varepsilon<|x|\le 2\varepsilon}(x)]x$ with $\varepsilon > 0$.

Exercise 9.7. Suppose that $(\mu_n)_{n\in\mathbb{N}}$ is a tight sequence in $\mathrm{ID}(\mathbb{R}^d)$ with Lévy measures ν_n such that $\int_{|x|\le 1} |x|\, \nu_n(dx) < \infty$ for all $n \in \mathbb{N}$, so that the drift γ_n^0, as defined in Definition 9.20, exists for all n.
(a) Show that $(\gamma_n^0)_{n\in\mathbb{N}}$ is not necessarily bounded.
(b) Suppose that $(\mu_n)_{n\in\mathbb{N}}$ converges weakly as $n \to \infty$ to some $\mu \in \mathrm{ID}(\mathbb{R}^d)$ for which the drift γ^0 exists. Does it follow that $\gamma_n^0 \to \gamma^0$ as $n \to \infty$?

Exercise 9.8. Let $(\mu_n)_{n\in\mathbb{N}}$ be a sequence in $\mathrm{ID}(\mathbb{R}^d)$ with characteristic triplets $(A_n, \nu_n, \gamma_n^c)_c$ with respect to some truncation function c (not necessarily continuous or subject to (9.35)).
(a) (General dimension) Define $\rho_n(dx) := (1 \wedge |x|^2)\nu_n(dx)$. Show that $(\mu_n)_{n\in\mathbb{N}}$ is tight if and only if $(A_n)_{n\in\mathbb{N}}$ and $(\gamma_n^c)_{n\in\mathbb{N}}$ are bounded and $(\rho_n)_{n\in\mathbb{N}}$ is tight and uniformly bounded.
Hint: Prove the claim first under the additional assumption that c is continuous and satisfies (9.35), and afterwards relax this condition.
(b) (Dimension 1) If $d = 1$, c satisfies (9.28), and $(\zeta_n, \gamma_n^c)_c$ is the characteristic pair of μ_n with respect to c, show that $(\mu_n)_{n\in\mathbb{N}}$ is tight if and only if $(\gamma_n^c)_{n\in\mathbb{N}}$ is bounded and $(\zeta_n)_{n\in\mathbb{N}}$ is tight and uniformly bounded.

10 Lévy processes

Abstract: Lévy processes are introduced in this chapter and some of their elementary properties are considered. In particular, the one-to-one correspondence between Lévy processes in law and infinitely divisible distributions is established. It is also shown that every Lévy process in law has a modification that is a Lévy process, thereby establishing the existence of many such processes.

The increments of Lévy processes play the same role in continuous time modeling as i. i. d. sequences do in discrete time. In particular, Lévy processes are often used as driving processes for stochastic differential equations and in the definition of continuous time moving average processes. To see the relation between Lévy processes and i. i. d. sequences, let $(Z_n)_{n\in\mathbb{N}}$ be an i. i. d. sequence of random vectors and define

$$S_n := \sum_{k=1}^{n} Z_k \quad \text{for } n \in \mathbb{N}_0.$$

Then $S = (S_n)_{n\in\mathbb{N}_0}$ is known as a *random walk*, and it follows immediately from the definition that $S_0 = 0$, that S has independent increments in the sense that

$$S_{t_1} - S_{t_0}, S_{t_2} - S_{t_1}, \ldots, S_{t_n} - S_{t_{n-1}}$$

are independent for all $n \in \mathbb{N}$ and $t_0, t_1, \ldots, t_n \in \mathbb{N}_0$ with $t_0 < t_1 < \cdots < t_n$, and that the increments of S are stationary in the sense that

$$S_{t+h} - S_t \stackrel{d}{=} S_h - S_0 = S_h \quad \forall\, t, h \in \mathbb{N}_0.$$

Lévy processes constitute the natural continuous time analogs of random walks, in the sense that they have stationary and independent increments and take the value 0 at time $t = 0$. The only additional requirement that we shall impose is that with probability one they have càdlàg paths. While Lévy processes are often defined only as processes indexed by $[0, \infty)$, we shall be interested also in two-sided Lévy processes indexed by $\mathbb{R}$ and include this case in our definition.

10.1 Definition, elementary properties and examples

Throughout the chapter we shall use the same notation as in Chapters 8 and 9. Further, unless specified otherwise, in this and the next chapters on Lévy processes the set I will always refer to either $[0, \infty)$ or $\mathbb{R}$.

https://doi.org/10.1515/9783111325033-010

In Section 8.1, we have already defined one-sided Lévy processes. Let us now state the definition in such a way as to include two-sided Lévy processes. Before we do that, let us define càdlàg and càglàd functions, càdlàg processes and jumps.

Definition 10.1 (Càdlàg, càglàd, and jumps). Let $I = [0, \infty)$ or $I = \mathbb{R}$.

(a) A function $f : I \to \mathbb{R}^d$ is called *càdlàg*, if it is right-continuous on I, and has finite left-limits at every $t \in \mathbb{R}$ when $I = \mathbb{R}$ and at every $t \in (0, \infty)$ when $I = [0, \infty)$. For a càdlàg function f, we denote the left-limit at $t \in (0, \infty)$ or $t \in \mathbb{R}$, respectively, by

$$f(t-) := \lim_{x \uparrow t} f(x)$$

and when $I = [0, \infty)$ we define additionally

$$f(0-) := f(0).$$

(The acronym *càdlàg* is the abbreviation for the French phrase *continu à droite, limites à gauche*). The *jump size* of the càdlàg function f at a point $t \in I$ is defined by

$$\Delta f(t) := f(t) - f(t-).$$

We say that *a jump of f at time t occurs* if $\Delta f(t) \neq 0$.

(b) A function $f : I \to \mathbb{R}^d$ is called *càglàd* (acronym for the French phrase *continu à gauche, limites à droite*), if it is left-continuous on I and has finite right-limits at every $t \in I$.

(c) An $\mathbb{R}^d$-valued stochastic process $L = (L_t)_{t \in I}$ defined on a probability space $(\Omega, \mathcal{F}, P)$ is called a *càdlàg process*, or simply said to be *càdlàg*, if the sample paths of L are almost surely càdlàg, i. e. if there is some set $\Omega_0 \in \mathcal{F}$ with $P(\Omega_0) = 1$ such that the sample path $I \to \mathbb{R}^d$, $u \mapsto L_u(\omega)$ is càdlàg for all $\omega \in \Omega_0$. If *every* path of L is càdlàg, i. e. if Ω_0 can be chosen to equal to Ω, then we say that L is a *strictly càdlàg process* or simply *strictly càdlàg*. For a càdlàg process $L = (L_t)_{t \in I}$, we define L_{t-} pathwise by $L_{t-}(\omega)$ as in (a) for $\omega \in \Omega_0$ and by $L_{t-}(\omega) := L_t(\omega)$ for $\omega \notin \Omega_0$. Then, the *jump size* ΔL_t *at* $t \in I$ is defined pathwise by $\Delta L_t(\omega) := L_t(\omega) - L_{t-}(\omega)$ for all $\omega \in \Omega$.

Observe that a càdlàg function f is continuous at a point $t \in I$ if and only if $\Delta f(t) \neq 0$, i. e., if no jump occurs at t. Also observe that formally the definition of L_{t-} and hence of ΔL_t in (c) above depends on the chosen set Ω_0. As we are usually interested only in properties holding almost surely or in distribution, this does not cause a problem. Observe that by definition, L_{t-} and hence ΔL_t are $\mathcal{F}$-measurable, i. e., random variables.

The definition of a Lévy process now is as follows:

Definition 10.2 (Lévy process). Let $I = [0, \infty)$ or $I = \mathbb{R}$. A stochastic process $L = (L_t)_{t \in I}$ on a probability space $(\Omega, \mathcal{F}, P)$ with $L_t(\omega) \in \mathbb{R}^d$ for each $\omega \in \Omega$ and $t \in I$ is called a *Lévy process* (in $\mathbb{R}^d$) or an $\mathbb{R}^d$-*valued Lévy process* if it has the following properties.

(a) The process L has *independent increments*, i. e.,

$$L_{t_1} - L_{t_0}, L_{t_2} - L_{t_1}, \dots, L_{t_n} - L_{t_{n-1}}$$

are independent for all $n \in \mathbb{N}$ and $t_0, \dots, t_n \in I$ with $t_0 < t_1 < \cdots < t_n$.

(b) The process L has *stationary increments*, i. e.,

$$L_{t+h} - L_t \overset{d}{=} L_h - L_0 \quad \forall\, t, h \in I.$$

(c) $L_0 = 0$ almost surely.

(d) The process L is càdlàg.

In the case $I = \mathbb{R}$, we refer to L as a *two-sided Lévy process*, while if $I = [0, \infty)$, we call it a *one-sided Lévy process.*

Remark 10.3. If $I = \mathbb{R}$ and condition (b) is satisfied when $t \in \mathbb{R}$ and $h \geq 0$, then it is also satisfied when $t \in \mathbb{R}$ and $h \leq 0$, since

$$L_{t+h} - L_t = -(L_t - L_{t-|h|}) \overset{d}{=} -(L_0 - L_{-|h|}) = L_h - L_0.$$

Example 10.4 (Deterministic Lévy process). The simplest example of a Lévy process is a deterministic linear drift. More precisely, for $\gamma \in \mathbb{R}^d$ define $L_t := \gamma t$, where $t \in [0, \infty)$ or $t \in \mathbb{R}$ respectively. Then it is easy to see that (L_t) defines a Lévy process in $\mathbb{R}^d$. It is called a *deterministic Lévy process with drift* γ. In this case, the characteristic function of L_t is

$$\phi_{L_t}(z) = \exp(\mathrm{i}t\langle \gamma, z\rangle), \quad z \in \mathbb{R}^d.$$

The notion of a deterministic Lévy process should not be confused with deterministic weakly stationary processes as considered in Definition 6.12.

Example 10.5 (Brownian motion, Poisson process). It follows immediately from Definitions 1.29 and 1.30 that both the one-dimensional standard Brownian motion and the Poisson process are one-dimensional Lévy processes (two-sided when indexed by $\mathbb{R}$, and one-sided when indexed by $[0, \infty)$).

Another example is the multivariate Brownian motion, defined as follows.

Definition 10.6 (Multivariate Brownian motion). Suppose that $A \in \mathbb{R}^{d\times d}$ is a non-negative definite symmetric real $d \times d$-matrix and that $\gamma \in \mathbb{R}^d$. Let $I = \mathbb{R}$ or $I = [0, \infty)$. Then a *d-dimensional Brownian motion with mean γ and covariance matrix A* is an $\mathbb{R}^d$-valued stochastic process $(B_t)_{t\in I}$ such that

(a) $B_0 = 0$,

(b) $B_{t_2} - B_{t_1}, B_{t_3} - B_{t_2}, \dots, B_{t_n} - B_{t_{n-1}}$, are independent for every $n \in \{2, 3, \dots\}$ and every $(t_1, t_2, \dots, t_n)^T \in I^n$ such that $t_1 < t_2 < \cdots < t_n$,

(c) $B_t - B_s$ has the (multivariate) normal distribution with mean $(t-s)\gamma$ and covariance matrix $(t-s)A$ for all $s, t \in I$ such that $t \geq s$, i. e.,

$$\mathcal{L}(B_t - B_s) = N((t-s)\gamma, (t-s)A).$$

(d) B has almost surely continuous sample paths.

If $A = I_d$, the identity matrix operating on $\mathbb{R}^d$, and $\gamma = 0$, then B is called (d-dimensional) *standard Brownian motion.*

If the process B satisfies conditions (a), (b) and (c) but not necessarily (d), we refer to it as a *Brownian motion in law* with mean γ and covariance matrix A.

Lemma 10.7 (Existence of Brownian motion). *For every non-negative definite symmetric matrix $A \in \mathbb{R}^{d\times d}$ and $\gamma \in \mathbb{R}^d$, a Brownian motion $B = (B_t)_{t\in I}$ with mean γ and covariance matrix A exists and is a Lévy process. The characteristic function of B_t, $t \in I$, is*

$$\phi_{B_t}(z) = \exp\left(t\,\mathrm{i}\langle \gamma, z\rangle - \frac{1}{2}|t|\langle z, Az\rangle \right) = \exp\left(t\,\mathrm{i}\gamma^T z - \frac{1}{2}|t| z^T Az \right), \quad z \in \mathbb{R}^d.$$

Proof. That Brownian motion is a Lévy process follows immediately from the definition. To show its existence, let $B^1 = (B_t^1)_{t\in I}, \dots, B^d = (B_t^d)_{t\in I}$ be independent one-dimensional standard Brownian motions, the existence of which was established in Section 1.4. Then it is easy to see that $B = (B^1, \dots, B^d)^T$ is a d-dimensional standard Brownian motion. To prove the existence of Brownian motion with given mean γ and covariance matrix A, let $W = (W_t)_{t\in I}$ be a d-dimensional standard Brownian motion. Let $\Sigma \in \mathbb{R}^{d\times d}$ be a non-negative definite symmetric real $(d \times d)$-matrix such that $\Sigma^2 = A$, which exists since A is symmetric and non-negative definite. Then it is easy to check that $B_t := \gamma t + \Sigma W_t$, $t \in I$, defines a d-dimensional Brownian motion with mean γ and covariance matrix A. The form of the characteristic function is immediate from the characteristic function of an $N(t\gamma, |t|A)$-distributed random vector. □

Our next example of a Lévy process is the compound Poisson process.

Definition 10.8 (Compound Poisson process). Suppose that $I = [0, \infty)$ or $I = \mathbb{R}$, and that $N = (N_t)_{t\in I}$ is the Poisson process with jump rate $a > 0$ constructed as in Theorem 1.31 (and restricted to $[0, \infty)$ when $I = [0, \infty)$). Let $(Y_i)_{i\in\mathbb{Z}}$ be an i. i. d. sequence of $\mathbb{R}^d$-valued random variables, independent of $(N_t)_{t\in I}$, all defined on the same probability space $(\Omega, \mathcal{F}, P)$, such that the distribution σ of Y_1 satisfies $\sigma(\{0\}) = 0$. Then the stochastic process $L = (L_t)_{t\in I}$, defined for $\omega \in \Omega$ by

$$L_t(\omega) = \begin{cases} \sum_{i=1}^{N_t(\omega)} Y_i(\omega), & t \geq 0, \\ -\sum_{i=N_t(\omega)+1}^{0} Y_i(\omega), & t < 0 \text{ if } I = \mathbb{R}, \end{cases}$$

is called a *compound Poisson process (in $\mathbb{R}^d$) with parameter* (or *jump rate*) $a > 0$ and *jump size distribution* σ.

In the definition above, we have been careful to use only Poisson processes that are constructed as in Theorem 1.31. Actually, as will be shown later in Theorem 12.18 and Remark 12.19(b), every Poisson process has a representation as in Theorem 1.31, so that we could also have omitted the mention of the explicit construction of N in Definition 10.8, but since we have not proved this fact yet, for the moment we can only consider underlying Poisson processes with a representation as in Theorem 1.31.

Obviously, every compound Poisson process in $\mathbb{R}$ with jump size distribution $\sigma = \delta_1$, the Dirac measure at 1, is a Poisson process by Theorem 1.31. The paths of a general compound Poisson process $L = (L_t)_{t\in I}$ in $\mathbb{R}^d$ are step functions with the same jump times as the underlying Poisson process N, but with jump size Y_i for the ith jump. In particular, L has almost surely càdlàg paths with $L_0 = 0$. The following theorem shows that every compound Poisson process is a Lévy process.

Theorem 10.9 (Compound Poisson processes are Lévy processes). *Suppose that $I = [0,\infty)$ or $I = \mathbb{R}$ and $L = (L_t)_{t\in I}$ is a compound Poisson process in $\mathbb{R}^d$ with parameter $a > 0$ and jump size distribution σ. Then L is a Lévy process, and the characteristic function of L_t, $t \in I$, is*

$$\mathbb{E}\,\mathrm{e}^{\mathrm{i}\langle z,L_t\rangle} = \begin{cases} \exp(ta(\widehat{\sigma}(z)-1)), & z \in \mathbb{R}^d, t \ge 0,\\ \exp(|t|a(\widehat{\sigma}(-z)-1)), & z \in \mathbb{R}^d, t < 0 \text{ if } I = \mathbb{R}, \end{cases}$$

so that $\mathcal{L}(L_t)$ is the compound Poisson distribution with parameter at and jump size distribution σ when $t > 0$, and a compound Poisson distribution with parameter $a|t|$ and jump size distribution σ' when $t < 0$, where σ' is the distribution defined by $\sigma'(B) := \sigma(-B)$ for $B \in \mathcal{B}_d$. In particular, the distribution of L_1 has the compound Poisson distribution with parameter a and jump size distribution σ.

We shall give the proof here for $I = [0,\infty)$ and, after Remark 10.18 below, for the case when $I = \mathbb{R}$.

Proof. Suppose that $I = [0,\infty)$. Let L, N and Y_i be as in Definition 10.8. Define

$$S_n := \sum_{i=1}^{n} Y_i, \quad \text{so that} \quad L_t = S_{N_t}.$$

Suppose that $B, B_1, B_2, \ldots \in \mathcal{B}_d$. Then

$$\begin{aligned} P(L_t \in B) &= P(S_{N_t} \in B)\\ &= \sum_{n=0}^{\infty} P(S_n \in B, N_t = n) = \sum_{n=0}^{\infty} P(N_t = n)P(S_n \in B). \end{aligned}$$

Using that $(S_n)_{n\in\mathbb{N}_0}$ has independent and stationary increments and that the Poisson process is a Lévy process, for $0 \le t_0 < t_1 < \cdots < t_k$ we have

$$\begin{aligned}
&P(L_{t_1} - L_{t_0} \in B_1, \dots, L_{t_k} - L_{t_{k-1}} \in B_k) \\
&= \sum_{n_0,\dots,n_k=0}^{\infty} P(N_{t_0} = n_0, N_{t_1} - N_{t_0} = n_1, \dots, N_{t_k} - N_{t_{k-1}} = n_k, \\
&\qquad S_{n_0+n_1} - S_{n_0} \in B_1, \dots, S_{n_0+\cdots+n_k} - S_{n_0+\cdots+n_{k-1}} \in B_k) \\
&= \sum_{n_0,\dots,n_k=0}^{\infty} P(N_{t_0} = n_0)\, P(N_{t_1} - N_{t_0} = n_1) \cdots P(N_{t_k} - N_{t_{k-1}} = n_k) \\
&\qquad \times P(S_{n_1} \in B_1) \cdots P(S_{n_k} \in B_k) \\
&= \left(\sum_{n_0=0}^{\infty} P(N_{t_0} = n_0) \right) \prod_{i=1}^{k} \left(\sum_{n_i=0}^{\infty} \underbrace{P(N_{t_i} - N_{t_{i-1}} = n_i)}_{=P(N_{t_i-t_{i-1}}=n_i)}\, P(S_{n_i} \in B_i) \right) \\
&= P(L_{t_1-t_0} \in B_1) \cdots P(L_{t_k-t_{k-1}} \in B_k).
\end{aligned}$$

Putting $k = 1$ in this formula, we see that L has stationary increments. From this, we conclude that for general $k \in \mathbb{N}$

$$\begin{aligned}
&P(L_{t_1} - L_{t_0} \in B_1, \dots, L_{t_k} - L_{t_{k-1}} \in B_k) \\
&= P(L_{t_1} - L_{t_0} \in B_1) \cdots P(L_{t_k} - L_{t_{k-1}} \in B_k),
\end{aligned}$$

which shows that L has independent increments. It is clear that L has almost surely càdlàg paths and that $L_0 = 0$, so that L is a Lévy process in $\mathbb{R}^d$. That L_t has the compound Poisson distribution with jump rate at and jump size distribution σ for $t > 0$ follows from $L_t = \sum_{i=1}^{N_t} Y_i$, since N_t is Poisson distributed with parameter at. The form of the characteristic function then follows from Theorem 8.7. □

Remark 10.10. An application of Example 9.19 shows that the jump rate a and the jump size distribution σ of a compound Poisson process $L = (L_t)_{t\in I}$ are uniquely determined by L, since by the previous theorem, L_1 is compound Poisson distributed with parameter a and jump size distribution σ.

Proposition 10.11 (Linear transformations for independent Lévy proc.). *The following results apply to independent Lévy processes, $L^{(1)}, \dots, L^{(n)}$, $n \in \mathbb{N}$, which are all one-sided or all two-sided.*

(a) *If $L^{(1)}, \dots, L^{(n)}$ are all in $\mathbb{R}^d$ and $C_1, \dots, C_n \in \mathbb{R}^{m\times d}$ then $L := C_1 L^{(1)} + \cdots + C_n L^{(n)}$ is a Lévy process in $\mathbb{R}^m$. In particular, $C_1 L^{(1)}$ is a Lévy process, and each component of any Lévy process is a one-dimensional Lévy process.*

(b) *If $L^{(1)}, \dots, L^{(n)}$ are in $\mathbb{R}^{d_1}, \dots, \mathbb{R}^{d_n}$, respectively, then L, defined by $L := ((L^{(1)})^T, \dots, (L^{(n)})^T)^T$ is a Lévy process in $\mathbb{R}^{d_1+\cdots+d_n}$.*

Proof. These results are easily established by checking that L satisfies the defining properties of a Lévy process in each case. □

In particular, the sum of a Brownian motion and an independent compound Poisson process is again a Lévy process.

Next, we shall consider elementary distributional properties of Lévy processes and introduce the concept of a Lévy process in law.

Definition 10.12 (Lévy process in law). Suppose that $I = [0,\infty)$ or $I = \mathbb{R}$. A stochastic process $L = (L_t)_{t\in I}$ on a probability space $(\Omega, \mathcal{F}, P)$ with $L_t(\omega) \in \mathbb{R}^d$ for each $\omega \in \Omega$ and $t \in I$ is called a *Lévy process in law*, if it satisfies conditions (a), (b) and (c) of Definition 10.2 and is *continuous in probability*, i. e.,

$$\lim_{s\to t; s\in I} P(|L_s - L_t| > \varepsilon) = 0 \quad \forall\, \varepsilon > 0, \forall\, t \in I.$$

Remark 10.13. Observe that d-dimensional Brownian motion B as defined in Definition 10.6 and the (one-dimensional) Poisson process N as defined in Definition 1.30 are indeed Lévy processes in law. The defining properties (a), (b) and (c) are clear. To see that Brownian motion is also continuous in probability, observe that $B_t - B_s$ is $N((t-s)\gamma, |t-s|A)$-distributed, hence $B_t - B_s$ converges in distribution to $N(0, 0A) = \delta_0$ as $s \to t$, hence also in probability to 0 by Slutsky's lemma. The same is true for a Poisson process N, since $|N_t - N_s|$, which has mean $\lambda|t-s|$, converges in distribution to δ_0 as $s \to t$.

Not only are both Brownian motion and the Poisson process Lévy processes in law, but every Lévy process is a Lévy process in law, as we shall now see:

Proposition 10.14 (Lévy processes are Lévy processes in law). *Suppose that $I = [0,\infty)$ or $I = \mathbb{R}$.*

(a) *Then every stochastic process $L = (L_t)_{t\in I}$ on $(\Omega, \mathcal{F}, P)$ which has stationary increments (i. e., satisfies property (b) of Definition* 10.2*) and is right-continuous in probability at 0, i. e., satisfies*

$$\lim_{s\to 0; s\geq 0} P(|L_s - L_0| > \varepsilon) = 0 \quad \forall\, \varepsilon > 0,$$

is necessarily continuous in probability.

(b) *Every Lévy process $L = (L_t)_{t\in I}$ is continuous in probability and is therefore a Lévy process in law. In particular, for every fixed $t \in I$, $L_{t-} = L_t$ almost surely, so that $\Delta L_t = 0$ almost surely for every fixed point $t \in I$.*

Proof. (a) It follows from the stationary increments property that $L_s - L_t \overset{d}{=} L_{s-t} - L_0$ for $s \geq t$ and that $L_s - L_t = -(L_t - L_s) \overset{d}{=} -(L_{t-s} - L_0)$ for $s \leq t$, so that $|L_s - L_t| \overset{d}{=} |L_{|t-s|} - L_0|$ for all $s, t \in I$. The right-continuity in probability of L at 0 then clearly implies continuity in probability of L.

(b) Let $(s_n)_{n\in\mathbb{N}}$ be a sequence in $[0,\infty)$ converging to 0. Since L has càdlàg paths, L is almost surely right-continuous at 0, so that L_{s_n} converges almost surely and hence in probability to 0 as $n\to\infty$. Since this is true for every such sequence $(s_n)_{n\in\mathbb{N}}$, it follows that $\lim_{s\to 0, s\geq 0} P(|L_s|>\varepsilon)=0$ for every $\varepsilon>0$. Since $L_0=0$ a. s., L is right-continuous in probability at 0 and is hence continuous in probability by (a).

Since L has càdlàg paths, the almost sure limit $L_{t-}=\lim_{s\uparrow t} L_s$ exists for every $t\in I$ (with $L_{0-}:=L_0$ if $I=[0,\infty)$). It is therefore also the limit in probability and by the uniqueness of such limits, $L_{t-}=L_t$ almost surely. □

Corollary 10.15 (Marginal distribution at negative times). *If $L=(L_t)_{t\in\mathbb{R}}$ is a two-sided Lévy process in law, then*

$$\mathcal{L}(L_{-t})=\mathcal{L}(-L_t)\quad\forall\, t\in\mathbb{R}.$$

If in addition, L is a two-sided Lévy process, then

$$\mathcal{L}(L_{-t})=\mathcal{L}(-L_t)=\mathcal{L}(-L_{t-}).$$

Proof. The first assertion follows from

$$\mathcal{L}(L_{-t})=\mathcal{L}(L_{-t}-L_0)=\mathcal{L}(L_0-L_t)=\mathcal{L}(-L_t),$$

and the second is clear from Proposition 10.14(b). □

The next results show that the finite-dimensional distributions of a Lévy process are completely determined by all one-dimensional distributions $\mathcal{L}(L_t)$ with $t\geq 0$. It will be proved later in Theorem 10.26 that even more is true, namely that they are characterized already by just $\mathcal{L}(L_1)$.

Proposition 10.16 (Finite dimensional distr. of Lévy processes in law). *Suppose that $I=[0,\infty)$ or $I=\mathbb{R}$, $L=(L_t)_{t\in I}$ is a Lévy process in law in $\mathbb{R}^d$, $m,n\in\mathbb{N}_0$ with $n+m>0$ and*

$$t_{-m}<t_{-m+1}<\cdots<t_{-1}<t_0:=0<t_1<\cdots<t_n$$

with $t_{-m},\dots,t_n\in I$. Define the function $G:\mathbb{R}^{n+m}\to\mathbb{R}^{n+m+1}$ by

$$\begin{aligned}&G(x_{-m},\dots,x_{-1},x_1,\dots,x_n)\\&\quad:=(-x_{-m}-\cdots-x_{-1},\dots,-x_{-2}-x_{-1},-x_{-1},0,x_1,x_1+x_2,\dots,x_1+\cdots+x_n).\end{aligned}$$

Then

$$(L_{t_{-m}},L_{t_{-m+1}},L_{t_{-1}},L_0,L_{t_1},\dots,L_{t_n})^T\overset{d}{=}G\Bigg(\bigotimes_{i=-m}^{n-1}\mathcal{L}(L_{t_{i+1}-t_i})\Bigg),\tag{10.1}$$

where $G(\mu)$ denotes the image measure of a measure μ under the mapping G and $\otimes$ the product measure. In particular, if $L' = (L'_t)_{t\in I}$ is another Lévy process in law such that

$$\mathcal{L}(L_t) = \mathcal{L}(L'_t) \quad \forall\, t \geq 0,$$

then L and L' are equal in law, *i. e., they have the same finite-dimensional distributions.*

Proof. Since L is a Lévy process in law, we have

$$\begin{aligned}\mathcal{L}((L_{t_{-m+1}} - L_{t_{-m}}, \dots, L_0 - L_{t_{-1}}, L_{t_1} - L_0, \dots, L_{t_n} - L_{t_{n-1}})^T) \\ = \bigotimes_{i=-m}^{n-1} \mathcal{L}(L_{t_{i+1}} - L_{t_i}) = \bigotimes_{i=-m}^{n-1} \mathcal{L}(L_{t_{i+1}-t_i}).\end{aligned}$$

The result then follows from the fact that $(L_{t_{-m}}, L_{t_{-m+1}}, \dots, L_{t_{-1}}, L_0, L_{t_1}, \dots, L_{t_n})^T$ is the image of $(L_{t_{-m+1}} - L_{t_{-m}}, \dots, L_0 - L_{t_{-1}}, L_{t_1} - L_0, \dots, L_{t_n} - L_{t_{n-1}})^T$ under G (almost surely). □

As already pointed out in Remark 1.32(a), it follows from Proposition 10.16 that the Poisson processes in law with parameter $\lambda > 0$ and Brownian motion in law with mean γ and covariance matrix A are unique in distribution. Similarly, by Proposition 10.16 and Theorem 10.9, the compound Poisson process with jump rate $a > 0$ and jump size distribution σ is unique in distribution.

The following result shows how two-sided Lévy processes can be obtained from one-sided Lévy processes. This construction seems very natural in view of Corollary 10.15.

Theorem 10.17 (Construction of two-sided Lévy processes). *Suppose that $L = (L_t)_{t\geq 0}$ is a one-sided Lévy process and let $(L'_t)_{t\geq 0}$ be an independent copy of L. For $t \in \mathbb{R}$ define*

$$L''_t := \begin{cases} L_t, & \text{if } t \geq 0, \\ -L'_{(-t)-}, & \text{if } t < 0. \end{cases}$$

Then $(L''_t)_{t\in\mathbb{R}}$ is a two-sided Lévy process that extends L to $\mathbb{R}$. It is unique in law in the sense that if L''' is another two-sided Lévy process that extends L to $\mathbb{R}$, then L'' and L''' have the same finite dimensional distributions.

Proof. With probability one $L''_0 = 0$ and L' has càdlàg paths, hence $(L'_{-t})_{t\in(-\infty,0]}$ has càglàd paths with probability 1, hence $(L'_{(-t)-})_{t\in(-\infty,0]}$ has càdlàg paths with probability 1 and consequently so does L''. To show that L'' has independent increments, suppose that $t_0 < t_1 < \cdots < t_n \in \mathbb{R}$. Without loss of generality, we can assume that there are some $j \in \{0, \dots, n\}$ such that $t_j = 0$. Then $L''_{t_{j+1}} - L''_{t_j}, \dots, L''_{t_n} - L''_{t_{n-1}}$ are independent, and since $L''_{t_{i+1}} - L''_{t_i} = -(L'_{(-t_{i+1})-} - L'_{(-t_i)-}) = L'_{-t_i} - L'_{-t_{i+1}}$ a. s. for $i < j$ by Proposition 10.14(b), it follows that $L''_{t_1} - L''_{t_0}, \dots, L''_{t_j} - L''_{t_{j-1}}$ are independent, and obviously independent of $L''_{t_{j+1}} - L''_{t_j}, \dots, L''_{t_n} - L''_{t_{n-1}}$ by the independence of L and L'. Hence L'' has independent increments. To show that L has stationary increments, suppose that $h \geq 0$ and $t \in \mathbb{R}$. If

$t \geq 0$, then $L''_{t+h} - L''_t = L_{t+h} - L_t \stackrel{d}{=} L_h - L_0 = L''_h - L''_0$ as required. If $t < 0$ and $t + h \leq 0$, then by Proposition 10.14(b)

$$L''_{t+h} - L''_t = L'_{(-t)-} - L'_{(-(t+h))-} \stackrel{d}{=} L'_{-t} - L'_{-(t+h)} \stackrel{d}{=} L'_h - L'_0 \stackrel{d}{=} L''_h - L''_0.$$

Finally, if $t < 0$ and $t + h > 0$, then

$$L''_{t+h} - L''_t = (L''_{t+h} - L''_0) + (L''_0 - L''_t) \stackrel{d}{=} (L_{t+h} - L_0) + (L'_{-t} - L'_0)$$

by the parts just shown. Observe also that $L_{t+h} - L_0$ and $L'_{-t} - L'_0$ are independent. Since $L'_{-t} - L'_0 \stackrel{d}{=} L'_h - L'_{t+h} \stackrel{d}{=} L_h - L_{t+h}$ and since $L_h - L_{t+h}$ is independent of $L_{t+h} - L_0$, we obtain

$$L''_{t+h} - L''_t \stackrel{d}{=} (L_{t+h} - L_0) + (L_h - L_{t+h}) = L_h - L_0.$$

With the aid of Remark 10.3 we conclude finally that L'' has stationary increments and is therefore a two-sided Lévy process extending L from $[0, \infty)$ to $\mathbb{R}$. The uniqueness assertion regarding the finite dimensional distributions is clear from Proposition 10.16. □

Remark 10.18. Theorem 10.17 can also be easily modified to construct two-sided Lévy processes *in law* from one-sided ones. More precisely, if $L = (L_t)_{t\geq 0}$ is a one-sided Lévy process in law and $L' = (L_t)_{t\geq 0}$ is an independent copy of L, then $L'' = (L''_t)_{t\in\mathbb{R}}$, defined by

$$L''_t := \begin{cases} L_t, & \text{if } t \geq 0, \\ -L'_{-t}, & \text{if } t < 0, \end{cases}$$

is a two-sided Lévy process in law which extends L to $\mathbb{R}$, and it is unique in law (see Exercise 10.1). The reason why $L''_t := -L'_{(-t)-}$, $t < 0$, in the statement of the theorem is replaced by $L''_t := -L'_{-t}$, $t < 0$, in this remark is that, without assuming any regularity of the paths of L, there is no guarantee of the existence of $L'_{(-t)-}$.

We can now complete the proof of Theorem 10.9 by dealing with the case when $I = \mathbb{R}$.

Proof of Theorem 10.9 *when* $I = \mathbb{R}$. Let $I = \mathbb{R}$. That $(L_t)_{t\geq 0}$, as constructed in the proof when $I = [0, \infty)$, is a one-sided Lévy process has already been shown, and the definition of L on $(-\infty, 0]$ in the statement of the theorem ensures that it coincides with the process on $(-\infty, 0]$ obtained by the construction of a two-sided Lévy processes as described in Theorem 10.17. Hence $(L_t)_{t\in\mathbb{R}}$ is a Lévy process. The formula for the characteristic function for $t < 0$ then follows from the already established formula when $I = [0, \infty)$ and Corollary 10.15, since for $t < 0$ we have $\mathcal{L}(L_t) = \mathcal{L}(-L_{-t})$ and hence

$$\mathbb{E}\, e^{i\langle z, L_t\rangle} = \mathbb{E}\, e^{i\langle -z, L_{-t}\rangle} = \exp(|t| a(\widehat{\sigma}(-z) - 1)).$$ □

Next we observe that the independent increment property implies an independence property for the σ-algebras generated by the increments, which correspond to the past and future up to time t.

Proposition 10.19 (Independence of generated σ-algebras). *Suppose that $I = [0,\infty)$ or $I = \mathbb{R}$ and that $L = (L_t)_{t\in I}$ is a Lévy process in law. For $t \in I$, define*

$$\mathcal{A}_t^p := \sigma(L_u - L_s : u, s \in I, s \le u \le t), \quad \text{and}$$
$$\mathcal{A}_t^f := \sigma(L_u - L_s : u, s \in I, t \le s \le u),$$

where $\sigma(\cdot)$ denotes the sigma-algebra generated by the random variables in the parentheses. Then

$$\mathcal{A}_t^p = \sigma(L_t - L_s : s \in I, s \le t), \quad \mathcal{A}_t^f = \sigma(L_s - L_t : s \in I, s \ge t),$$

and $\mathcal{A}_t^p$ and $\mathcal{A}_t^f$ are independent for each fixed $t \in I$. Further, for $t_0 < t_1 < \cdots < t_n \in I$, the σ-algebras $\sigma(L_u - L_s : u, s \le t_0), \sigma(L_u - L_s : u, s \in [t_0, t_1]), \ldots, \sigma(L_u - L_s : u, s, \in [t_{n-1}, t_n])$ and $\sigma(L_u - L_s : u, s \ge t_n)$ are independent.

Proof. That $\mathcal{A}_t^p = \sigma(L_t - L_s : s \in I, s \le t)$ follows readily from the fact that every increment $L_u - L_s$ with $s \le u \le t$ can be written as $(L_t - L_s) - (L_t - L_u)$, and similarly for $\mathcal{A}_t^f$. To see that $\mathcal{A}_t^p$ and $\mathcal{A}_t^f$ are actually independent for fixed t, denote by $\mathcal{G}_t^p$ the set of all finite intersections of sets of the form $\{L_u - L_t \in B\}$ with $u \le t$ and $B \in \mathcal{B}_d$, and by $\mathcal{G}_t^f$ the set of all finite intersections of sets of the form $\{L_s - L_t \in B\}$ with $s \ge t$ and $B \in \mathcal{B}_d$. Then $\mathcal{G}_t^p$ and $\mathcal{G}_t^f$ are independent, since for $u_n < \cdots < u_1 \le t \le s_1 < \cdots < s_m$ the random vectors $(L_{s_1} - L_t, \ldots, L_{s_m} - L_t)^T$ and $(L_t - L_{u_1}, \ldots, L_t - L_{u_n})^T$ are independent by the same argument that led to (10.1). Since $\mathcal{G}_t^p$ and $\mathcal{G}_t^f$ form generating systems of $\mathcal{A}_t^p$ and $\mathcal{A}_t^f$, respectively, which are closed under finite intersections, the independence of $\mathcal{A}_t^p$ and $\mathcal{A}_t^f$ follows. Finally, the independence of $\sigma(L_u - L_s : u, s \le t_0), \sigma(L_u - L_s : u, s \in [t_0, t_1]), \ldots, \sigma(L_u - L_s : u, s, \in [t_{n-1}, t_n]), \sigma(L_u - L_s : u, s \ge t_n)$ follows inductively from the independence of $\mathcal{A}_t^p$ and $\mathcal{A}_t^f$ for each t. □

We close this section with a generalization of the class of Lévy processes to that of the so-called additive processes. These are processes continuous in probability with all the properties of a Lévy process apart from the stationarity of the increments. More specifically, they are defined as follows.

Definition 10.20 (Additive process). Let $I = [0,\infty)$ or $I = \mathbb{R}$. A stochastic process $L = (L_t)_{t\in I}$ on a probability space $(\Omega, \mathcal{F}, P)$ with $L_t : \Omega \to \mathbb{R}^d$ is called an *additive process* (in $\mathbb{R}^d$), if

(a) The process L has *independent increments*, i. e.,

$$L_{t_1} - L_{t_0}, L_{t_2} - L_{t_1}, \ldots, L_{t_n} - L_{t_{n-1}}$$

are independent for all $n \in \mathbb{N}$ and $t_0, \ldots, t_n \in I$ with $t_0 < t_1 < \cdots < t_n$.

(b) The process L is continuous in probability.
(c) $L_0 = 0$ almost surely.
(d) The process has almost surely càdlàg paths on I.

If a process satisfies only conditions (a), (b) and (c), then we call it an *additive process in law*.

In contrast with Lévy processes, a process that satisfies conditions (a), (c) and (d) of Definition 10.20 is not automatically continuous in probability. For example, let $f : [0, \infty) \to \mathbb{R}^d$ be a càdlàg function with $f(0) = 0$ that has a discontinuity at $t_0 > 0$. Then $L = (L_t)_{t \geq 0}$, defined by $L_t(\omega) = f(t)$, satisfies conditions (a), (c) and (d), but is not continuous in probability, since for $0 < \varepsilon < |\Delta f(t_0)|$ we have $P(|L_{t_0} - L_s| > \varepsilon) \to 1$ as $s \uparrow t_0$. Hence, for additive processes, the requirement of continuity in probability does not follow from the other properties.

Remark 10.21. An inspection of the proof of Proposition 10.19 shows that it holds word for word for additive processes in law (in fact for all processes with independent increments). Moreover, every additive process L satisfies $\Delta L_t = 0$ a. s. for each fixed $t \in I$. This follows exactly as in the proof of the second part of Proposition 10.14(b).

10.2 Lévy processes in law and infinitely divisible distributions

In this section, we shall show that there is a one-to-one correspondence between the class of Lévy processes in law and the class of infinitely divisible distributions. We already observed in Proposition 8.2 that the marginal distributions of a one-sided Lévy process are infinitely divisible. This continues to hold for one- and two-sided Lévy processes in law:

Proposition 10.22. *Let $L = (L_t)_{t \in I}$ be a one- or two-sided Lévy process in law in $\mathbb{R}^d$. Then the distribution $\mathcal{L}(L_t)$ is infinitely divisible for all $t \in I$.*

Proof. The proof is word for word the same as that of Proposition 8.2. □

We now aim to establish the converse by showing how to construct, for an infinitely divisible distribution μ, a Lévy process (L_t) in law with $\mathcal{L}(L_1) = \mu$. We shall also see the connection between $\mathcal{L}(L_t)$ and $\mathcal{L}(L_1)$ for Lévy processes in law. We need the following result.

Lemma 10.23. *Suppose that $\mu \in \mathrm{ID}(\mathbb{R}^d)$ has characteristic function $\widehat{\mu}$ and characteristic triplet $(A, \nu, \gamma)_c$ with respect to some truncation function c. Then, for each $t \geq 0$, the function $z \mapsto \exp(t \operatorname{Log} \widehat{\mu}(z))$ is the characteristic function of an infinitely divisible distribution with characteristic triplet $(tA, t\nu, t\gamma)_c$.*

Proof. From

$$\operatorname{Log}\widehat{\mu}(z) = \mathrm{i}\langle \gamma, z\rangle - \frac{1}{2}\langle z, Az\rangle + \int_{\mathbb{R}^d} (\mathrm{e}^{\mathrm{i}\langle z,x\rangle} - 1 - \mathrm{i}\langle z, c(x)\rangle)\, \nu(\mathrm{d}x),$$

we immediately see that $\exp(t \operatorname{Log}\widehat{\mu}(z))$ is equal to

$$\exp\Big(\mathrm{i}t\langle \gamma, z\rangle - \frac{1}{2}\langle z, tAz\rangle + \int_{\mathbb{R}^d} (\mathrm{e}^{\mathrm{i}\langle z,x\rangle} - 1 - \mathrm{i}\langle z, c(x)\rangle)\, t\nu(\mathrm{d}x)\Big),$$

showing that $z \mapsto \exp(t \operatorname{Log} z)$ is the characteristic function of an infinitely divisible distribution with characteristic triplet $(tA, t\nu, t\gamma)_c$. □

Definition 10.24 (Distinguished t^{th} power). Let $\phi : \mathbb{R}^d \to \mathbb{C}$ be a continuous function with $\phi(0) = 1$ and $\phi(z) \neq 0$ for all $z \in \mathbb{R}^d$. Then, for each $t \geq 0$, the function $\phi^t : \mathbb{R}^d \to \mathbb{C}$, defined by

$$\phi^t(z) := \exp[t \operatorname{Log} \phi(z)], \quad z \in \mathbb{R}^d,$$

is called the *distinguished t^{th} power* of ϕ. If $\phi = \widehat{\mu}$ is the characteristic function of an infinitely divisible distribution μ, then the distribution with characteristic function ϕ^t is called the *distinguished t^{th} power* of μ and denoted by μ^{*t}. By Lemma 10.23, μ^{*t} is indeed a distribution and it is infinitely divisible.

By definition, for $n \in \mathbb{N}$, the distinguished $(1/n)^{\text{th}}$ power of $\mu \in \mathrm{ID}(\mathbb{R}^d)$ coincides with the nth convolution root, and the distinguished n^{th} power coincides with the n-fold convolution of μ with itself. Hence the notation μ^{*t} extends unambiguously the notations $\mu^{*1/n}$ and μ^{*n} that we have used previously. The next lemma is an immediate consequence of the definition.

Lemma 10.25. *Suppose that $\mu \in \mathrm{ID}(\mathbb{R}^d)$ and $s, t \geq 0$. Then $(\mu^{*t})^{*s} = (\mu^{*s})^{*t} = \mu^{*st}$, $\mu^{*0} = \delta_0$, and $\mu^{*r} \xrightarrow{w} \mu^{*t}$ as $r \to t$, $r \geq 0$.*

We come now to the one-to-one relation between infinitely divisible distributions and Lévy processes in law.

Theorem 10.26 (I. d. distributions and Lévy processes in law). *Suppose that $I = \mathbb{R}$ or $I = [0, \infty)$.*

(a) *If $L = (L_t)_{t\in I}$ is a Lévy process in law then $\mathcal{L}(L_t) \in \mathrm{ID}(\mathbb{R}^d)$ for all $t \in I$ and*

$$\mathcal{L}(L_t) = \begin{cases} [\mathcal{L}(L_1)]^{*t}, & t \geq 0, \\ [\mathcal{L}(-L_1)]^{*|t|}, & t < 0 \text{ if } I = \mathbb{R}. \end{cases}$$

(b) *If* $L = (L_t)_{t\in I}$ *and* $X = (X_t)_{t\in I}$ *are two Lévy processes in law such that* $\mathcal{L}(L_{t_0}) = \mathcal{L}(X_{t_0})$ *for some* $t_0 \in I \setminus \{0\}$ *then L and X are identical in law, i. e., they have the same finite dimensional distributions.*

(c) *For every* $\mu \in \mathrm{ID}(\mathbb{R}^d)$ *there exists a Lévy process in law* $L = (L_t)_{t\in I}$ *such that* $\mathcal{L}(L_1) = \mu$. *This Lévy process in law is unique in law by* (b).

Proof. (a) Let $L = (L_t)_{t\in I}$ be a Lévy process in law. By Proposition 10.22, we have $\mathcal{L}(L_t) \in \mathrm{ID}(\mathbb{R}^d)$ for all $t \in I$. Since $L_1 = (L_{1/n} - L_0) + (L_{2/n} - L_{1/n}) + \cdots + (L_{n/n} - L_{(n-1)/n})$, we have $\mathcal{L}(L_1) = [\mathcal{L}(L_{1/n})]^{*n}$ and hence $\mathcal{L}(L_{1/n}) = [\mathcal{L}(L_1)]^{*1/n}$ for $n \in \mathbb{N}$ by Theorem 8.26. Using the same argument and Lemma 10.25, we obtain

$$\mathcal{L}(L_{m/n}) = [\mathcal{L}(L_{1/n})]^{*m} = [(\mathcal{L}(L_1))^{*1/n}]^{*m} = [\mathcal{L}(L_1)]^{*m/n}$$

for $m, n \in \mathbb{N}$. For $t \in [0,\infty) \setminus \mathbb{Q}$ choose a sequence $(r_n)_{n\in\mathbb{N}}$ in $\mathbb{Q} \cap (0,\infty)$ such that $r_n \to t$ as $n \to \infty$. Then $\mathcal{L}(L_{r_n}) \xrightarrow{w} \mathcal{L}(L_t)$ by the continuity in probability of Lévy processes in law, and $\mathcal{L}(L_{r_n}) = [\mathcal{L}(L_1)]^{*r_n} \xrightarrow{w} [\mathcal{L}(L_1)]^{*t}$ by Lemma 10.25. Hence $\mathcal{L}(L_t) = [\mathcal{L}(L_1)]^{*t}$ for all $t \ge 0$. That $\mathcal{L}(L_t) = [\mathcal{L}(-L_1)]^{*|t|}$ for $t < 0$ follows from Corollary 10.15.

(b) Since $\mathcal{L}(L_{-t}) = \mathcal{L}(-L_t)$, for every two-sided Lévy process in law by Corollary 10.15, we can assume without loss of generality that $t_0 > 0$. It then follows, as in the proof of (a), that $\mathcal{L}(L_t) = [\mathcal{L}(L_{t_0})]^{*t/t_0} = [\mathcal{L}(X_{t_0})]^{*t/t_0} = \mathcal{L}(X_t)$ for each $t \ge 0$. Hence, by Proposition 10.16, L and X have the same finite-dimensional distributions.

(c) Suppose that $\mu \in \mathrm{ID}(\mathbb{R}^d)$ and define $\mathcal{T} := \{\mathbf{t} = (t_1,\dots,t_n)^T \in [0,\infty)^n : t_1 < \cdots < t_n, n \in \mathbb{N}\}$. For $\mathbf{t} = (t_1,\dots,t_n)^T \in \mathcal{T}$ define

$$\rho_{\mathbf{t}} := \mu^{*t_1} \otimes \mu^{*(t_2-t_1)} \otimes \cdots \otimes \mu^{*(t_n-t_{n-1})},$$

the function $f_n : (\mathbb{R}^d)^n \to (\mathbb{R}^d)^n$,

$$(x_1,\dots,x_d)^T \mapsto (x_1, x_1 + x_2, x_1 + x_2 + x_3, \dots, x_1 + \cdots + x_n)^T,$$

and $\mu_{\mathbf{t}}$ to be the image measure of $\rho_{\mathbf{t}}$ under the mapping f_n, i. e.

$$\mu_{\mathbf{t}} := f_n(\rho_{\mathbf{t}}) = \rho_{\mathbf{t}} \circ f_n^{-1}.$$

We want to show that $(\mu_{\mathbf{t}})_{\mathbf{t}\in\mathcal{T}}$ is a consistent family of finite-dimensional distributions in the sense of (1.6), i. e.,

$$\begin{aligned}&\widehat{\mu}_{(t_1,\dots,t_n)}(u_1,\dots,u_{j-1},0,u_{j+1},\dots,u_n)\\&\quad= \widehat{\mu}_{(t_1,\dots,t_{j-1},t_{j+1},\dots,t_n)}(u_1,\dots,u_{j-1},u_{j+1},\dots,u_n)\end{aligned} \tag{10.2}$$

for all $n \in \mathbb{N}$, $j \in \{1,\dots,n\}$, $(t_1,\dots,t_n)^T \in \mathcal{T}$ and $u_1,\dots,u_{j-1},u_{j+1},\dots,u_n \in \mathbb{R}^d$. To see this, suppose that $\mathbf{t} = (t_1,\dots,t_n)^T \in \mathcal{T}$ and that $(Z_1,\dots,Z_n)^T$ is a random vector with values in $(\mathbb{R}^d)^n$ and distribution $\rho_{\mathbf{t}}$. Define $t_0 := 0$. Then $Z_1, \dots, Z_n$ are independent with $\mathcal{L}(Z_k) = \mu^{*(t_k - t_{k-1})}$. For $t \geq 0$ denote the characteristic function of μ^{*t} by ϕ_t. Then $\mu_{\mathbf{t}} = \mathcal{L}(f_n((Z_1, \dots, Z_n)^T))$ so that

$$\widehat{\mu}_{\mathbf{t}}(u_1, \dots, u_n) = \mathbb{E}\exp\left(\mathrm{i}\sum_{m=1}^{n}\left\langle u_m, \sum_{k=1}^{m} Z_k\right\rangle\right)$$
$$= \mathbb{E}\exp\left(\mathrm{i}\sum_{k=1}^{n}\left\langle \sum_{m=k}^{n} u_m, Z_k\right\rangle\right) = \prod_{k=1}^{n}\phi_{t_k - t_{k-1}}\left(\sum_{m=k}^{n} u_m\right).$$

Suppose that $j \in \{1, \dots, n\}$ and $u_1, \dots, u_{j-1}, u_{j+1}, \dots, u_n \in \mathbb{R}^d$, and define $u_j := 0$ and $t_{n+1} := t_n$ (needed if $j = n$). Then

$$\widehat{\mu}_{\mathbf{t}}(u_1, \dots, u_{j-1}, 0, u_{j+1}, \dots, u_n)$$
$$= \phi_{t_j - t_{j-1}}\left(\sum_{m=j}^{n} u_m\right) \cdot \phi_{t_{j+1} - t_j}\left(\sum_{m=j+1}^{n} u_m\right) \cdot \prod_{\substack{k=1,\dots,n \\ k \neq j, j+1}} \phi_{t_k - t_{k-1}}\left(\sum_{m=k}^{n} u_m\right).$$

Since $\sum_{m=j}^{n} u_m = \sum_{m=j+1}^{n} u_m$ and

$$\phi_{t_j - t_{j-1}}(z) \cdot \phi_{t_{j+1} - t_j}(z)$$
$$= \exp[(t_j - t_{j-1}) \operatorname{Log} \widehat{\mu}(z)] \cdot \exp[(t_{j+1} - t_j) \operatorname{Log} \widehat{\mu}(z)]$$
$$= \exp[(t_{j+1} - t_{j-1}) \operatorname{Log} \widehat{\mu}(z)] = \phi_{t_{j+1} - t_{j-1}}(z)$$

for $z \in \mathbb{R}^d$, we conclude that

$$\widehat{\mu}_{\mathbf{t}}(u_1, \dots, u_{j-1}, 0, u_{j+1}, \dots, u_n)$$
$$= \phi_{t_{j+1} - t_{j-1}}\left(\sum_{m=j+1}^{n} u_m\right) \cdot \prod_{\substack{k=1,\dots,n \\ k \neq j, j+1}} \phi_{t_k - t_{k-1}}\left(\sum_{m=k}^{n} u_j\right)$$
$$= \widehat{\mu}_{(t_1, \dots, t_{j-1}, t_{j+1}, \dots, t_n)}(u_1, \dots, u_{j-1}, u_{j+1}, \dots, u_n),$$

which is (10.2). It then follows from Kolmogorov's consistency theorem (Theorem 1.13) that there exists a stochastic process $L = (L_t)_{t \geq 0}$ with $(\mu_{\mathbf{t}})_{\mathbf{t} \in \mathcal{T}}$ as its finite dimensional distributions. To see that L is a (one-sided) Lévy process in law with finite dimensional distributions $\mu_{\mathbf{t}}$, $\mathbf{t} = (t_1, \dots, t_n)^T \in \mathcal{T}$, denote by g_n the inverse function of f_n, so that

$$g_n(x_1, x_2, \dots, x_n) = (x_1, x_2 - x_1, x_3 - x_2, \dots, x_n - x_{n-1})^T, \quad x_1, \dots, x_n \in \mathbb{R}^d.$$

Then

$$(L_{t_1}, L_{t_2} - L_{t_1}, \dots, L_{t_n} - L_{t_{n-1}})^T = g_n(L_{t_1}, L_{t_2}, \dots, L_{t_n})$$
$$\stackrel{d}{=} g_n(\mu_{\mathbf{t}}) = g_n(f_n(\rho_{\mathbf{t}})) = \rho_{\mathbf{t}}.$$

From the definition of ρ_t, we see that $\mathcal{L}(L_{t_1}) = \mu^{*t_1}$. This implies that $\mathcal{L}(L_0) = \delta_0$ so that $L_0 = 0$ a. s., that $L_{t_1} - L_0, L_{t_2} - L_{t_1}, \ldots, L_{t_n} - L_{t_{n-1}}$ are independent, and that $\mathcal{L}(L_{t_2} - L_{t_1}) = \mu^{*(t_2-t_1)}$. Hence L has stationary and independent increments and $\mathcal{L}(L_t) = \mu^{*t}$ for all $t \geq 0$. Since $\mu^{*t} \xrightarrow{w} \mu^{*0} = \delta_0$ by Lemma 10.25, it follows that L_t converges in distribution, and hence in probability, to $L_0 = 0$ as $t \downarrow 0$. From Proposition 10.14(a) we conclude that L is continuous in probability, and is therefore a one-sided Lévy process in law with $\mathcal{L}(L_1) = \mu$. Using Remark 10.18, L can be extended to a two-sided Lévy process in law with $\mathcal{L}(L_1) = \mu$. □

Since infinitely divisible distributions are characterized in terms of their characteristic triplet by the Lévy–Khintchine-formula, the one-to-one-correspondence between infinitely divisible distributions and Lévy processes in law allows the definition of a characteristic triplet of Lévy processes (in law):

Definition 10.27. Suppose that $\mu \in \mathrm{ID}(\mathbb{R}^d)$ has characteristic exponent Ψ_μ and characteristic triplet $(A, \nu, \gamma)_c$ with respect to some fixed truncation function $c : \mathbb{R}^d \to \mathbb{R}^d$, and suppose that $I = [0, \infty)$ or $I = \mathbb{R}$. If $L = (L_t)_{t \in I}$ is a Lévy process or, more generally, a Lévy process in law, with $\mathcal{L}(L_1) = \mu$, then we call μ the *infinitely divisible distribution corresponding to L* and conversely we call L a *Lévy process (in law) corresponding to μ.* The triplet $(A, \nu, \gamma)_c$ is also called the *characteristic triplet of L with respect to c*, or the *Lévy–Khintchine triplet of L with respect to c*, or the *generating triplet of L with respect to c*, and we speak of A as the *Gaussian variance*, ν as the *Lévy measure* and γ as the *location parameter (with respect to c)* of L. When $c(x) = \mathbf{1}_D(x)x$ is the standard truncation function, we also speak of *the (standard) characteristic/generating/Lévy–Khintchine triplet* of L and simply write (A, ν, γ). Similarly, when the drift γ^0 of $\mathcal{L}(L_1)$ exists (i. e., when $\int_{|x|\leq 1} |x|\,\nu(dx) < \infty$), we call γ^0 the *drift of L*, and when the center γ^m of $\mathcal{L}(L_1)$ exists (i. e. when $\int_{|x|>1} |x|\,\nu(dx) < \infty$), we call γ^m the *center of L*. The characteristic exponent Ψ_μ of μ is also called the *characteristic exponent of L* and we denote it by $\Psi_L := \Psi_\mu = \Psi_{L_1}$. The representation

$$\begin{aligned} &\mathbb{E}\,\mathrm{e}^{\mathrm{i}\langle z, L_t\rangle} \\ &= \exp\Bigg[t\Bigg(\mathrm{i}\langle\gamma, z\rangle - \frac{1}{2}\langle z, Az\rangle + \int_{\mathbb{R}^d} (\mathrm{e}^{\mathrm{i}\langle z,x\rangle} - 1 - \mathrm{i}\langle z, c(x)\rangle)\,\nu(dx)\Bigg)\Bigg], \end{aligned} \tag{10.3}$$

for $z \in \mathbb{R}^d$ and $t \geq 0$ (together with $\mathbb{E}\,\mathrm{e}^{\mathrm{i}\langle z, L_t\rangle} = \mathbb{E}\,\mathrm{e}^{\mathrm{i}\langle -z, L_{-t}\rangle}$ for $t < 0$ when $I = \mathbb{R}$) is called the *Lévy–Khintchine representation of L with respect to c*; when $c(x) = \mathbf{1}_D(x)x$ it is called *the (standard) Lévy–Khintchine representation of L*. When $d = 1$, the *characteristic pair of L* is defined similarly as the characteristic pair of $\mathcal{L}(L_1)$.

Remark 10.28. (a) Since a Lévy process in law corresponding to an infinitely divisible distribution μ is unique in law, we also speak of *the* Lévy process in law corresponding to μ.

(b) When $I = \mathbb{R}$, the Lévy–Khintchine representation (10.3) can be written for all $t \in \mathbb{R}$ and $z \in \mathbb{R}^d$ in the compact form

$$\mathbb{E}\, e^{i\langle z, L_t\rangle} = \exp\Big[|t| \Big(i\langle \gamma, z\,\mathrm{sign}(t)\rangle - \frac{1}{2}\langle z, Az\rangle + \int_{\mathbb{R}^d} \big(e^{i\langle z\,\mathrm{sign}(t), x\rangle} - 1 - i\langle z\,\mathrm{sign}(t), c(x)\rangle\big)\,\nu(dx)\Big)\Big],$$

where $t \mapsto \mathrm{sign}(t) = \mathbf{1}_{(0,\infty)}(t) - \mathbf{1}_{(-\infty,0)}(t)$ denotes the sign function.

(c) It is straightforward to see from the proof of Theorem 10.26 that also to every $t_0 \in I \setminus \{0\}$ and $\mu \in \mathrm{ID}(\mathbb{R}^d)$ there exists a Lévy process in law $L = (L_t)_{t\in I}$, unique in law, such that $\mathcal{L}(L_{t_0}) = \mu$.

(d) It is sometimes convenient, in particular when working with stochastic integrals, to have the process defined on a complete probability space. Recall that a probability space $(\Omega, \mathcal{F}, P)$ is *complete* if $A \in \mathcal{F}$, $B \subset A$ and $P(A) = 0$ imply $B \in \mathcal{F}$. To every probability space $(\Omega, \mathcal{F}, P)$, there exists a smallest σ-algebra $\overline{\mathcal{F}} \supset \mathcal{F}$ on Ω and a probability measure $\overline{P}$ on $(\Omega, \overline{\mathcal{F}})$ that extends P and such that $(\Omega, \overline{\mathcal{F}}, \overline{P})$ is complete. The probability space $(\Omega, \overline{\mathcal{F}}, \overline{P})$ is called the *completion* of $(\Omega, \mathcal{F}, P)$ (see, e. g., Cohn [33, Section 1.5]). Since the (joint) distribution of random vectors defined on $(\Omega, \mathcal{F}, P)$ is unchanged when they are considered as random vectors on $(\Omega, \overline{\mathcal{F}}, \overline{P})$, every Lévy process (in law) on $(\Omega, \mathcal{F}, P)$ is also a Lévy process (in law) on the completion $(\Omega, \overline{\mathcal{F}}, \overline{P})$. In particular, it follows from Theorem 10.26 that to every $\mu \in \mathrm{ID}(\mathbb{R}^d)$ there exists a Lévy process in law $L = (L_t)_{t\in I}$ defined on a *complete* probability space such that $\mathcal{L}(L_1) = \mu$.

Example 10.29. If $L = (L_t)_{t\in I}$ and $X = (X_t)_{t\in I}$ are two independent Lévy processes (in law) in $\mathbb{R}^d$ with characteristic triplets $(A_L, \nu_L, \gamma_L)_c$ and $(A_X, \nu_X, \gamma_X)_c$ with respect to some fixed truncation function $c : \mathbb{R}^d \to \mathbb{R}^d$ then $L + X$ is a Lévy process (in law) with characteristic triplet $(A_L + A_X, \nu_L + \nu_X, \gamma_L + \gamma_X)_c$ and characteristic exponent $\Psi_L + \Psi_X$. This follows at once from Proposition 10.11(a) and Example 9.15.

Remark 10.30. Let $X = (X_t)_{t\in I}$ be a Lévy process (in law) in $\mathbb{R}^d$ with characteristic exponent Ψ_X and characteristic triplet $(A, \nu, \gamma^{c_1})_{c_1}$ for a given truncation function $c_1 : \mathbb{R}^d \to \mathbb{R}^d$. Suppose that $C \in \mathbb{R}^{m\times d}$, $b \in \mathbb{R}^m$ and $c_2 : \mathbb{R}^m \to \mathbb{R}^m$ is a fixed truncation function in $\mathbb{R}^m$. Then $Y := (bt + CX_t)_{t\in I}$ is a Lévy process (in law) in $\mathbb{R}^m$ with characteristic exponent $z \mapsto \Psi_Y(z) = ib^T z + \Psi_X(C^T z)$ and characteristic triplet $(A_Y, \nu_Y, \gamma_Y^{c_2})_{c_2}$ given by (9.31)–(9.33). Similarly, when the drift γ^0 of X exists, then Y has drift $b + C\gamma^0$, and if the center γ^m of X exists, then Y has center $b + C\gamma^m$. This follows immediately from Proposition 10.11(a), Theorem 9.25 and Corollary 9.26.

We next relate independence of the components of a Lévy process to properties of its characteristic triplet.

Theorem 10.31 (Lévy processes with independent components). *Let $X = (X_t = (X_t^1, \dots, X_t^d)^T)_{t \in I}$ be a stochastic process in $\mathbb{R}^d$, and suppose that $t_0 \in I$ with $t_0 \neq 0$. Then the following are equivalent:*
(i) *$X^1, \dots, X^d$ are independent Lévy processes (in law).*
(ii) *X is a Lévy process (in law) with $X^1, \dots, X^d$ independent.*
(iii) *X is a Lévy process (in law) with $X^1, \dots, X^d$ pairwise independent.*
(iv) *X is a Lévy process (in law) with $X_{t_0}^1, \dots, X_{t_0}^d$ independent.*
(v) *X is a Lévy process (in law) with $X_{t_0}^1, \dots, X_{t_0}^d$ pairwise independent.*
(vi) *X is a Lévy process (in law) with standard characteristic triplet (A, ν, γ), where A is diagonal and ν is supported on the axes.*

Proof. The implication (i) $\Longrightarrow$ (ii) follows from Proposition 10.11(b), the implications (ii) $\Longrightarrow$ (iii) $\Longrightarrow$ (v) are clear, and the implications (v) $\Longrightarrow$ (iv) $\Longrightarrow$ (vi) follow from Theorem 9.28. To show that (vi) implies (i), let X be a Lévy process (in law) with standard characteristic triplet (A, ν, γ) where A is diagonal and ν is supported on the axes. Since $\mathcal{L}((X_1^1, \dots, X_1^d)^T)$ has characteristic triplet (A, ν, γ), we know from Theorem 9.28 that $X_1^1, \dots, X_1^d$ are independent. Let $Y^1, \dots, Y^d$ be independent Lévy processes (in law) with $\mathcal{L}(Y_1^j) = \mathcal{L}(X_1^j)$ for all $j \in \{1, \dots, d\}$. Then by Proposition 10.11(b), $Y = (Y^1, \dots, Y^d)$ is a Lévy process (in law) with the same characteristic triplet as X and hence has the same finite dimensional distributions as X by Theorem 10.26. But since $Y^1, \dots, Y^d$ are independent, so are $X^1, \dots, X^d$, giving (ii) and hence (i). □

10.3 Existence of Lévy processes

In this section, we show that every Lévy process in law has a modification which has càdlàg paths and is therefore a Lévy process. Consequently, it follows that to every infinitely divisible distribution μ there exists a Lévy process $L = (L_t)_{t \in I}$, unique in distribution, with $\mathcal{L}(L_1) = \mu$. We follow the exposition of Itô [51, Section 1.3] in this section including, in particular, the proof for the broader class of additive processes. We first need to make precise the concepts of modification and indistinguishability for $\mathbb{R}^d$-valued stochastic processes.

Definition 10.32 (Modification; indistinguishable). Suppose that $J \subset \mathbb{R}$ and that $X = (X_t)_{t \in J}$ and $Y = (Y_t)_{t \in J}$ are two $\mathbb{R}^d$-valued stochastic processes on a probability space $(\Omega, \mathcal{F}, P)$. Then X and Y are called
(a) *modifications* or *versions of each other*, if

$$P(X_t = Y_t) = 1 \quad \forall \, t \in J,$$

(b) and *indistinguishable*, if there exists a set $\Omega_0 \in \mathcal{F}$ with $P(\Omega_0) = 1$ such that

$$X_t(\omega) = Y_t(\omega) \quad \forall\, t \in J \;\forall\, \omega \in \Omega_0,$$

i. e., if $[X_t = Y_t \;\forall\, t \in J]$ almost surely.

Remark 10.33. Suppose that $J \subset \mathbb{R}$ and that $X = (X_t)_{t\in J}$ and $Y = (Y_t)_{t\in J}$ are two $\mathbb{R}^d$-valued stochastic processes defined on the same probability space. It is obvious that if X and Y are indistinguishable, then they are modifications of each other, and if X and Y are modifications of each other, then they are identical in law (i. e., have the same finite dimensional distributions). Conversely, if J is countable and X and Y are modifications of each other, then it is easy to see that they are also indistinguishable. This is no longer true however if J is uncountable (see Exercise 10.4). However, if both X and Y are right-continuous (or left-continuous) and are modifications of each other, then they are indistinguishable, as seen in the next lemma.

Lemma 10.34. *Suppose that $I = [0,\infty)$ or $I = \mathbb{R}$, that $X = (X_t)_{t\in I}$ and $Y = (Y_t)_{t\in I}$ are two stochastic processes which are modifications of each other, and that X and Y both have almost surely right-continuous sample paths, or both have almost surely left-continuous sample paths. Then X and Y are indistinguishable.*

Proof. Define $J := I \cap \mathbb{Q}$. Then $(X_t)_{t\in J}$ and $(Y_t)_{t\in J}$ are modifications of each other and hence indistinguishable by the previous remark, so that there exists a set $\Omega_0 \in \mathcal{F}$ such that $P(\Omega_0) = 1$ and $X_t(\omega) = Y_t(\omega)$ for all $t \in J$ and $\omega \in \Omega$. If now both X and Y are almost surely right-continuous, then there exists $\Omega_1 \in \mathcal{F}$ with $P(\Omega_1) = 1$ such that $X_t(\omega) = \lim_{h\downarrow 0} X_{t+h}(\omega)$ and $Y_t(\omega) = \lim_{h\downarrow 0} Y_{t+h}(\omega)$ for all $\omega \in \Omega_1$ and $t \in I$. Hence $X_t(\omega) = Y_t(\omega)$ for all $t \in I$ and $\omega \in \Omega_0 \cap \Omega_1$, so that X and Y are indistinguishable. The same is true when both X and Y are left-continuous almost surely. □

The next lemma presents some properties of processes that are continuous in probability.

Lemma 10.35. *Suppose that $-\infty < \alpha < \beta < \infty$ and $X = (X_t)_{t\in[\alpha,\beta]}$ is an $\mathbb{R}^d$-valued stochastic process which is continuous in probability. Then for every $\delta > 0$ there exists $K = K(\delta) \in (0,\infty)$ such that*

$$P(|X_t| \geq K) \leq \delta \quad \forall\, t \in [\alpha,\beta]. \tag{10.4}$$

Moreover X is uniformly continuous in probability in the sense that for every $\varepsilon > 0$ and $\eta > 0$ there exist some $\gamma > 0$ such that

$$P(|X_s - X_t| \geq \varepsilon) \leq \eta \quad \forall\, s,t \in [\alpha,\beta] : |s-t| \leq \gamma. \tag{10.5}$$

Proof. Suppose that $\delta > 0$. Since X is $\mathbb{R}^d$-valued, for every $t \in [\alpha,\beta]$ there exists some $K_t \in (0,\infty)$ such $P(|X_t| \geq K_t/2) \leq \delta/2$. Since X is continuous in probability at t, there exists $\xi(t) > 0$ such that $P(|X_s - X_t| \geq K_t/2) \leq \delta/2$ for all $s \in (t-\xi(t), t+\xi(t)) \cap [\alpha,\beta]$.

Since $[\alpha, \beta]$ is compact, it can be covered by finitely many such intervals, say associated with $t_1, \dots, t_n$, and (10.4) follows with $K := \max\{K_{t_1}, \dots, K_{t_n}\}$. To see (10.5), suppose that $\varepsilon > 0$ and $\eta > 0$. By stochastic continuity, for each $t \in [\alpha, \beta]$ there exists $\zeta(t) > 0$ such that $P(|X_s - X_t| \geq \varepsilon/2) \leq \eta/2$ for all $s \in (t - \zeta(t), t + \zeta(t)) \cap [\alpha, \beta]$. Again, $[\alpha, \beta]$ can be covered by finitely many such intervals, say associated with $t_1 < \cdots < t_m$, and then $P(|X_s - X_v| \geq \varepsilon) \leq \eta$ whenever $s, v \in (t - \zeta(t_j), t + \zeta(t_j))$ for the same t_j. Now it is clearly possible to choose $\gamma > 0$ small enough to ensure that for each pair $s, v \in [\alpha, \beta]$ with $|s - v| \leq \gamma$ there exists some $j \in \{1, \dots, m\}$ with $s, v \in (t_j - \zeta(t_j), t_j + \zeta(t_j))$. This gives (10.5). □

Next we want to exploit the independent increments property. We start with Etemadi's maximal inequality for (finite) sums of independent summands:

Theorem 10.36 (Etemadi's maximal inequality). *Let $(Y_n)_{n\in\mathbb{N}}$ be a sequence of independent random vectors on $\mathbb{R}^d$ and define $S_n := \sum_{j=1}^n Y_j$ for $n \in \mathbb{N}_0$. Then*

$$P\Big(\max_{j\in\{0,\dots,n\}} |S_j| \geq u\Big) \leq 3 \max_{j\in\{1,\dots,n\}} P(|S_j| \geq u/3) \quad \forall\, u \geq 0\ \forall\, n \in \mathbb{N}.$$

Proof. For $j \in \{1, \dots, n\}$ and $r > 0$, define

$$A_j := \Big\{\max_{k\in\{1,\dots,j-1\}} |S_k| < 3r,\ |S_j| \geq 3r\Big\}.$$

Then $A_1, \dots, A_n$ are pairwise disjoint and

$$\bigcup_{j=1}^n A_j = \Big\{\max_{j\in\{1,\dots n\}} |S_j| \geq 3r\Big\} = \Big\{\max_{j\in\{0,\dots n\}} |S_j| \geq 3r\Big\}$$

(the latter inequality follows from the fact that $S_0 = 0$). Hence

$$\begin{aligned} P\Big(\max_{j\in\{0,\dots,n\}} |S_j| \geq 3r\Big) &\leq P(|S_n| \geq r) + P\Big(\bigcup_{j=1}^{n-1} A_j \cap \{|S_n| < r\}\Big) \\ &= P(|S_n| \geq r) + \sum_{j=1}^{n-1} P(A_j \cap \{|S_n| < r\}). \end{aligned}$$

However, since $A_j \cap \{|S_n| < r\} \subset A_j \cap \{|S_n - S_j| > 2r\}$ and since A_j and $\{|S_n - S_j| > 2r\}$ are independent, we further conclude that

$$\begin{aligned} &P\Big(\max_{j\in\{0,\dots,n\}} |S_j| \geq 3r\Big) \\ &\quad\leq P(|S_n| \geq r) + \sum_{j=1}^{n-1} P(A_j)P(|S_n - S_j| > 2r) \\ &\quad\leq P(|S_n| \geq r) + \max_{j\in\{1,\dots,n\}} P(|S_n - S_j| > 2r) \end{aligned}$$

$$\le P(|S_n| \ge r) + \max_{j\in\{1,\dots,n\}} (P(|S_n| > r) + P(|S_j| > r))$$
$$\le 3 \max_{j\in\{1,\dots,n\}} P(|S_j| \ge r).$$

Replacing $3r$ by u gives the inequality for $u > 0$, and for $u = 0$ it is trivial. □

Corollary 10.37. *Let $(Y_n)_{n\in\mathbb{N}}$ be a sequence of independent random vectors on $\mathbb{R}^d$ and define $S_n := \sum_{j=1}^n Y_j$ for $n \in \mathbb{N}_0$. Then*

$$P\Big(\max_{j,k\in\{0,\dots,n\}} |S_j - S_k| \ge u\Big) \le 3 \max_{j\in\{1,\dots,n\}} P(|S_j| \ge u/6) \quad \forall\, u \ge 0, n \in \mathbb{N}.$$

Proof. This is immediate from Theorem 10.36, since $\{\max_{j,k\in\{0,\dots,n\}} |S_j - S_k| \ge u\} \subset \{\max_{j\in\{0,\dots,n\}} |S_j| \ge u/2\}$. □

We also need the notion of upcrossings.

Definition 10.38 (Upcrossings). Let $J \subset \mathbb{R}$ be countable and let $X = (X_t)_{t\in J}$ be a real-valued stochastic process. Suppose that $a, b \in \mathbb{R}$ and $a < b$. Then the *number $U_{a,b}(X)$ of $[a,b]$-upcrossings of X over J* is defined as the supremum of all numbers $n \in \mathbb{N}_0$ for which there exist numbers $s_1 < t_1 < \cdots < s_n < t_n$, all in J, such that

$$X_{s_k} \le a \quad \text{and} \quad X_{t_k} \ge b \quad \forall\, k \in \{1,\dots,n\}.$$

It is easy to see that $U_{ab}((X_t)_{t\in K})$ is a random variable for all finite $K \subset J$, that $U_{ab}((X_t)_{t\in K}) \le U_{ab}((X_t)_{t\in K'})$ for $K \subset K' \subset J$, and that

$$U_{ab}\big((X_t)_{t\in J}\big) = \lim_{K\uparrow J, K \text{ finite}} U_{ab}\big((X_t)_{t\in K}\big),$$

from which it follows that $U_{ab}((X_t)_{t\in J})$ is an $\mathbb{N}_0 \cup \{\infty\}$-valued random variable for all countable J. The following result bounds the expected number of $[a,b]$-upcrossings of a (finite) sum with independent increments. Usually, Doob's upcrossing inequality is formulated for martingales (or submartingales) $X = (X_n)_{n\in\mathbb{N}_0}$ as $\mathbb{E}(U_{ab}((X_t)_{t=0,\dots,n})) \le \frac{\mathbb{E}((X_n-a)^+)}{b-a}$, but we need a variant of this result in order to deal with processes which do not necessarily have finite expectation.

Lemma 10.39 (Doob's upcrossing inequality). *Let $Y_1, \dots, Y_n$ be independent real-valued random variables and define $S_k := \sum_{j=1}^k Y_j$ for $k \in \{0,\dots,n\}$. Suppose that $a, b \in \mathbb{R}$, $a < b$ and*

$$P\Big(\max_{j,k\in\{0,\dots,n\}} |S_j - S_k| \ge b - a\Big) \le \delta < 1.$$

Then

$$P\big(U_{ab}\big((S_t)_{t=0,\dots,n}\big) \ge m\big) \le \delta^m \quad \forall\, m \in \mathbb{N}_0, \tag{10.6}$$

and, in particular,

$$\mathbb{E}(U_{ab}((S_t)_{t=0,\dots,n})) \le \frac{\delta}{1-\delta}. \tag{10.7}$$

Proof. We shall prove equation (10.6) by induction. It is trivially true for $m \in \{0,1\}$. We shall show that if it holds for some $m \ge 1$, then it also holds for $m+1$ and since it holds for $m = 1$, it must hold for all $m \in \{0,1,2,\dots\}$.

Suppose that $m \ge 1$ and define

$$\mathcal{T}_{2m} := \{(s_1,t_1,\dots,s_m,t_m)^T \in \{0,\dots,n\}^{2m} : s_1 < t_1 < \dots < s_m < t_m\}.$$

Then, for $(s_1,\dots,t_m)^T \in \mathcal{T}_{2m}$, define

$$\begin{aligned}
&A_{(s_1,\dots,t_m)}\\
&:= \{S_t > a\ \forall\, t \in \{0,\dots,s_1-1\},\ S_{s_1} \le a,\ S_t < b\ \forall\, t \in \{s_1+1,\dots,t_1-1\},\\
&\quad S_{t_1} \ge b,\ S_t > a\ \forall\, t \in \{t_1+1,\dots,s_2-1\},\ S_{s_2} \le a,\\
&\quad S_t < b\ \forall\, t \in \{s_2+1,\dots,t_2-1\},\ S_{t_2} \ge b,\ \dots,\\
&\quad S_{s_m} \le a,\ S_t < b\ \forall\, t \in \{s_m+1,\dots,t_m-1\},\ S_{t_m} \ge b\}.
\end{aligned}$$

Occurrence of the event $A_{(s_1,\dots,t_m)}$ means that s_1 is the first time when S is less than or equal to a, t_1 is the first time after s_1 when S is greater or equal than b, s_2 is the first time after t_1 when S is less than or equal to a, etc. It is then evident that

$$\{U_{ab}((S_t)_{t=0,\dots,n}) \ge m\} = \bigcup_{(s_1,\dots,t_m)^T \in \mathcal{T}_{2m}} A_{(s_1,\dots,t_m)},$$

where the sets in the union are pairwise disjoint. Hence

$$P(U_{ab}((S_t)_{t=0,\dots,n}) \ge m+1) = \sum_{(s_1,\dots,t_{m+1})^T \in \mathcal{T}_{2m+2}} P(A_{(s_1,\dots,t_{m+1})}). \tag{10.8}$$

Now let $(s_1',t_1',\dots,s_m',t_m')^T$ be a fixed vector in $\mathcal{T}_{2m}$. Then

$$\begin{aligned}
&\bigcup_{(s_1,\dots,t_{m+1})^T \in \mathcal{T}_{2m+2}: s_1=s_1',\dots,t_m=t_m'} A_{(s_1,\dots,t_{m+1})}\\
&\quad \subset A_{(s_1',\dots,t_m')} \cap \Big\{\max_{j,k\in\{t_m'+1,\dots,n\}} |S_j - S_k| \ge b-a\Big\}
\end{aligned}$$

where the sets on the right hand side are independent. Hence

$$\begin{aligned}
&\sum_{(s_1,\dots,t_{m+1})^T \in \mathcal{T}_{2m+2}: s_1=s_1',\dots,t_m=t_m'} P(A_{(s_1,\dots,t_{m+1})})\\
&\le P(A_{(s_1',\dots,t_m')})P\Big(\Big\{\max_{j,k\in\{t_m'+1,\dots,n\}} |S_j - S_k| \ge b-a\Big\}\Big) \le P(A_{(s_1',\dots,t_m')})\,\delta
\end{aligned}$$

by assumption. Summing over $(s'_1, \dots, t'_m)^T \in \{0, \dots, n\}^{2m}$ in (10.8) gives

$$P(U_{ab}((S_t)_{t=0,\dots,n}) \geq m+1) \leq P(U_{ab}((S_t)_{t=0,\dots,n}) \geq m)\,\delta,$$

completing the induction argument.

Equation (10.7) is then an immediate consequence of (10.6) since $\mathbb{E}(X) = \sum_{m=1}^{\infty} P(X \geq m)$ for every $\mathbb{N}_0$-valued random variable X. □

Corollary 10.40. *Let $Y_1, \dots, Y_n$ be independent real-valued random variables and define $S_k := \sum_{j=1}^{k} Y_j$ for $k \in \{0, \dots, n\}$. Suppose that $-\infty < a < b < \infty$ and*

$$P\left(|S_j| \geq \frac{b-a}{6}\right) \leq \frac{1}{6} \quad \forall j \in \{1, \dots, n\}.$$

Then

$$\mathbb{E}(U_{ab}((S_t)_{t=0,\dots,n}) \leq 1.$$

Proof. This follows at once from Corollary 10.37 and Lemma 10.39. □

The next result, due to Doob [35], is the main result of this section. It applies to additive processes in law and hence, in particular, to Lévy processes in law.

Theorem 10.41 (Doob's regularization result for additive proc. in law). *Let $I = [0, \infty)$ or $I = \mathbb{R}$ and let $L = (L_t)_{t \in I}$ be an $\mathbb{R}^d$-valued additive process defined on a probability space $(\Omega, \mathcal{F}, P)$. Then there exists a modification $X = (X_t)_{t \in I}$ of L on $(\Omega, \mathcal{F}, P)$ such that all paths of X are càdlàg. Moreover, every other modification X' of L on $(\Omega, \mathcal{F}, P)$ with càdlàg paths is indistinguishable from X and every modification X' of L is continuous in probability with independent increments.*

Proof. The uniqueness assertion is clear, since any other modification X' which is a modification of L must also be a modification of X, and by the right-continuity of the sample paths it follows from Lemma 10.34 that X and X' are indistinguishable. That every modification of L is also continuous in probability and has independent increments is also clear. It therefore remains only to show the existence of X. By considering each component separately, we can and shall assume without loss of generality that $d = 1$. Denote by $\mathbb{Q}_{\text{dyad}}$ the dyadic rational numbers, i. e.,

$$\mathbb{Q}_{\text{dyad}} := \{q \in \mathbb{Q} : \exists n \in \mathbb{N}, j \in \mathbb{Z} : q = 2^{-n} j\}. \tag{10.9}$$

Then $\mathbb{Q}_{\text{dyad}}$ is obviously dense in $\mathbb{R}$. Let $[\alpha, \beta] \subset I$ be compact with $\alpha, \beta \in \mathbb{Z}$. Since L is continuous in probability, by Lemma 10.35, there exists for every $\delta > 0$ some $K(\delta) < \infty$ such that $\sup_{t \in [\alpha,\beta]} P(|L_t - L_\alpha| \geq K(\delta)) \leq \delta$. Define

$$r_{j,n} := \alpha + j 2^{-n}(\beta - \alpha) \quad \text{for } n \in \mathbb{N} \text{ and } j \in \{0, 1, \dots, 2^n\}.$$

Since L has independent increments, we can write

$$L_{r_{k,n}} - L_a = \sum_{j=1}^{k} Y_{j,n} \quad \forall\, k \in \{0,\dots,2^n\}$$

with

$$Y_{j,n} := L_{r_{j,n}} - L_{r_{j-1,n}}, \quad j \in \{1,\dots,2^n\},$$

where $Y_{1,n},\dots,Y_{2^n,n}$ are independent. From Etemadi's maximal inequality (Theorem 10.36), we conclude that

$$P\Big(\max_{k\in\{0,\dots,2^n\}} |L_{r_{k,n}} - L_a| \ge 3K(\delta)\Big) \le 3 \max_{k\in\{1,\dots,2^n\}} P(|L_{r_{k,n}} - L_a| \ge K(\delta)) \le \delta.$$

Letting n tend to infinity, we conclude that $P(\sup_{t\in[\alpha,\beta]\cap\mathbb{Q}_{\text{dyad}}} |L_t - L_\alpha| \ge 3K(\delta)) \le \delta$. Letting $\delta \to 0$, then letting $\alpha \to -\infty$ if $I = \mathbb{R}$ (otherwise setting α equal to 0) and letting $\beta \to \infty$, we conclude that there is $\Omega_1 \in \mathcal{F}$ with $P(\Omega_1) = 1$ such that

$$\sup_{t\in[-N,N]\cap I\cap\mathbb{Q}_{\text{dyad}}} |L_t(\omega)| < \infty \quad \forall\, N \in \mathbb{N}, \omega \in \Omega_1. \tag{10.10}$$

Now let α and β be integers such that $\alpha < \beta$, $[\alpha,\beta] \subset I$ and $r_{j,n} = \alpha + j2^{-n}(\beta-\alpha)$ as above. Suppose that $a < b$ and $a, b \in \mathbb{Q}$. By equation (10.5) there exists $\gamma > 0$ (depending on a, b, α and β) such that $P(|L_{r_{k,n}} - L_{r_{j,n}}| \ge (b-a)/6) \le 1/6$ for all $n \in \mathbb{N}$ and $j, k \in \{0,\dots,2^n\}$ with $|r_{k,n} - r_{j,n}| \le \gamma$. From Corollary 10.40 we conclude that for each j the expected number of $[a,b]$-upcrossings of $(L_{r_{k,n}} - L_{r_{j,n}})_{k=j,\dots,2^n:2^{-n}(k-j)(\beta-\alpha)\le\gamma}$ is less than or equal to 1. Since $[\alpha,\beta]$ can be covered by $\lceil(\beta-\alpha)/\gamma\rceil$ intervals of length γ (here, $\lceil x\rceil$ denotes the smallest integer greater than or equal to x for $x \in \mathbb{R}$), it follows that the expected number of upcrossings of $(L_{r_{k,n}} - L_\alpha)_{k=0,\dots,2^n}$ is less than $2\lceil(\beta-\alpha)/\gamma\rceil$. Since $U_{ab}((L_{r_{k,n}} - L_\alpha)_{k=0,\dots,2^n})$ is non-decreasing in n, monotone convergence shows that $\mathbb{E}(U_{ab}((L_t - L_\alpha)_{t\in[\alpha,\beta]\cap\mathbb{Q}_{\text{dyad}}})) \le 2\lceil(\beta-\alpha)/\gamma\rceil$. In particular, $U_{ab}((L_t - L_\alpha)_{t\in[\alpha,\beta]\cap\mathbb{Q}_{\text{dyad}}})$ is almost surely finite. Hence there exists $\Omega^{a,b,\alpha,\beta} \in \mathcal{F}$ with $P(\Omega^{a,b,\alpha,\beta}) = 1$ such that $U_{ab}((L_t(\omega) - L_\alpha(\omega))_{t\in[\alpha,\beta]\cap\mathbb{Q}_{\text{dyad}}}) < \infty$ for all $\omega \in \Omega^{a,b,\alpha,\beta}$. Define

$$\Omega_0 := \Omega_1 \cap \bigcap_{a,b\in\mathbb{Q},\alpha,\beta\in\mathbb{Z},[\alpha,\beta]\subset I} \Omega^{a,b,\alpha,\beta}.$$

Then $\Omega_0 \in \mathcal{F}$, $P(\Omega_0) = 1$, the number of $[a,b]$-upcrossings of $(L_t(\omega))_{t\in I\cap\mathbb{Q}_{\text{dyad}}}$ over every compact subinterval of I is finite for all $a, b \in \mathbb{Q}$ and $\omega \in \Omega_0$, and (10.10) holds for $\omega \in \Omega_0$.

Now suppose that $\omega \in \Omega_0$. We claim that $\lim_{s>t,s\in\mathbb{Q}_{\text{dyad}},s\to t} L_s(\omega)$ exists for all $t \in I$. Suppose not. Then there is $t' \in I$ and a sequence $(s_n)_{n\in\mathbb{N}}$ in $\mathbb{Q}_{\text{dyad}}$ with $s_n > t'$ and $s_n \to t'$ such that $\liminf_{n\to\infty} L_{s_n}(\omega) < \limsup_{n\to\infty} L_{s_n}(\omega)$, since the lim sup and lim inf are both finite by (10.10). Hence there must exist rationals $a < b$ such that

$$-\infty < \liminf_{n\to\infty} L_{s_n}(\omega) < a < b < \limsup_{n\to\infty} L_{s_n}(\omega) < +\infty.$$

It follows that the number of $[a,b]$-upcrossings of $(L_s(\omega))_{s\in\mathbb{Q}_{\mathrm{dyad}},|s-t'|\le 1}$ is infinite, contradicting $\omega \in \Omega_0$.

Now define X_t for $t \in I$ by

$$X_t(\omega) := \begin{cases} \lim_{s>t, s\in\mathbb{Q}_{\mathrm{dyad}}, s\to t} L_s(\omega), & \omega \in \Omega_0, \\ 0, & \omega \notin \Omega_0. \end{cases}$$

Then $X = (X_t)_{t\in I}$ is a stochastic process and it is easy to see from the construction that $t \mapsto X_t(\omega)$ is right-continuous on I.

Next, suppose that $X_{t'-}(\omega)$ does not exist for some $t' \in I$. Then there exists a sequence $(s_n)_{n\in\mathbb{N}}$ in $I \cap (-\infty, t')$ converging to t' and rationals $a < b$ such that

$$-\infty < \liminf_{n\to\infty} X_{s_n}(\omega) < a < b < \limsup_{n\to\infty} X_{s_n}(\omega) < \infty$$

(using (10.10) again). By construction, this means that the number of $[a,b]$-upcrossings of $(L_s(\omega))_{s\in\mathbb{Q}_{\mathrm{dyad}},|s-t'|\le 1}$ must be infinite, contradicting the construction of Ω_0.

It follows that every path of X is càdlàg. Finally, to see that X is a modification of L, it suffices to observe that both X_t and L_t are the limits in probability of L_s for $s \downarrow t$ with $s \in \mathbb{Q}_{\mathrm{dyad}}$ (the latter by stochastic continuity of L) and hence must be equal almost surely. □

Observe that the above proof did not make use of the property $L_0 = 0$ of additive processes, but only used the stochastic continuity and independent increment properties.

Corollary 10.42 (Doob's regularization result for Lévy processes in law). *Suppose that $I = [0,\infty)$ or $I = \mathbb{R}$ and let $L = (L_t)_{t\in I}$ be an $\mathbb{R}^d$-valued Lévy process in law. Then there exists a modification $X = (X_t)_{t\in I}$ that is a Lévy process, and any other modification X' of L that is a Lévy process is indistinguishable from X.*

Proof. This is immediate from Theorem 10.41, on observing that the stationary increments property is preserved on going from L to X. □

The next result follows immediately from Theorem 10.26 and Corollary 10.42.

Corollary 10.43 (Existence of Lévy processes). *Let $I = \mathbb{R}$ or $I = [0,\infty)$. Then to every $\mu \in \mathrm{ID}(\mathbb{R}^d)$ and $t_0 \in I\setminus\{0\}$ there exists a Lévy process $L = (L_t)_{t\in I}$, unique in law, with $\mathcal{L}(L_{t_0}) = \mu$.*

Observe that the Lévy process of Corollary 10.43 can be constructed on a complete probability space by Remark 10.28(d). Specific examples of Lévy processes will be considered in Chapter 14.

10.4 Exercises

Exercise 10.1. Prove Remark 10.18.

Exercise 10.2. Let $(\Omega, \mathcal{F}, P)$ be a probability space and $\mathcal{G}$, $\mathcal{H}$, $\mathcal{A}$ be sub-sigma-algebras such that $\mathcal{G}$ and $\mathcal{H}$ are independent and such that $\mathcal{A}$ is P-trivial, i. e. such that $P(A) \in \{0,1\}$ for all $A \in \mathcal{A}$. Show that $\sigma(\mathcal{H}, \mathcal{A})$ and $\mathcal{G}$ are independent.

Exercise 10.3. Let $L = (L_t)_{t\in\mathbb{R}}$ be a Lévy process and define $\mathcal{A}_t^p$ and $\mathcal{A}_t^f$ as in Proposition 10.19. Further, denote by $\mathcal{C}_t^p$ be the σ-algebra that is generated by $\mathcal{A}_t^p$ and L_0, and by $\mathcal{C}_t^f$ the σ-algebra that is generated by $\mathcal{A}_t^f$ and L_0. Show that $\mathcal{C}_t^p$ and $\mathcal{C}_t^f$ are independent for each fixed $t \in \mathbb{R}$.

Exercise 10.4. Consider the probability space $(\Omega = [0,1], \mathcal{F} = \mathcal{B}_1, P = \lambda)$, where λ denotes Lebesgue measure on $[0,1]$. Define the stochastic processes $X = (X_t)_{t\in[0,1]}$ and $Y = (Y_t)_{t\in[0,1]}$ by $X_t(\omega) = 1$ if $t = \omega$ and $X_t(\omega) = 0$ otherwise, and $Y_t(\omega) := 0$ for all $t, \omega \in [0,1]$. Show that X and Y are modifications of each other, but not indistinguishable.

11 Distributional properties of Lévy processes and the strong law of large numbers

Abstract: In this chapter, we study some distributional properties of Lévy processes. In particular, we investigate their Markov property, their behavior under time reversal, and their moment properties in terms of the characteristic triplets. We also treat the strong law of large numbers for Lévy processes.

11.1 The Markov property and time reversal of Lévy processes

In this section, we establish the Markov property of Lévy processes and consider the behavior of Lévy processes under time reversal.

Theorem 11.1 (Markov property of Lévy processes). *Suppose that $I = [0, \infty)$ or $I = \mathbb{R}$, let $L = (L_t)_{t\in I}$ be a Lévy process (in law), and suppose that $t \in I$. Then the stochastic process $M = (M_s)_{s\geq 0}$, defined by $M_s = L_{t+s} - L_t$, is a one-sided Lévy process (in law) with the same finite dimensional distributions, and hence the same characteristic triplet, as $(L_t)_{t\geq 0}$. Moreover, M is independent of $(L_t - L_s)_{s\leq t}$.*

Proof. It is clear that $M_0 = 0$ and that M is càdlàg if L is Lévy (and continuous in probability if L is Lévy in law). Suppose that $0 \leq s_0 < s_1 < \cdots < s_n$. Then $M_{s_i} - M_{s_{i-1}} = L_{s_i+t} - L_{s_{i-1}+t}$, from which it follows that M has independent increments. Since

$$M_{s+h} - M_s = L_{t+s+h} - L_{t+s} \stackrel{d}{=} (L_{t+h} - L_t) = M_h$$

for $h \geq 0$, the increments of M are also stationary, showing that M is indeed a Lévy process (or Lévy in law if L is Lévy in law). Since $M_h = L_{t+h} - L_t \stackrel{d}{=} L_h$, M and $(L_s)_{s\geq 0}$ have the same distribution by Theorem 10.26(b) and hence the same characteristic triplet. That M is independent of $(L_t - L_s)_{s\leq t}$ follows from Proposition 10.19. □

Remark 11.2. If L is a two-sided Lévy process (in law) and $t \in I$, we can still define $M = (M_s)_{s\in\mathbb{R}}$ as a two-sided process through $M_s = L_{t+s} - L_t$. It then follows, as in the proof of Theorem 11.1, that M is a two-sided Lévy process with the same characteristic triplet as L. However, the independence of M and $(L_t - L_s)_{s\leq t}$ no longer holds. This is why, in the statement of the theorem, the process M is defined only as a one-sided process.

The Markov property says in particular that the process $(L_{s+t} - L_t)_{s\geq 0}$, which is obtained by looking into the *future* after time t, is again a Lévy process with the same distribution as L. What happens if we look into the *past* by considering $(L_{t-s} - L_t)_{s\geq 0}$? The answer is provided by the time-reversal property of Lévy processes, for which we have to modify the reversed process slightly in order to get càdlàg paths.

https://doi.org/10.1515/9783111325033-011

Definition 11.3. If $L = (L_s)_{s\in\mathbb{R}}$ is a two-sided càdlàg stochastic process, then for each $t \in \mathbb{R}$, the *(at t) time-reversed process* $\widetilde{L}^t = (\widetilde{L}^t_s)_{s\in\mathbb{R}}$ is defined by

$$\widetilde{L}^t_s := L_{(t-s)-} - L_{t-}.$$

If $L = (L_t)_{t\geq 0}$ is a one-sided càdlàg stochastic process, then for each $t > 0$, the *(at t) time-reversed process* $\widetilde{L}^t = (\widetilde{L}^t_s)_{s\in[0,t]}$ is defined by

$$\widetilde{L}^t_s := L_{(t-s)-} - L_{t-}, \quad s \in [0,t].$$

Theorem 11.4 (Time-reversal of Lévy processes). *The time-reversed process of a one- or two-sided càdlàg process L is again a càdlag stochastic process. If* $L = (L_t)_{t\in I}$ *is a Lévy process, then so is* $\widetilde{L}^t$ *for each* $t \in I$ *(restricted to* $[0,t]$ *when* $I = [0,\infty)$*) with the same finite dimensional distributions as* $-L$*.*

Proof. Since $s \mapsto L_s$ is càdlàg almost surely, $s \mapsto L_{t-s}$ is càglàd almost surely, from which it follows easily that $s \mapsto L_{(t-s)-} - L_{t-}$ is càdlàg almost surely, so that $\widetilde{L}^t$ is a càdlàg stochastic process. Now let L be a two-sided Lévy process, then $\widetilde{L}^t_0 = L_{t-} - L_{t-} = 0$. Suppose that $s_0 < s_1 < \dots < s_n$. Then

$$\widetilde{L}^t_{s_i} - \widetilde{L}^t_{s_{i-1}} = L_{(t-s_i)-} - L_{(t-s_{i-1})-} = \lim_{\varepsilon\downarrow 0}(L_{t-s_i-\varepsilon} - L_{t-s_{i-1}-\varepsilon}),$$

and since $L_{t-s_1-\varepsilon} - L_{t-s_0-\varepsilon}, \dots, L_{t-s_n-\varepsilon} - L_{t-s_{n-1}-\varepsilon}$ are independent for each ε, so is their limit as $\varepsilon \downarrow 0$ (see Exercise 11.1), showing that $\widetilde{L}^t$ has independent increments. Suppose that $h, s \in \mathbb{R}$. Then

$$\begin{aligned}\widetilde{L}^t_{s+h} - \widetilde{L}^t_s &= L_{(t-s-h)-} - L_{(t-s)-} \stackrel{d}{=} L_{t-s-h} - L_{t-s} \\ &\stackrel{d}{=} L_{t-h} - L_t \stackrel{d}{=} L_{(t-h)-} - L_{t-} = \widetilde{L}^t_h\end{aligned}$$

by Proposition 10.14(b) and the stationarity of the increments of L. Hence $\widetilde{L}^t$ has stationary increments and is therefore a Lévy process. Since $\widetilde{L}^t_h \stackrel{d}{=} L_{t-h} - L_t \stackrel{d}{=} L_{-h} \stackrel{d}{=} -L_h$, it has the same finite dimensional distributions as $-L$. This proves the claim for two-sided Lévy processes, and for a one-sided Lévy process it follows by extending it to a two-sided Lévy process (cf. Theorem 10.17) and then restricting the derived two-sided $\widetilde{L}^t$ to $[0,t]$. □

11.2 Moments of Lévy processes and infinitely divisible distributions

In this section, we characterize the existence of certain moments of infinitely divisible distributions and Lévy processes in terms of their Lévy measures. We again closely follow the exposition in Sato [79, Section 25]. We start with the following lemma, which shows in particular that all exponential moments of an infinitely divisible distribution on $\mathbb{R}^d$ are finite if the Lévy measure has bounded support. Furthermore if L is a Lévy process whose Lévy measure has bounded support, then $\sup_{|s|\le t}|L_s|$ also has all exponential moments finite for all $t > 0$. Here we interpret $\sup_{|s|\le t}|L_s|$ as the random variable

$$\omega \mapsto \begin{cases} \sup_{|s|\le t}|L_s|(\omega) = \sup_{|s|\le t, s/t\in\mathbb{Q}}|L_s(\omega)|, & \omega \in \Omega_0, \\ 0, & \omega \notin \Omega_0, \end{cases}$$

where $\Omega_0 \in \mathcal{F}$ with $P(\Omega_0) = 1$ is a set such that all paths $t \mapsto L_t(\omega)$ with $\omega \in \Omega_0$ are càdlàg. Interpreted in this way, $\sup_{|s|\le t}|L_s|$ is indeed a random variable.

Lemma 11.5 (Compactly supported LM and finite exponential moments). *Suppose that $\mu \in \mathrm{ID}(\mathbb{R}^d)$ with Lévy measure ν supported on a bounded set. Then all exponential moments of μ are finite, i. e., $\int_{\mathbb{R}^d} e^{b|x|}\,\mu(dx) < \infty$ for all $b > 0$. Moreover, if $L = (L_t)_{t\in I}$ is a Lévy process in $\mathbb{R}^d$ with Lévy measure supported on a bounded set, where $I = [0,\infty)$ or $I = \mathbb{R}$, then it is even true that all exponential moments of $\sup_{s\in I,|s|\le t}|L_s|$ are finite for all $t > 0$, i. e., that $\mathbb{E}\exp\{b\sup_{s\in I,|s|\le t}|L_s|\} < \infty$ for all $b, t > 0$.*

Proof. If $\mu = \mathcal{L}(X)$ with $X = (X_1,\dots,X_d)^T$, then the Lévy measure of each X_j has bounded support and $|X| \le |X_1| + \dots + |X_d|$, hence $\mathbb{E}\,e^{b|X|} \le \prod_{j=1}^d(\mathbb{E}\,e^{bd|X_j|})^{1/d}$ by Hölder's inequality. It therefore suffices to prove the first and second assertion for $d = 1$, so we shall assume from now on that $d = 1$.

To prove the first assertion, suppose that $\mu \in \mathrm{ID}(\mathbb{R})$ with Lévy measure ν supported on $[-a,a]$. Define the function $g : \mathbb{C} \to \mathbb{C}$ by $g(z) := \int_{[-a,a]}(e^{izx} - 1 - izx)\,\nu(dx)$. Observe that for each compact $K \subset \mathbb{C}$ there is a constant $C_K < \infty$ such that $|e^{izx} - 1 - izx| \le C_K x^2$ for all $z \in K$ and $x \in [-a,a]$, and that x^2 is ν-integrable over $[-a,a]$. It follows that g is well-defined for all $z \in \mathbb{C}$, and even continuous by the continuity lemma (a consequence of the dominated convergence theorem, see, e. g., Bauer [10, Lemma 16.1]). Let ρ be a simple closed curve in $\mathbb{C}$. By Fubini's theorem and Cauchy's theorem (see, e. g., Conway [34, Proposition IV.2.15]), we have for the complex path integral

$$\int_\rho g(z)\,dz = \int_{[-a,a]}\left(\int_\rho (e^{izx} - 1 - izx)\,dz\right)\nu(dx) = \int_{[-a,a]} 0\,\nu(dx) = 0.$$

Hence $g : \mathbb{C} \to \mathbb{C}$ is an entire function by Morera's theorem (see, e. g., Conway [34, Theorem IV.5.10]).

Now consider the truncation function $c(x) = x\mathbf{1}_{[-a,a]}(x)$ and denote the characteristic triplet of $\mu \in \mathrm{ID}(\mathbb{R})$ with respect to c by $(A,\nu,\gamma)_c$. Define $f(z) := \exp[-Az^2/2 + g(z) + i\gamma z]$ for $z \in \mathbb{C}$. Then f is an entire function and $f(z) = \widehat{\mu}(z)$ for $z \in \mathbb{R}$. Let X be a random variable with distribution μ. Then $\widehat{\mu}$ and f are infinitely often differentiable and $f(z) = \sum_{n=0}^\infty \frac{i^n a_n}{n!} z^n$ for all $z \in \mathbb{R}$, where $a_n := i^{-n} f^{(n)}(0)$, and the power series converges on the whole of $\mathbb{C}$. In particular, since f is n-times differentiable for every even $n \in \mathbb{N}$, it follows that $\mathbb{E}|X|^n < \infty$ for all even $n \in \mathbb{N}$ (see, e. g., Klenke [57, Theorem 15.34]), and hence that $\mathbb{E}|X|^n < \infty$ for all $n \in \mathbb{N}_0$. Then $\mathbb{E}X^n = (-\mathrm{i})^n f^{(n)}(0) = \alpha_n$. Let $\beta_n := \mathbb{E}|X|^n$ for $n \in \mathbb{N}_0$. Since $2|x| \le 1 + x^2$ for $x \in \mathbb{R}$, we have $|X|^{2m+1} \le (|X|^{2m} + |X|^{2m+2})/2$ for $m \in \mathbb{N}_0$, hence $\beta_{2m} = \alpha_{2m}$ and

$$\beta_{2m+1} = \mathbb{E}|X|^{2m+1} \le \frac{1}{2}(\mathbb{E}|X|^{2m} + \mathbb{E}|X|^{2m+2}) = \frac{\alpha_{2m} + \alpha_{2m+2}}{2}$$

for $m \in \mathbb{N}_0$. Since $\sum_{n=0}^{\infty} \frac{|\alpha_n z^n|}{n!} < \infty$ for all $z \in \mathbb{C}$ it follows also that $\sum_{n=0}^{\infty} \frac{\beta_n |z|^n}{n!} < \infty$ for all $z \in \mathbb{C}$. But for $b > 0$ we have

$$\int_{\mathbb{R}} \mathrm{e}^{b|x|}\,\mu(\mathrm{d}x) = \int_{\mathbb{R}} \sum_{n=0}^{\infty} \frac{b^n |x|^n}{n!}\,\mu(\mathrm{d}x) = \sum_{n=0}^{\infty} \frac{b^n}{n!}\mathbb{E}|X|^n = \sum_{n=0}^{\infty} \frac{\beta_n b^n}{n!} < \infty.$$

This shows that all exponential moments of μ are finite.

Now let $L = (L_t)_{t\in I}$ be a Lévy process in $\mathbb{R}$ with Lévy measure supported in $[-a, a]$. For $t > 0$ define $M := \sup_{0\le s\le t} |L_s| = \sup_{0\le s\le t, s/t\in\mathbb{Q}_{\mathrm{dyad}}} |L_s|$, where $\mathbb{Q}_{\mathrm{dyad}}$ denotes the dyadic rational numbers as in (10.9). By monotone convergence and Etemadi's maximal inequality (Theorem 10.36), we obtain for each $u \ge 0$

$$\begin{aligned} P(M > u) &= \mathbb{E}(\mathbf{1}_{(u,\infty)} \circ M) \\ &= \lim_{n\to\infty} \mathbb{E}(\mathbf{1}_{(u,\infty)} \circ \max\{|L_{2^{-n}jt}| : j = 0, \dots, 2^n\}) \\ &\le 3 \lim_{n\to\infty} \max_{j\in\{1,\dots,2^n\}} P(|L_{2^{-n}jt}| \ge u/3) \\ &= 3 \sup_{s\in[0,t], s/t\in\mathbb{Q}_{\mathrm{dyad}}} P(|L_s| \ge u/3). \end{aligned}$$

Now suppose that $b > 0$. Then, for each $w \ge 1$, it follows that

$$\begin{aligned} P(\mathrm{e}^{bM} > w) = P(M > (\log w)/b) &\le 3 \sup_{s\in[0,t], s/t\in\mathbb{Q}} P(|L_s| \ge (\log w)/(3b)) \\ &= 3 \sup_{s\in[0,t], s/t\in\mathbb{Q}} P(\mathrm{e}^{6b|L_s|} \ge w^2) \le 3w^{-2} \sup_{s\in[0,t], s/t\in\mathbb{Q}} \mathbb{E}\,\mathrm{e}^{6b|L_s|} \end{aligned}$$

by Markov's inequality. Integrating over $w \ge 1$, we see that $\mathbb{E}\,\mathrm{e}^{bM}$ is finite, provided we can show finiteness of $\sup_{s\in[0,t], s/t\in\mathbb{Q}} \mathbb{E}\,\mathrm{e}^{6b|L_s|}$. Observe that $\mathbb{E}\,\mathrm{e}^{6b|L_s|} < \infty$ for all $s \in [0, t]$ by the part already shown since $\mathcal{L}(L_s)$ has Lévy measure $s\nu$. In particular, $\mathbb{E}\,\mathrm{e}^{6bL_s} < \infty$ and $\mathbb{E}\,\mathrm{e}^{-6bL_s} < \infty$. But from the independent and stationary increments property of L, we obtain $\mathbb{E}\,\mathrm{e}^{6bL_{t/n}} = (\mathbb{E}\,\mathrm{e}^{6bL_t})^{1/n}$ for each $n \in \mathbb{N}$ and further $\mathbb{E}\,\mathrm{e}^{6bL_s} = (\mathbb{E}\,\mathrm{e}^{6bL_t})^{s/t}$ for $s \ge 0$ with $s/t \in \mathbb{Q}$. Hence $\sup_{0\le s\le t, s/t\in\mathbb{Q}} \mathbb{E}\,\mathrm{e}^{6bL_s} \le \sup_{0\le s\le t, s/t\in\mathbb{Q}} (\mathbb{E}\,\mathrm{e}^{6bL_t})^{s/t} < \infty$ and similarly $\sup_{0\le s\le t, s/t\in\mathbb{Q}} \mathbb{E}\,\mathrm{e}^{-6bL_s} < \infty$. But since $\mathrm{e}^{|y|} \le \mathrm{e}^y + \mathrm{e}^{-y}$ for $y \in \mathbb{R}$ this implies that $\sup_{0\le s\le t, s/t\in\mathbb{Q}} \mathbb{E}\,\mathrm{e}^{6b|L_s|} < \infty$, finishing the proof when $I = [0, \infty)$. The fact that $\mathbb{E}\exp(b \sup_{|s|\le t} |L_s|) < \infty$ when $I = \mathbb{R}$ follows by application of the preceding argument to $(L_s)_{s\ge0}$ and $(L_s)_{s\le0}$ separately. □

The following formula for general moments of compound Poisson distributions is useful.

Lemma 11.6 (Moments of compound Poisson distributions). *Let $\mu = \mathcal{L}(X)$ be a compound Poisson distribution on $\mathbb{R}^d$ with Lévy measure ν. Let $g : \mathbb{R}^d \to \mathbb{R}$ be a Borel measurable function such that $\mathbb{E}|g(X)| < \infty$ or $g \geq 0$. Then*

$$\mathbb{E}\, g(X) = \mathrm{e}^{-\nu(\mathbb{R}^d)} \sum_{n=0}^{\infty} \frac{1}{n!} \int_{\mathbb{R}^d} g(y)\, \nu^{*n}(\mathrm{d}y),$$

where ν^{*n} denotes the n-fold convolution of ν with itself.

Proof. Suppose first that $g \geq 0$. Since X has a compound Poisson distribution, there is a Poisson distributed random variable M with parameter $\lambda := \nu(\mathbb{R}^d) \in (0, \infty)$ and an i. i. d. sequence $(Y_j)_{j\in\mathbb{N}}$ with distribution $\nu/\nu(\mathbb{R}^d)$, independent of M, such that $\mu = \mathcal{L}(\sum_{j=1}^M Y_j)$. Conditioning on M, we obtain

$$\begin{aligned}
\mathbb{E}[g(X)] &= \sum_{n=0}^{\infty} P(M = n)\, \mathbb{E}(g(X)|M = n) = \sum_{n=0}^{\infty} \mathrm{e}^{-\lambda}\frac{\lambda^n}{n!} \mathbb{E}\left[g\left(\sum_{j=1}^{n} Y_j\right)\right] \\
&= \sum_{n=0}^{\infty} \mathrm{e}^{-\lambda}\frac{\lambda^n}{n!} \int_{\mathbb{R}^d} g(y) \left[\frac{\nu}{\lambda}\right]^{*n} (\mathrm{d}y) \\
&= \mathrm{e}^{-\lambda} \sum_{n=0}^{\infty} \frac{1}{n!} \int_{\mathbb{R}^d} g(y)\, \nu^{*n}(\mathrm{d}y).
\end{aligned}$$

If $\mathbb{E}|g(X)| < \infty$, the result is obtained by applying the same argument to the positive and negative parts of g. □

We are concerned also with finiteness of (possibly fractional) moments of the form $\mathbb{E}|X|^\alpha$ with $\alpha > 0$ and logarithmic moments of the form $\mathbb{E}(\log^+|X|)^\alpha$. These can be handled in a unified way by introducing submultiplicative functions.

Definition 11.7 (Submultiplicative and locally bounded functions). A function $g : \mathbb{R}^d \to \mathbb{R}$ is called *submultiplicative* if it is non-negative and if there exists a constant $a > 0$ such that

$$g(x + y) \leq a\, g(x)\, g(y) \quad \forall\, x, y \in \mathbb{R}^d. \tag{11.1}$$

The function g is *locally bounded* if it is bounded on every compact set.

Remark 11.8. It follows immediately from (11.1) that any submultiplicative function g satisfies

$$g(x_1 + \cdots + x_n) \leq a^{n-1} g(x_1) \cdots g(x_n)$$

for all $n \in \mathbb{N}$ and $x_1, \ldots, x_n \in \mathbb{R}^d$.

Example 11.9. The product of two submultiplicative functions is obviously submultiplicative. If g is submultiplicative, then so is $x \mapsto [g(x)]^\alpha$ for any $\alpha > 0$. Also, the function $x \mapsto e^{b|x|}$ is submultiplicative for each $b > 0$, and when $d = 1$ then the function $x \mapsto e^{bx}$ is submultiplicative for each $b \in \mathbb{R}$. It is easy to see that the function $x \mapsto (|x| \vee b)^\alpha$ is submultiplicative for any $\alpha, b > 0$. Since $|x+y|^\alpha \le |x|^\alpha + |y|^\alpha$ for $\alpha \in (0,1]$, the function $x \mapsto \exp(b|x|^\alpha)$ is submultiplicative for $\alpha \in (0,1]$ and $b > 0$. Consequently, as a product of two submultiplicative functions, so is $x \mapsto (|x|^\beta \vee 1)\exp(b|x|^\alpha)$ with $\alpha \in (0,1]$ and $b, \beta > 0$. Finally, the function $x \mapsto \log(|x| \vee e)$ is submultiplicative, as can be seen from the inequalities,

$$\begin{aligned}\log(|x+y| \vee e) &\le \log[(|x| \vee e) + (|y| \vee |e|)] \\ &\le \log 2 + \log[(|x| \vee e) \vee (|y| \vee e)] \\ &\le \log 2 + \log(|x| \vee e) + \log(|y| \vee e) \\ &\le 4 \log(|x| \vee e) \log(|y| \vee e)\end{aligned}$$

for $x, y \in \mathbb{R}^d$. As a consequence, the function $x \mapsto [\log(|x| \vee e)]^\alpha$ is submultiplicative for any $\alpha > 0$. Further examples of submultiplicative functions can be found in Sato [79, Proposition 25.4].

Lemma 11.10 (Exponential bound for submultiplicative functions). *If $g : \mathbb{R}^d \to [0,\infty)$ is submultiplicative and locally bounded, then there exist strictly positive a and b such that*

$$g(x) \le a e^{b|x|} \quad \forall\, x \in \mathbb{R}^d.$$

Proof. Let a be as in (11.1). Choose $\alpha > a^{-1}$ such that $\alpha \ge \sup_{|x|\le 1} g(x)$. For $n - 1 < |x| \le n$, we then have, by Remark 11.8,

$$g(x) \le a^{n-1}[g(x/n)]^n \le a^{n-1}\alpha^n \le \alpha(a\alpha)^{|x|}$$

from which the claim follows. □

We also need the following result.

Lemma 11.11. *If X and Y are two independent $\mathbb{R}^d$-valued random variables and $g : \mathbb{R}^d \to [0,\infty)$ is submultiplicative and Borel-measurable, then $\mathbb{E}\, g(X+Y) < \infty$ if and only if $\mathbb{E}\, g(X) < \infty$ and $\mathbb{E}\, g(Y) < \infty$.*

Proof. Let a be as in (11.1). If $\mathbb{E}\, g(X) < \infty$ and $\mathbb{E}\, g(Y) < \infty$, then

$$\mathbb{E}\, g(X+Y) \le \mathbb{E}\big(a g(X) g(Y)\big) = a\, \mathbb{E}\, g(X)\, \mathbb{E}\, g(Y) < \infty$$

by independence. Conversely, if $\mathbb{E}\, g(X+Y) < \infty$, then

$$\infty > \int_{(\mathbb{R}^d)^2} g(x+y)\,(P_X \otimes P_Y)(dx, dy) = \int_{\mathbb{R}^d}\Big(\int_{\mathbb{R}^d} g(x+y)\, P_X(dx)\Big) P_Y(dy).$$

It follows that there exists some $y_0 \in \mathbb{R}^d$ for which $\mathbb{E}\, g(X+y_0) = \int_{\mathbb{R}^d} g(x+y_0)\, P_X(\mathrm{d}x) < \infty$. It then follows from the definition (11.1) of the submultiplicativity of g that

$$\mathbb{E}\, g(X) \le \mathbb{E}(ag(X+y_0)g(-y_0)) = ag(-y_0)\,\mathbb{E}\, g(X+y_0) < \infty.$$

Similarly, $\mathbb{E}\, g(Y) < \infty$. □

The next theorem characterizes the finiteness of certain moments of Lévy processes, and hence of infinitely divisible distributions, in terms of the Lévy measure. For any Lévy measure ν and Borel set A we shall denote by $\nu_{|A}$ the measure $\nu_{|A}(\cdot) := \nu(A \cap \cdot)$.

Theorem 11.12 (Moment conditions for Lévy processes). *Suppose that $g : \mathbb{R}^d \to [0,\infty)$ is a submultiplicative, locally bounded Borel-measurable function, that $I = \mathbb{R}$ or $I = [0,\infty)$ and that $L = (L_t)_{t\in I}$ is a Lévy process in $\mathbb{R}^d$ with Lévy measure ν. Then the following are equivalent:*
(i) *L_t has* finite g-moment, *i. e., $\mathbb{E}[g(L_t)] < \infty$, for some $t > 0$.*
(ii) *L_t has finite g-moment for all $t > 0$.*
(iii) *$\nu_{|\{|x|>1\}}$ has finite g-moment, i. e., $\int_{x\in\mathbb{R}^d:|x|>1} g(x)\,\nu(\mathrm{d}x) < \infty$.*

If, in addition, g is radially symmetric, i. e., is of the form $g(x) = h(|x|)$ for some function $h : [0,\infty) \to [0,\infty)$ and if $I = \mathbb{R}$, then each of the following statements (iv) *and* (v) *is also equivalent to* (i)–(iii)*:*
(iv) *L_t has finite g-moment for some $t < 0$.*
(v) *L_t has finite g-moment for all $t \in I$.*

If g is not only radially symmetric, but has the form $g(x) = h(|x|)$, where $h : [0,\infty) \to [0,\infty)$ is non-decreasing and if $I = \mathbb{R}$ or $I = [0,\infty)$, then each of the following statements is also equivalent to (i)–(iii).
(vi) *$\sup_{|s|\le t} |L_s|$ has finite h-moment, i. e., $\mathbb{E}[h(\sup_{|s|\le t} |L_s|)] < \infty$, for some $t > 0$.*
(vii) *$\sup_{|s|\le t} |L_s|$ has finite h-moment for all $t > 0$.*

Proof. Suppose that g satisfies (11.1) with $a > 0$. The implication (ii) $\Longrightarrow$ (i) is clear. For the other implications, denote the standard characteristic triplet of L by (A, ν, γ). Let $X = (X_t)_{t\in I}$ be a compound Poisson process with Lévy measure $\nu_X := \nu_{|\{|x|>1\}}$ if $\nu_X \ne 0$ and the zero process otherwise, and let $Y = (Y_t)_{t\in I}$ be a Lévy process with characteristic triplet $(A, \nu_Y := \nu_{|\{|x|\le 1\}}, \gamma)$ such that X and Y are independent. It follows that $X+Y$ is a Lévy process with the same characteristic triplet and hence the same finite-dimensional distributions as L. Since both L and $X+Y$ are càdlàg, it follows also that $\sup_{|s|\le t} |X_s+Y_s| = \sup_{|s|\le t, s/t\in\mathbb{Q}} |X_s+Y_s|$ and $\sup_{|s|\le t} |L_s| = \sup_{s\le t, s/t\in\mathbb{Q}} |L_s|$ have the same distribution. Hence we can and shall assume that $L = X+Y$ with X and Y independent.

To prove that (i) $\Longrightarrow$ (iii) suppose that $\mathbb{E}\, g(L_t) < \infty$ for some $t > 0$. Since $L_t = X_t + Y_t$ where X_t and Y_t are independent, it follows from Lemma 11.11 that $\mathbb{E}\, g(X_t) < \infty$. But X_t has a compound Poisson distribution with Lévy measure $t\nu_X$ (or is the zero process when $\nu_X = 0$). Hence $\sum_{n=0}^{\infty} \frac{1}{n!} \int_{\mathbb{R}^d} g(x)\, t^n \nu_X^{*n}(\mathrm{d}x) < \infty$ by Lemma 11.6 and in particular $\int_{\mathbb{R}^d} g(x)\nu_X(\mathrm{d}x) < \infty$, which is (iii).

To show that (iii) $\Longrightarrow$ (ii), let ν_X have finite g-moment and suppose that $t > 0$. Then X_t has Lévy measure $t\nu_X$. From Remark 11.8, we conclude that

$$\int_{\mathbb{R}^d} g(y)\, \nu_X^{*n}(\mathrm{d}x) = \int_{\mathbb{R}^d} \cdots \int_{\mathbb{R}^d} g(x_1 + \cdots + x_n)\, \nu_X(\mathrm{d}x_1) \ldots \nu_X(\mathrm{d}x_n)$$
$$\le a^{n-1} \Big(\int_{\mathbb{R}^d} g(x)\, \nu_X(\mathrm{d}x) \Big)^n,$$

and hence, by Lemma 11.6 when $\nu_X \neq 0$,

$$\mathbb{E}\, g(X_t) = \mathrm{e}^{-t\nu_X(\mathbb{R}^d)} \sum_{n=0}^{\infty} \frac{t^n}{n!} \int_{\mathbb{R}^d} g(x)\, \nu_X^{*n}(\mathrm{d}x)$$
$$\le a^{-1} \mathrm{e}^{-t\nu_X(\mathbb{R}^d)} \sum_{n=0}^{\infty} \frac{(at)^n}{n!} \Big(\int_{\mathbb{R}^d} g(x)\, \nu_X(\mathrm{d}x) \Big)^n$$
$$= a^{-1} \mathrm{e}^{-t\nu_X(\mathbb{R}^d)} \exp\Big\{ at \int_{\mathbb{R}^d} g(x)\, \nu_X(\mathrm{d}x) \Big\} < \infty.$$

That $\mathbb{E}\, g(X_t) = g(0) < \infty$, when $\nu_X = 0$, is trivial. By the submultiplicativity of g, Lemma 11.10 and the independence of X_t and Y_t, we also have, for some strictly positive a and b,

$$\mathbb{E} g(L_t) \le \mathbb{E}[a g(X_t) g(Y_t)] = a\mathbb{E}[g(X_t)]\mathbb{E}[g(Y_t)] \le a\mathbb{E}[g(X_t)]\mathbb{E}[a\mathrm{e}^{b|Y_t|}]$$

and the latter is finite by Lemma 11.5. This shows (ii).

The equivalences of (i) and (iv) and of (ii) and (v) are clear when g is radially symmetric, since $\mathcal{L}(L_{-t}) = \mathcal{L}(-L_t)$ and hence $\mathcal{L}(g(L_{-t})) = \mathcal{L}(g(L_t))$ for $t \in \mathbb{R}$. That (vii) $\Longrightarrow$ (vi) $\Longrightarrow$ (i) is clear when $g(x) = h(|x|)$ with non-decreasing h, so it only remains to show that (iii) $\Longrightarrow$ (vii) in this case. Denote by $\mathbf{e}_1 = (1, 0, \ldots, 0)^T \in \mathbb{R}^d$ the first unit vector in $\mathbb{R}^d$ and observe that $h(u + v) = g((u + v)\mathbf{e}_1) \le a g(u\mathbf{e}_1) g(v\mathbf{e}_1) = a h(u) h(v)$ for $u, v \in [0, \infty)$. Suppose that $t > 0$. Then

$$\mathbb{E}\Big[h\Big(\sup_{|s|\le t} |L_s|\Big)\Big] \le \mathbb{E}\Big[h\Big(\sup_{|s|\le t} |X_s| + \sup_{|s|\le t} |Y_s|\Big)\Big] \le a\mathbb{E}h\Big(\sup_{|s|\le t} |X_s|\Big)\, \mathbb{E}h\Big(\sup_{|s|\le t} |Y_s|\Big).$$

As in the proof of Lemma 11.10, we have $h(u) \le a\mathrm{e}^{bu}$ for all $u \ge 0$ for some $a > 0$ and $b > 0$. Hence $\mathbb{E}h(\sup_{|s|\le t} |Y_s|) < \infty$ by Lemma 11.5. For the other factor, observe that by independence of $(X_s)_{s\ge 0}$ and $(X_s)_{s\le 0}$

$$\mathbb{E}\, h\Big(\sup_{|s|\le t} |X_s|\Big) \le a\, \mathbb{E}\, h\Big(\sup_{0\le s\le t} |X_s|\Big)\, \mathbb{E}\, h\Big(\sup_{-t\le s\le 0} |X_s|\Big) = a\Big(\mathbb{E}\, h\Big(\sup_{0\le s\le t} |X_s|\Big)\Big)^2$$

when $I = \mathbb{R}$, so that it suffices to show that $\mathbb{E}[h(\sup_{0\le s\le t}|X_t|)] < \infty$. This is clear when $\nu_X = 0$, so suppose that $\nu_X \neq 0$. Then for $s \geq 0$, $X_s = \sum_{j=1}^{N_s} Z_j$, where Z_j has distribution $\nu_X/\nu_X(\mathbb{R}^d)$ and $(Z_j)_{j\in\mathbb{N}}$ is an i. i. d. sequence independent of the Poisson process N with jump rate $\lambda := \nu_X(\mathbb{R}^d)$, as constructed in Theorem 1.31. Hence $h(\sup_{0\le s\le t}|X_s|) \leq h(\sum_{j=1}^{N_t}|Z_j|)$, where $\sum_{j=1}^{N_t}|Z_j|$ has a one-dimensional compound Poisson distribution with jump rate λ and jump-size distribution $\sigma = \xi(\nu_X)$, where $\xi : \mathbb{R}^d \to \mathbb{R}$ is defined by $\xi(x) := |x|$. Since

$$\int_{u>1} h(u)\,\lambda\sigma(du) = \lambda \int_{x\in\mathbb{R}^d:|x|>1} h(\xi(x))\,\nu_X(dx) = \lambda \int_{x\in\mathbb{R}^d:|x|>1} g(x)\,\nu_X(dx)$$

is finite, it follows as in the proof that (iii) $\Longrightarrow$ (i) that $\mathbb{E}[h(\sum_{j=1}^{N_t}|Z_j|)]$ is finite, completing the proof. □

Remark 11.13. (a) If L is a two-sided Lévy process and $g : \mathbb{R}^d \to [0,\infty)$ is submultiplicative, locally bounded and Borel-measurable, then the conditions (i)–(iii) in the preceding theorem are in general not equivalent to (iv) or (v). Indeed, if we define $\tilde{g} : \mathbb{R}^d \to [0,\infty)$ by $\tilde{g}(x) := g(-x)$, then $\tilde{g}$ is submultiplicative, locally bounded and Borel-measurable. Since $\mathcal{L}(L_t) = \mathcal{L}(-L_{-t})$, condition (iv) is then equivalent to
(iii′) $\nu_{|\{|x|>1\}}$ has finite $\tilde{g}$-moment.
It is easy to see that conditions (iii) and (iii′) can be quite different, for example by taking $g(x) := x \vee 1$, or alternatively $g(x) := (-x) \vee 1$. It is also clear that (v) is equivalent to both (i) and (iv), but may in general be stronger than (i) or (iv) alone.

(b) Let $g_1, g_2 : \mathbb{R}^d \to [0,\infty)$ be locally bounded Borel-measurable functions such that there exist $b_1, b_2 > 0$ and $T > 0$ such that

$$b_1 g_1(x) \leq g_2(x) \leq b_2 g_1(x) \quad \forall\, x \in \mathbb{R}^d : |x| \geq T.$$

Then L_t has finite g_1-moment if and only if it has finite g_2-moment, and similarly for $\nu_{|\{|x|>1\}}$. Hence it is possible to apply Theorem 11.12 to functions that are not necessarily submultiplicative, but just behave for large x like a submultiplicative function. For example, the function $x \mapsto \log^+|x|$, $x \in \mathbb{R}^d$, where $\log^+(y)$ for $y \in \mathbb{R}$ is defined by

$$\log^+(y) := \log(y \vee 1),$$

is not submultiplicative, but coincides with the submultiplicative function $x \mapsto \log(|x| \vee \mathrm{e})$ for large $|x|$, hence Theorem 11.12 can be applied to the function $x \mapsto \log^+|x|$ as well. Similarly, the function $\mathbb{R}^d \to [0,\infty)$, $x \mapsto |x|^\alpha$ is not submultiplicative for $\alpha > 0$, but agrees with the submultiplicative function $x \mapsto (|x| \vee 1)^\alpha$ for large x, so that Theorem 11.12 can be applied. Similar comments apply to the functions $x \mapsto |x|^\beta \exp(b|x|^\alpha)$ and $x \mapsto (|x|^\beta \vee 1)\exp(b|x|^\alpha)$ with $\alpha \in (0,1]$ and $b, \beta > 0$.

(c) Although submultiplicativity of the function g can be relaxed, as seen in (b) above, it cannot be omitted completely. An example where (i) and (ii) of Theorem 11.12 are not equivalent for a non-submultiplicative function is presented in Exercise 14.9, and an example where (iii) does not imply (i) is presented in Exercise 11.4.

Let us collect the most important cases again in the following corollary.

Corollary 11.14 (Moment conditions for Lévy processes). *Suppose that $I = [0,\infty)$ or $I = \mathbb{R}$ and let $L = (L_t)_{t\in I}$ be a Lévy process in $\mathbb{R}^d$ with Lévy measure ν.*

(a) *If $\alpha > 0$ then $\mathbb{E}|L_t|^\alpha < \infty$ for some $t \in I \setminus \{0\}$, if and only if $\mathbb{E}|L_t|^\alpha < \infty$ for all $t \in I$, or $\int_{x\in\mathbb{R}^d:|x|>1} |x|^\alpha\, \nu(\mathrm{d}x) < \infty$, or $\mathbb{E}[\sup_{|s|\le t} |L_s|]^\alpha < \infty$ for some (equivalently all) $t > 0$.*

(b) *If $\alpha > 0$ then $\mathbb{E}[\log^+ |L_t|]^\alpha < \infty$ for some $t \in I \setminus \{0\}$, if and only if $\mathbb{E}[\log^+ |L_t|]^\alpha < \infty$ for all $t \in I$, or $\int_{\mathbb{R}^d} [\log^+ |x|]^\alpha\, \nu(\mathrm{d}x) < \infty$, or $\mathbb{E}[\sup_{|s|\le t} \log^+ |L_s|]^\alpha < \infty$ for some (equivalently all) $t > 0$.*

(c) *If $\beta > 0$ and $\alpha \in (0,1]$, then $\mathbb{E}\, e^{\beta|L_t|^\alpha} < \infty$ for some $t \in I \setminus \{0\}$, if and only if $\mathbb{E}\, e^{\beta|L_t|^\alpha} < \infty$ for all $t \in I$, or $\int_{x\in\mathbb{R}^d:|x|>1} e^{\beta|x|^\alpha}\, \nu(\mathrm{d}x) < \infty$, or $\mathbb{E}[\sup_{|s|\le t} e^{\beta|L_s|^\alpha}] < \infty$ for some (equivalently all) $t > 0$.*

(d) *If $d = 1$ and $\beta \in \mathbb{R}$ then $\mathbb{E}\, e^{\beta L_t} < \infty$ for some $t > 0$, if and only if $\mathbb{E}\, e^{\beta L_t} < \infty$ for all $t > 0$, or $\int_{\mathbb{R}\setminus[-1,1]} e^{\beta x}\, \nu(\mathrm{d}x) < \infty$.*

Proof. Immediate from Theorem 11.12 and Remark 11.13(b). □

The next theorem expresses the expectation and covariance matrix of a Lévy process at time t in terms of the characteristic triplet, and shows in particular that both are linear in t for $t > 0$.

Theorem 11.15 (Expectation and covariance matrix of Lévy processes). *Suppose that $I = [0,\infty)$ or $I = \mathbb{R}$ and that $L = (L_t)_{t\in I}$ is a Lévy process in $\mathbb{R}^d$ with standard characteristic triplet (A, ν, γ).*

(a) *If $\int_{|x|>1} |x|\, \nu(\mathrm{d}x) < \infty$, then $\mathbb{E}|L_t| < \infty$ for all $t \in I$ and*

$$\mathbb{E}L_t = t\Big(\int_{|x|>1} x\, \nu(\mathrm{d}x) + \gamma\Big) = t\gamma_m \quad \forall\, t \in I,$$

where γ_m denotes the center of $\mathcal{L}(L_1)$. In particular, the center and the mean of an infinitely divisible distribution coincide.

(b) *If $\int_{|x|>1} |x|^2\, \nu(\mathrm{d}x) < \infty$, then $\mathbb{E}|L_t|^2 < \infty$ for all $t \in I$ and the covariance matrix $\mathrm{Cov}(L_t) \in \mathbb{R}^{d\times d}$ of L_t is given by*

$$\mathrm{Cov}(L_t) = |t|\Big(A + \int_{\mathbb{R}^d} xx^T\, \nu(\mathrm{d}x)\Big) \quad \forall\, t \in I.$$

Proof. Define $\mu := \mathcal{L}(L_t)$ for fixed $t \geq 0$ and write $L_t = (L_t^1, \ldots, L_t^d)^T$ and $A = (a_{jk})_{j,k=1,\ldots,d}$. Both conditions imply the existence of the center

$$(\gamma_m^1, \ldots, \gamma_m^d)^T := \gamma_m = \gamma + \int\limits_{|x|>1} x\,\nu(\mathrm{d}x)$$

(cf., (9.27)) of μ. Writing

$$\Psi_L(z) = \mathrm{i}\langle \gamma_m, z\rangle - \frac{1}{2}\sum_{j,k=1}^{d} a_{jk} z_j z_k + \int\limits_{\mathbb{R}^d} (\mathrm{e}^{\mathrm{i}\langle z,x\rangle} - 1 - \mathrm{i}\langle z, x\rangle)\,\nu(\mathrm{d}x), \quad z \in \mathbb{R}^d,$$

we have $\widehat{\mu}(z) = \exp(t\Psi_L(z))$ by (9.26).

(a) If $\int_{|x|>1} |x|\,\nu(\mathrm{d}x) < \infty$, then $\mathbb{E}|L_t| < \infty$ for $t \geq 0$ by Corollary 11.14 and we know that $\mathbb{E}(L_t^j) = -\mathrm{i}\frac{\partial}{\partial z_j}\widehat{\mu}(z)|_{z=0}$ for $j \in \{1, \ldots, d\}$ (see, e. g., Bauer [9, Theorem 8.3.3]). However,

$$\begin{aligned}\frac{\partial}{\partial z_j}\widehat{\mu}(z) &= \widehat{\mu}(z)\, t \frac{\partial}{\partial z_j}\Psi_L(z)\\ &= \widehat{\mu}(z)\, t\left(\mathrm{i}\gamma_m^j - \sum_{k=1}^{d} a_{jk} z_k + \int\limits_{\mathbb{R}^d} (\mathrm{e}^{\mathrm{i}\langle z,x\rangle} - 1)\mathrm{i}x_j\,\nu(\mathrm{d}x)\right)\end{aligned}$$

by the differentiation lemma (see, e. g., Bauer [10, Corollary 16.3]). In particular, $\mathbb{E}L_t^j = t\gamma_m^j$ and hence $\mathbb{E}L_t = t\gamma_m$ for $t > 0$. The same is true for $t < 0$, since $\mathcal{L}(L_t) = \mathcal{L}(-L_{-t})$.

(b) If $\int_{|x|>1} |x|^2\,\nu(\mathrm{d}x) < \infty$, then $\mathbb{E}|L_t|^2 < \infty$ for all $t \geq 0$ by Corollary 11.14 and we know that $\mathbb{E}(L_t^j L_t^k) = (-\mathrm{i})^2 \frac{\partial^2}{\partial z_k \partial z_j}\widehat{\mu}(z)|_{z=0}$ for $j, k \in \{1, \ldots, d\}$ (see, e. g., Bauer [9, Theorem 8.3.3]). Again, from the differentiation lemma, we obtain

$$\begin{aligned}&\frac{\partial^2}{\partial z_k \partial z_j}\widehat{\mu}(z)\\ &\quad = t\left(\frac{\partial}{\partial z_k}\widehat{\mu}(z)\right)\left[\mathrm{i}\gamma_m^j - \sum_{l=1}^{d} a_{jl} z_l + \int\limits_{\mathbb{R}^d} (\mathrm{e}^{\mathrm{i}\langle z,x\rangle} - 1)\mathrm{i}x_j\,\nu(\mathrm{d}x)\right]\\ &\qquad + t\widehat{\mu}(z)\left[-a_{jk} + \int\limits_{\mathbb{R}^d} (\mathrm{e}^{\mathrm{i}\langle z,x\rangle}\mathrm{i}^2 x_j x_k + (\mathrm{e}^{\mathrm{i}\langle z,x\rangle} - 1)\mathrm{i}\delta_{j,k})\,\nu(\mathrm{d}x)\right],\end{aligned}$$

where $\delta_{j,k} = 1$ if $j = k$ and 0 otherwise. Evaluating this at $z = 0$ gives

$$\mathbb{E}(L_t^j L_t^k) = t^2\gamma_m^k\gamma_m^j + t a_{jk} + t\int\limits_{\mathbb{R}^d} x_j x_k\,\nu(\mathrm{d}x),$$

so that $\mathrm{Cov}(L_t^j, L_t^k) = t a_{jk} + t\int_{\mathbb{R}^d} x_j x_k\,\nu(\mathrm{d}x)$, or in matrix form, $\mathrm{Cov}(L_t) = tA + t\int_{\mathbb{R}^d} xx^T\,\nu(\mathrm{d}x)$. The case $t < 0$ follows again from $\mathcal{L}(L_t) = \mathcal{L}(-L_{-t})$. □

Remark 11.16 (Moments of infinitely divisible distributions). The equivalence of (i) and (iii) in Theorem 11.12 immediately implies a corresponding statement for infinitely divisible distributions. Namely, if μ is an infinitely divisible distribution on $\mathbb{R}^d$ with Lévy measure ν and $g : \mathbb{R}^d \to [0, \infty)$ is a submultiplicative, locally bounded Borel-measurable function, then μ has finite g-moment (i. e., $\int_{\mathbb{R}^d} g(x)\,\mu(\mathrm{d}x) < \infty$) if and only if $\int_{x\in\mathbb{R}^d:|x|>1} g(x)\,\nu(\mathrm{d}x) < \infty$. The same is true for the corresponding statements in Corollary 11.14 and Theorem 11.15. In particular, $\int_{\mathbb{R}^d} |x|\,\mu(\mathrm{d}x) < \infty$ if and only if $\int_{|x|>1} |x|\,\nu(\mathrm{d}x) < \infty$, in which case $\int_{\mathbb{R}^d} x\,\mu(\mathrm{d}x) = \gamma_m$, the center of μ, and $\int_{\mathbb{R}^d} |x|^2\,\mu(\mathrm{d}x) < \infty$ if and only if $\int_{|x|>1} |x|^2\,\nu(\mathrm{d}x) < \infty$, in which case the covariance matrix of μ is given by $A + \int_{\mathbb{R}^d} xx^T\,\nu(\mathrm{d}x)$, where A denotes the Gaussian variance of μ.

As a corollary, we can characterize when a two-sided Lévy process is an orthogonal increment process in the sense of Definition 4.1.

Corollary 11.17 (Characterization of OIP property of Lévy processes). *Let $L = (L_t)_{t\in\mathbb{R}}$ be a two-sided $\mathbb{R}$-valued Lévy process with standard characteristic triplet (A, ν, γ). Then L is an orthogonal increment process if and only if $\int_{\mathbb{R}^d:|x|>1} |x|^2\,\nu(\mathrm{d}x) < \infty$ and $\gamma_m = 0$, where γ_m denotes the center of $\mathcal{L}(L_1)$. Under these conditions, the distribution function F associated with L according to Proposition 4.2 is given by $t \mapsto F(t) = t\,\mathrm{Var}(L_1) = t(A + \int_{\mathbb{R}} x^2\,\nu(\mathrm{d}x))$.*

Proof. By Corollary 11.14, L satisfies $\mathbb{E}|L_t|^2 < \infty$ for all $t \in \mathbb{R}$ if and only if $\int_{|x|>1} x^2\,\nu(\mathrm{d}x) < \infty$. By Theorem 11.15, $\mathbb{E}L_t = 0$ for all $t \in \mathbb{R}$ if and only if $\mathbb{E}L_t = t\gamma_m = 0$ for all $t \in \mathbb{R}$, which is equivalent to $\gamma_m = 0$. Condition (4.5) is automatically satisfied for a Lévy process with finite second moments, since then independence implies uncorrelatedness. The right-continuity in mean-square of L follows from Theorem 11.15, since

$$\|L_{t+\delta} - L_t\|^2 = \mathbb{E}(L_{t+\delta} - L_t)^2 = \mathbb{E}L_\delta^2 = \delta\,\mathbb{E}L_1^2 \to 0 \quad \text{as } \delta \downarrow 0.$$

Finally, (4.7) implies that the distribution function is $F(t) = \mathrm{sign}(t)\,\mathbb{E}|L_t|^2 = t\,\mathrm{Var}(L_1) = t(A + \int_{\mathbb{R}} x^2\,\nu(\mathrm{d}x))$. □

In Corollary 11.14(a), we saw that for a Lévy process L and some constant $\alpha > 0$, the finiteness of $\mathbb{E}[\sup_{|s|\le t} |L_s|]^\alpha$ is equivalent to the finiteness of $\mathbb{E}|L_t|^\alpha$. Later, when proving the Lévy–Itô decomposition in Section 13.1, we will need a bound for $\mathbb{E}[\sup_{|s|\le t} |L_s|]^2$ in terms of $\mathbb{E}|L_t|^2$ when the Lévy process has mean zero. Such a bound can be obtained from Doob's L^2-inequality for martingales. Since we do not wish to assume familiarity with that inequality, we shall next give a proof in the special case when the martingale is a sum of independent mean zero random vectors. This will be quite sufficient for our purposes.

Theorem 11.18 (Doob's L^α-inequality, special case). *Suppose that $\alpha > 1$, $(Y_n)_{n\in\mathbb{N}}$ is a sequence of independent random vectors in $\mathbb{R}^d$ with finite α-moment and mean 0 and let $S_n := \sum_{j=1}^n Y_j$ for $n \in \mathbb{N}$. Then*

$$\mathbb{E}\Big(\max_{j\in\{1,\dots,n\}} |S_j|\Big)^\alpha \le \Big(\frac{\alpha}{\alpha-1}\Big)^\alpha \mathbb{E}|S_n|^\alpha$$

for each $n \in \mathbb{N}$.

Proof. Let $\mathcal{F}_j = \sigma(Y_1,\dots,Y_j) = \sigma(S_1,\dots,S_j)$ be the σ-algebra generated by $Y_1,\dots,Y_j$. For $j \in \{1,\dots,n\}$ and $y \in \mathbb{R}^d$ we find by conditioning on $S_j = y$ and using the independence of S_j and $S_n - S_j$, that

$$\begin{aligned}\mathbb{E}(|S_n||S_j = y) &= \mathbb{E}(|S_j + (S_n - S_j)||S_j = y)\\ &= \mathbb{E}(|y + S_n - S_j|)\\ &\ge |\mathbb{E}(y + S_n - S_j)| = |y|,\end{aligned}$$

so that

$$\mathbb{E}(|S_n||\mathcal{F}_j) = \mathbb{E}(|S_n||S_j) \ge |S_j|$$

(this is the submartingale property of $(|S_n|)_{n\in\mathbb{N}}$). For fixed $n \in \mathbb{N}$ and $u > 0$, define the stopping time,

$$\tau := \inf\{j \in \{1,\dots,n\} : |S_j| > u\} \wedge n,$$

i. e., τ is the first time when $|S_j|$ is greater than u if this happens prior to time n, and $\tau = n$ otherwise. It is easily seen that τ and $|S_\tau|$ are random variables. Further, since

$$\{\tau = j\} = \{|S_1| \le u,\dots,|S_{j-1}| \le u, |S_j| > u\}$$

for $j < n$, the event $\{\tau = j\}$ is $\mathcal{F}_j$-measurable when $j < n$ and, by a similar argument, also when $j = n$. Hence, by the submartingale property of $(|S_n|)_{n\in\mathbb{N}}$ established above,

$$\begin{aligned}\mathbb{E}|S_\tau| = \mathbb{E}\Big(\sum_{j=1}^n |S_j|\mathbf{1}_{\{\tau=j\}}\Big) &= \sum_{j=1}^n \mathbb{E}(|S_j|\mathbf{1}_{\{\tau=j\}})\\ &\le \sum_{j=1}^n \mathbb{E}(\mathbb{E}(|S_n|\mathbf{1}_{\{\tau=j\}}|\mathcal{F}_j))\\ &= \sum_{j=1}^n \mathbb{E}(|S_n|\mathbf{1}_{\{\tau=j\}}) = \mathbb{E}|S_n|.\end{aligned}$$

Writing $S_n^* := \max\{|S_1|,\dots,|S_n|\}$, we hence obtain

$$\begin{aligned}\mathbb{E}|S_n| \ge \mathbb{E}|S_\tau| &= \mathbb{E}(|S_\tau|\mathbf{1}_{\{S_n^*>u\}}) + \mathbb{E}(|S_\tau|\mathbf{1}_{\{S_n^*\le u\}})\\ &\ge uP(S_n^* > u) + \mathbb{E}(|S_n|\mathbf{1}_{\{S_n^*\le u\}}),\end{aligned}$$

so that

$$\mathbb{E}(|S_n|\mathbf{1}_{\{S_n^*>u\}}) \geq uP(S_n^* > u)$$

(this is Doob's maximal martingale inequality in our special case). Using this inequality together with Fubini's theorem and Hölder's inequality (with $\beta > 1$ such that $1/\alpha + 1/\beta = 1$), we find that

$$\begin{aligned}
\mathbb{E}|S_n^*|^\alpha &= \alpha \int_0^\infty P(S_n^* > u)u^{\alpha-1}\,du \\
&\leq \alpha \int_0^\infty \mathbb{E}(|S_n|\mathbf{1}_{\{S_n^*>u\}})u^{\alpha-2}\,du \\
&= \alpha\,\mathbb{E}\Bigg(|S_n| \int_0^{S_n^*} u^{\alpha-2}\,du\Bigg) \\
&= \frac{\alpha}{\alpha-1}\mathbb{E}(|S_n|(S_n^*)^{\alpha-1}) \\
&\leq \frac{\alpha}{\alpha-1}(\mathbb{E}|S_n|^\alpha)^{1/\alpha}(\mathbb{E}|S_n^*|^{(\alpha-1)\beta})^{1/\beta} \\
&= \frac{\alpha}{\alpha-1}(\mathbb{E}|S_n|^\alpha)^{1/\alpha}(\mathbb{E}|S_n^*|^\alpha)^{1-1/\alpha}.
\end{aligned}$$

Since $\mathbb{E}|S_n^*|^\alpha$ is easily seen to be finite by assumption, dividing by $\mathbb{E}|S_n^*|^\alpha$ gives the desired result. □

An application of Theorem 11.18 yields the following corollary.

Corollary 11.19 (Doob's L^α-inequality for Lévy processes). *Suppose that $\alpha > 1$ and $L = (L_t)_{t\geq 0}$ is an $\mathbb{R}^d$-valued Lévy process with Gaussian variance $A = (a_{ij})_{i,j=1,\ldots,d}$, Lévy measure ν such that $\int_{|x|>1} |x|^\alpha\,\nu(dx) < \infty$, and such that L_1 has mean zero. Then*

$$\mathbb{E}\Big(\sup_{0\leq s\leq t} |L_s|\Big)^\alpha \leq \Big(\frac{\alpha}{\alpha-1}\Big)^\alpha \mathbb{E}|L_t|^\alpha$$

for each $t \geq 0$. In particular, when $\alpha = 2$,

$$\mathbb{E}\Big(\sup_{0\leq s\leq t} |L_s|\Big)^2 \leq 4t\Bigg(\sum_{j=1}^d a_{jj} + \int_{\mathbb{R}^d} |x|^2\,\nu(dx)\Bigg), \qquad t \geq 0.$$

Proof. Suppose that $t \geq 0$. Then

$$\mathbb{E}\Big(\max_{j=0,\ldots,2^n} |L_{j2^{-n}t}|\Big)^\alpha \leq \Big(\frac{\alpha}{\alpha-1}\Big)^\alpha \mathbb{E}|L_t|^\alpha$$

for each $n \in \mathbb{N}$ by Theorem 11.18. Since $\max_{j=0,\ldots,2^n} |L_{j2^{-n}t}|$ converges from below to $\sup_{0\leq s\leq t} |L_s| = \sup_{0\leq s\leq t, s/t\in\mathbb{Q}_{\text{dyad}}} |L_s|$ as $n \to \infty$, where $\mathbb{Q}_{\text{dyad}}$ denotes the dyadic rational numbers as in (10.9), the first inequality follows from monotone convergence. The second inequality when $\alpha = 2$ follows, since

$$\mathbb{E}|L_t|^2 = \mathbb{E}(L_t^T L_t) = \mathbb{E}(\mathrm{trace}(L_t L_t^T)) = \mathrm{trace}(\mathrm{Cov}(L_t))$$

and $\mathrm{Cov}(L_t) = t(A + \int_{\mathbb{R}^d} xx^T\, \nu(\mathrm{d}x))$ by Theorem 11.15(b). □

We next express the exponential moments in terms of the characteristic triplet in dimension 1. An extension to dimension d can be found in Sato [79, Theorem 25.17].

Theorem 11.20 (Exponential moments of LPs). *Let $L = (L_t)_{t\geq 0}$ be a one-sided real-valued Lévy process with characteristic exponent Ψ and characteristic triplet $(A, \nu, \gamma)_c$, with respect to some truncation function $c : \mathbb{R} \to \mathbb{R}$, so that $\Psi_L(z) = \mathrm{i}\gamma z - Az^2/2 + \int_{\mathbb{R}}(\mathrm{e}^{\mathrm{i}zx} - 1 - \mathrm{i}zc(x))\,\nu(\mathrm{d}x)$ for all $z \in \mathbb{R}$. Suppose that $b \in \mathbb{R}$ and $\int_{\mathbb{R}\setminus[-1,1]} \mathrm{e}^{bx}\,\nu(\mathrm{d}x) < \infty$, so that $\mathbb{E}\,\mathrm{e}^{bL_t}$ exists for each $t \geq 0$ by Corollary 11.14. Then the integral $\int_{\mathbb{R}}(\mathrm{e}^{bx} - 1 - bc(x))\,\nu(\mathrm{d}x)$ converges absolutely and, writing*

$$\Psi_L(-\mathrm{i}b) := \gamma b + Ab^2/2 + \int_{\mathbb{R}} (\mathrm{e}^{bx} - 1 - bc(x))\,\nu(\mathrm{d}x),$$

we can express $\mathbb{E}\,\mathrm{e}^{bL_t}$ as

$$\mathbb{E}\,\mathrm{e}^{bL_t} = \exp\{t\Psi_L(-\mathrm{i}b)\} \quad \forall\, t \geq 0.$$

Proof. It follows, as in the proof of Proposition 9.17, that the convergence of $\int_{\mathbb{R}}(\mathrm{e}^{bx} - 1 - bc(x))\,\nu(\mathrm{d}x)$ does not depend on the choice of the specific truncation function, and that different truncation functions lead to the same value of $\Psi_L(-\mathrm{i}b)$. Hence we can and shall restrict attention to the case when $x \mapsto c(x) = x\mathbf{1}_{[-1,1]}(x)$ is the standard truncation function.

As in the proof of Theorem 11.12, we can find two independent Lévy processes X and Y such that $L_t \overset{d}{=} X_t + Y_t$ for all $t \geq 0$, where X is a compound Poisson process with Lévy measure $\nu_X := \nu_{|\{|x|>1\}}$ if $\nu_X \neq 0$ and $X = 0$ otherwise, and Y is a Lévy process with characteristic triplet $(A, \nu_Y := \nu_{|\{|x|\leq 1\}}, \gamma)$. It is then easy to see that $\Psi_L(-\mathrm{i}b) = \Psi_X(-\mathrm{i}b) + \Psi_Y(-\mathrm{i}b)$, provided all are definable and, by the independence of X and Y, $\mathbb{E}\,\mathrm{e}^{bL_t} = \mathbb{E}\,\mathrm{e}^{bX_t+bY_t} = \mathbb{E}\,\mathrm{e}^{bX_t}\,\mathbb{E}\,\mathrm{e}^{bY_t}$. Hence it suffices to establish the result for X and for Y separately.

Let us start with Y. Define $g : \mathbb{C} \to \mathbb{C}$ by $g(z) := \int_{[-1,1]}(\mathrm{e}^{\mathrm{i}zx} - 1 - \mathrm{i}zx)\,\nu(\mathrm{d}x)$. As seen in the proof of Lemma 11.5, g is an entire function so that $f_t : \mathbb{C} \to \mathbb{C}$, $z \mapsto \exp[t(\mathrm{i}\gamma z - Az^2/2 + g(z))]$ is definable for all $z \in \mathbb{C}$ and $t \geq 0$. Moreover f_t is an entire function that coincides with $\mathbb{E}\,\mathrm{e}^{\mathrm{i}zY_t}$ for all $z \in \mathbb{R}$ and $f_t(-\mathrm{i}b) = \exp(t\Psi_Y(-\mathrm{i}b))$. On the other hand, since $|\mathrm{e}^{\mathrm{i}zY_t}| \leq \mathrm{e}^{|Y_t \Im z|}$ for $z \in \mathbb{C}$ and, since all exponential moments of Y_t exist by Lemma 11.5, we see that $h_t(z) := \mathbb{E}\,\mathrm{e}^{\mathrm{i}zY_t}$ exists for all $z \in \mathbb{C}$ and, as in the proof of Lemma 11.5, we can apply Cauchy's and Fubini's theorems to conclude that

$$\int_{\rho} h_t(z)\,dz = \int_{\mathbb{R}} \Big(\int_{\rho} e^{izx}\,dz \Big) P_{Y_t}(dx) = \int_{\mathbb{R}} 0\, P_{Y_t}(dx) = 0,$$

so that h_t is also an entire function by Morera's theorem. Since f_t and h_t coincide for $z \in \mathbb{R}$, they must coincide on $\mathbb{C}$, so that in particular $\mathbb{E}\, e^{bY_t} = h_t(-ib) = f_t(-ib) = \exp(t\Psi_Y(-ib))$.

Let us now turn to X. Since $\int_{|x|>1} e^{bx}\,\nu(dx) < \infty$, it follows also that $\mathbb{E}\, e^{bX_t} < \infty$ and that $\int_{|x|>1} |e^{bx} - 1|\,\nu(dx) < \infty$. Define $\lambda := \nu_X(\mathbb{R}^d)$ and assume that $\lambda > 0$, since otherwise $X_t = 0$. Writing $X_t = \sum_{j=1}^{N_t} Z_j$, where N is a Poisson process with jump rate λ, independent of the i. i. d. jumps $Z_j, j \in \mathbb{N}$, with $\mathcal{L}(Z_j) = \nu_X/\lambda$, we obtain from Lemma 11.6 that

$$\begin{aligned}
\mathbb{E}\, e^{bX_t} &= e^{-\lambda t} \sum_{n=0}^{\infty} \frac{(\lambda t)^n}{n!} \int_{\mathbb{R}^d} e^{by}\, (\nu_X/\lambda)^{*n}(dy) \\
&= e^{-\lambda t} \sum_{n=0}^{\infty} \frac{(\lambda t)^n}{n!}\, \mathbb{E}\, e^{b(Z_1+\dots+Z_n)} \\
&= e^{-\lambda t} \sum_{n=0}^{\infty} \frac{(\lambda t)^n}{n!} \Big(\int_{|y|>1} e^{by} \lambda^{-1}\, \nu(dy) \Big)^n \\
&= e^{-\lambda t} \exp\Big(t \int_{|y|>1} e^{by}\, \nu(dy) \Big) \\
&= \exp\Big(t \int_{|y|>1} (e^{by} - 1)\, \nu(dy) \Big) = \exp(t\Psi_X(-ib)),
\end{aligned}$$

finishing the proof. □

11.3 Strong law of large numbers and a further growth condition for Lévy processes

As an application of the results of the previous section, we shall now prove the strong law of large numbers for Lévy processes. For random walks, $S_n = \sum_{j=1}^n Z_j$, where $(Z_j)_{j\in\mathbb{N}}$ is a sequence of i. i. d. $\mathbb{R}^d$-valued random vectors with finite expectation $\mathbb{E}(Z_1)$, the strong law of large numbers states that $n^{-1}S_n$ converges almost surely to $\mathbb{E}(Z_1)$ as $n \to \infty$. A proof can be found in most textbooks on probability theory, see, e. g., Billingsley [16, Thm. 6.1]. The following is the corresponding result for Lévy processes:

Theorem 11.21 (Strong law of large numbers for Lévy processes). *Suppose that $I = [0, \infty)$ or $\mathbb{R}$ and that $L = (L_t)_{t\in I}$ is an $\mathbb{R}^d$-valued Lévy process with $\mathbb{E}|L_1| < \infty$. Then $t^{-1}L_t$ converges almost surely to $\mathbb{E}(L_1)$ as $|t| \to \infty$.*

Proof. For $n \in \mathbb{N}$, we can write $L_n = \sum_{j=1}^{n}(L_j - L_{j-1})$, where $(L_j - L_{j-1})_{j\in\mathbb{N}}$ is i. i. d. with expectation $\mathbb{E}(L_1)$. From the strong law of large numbers for random walks we obtain almost sure convergence of $n^{-1}L_n$ to $\mathbb{E}(L_1)$ as $n \to \infty$. Denote by $\lfloor t \rfloor$ the largest integer smaller than or equal to t for $t \geq 1$. Then

$$t^{-1}L_t = \frac{\lfloor t \rfloor}{t}\lfloor t \rfloor^{-1}(L_t - L_{\lfloor t \rfloor}) + \frac{\lfloor t \rfloor}{t}\lfloor t \rfloor^{-1}L_{\lfloor t \rfloor}.$$

Since $\lfloor t \rfloor/t$ converges to 1 as $t \to \infty$, the second term converges almost surely to $\mathbb{E}(L_1)$, as has just been shown, and it remains only to show that $\lfloor t \rfloor^{-1}(L_t - L_{\lfloor t \rfloor})$ converges almost surely to 0 as $t \to \infty$. To see the latter, define the random variables $Z_n := \sup_{s\in[n,n+1]} |L_s - L_n|$ for $n \in \mathbb{N}_0$. Since L has càdlàg paths, we can write $Z_n = \sup_{s\in[n,n+1]\cap\mathbb{Q}_{\text{dyad}}} |L_s - L_n|$, where $\mathbb{Q}_{\text{dyad}}$ denotes the dyadic rational numbers as in (10.9). Since L is a Lévy process, we conclude that $(Z_n)_{n\in\mathbb{N}_0}$ is i. i. d., and from Corollary 11.14(a) we conclude that $\mathbb{E}|Z_0| < \infty$. Hence $n^{-1}\sum_{j=0}^{n-1} Z_j$ converges almost surely to $\mathbb{E}(Z_0)$, again by the strong law of large numbers for random walks. Consequently, $n^{-1}\sum_{j=0}^{n} Z_j = \frac{n+1}{n}(n+1)^{-1}\sum_{j=0}^{n} Z_j$ also converges almost surely to $\mathbb{E}(Z_0)$, and subtracting this from $n^{-1}\sum_{j=0}^{n-1} Z_j$ shows that $n^{-1}Z_n$ converges almost surely to 0 as $n \to \infty$. Since $|L_t - L_{\lfloor t \rfloor}| \leq Z_{\lfloor t \rfloor}$ we get the desired almost sure convergence of $\lfloor t \rfloor^{-1}(L_t - L_{\lfloor t \rfloor})$ to 0 as $t \to \infty$. The proof for $t \to -\infty$ (when $I = \mathbb{R}$) is similar. □

The strong law of large numbers implies in particular that if L is a Lévy process with finite expectation, then L_t grows at most linearly in the sense that $\limsup_{|t|\to\infty} |L_t/t| < \infty$ almost surely. What can be said about the growth when a Lévy process does not have finite expectation? There is a connection between the growth of a Lévy process and the existence of certain of its moments. We shall restrict attention to the following special case, which shows that if a Lévy process has finite log-moment, then it grows more slowly than any exponential function. This result will be useful later in connection with Ornstein–Uhlenbeck processes. The corresponding result for random walks is also of interest.

Proposition 11.22 (Growth of Lévy processes with finite log-moment).
(a) *Let $(Y_n)_{n\in\mathbb{N}}$ be an i. i. d. sequence of $\mathbb{R}^d$-valued random vectors such that $\mathbb{E}\log^+|Y_1| < \infty$. Then $\lim_{n\to\infty} \mathrm{e}^{-\varepsilon n}\sum_{j=1}^{n} Y_j = 0$ a. s. for every $\varepsilon > 0$.*
(b) *Suppose that $I = [0,\infty)$ or $\mathbb{R}$ and that $L = (L_t)_{t\in I}$ is an $\mathbb{R}^d$-valued Lévy process such that $\mathbb{E}\log^+|L_1| < \infty$. Then $\lim_{|t|\to\infty} \mathrm{e}^{-\varepsilon|t|}L_t = 0$ a. s. for every $\varepsilon > 0$.*

Proof. (a) Suppose that $\varepsilon > 0$. Then

$$\sum_{j=1}^{\infty} P(|Y_j| > \mathrm{e}^{j\varepsilon/2}) = \sum_{j=1}^{\infty} P\Big(\frac{\log|Y_1|}{\varepsilon/2} > j\Big) < \infty$$

since $\mathbb{E}\log^+|Y_1| < \infty$, so that $P(|Y_j| > e^{j\varepsilon/2}$ infinitely often) $= 0$ by the Borel–Cantelli lemma. For each ω in the sample space, denote by $J(\omega)$ the set of all $j \in \mathbb{N}$ for which $|Y_j(\omega)| > e^{j\varepsilon/2}$. Then J is finite almost surely and we have the inequality,

$$\left|\sum_{j=1}^{n} Y_j(\omega)\right| \le \sum_{j\in J(\omega)} |Y_j(\omega)| + \sum_{j=1}^{n} e^{\varepsilon j/2} = \sum_{j\in J(\omega)} |Y_j(\omega)| + \frac{e^{\varepsilon(n+1)/2} - e^{\varepsilon/2}}{e^{\varepsilon/2} - 1}$$

for each $n \in \mathbb{N}$. Since J is finite with probability 1, this implies that $\lim_{n\to\infty} e^{-\varepsilon n}\sum_{j=1}^{n} Y_j = 0$ almost surely.

(b) As in the proof of Theorem 11.21, define $Z_n := \sup_{s\in[n,n+1]} |L_s - L_n|$ for $n \in \mathbb{N}_0$. Then $(Z_n)_{n\in\mathbb{N}_0}$ is i. i. d. By assumption, $\mathbb{E}\log^+|L_1| < \infty$, hence $\mathbb{E}\log^+|Z_0| < \infty$ by Corollary 11.14(b). For $s \ge 0$, denoting again by $\lfloor s\rfloor$ the largest integer smaller than or equal to s, we have the inequality, for every $\varepsilon > 0$,

$$e^{-\varepsilon s}|L_s| \le e^{-\varepsilon\lfloor s\rfloor}(Z_0 + \cdots + Z_{\lfloor s\rfloor}) = e^{\varepsilon} e^{-\varepsilon(\lfloor s\rfloor+1)}(Z_0 + \cdots + Z_{\lfloor s\rfloor}),$$

and the latter converges almost surely to zero as $s \to \infty$ by (a). A similar argument, when $I = \mathbb{R}$, establishes the limit in (b) as $t \to -\infty$. □

11.4 Exercises

Exercise 11.1. Let $X_{j,n}$, $n \in \mathbb{N}$, $j = 1,\dots,m$, be random vectors on a given probability space such that $X_{1,n},\dots,X_{m,n}$ are independent for each n. Assume that $X_{j,n}$ converges in probability to some random vector X_j as $n \to \infty$, for each $j \in \{1,\dots,m\}$. Show that also $X_1,\dots,X_m$ are independent.
Hint: Use characteristic functions.

Exercise 11.2. Show that the function $g(x) = e^{b|x|}$, $x \in \mathbb{R}^d$, is not submultiplicative when $b < 0$.

Exercise 11.3. Let $g : \mathbb{R}^d \to [0,\infty)$ be a submultiplicative function that is not identically zero. Show that g is strictly positive. Deduce that the function $g(x) = |x|^\alpha$, $x \in \mathbb{R}^d$, is not submultiplicative for $\alpha > 0$.

Exercise 11.4. Let $L = (L_t)_{t\ge0}$ be a real-valued Lévy process with standard characteristic triplet (A,ν,γ). Suppose that $\alpha > 1$, $\beta > 0$ and $\mathbb{E}\, e^{\beta|L_t|^\alpha} < \infty$ for some $t > 0$.
(a) Show that $\nu = 0$. Deduce that (iii) of Theorem 11.12 does not imply (i) for the function $g(x) = e^{\beta|x|^\alpha}$. In particular, g cannot be submultiplicative.
(b) Now let $\alpha > 2$, $\beta > 0$ and $\mathbb{E}\, e^{\beta|L_t|^\alpha} < \infty$ for some $t > 0$. Show that $A = 0$ so that $L_t = \gamma t$ in this case, i. e., is deterministic. In particular, non-deterministic Lévy processes cannot have arbitrarily high moments.

Exercise 11.5 (Mean-square law of large numbers for Lévy processes). Let $L = (L_t)_{t\in I}$ be a one- or two-sided Lévy process in $\mathbb{R}^d$ with $\mathbb{E}|L_1|^2 < \infty$. Show that $t^{-1}L_t$ converges in $\mathcal{S}$, i. e., in mean-square, to $\mathbb{E}(L_1)$ as $|t| \to \infty$.

Exercise 11.6. (a) Let $\beta > 1$ and $(Y_n)_{n\in\mathbb{N}}$ be an i. i. d. sequence of $\mathbb{R}^d$-valued random vectors such that $\mathbb{E}|Y_n|^{1/\beta} < \infty$. Show that $n^{-\beta}\sum_{j=1}^n Y_j$ converges almost surely to 0 as $n \to \infty$.
Hint: For each $K > 0$ show that $\lim_{n\to\infty} n^{-\beta}\sum_{j=1}^n |Y_j|\mathbf{1}_{|Y_j|\le K} = 0$ and
$\limsup_{n\to\infty} n^{-\beta}\sum_{j=1}^n |Y_j|\mathbf{1}_{|Y_j|>K} \le [\mathbb{E}(|Y_1|^{1/\beta}\mathbf{1}_{|Y_1|>K})]^\beta$ almost surely.

(b) Let $\beta > 1$ and $L = (L_t)_{t\in I}$ be a one- or two-sided Lévy process in $\mathbb{R}^d$ with $\mathbb{E}|L_1|^{1/\beta} < \infty$. Show that $|t|^{-\beta}L_t$ converges almost surely to 0 as $|t| \to \infty$.

12 Lévy processes as random elements and their jump structure

Abstract: The study of sample path properties of Lévy processes is initiated in this chapter by regarding the sample paths of a d-dimensional Lévy process on a subinterval J of $\mathbb{R}$ as random elements of the space $D(J; \mathbb{R}^d)$ of $\mathbb{R}^d$-valued càdlàg functions on J with sigma-algebra generated by the coordinate projections. By modifying a càdlàg process (i. e., a stochastic process whose sample paths are only *almost surely* càdlàg) on a null set, we can always find a version that is strictly càdlàg and is indistinguishable from the original one, and by a slight abuse of language we shall sometimes also view càdlàg processes as random elements in $D(J; \mathbb{R}^d)$. We characterize both the Poisson process and Brownian motion in terms of their sample path behavior and consider in detail the jump structure of a general Lévy process and its connection with the Lévy measure. These findings lay the ground for the derivation of the fundamental Lévy–Itô path decomposition, which will be treated in the next chapter.

12.1 Lévy processes as random elements in the space of càdlàg functions

It is sometimes convenient to view the sample paths of stochastic processes as random elements, and in doing so we shall restrict attention to processes for which *all* paths are càdlàg. For that purpose, let $J \subset \mathbb{R}$ be a real interval (not necessarily restricted to be $\mathbb{R}$ or $[0, \infty)$). As in Definition 10.1, we call a function $f : J \to \mathbb{R}^d$ a *càdlàg function* if it is right-continuous on J and the left-limits $f(t-) := \lim_{u \uparrow t} f(u)$ exist and are finite at every $t \in J \setminus \{\inf J\}$. Also, if $t_J := \inf J \in J$, define $f(t_J-) := f(t_J)$. Denote the *space of* $\mathbb{R}^d$*-valued càdlàg functions on* J by

$$D(J; \mathbb{R}^d) := \{f : J \to \mathbb{R}^d : f \text{ càdlàg}\},$$

and for a càdlàg function f define the *jump size* $\Delta f(t)$ of f at t by

$$\Delta f(t) := f(t) - f(t-).$$

Similarly, a function $f : J \to \mathbb{R}^d$ is called *càglàd* if it is left-continuous on J and the right-limits $f(t+) := \lim_{u \downarrow t} f(u)$ exist and are finite at every $t \in J \setminus \{\sup J\}$. Let $X = (X_t)_{t \in J}$ be a family of $\mathbb{R}^d$-valued mappings on a probability space $(\Omega, \mathcal{F}, P)$ (a priori not necessarily a stochastic process). As in Definition 10.1(c), we shall say that X is *strictly càdlàg* if *all* paths $(X_t(\omega))_{t \in J}$ are càdlàg, i. e. are in $D(J; \mathbb{R}^d)$, and we call X *càdlàg* if the sample paths of L are only almost surely càdlàg. Similarly, X is called *strictly càglàd* if *all* paths are

https://doi.org/10.1515/9783111325033-012

càglàd, and *càglàd* if the sample paths of L are only almost surely càglàd. In the strictly càdlàg case, we can view X as a mapping $X : \Omega \to D(J; \mathbb{R}^d)$.

The space of càdlàg functions on $[0, 1]$ or on $[0, \infty)$ is usually topologised by the Skorohod-topology, based on a metric which makes it a separable and complete metric space (cf. Billingsley [17, Sections 12 and 16]). For our purposes, when we later prove the Lévy–Itô decomposition it is enough to know that when J is a compact interval, then $D(J; \mathbb{R}^d)$ will be a subspace of $L^\infty(J; \mathbb{R}^d)$, the space of bounded functions from J to $\mathbb{R}^d$, endowed with the supremum norm $\|\cdot\|_\infty$, and that $D(J; \mathbb{R}^d)$ will be complete with respect to the supremum norm.

Lemma 12.1 ($D(J; \mathbb{R}^d)$ is a Banach space for compact J). *Let J be a compact interval. Then $D(J; \mathbb{R}^d)$ is a closed subspace of $L^\infty(J; \mathbb{R}^d)$ endowed with the supremum norm $\|\cdot\|_\infty$. In particular, $(D(J; \mathbb{R}^d), \|\cdot\|_\infty)$ is a Banach space.*

Proof. That $D(J; \mathbb{R}^d)$ is a vector space is clear, and since càdlàg functions on compact intervals must be bounded, it is also a subspace of $L^\infty(J; \mathbb{R}^d)$. To show that $D(J; \mathbb{R}^d)$ is closed, let $(f_n)_{n\in\mathbb{N}}$ be a sequence in $D(J; \mathbb{R}^d)$, converging to $f \in L^\infty(J; \mathbb{R}^d)$ with respect to the supremum norm. We need to show that f is càdlàg, which we shall do by contradiction. Suppose that x_0 is a point in J other than the left endpoint, and suppose that f does not have a finite left limit $f(x_0-)$ at x_0. Then there would be two sequences $(u_n)_{n\in\mathbb{N}}$ and $(v_n)_{n\in\mathbb{N}}$ in J with $u_n, v_n < x_0$, $\lim_{n\to\infty} u_n = \lim_{n\to\infty} v_n = x_0$ and some $\delta > 0$ such that

$$|f(v_n) - f(u_n)| > \delta \quad \forall\, n \in \mathbb{N}.$$

Since $(f_m)_{m\in\mathbb{N}}$ converges uniformly to f by assumption, there is $m_0 \in \mathbb{N}$ such that

$$|f_{m_0}(x) - f(x)| < \delta/4 \quad \forall\, x \in J.$$

We conclude that

$$\begin{aligned}
&|f_{m_0}(u_n) - f_{m_0}(v_n)| \\
&\quad \geq |f(u_n) - f(v_n)| - |f(u_n) - f_{m_0}(u_n)| - |f(v_n) - f_{m_0}(v_n)| \\
&\quad > \delta - 2\frac{\delta}{4} = \frac{\delta}{2}.
\end{aligned}$$

However, this shows that $f_{m_0}(x_0-)$ does not exist, contradicting $f_{m_0} \in D(J; \mathbb{R}^d)$. Similarly, we can show that f is right-continuous at $x_0 \in J$, by considering sequences $(u_n)_{n\in\mathbb{N}}$ and $(v_n)_{n\in\mathbb{N}}$ in J converging to x_0 with $u_n, v_n \geq x_0$. Hence f is càdlàg and $D(J; \mathbb{R}^d)$ is a closed subspace of $L^\infty(J; \mathbb{R}^d)$. Since $L^\infty(J; \mathbb{R}^d)$ is complete, so is $D(J; \mathbb{R}^d)$ as a closed subspace. □

We shall endow the set $D(J; \mathbb{R}^d)$ with the following σ-algebra.

Notation 12.2. Let J be a real interval and for each $t \in J$ denote by $\pi_t : D(J; \mathbb{R}^d) \to \mathbb{R}^d$ the projection onto the tth coordinate, i. e., $\pi_t(f) = f_t$ for $f \in D(J; \mathbb{R}^d)$. Then we denote

$$\mathcal{F}_{D(J;\mathbb{R}^d)} := \sigma(\pi_t : t \in J),$$

which is the smallest σ-algebra on $D(J;\mathbb{R}^d)$ making all projections π_t Borel-measurable.

It can be shown that $\mathcal{F}_{D(J;\mathbb{R}^d)}$ (when $J = [0,1]$ or $J = [0,\infty)$) coincides with the Borel σ-algebra generated by the open sets of $D(J;\mathbb{R}^d)$ with respect to the Skorohod topology (cf. Billingsley [17, Theorems 12.5 and 16.6]). This result can then be used to identify various sets of importance belonging to $\mathcal{F}_{D(J;\mathbb{R}^d)}$, but we shall not use it here.

The following lemma shows that a strictly càdlàg stochastic process can also be viewed as a random element in $(D(J;\mathbb{R}^d), \mathcal{F}_{D(J;\mathbb{R}^d)})$.

Lemma 12.3 (Strictly càdlàg processes as random elements in $D(J;\mathbb{R}^d)$). *Let J be a real interval, $(\Omega, \mathcal{F}, P)$ a probability space and $X = (X_t)_{t\in J}$ a family of functions $X_t : \Omega \to \mathbb{R}^d$ such that X is strictly càdlàg. Then X is a stochastic process if and only if X regarded as a function $\Omega \to D(J;\mathbb{R}^d)$, $\omega \mapsto (X_t(\omega))_{t\in J}$ is $\mathcal{F} \to \mathcal{F}_{D(J;\mathbb{R}^d)}$-measurable.*

Proof. If $X : \Omega \to D(J;\mathbb{R}^d)$ is $\mathcal{F} \to \mathcal{F}_{D(J;\mathbb{R}^d)}$-measurable then, since $\pi_t : D(J;\mathbb{R}^d) \to \mathbb{R}^d$ is $\mathcal{F}_{D(J;\mathbb{R}^d)} \to \mathcal{B}_d$-measurable, it follows that $X_t = \pi_t \circ X$ is $\mathcal{F} \to \mathcal{B}_d$-measurable so that X is a stochastic process. Conversely, if each X_t is $\mathcal{F} \to \mathcal{B}_d$-measurable, then for each $A \in \mathcal{B}_d$ we have $X^{-1}(\pi_t^{-1}(A)) = (\pi_t \circ X)^{-1}(A) = X_t^{-1}(A) \in \mathcal{F}$. But by definition, $\{\pi_t^{-1}(A) : t \in J, A \in \mathcal{B}_d\}$ generates $\mathcal{F}_{D(J;\mathbb{R}^d)}$, so that X is $\mathcal{F} \to \mathcal{F}_{D(J;\mathbb{R}^d)}$-measurable. □

Having identified a strictly càdlàg process X as a random element, we can also speak of the *distribution* or *law* P_X or $\mathcal{L}(X)$ of X as the image measure of P under X, i. e., $\mathcal{L}(X) = P_X$ is the probability measure on $(D(J;\mathbb{R}^d), \mathcal{F}_{D(J;\mathbb{R}^d)})$ such that

$$(\mathcal{L}(X))(A) = P_X(A) = P(X^{-1}(A)) \quad \forall\, A \in \mathcal{F}_{D(J;\mathbb{R}^d)}.$$

By modifying a càdlàg process (i. e., a stochastic process whose sample paths are only almost surely càdlàg) on a null set we can always find a version which is strictly càdlàg and is indistinguishable from the original one, and by a slight abuse of language we shall sometimes also view càdlàg processes as random elements in $D(J;\mathbb{R}^d)$.

The next result says that two (strictly) càdlàg stochastic processes have the same law if and only if they have the same finite dimensional distributions.

Proposition 12.4 (Finite dimensional distributions and equality in law). *Let J be a real interval, $(\Omega, \mathcal{F}, P)$ a probability space and $X = (X_t)_{t\in J}$ and $Y = (Y_t)_{t\in J}$ be two $\mathbb{R}^d$-valued (strictly) càdlàg stochastic processes. Then X and Y have the same distribution (as random elements in $D(J;\mathbb{R}^d)$) if and only if they have the same finite dimensional distributions.*

Proof. Observe that

$$\mathcal{C} := \{(\pi_{t_1}, \dots, \pi_{t_n})^{-1}(B) : n \in \mathbb{N}, B \in \mathcal{B}_{dn}, t_1, \dots, t_d \in J\}$$

is a generator of $\mathcal{F}_{D(J;\mathbb{R}^d)}$ and is closed under finite intersections. By the uniqueness theorem for measures, it therefore follows that P_X and P_Y are equal if and only if P_X and P_Y agree on $\mathcal{C}$. However, since $(X_{t_1}, \dots, X_{t_n}) = (\pi_{t_1}, \dots, \pi_{t_n}) \circ X$ we have $P_X((\pi_{t_1}, \dots, \pi_{t_n})^{-1}(B)) = P_{(X_{t_1},\dots,X_{t_n})}(B)$ for $B \in \mathcal{B}_{dn}$, so that equality of P_X and P_Y on $\mathcal{C}$ just means that X and Y have the same finite dimensional distributions. □

Remark 12.5. Proposition 12.4 applies in particular to Lévy processes. Previously in Theorem 10.26(b) we had only shown that two Lévy processes $X = (X_t)_{t\in I}$ and $Y = (Y_t)_{t\in I}$ with $\mathcal{L}(X_t) = \mathcal{L}(Y_t)$ for some $t \neq 0$ are equal in law in the sense of finite dimensional distributions, but now we know that they are even unique in distribution when viewed as random elements in $(D(I; \mathbb{R}^d), \mathcal{F}_{D(I;\mathbb{R}^d)})$. In particular, when $X = (X_t)_{t\geq 0}$ is a Lévy process, it follows from the Markov property in Theorem 11.1 that $(X_t)_{t\geq 0}$ and $(X_{t+s} - X_s)_{t\geq 0}$ have the same distribution when viewed as random elements in $(D([0,\infty); \mathbb{R}^d), \mathcal{F}_{D([0,\infty);\mathbb{R}^d)})$.

It is easily seen that on each fixed compact interval, a càdlàg function can only have finitely many jumps of absolute size exceeding a given constant $\varepsilon > 0$, since otherwise the set of jump times where the jumps exceed ε in absolute size would have an accumulation point in the compact interval, and at this accumulation point the function cannot have a finite left-limit and simultaneously be right-continuous, contradicting the càdlàg property. In particular, the number of jump times is countable. We would like to show that for strictly càdlàg processes, the time of the k'th jump with jump size in a fixed Borel set that is bounded away from zero is a random variable, and also that the corresponding jump size is a random variable. For the latter we need the following lemma:

Lemma 12.6. *Let J be a real interval.*

(a) *Let $(X_t)_{t\in J}$ be a strictly càdlàg or strictly càglàd process on a probability space $(\Omega, \mathcal{F}, P)$ with values in $\mathbb{R}^d$. Let $T : \Omega \to J$ be a random variable taking values in J. Then $X_T : \Omega \to \mathbb{R}^d$, defined by $X_T(\omega) := X_{T(\omega)}(\omega)$, is a random vector, i. e., $\mathcal{F}$-measurable.*

(b) *Let $X = (X_t)_{t\in J}$ be an $\mathbb{R}^d$-valued strictly càdlàg stochastic process on a probability space $(\Omega, \mathcal{F}, P)$ and $T : \Omega \to J$ a random variable. Then $(X_{t-})_{t\in J}$ is a strictly càglàd stochastic process, $(\Delta X_t)_{t\in J}$ is a stochastic process, and X_{T-} and ΔX_T are random vectors.*

Proof. (a) We shall assume that X is strictly càdlàg, the proof for strictly càglàd processes being similar. Suppose first that T takes only countably many values, $(t_i)_{i\in L}$ for a countable index set L. Then for each Borel set B in $\mathbb{R}^d$ we have

$$\{X_T \in B\} = \bigcup_{i\in L} \{X_T \in B, T = t_i\} = \bigcup_{i\in L} \{X_{t_i} \in B, T = t_i\} \in \mathcal{F},$$

since T and X_{t_i} are random vectors. Now let $T : \Omega \to J$ be a general random variable. If the right endpoint r_J of J is finite, denote $\widetilde{J} := J \cup [r_J, r_J + 1]$, otherwise denote $\widetilde{J} := J$. For $k \in \mathbb{N}$ define $T_k : \Omega \to \widetilde{J}$ by $T_k(\omega) := i2^{-k}$, where i is the unique integer such that $T(\omega) \in [(i-1)2^{-k}, i2^{-k})$. Then each T_k is a random variable with values in $\widetilde{J}$, and decreases pointwise to T as $k \to \infty$. Since X_{T_k} is a random vector by the part already shown, and since X is strictly càdlàg, it follows that X_{T_k} converges pointwise to X_T as $k \to \infty$, showing that X_T is also $\mathcal{F}$-measurable.

(b) It is clear that $(X_{t-})_{t\in J}$ is a strictly càglàd process, since X is strictly càdlàg. Since $\Delta X_t = X_t - X_{t-}$, $(\Delta X_t)_{t\in I}$ must also be a stochastic process. X_{T-} is a random vector by (a), and ΔX_T is a random vector since $\Delta X_T = X_T - X_{T-}$. □

In the following, we call a function $f : [0,\infty) \to \mathbb{R}^d$ a *càdlàg step function* if it is càdlàg, has only finitely many jumps on each compact interval and is constant between two consecutive jumps. Also, by $\overline{B}$ we denote the topological closure of a set $B \subset \mathbb{R}^d$.

Proposition 12.7. *Let $X = (X_t)_{t\ge 0}$ be an $\mathbb{R}^d$-valued strictly càdlàg stochastic process on a probability space $(\Omega, \mathcal{F}, P)$.*

(a) *Suppose that $B \in \mathcal{B}_d$ is bounded away from zero, i. e., that $0 \notin \overline{B}$. For each $k \in \mathbb{N}$, denote by T_k^B the time of the kth jump of X with jump size in B, with the convention that $T_k^B = +\infty$ if no k^{th} jump exists, i. e.,*

$$T_1^B := \inf\{t > 0 : \Delta X_t \in B\} \quad \textit{and} \quad T_k^B := \inf\{t > T_{k-1}^B : \Delta X_t \in B\} \quad \textit{for } k \ge 2.$$

Define also

$$Y_k^B := \begin{cases} \Delta X_{T_k^B}, & \textit{if } T_k^B < \infty, \\ 0, & \textit{if } T_k^B = \infty. \end{cases}$$

Then (T_k^B, Y_k^B) is a $[0,\infty] \times \mathbb{R}^d$-valued random vector for each $k \in \mathbb{N}$.

(b) *Suppose that for each $\omega \in \Omega$, $t \mapsto X_t(\omega)$ has only finitely many jumps on compact intervals, and denote the time of the kth jump of X by T_k, so that $T_1 := \inf\{t > 0 : \Delta X_t \neq 0\}$ and $T_k := \inf\{t > T_{k-1} : \Delta X_t \neq 0\}$ for $k \ge 2$. Define $Y_k := \Delta X_{T_k}$ if $T_k < \infty$ and $Y_k := 0$ otherwise. Then each (T_k, Y_k) is a $[0,\infty] \times \mathbb{R}^d$-valued random vector.*

(c) *The set of all $\omega \in \Omega$ for which $t \mapsto X_t(\omega)$ has only finitely many jumps on compact intervals is in $\mathcal{F}$.*

(d) *The set of all $\omega \in \Omega$ for which $t \mapsto X_t(\omega)$ is a càdlàg step function is in $\mathcal{F}$.*

(e) *For $d = 1$, the set of all $\omega \in \Omega$ for which $t \mapsto X_t(\omega)$ is a càdlàg step function with only jumps of size 1 is in $\mathcal{F}$.*

Proof. (a) For ease of notation, we write T_k instead of T_k^B and Y_k instead of Y_k^B in this part of the proof.

(i) First consider the case when $B = \{x \in \mathbb{R}^d : |x| > r\}$ for some $r > 0$. Since X is strictly càdlàg, it is easy to see that for every $t \in [0,\infty)$, we have

$$\{T_1 < t\} = \bigcup_{l\in\mathbb{N}} \bigcap_{m\in\mathbb{N}} \bigcup_{\substack{p,q\in\mathbb{Q},\\ 0<p<q<t,\\ |q-p|\le 1/m}} \left\{|X_q - X_p| \ge r + \frac{1}{l}\right\}.$$

Consequently, $\{T_1 < t\} \in \mathcal{F}$ for every $t \in [0,\infty)$, showing that T_1 is a (numerical) random variable. Since

$$\{T_k < t\} = \bigcup_{l\in\mathbb{N}} \bigcap_{m\in\mathbb{N}} \bigcup_{\substack{p,q\in\mathbb{Q},\\ 0<p<q<t,\\ |q-p|\le 1/m}} \left(\left\{|X_q - X_p| \ge r + \frac{1}{l}\right\} \cap \{p > T_{k-1}\}\right)$$

for $k \ge 2$ and $t \in [0,\infty)$, we inductively see that each T_k is a random variable. Now define

$$\widetilde{T}_k := \begin{cases} T_k, & \text{if } T_k \in (0,\infty), \\ 0, & \text{if } T_k = +\infty, \end{cases}$$

and observe that $Y_k = \Delta X_{\widetilde{T}_k}$ for each k (since $\Delta X_0 = 0$ by Definition 10.1). Hence Y_k is a random vector by Lemma 12.6 and it follows that (T_k, Y_k) is a $[0,\infty]\times\mathbb{R}^d$-valued random vector.

(ii) Now consider the general case when $B \in \mathcal{B}_d$ and is bounded away from 0. Choose $r > 0$ such that $B \subset \{x \in \mathbb{R}^d : |x| > r\}$, denote by S_k the time of the kth jump of X with size in $\{x \in \mathbb{R}^d : |x| > r\}$, and define

$$Z_k := \begin{cases} \Delta X_{S_k}, & \text{if } S_k < \infty, \\ 0, & \text{if } S_k = \infty. \end{cases}$$

Then $T_k \ge S_k$ and for $k \le m$ we have $T_k = S_m < \infty$ if and only if there are exactly k indices $l_1 < \cdots < l_k$ in $\mathbb{N}$ with $l_k = m$, $Z_{l_1},\dots,Z_{l_k} \in B$ and $Z_s \notin B$ for all $s \in \{1,\dots,m\} \setminus \{l_1,\dots,l_k\}$. In other words, defining $A_{k,m}$ as

$$\bigcup_{\substack{(l_1,\dots,l_k)\in\mathbb{N}^k,\\ l_1<\dots<l_k=m}} \{Z_{l_1},\dots,Z_{l_k} \in B, Z_s \notin B\ \forall\, s \in \{1,\dots,m\} \setminus \{l_1,\dots,l_k\}\}$$

for $k \le m$, we have

$$\widetilde{T}_k := \left\{\begin{matrix} T_k, & \text{if } T_k \in (0,\infty) \\ 0, & \text{if } T_k = +\infty \end{matrix}\right\} = \sum_{m=k}^{\infty} S_m \mathbf{1}_{A_{k,m}},$$

and this is measurable since $A_{k,m} \in \mathcal{F}$ by part (i). Hence also T_k and Y_k are measurable, the latter again by Lemma 12.6.

(b) Define $B_1 := \{x \in \mathbb{R}^d : |x| > 1\}$ and $B_m := \{x \in \mathbb{R}^d : 1/m < |x| \le 1/(m-1)\}$ for $m \ge 2$. Then $T_1 = \inf\{T_j^{B_m} : j, m \in \mathbb{N}\}$ and is hence a random variable. Moreover,

$T_k = \inf\{T_j^{B_m} : j, m \in \mathbb{N}, T_j^{B_m} > T_{k-1}\}$ for each $k \geq 2$ and inductively it follows that each T_k is a random variable. That Y_k is also a random vector follows from Lemma 12.6 as in part (a) above.

(c) With B_m as above, the event that X has only finitely many jumps on compact intervals can be expressed as

$$\bigcap_{M=1}^{\infty} \left\{ \sum_{j,m=1}^{\infty} \mathbf{1}_{[0,M]} \circ T_j^{B_m} < \infty \right\}, \tag{12.1}$$

which is measurable by part (a).

(d) Define $R_1 := \inf\{t > 0 : X_t \neq X_0\}$ and, for $n \geq 2$, $R_n := \inf\{t > R_{n-1} : X_t \neq X_{R_{n-1}}\}$ if $R_{n-1} < \infty$, and $R_n := +\infty$ otherwise. Since $\{R_1 \geq t\} = \bigcap_{q \in \mathbb{Q} \cap [0,t)} \{X_q = X_0\}$ for $t \in [0, \infty)$ it follows that R_1 is a random variable. Hence also $\widetilde{R}_1 := R_1 \mathbf{1}_{[0,\infty)}(R_1)$ is a $[0, \infty)$-valued random variable and it follows from Lemma 12.6 that $X_{\widetilde{R}_1}$ is a random vector. For $t \in [0, \infty)$, we then have

$$\{R_2 \geq t\} = \{R_1 = \infty\} \cup \bigcap_{q \in \mathbb{Q} \cap [0,t)} ((\{X_q = X_{\widetilde{R}_1}\} \cap \{q \geq R_1\}) \cup \{q < R_1\})$$

from which it follows that R_2 is also a random variable. An inductive argument then shows that all R_n are (numerical) random variables. It may happen that $R_n = R_{n-1}$ as, for example, if $X_t(\omega) = t$ (then $R_1(\omega) = R_2(\omega) = \cdots = 0$). The event that $t \mapsto X_t(\omega)$ is a step function is the same as the event that $\lim_{n\to\infty} R_n(\omega) = \infty$. Hence

$$\{\omega \in \Omega : X_\cdot(\omega) \text{ is a step function}\} = \bigcap_{m=1}^{\infty} \bigcup_{n=1}^{\infty} \{R_n > m\} \in \mathcal{F}.$$

(e) Let $B_n := (-\frac{1}{n}, -\frac{1}{n+1}] \cup [\frac{1}{n+1}, \frac{1}{n})$ for $n \in \mathbb{N}$ and $B_0 := (-\infty, -1] \cup (1, \infty)$, so that $\bigcup_{n=0}^{\infty} B_n = \mathbb{R} \setminus \{0, 1\}$ and each B_n is bounded away from zero. Let $A := \{\omega \in \Omega : X_\cdot(\omega) \text{ is a step function}\}$. Then the set of all $\omega \in \Omega$, for which $X_\cdot(\omega)$ is a step function with only jumps of size 1 can be described as $A \cap \bigcap_{n=0}^{\infty} \{T_1^{B_n} = +\infty\}$, which is in $\mathcal{F}$ by parts (a) and (d). □

Proposition 12.7 enables us to identify some important sets and functions that are measurable with respect to $\mathcal{F}_{D([0,\infty);\mathbb{R}^d)}$.

Corollary 12.8 ($\mathcal{F}_{D([0,\infty);\mathbb{R}^d)}$-meas. functionals and sets in $D([0,\infty); \mathbb{R}^d)$).

(a) *Let $B \in \mathcal{B}_d$ be bounded away from zero, and for each càdlàg function $f \in D([0,\infty); \mathbb{R}^d)$ and $k \in \mathbb{N}$ denote by $\xi_k^B(f)$ the time of the kth jump of f with jump size in B, i. e.,*

$$\xi_1^B(f) := \inf\{t > 0 : \Delta f_t \in B\} \quad \textit{and} \quad \xi_k^B(f) := \inf\{t > \xi_{k-1}^B : \Delta f_t \in B\}, \quad k \geq 2.$$

Further, define $\zeta_k^B(f)$ *as* $\Delta f_{\xi_k^B}$ *if* $\xi_k^B < \infty$ *and* $\zeta_k^B = 0$ *if* $\xi_k^B = \infty$. *Then each of the functions* $(\xi_k^B, \zeta_k^B) : D([0,\infty); \mathbb{R}^d) \to [0,\infty] \times \mathbb{R}^d, f \mapsto (\xi_k^B(f), \zeta_k^B(f))$ *is* $\mathcal{F}_{D([0,\infty);\mathbb{R}^d)} \to \mathcal{B}([0,\infty] \times \mathbb{R}^d)$*-measurable.*

(b) *The set of all* $\mathbb{R}^d$*-valued càdlàg functions which have only finitely many jumps on compact intervals is an element of* $\mathcal{F}_{D([0,\infty);\mathbb{R}^d)}$.

(c) *The set of all* $\mathbb{R}^d$*-valued càdlàg step functions is an element of* $\mathcal{F}_{D([0,\infty);\mathbb{R}^d)}$.

(d) *The set of all* $\mathbb{R}$*-valued càdlàg step functions with only jumps of size 1 is an element of* $\mathcal{F}_{D([0,\infty);\mathbb{R})}$.

Proof. This follows immediately from Proposition 12.7 by considering the (canonical) stochastic process $X = (X_t)_{t\geq 0}$ on $(D([0,\infty); \mathbb{R}^d), \mathcal{F}_{D([0,\infty);\mathbb{R}^d)})$ defined by $X_t := \pi_t$, so that X is the identity on $D([0,\infty); \mathbb{R}^d)$ (the probability measure is irrelevant, since we are making assertions only about measurability). □

Another consequence of Proposition 12.7 is the following, which shows that suitable functionals of jumps of càdlàg processes give again stochastic processes with the desired measurability.

Corollary 12.9. *Let* $X = (X_t)_{t\geq 0}$ *be an* $\mathbb{R}^d$*-valued strictly càdlàg process and let* $f : [0,\infty) \times \mathbb{R}^d \to \mathbb{R}^m$ *be Borel-measurable with* $f(s,0) = 0$ *for all* $s \in [0,\infty)$. *Define* $B_1 := \{x \in \mathbb{R}^d : |x| > 1\}$ *and* $B_m := \{x \in \mathbb{R}^d : 1/m < |x| \leq 1/(m-1)\}$ *for* $m \geq 2$. *Then for* $0 \leq u < v < \infty$, *with the notations used in Proposition 12.7,*

$$\sum_{u<s\leq v} |f(s, \Delta X_s)| = \sum_{m,j=1}^{\infty} |f(T_j^{B_m}, Y_j^{B_m})| \cdot \mathbf{1}_{(u,v]}(T_j^{B_m})$$

is a $\sigma(X_s - X_u : s \in (u,v])$*-measurable* $[0,\infty]$*-valued random variable. If, additionally,* $C_{u,v} := \sum_{u<s\leq v} |f(s, \Delta X_s)| < \infty$ *almost surely, then*

$$\mathbf{1}_{[0,\infty)}(C_{u,v}) \sum_{u<s\leq v} f(s, \Delta X_s) = \mathbf{1}_{[0,\infty)}(C_{u,v}) \sum_{m,j=1}^{\infty} f(T_j^{B_m}, Y_j^{B_m}) \mathbf{1}_{(u,v]}(T_j^{B_m})$$

is a $\sigma(X_s - X_u : s \in (u,v])$*-measurable* $\mathbb{R}^d$*-valued random vector. (By modifying* X *on a set of measure zero if necessary, we can assume in the latter assertion that* $C_{u,v} < \infty$ *everywhere and hence omit the factor* $\mathbf{1}_{[0,\infty)}(C_{u,v})$.)

Proof. Let $\mathcal{G} := \sigma(X_s - X_u : s \in (u,v])$. That $\sum_{u<s\leq v} |f(s, \Delta X_s)|$ is equal to $\sum_{m,j=1}^{\infty} |f(T_j^{B_m}, Y_j^{B_m})| \mathbf{1}_{(u,v]}(T_j^{B_m})$ is clear, since $f(s,0) = 0$ and the right-hand side of this equation is just a specific enumeration of the considered jumps. To show that it is $\mathcal{G}$-measurable, define the strictly càdlàg process $Z = (Z_t)_{t\geq 0}$ by $Z_t = 0$ for $t \in [0,u]$, $Z_t = X_t - X_u$ for $t \in (u,v]$, and $Z_t = X_v - X_u$ for $t > v$. Then $\Delta Z_t = \Delta X_t$ when $t \in (u,v]$ and $\Delta Z_t = 0$ otherwise. Further, each Z_t is $\mathcal{G}$-measurable, hence Z can be viewed as a stochastic process on the probability space $(\Omega, \mathcal{G}, P_{|\mathcal{G}})$. Denoting the time and size of the kth jump of Z with jump size in B_m by $T_{k,Z}^{B_m}$ and $Y_{k,Z}^{B_m}$, respectively, we then obviously have

$$\sum_{u<s\leq v} |f(s,\Delta X_s)| = \sum_{s\geq 0} |f(s,\Delta Z_s)| = \sum_{m,j=1}^{\infty} |f(T_{j,Z}^{B_m}, Y_{j,Z}^{B_m})|,$$

and the latter is $\mathcal{G}$-measurable by Proposition 12.7(a). A similar argument holds for $\sum_{u<s\leq v} f(s,\Delta X_s)$, when it converges absolutely almost surely. □

Example 12.10. Let X be an $\mathbb{R}^d$-valued compound Poisson process on $[0,\infty)$. Then almost surely, it has only finitely many jumps on compact intervals, hence $\sum_{u<s\leq v} f(s,\Delta X_s)$ converges almost surely for each Borel-measurable function $f : [0,\infty)\times\mathbb{R}^d$ with $f(s,0) = 0$ and is $\sigma(X_s - X_u : s \in (u,v])$-measurable.

Remark 12.11. Corollary 12.9 can easily be extended to strictly càdlàg processes defined on J, where $J = \mathbb{R}$ (or a subinterval of $\mathbb{R}$) in the sense that $\sum_{u<s\leq v} |f(s,\Delta X_s)|$ and $\mathbf{1}_{[0,\infty)}(C_{u,v}) \sum_{u<s\leq v} f(s,\Delta X_s)$ are $\sigma(X_s - X_u : s \in (u,v])$-measurable for $u < v$ in J. This can be done as in the proof of Corollary 12.9, by introducing Z and counting only the jumps of Z which appear from time u onwards.

Having identified some mappings and sets that are measurable with respect to $\mathcal{F}_{D([0,\infty);\mathbb{R}^d)}$, we can use them to show that certain functionals of stochastic processes have the same distribution, provided the stochastic processes have the same finite dimensional distributions.

Corollary 12.12 (Functionals of càdlàg processes with same distribution). *Let $X = (X_t)_{t\geq 0}$ and $Z = (Z_t)_{t\geq 0}$ be two $\mathbb{R}^d$-valued strictly càdlàg stochastic processes (not necessarily defined on the same probability space) that have the same finite dimensional distributions.*

(a) *For each $m \in \mathbb{N}$ let $B_m \in \mathcal{B}_d$ be bounded away from zero and, as in Proposition* 12.7(a), *denote the time of the kth jump of X with jump size in B_m by $T_{k,X}^{B_m}$ and the corresponding jump size by $Y_{k,X}^{B_m}$, with $Y_{k,X}^{B_m} := 0$ if $T_{k,X}^{B_m} = +\infty$. Similarly, define $T_{k,Z}^{B_m}$ and $Y_{k,Z}^{B_m}$ for the corresponding quantities associated with Z. Then $(T_{k,X}^{B_m}, Y_{k,X}^{B_m})_{k,m\in\mathbb{N}}$ and $(T_{k,Z}^{B_m}, Y_{k,Z}^{B_m})_{k,m\in\mathbb{N}}$ have the same finite dimensional distributions.*

(b) *If almost surely all paths of X have only finitely many jumps on compact intervals, then almost surely all paths of Z have only finitely many jumps on compact intervals. In that case if X and Z are modified (if necessary) on a set of probability* 0 *so as to make all paths of X and Z have only finitely many jumps on compact intervals, then the sequences $(T_{k,X}, Y_{k,X})_{k\in\mathbb{N}}$ and $(T_{k,Z}, Y_{k,Z})_{k\in\mathbb{N}}$, defined as in Proposition* 12.7(b), *will have the same finite dimensional distributions.*

(c) *If almost surely all paths of X are càdlàg step functions, then almost surely all paths of Z are càdlàg step functions.*

(d) *If $d = 1$ and almost surely all paths of X are càdlàg step functions with jump size* 1, *then almost surely all paths of Z are càdlàg step functions with jump size* 1.

Proof. (a) With $\xi_k^{B_m}$ and $\zeta_k^{B_m}$ as in Corollary 12.8, this follows from the facts that $(T_{k,X}^{B_m}, Y_{k,X}^{B_m}) = (\xi_k^{B_m}, \zeta_k^{B_m}) \circ X$, $(T_{k,Z}^{B_m}, Y_{k,Z}^{B_m}) = (\xi_k^{B_m}, \zeta_k^{B_m}) \circ Z$, each $(\xi_k^{B_m}, \zeta_k^{B_m})$ is $\mathcal{F}_{D([0,\infty);\mathbb{R}^d)}$-measurable and X and Y have the same distribution as elements in $(D([0,\infty);\mathbb{R}^d), \mathcal{F}_{D([0,\infty);\mathbb{R}^d)})$.

(b) Define $B_1 := \{x \in \mathbb{R}^d : |x| > 1\}$ and $B_m := \{x \in \mathbb{R}^d : 1/m < |x| \le 1/(m-1)\}$ for $m \ge 2$. Since $(T_{j,X}^{B_m})_{j,m\in\mathbb{N}}$ and $(T_{j,Z}^{B_m})_{j,m\in\mathbb{N}}$ have the same finite dimensional distributions by (a), the sets $\bigcap_{M=1}^{\infty}\{\sum_{j,m=1}^{\infty} \mathbf{1}_{[0,M]} \circ T_{j,X}^{B_m} < \infty\}$ and $\bigcap_{M=1}^{\infty}\{\sum_{j,m=1}^{\infty} \mathbf{1}_{[0,M]} \circ T_{j,Z}^{B_m} < \infty\}$ have the same probability and hence by (12.1) the paths of Z almost surely have only finitely many jumps on compact intervals. In that case, after modifying X and Z if necessary, $(T_{k,X}, Y_{k,X})$ and $(T_{k,Z}, Y_{k,Z})$ are random vectors. To see that $(T_{k,X}, Y_{k,X})_{k\in\mathbb{N}}$ and $(T_{k,Z}, Y_{k,Z})_{k\in\mathbb{N}}$ have the same finite dimensional distributions, denote by A the set of all $f \in D([0,\infty);\mathbb{R}^d)$ that have only finitely many jumps on compact intervals, which is in $\mathcal{F}_{D([0,\infty);\mathbb{R}^d)}$ by Corollary 12.8. Hence $\eta = (\eta_t)_{t\ge 0} : D([0,\infty);\mathbb{R}^d) \to D([0,\infty);\mathbb{R}^d)$, defined by $\eta(f) = f$ if $f \in A$ and $\eta(f) = \mathbf{0}$ if $f \notin A$, where $\mathbf{0} \in D([0,\infty);\mathbb{R}^d)$ is the null-function, is $\mathcal{F}_{D([0,\infty);\mathbb{R}^d)} \to \mathcal{F}_{D([0,\infty);\mathbb{R}^d)}$-measurable. Let $\xi_{k,\eta}$ be the time of the kth jump of η and denote by $\zeta_{k,\eta}$ the corresponding size of the jump, again with the convention that $\zeta_{k,\eta} := 0$ if $\xi_{k,\eta} = +\infty$. By Proposition 12.7(b), each $(\xi_{k,\eta}, \zeta_{k,\eta})$ is $\mathcal{F}_{D([0,\infty);\mathbb{R}^d)} \to \mathcal{B}([0,\infty] \times \mathbb{R}^d)$-measurable. Coming back to the (possibly modified) processes X and Z, we observe that $(T_{k,X}, Y_{k,X}) = (\xi_{k,\eta}, \zeta_{k,\eta}) \circ X$ and $(T_{k,Z}, Y_{k,Z}) = (\xi_{k,\eta}, \zeta_{k,\eta}) \circ Z$, and hence $(T_{k,X}, Y_{k,X})_{k\in\mathbb{N}}$ and $(T_{k,Z}, Y_{k,Z})_{k\in\mathbb{N}}$ have the same finite dimensional distributions, since $\mathcal{L}(X) = \mathcal{L}(Z)$.

(c) Denote by $S \subset D([0,\infty);\mathbb{R}^d)$ the set of step functions. Then $S \in \mathcal{F}_{D([0,\infty);\mathbb{R}^d)}$ by Corollary 12.8(b), and $P_X(S) = P(X \in S) = 1$. Since X and Z have the same distribution as random elements in $D([0,\infty);\mathbb{R}^d)$, it follows that Z also takes on values in S with probability 1. The proof for (d) is completely analogous. □

We can now show that from a given Lévy process we can construct new Lévy processes as certain functionals of the jumps.

Theorem 12.13 (Lévy processes constructed from jumps of LPs). *Suppose that $I = [0,\infty)$ or $I = \mathbb{R}$ and let $L = (L_t)_{t\in I}$ be an $\mathbb{R}^d$-valued Lévy process. Let $f : \mathbb{R}^d \to \mathbb{R}^m$ be a Borel-measurable function with $f(0) = 0$ and assume that $\sum_{u<s\le v} |f(\Delta L_s)| < \infty$ almost surely for each $u < v$ in I. By modifying L on a null-set, we can assume without loss of generality that $\sum_{u<s\le v} |f(\Delta L_s)| < \infty$ everywhere for each $u < v$ in I. Then $M = (M_t)_{t\in I}$, defined by*

$$M_t = \begin{cases} \sum_{0<s\le t} f(\Delta L_s), & t \ge 0, \\ -\sum_{t<s\le 0} f(\Delta L_s), & t < 0, \end{cases}$$

is an $\mathbb{R}^m$-valued Lèvy process that is strictly càdlàg.

The characteristic triplet of the process M will be determined later in Corollary 12.37.

Proof. Since $\sum_{0<s\le t} |f(\Delta L_s(\omega))|$, $t \in [0,\infty)$, is a non-decreasing function of t for each ω, $\sum_{0<s\le t} |f(\Delta L_s(\omega))| < \infty$ implies $\sum_{0<s\le t'} |f(\Delta L_s(\omega))| < \infty$ for each $t' \in [0,t]$. Letting t tend to infinity along the integers, we see that there is a null-set such that $\sum_{0<s\le t} f(\Delta L_s(\omega))$ converges absolutely for all $t \ge 0$ and all ω outside this null-set; similar reasoning applies when $t < 0$.

Clearly $M_0 = 0$ and M is strictly càdlàg. To see that M has independent increments, let $\mathcal{G}_{u,v} := \sigma(L_s - L_u : s \in (u,v])$ for $u < v$, and consider the times $t_0 < t_1 < \cdots < t_n$ in I. By Proposition 10.19, $\mathcal{G}_{t_0,t_1}, \mathcal{G}_{t_1,t_2}, \ldots, \mathcal{G}_{t_{n-1},t_n}$ are independent, and since $M_{t_i} - M_{t_{i-1}} = \sum_{t_{i-1}<s\le t_i} f(\Delta L_s)$ is $\mathcal{G}_{t_{i-1},t_i}$-measurable by Corollary 12.9 and Remark 12.11, it follows that M has independent increments.

To show that M has stationary increments, suppose that $t \in I$. By the Markov property in Theorem 11.1, the process $Z = (Z_s := (L_{t+s} - L_t))_{s\ge 0}$ is a Lévy process with the same distribution as $(L_s)_{s\ge 0}$ and $\Delta Z_s = \Delta L_{t+s}$ for $s > 0$. Define B_m for $m \in \mathbb{N}$ as in Corollary 12.9, and denote by $T^{B_m}_{j,L}$ and $Y^{B_m}_{j,L}$ the time and size of the jth jump of $(L_s)_{s\ge 0}$ with jump size in B_m, and similarly for Z. Then, by Corollary 12.9, we have for $h \ge 0$

$$M_{t+h} - M_t = \sum_{t<s\le t+h} f(\Delta L_s) = \sum_{0<s\le h} f(\Delta Z_s) = \sum_{m,j=1}^{\infty} f(Y^{B_m}_{j,Z}) \mathbf{1}_{(0,h]}(T^{B_m}_{j,Z})$$

and

$$M_h - M_0 = \sum_{0<s\le h} f(\Delta L_s) = \sum_{m,j=1}^{\infty} f(Y^{B_m}_{j,L}) \mathbf{1}_{(0,h]}(T^{B_m}_{j,L}).$$

But since $(T^{B_m}_{k,Z}, Y^{B_m}_{k,Z})_{k,m\in\mathbb{N}}$ and $(T^{B_m}_{k,L}, Y^{B_m}_{k,L})_{k,m\in\mathbb{N}}$ have the same finite-dimensional distributions by Corollary 12.12, it follows that also $M_{t+h} - M_t$ and $M_h - M_0 = M_h$ have the same distribution for $h \ge 0$. Hence M has stationary increments and consequently is a Lévy process. □

Example 12.14. Let L be an $\mathbb{R}^d$-valued compound Poisson process on $[0,\infty)$, and let $f : \mathbb{R}^d \to \mathbb{R}^m$ be a Borel-measurable function with $f(0) = 0$. Then $M := (\sum_{0<s\le t} f(\Delta L_s))_{t\ge 0}$ is a Lévy process, and since its sample paths are almost surely càdlàg step functions, it will follow from Corollary 12.27 below that it also is a compound Poisson process or the zero-process. The fact that M is a compound Poisson process when $f(x) \ne 0$ for $x \ne 0$ can also be seen more directly be using the representation $L_t = \sum_{i=1}^{N_t} Y_i$ as in Definition 10.8, which results in $M_t = \sum_{i=1}^{N_t} f(Y_i)$.

Theorem 12.13 also has a version for additive processes:

Proposition 12.15 (Additive processes constructed from jumps). *Let $L = (L_t)_{t\ge 0}$ be an $\mathbb{R}^d$-valued additive process. Let $f : [0,\infty) \times \mathbb{R}^d \to \mathbb{R}^m$ be a Borel-measurable function such that $f(s,0) = 0$ for all $s \ge 0$ and $\sum_{0<s\le t} |f(s,\Delta L_s)| < \infty$ almost surely for each $t \ge 0$. By modifying L on a null-set, we can assume without loss of generality that $\sum_{0<s\le t} |f(s,\Delta L_s)| < \infty$ everywhere for each t. Then $M = (M_t)_{t\ge 0}$, defined by*

$$M_t = \sum_{0<s\le t} f(s, \Delta L_s), \quad t \ge 0,$$

is an $\mathbb{R}^m$*-valued additive process which is strictly càdlàg.*

Proof. The facts that $M = 0$, M is strictly càdlàg and with independent increments are established by a completely analogous proof to that of Theorem 12.13, this time using Remark 10.21 instead of Proposition 10.19. To see that M is continuous in probability, observe that $\Delta L_t = 0$ a. s. for each fixed $t \ge 0$ by Remark 10.21. Hence, $\Delta M_t = f(t, \Delta L_t) = 0$ a. s., so that $M_{t-} = M_t$ almost surely. The càdlàg property of M then shows that M is continuous in probability at t and hence continuous in probability everywhere. □

12.2 Characterizations of the Poisson process and of Brownian motion

In this section, we characterize the Poisson process and Brownian motion in terms of their path behaviour. We first extend the definition of the Poisson distribution. Recall that a Poisson distribution μ with parameter (mean) $\alpha \in (0, \infty)$ is given by $\mu(\{k\}) = \mathrm{e}^{-\alpha}\alpha^k/k!$ for $k \in \mathbb{N}_0$. We extend this to $\alpha \in [0, \infty]$:

Definition 12.16 (Poisson distribution with parameter 0 or $+\infty$). A distribution μ on $\overline{\mathbb{R}}$ is called a *Poisson distribution with parameter (mean)* 0 if $\mu = \delta_0$, the Dirac measure at 0. It is called a *Poisson distribution with parameter (mean)* $+\infty$ if $\mu = \delta_{+\infty}$, the Dirac measure at $+\infty$. When $\alpha \in [0, \infty)$, we speak of a *Poisson distribution with finite parameter*.

We start with a result for additive processes whose sample paths are almost surely càdlàg step functions of size 1.

Proposition 12.17. *Let* $L = (L_t)_{t\ge 0}$ *be an* $\mathbb{R}$*-valued additive process whose sample paths are almost surely non-decreasing càdlàg step functions with jump size* 1*. Then each increment* $L_t - L_s$ *of* L *is Poisson distributed with finite parameter for every* $t \ge s \ge 0$.

Proof. Since, together with L, $(L_{s+t} - L_s)_{t\ge 0}$ is an additive process with paths which are almost surely non-decreasing càdlàg step functions of jump size 1, we can assume without loss of generality that $s = 0$ and, for notational simplicity, that $t = 1$ (the case $t = 0$ being trivial). Since L has non-decreasing càdlàg sample paths and $L_0 = 0$, we must have $L_1 \ge 0$ almost surely. For each $n \in \mathbb{N}$ we have $L_1 = \sum_{j=1}^{2^n}(L_{j2^{-n}} - L_{(j-1)2^{-n}})$, where the summands are independent with values almost surely in $\mathbb{N}_0$. Define, for $n \in \mathbb{N}$ and $j \in \{1, \dots, 2^n\}$,

$$Z_{j,n} := (L_{j2^{-n}} - L_{(j-1)2^{-n}}) \wedge 1, \quad p_{j,n} := P(Z_{j,n} = 1),$$
$$Z_n := \sum_{j=1}^{2^n} Z_{j,n} \quad \text{and} \quad p_n := \sum_{j=1}^{2^n} p_{j,n}.$$

Then $Z_{j,n}, j = 1, \dots, 2^n$, are independent and Bernoulli distributed with parameter $p_{j,n}$. Hence Z_n has expectation p_n. Since the sample paths of L are almost surely step functions of jump size 1, and since L_1 is finite, L can have only finitely many jumps (of size 1) in $[0,1]$, and it follows that Z_n converges almost surely to L_1. It is also clear that the sequence $(Z_n)_{n\in\mathbb{N}}$ is non-decreasing. By monotone convergence it follows that $p_n = \mathbb{E}\, Z_n$ converges to $p := \mathbb{E}\, L_1 \in [0,\infty]$ as $n \to \infty$. Using dominated convergence, we find that for every $\beta > 0$ (observe that $0 \le L_1 < \infty$),

$$0 < \mathbb{E}\,\mathrm{e}^{-\beta L_1} = \lim_{n\to\infty} \mathbb{E}\,\mathrm{e}^{-\beta Z_n} = \lim_{n\to\infty} \prod_{j=1}^{2^n} \mathbb{E}\,\mathrm{e}^{-\beta Z_{j,n}}.$$

Since

$$\mathbb{E}\,\mathrm{e}^{-\beta Z_{j,n}} = (1 - p_{j,n}) + p_{j,n}\mathrm{e}^{-\beta} = 1 - p_{j,n}(1 - \mathrm{e}^{-\beta}) \le \mathrm{e}^{-p_{j,n}(1-\mathrm{e}^{-\beta})},$$

we conclude that

$$0 < \mathbb{E}\,\mathrm{e}^{-\beta L_1} = \lim_{n\to\infty} \prod_{j=1}^{2^n} (1 - p_{j,n}(1 - \mathrm{e}^{-\beta})) \le \liminf_{n\to\infty} \mathrm{e}^{-p_n(1-\mathrm{e}^{-\beta})} = \mathrm{e}^{-p(1-\mathrm{e}^{-\beta})}$$

for every β. This implies that $p < \infty$. Similarly, we obtain

$$\mathbb{E}\,\mathrm{e}^{\mathrm{i}\beta L_1} = \lim_{n\to\infty} \prod_{j=1}^{2^n} (1 - p_{j,n}(1 - \mathrm{e}^{\mathrm{i}\beta})) \quad \forall\, \beta \in \mathbb{R}.$$

If we can identify the right-hand side of this equation as the characteristic function of a Poisson distribution with parameter p, we are done. To do so, observe that by the uniform continuity in probability of L on $[0,1]$ (Lemma 10.35), we have

$$\lim_{n\to\infty} \max_{j=1,\dots,2^n} p_{j,n} \le \lim_{n\to\infty} \sup_{0\le s,t\le 1, |t-s|\le 2^{-n}} P(|L_t - L_s| \ge 1) = 0.$$

By the logarithmic series (8.2), there is $C \in (0,\infty)$ such that $|\log(1-z) + z| \le C|z|^2$ for all $z \in \mathbb{C}$ with $|z| \le 1/2$. Hence

$$\begin{aligned}
&\left|\sum_{j=1}^{2^n} \log(1 - p_{j,n}(1 - \mathrm{e}^{\mathrm{i}\beta})) + p_n(1 - \mathrm{e}^{\mathrm{i}\beta})\right| \\
&\quad \le \sum_{j=1}^{2^n} |\log(1 - p_{j,n}(1 - \mathrm{e}^{\mathrm{i}\beta})) + p_{j,n}(1 - \mathrm{e}^{\mathrm{i}\beta})| \\
&\quad \le 4C \sum_{j=1}^{2^n} p_{j,n}^2 \le 4Cp \max_{j=1,\dots,2^n} p_{j,n} \to 0 \quad \text{as } n \to \infty.
\end{aligned}$$

Since p_n converges to p, we conclude that

$$\mathbb{E}\,e^{i\beta L_1} = \exp\left(\lim_{n\to\infty}\sum_{j=1}^{2^n}\log(1-p_{j,n}(1-e^{i\beta}))\right) = \exp(-p(1-e^{i\beta}))$$

for $\beta \in \mathbb{R}$, which is the characteristic function of a Poisson distribution (cf., Theorem 8.7), as claimed. □

Poisson processes were defined in Definition 1.30, and there we mentioned that some of the conditions in the definition can be relaxed. Further, in Theorem 1.31 we gave an explicit method for constructing a Poisson process. The following theorem shows that *every* Poisson process can be constructed in this way, and that Poisson processes are exactly the Lévy processes whose paths are càdlàg step functions with jump size 1. In the proof, we will take full advantage of the machinery developed in the previous section regarding the distribution of the jump times.

Theorem 12.18 (Characterization of the Poisson process). *Suppose that $L = (L_t)_{t\geq 0}$ is an $\mathbb{R}$-valued stochastic process. Then the following are equivalent:*
(i) *L is a (one-sided) Poisson process.*
(ii) *L is a Lévy process such that L_t is Poisson distributed with parameter λt for some $t > 0$ and some $\lambda \in (0,\infty)$.*
(iii) *L is a non-zero Lévy process whose sample paths are almost surely non-decreasing step functions with jump size 1.*
(iv) *There exists an i. i. d. sequence $(S_n)_{n\in\mathbb{N}}$ of exponentially distributed random variables with parameter $\lambda' \in (0,\infty)$, i. e., with probability density $\lambda' e^{\lambda' t}\mathbf{1}_{[0,\infty)}(t)$, $t \in \mathbb{R}$, such that almost surely,*

$$L_t = \sup\left\{n \in \mathbb{N}_0 : \sum_{k=1}^{n} S_k \leq t\right\} \quad \forall\, t \geq 0.$$

(v) *L is a Lévy process with Gaussian variance 0, drift 0 and Lévy measure $\lambda''\delta_1$ for some $\lambda'' \in (0,\infty)$.*

If any one of the above conditions holds, then the differences between consecutive jump times of the one-sided Poisson process are i. i. d. exponentially distributed and the parameters λ, λ' and λ'' for which (ii), (iv) *and* (v) *hold satisfy $\lambda = \lambda' = \lambda''$.*

Proof. It is clear that (i) $\Longrightarrow$ (ii) and (i) $\Longrightarrow$ (iii), and that (iv) implies (i) is clear from Theorem 1.31. It follows from Proposition 12.17 that (iii) $\Longrightarrow$ (ii) and, from Theorem 1.31, that (iv) implies (i) and hence also (ii) with $\lambda' = \lambda$. Further, the equivalence of (ii) and (v) with $\lambda = \lambda''$ follows from Theorem 10.26(a) and Examples 9.16(c) and 9.21. To complete the proof, it therefore remains only to show that (ii) $\Longrightarrow$ (iv). So let L be a Lévy process such that L_{t_0} is Poisson distributed with parameter $\lambda \in (0,\infty)$ for some $t_0 > 0$.

Let $(S'_n)_{n\in\mathbb{N}}$ be an i. i. d. sequence of exponentially distributed random variables with parameter λ, possibly on a different probability space, and define the process $L' = (L'_t)_{t\geq 0}$ by $L'_t = \sup\{n \in \mathbb{N}_0 : \sum_{k=1}^n S'_k \leq t\}$. By the implication (iv) $\Longrightarrow$ (ii) already proved, L' is a Lévy process and L'_{t_0} is Poisson distributed with parameter λt_0. By modifying L and L' on null sets if necessary, we can assume that both are strictly càdlàg. Since $\mathcal{L}(L'_{t_0}) = \mathcal{L}(L_{t_0})$, it follows from Theorem 10.26(b) that L' and L have the same finite-dimensional distributions. Since the paths of L' are almost surely step functions with jumps of size 1, so are the paths of L by Corollary 12.12(d). Denote by $T_{k,L}$ the time of the kth jump of L and by $T_{k,L'}$ the time of the kth jump of L', and define $T_{0,L} := 0$ and $T_{0,L'} := 0$. Then $T_{k,L'} = \sum_{j=1}^k S'_j$ for all $k \in \mathbb{N}_0$, and it follows from Corollary 12.12(b) that $(T_{k,L})_{k\in\mathbb{N}}$ and $(T'_{k,L})_{k\in\mathbb{N}}$ have the same finite dimensional distributions. Hence both $(S'_k = T'_k - T'_{k-1})_{k\in\mathbb{N}}$ and $(S_k := T_{k,L} - T_{k-1,L})_{k\in\mathbb{N}}$ are i. i. d. exponentially distributed with parameter λ. However, since the paths of L are almost surely step functions with jump size 1, and the $T_{k,L}$ are the jump times, it follows that L can be represented as

$$L_t = \sup\{n \in \mathbb{N}_0 : T_{n,L} \leq t\} = \sup\left\{n \in \mathbb{N}_0 : \sum_{k=1}^n S_k \leq t\right\}.$$

This establishes (iv) and shows that the waiting times between two consecutive jumps of L are i. i. d. exponentially distributed with parameter λ. □

Remark 12.19. (a) Results similar to those of Theorem 12.18 hold also for two-sided Lévy processes. Thus, $L = (L_t)_{t\in\mathbb{R}}$ is a two-sided Poisson process if and only if it is a Lévy process and L_{t_0} is Poisson distributed with non-zero finite parameter for some $t_0 > 0$, or equivalently if and only if it is a non-zero Lévy process with sample paths which are almost surely non-decreasing càdlàg step functions with jump size 1, or equivalently, if and only if it has a representation as in Theorem 1.31. This follows simply by considering the independent processes $(L_t)_{t\geq 0}$ and $(-L_{(-t)-})_{t\geq 0}$ separately, and applying Theorem 12.18 to both of them.

(b) From Theorem 12.18 and (a) above we conclude that in the definition of a compound Poisson process (Definition 10.8), the requirement that the Poisson process N is constructed as in Theorem 1.31 can be omitted, since we have seen that every Poisson process has such a representation.

An interesting consequence of the previous results is that we can describe the distribution of the waiting times between consecutive jumps of a two-sided Poisson process prior to a specified time and also after it.

Corollary 12.20 (Waiting times for jumps before and after a given time). *Let $L = (L_t)_{t\in\mathbb{R}}$ be a two-sided Poisson process with jump rate $\lambda > 0$. For fixed $t \in \mathbb{R}$ denote by $T_1^t < T_2^t < \cdots$ the jump times of L after time t in increasing order, and by $T_{-1}^t > T_{-2}^t > \cdots$ the jump times of L prior to t, in decreasing order. Also, define $T_0^t := t$. Then $(T_m^t - T_{m-1}^t)_{m\in\mathbb{Z}}$ is an i. i. d. sequence of exponentially distributed random variables with parameter λ.*

Proof. For fixed $t \in \mathbb{R}$, define the process $M^t = (M_s^t)_{s\in\mathbb{R}}$ by $M_s^t := L_{t+s} - L_t$. It then follows from Remark 11.2 that M^t is a Lévy process with the same characteristic triplet as L, hence a Poisson process with jump rate λ. Denote by $0 < \Gamma_1^t < \Gamma_2^t < \cdots$ the jump times of M^t after time 0 and by $0 > \Gamma_{-1}^t > \Gamma_{-2}^t > \cdots$ the jump times of M^t prior to 0. Also, denote $\Gamma_0^t := 0$. By Theorem 12.18 and Remark 12.19(a), $(\Gamma_n^t - \Gamma_{n-1}^t)_{n\in\mathbb{N}}$ and $(\Gamma_n^t - \Gamma_{n-1}^t)_{n\in-\mathbb{N}_0}$ are two i. i. d. sequences that are exponentially distributed with parameter λ. The two sequences are independent, since $(M_s^t)_{s\geq 0}$ and $(M_s^t)_{s\leq 0}$ are independent by Proposition 10.19. The claim now follows from the observation that $T_m^t = \Gamma_m^t + t$ for every $m \in \mathbb{Z}$. □

Observe that in Corollary 12.20 we numbered the jump times before time t differently than in Theorem 1.31 (for $t = 0$), since there we denoted the jump times prior to 0 by $\Gamma_{-2} < \Gamma_{-1} < \Gamma_0 < 0$, in particular the last jump time before time 0 was labeled as Γ_0 and not as Γ_{-1}. This was convenient there since it allowed the definition of N_t as $N_t = \sup\{n \in \mathbb{Z} : \Gamma_n \leq t\}$ without having to distinguish between $t \geq 0$ and $t < 0$, while the notion used in Corollary 12.20 is more adapted for the i. i. d. property of the waiting times prior and after time t. The reader should keep this in mind.

Since (almost surely) no jump occurs at the fixed time t, one may wonder why the time between the last jump before time t and the first jump after time t is not exponentially distributed, but the independent sum of two exponentially distributed random variables with parameter λ. After all, it is the difference of two consecutive jumps. The explanation is that for labelling the jump times we need a reference point, this can be either 0 or a general t. (Suppose that you would only see all paths of a two-sided Poisson process. You would see all jumps, but without reference point, how would you say which is the first jump, etc.?) And when then having chosen the reference point t and using the notation as in Corollary 12.20, we actually make a condition, namely that t lies between T_{-1}^t and T_1^t, while such a condition does not apply to the other waiting times between consecutive jumps. This is the reason why the distribution of $T_1^t - T_{-1}^t$ is different from the distribution of $T_n^t - T_{n-1}^t$ for $n \in \{\ldots, -2, -1, 2, 3, \ldots\}$.

While we defined a Poisson process as a Lévy process that has a Poisson distribution at all strictly positive times and whose sample paths are step functions with jump size 1 (Definition 1.30), and then showed that every Poisson process can be represented as in Theorem 1.31, for the compound Poisson process we chose a different route and immediately defined it in terms of the representation given in Definition 10.8. The following theorem shows that it would also have been possible to take the same route as for the Poisson process, i. e., to define a compound Poisson process as a Lévy process with compound Poisson distribution at some strictly positive time.

Theorem 12.21 (First characterization of compound Poisson processes). *Suppose that $L = (L_t)_{t\geq 0}$ is an $\mathbb{R}^d$-valued stochastic process, that $a > 0$ and that σ is a probability distribution on $\mathbb{R}^d$ with $\sigma(\{0\}) = 0$. Then the following are equivalent.*

(i) *L is a compound Poisson process with parameter a and jump size distribution σ.*

(ii) *L is a Lévy process and there exists some $t_0 > 0$ such that L_{t_0} has a compound Poisson distribution with parameter at_0 and jump size distribution σ.*

(iii) *L is a Lévy process with Gaussian variance* 0, *Lévy measure $a\sigma$ and drift* 0.

Proof. The equivalence of (ii) and (iii) is clear from Theorem 10.26(a) together with Examples 9.16(b) and 9.21. That (i) implies (ii) follows readily from Theorem 10.9. For the converse, suppose that L is a Lévy process such that L_{t_0} is compound Poisson distributed with parameter at_0 and jump size distribution σ, and let L' be a compound Poisson process with parameter a and jump size distribution σ. Then L and L' have the same finite dimensional distributions by Theorem 10.26(b). Hence it follows from Corollary 12.12(c) and (b), that both L' and L have only paths that are càdlàg step functions, and that the sequences $(T_{k,L}, Y_{k,L})_{k\in\mathbb{N}}$ and $(T_{k,L'}, Y_{k,L'})_{k\in\mathbb{N}}$ of jump times and sizes of L and L', respectively, have the same finite dimensional distributions. But then, as in the proof of Theorem 12.18, it follows that the jump sizes of L are i. i. d. with the distribution $\mathcal{L}(Y_{1,L'}) = \sigma$ and are independent of the jump times $(T_{k,L'})_{k\in\mathbb{N}}$, which are the jump times of a Poisson process with parameter a. This gives (i). □

We shall show later in Corollary 12.27 that compound Poisson processes can also be characterized as the Lévy processes whose paths are almost surely càdlàg step functions.

We now turn to the characterization of Brownian motion as a Lévy process with continuous sample paths, making use of the following preliminary remark.

Remark 12.22. If T and U are two independent $\mathbb{R}^d$-valued random variables such that T is absolutely continuous (with respect to Lebesgue measure), then $U \neq T$ a. s. since

$$P(T = U) = \int_{\mathbb{R}^d} \Big(\int_{\mathbb{R}^d} \mathbf{1}_{\{u\}}(t)\, P_T(\mathrm{d}t) \Big) P_U(\mathrm{d}u) = \int_{\mathbb{R}^d} 0\, P_U(\mathrm{d}u) = 0.$$

Theorem 12.23 (Characterization of Brownian motion). *For an $\mathbb{R}^d$-valued stochastic process $L = (L_t)_{t\geq 0}$, the following are equivalent.*

(i) *L is a Brownian motion.*

(ii) *L is a Lévy process with zero Lévy measure.*

(iii) *L is a Lévy process with almost surely continuous sample paths.*

Proof. It is clear that (i) implies (iii).

To show that (iii) implies (ii), we proceed indirectly, denoting the (standard) characteristic triplet of L by (A, ν, γ) and supposing that the Lévy measure ν of L is non-zero. Then we can find a set $B \in \mathcal{B}_d$ that is bounded away from zero with $\nu(B) > 0$. Now let X and Z be two independent Lévy processes (possibly defined on another probability space), where X is a compound Poisson process with Lévy measure $\nu_{|B}$, and Z has characteristic triplet $(A, \nu_{|\mathbb{R}^d\setminus B}, \gamma - \int_{|x|\leq 1} x\mathbf{1}_B(x)\, \nu(\mathrm{d}x))$. Then $X + Z$ has characteristic triplet (A, ν, γ) and hence the same distribution as L. Hence, by Corollary 12.12(a), L has a jump with positive probability if and only if $X+Z$ has, so it suffices to show that $X+Z$ has jumps with positive probability. To see this, observe that, as a compound Poisson process, X has jumps with probability 1, and its jump times $T_{1,X}, T_{2,X}, \dots$ are all gamma-distributed and therefore absolutely continuous. Hence, by Remark 12.22, $P(T_{k,X} = T_{j,Z}^C) = 0$ for all $k, j \in \mathbb{N}$ and for all $C \in \mathcal{B}_d$ bounded away from zero, where $T_{j,Z}^C$ denotes the time of the jth jump of Z with jump size in C. This shows that almost surely X and Z have no simultaneous jumps, so that $P(\Delta(X+Z)_{T_{1,X}} = \Delta X_{T_{1,X}}) = 1$, showing that $X + Z$ and hence L have jumps with probability 1.

To show that (ii) implies (i), let L be a Lévy process with characteristic triplet $(A, 0, \gamma)$ and let L' be a Brownian motion with mean γ and covariance matrix A. Then L and L' are identical in law, so that their jump times have the same distributions by Corollary 12.12. But L' almost surely has no jumps, which implies that L almost surely has no jumps, in other words the sample paths of L are almost surely continuous. □

Remark 12.24. (a) Theorems 12.21 and 12.23 extend word for word, by the same proofs, to two-sided Lévy processes.

(b) It is also possible to show that an additive process has almost surely continuous sample paths if and only if all its increments are normally distributed. Itô [51, Theorem 1.4.1(i)] gives a proof in the one-dimensional case which transfers easily to the multivariate case, but we shall not need the result in this generality.

12.3 The jump structure of Lévy processes

In this section, we investigate the probabilistic structure of the jumps of a given Lévy process L. It turns out that this is very well described in terms of the Lévy measure ν, and in particular it will be shown that $\nu(C)$ for $C \in \mathcal{B}(\mathbb{R}^d \setminus \{0\})$ is the expected number of jumps of L in the interval $[0,1]$ with jump size in C. However, much more is true. The importance of the Lévy measure for describing the jumps can already be seen from Theorem 12.23, according to which L has jumps if and only if the Lévy measure ν of L is non-zero. Throughout this section, we use the following notation to denote the number of jumps.

Notation 12.25. Suppose that $I = [0,\infty)$ or $I = \mathbb{R}$ and let $L = (L_t)_{t\in I}$ be an $\mathbb{R}^d$-valued Lévy process on a probability space $(\Omega, \mathcal{F}, P)$. Choose $\Omega_0 \in \mathcal{F}$ such that $P(\Omega_0) = 1$ and, for all $\omega \in \Omega_0$, $t \mapsto L_t(\omega)$ is càdlàg and $L_0(\omega) = L_{0-}(\omega) = 0$.

Then for each $B \in \mathcal{B}(I \times (\mathbb{R}^d \setminus \{0\}))$ (the Borel-σ-algebra on $I \times (\mathbb{R}^d \setminus \{0\})$) define

$$N(B,\omega) = N^L(B,\omega) := \begin{cases} \#\{s \in I : (s, \Delta L_s) \in B\}, & \omega \in \Omega_0, \\ 0, & \omega \notin \Omega_0, \end{cases}$$

the number of points $(s, \Delta L_s)$ falling in B, where s is a jump time and ΔL_s is the corresponding jump size. Since

$$N(B,\omega) = \sum_{0<s<\infty} \mathbf{1}_B(s, \Delta L_s(\omega)), \tag{12.2}$$

when $I = [0, \infty)$, $N(B) : \Omega \to \mathbb{N}_0 \cup \{\infty\}$ is a random variable by Corollary 12.9. A similar statement is true when $I = \mathbb{R}$.

As a first result, we show that the number of jumps of L on compact intervals is finite almost surely if and only if L has finite Lévy measure, and that otherwise the number is almost surely infinite. This is made precise in the following theorem.

Theorem 12.26 (Number of jumps of a Lévy process). *Suppose that $I = [0, \infty)$ or $I = \mathbb{R}$ and let $L = (L_t)_{t\in I}$ be an $\mathbb{R}^d$-valued Lévy process with Lévy measure ν. Suppose that $J \subset I$ is a non-degenerate bounded interval and that N is defined as in* (12.2).

(a) *If $\nu(\mathbb{R}^d) = +\infty$, then $N(J \times (\mathbb{R}^d \setminus \{0\})) = +\infty$ almost surely, i. e., almost surely L has infinitely many jumps in J.*
(b) *If $\nu(\mathbb{R}^d) < \infty$, then $N(J \times (\mathbb{R}^d \setminus \{0\})) < \infty$ almost surely, i. e., almost surely the number of jumps of L in J is finite.*
(c) *If $I = [0, \infty)$ and $0 < \nu(\mathbb{R}^d) < \infty$, then almost surely, L has countably but infinitely many jumps in $(0, \infty)$, which can be enumerated by the jump times $T_{k,L}$, $k \in \mathbb{N}$, in increasing order. In this case, $T_{1,L}, T_{2,L} - T_{1,L}, T_{3,L} - T_{2,L}, \dots$ are i. i. d. exponentially distributed with parameter $\nu(\mathbb{R}^d)$.*

Proof. Denote the standard characteristic triplet of L by (A, ν, γ).

(a) Suppose that $\nu(\mathbb{R}^d) = \infty$. Defining $C_m := \{x \in \mathbb{R}^d : |x| > 1/m\}$ for $m \in \mathbb{N}$, it follows that $\nu(C_m) < \infty$ but $\nu(C_m) \uparrow \infty$ as $m \to \infty$. For $m \in \mathbb{N}$ such that $\nu(C_m) > 0$, let X^m and Z^m be two independent Lévy processes such that X^m is a compound Poisson process with Lévy measure $\nu_{|C_m}$, and Z^m has characteristic triplet $(A, \nu_{|\mathbb{R}^d\setminus C_m}, \gamma - \int_{|x|\le 1} x \mathbf{1}_{C_m}(x)\,\nu(\mathrm{d}x))$. As in the proof of Theorem 12.23, $X^m + Z^m$ has the same distribution as L and, almost surely, X^m and Z^m have no simultaneous jumps. Hence, almost surely,

$$N^L(J \times (\mathbb{R}^d \setminus \{0\})) = N^{X^m+Z^m}(J \times (\mathbb{R}^d \setminus \{0\})) \ge N^{X^m}(J \times (\mathbb{R}^d \setminus \{0\})).$$

However, by Definition 10.8, the number of jumps of the compound Poisson process X^m in the interval J is Poisson distributed with mean $a_m := \nu(C_m)\lambda^1(J)$, where λ^1 denotes one-dimensional Lebesgue-measure. For given $M \in \mathbb{N}$, we then conclude that

$$P(N^{X^m}(J \times (\mathbb{R}^d \setminus \{0\})) > M) = 1 - \sum_{k=0}^{M} \mathrm{e}^{-a_m} a_m^k/k! \to 1 \quad \text{as } m \to \infty$$

since $a_m \to \nu(\mathbb{R}^d)\lambda^1(J) = \infty$ as $m \to \infty$. This shows that, almost surely, L has infinitely many jumps in the interval J.

(b, c) Suppose that $\nu(\mathbb{R}^d) < \infty$. If $\nu(\mathbb{R}^d) = 0$, then L is a Brownian motion by Theorem 12.23 and Remark 12.24 and the result is clear. Assume therefore that $0 < \nu(\mathbb{R}^d) < \infty$. Let X and Z be independent Lévy processes such that X is a compound Poisson process with Lévy measure ν, Z is a Brownian motion, and $X + Z$ has the same distribution as L. Then Z has no jumps and X has only finitely many jumps on bounded intervals. Hence

L also has only finitely many jumps on bounded intervals by Corollary 12.12(b), showing (b). That the jump times of L have the same finite dimensional distributions as those of $X + Z$ and hence of X when $I = [0, \infty)$ also follows from Corollary 12.12(b), which gives (c) by the definition of compound Poisson processes. □

As a byproduct of the above, we can characterize compound Poisson processes as processes whose sample paths are almost surely step functions.

Corollary 12.27 (2nd characterization of compound Poisson processes). *Suppose that* $I = [0, \infty)$ *or* $I = \mathbb{R}$ *and let* $L = (L_t)_{t \in I}$ *be an* $\mathbb{R}^d$*-valued process. Then the following are equivalent.*

(i) *L is a compound Poisson process.*
(ii) *L is a non-zero Lévy process whose sample paths are almost surely step functions.*
(iii) *L is a Lévy process with Gaussian variance* 0*, non-zero but finite Lévy measure, and drift* 0.

Proof. The equivalence of (i) and (iii) follows from Theorem 12.21 and Remark 12.24 and the fact that (i) implies (ii) is clear by the definition of a compound Poisson process. It remains only to show that (ii) implies (iii).

Assuming that (ii) holds, the sample paths of L are almost surely step functions, L can only have finitely many jumps in finite time and hence $\nu(\mathbb{R}^d) < \infty$ by Theorem 12.26. Also, $\nu(\mathbb{R}^d) > 0$, as otherwise L would be a Brownian motion by Theorem 12.23. Since the only continuous step function is the zero-function, this would imply that L is the zero-process, contradicting (ii). Now let X be a compound Poisson process with Lévy measure ν, and let Z be an independent Brownian motion with the same Gaussian variance A and the same drift γ_L as L. Then $X + Z$ has the same distribution as L. Since $t^{-1/2}L_t$ converges almost surely to 0 as $t \to 0$ by the step function assumption and the fact that $L_0 = 0$ almost surely, $t^{-1/2}X_t + t^{-1/2}Z_t$ must converge in distribution to 0 as $t \to 0$. But $t^{-1/2}X_t$ converges almost surely to 0 since X is a compound Poisson process, and $t^{-1/2}Z_t \overset{d}{=} N(t^{1/2}\gamma_L, A) \overset{d}{\to} N(0, A)$. This implies that $A = 0$. Hence L is equal in distribution to a Poisson process plus a linear drift $\gamma_L t$, and it is clear that the sample paths can only be step functions if also $\gamma_L = 0$. This completes the proof that (ii) implies (iii). □

An important feature of the jump structure of a Lévy process is the independence of the random variables $N(C)$ and $N(C')$ for disjoint sets $C, C' \in \mathcal{B}(I \times (\mathbb{R}^d \setminus \{0\}))$. A key tool for proving this is the following so-called *Fundamental Lemma* of Itô [51, p. 58]. We follow Itô's proof closely. To keep notation simple, in the following we write (X, Y) rather than $(X^T, Y)^T$.

Lemma 12.28 (Fundamental Lemma). *Let* $(X, Y) = (X_t, Y_t)_{t \geq 0}$ *be an* $\mathbb{R}^d \times \mathbb{R}$ *valued additive process and suppose that almost surely the sample paths of Y are step functions with jump size* 1 *and X and Y have no simultaneous jumps, i. e.,*

$$P(|\Delta X_t| \cdot |\Delta Y_t| = 0 \;\forall\, t \geq 0) = 1.$$

Then the two processes X and Y are independent.

Proof. By modifying (X, Y) on a null-set, we can assume without loss of generality that we are working with a strictly càdlàg version. Decomposing the jumps of X and Y according to their respective sizes as in Corollary 12.9, it is easily seen that $\{|\Delta X_t| \cdot |\Delta Y_t| = 0 \;\forall\, t \geq 0)$ is indeed an event, i. e. a set belonging to the underlying σ-algebra.

We first show that $X_t - X_s$ and $Y_t - Y_s$ are independent for given $0 \leq s < t$. For that purpose, define

$$U_{n,k} := X_{s+(t-s)k/n} - X_{s+(t-s)(k-1)/n} \quad \text{and}$$
$$V_{n,k} := Y_{s+(t-s)k/n} - Y_{s+(t-s)(k-1)/n}$$

for $n \in \mathbb{N}$ and $k = 1, \dots, n$, so that

$$X_t - X_s = \sum_{k=1}^{n} U_{n,k} \quad \text{and} \quad Z := Y_t - Y_s = \sum_{k=1}^{n} V_{n,k}.$$

Define also

$$U_n := \sum_{k=1}^{n} \mathbf{1}_{\{0\}}(V_{n,k})\, U_{n,k} = \sum_{k=1,\dots,n:\, V_{n,k}=0} U_{n,k} \quad \text{for } n \in \mathbb{N}.$$

By Proposition 12.17, $Z = Y_t - Y_s$ and all $V_{n,k}$ are Poisson distributed with finite parameter and are consequently integer-valued. Write $v = (v_1, \dots, v_n)^T$ for $v \in \mathbb{N}_0^n$. Then for any $p \in \mathbb{N}_0$ and $z \in \mathbb{R}^d$,

$$\begin{aligned}
&\mathbb{E}(\mathrm{e}^{\mathrm{i}z^T U_n} \mathbf{1}_{\{Z=p\}})\, P(Z = 0) \\
&= \sum_{\substack{v \in \mathbb{N}_0^n, \\ \sum_{j=1}^n v_j = p}} \mathbb{E}\Big(\prod_{\substack{k=1,\dots,n, \\ v_k=0}} \mathrm{e}^{\mathrm{i}z^T U_{n,k}} \cdot \mathbf{1}_{\{V_{n,1}=v_1\}} \cdots \mathbf{1}_{\{V_{n,n}=v_n\}} \Big) \\
&\qquad \times P(V_{n,1} = \cdots = V_{n,n} = 0) \\
&= \sum_{\substack{v \in \mathbb{N}_0^n, \\ \sum_{j=1}^n v_j = p}} \Big(\prod_{\substack{k=1,\dots,n, \\ v_k=0}} \mathbb{E}(\mathrm{e}^{\mathrm{i}z^T U_{n,k}} \mathbf{1}_{\{V_{n,k}=0\}}) \prod_{\substack{k=1,\dots,n, \\ v_k \geq 1}} P(V_{n,k} = v_k) \Big) \\
&\qquad \times \prod_{k=1}^{n} P(V_{n,k} = 0),
\end{aligned}$$

since (X, Y) is an additive process. Since trivially

$$\prod_{\substack{k=1,\dots,n,\\ v_k\geq 1}} P(V_{n,k}=v_k)\prod_{k=1}^{n} P(V_{n,k}=0) = \prod_{\substack{k=1,\dots,n,\\ v_k\geq 1}} P(V_{n,k}=0)\prod_{k=1}^{n} P(V_{n,k}=v_k)$$

for given $v=(v_1,\dots,v_n)^T \in \mathbb{N}_0^n$, $\mathbb{E}(\mathrm{e}^{\mathrm{i}z^T U_n}\mathbf{1}_{\{Z=p\}})\,P(Z=0)$ can be rewritten as

$$\sum_{\substack{v\in\mathbb{N}_0^n,\\ \sum_{j=1}^n v_j=p}}\ \prod_{\substack{k=1,\dots,n,\\ v_k=0}} \mathbb{E}(\mathrm{e}^{\mathrm{i}z^T U_{n,k}}\mathbf{1}_{\{V_{n,k}=0\}}) \prod_{\substack{k=1,\dots,n,\\ v_k\geq 1}} P(V_{n,k}=0)\prod_{k=1}^{n} P(V_{n,k}=v_k). \tag{12.3}$$

Again, using the additivity of (X,Y), we find that

$$\begin{aligned}
&\mathbb{E}(\mathrm{e}^{\mathrm{i}z^T U_n}\mathbf{1}_{\{Z=0\}})\,P(Z=p)\\
&\quad=\prod_{k=1}^{n}\mathbb{E}(\mathrm{e}^{\mathrm{i}z^T U_{n,k}}\mathbf{1}_{\{V_{n,k}=0\}}) \sum_{\substack{v\in\mathbb{N}_0^n,\\ \sum_{j=1}^n v_j=p}}\ \prod_{k=1}^{n} P(V_{n,k}=v_k).
\end{aligned}$$

Subtracting this from (12.3) and observing that $|\mathbb{E}(\mathrm{e}^{\mathrm{i}z^T U_{n,k}}\mathbf{1}_{\{V_{n,k}=0\}})|\leq 1$, we obtain

$$\begin{aligned}
&\left|\mathbb{E}(\mathrm{e}^{\mathrm{i}z^T U_n}\mathbf{1}_{\{Z=p\}})\,P(Z=0)-\mathbb{E}(\mathrm{e}^{\mathrm{i}z^T U_n}\mathbf{1}_{\{Z=0\}})\,P(Z=p)\right|\\
&\quad\leq \sum_{\substack{v\in\mathbb{N}_0^n,\\ \sum_{j=1}^n v_j=p}}\left|\prod_{\substack{k=1,\dots,n,\\ v_k\geq 1}} P(V_{n,k}=0)-\prod_{\substack{k=1,\dots,n,\\ v_k\geq 1}}\mathbb{E}(\mathrm{e}^{\mathrm{i}z^T U_{n,k}}\mathbf{1}_{\{V_{n,k}=0\}})\right|\\
&\qquad\qquad\times\prod_{k=1}^{n} P(V_{n,k}=v_k).
\end{aligned}$$

Using the well-known inequality that $|u_1\cdots u_m - w_1\cdots w_m|\leq\sum_{k=1}^m|u_k-w_k|$ for complex numbers u_k, w_k of modulus at most 1 (see, e. g., Billingsley [16, Section 27, Lemma 1, p. 358]), the latter can be further bounded above as

$$\begin{aligned}
&\sum_{\substack{v\in\mathbb{N}_0^n,\\ \sum_{j=1}^n v_j=p}}\ \sum_{\substack{k=1,\dots,n,\\ v_k\geq 1}}|P(V_{n,k}=0)-\mathbb{E}(\mathrm{e}^{\mathrm{i}z^T U_{n,k}}\mathbf{1}_{\{V_{n,k}=0\}})|\prod_{k=1}^{n} P(V_{n,k}=v_k)\\
&\quad\leq\sum_{\substack{v\in\mathbb{N}_0^n,\\ \sum_{j=1}^n v_j=p}}\ \sum_{\substack{k=1,\dots,n,\\ v_k\geq 1}}\mathbb{E}(|1-\mathrm{e}^{\mathrm{i}z^T U_{n,k}}|)\prod_{k=1}^{n} P(V_{n,k}=v_k)\\
&\quad\leq p\sup_{k=1,\dots,n}\mathbb{E}(|1-\mathrm{e}^{\mathrm{i}z^T U_{n,k}}|)\sum_{\substack{v\in\mathbb{N}_0^n,\\ \sum_{j=1}^n v_j=p}} P(V_{n,1}=v_1,\dots,V_{n,n}=v_n)\\
&\quad\leq p\sup_{k=1,\dots,n}\mathbb{E}(|1-\mathrm{e}^{\mathrm{i}z^T U_{n,k}}|).
\end{aligned}$$

Since X is continuous in probability, it is uniformly continuous in probability on $[s,t]$ by Lemma 10.35. In particular, for any $\varepsilon \in (0,1)$ there is $n(\varepsilon) \in \mathbb{N}$ such that

$$P(|U_{n,k}| \geq \varepsilon) \leq \varepsilon \quad \forall\, n \geq n(\varepsilon), k = 1, \ldots, n.$$

For $n \geq n(\varepsilon)$, this gives

$$\sup_{k=1,\ldots,n} \mathbb{E}(|1 - \mathrm{e}^{\mathrm{i}z^T U_{n,k}}|) \leq 2\varepsilon + \sup_{x \in \mathbb{R}^d : |x| \leq \varepsilon} |1 - \mathrm{e}^{\mathrm{i}z^T x}|.$$

Letting $n \to \infty$ and then $\varepsilon \downarrow 0$, the above bounds imply that

$$\lim_{n\to\infty} |\mathbb{E}(\mathrm{e}^{\mathrm{i}z^T U_n} \mathbf{1}_{\{Z=p\}})\, P(Z=0) - \mathbb{E}(\mathrm{e}^{\mathrm{i}z^T U_n} \mathbf{1}_{\{Z=0\}})\, P(Z=p)| = 0. \tag{12.4}$$

Since the sample paths of Y are step functions, the number of jumps of Y in $[s,t]$ is almost surely finite. Denoting the jump times of Y by $T_{k,Y}$, the assumption of no simultaneous jumps implies that the sample paths of X are continuous at every $T_{k,Y}$ (almost surely), which implies that

$$X_t - X_s - U_n = \sum_{\substack{k=1,\ldots,n,\\ V_{n,k} \neq 0}} (X_{s+(t-s)k/n} - X_{s+(t-s)(k-1)/n}) \to 0 \quad \text{a. s. as } n \to \infty.$$

Hence, for any $q \in \mathbb{N}_0$, $\mathbb{E}(\mathrm{e}^{\mathrm{i}z^T U_n} \mathbf{1}_{\{Z=q\}})$ converges to $\mathbb{E}(\mathrm{e}^{\mathrm{i}z^T (X_t - X_s)} \mathbf{1}_{\{Z=q\}})$ as $n \to \infty$, and inserting this in (12.4) gives

$$\begin{aligned} &\mathbb{E}(\mathrm{e}^{\mathrm{i}z^T(X_t-X_s)} \mathbf{1}_{\{Y_t - Y_s = p\}}) P(Y_t - Y_s = 0) \\ &\quad = \mathbb{E}(\mathrm{e}^{\mathrm{i}z^T(X_t-X_s)} \mathbf{1}_{\{Y_t - Y_s = 0\}}) P(Y_t - Y_s = p). \end{aligned} \tag{12.5}$$

Summing this equation over all $p \in \mathbb{N}_0$ leads to

$$\mathbb{E}(\mathrm{e}^{\mathrm{i}z^T(X_t-X_s)}) P(Y_t - Y_s = 0) = \mathbb{E}(\mathrm{e}^{\mathrm{i}z^T(X_t-X_s)} \mathbf{1}_{\{Y_t - Y_s = 0\}}).$$

Inserting this again into (12.5) gives

$$\begin{aligned} &\mathbb{E}(\mathrm{e}^{\mathrm{i}z^T(X_t-X_s)} \mathbf{1}_{\{Y_t - Y_s = p\}}) P(Y_t - Y_s = 0) \\ &\quad = \mathbb{E}(\mathrm{e}^{\mathrm{i}z^T(X_t-X_s)}) P(Y_t - Y_s = 0) P(Y_t - Y_s = p). \end{aligned}$$

Since $Y_t - Y_s$ is Poisson distributed with finite parameter by Proposition 12.17, we can divide the above equation by $P(Y_t - Y_s = 0) > 0$, yielding

$$\mathbb{E}(\mathrm{e}^{\mathrm{i}z^T(X_t-X_s)} \mathbf{1}_{\{Y_t - Y_s = p\}}) = \mathbb{E}(\mathrm{e}^{\mathrm{i}z^T(X_t-X_s)}) P(Y_t - Y_s = p) \quad \forall\, p \in \mathbb{N}_0.$$

For $w \in \mathbb{R}$, this gives

$$\mathbb{E}(\mathrm{e}^{\mathrm{i}z^T(X_t-X_s)}\mathrm{e}^{\mathrm{i}w(Y_t-Y_s)}) = \sum_{p=0}^{\infty} \mathbb{E}(\mathrm{e}^{\mathrm{i}z^T(X_t-X_s)}\mathrm{e}^{\mathrm{i}wp}\mathbf{1}_{\{Y_t-Y_s=p\}})$$
$$= \mathbb{E}(\mathrm{e}^{\mathrm{i}z^T(X_t-X_s)})\,\mathbb{E}(\mathrm{e}^{\mathrm{i}w(Y_t-Y_s)}),$$

showing the independence of $X_t - X_s$ and $Y_t - Y_s$ for arbitrary $0 \leq s < t$.

Now suppose that $0 = t_0 < t_1 < t_2 < \cdots < t_n$. Since (X, Y) is an additive process, $(X_{t_k} - X_{t_{k-1}}, Y_{t_k} - Y_{t_{k-1}})$, $k = 1, \ldots, n$, and (X_0, Y_0) are independent (observe that $(X_0, Y_0) = (0, 0)$ almost surely), and consequently so are

$$(X_{t_k} - X_{t_{k-1}})_{k=1,\ldots,n},\ X_0,\ (Y_{t_k} - Y_{t_{k-1}})_{k=1,\ldots,n},\ Y_0,$$

by the previously established independence. It follows that the σ-algebra generated by X_0 and $(X_{t_k} - X_{t_{k-1}})_{k=1,\ldots,n}$, $n \in \mathbb{N}$, is independent of the corresponding σ-algebra for Y, showing the independence of X and Y. □

We can now show that processes constructed from jumps of a Lévy process with jump sizes in appropriate disjoint regions give rise to independent processes, and that they are even independent from the original process after subtracting the corresponding jumps. More precisely, we have the following theorem, where again we write $(M^1, \ldots, M^k, R)$ rather than the more cumbersome $((M^1)^T, \ldots, (M^k)^T, R^T)^T$. A similar notation will be used also in the future without further mention.

Theorem 12.29 (Indep. processes constructed from different jumps). *Suppose that $I = [0, \infty)$ or $I = \mathbb{R}$ and $L = (L_t)_{t\in I}$ is an $\mathbb{R}^d$-valued Lévy process.*

(a) *Let $B_1, \ldots, B_k \in \mathcal{B}(I \times \mathbb{R}^d)$ be pairwise disjoint sets which are bounded away from zero in space, i. e., such that there is some $a > 0$ such that $B_j \subset I \times \{u \in \mathbb{R}^d : |u| > a\}$ for all $j = 1, \ldots, k$. Let $f_j : I \times \mathbb{R}^d \to \mathbb{R}^{m_j}$ be Borel-measurable functions for each $j \in \{1, \ldots, k\}$. Consider the $\mathbb{R}^{m_j}$-valued processes $M^j = (M^j_t)_{t\in I}$, defined for $j \in \{1, \ldots, k\}$ by*

$$M^j_t = \begin{cases} \sum_{0<s\leq t,(s,\Delta L_s)\in B_j} f_j(s, \Delta L_s), & t \geq 0, \\ -\sum_{t<s\leq 0,(s,\Delta L_s)\in B_j} f_j(s, \Delta L_s), & t < 0 \text{ (when } I = \mathbb{R}), \end{cases}$$

and the $\mathbb{R}^d$-valued process $R = (R_t)_{t\in I}$ defined by

$$R_t = \begin{cases} L_t - \sum_{0<s\leq t,(s,\Delta L_s)\in\bigcup_{j=1}^k B_j} \Delta L_s, & t \geq 0, \\ L_t + \sum_{t<s\leq 0,(s,\Delta L_s)\in\bigcup_{j=1}^k B_j} \Delta L_s, & t < 0 \text{ (when } I = \mathbb{R}), \end{cases}$$

i. e., the process which arises from L by subtracting the jumps that fall into $B_1 \cup \cdots \cup B_k$. Then $(M^1, \ldots, M^k, R)$ is an additive process, and the component processes $M^1, \ldots, M^k$ and R are independent.

(b) *Let $C_1,\dots,C_k \in \mathcal{B}_d$ be pairwise disjoint and bounded away from zero and $f_j : \mathbb{R}^d \to \mathbb{R}^{m_j}$ be Borel-measurable functions, $j = 1,\dots,k$. Define the processes $M^j = (M_t^j)_{t\in I}$ and $R = (R_t)_{t\in I}$ by*

$$M_t^j = \sum_{\substack{0<s\le t\\ \Delta L_s\in C_j}} f_j(\Delta L_s) \quad \text{and} \quad R_t = L_t - \sum_{\substack{0<s\le t\\ \Delta L_s\in\bigcup_{j=1}^k C_k}} \Delta L_s$$

for $t \ge 0$, and adjusted as in (a) for $t < 0$ when $I = \mathbb{R}$. Then $(M^1,\dots,M^k,R)$ is a Lévy process, and the component processes $M^1,\dots,M^k,R$ are independent.

Since the space component of the sets under consideration is bounded away from zero, and since the time horizon is bounded for each fixed t, it follows from the càdlàg property that the sums defining M^j and R above are over finite sets almost surely and hence converge absolutely almost surely. In Corollary 12.39 below, we will relax the condition in (b) that the sets are bounded away from zero by replacing it by a more general integration condition. Standard choices for the function f_j in (b) are $f_j(x) = x$ and $f_j = \mathbf{1}_{\mathbb{R}^d\setminus\{0\}}$ leading, for $t \ge 0$, to $M_t^j = \sum_{0<s\le t,\Delta L_s\in C_j} \Delta L_s$ and $M_t^j = \#\{s \in (0,t] : \Delta L_s \in C_j\} = N^L((0,t]\times C_j)$, respectively.

Proof. (a) Consider first the one-sided case when $I = [0,\infty)$. Since the assertions are not affected if L is changed on a null-set, we can assume without loss of generality that $L_0 = 0$ everywhere and that L is strictly càdlàg.

Step 1: First, we shall show that $(M^1,\dots,M^k,R)$ is an additive process. The process is clearly well defined since the number of jumps in each B_j is finite. It is also clear that $M_0^j = 0$, $R_0 = 0$ and the paths of $(M^1,\dots,M^k,R)$ are càdlàg. Define $g : [0,\infty)\times\mathbb{R}^d \to \mathbb{R}^d$ by $g(s,x) := x$ and the process $S = (S_t)_{t\ge0}$ by

$$S_t := \sum_{\substack{0<s\le t,\\ (s,\Delta L_s)\in\bigcup_{j=1}^k B_j}} \Delta L_s = \sum_{\substack{0<s\le t,\\ (s,\Delta L_s)\in\bigcup_{j=1}^k B_j}} g(s,\Delta L_s), \quad t \ge 0,$$

so that $R_t = L_t - S_t$. Define $\overline{f} : [0,\infty)\times\mathbb{R}^d \to \mathbb{R}^{m_1}\times\cdots\times\mathbb{R}^{m_k}\times\mathbb{R}^d$ by

$$\overline{f} := (f_1\mathbf{1}_{B_1},\dots,f_k\mathbf{1}_{B_k},g\mathbf{1}_{B_1\cup\cdots\cup B_k}),$$

so that

$$(M_t^1,\dots,M_t^k,S_t) = \sum_{0<s\le t} \overline{f}(s,\Delta L_s), \quad t \ge 0.$$

Then $(M^1,\dots,M^k,S)$ is an additive process by Proposition 12.15 and is therefore continuous in probability. Since $R = L - S$, $(M^1,\dots,M^k,R = L - S)$ is also continuous in probability. To see that it has independent increments, define $\mathcal{G}_{u,v} := \sigma(L_s - L_u : s \in (u,v])$ for

$0 \le u < v$. It then follows from Corollary 12.9 that $(M_v^1 - M_u^1, \dots, M_v^k - M_u^k, S_v - S_u)$, and hence $(M_v^1 - M_u^1, \dots, M_v^k - M_u^k, R_v - R_u)$, is $\mathcal{G}_{u,v}$ measurable. Since $\mathcal{G}_{0,t_1}, \mathcal{G}_{t_1,t_2}, \dots, \mathcal{G}_{t_{n-1},t_n}$ are independent for $0 < t_1 < \dots < t_n$ by Remark 10.21, it follows that $(M^1, \dots, M^k, R)$ has independent increments and is hence an additive process. As a byproduct, since $L_0 = 0$ everywhere, we have also shown that $(M^1, \dots, M^k, S)$ and $(M^1, \dots, M^k, R)$ are adapted to the natural filtration of L, i. e., that $(M_t^1, \dots, M_t^k, S_t)$ and $(M_t^1, \dots, M_t^k, R_t)$ are $\sigma(L_u : u \in [0, t])$-measurable for each $t \ge 0$.

Step 2: Next, we show that $M^1, \dots, M^k, R$ are independent in the special case when $f_j = \mathbf{1}_{I \times (\mathbb{R}^d \setminus \{0\})}$ for $j = 1, \dots, k$. In that case,

$$M_t^j = \#\{s \in (0, t] : (s, \Delta L_s) \in B_j\} = N^L(((0, t] \times \mathbb{R}^d) \cap B_j).$$

Obviously, the sample paths of M^j are step functions with jump size 1. We shall establish the claim by induction on k.

When $k = 1$, we have $R = L - S$. Obviously, M^1 and R have no simultaneous jumps and, since (M^1, R) is an additive process, it follows from Lemma 12.28 that M^1 and R are independent.

For the induction step, suppose that $k \ge 2$, define

$$R_t^{k-1} := L_t - \sum_{0 < s \le t, (s, \Delta L_s) \in B_1 \cup \dots B_{k-1}} \Delta L_s, \quad t \ge 0,$$

and assume that $M^1, \dots, M^{k-1}$ and $R^{k-1} = (R_t^{k-1})_{t \ge 0}$ are independent. Since $B_1, \dots, B_k$ are pairwise disjoint, we have $(s, \Delta L_s) \in B_k$ if and only if $(s, \Delta R_s^{k-1}) \in B_k$, in which case $(s, \Delta L_s) = (s, \Delta R_s^{k-1})$. It follows that M^k and $R = R^k$ can also be written as

$$\begin{aligned} M_t^k &= \sum_{0 < s \le t, (s, \Delta R_s^{k-1}) \in B_k} 1 \quad \text{and} \\ R_t &= R_t^{k-1} - \sum_{0 < s \le t, (s, \Delta R_s^{k-1}) \in B_k} \Delta R_s^{k-1}, \quad t \ge 0. \end{aligned}$$

By the observation made in Step 1, M^k and R are hence measurable with respect to the natural filtration of R^{k-1}. Since $M^1, \dots, M^{k-1}$ and R^{k-1} are independent, this shows independence of $M^1, \dots, M^{k-1}$ and (M^k, R). But as M^k and R have no simultaneous jumps, they are themselves independent, again by Lemma 12.28. Hence we obtain the independence of $M^1, \dots, M^k, R$, finishing the induction step.

Step 3: Now we show the independence of $M^1, \dots, M^k, R$ for general Borel-measurable functions f_j. Replacing f_j by $f_j \mathbf{1}_{B_j}$ obviously leads to the same processes, so we can assume without loss of generality that $f_j(s, x) = 0$ for $(s, x) \notin B_j$. For each $j \in \{1, \dots, k\}$, we can find a sequence of finite partitions of B_j into Borel sets,

$$B_j = B_{j,n,1} \cup \dots \cup B_{j,n,p(n)}, \quad n \in \mathbb{N}$$

(with pairwise disjoint sets for each fixed n), and constants $c_{j,n,l} \in \mathbb{R}^{m_j}$, $n \in \mathbb{N}$, $l = 1, \dots, p(n)$, such that

$$f_{j,n} := \sum_{l=1}^{p(n)} c_{j,n,l} \mathbf{1}_{B_j,n,l}$$

converges pointwise to f_j as $n \to \infty$; this follows since f_j is Borel-measurable, as can be seen by considering the components and their positive and negative parts separately. Defining

$$M_t^{j,n,l} := \#\{s \in (0,t] : (s, \Delta L_s) \in B_{j,n,l}\}, \quad t \geq 0, n \in \mathbb{N}, l \in \{1, \dots, p(n)\},$$

it follows that for fixed $n \in \mathbb{N}$ the processes R and $M^{j,n,l}$ with $j \in \{1, \dots, k\}$ and $l \in \{1, \dots, p(n)\}$ are all independent by Step 2. Defining

$$M_t^{j,n} := \sum_{0<s\leq t, (s,\Delta L_s) \in B_j} f_{j,n}(s, \Delta L_s) = \sum_{l=1}^{p(n)} c_{j,n,t} M_t^{j,n,l}, \quad t \geq 0,$$

we conclude that the processes $M^{1,n}, \dots, M^{k,n}, R$ are also independent for each fixed $n \in \mathbb{N}$. However, $M_t^{j,n}$ converges pointwise to M_t^j as $n \to \infty$ (observe that $(s, \Delta L_s) \in B_j$ only for finitely many $s \leq t$). Since probability limits of independent random vectors remain independent (Exercise 11.1), it follows that also $M^1, \dots, M^k, R$ are independent. This establishes the claim when $I = [0, \infty)$.

Step 4: Now suppose that $I = \mathbb{R}$. Again, since the assertions are not affected if L is changed on a null-set, we can assume without loss of generality that $L_{0-} = L_0 = 0$ everywhere and that L is strictly càdlàg. Decompose B_j into $B_j = (((0,\infty)\times\mathbb{R}^d)\cap B_j)\cup(((-\infty,0]\times \mathbb{R}^d) \cap B_j) =: B_j^+ \cup B_j^-$, and define $M_t^{j,+} := \sum_{0<s\leq t,(s,\Delta L_s)\in B_j^+} f_j(s, \Delta L_s)$ for $t \geq 0$, and R_t^+ similarly. By Step 3, $M^{1,+}, \dots, M^{k,+}, R^+$ are independent and, by Step 1, are measurable with respect to $\sigma(L_s : s \geq 0)$. Applying a similar argument to the negative time axis, we get independent processes $M^{1,-}, \dots, M^{k,-}, R^-$, which are $\sigma(L_s : s \leq 0)$-measurable. Using the independence of $(L_s)_{s\geq 0}$ and $(L_s)_{s\leq 0}$, we then see from Steps 1 and 3 that $(M^1, \dots, M^k, R)$ is an additive process with independent components.

(b) By an argument similar to that in Step 4 of (a), we may restrict attention to the one-sided case. Defining $B_k := I \times C_k$ and $S_t := \sum_{0<s\leq t, \Delta L_s \in C_1\cup\cdots\cup C_k} \Delta L_s$ the processes $M^1, \dots, M^k, S, R$ reduce to the corresponding processes appearing in (a) and its proof. Hence $(M^1, \dots, M^k, S)$ and $(M^1, \dots, M^k, R)$ are additive processes and we only have to show that they have stationary increments. However, this follows as in the proof of Theorem 12.13 and Corollary 12.12, on observing that for each $h \geq 0$, $(M_t^1, \dots, M_t^k, S_t, L_t)$ and $(M_{t+h}^1 - M_h^1, \dots, M_{t+h}^k - M_h^k, S_{t+h} - S_h, L_{t+h} - L_h)$ can be written as the same deterministic functional applied to the processes L and $(L_{t+h} - L_h)_{t\geq 0}$, and that L and

$(L_{t+h} - L_h)_{t\geq 0}$ have the same distribution by the Markov property in Theorem 11.1. It follows that $(M^1, \dots, M^k, S, L)$ and hence $(M^1, \dots, M^k, R = L - S)$ is a Lévy process. □

Remark 12.30. It is worth noting that if $I = [0, \infty)$, $L_0(\omega) = 0$ for all ω and L is strictly càdlàg, then the process $(M^1, \dots, M^k, R)$ defined in (a) or (b) of Theorem 12.29 is adapted to the natural filtration of L, as shown in Step 1 of the proof.

The next result is of fundamental importance in the analysis of the jump structure of Lévy processes and shows in particular that the Lévy measure specifies the expected number of jumps of a Lévy process.

Theorem 12.31 (Lévy measure as the expected number of jumps). *Suppose that $I = [0, \infty)$ or $I = \mathbb{R}$ and that $L = (L_t)_{t\in I}$ is an $\mathbb{R}^d$-valued Lévy process with Lévy measure ν. Define $N = N^L$ as in Notation 12.25 and denote by λ^1 Lebesgue measure on $(\mathbb{R}, \mathcal{B}_1)$. Then*

$$\mathbb{E}(N(B)) = (\lambda^1 \otimes \nu)(B) \quad \forall\, B \in \mathcal{B}(I \times (\mathbb{R}^d \setminus \{0\})). \tag{12.6}$$

In particular (for a modification that is strictly càdlàg)

$$\nu(C) = \frac{\mathbb{E}(\#\{s \in (u,t] : \Delta L_s \in C\})}{t-u} \quad \forall\, u, t \in I \text{ with } u < t, C \in \mathcal{B}(\mathbb{R}^d \setminus \{0\}), \tag{12.7}$$

i. e., $(t-u)\nu(C)$ is the expected number of jumps with jump size in C that occur in the interval $(u, t]$, so that

$$\nu(C) = \mathbb{E}(\#\{s \in (0,1] : \Delta L_s \in C\}) \quad \forall\, C \in \mathcal{B}(\mathbb{R}^d \setminus \{0\}). \tag{12.8}$$

Proof. Choose a strictly cadlàg version with $L_0 = L_{0-} = 0$ everywhere, and define N for $B \in \mathcal{B}(I \times (\mathbb{R}^d \setminus \{0\}))$ as in Notation 12.25. Observe that (12.7) and hence (12.8) are immediate consequences of (12.6) with $B = (u,t] \times C$ for $C \in \mathcal{B}(\mathbb{R}^d \setminus \{0\})$.

To show (12.6), first suppose that $u < t$, $u, t \in I$ and $C \in \mathcal{B}_d$ with $0 \notin \overline{C}$. Define

$$X_t := \sum_{0<s\leq t, \Delta L_s \in C} \Delta L_s, \quad t \geq 0$$

(with a corresponding two-sided version if $I = \mathbb{R}$) and $Y_t := L_t - X_t$. Then $X = (X_t)_{t\in I}$ and $Y = (Y_t)_{t\in I}$ are independent Lévy processes by Theorem 12.29(b). Denote their Lévy measures by ν_X and ν_Y, respectively. Then $\nu_X + \nu_Y = \nu$ by Example 9.15. Since the sample paths of X are obviously almost surely step functions, it follows from Corollary 12.27 that X is either a compound Poisson process, in which case $0 < \nu_X(\mathbb{R}^d) < \infty$, or else it is the zero process, in which case $\nu_X(\mathbb{R}^d) = 0$. In both cases, the number of jumps of X in the interval $(u,t]$ is Poisson distributed with parameter $(t-u)\nu_X(\mathbb{R}^d) < \infty$, and the same is true for the closed interval $[u,t]$. Since X jumps when and only when L has a jump with jump size in C, in which case the jump sizes are the same, we find that

$$\mathbb{E}(N^L([u,t] \times C)) = \mathbb{E}(N^X([u,t] \times C)) = (t-u)\nu_X(\mathbb{R}^d).$$

If X is the zero process, then $\nu_X(\mathbb{R}^d) = 0 \leq \nu(C)$. If X is a compound Poisson process, then all of its jump sizes lie in C. Denoting the jump size distribution of X by σ, it follows that $\sigma(C) = 1$ and $\sigma(\mathbb{R}^d \setminus C) = 0$. However, since $\nu_X = \nu_X(\mathbb{R}^d)\sigma$, by Theorem 12.21, we conclude

$$\mathbb{E}(N^L([u,t] \times C)) = (t-u)\nu_X(\mathbb{R}^d) = (t-u)\nu_X(C) \leq (t-u)\nu(C),$$

the latter inequality holding since $\nu_X + \nu_Y = \nu$. To show the converse inequality, we use the same trick that was used to show the implication (iii) $\Longrightarrow$ (ii) in the proof of Theorem 12.23. Denote the characteristic triplet of L by (A, ν, γ), let U be a compound Poisson process with Lévy measure $\nu_{|C}$ (or the zero process if $\nu(C) = 0$), and let V be a Lévy process with characteristic triplet $(A, \nu_{|\mathbb{R}\setminus C}, \gamma - \int_{\{|x|\leq 1\}} x\mathbf{1}_C(x)\,\nu(dx))$, independent of U. Then $U + V$ has the same characteristic triplet and hence the same distribution as L. By Remark 12.22, U and V almost surely have no simultaneous jumps. In particular, with probability 1, a jump of U is not cancelled or altered in size by a jump of V so that, by Corollary 12.12(a),

$$\begin{aligned}\mathbb{E}(N^L([u,t] \times C)) &= \mathbb{E}(N^{U+V}([u,t] \times C)) \\ &\geq \mathbb{E}(N^U([u,t] \times C)) = (t-u)\nu_U(\mathbb{R}^d) = (t-u)\nu(C),\end{aligned}$$

where we again used the fact that the number of jumps of the compound Poisson process U during the interval $[u,t]$ is Poisson distributed with parameter $(u-t)\nu_U(\mathbb{R}^d)$. Together with the reverse inequality already established, this implies that

$$\mathbb{E}(N^L([u,t] \times C)) = (t-u)\nu(C) = (\lambda^1 \otimes \nu)([u,t] \times C)$$

provided $C \in \mathcal{B}_d$ with $0 \notin \overline{C}$.

To see that (12.6) holds for general $B \in \mathcal{B}(I \times (\mathbb{R}^d \setminus \{0\}))$, observe first that $\mathbb{E}(N(\emptyset)) = 0$ and that for any sequence $(B_n)_{n\in\mathbb{N}}$ of pairwise disjoint sets in $\mathcal{B}(I \times (\mathbb{R}^d \setminus \{0\}))$, we obviously have $N^L(\bigcup_{n=1}^\infty B_n) = \sum_{n=1}^\infty N^L(B_n)$, so that by monotone convergence, $\mathbb{E}(N^L(\bigcup_{n=1}^\infty B_n)) = \sum_{n=1}^\infty \mathbb{E}(N^L(B_n))$. The mapping $B \mapsto \mathbb{E}(N^L(B))$ therefore defines a measure on $\mathcal{B}(I \times (\mathbb{R}^d \setminus \{0\}))$. Since $\mathcal{B}(I \times (\mathbb{R}^d \setminus \{0\}))$ is generated by all sets of the form $[u,t] \times C$ with $C \in \mathcal{B}(\mathbb{R}^d \setminus \{0\})$, and since this generator is moreover closed under finite intersections, it follows from the cases already treated that $\mathbb{E}(N^L(\cdot))$ is equal to $\lambda^1 \otimes \nu$, completing the proof of (12.6). □

A typical application of Theorem 12.31 is the following:

Example 12.32. Let $L = (L_t)_{t\in I}$ be an $\mathbb{R}^d$-valued Lévy process with Lévy measure ν_L and suppose that $f : \mathbb{R}^d \to \mathbb{R}^m$ is Borel-measurable with $f(0) = 0$. Assume that $\sum_{u<s\leq v} |f(\Delta L_s)| < \infty$ almost surely for all $u < v$ in I, and define the Lévy process M as in Theorem 12.13. Then $\Delta M_t = f(\Delta L_t)$. Denote the Lévy measure of M by ν_M. Then for $B \in \mathcal{B}_m$ we obtain from equation (12.8)

$$\begin{aligned}\nu_M(B) &= \mathbb{E}(\#\{s \in (0,1] : \Delta M_s \in B \setminus \{0\}\}) \\ &= \mathbb{E}(\#\{s \in (0,1] : f(\Delta L_s) \in B \setminus \{0\}\}) \\ &= \mathbb{E}(\#\{s \in (0,1] : \Delta L_s \in f^{-1}(B \setminus \{0\})) \\ &= \nu_L(f^{-1}(B \setminus \{0\})) = f(\nu_L)(B \setminus \{0\}),\end{aligned}$$

where $f(\nu_L)$ denotes the image measure of ν_L under the mapping f. It follows that the Lévy measure of M is given by

$$\nu_M(\mathrm{d}y) = \mathbf{1}_{\mathbb{R}^m \setminus \{0\}}(y) f(\nu_L)(\mathrm{d}y), \quad y \in \mathbb{R}^m.$$

In Corollary 12.37, we shall rederive this formula using a slightly different approach, which also yields the characteristic triplet of M and a characterization of the conditions under which the above sums converge absolutely with probability 1.

Another application of Theorem 12.31 is in the proof that two independent Lévy processes almost surely have no simultaneous jumps. We have already shown this when one of them is a compound Poisson process (cf., the proof of Theorem 12.31 using Remark 12.22), but the following corollary gives the result with no such restriction.

Corollary 12.33 (Jumps of independent Lévy processes). *Let $L = (L_t)_{t \in I}$ and $X = (X_t)_{t \in I}$ be two Lévy processes in $\mathbb{R}^d$ and $\mathbb{R}^m$, respectively, and suppose that they are independent. Then almost surely L and X do not jump simultaneously, i. e., the probability that there exists a time $t \in I$ such that $|\Delta L_t| > 0$ and $|\Delta X_t| > 0$ is zero.*

Proof. Since L and X are independent, so are L_1 and X_1. By Theorem 9.28 (more precisely by Exercise 9.3), the Lévy measure ν of (L, X) is supported on $B := (\mathbb{R}^d \times \{0\}) \cup (\{0\} \times \mathbb{R}^m)$. Hence $\nu(\mathbb{R}^{d+m} \setminus B) = 0$ and consequently the expected number of jumps of (L, X) that fall into $\mathbb{R}^{d+m} \setminus B$ in the time intervals $[u, v]$ and therefore in all of I is 0. Hence, almost surely, no jumps of jump size in $\mathbb{R}^{d+m} \setminus B$ occur, equivalently, almost surely L and X have no simultaneous jumps. □

Remark 12.34. Theorem 12.31 allows us to add a seventh characterization of the independence of $X^1, \ldots, X^d$ in Theorem 10.31, since the condition that ν is supported on the axes in (vi) can be replaced by the condition that X^i and X^j almost surely have no simultaneous jumps for all $i \neq j$. In the context of Lévy processes, this gives a generalization of Lemma 12.28.

The following result summarizes some further important properties of the number of jumps falling into certain sets, showing that N^L is a Poisson random measure in the sense of Definition 12.40 below (see, Corollary 12.41). Recall Definition 12.16 of the Poisson distribution with parameter $\alpha \in [0, \infty]$.

Theorem 12.35 (Poisson random measure property of N). *Let I, L, $N = N^L$, ν and λ^1 be as in Theorem 12.31. Then the following statements are true.*

(a) *$N^L(B)$ is Poisson distributed with parameter $(\lambda^1 \otimes \nu)(B)$ for $B \in \mathcal{B}(I \times (\mathbb{R}^d \setminus \{0\}))$.*
(b) *If $B_1, \dots, B_k \in \mathcal{B}(I \times (\mathbb{R}^d \setminus \{0\}))$ are pairwise disjoint, then $N^L(B_1), \dots, N^L(B_k)$ are independent.*
(c) *For every $\omega \in \Omega$ (the underlying sample space of L), $B \mapsto N^L(B, \omega)$ is an $\mathbb{N}_0 \cup \{\infty\}$-valued measure on $(I \times \mathbb{R}^d \setminus \{0\}, \mathcal{B}(I \times (\mathbb{R}^d \setminus \{0\})))$.*

Proof. Condition (c) is obvious, since N^L counts the number of jumps of L that occur in B, so that $N^L(\cdot, \omega)$ is actually a counting measure for fixed $\omega \in \Omega$. Let us prove (b). So let $B_1, \dots, B_k \in \mathcal{B}(I \times \mathbb{R}^d \setminus \{0\})$ be pairwise disjoint. Define

$$B_{j,n} := B_j \cap ((-n, n] \times \{x \in \mathbb{R}^d : |x| \geq 1/n\}), \quad n \in \mathbb{N},\ j \in \{1, \dots, k\}.$$

Choosing $f_j = \mathbf{1}_{I \times (\mathbb{R}^d \setminus \{0\})}$ and $|t| > n$, it follows from Theorem 12.29 that $N^L(B_{1,n}), \dots, N^L(B_{k,n})$ are independent for each fixed $n \in \mathbb{N}$. Since $(B_{j,n})_{n \in \mathbb{N}}$ is an increasing sequence with limit B_j and since $N^L(\omega, \cdot)$ is a measure, it follows that $N^L(B_{j,n})$ converges to $N^L(B_j)$ almost surely (even everywhere) as $n \to \infty$. Since independence is preserved under convergence in probability (Exercise 11.1), we conclude that $N^L(B_1), \dots, N^L(B_k)$ are also independent.

For the proof of (a), suppose that $B \in \mathcal{B}(I \times (\mathbb{R}^d \setminus \{0\}))$. We already know that $N^L(B)$ is $\mathbb{N}_0 \cup \{\infty\}$-valued and that it has expectation $(\lambda^1 \otimes \nu)(B)$ by Theorem 12.31, so that we only have to show that it is Poisson distributed. Using an approach similar to that in the proof of (b), define

$$B_n := B \cap ((-n, n] \times \{x \in \mathbb{R}^d : |x| \geq 1/n\}), \quad n \in \mathbb{N},$$

and the process $M^n = (M_t^n)_{t \in I}$ by

$$M_t^n := \sum_{0 < s \leq t, (s, \Delta L_s) \in B_n} 1 = N^L(B_n \cap ((0, t] \times (\mathbb{R}^d \setminus \{0\})), \quad t \geq 0,$$

with similar definitions, when $t < 0$ (if $I = \mathbb{R}$). By Theorem 12.29(a), M^n is an additive process. Since its sample paths are obviously non-decreasing step functions with jump size 1, it follows from Proposition 12.17 that M_t^n is Poisson distributed with finite parameter for every $t \geq 0$. A similar argument applies if $t < 0$. Since $N^L(B_n) = M_t^n - M_{-t}^n$ for $t > n$ with M_t^n and M_{-t}^n being independent, we see that $N^L(B_n)$ is Poisson distributed, and by Theorem 12.31 it has parameter $\alpha_n := (\lambda^1 \otimes \nu)(B_n)$. Since B_n increases from below to B as $n \to \infty$, it follows that $N(B_n)$ is an increasing sequence that converges to $N(B)$. Hence $\alpha_n := \mathbb{E}(N^L(B_n))$ converges to $\alpha := \mathbb{E}(N^L(B)) \in [0, \infty]$ as $n \to \infty$ by monotone convergence. Since $N(B_n)$ is Poisson distributed with parameter α_n, it is easy to see that the limit $N(B)$ is Poisson distributed with parameter α. □

In Theorem 12.13, we constructed a new Lévy process from a given Lévy process, by setting $M_t = \sum_{0<s\leq t} f(\Delta L_s)$ and, in Proposition 12.15, a new additive process from a given

additive process by setting $M_t = \sum_{0<s\le t} f(s, \Delta L_s)$, the latter construction being more general. We assumed that the corresponding sums converge absolutely almost surely, but did not give conditions under which this occurs, except when L is a compound Poisson process. We are now in a position to characterize, for a general Lévy process, the almost sure absolute convergence of the corresponding sum in terms of the Lévy measure, and to determine the characteristic triplet of the distribution of the sum.

Theorem 12.36 (Summability of jumps of Lévy processes). *Suppose that $I = [0,\infty)$ or $I = \mathbb{R}$, let $L = (L_t)_{t\in I}$ be an $\mathbb{R}^d$-valued Lévy process with Lévy measure ν, and denote 1-dimensional Lebesgue measure by λ^1.*

(a) *Let $g : I \times \mathbb{R}^d \to [0,\infty)$ be Borel-measurable with $g(s,0) = 0$ for all $s \in I$. Then the following are equivalent:*
   (i) *The sum $\sum_{s\in I} g(s, \Delta L_s)$ is finite almost surely.*
   (ii) *The sum $\sum_{s\in I} g(s, \Delta L_s)$ is finite with positive probability.*
   (iii) $\int_I \int_{\mathbb{R}^d} (1 \wedge g(s,y))\, \nu(\mathrm{d}y)\, \mathrm{d}s < \infty$.

(b) *Let $f : I \times \mathbb{R}^d \to \mathbb{R}^m$ be Borel-measurable with $f(s,0) = 0$ for all $s \in I$. Assume that $\sum_{s\in I} f(s, \Delta L_s)$ converges absolutely almost surely, i.e., that $\int_I \int_{\mathbb{R}^d} (1 \wedge |f(s,y)|)\, \nu(\mathrm{d}y)\, \mathrm{d}s < \infty$ by (a). Then the random variable*

$$X := \sum_{s\in I} f(s, \Delta L_s)$$

is infinitely divisible with Gaussian variance 0, Lévy measure $\nu_f(\mathrm{d}x) = f(\lambda^1 \otimes \nu)(\mathrm{d}x)\, \mathbf{1}_{\mathbb{R}^m\setminus\{0\}}(x)$ (the image measure of $\lambda^1 \otimes \nu$ under f restricted to $\mathbb{R}^m\setminus\{0\}$) satisfying $\int_{\mathbb{R}^m} (|x| \wedge 1)\, \nu_f(\mathrm{d}x) < \infty$, and with drift 0. The characteristic function of X is given by

$$\begin{aligned}\phi_X(t) &= \exp\left\{\int_I \int_{\mathbb{R}^d} (\mathrm{e}^{\mathrm{i}\langle f(s,y),t\rangle} - 1)\, \nu(\mathrm{d}y)\, \mathrm{d}s\right\} \\ &= \exp\left\{\int_{\mathbb{R}^m} (\mathrm{e}^{\mathrm{i}\langle x,t\rangle} - 1)\, \nu_f(\mathrm{d}x)\right\} \end{aligned} \tag{12.9}$$

for $t \in \mathbb{R}^m$.

Proof. (a) Let $N = N^L$ be defined as in Notation 12.25 and define

$$\rho := \lambda^1 \otimes \nu \quad \text{and} \quad X := \sum_{s\in I} g(s, \Delta L_s).$$

Let us first derive the Laplace transform of X at $t > 0$, i.e., $\mathrm{Lap}_X(t) := \mathbb{E}\, \mathrm{e}^{-tX}$.

Consider first the case when $g = c\mathbf{1}_B$ with $c \in [0,\infty)$ and $B \in \mathcal{B}(I \times (\mathbb{R} \setminus \{0\}))$. Recall that $N(B)$ is Poisson distributed with parameter $(\lambda^1 \otimes \nu)(B) = \rho(B)$ by Theorem 12.35. Then if $\rho(B) < \infty$, we obtain for $t > 0$

$$\begin{aligned}\mathrm{Lap}_X(t) &= \sum_{k=0}^{\infty} \mathrm{e}^{-ckt} P(N(B)=k) = \sum_{k=0}^{\infty} \mathrm{e}^{-ckt} \frac{(\rho(B))^k}{k!} \mathrm{e}^{-\rho(B)} \\ &= \exp\{\rho(B)(\mathrm{e}^{-ct}-1)\},\end{aligned}$$

and the same holds when $\rho(B) = \infty$ in which case $X = c \cdot \infty$.

Now if $g = \sum_{j=1}^{n} c_j \mathbf{1}_{B_j}$ is an elementary function with $c_j \in [0,\infty)$ and $B_1,\dots,B_n \in \mathcal{B}(I \times (\mathbb{R}^d \setminus \{0\}))$ pairwise disjoint, then by the independence of $c_j N(B_j)$, $j = 1,\dots,n$, where $N(B_j) = \sum_{s\in I} \mathbf{1}_{B_j}(s, \Delta L_s)$ (Theorem 12.35),

$$\begin{aligned}\mathrm{Lap}_X(t) &= \prod_{j=1}^{n} \mathrm{Lap}_{c_j N(B_j)}(t) = \exp\left\{\sum_{j=1}^{n} \rho(B_j)(\mathrm{e}^{-c_j t}-1)\right\} \\ &= \exp\left\{\int_{I\times\mathbb{R}^d} (\mathrm{e}^{-g(s,y)t}-1)\,\rho(\mathrm{d}(s,y))\right\}.\end{aligned}$$

For general measurable $g : I \times \mathbb{R}^d \to [0,\infty)$ with $g(s,0) = 0$ choose a sequence of elementary functions g_n converging from below to g. Then $X_n := \sum_{s\in I} g_n(s,\Delta L_s)$ converges from below to $X = \sum_{s\in I} g(s,\Delta L_s)$ and hence $\mathrm{Lap}_{X_n}(t)$ converges for $t > 0$ to $\mathrm{Lap}_X(t)$ as $n \to \infty$ by dominated convergence. On the other hand, $0 \le 1 - \mathrm{e}^{-g_n(s,y)t}$ converges from below to $1-\mathrm{e}^{-g(s,y)t}$ and hence $\int_{I\times\mathbb{R}^d} (\mathrm{e}^{-g_n(s,y)t}-1)\,\rho(\mathrm{d}(s,y))$ converges to $\int_{I\times\mathbb{R}^d} (\mathrm{e}^{-g(s,y)t}-1)\,\rho(\mathrm{d}(s,y))$ by monotone convergence, so that

$$\mathrm{Lap}_X(t) = \exp\left\{\int_{I\times\mathbb{R}^d} (\mathrm{e}^{-g(s,y)t}-1)\,\rho(\mathrm{d}(s,y))\right\}, \quad t > 0, \tag{12.10}$$

in the general case also.

We next show the equivalence of (ii) and (iii). Since $\mathrm{Lap}_X(1) = \int_{\{X<\infty\}} \mathrm{e}^{-x} P_X(\mathrm{d}x)$, we have $P(X < \infty) > 0$ if and only if $\mathrm{Lap}_X(1) > 0$. Hence, by (12.10), (ii) is equivalent to $\int_{I\times\mathbb{R}^d} (1-\mathrm{e}^{-g(s,y)})\,\rho(\mathrm{d}(s,y)) < \infty$. But since $1-\mathrm{e}^{-1} \le 1-\mathrm{e}^{-g(s,y)} \le 1$ when $g(s,y) > 1$ we have $\int_{\{g(s,y)>1\}} (1-\mathrm{e}^{-g(s,y)})\,\rho(\mathrm{d}(s,y)) < \infty$ if and only if $\int_{\{g(s,y)>1\}} \rho(\mathrm{d}(s,y)) < \infty$. For the integral over $\{0 \le g(s,y) \le 1\}$, observe that there are constants $C_1, C_2 \in (0,\infty)$ such that $C_1 x \le 1-\mathrm{e}^{-x} \le C_2 x$ for all $x \in [0,1]$ since $\lim_{x\to 0} \frac{1-\mathrm{e}^{-x}}{x} = 1$. Hence $C_1 g(s,y) \le 1-\mathrm{e}^{-g(s,y)} \le C_2 g(s,y)$ for $0 \le g(s,y) \le 1$ and we conclude that $\int_{\{g(s,y)\le 1\}} (1-\mathrm{e}^{-g(s,y)})\,\rho(\mathrm{d}(s,y)) < \infty$ if and only if $\int_{\{g(s,y)\le 1\}} g(s,y)\,\rho(\mathrm{d}(s,y)) < \infty$. Summing up, we see that (ii) is equivalent to $\int_{I\times\mathbb{R}^d} (1 \wedge g(s,y))\,\rho(\mathrm{d}(s,y)) < \infty$, which is (iii).

For the equivalence of (i) and (ii), let $(B_n)_{n\in\mathbb{N}}$ be a sequence of pairwise disjoint sets in $\mathcal{B}(I \times (\mathbb{R}^d \setminus \{0\}))$ such that $\bigcup_{n=1}^{\infty} B_n = I \times (\mathbb{R}^d \setminus \{0\})$ and each B_n is contained in $(-n,n] \times \{y \in \mathbb{R}^d : |y| > 1/n\}$. It is obvious that such a sequence exists. Then $N(B_n)$ is finite almost surely and hence $X_n = \sum_{s\in I} \mathbf{1}_{B_n}(s,\Delta L_s) g(s,\Delta L_s) < \infty$ almost surely. Since $X = \sum_{n=1}^{\infty} X_n$ we conclude that the event $\{X < \infty\}$ is measurable with respect to the tail-σ-algebra of $(X_n)_{n\in\mathbb{N}}$ (after modifying X_n so that $X_n < \infty$ everywhere, if necessary).

However, $(X_n)_{n\in\mathbb{N}}$ is an independent sequence by Theorem 12.29(a) and hence the tail-σ-algebra of $(X_n)_{n\in\mathbb{N}}$ is P-trivial by Kolmogorov's 0-1-law, so that $P(\{X < \infty\}) \in \{0,1\}$. This gives the equivalence of (i) and (ii).

(b) Suppose that $\rho = \lambda^1 \otimes \nu$ and $N = N^L$ as in the proof of (a). For $c \in \mathbb{R}^m$ and $B \in \mathcal{B}(I \times (\mathbb{R}^d \setminus \{0\}))$ with $\rho(B) < \infty$, we obtain the characteristic function $\phi_{cN(B)}$ as

$$\phi_{cN(B)}(t) = \sum_{k=0}^{\infty} \mathrm{e}^{\mathrm{i}\langle ck,t\rangle} P(N(B) = k) = \exp\{\rho(B)(\mathrm{e}^{\mathrm{i}\langle c,t\rangle} - 1)\}, \quad t \in \mathbb{R}^m,$$

with a similar calculation as in (a) for the Laplace transform. Also, by arguments similar to those in (a), if f is an elementary function of the form $f = \sum_{j=1}^n c_j \mathbf{1}_{B_j}$ with $c_1, \ldots, c_n \in \mathbb{R}^m$ and pairwise disjoint $B_1, \ldots, B_n \in \mathcal{B}(I \times (\mathbb{R}^d \setminus \{0\}))$ with $\rho(B_j) < \infty$, we see that

$$\phi_{\sum f(s,\Delta L_s)}(t) = \exp\Big\{\int_{I\times\mathbb{R}^d} (\mathrm{e}^{\mathrm{i}\langle f(s,y),t\rangle} - 1)\,\rho(\mathrm{d}(s,y))\Big\}, \quad t \in \mathbb{R}^m. \tag{12.11}$$

Now let $f : I \times \mathbb{R}^d \to \mathbb{R}^m$ be Borel–measurable with $f(s,0) = 0$ and choose a sequence $(f_n)_{n\in\mathbb{N}}$ of elementary functions of the above form such that f_n converges everywhere to f as $n \to \infty$ and such that $|f_n| \le |f|$ for all $n \in \mathbb{N}$. Defining

$$X_n := \sum_{s\in I} f_n(s,\Delta L_s) \quad \text{and} \quad X := \sum_{s\in I} f(s,\Delta L_s),$$

we conclude that X_n converges almost surely to X as $n \to \infty$ by dominated convergence. Hence, ϕ_{X_n} converges pointwise to ϕ_X as $n \to \infty$. On the other hand, there exists a constant C such that

$$\big|\mathrm{e}^{\mathrm{i}\langle f_n(s,y),t\rangle} - 1\big| \le C\big|f_n(s,y)\big|\,|t| \wedge 2 \le C\big|f(s,y)\big|\,|t| \wedge 2$$

(see the proof of Proposition 9.18(a)), and since $\int_{I\times\mathbb{R}^m} C|f(s,y)|\,|t| \wedge 2\,\rho(\mathrm{d}(s,y)) < \infty$ by assumption and part (a), an application of the dominated convergence theorem to (12.11) shows that $\phi_{X_n}(t)$ converges to $\exp\{\int_{I\times\mathbb{R}^d} (\mathrm{e}^{\mathrm{i}\langle f(s,y),t\rangle} - 1)\,\rho(\mathrm{d}(s,y))\}$. This gives the first equality in (12.9). Using the transformation theorem for integration with respect to image measures, we can rewrite $\phi_X(t)$ as

$$\exp\Big(\int_{\mathbb{R}^m} (\mathrm{e}^{\mathrm{i}\langle x,t\rangle} - 1) f(\rho)(\mathrm{d}x)\Big) = \exp\Big(\int_{\mathbb{R}^m\setminus\{0\}} (\mathrm{e}^{\mathrm{i}\langle x,t\rangle} - 1) f(\rho)(\mathrm{d}x)\Big)$$

which is the second equality. The measure ν_f obviously satisfies $\nu_f(\{0\}) = 0$ and

$$\int_{\mathbb{R}^m} (|x| \wedge 1) f(\rho)(\mathrm{d}x) = \int_{I\times\mathbb{R}^d} (\big|f(s,y)\big| \wedge 1)\rho(\mathrm{d}(s,y)) < \infty$$

by (a), so that ν_f is a Lévy measure integrating $1 \wedge |x|$. The fact that X is infinitely divisible with the given characteristic triplet now follows from equation (12.9) together with Proposition 9.18(a). □

When the functions g and f depend mainly on $y \in \mathbb{R}^d$ and on $s \in I$ only through an indicator function of an interval, the previous result simplifies as follows.

Corollary 12.37 (Lévy processes constructed from jumps). *Suppose that $I = [0, \infty)$ or $I = \mathbb{R}$ and let $L = (L_t)_{t\in I}$ be an $\mathbb{R}^d$-valued Lévy process with Lévy measure ν.*

(a) *Let $g : \mathbb{R}^d \to [0, \infty)$ be Borel-measurable with $g(0) = 0$ and let $J \subset I$ be a bounded non-degenerate interval. Then the following are equivalent.*
  (i) *The sum $\sum_{s\in J} g(\Delta L_s)$ is finite almost surely.*
  (ii) *The sum $\sum_{s\in J} g(\Delta L_s)$ is finite with positive probability.*
  (iii) $\int_{\mathbb{R}^d} (1 \wedge g(y))\, \nu(\mathrm{d}y) < \infty$.

In particular, if $p \in (0, \infty)$, $J \subset I$ and J is bounded and non-degenerate, then

$$\sum_{s\in J} |\Delta L_s|^p < \infty \quad a.s. \quad \iff \quad \int_{|y|\le 1} |y|^p\, \nu(\mathrm{d}y) < \infty. \tag{12.12}$$

(b) *Let $f : \mathbb{R}^d \to \mathbb{R}^m$ be Borel-measurable with $f(0) = 0$. If $\sum_{s\in J} f(\Delta L_s)$ converges almost surely absolutely for each bounded interval $J \subset I$, i. e., if $\int_{\mathbb{R}^d} (1 \wedge |f(y)|)\, \nu(\mathrm{d}y) < \infty$ (by (a)), then, after modifying L on a null-set so that the sums are absolutely convergent everywhere, the process $M = (M_t)_{t\in I}$ defined by*

$$M_t := \begin{cases} \sum_{0<s\le t} f(\Delta L_s), & t \ge 0, \\ -\sum_{t<s\le 0} f(\Delta L_s), & t < 0, \end{cases}$$

is an $\mathbb{R}^m$-valued Lévy process (cf. Theorem 12.13). Further, its Gaussian variance is 0, its Lévy measure $\nu_f(\mathrm{d}x) = f(\nu)(\mathrm{d}x)\mathbf{1}_{\mathbb{R}^m\setminus\{0\}}(x)$ satisfies $\int_{\mathbb{R}^m} (|x| \wedge 1)\, \nu_f(\mathrm{d}x) < \infty$, and the drift of M is 0.

Proof. This follows immediately from Theorem 12.36, applied to the functions $\overline{g} : I \times \mathbb{R}^d \to [0, \infty)$, $\overline{g}(s, y) = g(y)\mathbf{1}_J(s)$, and $\overline{f} : I \times \mathbb{R}^d \to \mathbb{R}^m$, $\overline{f}(s, y) = f(y)\mathbf{1}_J(s)$, respectively. Equation (12.12) is then clear. □

Applying this corollary to the function $f(x) = \mathbf{1}_B(x)x$, we immediately obtain the following result.

Corollary 12.38. *Suppose that $I = [0, \infty)$ or $I = \mathbb{R}$ and let $L = (L_t)_{t\in I}$ be an $\mathbb{R}^d$-valued Lévy process with Lévy measure ν. Suppose also that $B \in \mathcal{B}_d$, $0 \notin B$ and $\int_{B\cap\{x\in\mathbb{R}^d : |x|\le 1\}} |x|\, \nu(\mathrm{d}x) < \infty$. Then $\sum_{s\in J, \Delta L_s\in B} |\Delta L_s| < \infty$ almost surely for each bounded interval J, and, after modifying L on a null-set so that the sums are absolutely convergent everywhere, the Lévy process $M = (M_t)_{t\in I}$ defined by*

$$M_t := \begin{cases} \sum_{0<s\le t, \Delta L_s \in B} \Delta L_s, & t \ge 0, \\ -\sum_{t<s\le 0, \Delta L_s\in B} \Delta L_s, & t<0, \end{cases}$$

has Gaussian variance 0, *Lévy measure* $\nu_M(\mathrm{d}x) = \mathbf{1}_B(x)\,\nu(\mathrm{d}x)$, *and drift* 0.

With the aid of Corollary 12.37, it is now easy to give an extension of Theorem 12.29 to pairwise disjoint sets which are not necessarily bounded away from zero, but such that a suitable integration condition is satisfied. We confine ourselves to the extension of part (b) of Theorem 12.29.

Corollary 12.39 (Indep. processes constructed from different jumps). *Let* $I = [0, \infty)$ *or* $I = \mathbb{R}$ *and let* $L = (L_t)_{t\in I}$ *be an* $\mathbb{R}^d$*-valued Lévy process with Lévy measure* ν*. Suppose also that the sets* $C_1, \dots, C_k \in \mathcal{B}_d$ *are pairwise disjoint and that* $f_j : \mathbb{R}^d \to \mathbb{R}^{m_j}$ *are Borel-measurable functions such that* $f_j(0) = 0$ *and* $\int_{C_j} (1 \wedge |f_j(x)|)\,\nu(\mathrm{d}x) < \infty$ *for all* $j = 1, \dots, k$*. Define* M^j *as in Theorem* 12.29(b)*. Then* $(M^1, \dots, M^k)$ *is a Lévy process and* $M^1, \dots, M^k$ *are independent. If additionally* $\int_{\{|x|\le 1\}\cap\bigcup_{j=1}^k C_j} |x|\,\nu(\mathrm{d}x) < \infty$*, define* R *as in Theorem* 12.29(b)*. Then* $(M^1, \dots, M^k, R)$ *is a Lévy process and* $M^1, \dots, M^k, R$ *are independent.*

Proof. By Corollary 12.37, the sums defining M^j converge absolutely almost surely, and by Theorem 12.13, $(M^1, \dots, M^k)$ is a Lévy process. Under the additional assumption that $\int_{\{|x|\le 1\}\cap\bigcup_{j=1}^k C_k} |x|\,\nu(\mathrm{d}x) < \infty$, define $S_t := \sum_{0<s\le t, \Delta L_s \in \bigcup_{j=1}^k C_j} \Delta L_s$ for $t \ge 0$, with a similar definition if $t < 0$. Again by Corollary 12.37, the sum defining S converges absolutely almost surely, and $(M^1, \dots, M^k, S)$ is a Lévy process by Theorem 12.13. By the same arguments as in the proof of Theorem 12.29(a)-Step 1 and (b), we conclude that $(M^1, \dots, M^k, R)$ is also a Lévy process. For $n \in \mathbb{N}$, define $M^{j,n}$ and R^n by the corresponding expressions with C_j replaced by $C_j \cap (\mathbb{R}^d \setminus [-n^{-1}, n^{-1}]^d)$. By Theorem 12.29(b), $M^{1,n}, \dots, M^{k,n}, R^n$ are independent for each fixed n. Since $M_t^{k,n}$ converges almost surely to M_t^k as $n \to \infty$, and R_t^n converges almost surely to R_t provided the sum defining R converges, almost surely absolutely on compact sets, in which case we shall say that R *exists*, it follows that $M_t^1, \dots, M_t^k$ and R_t are also independent by Exercise 11.1. Applying the same argument to the finite dimensional distributions of the M^j and of R, we see that $M^1, \dots, M^k$ (and R, if it exists) are independent, finishing the proof. It is also possible to derive the independence of $M^1, \dots, M^k$ (and R, if it exists) from the fact that, by construction, no two of them have simultaneous jumps, so that their Lévy measures must be concentrated on the corresponding axes by Theorem 12.31. Since the Gaussian variances of $M^1, \dots, M^k$ are zero by Theorem 12.36(b), the Gaussian variance of $(M^1, \dots, M^k, R)$ must be a block diagonal matrix (if R exists), and independence follows from Theorem 10.31(vi) (and Exercise 9.3). See also Remark 12.34. □

We end this section with the definition of a Poisson random measure and some related comments.

Definition 12.40 (Poisson random measure). Let $(\Theta, \mathcal{A}, \rho)$ be a σ-finite measure space. A family of $\mathbb{N}_0 \cup \{+\infty\}$-valued random variables $(M(B) : B \in \mathcal{A})$ on a probability space $(\Omega, \mathcal{F}, P)$ is called a *Poisson random measure on* $(\Theta, \mathcal{A})$ *with mean measure* (or *intensity measure*) ρ, if

(a) for every $B \in \mathcal{A}$, $M(B, \cdot)$ is a Poisson distributed random variable on $(\Omega, \mathcal{F}, P)$ with parameter $\rho(B)$,

(b) whenever $B_1, \dots, B_k \in \mathcal{A}$ are pairwise disjoint, the random variables $M(B_1), \dots, M(B_k)$ are independent, and

(c) for every $\omega \in \Omega$, $M(\cdot, \omega) : \mathcal{A} \to \mathbb{N}_0 \cup \{+\infty\}$, $B \mapsto M(B, \omega)$, is a measure on $(\Theta, \mathcal{A})$.

If the σ-algebra $\mathcal{A}$ is clear, we simply speak of a *Poisson random measure on* Θ.

With this notation, Theorem 12.35 can be rephrased as follows.

Corollary 12.41. *Let I, L, $N = N^L$, ν and λ^1 be as in Theorem* 12.31. *Then $N = N^L$ is a Poisson random measure on $I \times (\mathbb{R}^d \setminus \{0\})$ with mean measure $\lambda^1 \otimes \nu$.*

In most books, Theorem 12.35 would be stated in the form of Corollary 12.41, but we prefer to state it explicitly in order to keep the notation simple.

If M is a Poisson random measure on $(\Omega, \mathcal{F}, P)$ then, by part (c) of Definition 12.40, for each ω we can integrate deterministic functions with respect to the measure $M(\cdot, \omega)$ on $(\Theta, \mathcal{A})$. For the Poisson random measure $N = N^L$ of Corollary 12.41 this leads to the familiar sums of jumps, as shown in the following lemma.

Lemma 12.42 (Integration with respect to N^L). *Suppose that $I = [0, \infty)$ or $I = \mathbb{R}$ and let $L = (L_t)_{t \in I}$ be an $\mathbb{R}^d$-valued Lévy process on I which is strictly càdlàg and satisfies $L_0 = L_{0-} = 0$ everywhere. Let $N = N^L$ be defined as in Notation* 12.25.

(a) *Let $g : I \times (\mathbb{R}^d \setminus \{0\}) \to [0, \infty]$ be Borel-measurable. Then for each ω in the sample space Ω of L,*

$$\int\limits_{I \times (\mathbb{R}^d \setminus \{0\})} g(s,x)\, N^L(\mathrm{d}(s,x), \omega) = \sum_{s \in I, \Delta L_s(\omega) \neq 0} g(s, \Delta L_s(\omega)), \tag{12.13}$$

where the left-hand side denotes integration with respect to the measure $N^L(\cdot, \omega)$ for each fixed $\omega \in \Omega$. Moreover $\int_{I \times (\mathbb{R}^d \setminus \{0\})} g(s,x)\, N^L(\mathrm{d}(s,x), \cdot)$ is a $[0, \infty]$-valued random variable on $(\Omega, \mathcal{F}, P)$.

(b) *Let $f : I \times (\mathbb{R}^d \setminus \{0\}) \to \mathbb{R}^m$ be Borel-measurable, define*

$$\Omega_0 := \left\{ \omega \in \Omega : \sum_{s \in I, \Delta L_s(\omega) \neq 0} |f(s, \Delta L_s(\omega))| < \infty \right\}$$

and assume that $P(\Omega_0) = 1$. Then

$$\mathbf{1}_{\Omega_0}(\omega) \int\limits_{I\times(\mathbb{R}^d\setminus\{0\})} f(s,x)\, N^L(\mathrm{d}(s,x),\omega) = \mathbf{1}_{\Omega_0}(\omega) \sum_{s\in I,\Delta L_s(\omega)\neq 0} f(s,\Delta L_s(\omega)),$$

and both expressions are measurable in ω.

Proof. (a) If $g = \sum_{i=1}^n c_i \mathbf{1}_{B_i}$ is an elementary function with $c_i \geq 0$ and the sets $B_i \in \mathcal{B}(I \times (\mathbb{R}^d \setminus \{0\}))$ are pairwise disjoint, then

$$\begin{aligned}\int\limits_{I\times(\mathbb{R}^d\setminus\{0\})} g(s,x)\, N^L(\mathrm{d}(s,x),\omega) &= \sum_{i=1}^n c_i\, N^L(B_i,\omega) = \sum_{i=1}^n c_i \sum_{(s,\Delta L_s(\omega))\in B_i} 1\\ &= \sum_{s\in I,\Delta L_s(\omega)\neq 0} g(s,\Delta L_s(\omega)),\end{aligned}$$

so that (12.13) holds and $\sum_{i=1}^n c_i\, N^L(B_i)$ is a random variable. For general Borel-measurable $g \geq 0$, we can find a sequence $(g_n)_{n\in\mathbb{N}}$ of elementary functions converging from below to g. Then (12.13) holds for each g_n. Taking the limit of each side as $n \to \infty$ shows that it holds also for g and that the limiting integral is a measurable function of ω.

(b) This is immediate from (a) by considering the components of f separately and decomposing them into positive and negative parts. Observe that $\Omega_0 \in \mathcal{F}$, the underlying σ-algebra on Ω, by (a). □

Remark 12.43. Let M be a Poisson random measure on $(\Theta,\mathcal{A})$ with intensity measure ρ and let $g : \Theta \to [0,\infty)$ and $f : \Theta \to \mathbb{R}^m$ be Borel-measurable. Then $\omega \mapsto \int_\Theta g(y)\, M(\mathrm{d}y,\omega)$ is a random variable and the function defined as $\omega \mapsto \int_\Theta f(y)\, M(\mathrm{d}y,\omega)$ if $\int_\Theta |f(y)|\, M(\mathrm{d}y,\omega) < \infty$ and as $\omega \mapsto 0$ otherwise, is a random vector. This can be shown by an argument completely analogous to the proof of Lemma 12.42. For such random vectors, one usually suppresses the dependence on Ω and simply writes $\int_\Theta f(y)\, M(\mathrm{d}y)$. As in Theorem 12.36, one can then show that $\int_\Theta |f(y)|\, M(\mathrm{d}y) < \infty$ almost surely if and only if $\int_\Theta (1 \wedge |f(y)|)\, \rho(\mathrm{d}y) < \infty$, in which case $\int_\Theta f(y)\, M(\mathrm{d}y)$ has characteristic function $\exp\{\int_\Theta (\mathrm{e}^{\mathrm{i}\langle f(y),t\rangle} - 1)\, \rho(\mathrm{d}y)\}$, cf. Kyprianou [61, Thm. 2.7].

12.4 Exercises

Exercise 12.1. Let $L = (L_t)_{t\in I}$ be a Lévy process in $\mathbb{R}^d$ defined on a probability space $(\Omega,\mathcal{F},P)$ and assume that L is strictly càdlàg and that $L_0 = 0$ everywhere. Consider the following sets

$$\begin{aligned}A_1 &:= \{\omega \in \Omega : L_.(\omega) \text{ is continuous}\},\\ A_2 &:= \{\omega \in \Omega : L_.(\omega) \text{ is piecewise continuous}\},\\ A_3 &:= \{\omega \in \Omega : L_.(\omega) \text{ is a step function}\},\\ A_4 &:= \{\omega \in \Omega : L_.(\omega) \text{ is a step function with jumps of modulus 4}\}.\end{aligned}$$

Show that $P(A_1), P(A_2), P(A_3), P(A_4) \in \{0,1\}$.

Exercise 12.2. Let $L = (L_t)_{t\geq 0}$ be a Lévy process in $\mathbb{R}^d$.
(a) Show that $M_t := \sum_{0<s\leq t} |\Delta L_s|^2 < \infty$ almost surely for every $t \geq 0$.
(b) Show that M is a Lévy process with non-decreasing sample paths and calculate the characteristic triplet of M in terms of the Lévy measure of L.

Exercise 12.3 (See [58]). Let $L = (L_t)_{t\geq 0}$ be a Lévy process in $\mathbb{R}$ and let $\gamma \in \mathbb{R}$ and $b > 0$. Show that $\sum_{0<s\leq t} \log(1 + b|\Delta L_s|^2) < \infty$ almost surely and that $M = (M_t)_{t\geq 0}$, defined by $M_t = \gamma t - \sum_{0<s\leq t} \log(1 + b(\Delta L_s)^2)$ is a Lévy process. Denoting the Lévy measures of L and M by ν_L and ν_M, respectively, show that $\nu_M([0,\infty)) = 0$ and that

$$\nu_M((-\infty, -x]) = \nu_L(\{y \in \mathbb{R} : |y| \geq \sqrt{(e^x - 1)/b}\}).$$

What is the drift of M?

Exercise 12.4 (See [63]). Let $(X_t, Y_t)_{t\geq 0}$ be a bivariate Lévy process. Show that $\sum_{0<s\leq t} |(e^{-\Delta X_s} - 1)\Delta \eta_s| < \infty$ almost surely and that the process $M = (M_t)_{t\geq 0}$ defined by $M_t := Y_t + \sum_{0<s\leq t}(e^{-\Delta X_s} - 1)\Delta Y_s$ is a Lévy process. Determine the characteristic triplet of M in terms of the characteristic triplet of (X, Y).

13 The Lévy–Itô decomposition of Lévy processes and consequences

Abstract: In this chapter, we continue the study of sample path properties of Lévy processes initiated in the last chapter. We derive the fundamental Lévy–Itô path decomposition and treat Lévy processes of finite variation and subordinators. As a consequence, in Corollary 13.16 we show that every two-sided $\mathbb{R}$-valued Lévy process can be written as an independent sum of two Lévy processes, one of which is of finite variation and the other an orthogonal increment process. This will play a key role in Chapter 15 when defining the integral with respect to a Lévy process. Laplace transforms are also introduced, along with a characterization of infinitely divisible distributions concentrated on the positive half-line.

13.1 The Lévy–Itô decomposition of a Lévy process

In this section, we prove the famous Lévy–Itô decomposition, which states that every Lévy process can be written as a compensated sum of its jumps and an independent Brownian motion. However, we first give a simplified version in which the Lévy measure integrates $|x|$ in a neighourhood of zero so that the sum of jumps converges. This version follows easily from the considerations of the last chapter.

Theorem 13.1 (Simplified Lévy–Itô decomp. when $\int_{|x|\le 1}|x|\,\nu(\mathrm{d}x)<\infty$). *Suppose that $I = [0,\infty)$ or $I = \mathbb{R}$ and that $L = (L_t)_{t\in I}$ is an $\mathbb{R}^d$-valued Lévy process with characteristic triplet $(A,\nu,\gamma^0)_0$ with respect to the extended truncation function $c(x) = 0$, $x \in \mathbb{R}^d$. Suppose also that $\int_{|x|\le 1}|x|\,\nu(\mathrm{d}x) < \infty$. Then, almost surely, $\sum_{0<s\le t}\Delta L_s$ and $\sum_{t<s\le 0}\Delta L_s$ (when $t < 0$ and $I = \mathbb{R}$) converge absolutely for each $t \in I$ and, after modifying L on a null-set so that the sums are absolutely convergent everywhere, the process $X = (X_t)_{t\in I}$, defined by*

$$X_t = \begin{cases} \sum_{0<s\le t}\Delta L_s, & t \ge 0,\\ -\sum_{t<s\le 0}\Delta L_s, & t < 0 \text{ (when } I = \mathbb{R}) \end{cases}$$

is a Lévy process in $\mathbb{R}^d$ with characteristic triplet $(0,\nu,0)_0$. Moreover, the process $B = (B_t)_{t\in I}$, defined by

$$B_t := L_t - X_t - \gamma^0 t, \quad t \in I,$$

is a Lévy process with characteristic triplet $(A,0,0)_0$ (i. e., a Brownian motion with covariance matrix A and drift 0), and it is independent of X. Hence L is the sum of the independent processes X, B and the deterministic Lévy process $(t \mapsto \gamma^0 t)$.

https://doi.org/10.1515/9783111325033-013

Proof. By Corollary 12.38, the sum defining X_t converges almost surely for each $t \in I$, and even everywhere after modifying L appropriately on a null-set. By Corollaries 12.38 and 12.39, X and $L - X$ are independent Lévy processes, and the characteristic triplet of X is $(0, \nu, 0)_0$. Since $L_1 = X_1 + (L_1 - X_1)$ and since, by the independence of $L - X$ and X, their characteristic triplets add to that of L, the characteristic triplet of $L - X$ must be $(A, 0, \gamma^0)_0$. By Theorem 12.23, $L - X$ must be a Brownian motion, and subtracting its drift, we get the desired decomposition into independent Lévy processes. □

Theorem 13.1 says that every Lévy process L whose Lévy measure ν satisfies $\int_{|x|\le 1} |x|\,\nu(\mathrm{d}x) < \infty$ can be decomposed into a "jump part" X and a continuous part $L - X = (B_t + \gamma^0 t)_{t\in I}$. But what happens if a Lévy process L is such that its Lévy measure does not integrate $x \mapsto |x|$ in a neighborhood of the origin? Then the sum $\sum_{0<s\le t} \Delta L_s$ will not converge absolutely for $t > 0$ by (12.12). In order to overcome this difficulty, we can consider limits such as $\lim_{\varepsilon\downarrow 0} \sum_{0<s\le t, |\Delta L_s|>\varepsilon} \Delta L_s$. But this limit will also not exist in general, although convergence can be established if the mean is first subtracted (provided it exists). To motivate the argument, suppose again that $I = [0, \infty)$ or $I = \mathbb{R}$ and that L is a general Lévy process $L = (L_t)_{t\in I}$. For $\varepsilon > 0$, define

$$L_t^\varepsilon := \begin{cases} \sum_{0<s\le t, |\Delta L_s|>\varepsilon} \Delta L_s, & t \ge 0, \\ -\sum_{t<s\le 0, |\Delta L_s|>\varepsilon} \Delta L_s, & t < 0 \text{ (when } I = \mathbb{R}). \end{cases} \tag{13.1}$$

Then L^ε is a Lévy process. Decomposing it further into the jumps that are larger than 1 and those that are between ε and 1 in absolute size, we have for $0 < \varepsilon \le 1$ and $t \ge 0$,

$$L_t^\varepsilon = \sum_{0<s\le t, \varepsilon<|\Delta L_s|\le 1} \Delta L_s + \sum_{0<s\le t, |\Delta L_s|>1} \Delta L_s =: L_t^{\varepsilon,b} + L_t^\sharp, \quad \text{say},$$

with a similar decomposition for $t < 0$ when $I = \mathbb{R}$. The sums defining $L^{\varepsilon,b}$ and $L^\sharp$ converge absolutely almost surely and for each fixed $\varepsilon \in (0, 1]$ give rise to independent Lévy processes by Theorem 12.29(b). The process $L^\sharp$ is a compound Poisson process including the "big jumps" and does not change as $\varepsilon \downarrow 0$, but its mean may not exist. The first process $L^{\varepsilon,b}$ is a compound Poisson process with Lévy measure $\nu_{|\{x\in\mathbb{R}^d : \varepsilon<|x|\le 1\}}$ (cf., Corollary 12.38). If $\gamma^{\varepsilon,b}$ denotes its location parameter with respect to the standard truncation function, then $\gamma^{\varepsilon,b} = \int_{\varepsilon<|x|\le 1} x\,\nu(\mathrm{d}x)$ by Example 9.16(b) and hence its expectation is $\mathbb{E}L_t^{\varepsilon,b} = t\int_{\varepsilon<|x|\le 1} x\,\nu(\mathrm{d}x)$ by Theorem 11.15. Now let $(\varepsilon_n)_{n\in\mathbb{N}_0}$ be a sequence of real numbers such that $1 = \varepsilon_0 \ge \varepsilon_1 \ge \cdots > 0$ and such that $\varepsilon_n \downarrow 0$ as $n \to \infty$. Defining

$$M_t^j := \sum_{0<s\le t, \varepsilon_j<|\Delta L_s|\le\varepsilon_{j-1}} \Delta L_s - t \int_{\varepsilon_j<|x|\le\varepsilon_{j-1}} x\,\nu(\mathrm{d}x), \quad t > 0, j \in \mathbb{N}, \tag{13.2}$$

with a similar definition for $t < 0$ (when $I = \mathbb{R}$), we have

$$L_t^{\varepsilon_n,b} - \mathbb{E}(L_t^{\varepsilon_n,b}) = L_t^{\varepsilon_n,b} - t \int\limits_{\varepsilon_n<|x|\le 1} x\,\nu(\mathrm{d}x) = \sum_{j=1}^{n} M_t^j, \quad t \in I. \tag{13.3}$$

The processes M^j are compound Poisson processes with drifts (or possibly the zero process) such that each M^j has mean zero, and they are independent by Theorem 12.29(b). Since the Lévy measure of M^j is $\nu_{|\{x\in\mathbb{R}^d:\varepsilon_j<|x|\le\varepsilon_{j-1}\}}$, the covariance matrix of M_t^j is given by $|t|\int_{x\in\mathbb{R}^d:\varepsilon_j<|x|\le\varepsilon_{j-1}} xx^T\,\nu(\mathrm{d}x)$ by Theorem 11.15(b). Writing $M_t^j = (M_t^{j,1},\dots,M_t^{j,d})^T$, the random variable $M_t^{j,i}$ has expectation 0 and variance $|t|\int_{x\in\mathbb{R}^d:\varepsilon_j<|x|\le\varepsilon_{j-1}} x_i^2\,\nu(\mathrm{d}x)$ for each $i\in\{1,\dots,d\}$. Using the independence of $\{M_t^{j,i}, j\in\mathbb{N}\}$, we see that

$$\begin{aligned}\mathbb{E}\Bigg(\sum_{j=m+1}^{n} M_t^{j,i}\Bigg)^2 &= \operatorname{Var}\Bigg(\sum_{j=m+1}^{n} M_t^{j,i}\Bigg) = \sum_{j=m+1}^{n}\operatorname{Var}(M_t^{j,i})\\ &= |t|\sum_{j=m+1}^{n}\int\limits_{x\in\mathbb{R}^d:\varepsilon_j<|x|\le\varepsilon_{j-1}} x_i^2\,\nu(\mathrm{d}x)\\ &= |t|\int\limits_{x\in\mathbb{R}^d:\varepsilon_n<|x|\le\varepsilon_m} x_i^2\,\nu(\mathrm{d}x)\end{aligned}$$

for $n\ge m$. Since $\int_{|x|\le 1} x_i^2\,\nu(\mathrm{d}x)<\infty$, it follows that $(\sum_{j=1}^n M_t^{i,j})_{n\in\mathbb{N}}$ is a Cauchy sequence in $\mathcal{S}$, which therefore converges in $\mathcal{S}$ to a random variable. Hence, by (13.3), $L_t^{\varepsilon_n,b}-\mathbb{E}(L_t^{\varepsilon_n,b})$ converges in mean square and hence in probability to some random vector as $n\to\infty$ for each fixed t. Hence also $L_t^{\varepsilon_n} - \int_{\varepsilon_n<|x|\le 1} x\,\nu(\mathrm{d}x)$ converges in probability to a random variable. Actually, the convergence is also almost sure since sums with independent increments converge almost surely if and only if they converge in distribution, cf. Chow and Teicher [32, Thm. 8.3.4], or Theorem 16.9 below. Soon, however, we shall prove much more without reference to this theorem. We shall show in fact that the convergence is almost surely uniform in t on compact subsets of I, and that we do not even need to restrict to convergence along a subsequence, in other words $\lim_{\varepsilon\downarrow 0}(L_t^{\varepsilon} - t\int_{\varepsilon<|x|\le 1} x\,\nu(\mathrm{d}x))$ exists almost surely uniformly on compact subsets of I. The resulting Lévy process and an independent Brownian motion are the components of the Lévy–Itô decomposition. To prove almost sure uniform convergence on compacts along a fixed sequence $(\varepsilon_n)_{n\in\mathbb{N}}$ as above, a key result is the following version of Etemadi's inequality (cf., Theorem 10.36) for independent càdlàg processes with the supremum norm.

Lemma 13.2 (Etemadi's maximal inequality for càdlàg processes). *Suppose that $T>0$ and $M^n=(M_t^n)_{t\in[0,T]}$, $n\in\mathbb{N}$, is an independent sequence of $\mathbb{R}^d$-valued strictly càdlàg processes on $[0,T]$, and define $S^n := \sum_{j=1}^n M^j$ for $n\in\mathbb{N}_0$. Denote by $\|\cdot\|_\infty$ the supremum norm on $[0,T]$, i. e. $\|S^n(\omega)\|_\infty = \sup_{t\in[0,T]}|S^n(\omega)|$. Then*

$$P\Big(\max_{j\in\{0,\dots,n\}}\|S^j\|_\infty\ge u\Big)\le 3\max_{j\in\{1,\dots,n\}} P(\|S^j\|_\infty\ge u/3)\quad \forall\, u\ge 0, n\in\mathbb{N}.$$

Proof. The proof is word for word the same as that of Etemadi's maximal inequality, Theorem 10.36, with Euclidean norms replaced throughout by supremum norms, e. g. $|S_j|$ by $\|S^j\|_\infty$. □

We can now state and prove the famous Lévy–Itô decomposition in full generality.

Theorem 13.3 (Lévy–Itô decomposition). *Suppose that $I = [0,\infty)$ or $I = \mathbb{R}$ and let $L = (L_t)_{t\in I}$ be an $\mathbb{R}^d$-valued Lévy process with standard characteristic triplet (A,ν,γ). For each $\varepsilon \in (0,1]$, define $L^\varepsilon := (L_t^\varepsilon)_{t\in I}$ by* (13.1) *and $X^\varepsilon := (X_t^\varepsilon)_{t\in I}$ by*

$$X_t^\varepsilon = L_t^\varepsilon - t \int\limits_{\varepsilon<|x|\le 1} x\,\nu(\mathrm{d}x)$$

$$= \begin{cases} \sum_{0<s\le t, |\Delta L_s|>\varepsilon} \Delta L_s - t\int_{\varepsilon<|x|\le 1} x\,\nu(\mathrm{d}x), & t \ge 0, \\ -\sum_{t<s\le 0, |\Delta L_s|>\varepsilon} \Delta L_s - t\int_{\varepsilon<|x|\le 1} x\,\nu(\mathrm{d}x), & t < 0 \text{ (when } I = \mathbb{R}). \end{cases}$$

(a) *Then as $\varepsilon \downarrow 0$, X^ε converges almost surely uniformly on compact subsets of I to a Lévy process X with standard characteristic triplet $(0,\nu,0)$. Further, the process $B = (B_t)_{t\in I}$, defined by*

$$B_t := L_t - X_t - \gamma t, \quad t \in I,$$

is a Lévy process with standard characteristic triplet $(A,0,0)$ (i. e., a Brownian motion with covariance matrix A and drift 0), and it is independent of X, hence L is an independent sum of X, B and the deterministic Lévy process $(t \mapsto \gamma t)$.

(b) *The process $X = (X_t)_{t\in I}$ can be further decomposed into the two independent Lévy processes $Y = (Y_t)_{t\in I}$ and $Z = (Z_t)_{t\in I}$, where*

$$Z_t = \begin{cases} \sum_{0<s\le t, |\Delta L_s|>1} \Delta L_s, & t \ge 0, \\ -\sum_{t<s\le 0, |\Delta L_s|>1} \Delta L_s, & t < 0 \quad (\text{when } I = \mathbb{R}) \end{cases}$$

and

$$Y_t = X_t - Z_t$$

$$= \lim_{\varepsilon\downarrow 0} \left(\begin{cases} \sum_{0<s\le t, \varepsilon<|\Delta L_s|\le 1} \Delta L_s - t\int_{\varepsilon<|x|\le 1} x\,\nu(\mathrm{d}x), & t \ge 0, \\ -\sum_{t<s\le 0, \varepsilon<|\Delta L_s|\le 1} \Delta L_s - t\int_{\varepsilon<|x|\le 1} x\,\nu(\mathrm{d}x), & t < 0 \end{cases} \right).$$

The convergence in the definition of Y is almost surely uniform on compacts, Z is a compound Poisson process with standard characteristic triplet $(0, \nu_{|\{x\in\mathbb{R}^d : |x|>1\}}, 0)$, Y has standard characteristic triplet $(0, \nu_{|\{x\in\mathbb{R}^d : |x|\le 1\}}, 0)$, all moments of Y exist and Y has mean 0. In particular, L can be written as an independent sum of the four Lévy processes Y, Z, B and $(t \mapsto \gamma t)$.

Proof. (a) We shall follow the proof given in Sato [79, Section 20]. By modifying L on a null-set if necessary, we shall assume throughout that L is strictly càdlàg.

Step 1: We first show that X^ε converges almost surely uniformly on compacts to a càdlàg process along any given sequence $(\varepsilon_n)_{n\in\mathbb{N}_0}$ with $1 = \varepsilon_0 \geq \varepsilon_1 \geq \ldots > 0$ and $\varepsilon_n \downarrow 0$. By considering the positive time axis and the negative time axis separately (when $I = \mathbb{R}$), it suffices to prove the result for the positive time axis only. We shall suppose therefore that $T \in \mathbb{N}$ and show that almost surely, $X_t^{\varepsilon_n}$ converges uniformly to a càdlàg process on $[0, T]$ as $n \to \infty$. Since the sum involving the jumps greater than 1 in absolute size (denoted by $L_t^\sharp$ in the motivation) does not change as $\varepsilon_n \to 0$, it suffices to show the uniform convergence of

$$S_t^n := L_t^{\varepsilon_n,\flat} - t \int_{\varepsilon_n<|x|\leq 1} x\,\nu(\mathrm{d}x) := \sum_{0<s\leq t,\varepsilon_n<|\Delta L_s|\leq 1} \Delta L_s - t \int_{\varepsilon_n<|x|\leq 1} x\,\nu(\mathrm{d}x)$$

(cf., the motivating argument prior to Lemma 13.2). Defining $M^j = (M_t^j)$ as in (13.2), we have seen in (13.3) that $S_t^n = \sum_{j=1}^n M_t^j$, so that we have to show almost sure uniform convergence of $\sum_{j=1}^n M^j$ on $[0, T]$ as $n \to \infty$. Recall that M^j, $j \in \mathbb{N}$, are independent Lévy processes with mean zero, Gaussian variance 0 and Lévy measure $\nu_{|\{x\in\mathbb{R}^d:\varepsilon_j<|x|\leq\varepsilon_{j-1}\}}$. Writing $\|\cdot\|_\infty$ for the supremum norm on $[0, T]$, we conclude from Lemma 13.2 and Markov's inequality that for each $m > n$ and $u > 0$,

$$\begin{aligned}
&P\Big(\max_{k,l\in\{n,\ldots,m\}} \|S^k - S^l\|_\infty > u\Big) \\
&\leq P\Big(\max_{k\in\{n,\ldots,m\}} \|S^k - S^n\|_\infty \geq u/2\Big) \\
&= P\left(\max_{k\in\{n,\ldots,m\}} \left\|\sum_{j=n+1}^{k} M^j\right\|_\infty \geq u/2\right) \\
&\leq 3 \max_{k\in\{n,\ldots,m\}} P\left(\left\|\sum_{j=n+1}^{k} M^j\right\|_\infty \geq u/6\right) \\
&\leq \frac{108}{u^2} \max_{k\in\{n,\ldots,m\}} \mathbb{E}\left\|\sum_{j=n+1}^{k} M^j\right\|_\infty^2 .
\end{aligned}$$

Since $\mathbf{1}_{(u,\infty)} \circ \max_{k,l\in\{n,\ldots,m\}} \|S^k - S^l\|_\infty$ converges from below to $\mathbf{1}_{(u,\infty)} \circ \sup_{k,l\geq n} \|S^k - S^l\|_\infty$ as $m \to \infty$, we conclude from monotone convergence that

$$P\Big(\sup_{k,l\geq n} \|S^k - S^l\|_\infty > u\Big) \leq \frac{108}{u^2} \sup_{k\geq n} \mathbb{E}\left\|\sum_{j=n+1}^{k} M^j\right\|_\infty^2 .$$

But $\sum_{j=n+1}^k M^j$ is a Lévy process with Gaussian variance 0, mean 0 and Lévy measure $\nu_{|\{x\in\mathbb{R}^d:\varepsilon_k<|x|\leq\varepsilon_n\}}$. Hence

$$\mathbb{E}\left\|\sum_{j=n+1}^{k} M^j\right\|_\infty^2 \le 4T \int_{\varepsilon_k<|x|\le\varepsilon_n} |x|^2\,\nu(dx) \le 4T \int_{|x|\le\varepsilon_n} |x|^2\,\nu(dx)$$

by Corollary 11.19. However, the right-hand side converges to 0 as $n \to \infty$, since $\varepsilon_n \downarrow 0$, so we see that $\sup_{k,l\ge n} \|S^k - S^l\|_\infty$ converges in probability to 0 as $n \to \infty$. Since it is a non-increasing sequence, it must also converge almost surely to 0. But this shows that almost surely, $(S^n)_{n\in\mathbb{N}}$ is a Cauchy sequence in $(D([0,T];\mathbb{R}^d), \|\cdot\|_\infty)$ and hence converges almost surely uniformly on $[0,T]$ by the completeness of $(D([0,T];\mathbb{R}^d), \|\cdot\|_\infty)$ (cf., Lemma 12.1). It is clear that the limit is a stochastic process (if suitably defined for ω, for which $S^n(\omega)$ does not converge uniformly). We have thus shown that X^{ε_n} converges almost surely uniformly on compact subsets of I to a stochastic process.

Step 2: We now show that X^ε converges almost surely uniformly on compact sets as $\varepsilon \downarrow 0$. This does not follow trivially from Step 1, since the exceptional null-set on which X^{ε_n} does not converge uniformly on compacts depends on the specific sequence $(\varepsilon_n)_{n\in\mathbb{N}_0}$. Since there are uncountably many sequences converging to 0, we cannot conclude directly from Step 1 that X^ε also converges almost surely uniformly on compacts as $\varepsilon \downarrow 0$. With some more work, we can, however, deduce it from Step 1. As in Step 1, it suffices to consider the positive axis only. We shall suppose therefore that $T > 0$ and show that $X^\varepsilon = (X_t^\varepsilon)$ converges almost surely uniformly in $t \in [0,T]$ to a càdlàg process. Defining

$$S_t^\varepsilon := X_t^\varepsilon - \sum_{0<s\le t, |\Delta L_s|>1} \Delta L_s = \sum_{0<s\le t, \varepsilon<|\Delta L_s|\le 1} \Delta L_s - t \int_{\varepsilon<|x|\le 1} x\,\nu(dx)$$

for $0 < \varepsilon \le 1$, this is again equivalent to showing almost sure uniform convergence of $S^\varepsilon = (S_t^\varepsilon)$ in $t \in [0,T]$ as $\varepsilon \downarrow 0$.

Denote again by $\|\cdot\|_\infty$ the supremum norm on $[0,T]$ and write

$$|||S^\varepsilon||| := \|S^\varepsilon\|_\infty \wedge 1.$$

Choose a non-increasing sequence $(\varepsilon_n)_{n\in\mathbb{N}}$ of real numbers in $(0,1)$ converging to 0. By Step 1, S^{ε_n} converges almost surely uniformly on $[0,T]$ to a càdlàg process. Hence it is enough to show that there exists a subsequence $(\varepsilon_{n_k})_{k\in\mathbb{N}}$ such that

$$\lim_{k\to\infty} \sup_{\delta\in(0,\varepsilon_{n_k})} |||S^\delta - S^{\varepsilon_{n_k}}||| = 0 \quad \text{almost surely.}$$

Observe that for given $\delta \in (0,\varepsilon_n)$ and $\delta' \in (\delta,\varepsilon_n) \cap \mathbb{Q}$, we have

$$|||S^{\delta'}(\omega) - S^\delta(\omega)||| \le \sum_{0<s\le T, \delta<|\Delta L_s(\omega)|\le\delta'} |\Delta L_s(\omega)| + T \int_{\delta<|x|\le\delta'} |x|\,\nu(dx)$$

for all ω in the sample space Ω. Since we assumed that L is strictly càdlàg, it only has finitely many jumps of absolute size in $[\delta,1]$ for every $\omega \in \Omega$, and we conclude that $\lim_{\delta'\downarrow\delta,\delta'\in\mathbb{Q}} |||S^{\delta'}(\omega) - S^{\delta}(\omega)||| = 0$ for each $\omega \in \Omega$. Hence, $\sup_{\delta\in(0,\varepsilon_n)} |||S^{\delta} - S^{\varepsilon_n}||| = \sup_{\delta\in\mathbb{Q}\cap(0,\varepsilon_n)} |||S^{\delta} - S^{\varepsilon_n}|||$ and it is enough to show that there exists a subsequence $(\varepsilon_{n_k})_{k\in\mathbb{N}}$ such that

$$\lim_{k\to\infty} \sup_{\delta\in\mathbb{Q}\cap(0,\varepsilon_{n_k})} |||S^{\delta} - S^{\varepsilon_{n_k}}||| = 0 \quad \text{almost surely.} \tag{13.4}$$

Let $(\delta_m)_{m\in\mathbb{N}}$ be an enumeration of $\mathbb{Q} \cap (0,1)$. Since for each fixed $n \in \mathbb{N}$, $\max_{j\in\{1,\ldots,m\},\delta_j<\varepsilon_n} |||S^{\delta_j} - S^{\varepsilon_n}|||$ converges everywhere from below to $\sup_{\delta\in\mathbb{Q}\cap(0,\varepsilon_n)} |||S^{\delta} - S^{\varepsilon_n}|||$ as $m \to \infty$, it follows that for each $n \in \mathbb{N}$ there exist finitely many points $a_{j,n} \in \mathbb{Q}\cap(0,\varepsilon_n)$, $j \le N(n)$, such that

$$P\Big(\sup_{\delta\in\mathbb{Q}\cap(0,\varepsilon_n)} |||S^{\delta} - S^{\varepsilon_n}||| - \max_{j=1,\ldots,N(n)} |||S^{a_{j,n}} - S^{\varepsilon_n}||| > \frac{1}{n}\Big) < \frac{1}{n}.$$

Let $a_1, a_2, \ldots$ be the rearrangement of all points $a_{j,n}$ with $j = 1,\ldots,N(n)$ and $n \in \mathbb{N}$ in decreasing order. Further, let $J(n)$ be the (countably infinite) set of all indices $j \in \mathbb{N}$ for which $a_j \in (0,\varepsilon_n)$. Since

$$\max_{j=1,\ldots,N(n)} |||S^{a_{j,n}} - S^{\varepsilon_n}||| \le \sup_{j\in J(n)} |||S^{a_j} - S^{\varepsilon_n}||| \le \sup_{\delta\in\mathbb{Q}\cap(0,\varepsilon_n)} |||S^{\delta} - S^{\varepsilon_n}|||,$$

we also have

$$P\Big(\sup_{\delta\in\mathbb{Q}\cap(0,\varepsilon_n)} |||S^{\delta} - S^{\varepsilon_n}||| - \sup_{j\in J(n)} |||S^{a_j} - S^{\varepsilon_n}||| > \frac{1}{n}\Big) < \frac{1}{n}.$$

This shows that

$$\sup_{\delta\in\mathbb{Q}\cap(0,\varepsilon_n)} |||S^{\delta} - S^{\varepsilon_n}||| - \sup_{j\in J(n)} |||S^{a_j} - S^{\varepsilon_n}|||$$

converges in probability to 0 as $n \to \infty$. Hence there exists a subsequence $(\varepsilon_{n_k})_{k\in\mathbb{N}}$ of $(\varepsilon_n)_{n\in\mathbb{N}}$ such that

$$\sup_{\delta\in\mathbb{Q}\cap(0,\varepsilon_{n_k})} |||S^{\delta} - S^{\varepsilon_{n_k}}||| - \sup_{j\in J(n_k)} |||S^{a_j} - S^{\varepsilon_{n_k}}|||$$

converges almost surely to 0 as $k \to \infty$. Denote $j_k := \min J(n_k)$ (so that $J(n_k) = \{j_k, j_k + 1, \ldots\}$) and observe that

$$\sup_{j\in J(n_k)} |||S^{a_j} - S^{\varepsilon_{n_k}}||| \le \sup_{j\in J(n_k)} |||S^{a_j} - S^{a_{j_k}}||| + |||S^{a_{j_k}} - S^{\varepsilon_{n_k}}|||.$$

However, since $S^{a_{j_k}}$ converges almost surely uniformly as $k \to \infty$ by Step 1, it is almost surely a Cauchy sequence in $(D([0,T];\mathbb{R}^d), \|\cdot\|_\infty)$, from which we conclude that

$\sup_{j\in J(n_k)} |||S^{a_j} - S^{a_{j_k}}|||$ converges almost surely to 0 as $k \to \infty$. Further, $S^{a_{j_k}}$ and $S^{\varepsilon_{n_k}}$ converge almost surely uniformly by Step 1, and by considering $S^{a_{j_1}}, S^{\varepsilon_{n_1}}, S^{a_{j_2}}, S^{\varepsilon_{n_2}}, \ldots$ (more precisely, a suitable subsequence of it to make the indices non-increasing) we conclude from Step 1 that the limits of both sequences must be the same, so that $|||S^{a_{j_k}} - S^{\varepsilon_{n_k}}|||$ converges almost surely to 0 as $k \to \infty$. Hence $\sup_{\delta\in\mathbb{Q}\cap(0,\varepsilon_{n_k})} |||S^{\delta} - S^{\varepsilon_{n_k}}|||$ converges almost surely to 0 as $k \to \infty$, which is (13.4) and hence the desired uniform convergence.

Step 3: Denote the limit process of X^ε by X. We have to show that X and $L-X$ are independent Lévy processes with characteristic triplets $(0,\nu,0)$ and $(A,0,\gamma)$, respectively.

Since each X^ε has càdlàg paths a. s., so also has X, its uniform limit on compacts. Hence $X-L$ also has càdlàg paths almost surely. It is clear that $X_0 = L_0 - X_0 = 0$ almost surely. By Theorem 12.29(b) we know that X^ε and $L-X^\varepsilon$ are independent Lévy processes. In particular, for each $t,h\in I$, we have $X^\varepsilon_{t+h} - X^\varepsilon_t \stackrel{d}{=} X^\varepsilon_h - X^\varepsilon_0$. Since X^ε_s converges almost surely to X_s for all $s\in I$, we conclude also that $X_{t+h} - X_t \stackrel{d}{=} X_h - X_0$, so that X has stationary increments. A similar argument applies to $L-X$. To see the independence of the increments, suppose that $t_0 < t_1 < \cdots < t_n \in I$. Then $X^\varepsilon_{t_1} - X^\varepsilon_{t_0}, X^\varepsilon_{t_2} - X^\varepsilon_{t_1}, \ldots, X^\varepsilon_{t_n} - X^\varepsilon_{t_{n-1}}$ are independent. Since $X^\varepsilon_{t_i} - X^\varepsilon_{t_{i-1}}$ converges almost surely to $X_{t_i} - X_{t_{i-1}}$ for each i, it follows from Exercise 11.1 that also $X_{t_1} - X_{t_0}, X_{t_2} - X_{t_1}, \ldots, X_{t_n} - X_{t_{n-1}}$ are independent. Hence X has independent increments and a similar argument applies to $L-X$. Applying the same argument to the finite dimensional distributions, we also see that X and $L-X$ are independent by the independence of X^ε and $L-X^\varepsilon$. Hence X and $L-X$ are independent Lévy processes.

To obtain the characteristic triplets, observe that X^ε is a compound Poisson process with the added linear drift $-\int_{\varepsilon<|x|\le 1} x\,\nu(dx)$. By Corollary 12.38 and equation (9.24), the standard characteristic triplet of X^ε is given by $(A^\varepsilon = 0, \nu^\varepsilon = \nu|_{\{x\in\mathbb{R}^d:|x|>\varepsilon\}}, \gamma^\varepsilon = 0)$. Since characteristic triplets of independent Lévy processes add to give that of the sum, the standard characteristic triplet of $L-X^\varepsilon$ is $(A - A^\varepsilon = A, \nu - \nu^\varepsilon = \nu|_{\{x\in\mathbb{R}^d:|x|\le\varepsilon\}}, \gamma)$. Denote by A^X and A^{L-X} the Gaussian variances of X and $L-X$ and by ν^X and ν^{L-X} their Lévy measures, respectively. Consider the truncation function

$$c(x) = \frac{1}{1+|x|^2}x, \quad x\in\mathbb{R}^d,$$

which is continuous and satisfies (9.35). Observe that $L_1 - X_1^\varepsilon$ converges in distribution to $L_1 - X_1$ as $\varepsilon \downarrow 0$. From equation (9.37) in Theorem 9.31, we have

$$\lim_{\varepsilon\downarrow 0} \int_{\mathbb{R}^d} f(x)\,(\nu - \nu^\varepsilon)(dx) = \int_{\mathbb{R}^d} f(x)\,\nu^{L-X}(dx)$$

for every bounded continuous function $f:\mathbb{R}^d \to \mathbb{R}$ vanishing in a neighbourhood of 0. However, since $\nu-\nu^\varepsilon = \nu|_{\{x\in\mathbb{R}^d:|x|\le\varepsilon\}}$, the limit on the left-hand side must be zero, implying that $\nu^{L-X} = 0$. From (9.36), we then obtain

$$A^{L-X} = A^{L-X} + \int_{\mathbb{R}^d} c(x)c(x)^T \, \nu^{L-X}(\mathrm{d}x)$$
$$= \lim_{\varepsilon \downarrow 0}\Big(A + \int_{|x|\le\varepsilon} c(x)c(x)^T \, \nu(\mathrm{d}x)\Big) = A,$$

the latter by dominated convergence since $c(x)c(x)^T$ is integrable with respect to ν in a neighbourhood of zero. Thus the Gaussian variance of $L - X$ is A and its Lévy measure is the zero measure. Since the characteristic triplets of X and $L - X$ add up to the characteristic triplet of L, it follows that the Gaussian variance of X is zero and that ν is the Lévy measure of X.

It remains to determine the location parameters of X and $L - X$. Denote by γ^X the location parameter of X with respect to the standard truncation function and by γ_c^X its location parameter with respect to c. Similarly, denote by γ_c^ε the location parameter of X^ε with respect to c. Then

$$\gamma_c = \gamma^X + \int_{\mathbb{R}^d} (c(x) - x\mathbf{1}_{|x|\le1})\, \nu(\mathrm{d}x) \quad \text{and}$$
$$\gamma_c^\varepsilon = \int_{|x|>\varepsilon} (c(x) - x\mathbf{1}_{|x|\le1})\, \nu(\mathrm{d}x),$$

by Proposition 9.17. Since $X_1^\varepsilon \xrightarrow{d} X_1$, $\lim_{\varepsilon\downarrow0} \gamma_c^\varepsilon = \gamma_c$ by equation (9.38) and we conclude that

$$\gamma_X = \lim_{\varepsilon \downarrow 0}\Big(- \int_{|x|\le\varepsilon} (c(x) - x\mathbf{1}_{|x|\le1})\, \nu(\mathrm{d}x)\Big) = 0.$$

Again, since the characteristic triplets of X_1 and $L_1 - X_1$ add to that of their sum, the location parameter of $L - X$ with respect to the standard truncation function is γ.

(b) Since the jumps of X and L coincide, we can also write $Z_t = \sum_{0<s\le t, |\Delta X_s|>1} \Delta X_s$ when $t \ge 0$, with a similar expression when $t < 0$, and replace the jumps of L in the definition of Y by those of X. An argument like that in (a) shows that Z and $Y = X - Z$ are independent Lévy processes, and the almost sure uniform convergence is evident from the definition of Y. That Z is a compound Poisson process with the given characteristic triplet is clear from Corollary 12.38 and equation (9.24), and the characteristic triplet of Y follows from that of X and Z, since those of Y and Z add up to that of X. That Y has mean zero then follows from Theorem 11.15(a), and that it has moments of all orders from Corollary 11.14(a).

Finally, the independence of Y, Z, B follows from the independence of X and B and the fact that $(Z_t)_{t\ge0}$ and hence $(Y_t)_{t\ge0}$ are adapted to the natural filtration of $(X_t)_{t\ge0}$ by Remark 12.30 (after possibly modifying the processes on a null-set, which does not influence independence). A similar result holds for the negative axis. □

We have seen in Theorem 13.3 that

$$X_t^\varepsilon = L_t^\varepsilon - t \int_{\varepsilon<|x|\le 1} x\,\nu(\mathrm{d}x) = L_t^\varepsilon - t \int_{|x|>\varepsilon} x\mathbf{1}_{|x|\le 1}\,\nu(\mathrm{d}x)$$

converges almost surely uniformly on compacts as $\varepsilon \downarrow 0$, whether or not L_t^ε converges as $\varepsilon \downarrow 0$. There is nothing special about the use of the standard truncation function $x \mapsto x\mathbf{1}_{|x|\le 1}$, $x \in \mathbb{R}^d$, when “compensating” the sum of jumps, one can use any truncation function in the same way, as shown in the following corollary.

Corollary 13.4 (Lévy–Itô decomposition with general truncation funct.). *Suppose that $c : \mathbb{R}^d \to \mathbb{R}^d$ is a fixed truncation function, $I = [0,\infty)$ or $I = \mathbb{R}$ and that $L = (L_t)_{t\in I}$ is an $\mathbb{R}^d$-valued Lévy process with characteristic triplet $(A, \nu, \gamma^c)_c$ with respect to c. For each $\varepsilon \in (0,1]$, define $L^\varepsilon = (L_t^\varepsilon)_{t\in I}$ by (13.1). Then*

$$L_t^\varepsilon - t \int_{|x|>\varepsilon} c(x)\,\nu(\mathrm{d}x)$$

converges almost surely on compact subsets of I to a Lévy process $\widetilde{X}$ with characteristic triplet $(0, \nu, 0)_c$ with respect to c. Moreover, the process $B = (B_t)_{t\in I}$, defined by

$$B_t := L_t - \widetilde{X}_t - \gamma^c t, \quad t \in I,$$

is a Lévy process with characteristic triplet $(A, 0, 0)_c$ with respect to c (i. e., a Brownian motion with covariance matrix A and drift 0) and is independent of $\widetilde{X}$. Hence L is the sum of the independent components, $\widetilde{X}$, B and the deterministic Lévy process $(t \mapsto \gamma^c t)$.

Proof. With X^ε and X defined as in Theorem 13.3, we can write

$$L_t^\varepsilon - t \int_{|x|>\varepsilon} c(x)\,\nu(\mathrm{d}x) = X_t^\varepsilon + t \int_{|x|>\varepsilon} (x\mathbf{1}_{|x|\le 1} - c(x))\,\nu(\mathrm{d}x).$$

As $\varepsilon \downarrow 0$ the first term on the right converges almost surely uniformly on compact subsets of I to X_t by Theorem 13.3, while the second term converges uniformly in t on compact subsets of I to $t \int_{\mathbb{R}^d} (x\mathbf{1}_{|x|\le 1} - c(x))\,\nu(\mathrm{d}x)$. Since c is a truncation function the convergence of the integral follows from (9.3). The limit process $\widetilde{X}$ therefore satisfies

$$\widetilde{X}_t = X_t + t \int_{\mathbb{R}^d} (x\mathbf{1}_{|x|\le 1} - c(x))\,\nu(\mathrm{d}x).$$

By Theorem 13.3, the standard characteristic triplet of $\widetilde{X}$ is given by $(0, \nu, \int_{\mathbb{R}^d} (x\mathbf{1}_{|x|\le 1} - c(x))\,\nu(\mathrm{d}x))$ and hence, by Proposition 9.17, its characteristic triplet with respect to c is given by $(0, \nu, 0)_c$. The independence of $\widetilde{X}$ and the Lévy process B follows again from

Theorem 13.3, and the remaining assertions are clear since characteristic triplets of independent Lévy processes add to that of their sum. □

Of course, as was done in Theorem 13.3(b), it is possible to further decompose the process $\widetilde{X}$ above into a compound Poisson process and an independent Lévy process having compactly supported Lévy measure.

Remark 13.5. From Corollary 12.37, we see that $\sum_{0<s\leq t} \Delta L_s$ does not converge absolutely unless $\int_{|x|\leq 1} |x|\,\nu(dx) < \infty$. Nevertheless, in the setting of Corollary 13.4, the limit as $\varepsilon \downarrow 0$ of $\widetilde{X}_t^\varepsilon = \sum_{0<s\leq t, |\Delta L_s|>\varepsilon} \Delta L_s - t\int_{|x|>\varepsilon} c(x)\,\nu(dx)$ exists. Since subtracting a deterministic term leads to convergence, one speaks of $\widetilde{X}$ also as a *compensated sum of jumps with respect to c*.

Definition 13.6. For a Lévy process $L = (L_t)_{t\in I}$ with characteristic triplet $(A, \nu, \gamma^c)_c$ with respect to some truncation function c, the processes $B, \widetilde{X}$ and $(\gamma^c t)_{t\in I}$ in Corollary 13.4 are called the *Gaussian* or *Brownian motion*, the *jump*, and the *linear deterministic components* of L with respect to c, respectively. Observe that B does not depend on the specific truncation function c, while $\widetilde{X}$ and the deterministic term do.

13.2 Lévy processes of finite variation

In this section, we characterize the Lévy processes which are of finite variation. We begin by defining what it means to be a function of finite variation.

Definition 13.7 (Total variation, finite variation).

(a) Suppose that $[a, b] \subset \mathbb{R}$ is a compact interval. Then a *partition* π of $[a, b]$ is a finite sequence

$$\pi = (t_0, t_1, \dots, t_n)$$

of points such that

$$a = t_0 < t_1 < \dots < t_n = b$$

with $n \in \mathbb{N}_0$. The set of all partitions of $[a, b]$ will be denoted by $\Pi_{[a,b]}$. Now if $g : [a, b] \to \mathbb{R}^d$ is an $\mathbb{R}^d$-valued function on $[a, b]$ then

$$\mathrm{TV}_{[a,b]}(g) := \sup_{(t_0,\dots,t_n)\in\Pi_{[a,b]}} \sum_{j=1}^{n} |g(t_j) - g(t_{j-1})| \in [0, \infty]$$

is called the *total variation* of g on $[a, b]$.

(b) If $I = [0, \infty)$ or $I = \mathbb{R}$ then the function $g : I \to \mathbb{R}$ is said to be of *finite variation* if

$$\mathrm{TV}_{[a,b]}(g) < \infty \quad \text{for all compact subintervals } [a, b] \text{ of } I.$$

If $a = b$, then $\mathrm{TV}_{[a,a]}(g) = 0$, with the usual convention that the empty sum $\sum_{j=1}^{0}$ is interpreted as 0.

Remark 13.8. Let $I = [0, \infty)$ or $I = \mathbb{R}$.

(a) It follows easily from the definition that $\mathrm{TV}_{[c,d]}(g) \leq \mathrm{TV}_{[a,b]}(g)$ whenever $[c, d] \subset [a, b] \subset I$ and $g : I \to \mathbb{R}^d$.

(b) If g_1 and g_2 are two functions, $g_1 : I \to \mathbb{R}^d$ and $g_2 : I \to \mathbb{R}^d$, and $\lambda \in \mathbb{R}$ then

$$\mathrm{TV}_{[a,b]}(g_1 + g_2) \leq \mathrm{TV}_{[a,b]}(g_1) + \mathrm{TV}_{[a,b]}(g_2)$$

for all $[a, b] \subset I$ by the triangle inequality and $\mathrm{TV}_{[a,b]}(\lambda g_1) = |\lambda| \,\mathrm{TV}_{[a,b]}(g_1)$. Consequently, when g_1 and g_2 are of finite variation and $\lambda \in \mathbb{R}$, then $g_1 + g_2$ and λg_1 are also of finite variation, so that the set of all $\mathbb{R}^d$-valued functions of finite variation on I forms a vector space over $\mathbb{R}$ with the usual addition of functions and multiplication by scalars.

(c) It is easily seen that a function $g = (g_1, \ldots, g_d)^T : I \to \mathbb{R}^d$ is of finite variation if and only if all of its components $g_1, \ldots, g_d$ are of finite variation.

(d) Some authors say that $g : I \to \mathbb{R}$ is of finite variation or of "bounded variation" if additionally $\lim_{a\to\infty} \mathrm{TV}_{[0,a]}(g) < \infty$ and (when $I = \mathbb{R}$) $\lim_{a\to-\infty} \mathrm{TV}_{[a,0]}(g) < \infty$. With such nomenclature, our definition would fit in as being "locally of finite variation" or "locally of bounded variation". In the following, by functions of finite variation we shall always mean functions as defined in Definition 13.7(b).

We shall now give some examples of functions of finite variation.

Example 13.9. Suppose that $I = [0, \infty)$ or $I = \mathbb{R}$ and let g be a function, $g : I \to \mathbb{R}$.

(a) If g can be written as the difference $g_1 - g_2$ of two non-decreasing functions g_1 and g_2, then g is of finite variation. To see this, observe that for $[a, b] \subset \mathbb{R}$ and $\pi = (t_0, t_1, \ldots, t_n) \in \Pi_{[a,b]}$ we have

$$\sum_{j=1}^{n} |g_1(t_j) - g_1(t_{j-1})| = \sum_{j=1}^{n} (g_1(t_j) - g_1(t_{j-1})) = g_1(b) - g_1(a),$$

so that

$$\mathrm{TV}_{[a,b]}(g_1) = g_1(b) - g_1(a) < \infty.$$

Hence g_1 is of finite variation and similarly so is g_2, so that g is of finite variation by Remark 13.8(b). It can be shown that a real-valued function $g : I \to \mathbb{R}$ is of finite variation if and only if it can be represented as the difference of two non-decreasing functions (see, e. g., Cohn [33, Prop. 4.4.2]), but we will not need this characterization.

(b) Suppose that $g \in C^1(I; \mathbb{R})$, i. e., that $g : I \to \mathbb{R}$ is a differentiable function with continuous derivative. Then g is of finite variation. To see this, observe that for any compact interval $[a, b] \subset I$ and any partition $\pi = (t_0, \ldots, t_n) \in \Pi_{[a,b]}$, we have

$$\sum_{j=1}^{n}|g(t_j)-g(t_{j-1})| = \sum_{j=1}^{n}\left|\int_{t_{j-1}}^{t_j} g'(t)\,\mathrm{d}t\right| \leq \sum_{j=1}^{n}\int_{t_{j-1}}^{t_j}|g'(t)|\,\mathrm{d}t = \int_a^b |g'(t)|\,\mathrm{d}t.$$

Hence

$$\mathrm{TV}_{[a,b]}(g) \leq \int_a^b |g'(t)|\,\mathrm{d}t \leq (b-a)\sup_{t\in[a,b]}|g'(t)| < \infty,$$

showing that g is of finite variation.

While continuous differentiability is sufficient for a function to be of finite variation, continuity alone is not sufficient. A standard example is given by $g : [0,\infty) \to \mathbb{R}$ with $g(0) := 0$ and $g(x) := x\sin(x^{-1})$ for $x > 0$, see Exercise 13.3. Another standard example is a typical path of standard Brownian motion, as we shall show in Lemma 13.14 below.

Of particular importance for us is the following example, which is a "pure jump function" of finite variation:

Example 13.10. Suppose that $I = [0,\infty)$ or $I = \mathbb{R}$ and let J be a countable subset of I and $(c_j)_{j\in J}$ be a finite or infinite sequence in $\mathbb{R}^d$ such that $\sum_{j\in J\cap K}|c_j| < \infty$ for every compact subset K of I. Define the function $g : I \to \mathbb{R}^d$ by

$$g(t) := \begin{cases} \sum_{j\in J, 0<j\leq t} c_j, & t \geq 0, \\ -\sum_{j\in J, t<j\leq 0} c_j, & t < 0 \text{ (when } I = \mathbb{R}). \end{cases}$$

Then g is càdlàg and of finite variation and for $t \neq 0$ it satisfies

$$\Delta g(t) = \begin{cases} c_t, & t \in J, \\ 0, & t \notin J. \end{cases} \tag{13.5}$$

Further, $\Delta g(0) = c_0$, when $I = \mathbb{R}$ and $0 \in J$, and $\Delta g(0) = 0$ otherwise.

Proof. By considering the components of g separately, by Remark 13.8(c), we can assume that $d = 1$. Define

$$g_1(t) := \begin{cases} \sum_{j\in J, 0<j\leq t, c_j>0} c_j, & t \geq 0 \\ -\sum_{j\in J, t<j\leq 0, c_j>0} c_j, & t < 0, \end{cases}$$

and

$$g_2(t) := \begin{cases} \sum_{j\in J, 0<j\leq t, c_j<0} |c_j|, & t \geq 0, \\ -\sum_{j\in J, t<j\leq 0, c_j<0} |c_j|, & t < 0. \end{cases}$$

Then g_1 and g_2 are well defined, càdlàg and non-decreasing, and hence $g = g_1 - g_2$ is càdlàg and of finite variation by Example 13.9(a). For $t > 0$ we also have $g(t-) = \sum_{j\in J, 0<j<t} c_j$ so that $\Delta g(t) = g(t) - g(t-)$ is equal to c_t when $t \in J$ and equal to zero when $t \notin J$. The proof for $t \leq 0$ is similar – when $t = 0$ and $I = [0, \infty)$, one has to be careful, since then $\Delta g(0) = 0$ by definition (even when $0 \in J$). □

In the example above, we see from (13.5) that

$$\sum_{0\neq t\in K, \Delta g(t)\neq 0} |\Delta g(t)| = \sum_{0\neq j\in K\cap J} |c_j| < \infty$$

for every compact subset K of I, so that the jump sizes of the function g there are absolutely summable over compact sets. It turns out that absolute summability of the jump sizes over compacts is actually a necessary condition for a function to be of finite variation:

Lemma 13.11 (Finite variation; absolutely summable jumps). *Suppose that $I = [0, \infty)$ or $I = \mathbb{R}$ and let $g : I \to \mathbb{R}^d$ be a càdlàg function. Then the number of jumps of g on I (i. e., the set of $t \in I$ for which $\Delta g(t) \neq 0$) is countable. If g is additionally of finite variation, then the jump sizes of g are absolutely summable on compacts, i. e., for every compact subset $K \subset I$,*

$$\sum_{s\in K, \Delta g(s)\neq 0} |\Delta g(s)| < \infty.$$

Proof. If $[a, b] \subset I$ and $n \in \mathbb{N}$, denote by $S_{a,b,n}$ the set of points in $[a, b]$ in which the càdlàg function g has a jump of absolute size $\geq 1/n$. Then $S_{a,b,n}$ must be finite, since otherwise it would have an accumulation point in $[a, b]$, and at this accumulation point the function g cannot have a finite left-limit and simultaneously be right-continuous, contradicting the càdlàg property (we have already used this argument before the statement of Lemma 12.6, but repeated it here for convenience). It follows that the number of jumps on I of absolute size $\geq 1/n$ must be countable, and by taking the union over $n \in \mathbb{N}$, we see that the number of all (non-zero) jumps of g on I is countable.

Now let g be a càdlàg function for which there is some compact set $K \subset I$ such that $\sum_{s\in K, \Delta g(s)\neq 0} |\Delta g(s)| = \infty$. By enlarging K if necessary we can assume that $K = [a, b] \subset I$ is an interval. We shall show that g cannot then be of finite variation. Since the sum of absolute jump sizes of jumps occuring in $[a, b]$ is infinite, so is the sum of absolute jump sizes occuring in (a, b), and we conclude that

$$\lim_{n\to\infty} \sum_{s\in(a,b), |\Delta g(s)|>1/n} |\Delta g(s)| = \sum_{s\in(a,b), \Delta g(s)\neq 0} |\Delta g(s)| = \infty.$$

If K is an arbitrary number such that $K > 0$, then there is some $n \in \mathbb{N}$ such that $\sum_{s\in(a,b), |\Delta g(s)|>1/n} |\Delta g(s)| > K + 1$. Denote by $s_1 < s_2 < \cdots < s_m$ the points in (a, b) at which a jump of absolute size greater or equal than $1/n$ occurs (the number m is finite as seen above), and define $s_0 := a$ and $s_{m+1} := b$. Let $\delta := \min\{s_{j+1} - s_j : j = 0, \ldots, m\}$ and for each $j \in \{1, \ldots, m\}$ choose $t_j^+ \in (s_j, s_j + \delta/2)$ and $t_j^- \in (s_j - \delta/2, s_j)$ such that

$$|g(t_j^+) - g(s_j)| < \frac{1}{2m} \quad \text{and} \quad |g(t_j^-) - g(s_j-)| < \frac{1}{2m} \quad \forall j \in \{1, \ldots, m\};$$

this is possible since g is càdlàg. Then

$$a < t_1^- < t_1^+ < t_2^- < t_2^+ < t_3^- < \cdots < t_m^- < t_m^+ < b,$$

and by looking at this particular partition, we obtain

$$\mathrm{TV}_{[a,b]}(g) \geq \sum_{j=1}^{m} |g(t_j^+) - g(t_j^-)|.$$

However,

$$\begin{aligned} |g(t_j^+) - g(t_j^-)| &\geq |g(s_j) - g(s_j-)| - |g(t_j^+) - g(s_j)| - |g(t_j^-) - g(s_j-)| \\ &\geq |\Delta g(s_j)| - \frac{1}{m}, \end{aligned}$$

and hence

$$\mathrm{TV}_{[a,b]}(g) \geq \sum_{j=1}^{m} (|\Delta g(s_j)| - 1/m) \geq (K+1) - 1 = K.$$

Since $K > 0$ was arbitrary, we conclude that $\mathrm{TV}_{[a,b]}(g) = \infty$, so that g cannot be of finite variation. □

Just as we defined càdlàg processes in Definition 10.1(c) as stochastic processes whose sample paths are càdlàg with probability 1, we now define finite variation processes by the corresponding property:

Definition 13.12 (Stochastic process of finite variation). Suppose that $I = [0, \infty)$ or $I = \mathbb{R}$. An $\mathbb{R}^d$-valued stochastic process $(L_t)_{t \in I}$, defined on a probability space $(\Omega, \mathcal{F}, P)$, is called a *finite variation process* or we say that *L is of finite variation*, if the sample paths of L are almost surely of finite variation, i. e., if there is some $\Omega_0 \in \mathcal{F}$ with $P(\Omega_0) = 1$ such that the sample path $I \to \mathbb{R}^d$, $u \mapsto L_u(\omega)$ is of finite variation for all $\omega \in \Omega_0$.

Example 13.13. The processes M constructed in Theorem 12.13 and Proposition 12.15 are of finite variation, since their sample paths are of the required form, as shown in Example 13.10.

Since the sample paths of a compound Poisson process are step functions (with only finitely many jumps on compact time intervals), a compound Poisson process is a finite variation process by Example 13.10. Also, a linear term $t \mapsto t\gamma$ is obviously of finite variation, and since the set of $\mathbb{R}^d$-valued functions of finite variation form a real vector space by Remark 13.8(b), compound Poisson processes with drift are also of finite variation. This means that every Lévy process in $\mathbb{R}^d$ with standard characteristic triplet $(0, \nu, \gamma)$ such that ν is finite and $\gamma \in \mathbb{R}^d$ is of finite variation.

The question now arises, are there other Lévy processes of finite variation? In particular, can the Lévy measure of a finite variation Lévy process be infinite? The answer to both questions is yes. In fact we shall show, in Theorem 13.15 below, that a Lévy process is of finite variation if and only if its Gaussian variance is 0 and its Lévy measure ν is such that $\int_{|x|\leq 1} |x|\,\nu(dx) < \infty$. Before doing so, we shall treat the case of one-dimensional Brownian motion.

Lemma 13.14 (Brownian motion is not of finite variation). *Suppose that $I = [0, \infty)$ or $I = \mathbb{R}$ and let $B = (B_t)_{t\in I}$ be a one-dimensional Brownian motion with variance $\sigma^2 > 0$ and drift $\gamma \in \mathbb{R}$. Then for any two points $a, b \in \mathbb{R}$ with $a < b$, the total variation of the sample paths of B on $[a, b]$ is almost surely infinite.*

Proof. Consider first the case when $\gamma = 0$. Suppose that $a < b$ and define

$$Y_n := \sum_{i=1}^{2^n} |B_{a+i2^{-n}(b-a)} - B_{a+(i-1)2^{-n}(b-a)}|.$$

Then $\mathrm{TV}_{[a,b]}(B) \geq Y_n$ for each $n \in \mathbb{N}$. Let X be a normally distributed random variable with mean 0 and variance σ^2. Then

$$|B_{a+i2^{-n}(b-a)} - B_{a+(i-1)2^{-n}(b-a)}| \stackrel{d}{=} |B_{2^{-n}(b-a)}| \stackrel{d}{=} 2^{-n/2}\sqrt{b-a}\,|X|$$

and hence

$$\mathbb{E}(Y_n) = 2^n \mathbb{E}|2^{-n/2}\sqrt{b-a}\,X| = 2^{n/2}\sqrt{b-a}\,\mathbb{E}|X|$$

and

$$\mathrm{Var}(Y_n) = 2^n \,\mathrm{Var}|2^{-n/2}\sqrt{b-a}\,X| = (b-a)\,\mathrm{Var}\,|X|.$$

Using Chebyshev's inequality, we conclude that

$$\begin{aligned}
&\sum_{n=1}^{\infty} P\Big(Y_n \leq \frac{1}{2}2^{n/2}\sqrt{b-a}\,\mathbb{E}|X|\Big) \\
&\quad = \sum_{n=1}^{\infty} P\Big(Y_n - \mathbb{E}(Y_n) \leq -\frac{1}{2}2^{n/2}\sqrt{b-a}\,\mathbb{E}|X|\Big) \\
&\quad \leq \sum_{n=1}^{\infty} P\Big(|Y_n - \mathbb{E}(Y_n)| \geq \frac{1}{2}2^{n/2}\sqrt{b-a}\,\mathbb{E}|X|\Big) \\
&\quad \leq \sum_{n=1}^{\infty} \frac{(b-a)\,\mathrm{Var}\,|X|}{2^{n-2}(b-a)\,(\mathbb{E}|X|)^2} < \infty.
\end{aligned}$$

Consequently, by the Borel–Cantelli lemma, $Y_n \le \frac{1}{2}2^{n/2}\sqrt{b-a}\,\mathbb{E}|X|$ infinitely often with probability 0 so that $Y_n > \frac{1}{2}2^{n/2}\sqrt{b-a}\,\mathbb{E}|X|$ infinitely often with probability 1. However, this implies that $\limsup_{n\to\infty} Y_n = +\infty$ almost surely and hence that $\mathrm{TV}_{[a,b]}(B) = +\infty$ almost surely.

If the drift γ of B is not zero, consider $Y_t := B_t - \gamma_t$. Then the sample paths of Y almost surely have infinite variation and $t \mapsto \gamma t$ has finite variation on $[a,b]$. By the vector space property of the functions of finite variation, we conclude that B also has infinite variation almost surely. □

We come now to the aforementioned characterization of those Lévy processes which have finite variation. The results include a zero-one-law, which states that the sample paths are of finite variation with probability 1 if and only if they are of finite variation with positive probability. However, of more importance is the very nice characterization in terms of the characteristic triplet.

Theorem 13.15 (Characterization of finite variation of Lévy processes). *Suppose that $I = [0,\infty)$ or $I = \mathbb{R}$ and let $L = (L_t)_{t\in\mathbb{R}}$ be an $\mathbb{R}^d$-valued Lévy process with Lévy measure ν and Gaussian variance A. Then the following statements are equivalent.*

(i) *L is a finite variation process.*

(ii) *The sample paths of L are of finite variation with positive probability.*

(iii) *There exist $a,b \in I$ with $a < b$ such that the total variation on $[a,b]$ of the sample paths of L is finite with positive probability.*

(iv) *$A = 0$ and $\int_{|x|\le 1} |x|\,\nu(dx) < \infty$.*

Proof. We start by showing that (iv) implies (i). Suppose therefore that $A = 0$ and $\int_{|x|\le 1} |x|\,\nu(dx) < \infty$. Denoting by γ^0 the drift of L, the simplified Lévy–Itô decomposition in Theorem 13.1 gives $L_t = \gamma^0 t + X_t$ with

$$X_t = \begin{cases} \sum_{0<s\le t} \Delta L_s, & t \ge 0, \\ -\sum_{t<s\le 0} \Delta L_s, & t < 0 \text{ (when } I = \mathbb{R}), \end{cases} \tag{13.6}$$

where the sums converge almost surely absolutely. Then $t \mapsto \gamma^0 t$ is of finite variation and by Example 13.10, X is also of finite variation almost surely. Hence L is also of finite variation almost surely, proving (i).

The facts that (i) implies (ii) and that (ii) implies (iii) are clear. We now prove that (iii) implies (iv). Suppose therefore that the total variation of the sample paths of L on $[a,b]$ is finite with positive probability. Lemma 13.11 then implies that $\sum_{s\in[a,b]} |\Delta L_s| < \infty$ with positive probability. From Corollary 12.37(a), we then conclude that $\int_{\mathbb{R}^d} (1 \wedge |x|)\,\nu(dx) < \infty$, i. e., $\int_{|x|\le 1} |x|\,\nu(dx) < \infty$. This implies, as in the already proved implication (iv) $\Longrightarrow$ (i), that X defined by (13.6) is a finite variation process. Hence the total variation of $B = (B_t)_{t\in I}$, defined by $B_t = L_t - X_t - \gamma^0 t$, is finite on $[a,b]$ with positive probability. Writing $B = (B_t^1, \dots, B_t^d)^T$, the same is true for the components. But by the simplified

Lévy–Itô decomposition in Theorem 13.1, B is a Brownian motion with covariance matrix $A = (a_{ij})_{i,j=1,\dots,d}$ and hence $B^1, \dots, B^d$ are Brownian motions with variances $a_{11}, \dots, a_{dd}$, respectively. Lemma 13.14 then implies that $a_{11} = \dots = a_{dd} = 0$ so that $A = 0$, finishing the proof of the implication (iii) $\Longrightarrow$ (iv). □

The Lévy–Itô decomposition shows us that every two-sided $\mathbb{R}$-valued Lévy process can be decomposed into an independent sum of two Lévy processes, one of which is an orthogonal increment process, while the other is of finite variation. Due to the importance of this decomposition, we state it below explicitly.

Corollary 13.16 (Two-sided LP as sum of OIP and finite variation proc.). *Let $L = (L_t)_{t\in\mathbb{R}}$ be a two-sided $\mathbb{R}$-valued Lévy process with Lévy measure ν and Gaussian variance A. Then L can be decomposed into the sum of three independent processes, namely a compound Poisson process Z with Lévy measure $\nu_{|\{|x|>1\}}$, a Lévy process M with Lévy measure $\nu_{|\{|x|\le 1\}}$, Gaussian variance A, mean 0 and finite moments of all orders, and a deterministic Lévy process $t \mapsto t\gamma'$ for some $\gamma' \in \mathbb{R}^d$. In particular, M is an orthogonal increment process in the sense of Definition 4.1 and L is the sum of an orthogonal increment process and an independent finite variation process. The distribution function of M is given by $t \mapsto t(A + \int_{|x|\le 1} x^2\,\nu(dx))$, $t \in \mathbb{R}$.*

Proof. The decomposition follows immediately from Theorem 13.3(b) with Z as defined there and $M := B + Y$. That M is an orthogonal increment process along with the form of its distribution function follows from Corollary 11.17. The finite variation process is $(Z_t + \gamma' t)_{t\in\mathbb{R}}$. □

Corollary 13.16 could have been derived earlier from Corollaries 12.38, 12.39 and 11.17 without specific use of the Lévy–Itô decomposition. We delayed its statement till now however since it fits into the framework of the current and the previous section and is generally regarded as a consequence of the Lévy–Itô decomposition.

Corollary 13.16 will be a key tool when introducing integration of deterministic functions with respect to Lévy processes in Chapter 15.

13.3 Subordinators

In this section, we discuss subordinators, defined as follows.

Definition 13.17 (Subordinator). Suppose that $I = [0, \infty)$ or $I = \mathbb{R}$. A *subordinator* is an $\mathbb{R}$-valued Lévy process $L = (L_t)_{t\in I}$ with sample paths that are almost surely non-decreasing functions.

We restrict attention here to dimension 1, although subordinators with increments in $[0, \infty)^d$ could also be considered. Since non-decreasing functions are of finite variation, we can now easily derive various characterizations of subordinators using our findings from the previous sections.

Theorem 13.18 (Characterization of subordinators). *Suppose that $I = [0, \infty)$ or $I = \mathbb{R}$ and let $L = (L_t)_{t\in I}$ be an $\mathbb{R}$-valued Lévy process. Then the following are equivalent.*

(i) *L is a subordinator.*

(ii) *The Gaussian variance of L is zero, the Lévy measure ν of L satisfies $\nu((-\infty, 0]) = 0$ and $\int_0^1 x\,\nu(\mathrm{d}x) < \infty$, and the drift γ^0 of L is non-negative.*

(iii) *There exists $t_0 > 0$ such that $L_{t_0} \geq 0$ almost surely.*

(iv) *For all $t \geq 0$, $L_t \geq 0$ almost surely.*

Proof. We shall show first that (i) implies (ii). If L is a subordinator, then it has non-decreasing sample paths, hence its paths are almost surely of finite variation. By Theorem 13.15, the Gaussian variance of L must be zero and the Lévy measure ν must satisfy $\int_{-1}^1 |x|\,\nu(\mathrm{d}x) < \infty$ so that, in particular, L has a drift, which we denote by γ^0. By the simplified Lévy–Itô decomposition (Theorem 13.1), we can write, almost surely, $L_t = \gamma^0 t + \sum_{0<s\leq t} \Delta L_s$ for all $t \geq 0$. It is clear that a non-decreasing càdlàg function cannot jump downwards, so that almost surely, no negative jump occurs. Since $\nu((-\infty, 0))$ is the expected number of negative jumps that occur until time 1 by Theorem 12.31, we conclude that $\nu((-\infty, 0)) = 0$ and hence $\nu((-\infty, 0]) = 0$. To see that γ^0 is non-negative, observe that a non-decreasing càdlàg function remains càdlàg and non-decreasing if a jump of it is removed. Hence also finitely many jumps can be removed, and it follows that $t \mapsto \gamma^0 t + \sum_{0<s\leq t, 0<\Delta L_s \leq 1/n} \Delta L_s$ is almost surely non-decreasing on $[0, 1]$ for each $n \in \mathbb{N}$. Since pointwise limits of non-decreasing functions remain non-decreasing, letting $n \to \infty$ we conclude that also $t \mapsto \gamma^0 t$ is non-decreasing on $[0, 1]$, so that $\gamma^0 \geq 0$.

To see that (ii) implies (i), let the characteristic triplet of L be as described in (ii). By the simplified Lévy–Itô decomposition in Theorem 13.1, we see that almost surely, $L_t = \gamma^0 t + \sum_{0<s\leq t} \Delta L_s$ for $t \geq 0$ and a similar expression holds for $t < 0$, when $I = \mathbb{R}$. Since $\nu((-\infty, 0]) = 0$, the expected number of jumps with negative jump size that appear in I is zero by Theorem 12.31, i. e., almost surely they do not occur. Since also $\gamma^0 \geq 0$, the above representation shows that the sample paths of L are almost surely non-decreasing.

That (i) implies (iv) is clear, since $L_t \geq L_0 = 0$ almost surely by (i) for $t \geq 0$. Also, (iv) obviously implies (iii).

To show that (iii) implies (ii), suppose that $L_{t_0} \geq 0$ almost surely for some $t_0 > 0$. We claim then that $L_{t_0/2} \geq 0$ almost surely for if this were not the case, then $P(L_{t_0/2} < 0) = P(L_{t_0} - L_{t_0/2} < 0) > 0$ and hence also

$$P(L_{t_0} < 0) \geq P(L_{t_0/2} < 0)\, P(L_{t_0} - L_{t_0/2} < 0) > 0,$$

since $L_{t_0} = L_{t_0/2} + (L_{t_0} - L_{t_0/2})$ where $L_{t_0/2}$ and $L_{t_0} - L_{t_0/2}$ are independent. Hence $L_{t_0/2} \geq 0$ a. s. By iterating this argument, we find that $L_{t_0/2^n} \geq 0$ almost surely for all $n \in \mathbb{N}$. Since L has stationary increments and since the sum of two non-negative random variables is again non-negative, it follows that $L_{qt_0} \geq 0$ for all $q \in \mathbb{Q}_{\mathrm{dyad}}$ with $q > 0$, where $\mathbb{Q}_{\mathrm{dyad}}$ denotes the dyadic rational numbers given by (10.9). Using the stationary increments of L again, we see that for each pair $0 \leq q_1 < q_2$ of dyadic rational numbers we have

$L_{q_2 t_0} - L_{q_1 t_0} \geq 0$ almost surely. Since the set of all such pairs is countable, we have almost surely that $L_{q_2 t_0} - L_{q_1 t_0} \geq 0$ for all $0 \leq q_1 < q_2 \in \mathbb{Q}_{\text{dyad}}$. But if for a càdlàg function $f : [0,\infty) \to \mathbb{R}$ we have $f(q_2 t_0) \geq f(q_1 t_0)$ for all $0 \leq q_1 < q_2 \in \mathbb{Q}_{\text{dyad}}$, then $f(t) - f(s) \geq 0$ for all $0 \leq s \leq t$ by right-continuity of f and the density of $\mathbb{Q}_{\text{dyad}}$ in $\mathbb{R}$. Since the paths of L are càdlàg almost surely, this shows that $(L_t)_{t\geq 0}$ is a subordinator. This is (i) when $I = [0,\infty)$ and we have already shown that for $(L_t)_{t\geq 0}$, (i) implies (ii). □

Remark 13.19. (a) The preceding theorem can easily be generalized to higher dimensions. To do so let $L = (L_t)_{t\in I}$ be an $\mathbb{R}^d$-valued Lévy process with Gaussian variance A and Lévy measure ν, and write $L_t = (L_t^1, \ldots, L_t^d)^T$. Then all components $L^1, \ldots, L^d$ of L are subordinators if and only if $A = 0$, $\nu(\mathbb{R}^d \setminus [0,\infty)^d) = 0$, $\int_{[0,1]^d} |x| \, \nu(dx) < \infty$ and the drift γ^0 of L lies in $[0,\infty)^d$. This is also equivalent to the fact that there exists $t_0 > 0$ such that $L_{t_0} \in [0,\infty)^d$ almost surely. This can be proved by arguments completely analogous to those in the proof of Theorem 13.18, or alternatively by using Theorem 13.18 in conjunction with Theorem 9.25 and Corollary 9.26. Sometimes $\mathbb{R}^d$-valued Lévy processes whose components are all subordinators are also called subordinators, or more specifically $\mathbb{R}^d$*-valued subordinators.*

(b) Using Theorems 13.15 and 13.18 together with Corollary 12.39, it is easily seen that an $\mathbb{R}$-valued Lévy process is of finite variation if and only if it can be written as the sum of two independent subordinators, see Exercise 13.6.

From Theorem 13.18, we can now easily characterize those infinitely divisible distributions μ that are concentrated on $[0,\infty)$, or more generally those which have support, denoted by $\operatorname{supp} \mu$, which is bounded from below.

Corollary 13.20 (Infinitely divisible distributions bounded from below). *Let μ be an infinitely divisible distribution on $\mathbb{R}$ with Gaussian variance A and Lévy measure ν.*

(a) *Then $\mu((-\infty,0)) = 0$, i. e., $\operatorname{supp}\mu \subset [0,\infty)$, if and only if $A = 0$, $\nu((-\infty,0)) = 0$, $\int_0^1 x\,\nu(dx) < \infty$, and the drift of μ is non-negative.*

(b) *Suppose that $K \in \mathbb{R}$. Then $\mu((-\infty,K)) = 0$, i. e., $\operatorname{supp}\mu \subset [K,\infty)$, if and only if $A = 0$, $\nu((-\infty,0)) = 0$, $\int_0^1 x\,\nu(dx) < \infty$, and the drift of μ is greater than or equal to K.*

(c) *μ is bounded from below if and only if $A = 0$, $\nu((-\infty,0)) = 0$ and $\int_0^1 x\,\nu(dx) < \infty$.*

Proof. For (a), let $L = (L_t)_{t\geq 0}$ be a Lévy process with $\mathcal{L}(L_1) = \mu$. The assertion then follows from the equivalence of (ii), (iii) and (iv) in Theorem 13.18 with $t_0 = 1$. Part (b) then follows directly from (a) and the facts that $\operatorname{supp}\mu \subset [K,\infty)$ if and only if $\operatorname{supp}(\mu * \delta_{-K}) \subset [0,\infty)$ and that μ and $\mu * \delta_{-K}$ share the same Gaussian variance and Lévy measure and that their drifts differ by K. Part (c) is an immediate consequence of (b). □

Corollary 13.20 characterizes non-negativity of an infinitely divisible random variable. It is also possible to characterize infinite divisibility among all non-negative random variables. This can of course be done in terms of a Lévy–Khintchine representation for the characteristic function, but for non-negative random variables it is more common to work with the Laplace transform rather than the characteristic function.

Definition 13.21 (Laplace transform; Laplace exponent). Let μ be a probability distribution supported in $[0,\infty)$ and X be a random variable with distribution μ. Then the *Laplace transform of* μ, also called the *Laplace transform of* X, and denoted by Lap_μ or equivalently Lap_X is defined by

$$\mathrm{Lap}_\mu(u) = \mathrm{Lap}_X(u) := \int_{[0,\infty)} \mathrm{e}^{-ux}\,\mu(\mathrm{d}x) = \mathbb{E}\,\mathrm{e}^{-uX}, \quad u \in [0,\infty).$$

The *Laplace exponent of* μ, also called the *Laplace exponent of* X, and denoted by Φ_μ or Φ_X, is defined as the negative of the logarithm of the Laplace transform, i. e., by

$$\Phi_\mu(u) := -\log(\mathrm{Lap}_\mu(u)) = -\log(\mathbb{E}\,\mathrm{e}^{-uX}) \quad \forall\, u \geq 0.$$

It is the unique $[0,\infty)$-valued function such that $\mathrm{Lap}_\mu(u) = \mathrm{e}^{-\Phi_\mu(u)}$ for all $u \geq 0$.

A distribution on $[0,\infty)$ is uniquely determined by its Laplace transform and hence by its Laplace exponent (see, e. g., Feller [44, Chapter XIII] for this and other basic facts about the Laplace transform). We have already encountered the Laplace transform in the proof of Theorem 12.36(a), and we shall now use our findings there to characterize infinite divisibility of a non-negative random variable in terms of its Laplace transform.

Theorem 13.22 (Infinite divisibility of non-negative distributions). *Let μ be a probability distribution on $[0,\infty)$, i. e., with $\operatorname{supp}\mu \subset [0,\infty)$. Then μ is infinitely divisible if and only if there exist $\gamma^0 \geq 0$ and a measure ν on $(0,\infty)$ satisfying $\int_{(0,\infty)} (x \wedge 1)\,\nu(\mathrm{d}x) < \infty$ such that*

$$\mathrm{Lap}_\mu(u) = \exp\Big\{-\gamma^0 u - \int_{(0,\infty)} (1-\mathrm{e}^{-ux})\,\nu(\mathrm{d}x)\Big\} \quad \forall\, u \geq 0, \tag{13.7}$$

i. e., such that the Laplace exponent has the representation

$$\Phi_\mu(u) = \gamma^0 u + \int_{(0,\infty)} (1-\mathrm{e}^{-ux})\,\nu(\mathrm{d}x) \quad \forall\, u \geq 0.$$

In that case, the Gaussian variance of μ is 0, its Lévy measure is given by $\mathbf{1}_{(0,\infty)}(x)\,\nu(\mathrm{d}x)$ (i. e., the extension of ν to $\mathbb{R}$ by setting it to 0 on $(-\infty,0]$) and the drift of μ is γ^0.

Proof. Suppose first that μ with $\operatorname{supp}\mu \subset [0,\infty)$ is infinitely divisible. Denote its Lévy measure by ν' and its drift by γ^0. By Corollary 13.20,the Gaussian variance of μ is zero, $\gamma^0 \geq 0$, $\nu'((-\infty,0]) = 0$ and $\int_0^1 x\,\nu'(\mathrm{d}x) < \infty$. Let $L = (L_t)_{t\geq 0}$ be a Lévy process such that $\mathcal{L}(L_1) = \mu$. Defining $X := \sum_{0<s\leq 1} \Delta L_s$, we have $L_1 = \gamma^0 + X$ by Theorem 13.1. Since almost surely, the jump sizes of L are non-negative, we can also write $X = \sum_{0<s\leq 1} |\Delta L_s|$. Defining

$g : [0,\infty) \times \mathbb{R} \to [0,\infty)$ by $g(s,x) = |x|\mathbf{1}_{(0,1]}(s)$ we have $X = \sum_{s\in[0,\infty)} g(s,\Delta L_s)$. Hence we can apply equation (12.10), according to which the Laplace transform of X is given for $u > 0$ by

$$\mathrm{Lap}_X(u) = \exp\Big\{\int_{\mathbb{R}}\int_{[0,\infty)} (\mathrm{e}^{-u|y|\mathbf{1}_{(0,1]}(s)} - 1)\,\mathrm{d}s\,\nu'(\mathrm{d}y)\Big\}$$

$$= \exp\Big\{\int_{(0,\infty)} (\mathrm{e}^{-yu} - 1)\,\nu'(\mathrm{d}y)\Big\}.$$

Setting $\nu := \nu'_{|(0,\infty)}$ and observing that

$$\mathrm{Lap}_\mu(u) = \mathrm{Lap}_{\gamma^0+X}(u) = \mathrm{e}^{-\gamma^0 u}\cdot \mathrm{Lap}_X(u),$$

we obtain (13.7) and the relation with the characteristic triplet of μ. If conversely a probability distribution μ on $[0,\infty)$ has a Laplace transform given by (13.7) with $\gamma^0 \geq 0$ and measure ν as stated, let μ' be an infinitely divisible distribution with Gaussian variance 0, Lévy measure ν' as given (the extension of ν to $\mathbb{R}$) and drift $\gamma^0 \geq 0$. Then $\mathrm{supp}(\mu') \subset [0,\infty)$ by Corollary 13.20 and by the part already proved, the Laplace transform of μ' is also given by the right-hand side of (13.7). Since the Laplace transform determines the distribution, we have $\mu = \mu'$, showing that μ is infinitely divisible. □

Example 13.23 (Gamma distribution). Suppose that μ is the gamma distribution (on $\mathbb{R}$) with parameters $\alpha,\beta > 0$, i. e., that μ has Lebesgue density $x \mapsto \frac{\beta^\alpha}{\Gamma(\alpha)}x^{\alpha-1}\mathrm{e}^{-\beta x}\mathbf{1}_{(0,\infty)}(x)$, where

$$\Gamma(w) = \int_0^\infty t^{w-1}\mathrm{e}^{-t}\,\mathrm{d}t, \quad w \in \mathbb{C} : \Re w > 0,$$

denotes the gamma function. For $\alpha = 1$ the gamma distribution reduces to the exponential distribution with parameter β. Using the substitution $v = (\beta + u)x$, the Laplace transform of the gamma distribution is calculated for $u \geq 0$ as

$$\mathrm{Lap}_\mu(u) = \frac{\beta^\alpha}{\Gamma(\alpha)}\int_0^\infty x^{\alpha-1}\mathrm{e}^{-(\beta+u)x}\,\mathrm{d}x$$

$$= \frac{\beta^\alpha}{\Gamma(\alpha)}(\beta+u)^{-\alpha}\int_0^\infty v^{\alpha-1}\mathrm{e}^{-v}\,\mathrm{d}v = \Big(\frac{\beta}{\beta+u}\Big)^\alpha,$$

so that the Laplace exponent is given by

$$\Phi_\mu(u) = \alpha\log(1 + u/\beta), \quad u \geq 0.$$

Arguing as in Example 9.16(e), an application of the fundamental theorem of calculus and Tonelli's theorem yields

$$\begin{aligned}\Phi_\mu(u) &= \alpha \int_0^u \frac{\mathrm{d}}{\mathrm{d}t} \log(1+t/\beta)\,\mathrm{d}t = \alpha \int_0^u \frac{1}{\beta+t}\,\mathrm{d}t \\ &= \alpha \int_0^u \int_0^\infty \mathrm{e}^{-(\beta+t)x}\,\mathrm{d}x\,\mathrm{d}t = \alpha \int_0^\infty \mathrm{e}^{-\beta x} \int_0^u \mathrm{e}^{-tx}\,\mathrm{d}t\,\mathrm{d}x \\ &= \int_0^\infty (1-\mathrm{e}^{-ux})\,\alpha \frac{\mathrm{e}^{-\beta x}}{x}\,\mathrm{d}x, \quad u \geq 0.\end{aligned}$$

An application of Theorem 13.22 shows that μ is infinitely divisible with Gaussian variance 0, drift 0 and Lévy measure having Lévy density $x \mapsto \alpha x^{-1}\mathrm{e}^{-\beta x}\mathbf{1}_{(0,\infty)}(x)$. For $\alpha = 1$ this reduces to the characteristic triplet of the exponential distribution found in Example 9.16(e). Since the gamma distribution with parameters $n/2$ and $\beta = 1/2$ with $n \in \mathbb{N}$ is the chi-squared-distribution with n degrees of freedom, it follows that the chi-squared-distribution is infinitely divisible.

Since for a Lévy process $L = (L_t)_{t\geq 0}$, the characteristic triplet of L_t is t times the characteristic triplet of L_1, Theorem 13.22 immediately implies that the Laplace transform of a subordinator L at any time $t \geq 0$ has the following form.

Corollary 13.24 (Laplace transform of subordinator). *Let $L = (L_t)_{t\geq 0}$ be a subordinator with Lévy measure ν and drift γ^0. Then for each $t \geq 0$, the Laplace transform of L_t is given by*

$$\mathrm{Lap}_{L_t}(u) = \exp\Big\{-\gamma^0 ut - t \int_{(0,\infty)} (1-\mathrm{e}^{-ux})\,\nu(\mathrm{d}x)\Big\} \quad \forall\, u \geq 0.$$

Since the distribution of a subordinator is uniquely determined by its distribution at time 1, we make the following definition, just as we did it for the characteristic exponent of a Lévy process.

Definition 13.25 (Laplace exponent of subordinator). Suppose that $I = [0,\infty)$ or $I = \mathbb{R}$ and let $L = (L_t)_{t\in I}$ be an $\mathbb{R}$-valued subordinator. Then the Laplace exponent of L_1 is also called the *Laplace exponent of* L, and also denoted by $\Phi_L := \Phi_{L_1}$.

Subordinators are widely used for the modeling of non-negative processes and in particular for the modeling of the stochastic volatility processes, which play an important role in financial mathematics.

Now let Ψ_X be the characteristic exponent of some infinitely divisible distribution $\mu = \mathcal{L}(X)$ on $[0,\infty)$, so that $\Psi_X(z) = \mathrm{i}\gamma^0 z + \int_{(0,\infty)} (\mathrm{e}^{\mathrm{i}zx} - 1)\,\nu(\mathrm{d}x)$ for $z \in \mathbb{R}$, where γ^0 and ν denote the drift and Lévy measure of μ, respectively. Then $\mathbb{E}\,\mathrm{e}^{\mathrm{i}zX} = \mathrm{e}^{\Psi_X(z)}$ for

$z \in \mathbb{R}$ by the definition of the characteristic exponent. Inserting formally the purely imaginary number $z = iu$ with $u \geq 0$ in Ψ_X instead of a real number, equation (13.7) means nothing other than $\mathbb{E}\, e^{-uX} = e^{\Psi_X(iu)}$ for all $u \geq 0$, i. e., that $\Phi_X(u) = -\Psi_X(iu)$. This can be formalized and extended to hold for all complex numbers in the upper half plane, as shown in Proposition 13.26.

Proposition 13.26 (Relation between characteristic/Laplace exponent). *Let* $\mu = \mathcal{L}(X)$ *be an infinitely divisible distribution on* $[0, \infty)$ *with Lévy measure* ν *and drift* γ^0. *Then* $\mathbb{E}|e^{izX}|$ *and* $\int_{(0,\infty)} |e^{izx} - 1|\, \nu(dx)$ *are finite for each* $z \in \mathbb{C}$ *with* $\Im z \geq 0$ *and*

$$\mathbb{E}\, e^{izX} = \exp\Big\{ i\gamma^0 z + \int_{(0,\infty)} (e^{izx} - 1)\, \nu(dx) \Big\} \quad \forall\, z \in \mathbb{C} : \Im z \geq 0. \tag{13.8}$$

Extending the characteristic exponent $z \mapsto \Psi_X(z)$, $z \in \mathbb{R}$, *of* μ *to the complex upper half-plane by setting*

$$\Psi_X(z) := i\gamma^0 z + \int_{(0,\infty)} (e^{izx} - 1)\, \nu(dx) \quad \forall\, z \in \mathbb{C} : \Im(z) \geq 0, \tag{13.9}$$

equation (13.8) *can be rewritten as*

$$\mathbb{E}\, e^{izX} = e^{\Psi_X(z)} \quad \forall\, z \in \mathbb{C} : \Im z \geq 0. \tag{13.10}$$

In particular, for $z = iu$ *with* $u \in [0, \infty)$, *we have*

$$\mathbb{E}\, e^{-uX} = e^{\Psi_X(iu)} = \exp\Big\{ -\gamma^0 u - \int_{(0,\infty)} (1 - e^{-ux})\, \nu(dx) \Big\} \quad \forall\, u \in [0, \infty), \tag{13.11}$$

i. e.,

$$\Phi_X(u) = -\Psi_X(iu) \quad \forall\, u \in [0, \infty). \tag{13.12}$$

Proof. Denote by $H := \{z \in \mathbb{C} : \Im z > 0\}$ the open upper complex half-plane and by $\overline{H} := \{z \in \mathbb{C} : \Im z\}$ the closed upper complex half-plane. That $\mathbb{E}|e^{izX}|$ is finite for $z \in \overline{H}$ is clear, since the support of X is contained in $[0, \infty)$. Similarly, since the support of ν is contained in $[0, \infty)$ and since $\int_{(0,1]} x\, \nu(dx) < \infty$ by Corollary 13.20(a), it follows easily that $\int_{(0,\infty)} |e^{izx} - 1|\, \nu(dx)$ is finite for each $z \in \overline{H}$. Hence we can extend Ψ_X to $\overline{H}$ by (13.9). The function $z \mapsto (e^{izx} - 1)$ on $\overline{H}$ is continuous in z and for each $z_0 \in \overline{H}$ we can find an open neighborhood U_{z_0} of z_0 such that $\int_{(0,\infty)} \sup_{z \in U_{z_0} \cap \overline{H}} |e^{izx} - 1|\, \nu(dx) < \infty$. The continuity lemma (an easy consequence of the dominated convergence theorem, see, e. g., Bauer [10, Lemma 16.1]) then shows that $z \mapsto \int_{(0,\infty)} (e^{izx} - 1)\, \nu(dx)$ is continuous on $\overline{H}$. Next, write $z = u + iv$ for $z \in H$ (the open half-plane), and observe that both

$\Re(\mathrm{e}^{\mathrm{i}(u+\mathrm{i}v)x} - 1)$ and $\Im(\mathrm{e}^{\mathrm{i}(u+\mathrm{i}v)x} - 1)$ are partially differentiable with respect to u and v for fixed $x > 0$ and that the derivatives satisfy the Cauchy–Riemann equations, since $z \mapsto \mathrm{e}^{\mathrm{i}zx}$ is actually holomorphic on H (e. g., Conway [34, Eq. III.2.24]). Again, it is easily seen that for each $z_0 \in H$ there is an open neighbourhood U_{z_0} of z_0 in H such that $\int_{(0,\infty)} \sup_{u+\mathrm{i}v\in U_{z_0}} |\frac{\partial}{\partial u}\Re(\mathrm{e}^{\mathrm{i}(u+\mathrm{i}v)x} - 1)|\, \nu(\mathrm{d}x) < \infty$, and similarly for the imaginary part and the partial derivatives with respect to v. It follows from the differentiation lemma (again an easy consequence of the dominated convergence theorem, see, e. g., Bauer [10, Cor. 16.3]) that $\int_{(0,\infty)} \Re(\mathrm{e}^{\mathrm{i}(u+\mathrm{i}v)x} - 1)\, \nu(\mathrm{d}x)$ is partially differentiable with respect to u and v and that the partial differential operators can be taken under the integral sign. The same is true for the imaginary part. Using the continuity lemma again, it is easy to see that these partial derivatives are continuous on H. Since $u + \mathrm{i}v \mapsto \mathrm{e}^{(u+\mathrm{i}v)x} - 1$ satisfies the Cauchy–Riemann equations for fixed $x \in (0, \infty)$, taking the partial derivatives under the integral shows that the same is true for $\int_{(0,\infty)} (\mathrm{e}^{\mathrm{i}(u+\mathrm{i}v)x} - 1)\, \nu(\mathrm{d}x)$, and since the partial derivatives are continuous we conclude that $z \mapsto \int_{(0,\infty)} (\mathrm{e}^{\mathrm{i}zx} - 1)$ is holomorphic on H (e. g., Conway [34, Thm. III.2.29]). (We have in effect just given proof of a differentiation lemma for integrals depending in a holomorphic way on a complex parameter. Although this result is standard and well known, it has turned out to be very difficult to give a textbook reference for this result. The only textbook reference we are aware of is by Elstrodt [42, Satz IV.5.8], which unfortunately is not available in English. See also the article [67] by Mattner for a version of the differentiation lemma for holomorphic parameter-dependent integrals.)

We have seen that Ψ_X and hence e^{Ψ_X} are continuous on $\overline{H}$ and holomorphic on H. By the same argument, it follows that $z \mapsto \mathbb{E}\,\mathrm{e}^{\mathrm{i}zX}$ is continuous on $\overline{H}$ and holomorphic on H. However, for $z \in \mathbb{R}$ we have $\mathbb{E}\,\mathrm{e}^{\mathrm{i}zX} = \mathrm{e}^{\Psi_X(z)}$, so that $z \mapsto \mathbb{E}\,\mathrm{e}^{\mathrm{i}zX}$ and $z \mapsto \mathrm{e}^{\Psi_X(z)}$ are both continuous on $\overline{H}$, holomorphic on H and agree on the boundary $\mathbb{R}$ of the upper half plane. It follows then, from a well-known property of holomorphic functions (see, e. g., Kawata [56, Cor. 1.14.1]) that both functions coincide on $\overline{H}$, i. e. that (13.10) and hence (13.8) hold. Equations (13.11) and (13.12) are then immediate consequences of (13.8) and (13.10). □

Remark 13.27. Let $H := \{z \in \mathbb{C} : \Im(z) > 0\}$, $\overline{H} := \{z \in \mathbb{C} : \Im(z) \geq 0\}$ and $f, g : \overline{H} \to \mathbb{C}$ be continuous on $\overline{H}$ and holomorphic on H. The proof of Proposition 13.26 used the fact that if f and g agree on the real line, then they agree everywhere on $\overline{H}$ (cf. [56, Cor. 1.14.1]). Similarly, if f and g agree on $\{\mathrm{i}u : u > 0\}$, or more generally on a set $S \subset H$ which has an accumulation point in H, then they agree on H by the identity theorem for analytic functions (e. g., Conway [34, Cor. IV.3.8]), hence they agree also on $\overline{H}$ by continuity. Now let $\mu = \mathcal{L}(X)$ be a probability distribution on $[0, \infty)$ with characteristic function $\widehat{\mu} : \mathbb{R} \to \mathbb{C}$ and Laplace transform $\mathrm{Lap}_\mu : [0, \infty) \to \mathbb{R}$. As seen in the proof of Proposition 13.26, the function $f : \overline{H} \to \mathbb{C}$, $z \mapsto \mathbb{E}\,\mathrm{e}^{\mathrm{i}Xz}$ is continuous on $\overline{H}$ and holomorphic on H and agrees with $\widehat{\mu}$ on the real line, while $[0, \infty) \to \mathbb{R}$, $u \mapsto f(\mathrm{i}u)$ agrees with Lap_μ. It follows that if $g : \overline{H} \to \mathbb{C}$ is some continuous function which agrees with $\widehat{\mu}$ on the real line and is holomorphic on H, then $[0, \infty) \to \mathbb{R}$, $u \mapsto g(\mathrm{i}u)$ equals the Laplace transform of μ.

Conversely, if $g : \overline{H} \to \mathbb{C}$ is continuous, holomorphic on H and the function $u \mapsto g(\mathrm{i}u)$ on $[0, \infty)$ agrees with the Laplace transform of μ, then g agrees with $\widehat{\mu}$ on the real line. Hence, in the case where one has the Laplace transform of a distribution on $[0, \infty)$ for which it is easy to guess an analytic extension yields the characteristic function, and vice versa.

Example 13.28 (Gamma distribution). In Example 13.23, we found that the Laplace transform of the gamma distribution μ with parameters $\alpha, \beta > 0$ is given by $\mathrm{Lap}_\mu(u) = \exp(-\alpha \log(1 + u/\beta))$ for $u \geq 0$. Now define $g : \overline{H} \to \mathbb{C}$ by $g(z) := \exp(-\alpha \log(1 - \mathrm{i}z/\beta))$, where log denotes the principal branch of the complex logarithm. This is continuous on $\overline{H}$ and holomorphic on H and $g(\mathrm{i}u)$ agrees with $\mathrm{Lap}_\mu(u)$ for $u \geq 0$. By Remark 13.27, the characteristic function of μ is given by

$$\widehat{\mu}(z) = \exp[-\alpha \log(1 - \mathrm{i}z\beta^{-1})], \quad z \in \mathbb{R}.$$

13.4 Exercises

Exercise 13.1. In Step 2 of the proof of Theorem 13.3, we used

$$|||S^\varepsilon||| := 1 \wedge \sup_{t \in [0,T]} |S_t^\varepsilon|$$

rather than the supremum norm $\|\cdot\|_\infty$ on $[0, T]$. The advantage of using $|||\cdot|||$ is that it is bounded (by 1). Where in the proof of Step 2 was it important to work with $|||\cdot|||$ rather than $\|\cdot\|$?

Exercise 13.2. Let $L = (L_t)_{t\geq 0}$ be an $\mathbb{R}^d$-valued Lévy process with standard characteristic triplet (A, ν, γ). Let $B \in \mathcal{B}_d$ with $0 \notin B$. Show that almost surely, $\sum_{0<s\leq t, |\Delta L_s|>\varepsilon, \Delta L_s \in B} \Delta L_s - t \int_{B \cap \{\varepsilon < |x| \leq 1\}} x\, \nu(\mathrm{d}x)$ converges uniformly on compact subsets of $[0, \infty)$ to a Lévy process X^B. Show further that X and $L - X^B$ are independent and determine the standard characteristic triplets of X and $L - X^B$.

Exercise 13.3. Let $g : [0, \infty) \to \mathbb{R}$ be given by $g(0) = 0$ and $g(x) = x \sin(x^{-1})$ for $x > 0$. Show that g is continuous but not of finite variation.
Hint: For $n \geq 3$, consider the partition $\pi_n = (t_{0,n}, \dots, t_{n,n})$ of $[0, 1]$, defined by $t_{0,n} = 0$, $t_{n,n} = 1$ and

$$t_{j,n} := \pi^{-1}((n - j) + 1/2)^{-1}, \quad j \in \{1, \dots, n - 1\}.$$

Exercise 13.4. Suppose that $a < b \in \mathbb{R}$ and $g : [a, b] \to \mathbb{R}$ is a càdlàg function. Denote by $\widetilde{\Pi}_{[a,b]}$ the set of all partitions $\pi = (t_0, t_1, \dots, t_n)$ of $[a, b]$, which are dyadic in the sense that

$$\{t_0, \dots, t_n\} \subset \{a + k2^{-m}(b - a) : k, m \in \mathbb{N}_0\}.$$

Show that

$$\mathrm{TV}_{[a,b]}(g) = \sup_{(t_0,\dots,t_n)\in\widetilde{\Pi}_{[a,b]}} \sum_{j=1}^{n} |g_{t_j} - g_{t_{j-1}}|.$$

Exercise 13.5. Prove Remark 13.19(a).

Exercise 13.6. Prove Remark 13.19(b), i. e., show that an $\mathbb{R}$-valued Lévy process is of finite variation if and only if it can be written as the difference of two independent subordinators.

14 Examples of Lévy processes

Abstract: So far we have encountered Brownian motion, Poisson processes and compound Poisson processes as our main examples of Lévy processes. However, we know from Theorem 10.26 that there is a one-to-one correspondence between Lévy processes in law and infinitely divisible distributions, and from Corollary 10.42 we know that every Lévy process in law has a modification that is a Lévy process. This means that the class of Lévy processes consists of many more processes than simply linear combinations of independent Brownian motion and compound Poisson processes. In this chapter, we define and derive properties of some important further examples of Lévy processes in this large class, including stable Lévy processes, the gamma process, the variance-gamma process and the inverse Gaussian Lévy process.

14.1 Stable distributions and stable Lévy processes

In this section, we treat an important class of infinitely divisible distributions and the associated Lévy processes, namely the stable distributions and stable Lévy processes. We define stable distributions as infinitely divisible distributions with a certain characteristic triplet. To prepare, we need the following lemma.

Lemma 14.1. *Suppose that $\alpha \in (0,2)$ and ρ is a finite measure on $(S^{d-1}, \mathcal{B}(S^{d-1}))$, where $S^{d-1} = \{x \in \mathbb{R}^d : |x| = 1\}$ denotes the unit sphere in $\mathbb{R}^d$. Then*

$$\nu(B) := \int_{S^{d-1}} \int_0^\infty \mathbf{1}_B(r\xi) r^{-1-\alpha} \, \mathrm{d}r \, \rho(\mathrm{d}\xi) \quad \forall\, B \in \mathcal{B}_d \tag{14.1}$$

defines a Lévy measure on $\mathbb{R}^d$. If $f : \mathbb{R}^d \to [0,\infty]$ is a non-negative Borel-measurable function, we have the identity

$$\int_{\mathbb{R}^d} f(x)\, \nu(\mathrm{d}x) = \int_{S^{d-1}} \int_0^\infty f(r\xi)\, r^{-1-\alpha} \, \mathrm{d}r \, \rho(\mathrm{d}\xi). \tag{14.2}$$

Furthermore, if a Lévy measure ν on $\mathbb{R}^d$ has a representation as in (14.1) *with some $\alpha \in (0,2)$ and some finite non-zero measure ρ on S^{d-1}, then α and ρ are uniquely determined by ν.*

Proof. Let ρ be a finite measure on $(S^{d-1}, \mathcal{B}(S^{d-1}))$ and define $\nu(B)$ for $B \in \mathcal{B}_d$ by (14.1). It is clear that $\nu(\emptyset) = 0$ and that ν is σ-additive, so that $\nu : \mathcal{B}_d \to [0,\infty]$ is indeed a measure, and it obviously satisfies $\nu(\{0\}) = 0$. To see that it is a Lévy measure, let $c_1, \dots, c_n \geq 0$ and $B_1, \dots, B_n \in \mathcal{B}_d$. Then

https://doi.org/10.1515/9783111325033-014

$$\int_{\mathbb{R}^d} \sum_{j=1}^{n} c_j \mathbf{1}_{B_j}(x)\, \nu(\mathrm{d}x) = \int_{S^{d-1}} \int_0^\infty \sum_{j=1}^{n} c_j \mathbf{1}_{B_j}(r\xi) r^{-1-\alpha}\, \mathrm{d}r\, \rho(\mathrm{d}\xi)$$

by (14.1), and since every Borel measurable function $f : \mathbb{R}^d \to [0, \infty]$ can be approximated from below by a non-decreasing sequence of elementary functions, we conclude that $\int_{\mathbb{R}^d} f(x)\nu(\mathrm{d}x) = \int_{S^{d-1}} \int_0^\infty f(r\xi)r^{-1-\alpha}\, \mathrm{d}r\, \rho(\mathrm{d}\xi)$ for such f, which is (14.2). Taking in particular $f(x) = 1 \wedge |x|^2$, we see that

$$\int_{\mathbb{R}^d} (1 \wedge |x|^2)\, \nu(\mathrm{d}x) = \int_{S^{d-1}} \left(\int_0^1 r^{1-\alpha}\, \mathrm{d}r + \int_1^\infty r^{-1-\alpha}\, \mathrm{d}r \right) \rho(\mathrm{d}\xi) < \infty$$

so that ν is a Lévy measure.

It remains to show uniqueness of ρ and α in the representation (14.1), provided ρ is non-zero. To see that α is uniquely determined by ν, an application of (14.2) to $f(x) = |x|^\beta \mathbf{1}_{(0,1]}(|x|)$ with $\beta > 0$ yields

$$\int_{|x|\le 1} |x|^\beta\, \nu(\mathrm{d}x) = \int_{S^{d-1}} \int_0^1 r^{\beta-1-\alpha}\, \mathrm{d}r\, \rho(\mathrm{d}\xi) = \rho(S^{d-1}) \int_0^1 r^{\beta-1-\alpha}\, \mathrm{d}r.$$

Since $0 < \rho(S^{d-1}) < \infty$ and $\int_0^1 r^{\beta-1-\alpha}\, \mathrm{d}r$ is finite if and only if $\beta > \alpha$, we obtain

$$\alpha = \inf\left\{ \beta > 0 : \int_{|x|\le 1} |x|^\beta\, \nu(\mathrm{d}x) < \infty \right\},$$

showing that α is uniquely determined by ν. To see the uniqueness of ρ, observe from (14.1), that for $C \in \mathcal{B}(S^{d-1})$,

$$\nu(\{x \in \mathbb{R}^d : |x| > 1, |x|^{-1}x \in C\}) = \int_C \int_1^\infty r^{-1-\alpha}\, \mathrm{d}r\, \rho(\mathrm{d}\xi) = \alpha^{-1}\rho(C),$$

showing that ρ is uniquely determined by ν. □

In (14.1), the Lévy measure ν is decomposed into a *spherical part* ρ and a *radial part*. Also observe that (14.2) naturally extends to integrable functions $f : \mathbb{R}^d \to [-\infty, \infty]$ that are not necessarily non-negative.

We can now define stable distributions:

Definition 14.2 (Stable distribution; index of stability). Let μ be a probability distribution on $\mathbb{R}^d$. Then μ is called a *stable distribution*, if it is a Gaussian distribution, or if it is infinitely divisible with zero Gaussian variance and Lévy measure ν of the form (14.1) with some $\alpha \in (0, 2)$ and some finite measure ρ on $(S^{d-1}, \mathcal{B}(S^{d-1}))$. In the latter case, the probability distribution μ is called *α-stable* (with $\alpha \in (0,2)$), and Gaussian distributions are also called *2-stable*. If μ is a stable distribution that is not a Dirac measure, then the unique $\alpha \in (0,2]$ such that μ is α-stable is called the *index of stability of* μ. A random vector X is called *stable* if $\mathcal{L}(X)$ is stable, and α-stability and the index of stability of X are defined likewise.

The following results shed some light on the name *stable distribution*:

Proposition 14.3 (I. i. d. sums of stable random vectors). *Suppose that $\alpha \in (0,2]$ and μ is an α-stable distribution on $\mathbb{R}^d$. Let $(X_n)_{n\in\mathbb{N}}$ be an i. i. d. sequence with distribution μ. Then for each $n \in \mathbb{N}$ there is a constant $d_n \in \mathbb{R}^d$ such that*

$$X_1 + \cdots + X_n \stackrel{d}{=} n^{1/\alpha} X_1 + d_n. \tag{14.3}$$

Proof. Consider first the case $\alpha = 2$. Let $\mu = N(\gamma, A)$ be the normal distribution on $\mathbb{R}^d$ with mean γ and covariance matrix A. Then $X_1 + \cdots + X_n$ is $N(n\gamma, nA)$ distributed, while $n^{1/2}X_1 + d_n$ for $d_n \in \mathbb{R}^d$ is $N(n^{1/2}\gamma + d_n, nA)$ distributed. Choosing $d_n = (n - n^{1/2})\gamma$, we obtain (14.3).

Now suppose that $\alpha \in (0,2)$ and that μ has Lévy measure (14.1) with some finite measure ρ on S^{d-1}. Then $X_1 + \cdots + X_n$ has Gaussian variance 0 and Lévy measure $n\nu$. On the other hand, by Theorem 9.25, $n^{1/\alpha}X_1$ has Gaussian variance 0 and Lévy measure $\nu'(B) = \nu(n^{-1/\alpha}B)$, $B \in \mathcal{B}_d$. Using the specific form of ν, from (14.1) and the substitution $s = n^{1/\alpha}r$, we find that

$$\begin{aligned}\nu'(B) &= \int_{S^{d-1}} \int_0^\infty \mathbf{1}_B(n^{1/\alpha} r\xi)\, r^{-1-\alpha}\, \mathrm{d}r\, \rho(\mathrm{d}\xi) \\ &= \int_{S^{d-1}} \int_0^\infty \mathbf{1}_B(s\xi) n^{1+1/\alpha} s^{-1-\alpha} n^{-1/\alpha}\, \mathrm{d}s\, \rho(\mathrm{d}\xi) = n\nu(B).\end{aligned}$$

Hence $X_1 + \cdots + X_n$ and $n^{1/\alpha}X_1$ have the same Gaussian variances and the same Lévy measures, and since a shift of size d_n in a distribution changes the characteristic triplet only by adding d_n to the location parameter, we see that there must be $d_n \in \mathbb{R}^d$ such that (14.3) is satisfied. □

Remark 14.4. Proposition 14.3 shows that a stable distribution μ is *stable* in the sense that n-fold convolutions μ^{*n} of μ with itself are again of the same type, in the sense that μ^{*n} is equal to a scaled and shifted version of μ. This is a kind of stability property of such distributions, which explains the name. As a converse to Proposition 14.3, it can be shown that if μ is a probability distribution on $\mathbb{R}^d$ such that for each $n \in \mathbb{N}$ there exist $c_n \in (0,\infty)$ and $d_n \in \mathbb{R}^d$ such that $X_1 + \cdots + X_n \stackrel{d}{=} c_n X_1 + d_n$ (where $(X_m)_{m\in\mathbb{N}}$ is i. i. d. with distribution μ), then μ must be stable in the sense of Definition 14.2; see, e. g., Kuelbs [60, Lemma 2.2] for a proof.

From Proposition 14.3, we see that stable distributions occur as limit distributions of suitably scaled and shifted random walks. More precisely, we have the following corollary.

Corollary 14.5 (Stable distributions as limit distributions). *Suppose that $\alpha \in (0,2]$ and μ is an α-stable distribution on $\mathbb{R}^d$. Then there exists an i. i. d. sequence $(X_n)_{n\in\mathbb{N}}$ of $\mathbb{R}^d$-valued random vectors and an $\mathbb{R}^d$-valued sequence $(b_n)_{n\in\mathbb{N}}$ such that*

$$n^{-1/\alpha}\left(\sum_{i=1}^{n} X_i - b_n\right) \xrightarrow{d} \mu, \quad n \to \infty.$$

Proof. For a given α-stable distribution μ, choose an i. i. d. sequence $(X_n)_{n\in\mathbb{N}}$ with distribution μ. Let $d_n \in \mathbb{R}^d$ be such that (14.3) is satisfied. Then $n^{-1/\alpha}(X_1 + \cdots + X_n - d_n) \stackrel{d}{=} X_1$, which clearly converges in distribution to μ as $n \to \infty$. □

Remark 14.6. As for Proposition 14.3, there also exists a converse to Corollary 14.5. Namely, it can be shown that if for an i. i. d. sequence $(X_n)_{n\in\mathbb{N}}$ there are $a_n \in (0,\infty)$ and $b_n \in \mathbb{R}^d$ such that $(\sum_{i=1}^{n} X_i - b_n)/a_n$ converges in distribution to some probability distribution μ as $n \to \infty$, then μ must be stable, see Sato [79, Thm. 15.7] for a proof. In that sense, stable distributions are the only distributions that appear as limit distributions of suitably scaled and shifted random walks, explaining the importance of stable distributions.

From Definition 14.2, we obtain the following lemma.

Lemma 14.7 (Existence of moments and drift of α-stable distributions). *Suppose that $\alpha \in (0,2)$, $\beta > 0$, that μ is a non-degenerate α-stable distribution with Lévy measure ν and that X is a random vector with distribution μ. Then $\int_{|x|\le 1} |x|^\beta\, \nu(\mathrm{d}x) < \infty$ if and only if $\beta > \alpha$, and $\mathbb{E}|X|^\beta < \infty$ if and only if $\beta < \alpha$. In particular, $\int_{|x|\le 1} |x|\, \nu(\mathrm{d}x) < \infty$ (so that the drift of μ exists) if and only if $\alpha < 1$, and $\mathbb{E}|X| < \infty$ if and only if $\alpha > 1$. The second moment of a non-Gaussian stable distribution is always infinite.*

Proof. It follows immediately from (14.1) and (14.2) that $\int_{|x|\le 1} |x|^\beta\, \nu(\mathrm{d}x) < \infty$ if and only if $\beta > \alpha$ and that $\int_{|x|>1} |x|^\beta\, \nu(\mathrm{d}x) < \infty$ if and only if $\beta < \alpha$. Hence, the drift of μ exists if and only if $\alpha < 1$, and $\mathbb{E}|X|^\beta < \infty$ if and only if $\beta < \alpha$, the latter by Corollary 11.14. The last assertion is clear, since the only non-Gaussian stable distributions are non-degenerate and have index of stability less then 2. □

By definition, all α-stable distributions with $\alpha < 2$ are parameterized by a finite Borel measure ρ on S^{d-1} and a location parameter $\gamma \in \mathbb{R}^d$ (the case $\rho = 0$ corresponds to a Dirac measure). In dimension 1 this becomes particularly simple, since then $S^0 = \{-1, 1\} \subset \mathbb{R}$ and the only finite measures ρ on S^0 must be of the form $\rho = p_{-1}\delta_{-1} + p_1\delta_1$ with $p_{-1}, p_1 \in [0,\infty)$. This shows that a one-dimensional distribution is α-stable (with $\alpha < 2$) if and only if it is infinitely divisible, with Gaussian variance 0 and Lévy measure ν satisfying

$$\nu(B) = p_{-1} \int_{B\cap(-\infty,0)} |x|^{-1-\alpha}\, \mathrm{d}x + p_1 \int_{B\cap(0,\infty)} x^{-1-\alpha}\, \mathrm{d}x,$$

or in other words, ν has the Lévy density

$$\nu(\mathrm{d}x) = |x|^{-1-\alpha}(p_{-1}\mathbf{1}_{(-\infty,0)}(x) + p_1\mathbf{1}_{(0,\infty)}(x))\, \mathrm{d}x.$$

With some tedious calculations, this can be rewritten in the following form.

Theorem 14.8 (Characteristic function of α-stable distribution in $\mathbb{R}$). *If $d = 1$, $\alpha \in (0,2)$ and μ is a non-degenerate α-stable distribution on $\mathbb{R}$, there exist unique constants $\sigma > 0$, $\beta \in [-1,1]$ and $\tau \in \mathbb{R}$ such that the characteristic function $\widehat{\mu}$ of μ is given for $z \in \mathbb{R}$ by*

$$\widehat{\mu}(z) = \exp\left(-\sigma^\alpha |z|^\alpha \left(1 - \mathrm{i}\beta \tan\frac{\pi\alpha}{2}\, \mathrm{sign}(z)\right) + \mathrm{i}\tau z\right), \quad \text{if } \alpha \neq 1, \tag{14.4}$$

and

$$\widehat{\mu}(z) = \exp\left(-\sigma|z|\left(1 + \mathrm{i}\beta\frac{2}{\pi}\, \mathrm{sign}(z)\log|z|\right) + \mathrm{i}\tau z\right), \quad \text{if } \alpha = 1. \tag{14.5}$$

Conversely, for every $\sigma > 0$, $\beta \in [-1,1]$ and $\tau \in \mathbb{R}$, the right-hand sides of (14.4) *and* (14.5) *define the characteristic functions of non-degenerate α-stable distributions with $\alpha \neq 1$ and $\alpha = 1$, respectively.*

Proof. We split the proof into four parts. First, we derive a specific representation of the gamma function $\Gamma(\beta)$ for $\beta \in (0,1)$. Then we will prove the statements one by one when $\alpha \in (0,1)$, $\alpha \in (1,2)$ and finally when $\alpha = 1$.

(i) Recall that for $\beta > 0$ the gamma function is given by $\Gamma(\beta) = \int_0^\infty t^{\beta-1}\mathrm{e}^{-t}\, \mathrm{d}t$. We claim that

$$\lim_{T\to\infty} \int_0^T \mathrm{e}^{\mathrm{i}t} t^{\beta-1}\, \mathrm{d}t = \mathrm{e}^{\mathrm{i}\beta\pi/2}\Gamma(\beta), \tag{14.6}$$

whenever $\beta \in (0,1)$. To see this, denote by log the principal branch of the logarithm and consider the holomorphic function

$$g(z) := \exp(-z + (\beta - 1)\log z), \quad z \in \mathbb{C} \setminus \{x \in \mathbb{R} : x \le 0\}.$$

For $0 < \delta < T < \infty$, consider the complex paths $\gamma_{1,\delta}, \gamma_{2,\delta,T}, \gamma_{3,T}, \gamma_{4,\delta,T}$, defined by

$$\begin{aligned}
\gamma_{1,\delta} &: [0,\pi/2] \to \mathbb{C}, \quad t \mapsto \delta\mathrm{e}^{\mathrm{i}t}, \\
\gamma_{2,\delta,T} &: [\delta,T] \to \mathbb{C}, \quad t \mapsto \mathrm{i}t, \\
\gamma_{3,T} &: [0,\pi/2] \to \mathbb{C}, \quad t \mapsto T\mathrm{e}^{\mathrm{i}(\pi/2-t)} \quad \text{and} \\
\gamma_{4,\delta,T} &: [0,T-\delta] \to \mathbb{C}, \quad t \mapsto T - t.
\end{aligned}$$

These four paths together describe a closed curve in $\mathbb{C} \setminus \{x \in \mathbb{R} : x \le 0\}$ so that, by Cauchy's integral theorem,

$$\int_{\gamma_{1,\delta}} g(z)\,\mathrm{d}z + \int_{\gamma_{2,\delta,T}} g(z)\,\mathrm{d}z + \int_{\gamma_{3,T}} g(z)\,\mathrm{d}z + \int_{\gamma_{4,\delta,T}} g(z)\,\mathrm{d}z = 0. \tag{14.7}$$

Since

$$\int_{\gamma_{1,\delta}} g(z)\,\mathrm{d}z = \int_0^{\pi/2} g(\delta \mathrm{e}^{\mathrm{i}t})\gamma_{1,\delta}'(t)\,\mathrm{d}t = \int_0^{\pi/2} \mathrm{e}^{-\delta \mathrm{e}^{\mathrm{i}t}} \delta^{\beta-1}\mathrm{e}^{(\beta-1)\mathrm{i}t}\mathrm{i}\delta\mathrm{e}^{\mathrm{i}t}\,\mathrm{d}t$$

and $\beta > 0$, we conclude that $\lim_{\delta\to 0}\int_{\gamma_{1,\delta}} g(z)\,\mathrm{d}z = 0$. Similarly,

$$\begin{aligned}\left|\int_{\gamma_{3,T}} g(z)\,\mathrm{d}z\right| &= \left|\int_0^{\pi/2} \mathrm{e}^{-T\mathrm{e}^{\mathrm{i}(\pi/2-t)}} T^{\beta-1}\mathrm{e}^{(\beta-1)\mathrm{i}(\pi/2-t)}(-\mathrm{i}T)\mathrm{e}^{\mathrm{i}(\pi/2-t)}\,\mathrm{d}t\right| \\ &\le T^\beta \int_0^{\pi/2} \mathrm{e}^{-T\cos(\pi/2-t)}\,\mathrm{d}t = T^\beta \int_0^{\pi/2} \mathrm{e}^{-T\sin(t)}\,\mathrm{d}t.\end{aligned}$$

Let $c > 0$ be a constant such that $\sin(t) \ge ct$ for all $t \in [0, \pi/2]$. Then

$$\left|\int_{\gamma_{3,T}} g(z)\,\mathrm{d}z\right| \le T^\beta \int_0^{\pi/2} \mathrm{e}^{-cTt}\,\mathrm{d}t = T^{\beta-1}c^{-1}(1 - \mathrm{e}^{-cT\pi/2})$$

which tends to 0 as $T \to \infty$ since $\beta < 1$. Since $\log(\mathrm{i}t)$ for $t > 0$ is given by $\log(t) + \mathrm{i}\pi/2$, an application of (14.7) now gives

$$\begin{aligned}0 &= \lim_{T\to\infty}\lim_{\delta\to 0}\int_{\gamma_{2,\delta,T}} g(z)\,\mathrm{d}z + \lim_{T\to\infty}\lim_{\delta\to 0}\int_{\gamma_{4,\delta,T}} g(z)\,\mathrm{d}z \\ &= \lim_{T\to\infty}\lim_{\delta\to 0}\mathrm{i}\,\mathrm{e}^{(\beta-1)\mathrm{i}\pi/2}\int_\delta^T \mathrm{e}^{-\mathrm{i}t}t^{\beta-1}\,\mathrm{d}t - \lim_{T\to\infty}\lim_{\delta\to 0}\int_0^{T-\delta} \mathrm{e}^{t-T}(T-t)^{\beta-1}\,\mathrm{d}t \\ &= \mathrm{e}^{\beta\mathrm{i}\pi/2}\lim_{T\to\infty}\int_0^T \mathrm{e}^{-\mathrm{i}t}t^{\beta-1}\,\mathrm{d}t - \lim_{T\to\infty}\int_0^T \mathrm{e}^{t-T}(T-t)^{\beta-1}\,\mathrm{d}t.\end{aligned}$$

The substitution $u = T - t$ results in $\lim_{T\to\infty}\int_0^T \mathrm{e}^{-\mathrm{i}t}t^{\beta-1}\,\mathrm{d}t = \mathrm{e}^{-\beta\mathrm{i}\pi/2}\Gamma(\beta)$, the complex conjugate of which is (14.6).

(ii) Now suppose that $\alpha \in (0,1)$. Partial integration and (14.6) then lead to

$$\begin{aligned}\int_0^\infty (\mathrm{e}^{\mathrm{i}t}-1)t^{-\alpha-1}\,\mathrm{d}t &= \lim_{T\to\infty}\left(\frac{(\mathrm{e}^{\mathrm{i}T}-1)T^{-\alpha}}{-\alpha}+\int_0^T \mathrm{i}\mathrm{e}^{\mathrm{i}t}\frac{t^{-\alpha}}{\alpha}\,\mathrm{d}t\right)\\ &= \mathrm{i}\alpha^{-1}\mathrm{e}^{\mathrm{i}(1-\alpha)\pi/2}\Gamma(1-\alpha)\\ &= -\mathrm{e}^{-\mathrm{i}\alpha\pi/2}\frac{\Gamma(1-\alpha)}{\alpha}.\end{aligned} \tag{14.8}$$

(Extending the gamma function analytically to $\mathbb{C}\setminus\{0,-1,-2,\dots\}$, the right-hand side of this equation can be written more elegantly as $\mathrm{e}^{-\mathrm{i}\alpha\pi/2}\Gamma(-\alpha)$, since the gamma function satisfies $w\,\Gamma(w)=\Gamma(w+1)$ for all $w\in\mathbb{C}\setminus(-\mathbb{N}_0)$.) The above equation leads to

$$\int_0^\infty (\mathrm{e}^{\mathrm{i}zt}-1)t^{-\alpha-1}\,\mathrm{d}t = \begin{cases} -z^\alpha\frac{\Gamma(1-\alpha)}{\alpha}(\cos\frac{\alpha\pi}{2}-\mathrm{i}\sin\frac{\alpha\pi}{2}), & z>0,\\ -|z|^\alpha\frac{\Gamma(1-\alpha)}{\alpha}(\cos\frac{\alpha\pi}{2}+\mathrm{i}\sin\frac{\alpha\pi}{2}), & z<0.\end{cases}$$

Since a non-degenerate distribution μ is α-stable if and only if is infinitely divisible, with Gaussian variance 0 and with Lévy measure having density $x\mapsto(p_{-1}\mathbf{1}_{(-\infty,0)}(x)+p_1\mathbf{1}_{(0,\infty)}(x))|x|^{-\alpha-1}$, where $p_{-1},p_1\geq 0$ and $p_{-1}+p_1>0$ (see the discussion preceding this theorem), and since our assumption that $\alpha<1$ implies that $\int_{-1}^1|x|\,\nu(\mathrm{d}x)<\infty$, we see from Proposition 9.18(a) that μ is α-stable if and only if there exist $\gamma_0\in\mathbb{R}$, $p_{-1}\geq 0$, $p_1\geq 0$ with $p_{-1}+p_1>0$ such that $\widehat{\mu}(z)$ is equal to

$$\begin{aligned}&= \exp\left\{\mathrm{i}\gamma_0 z+\int_{\mathbb{R}}(\mathrm{e}^{\mathrm{i}zt}-1)|t|^{-\alpha-1}(p_{-1}\mathbf{1}_{(-\infty,0)}(t)+p_1\mathbf{1}_{(0,\infty)}(t))\,\mathrm{d}t\right\}\\ &= \exp\left\{\mathrm{i}\gamma_0 z-|z|^\alpha\frac{\Gamma(1-\alpha)}{\alpha}\left(p_1\left(\cos\frac{\alpha\pi}{2}-\mathrm{i}\,\sin\frac{\alpha\pi}{2}\,\mathrm{sign}(z)\right)\right)\right.\\ &\qquad\left.-|z|^\alpha\frac{\Gamma(1-\alpha)}{\alpha}\left(p_{-1}\left(\cos\frac{\alpha\pi}{2}+\mathrm{i}\,\sin\frac{\alpha\pi}{2}\,\mathrm{sign}(z)\right)\right)\right\}\\ &= \exp\left\{\mathrm{i}\gamma_0 z-|z|^\alpha\frac{\Gamma(1-\alpha)}{\alpha}(p_1+p_{-1})\cos\frac{\alpha\pi}{2}\left(1-\mathrm{i}\frac{p_1-p_{-1}}{p_1+p_{-1}}\tan\frac{\alpha\pi}{2}\,\mathrm{sign}(z)\right)\right\}.\end{aligned}$$

This corresponds to the representation (14.4) with

$$\tau:=\gamma_0,\quad \sigma^\alpha:=\frac{\Gamma(1-\alpha)}{\alpha}(p_1+p_{-1})\cos\frac{\alpha\pi}{2},\quad \beta:=\frac{p_1-p_{-1}}{p_1+p_{-1}}\in[-1,1]. \tag{14.9}$$

Conversely, for given $\tau\in\mathbb{R}$, $\sigma>0$ and $\beta\in[-1,1]$ it is easily seen that there exist unique $\gamma_0\in\mathbb{R}$, $p_1\geq 0$, $p_{-1}\geq 0$ with $p_1+p_2>0$ such that (14.9) is satisfied, thus showing the converse assertion. The uniqueness of τ, σ and β is also clear from the converse direction, since different triplets (τ,σ,β) and (τ',σ',β') lead to different triplets (γ_0,p_1,p_{-1}) by (14.9).

(iii) Now suppose that $\alpha \in (1, 2)$. Using partial integration again and applying (14.8) to $\alpha - 1 \in (0, 1)$ results in

$$\begin{aligned}
&\int_0^\infty (\mathrm{e}^{\mathrm{i}t} - 1 - \mathrm{i}t)t^{-\alpha-1}\,\mathrm{d}t \\
&\quad = \lim_{T\to\infty}\left(\frac{(\mathrm{e}^{\mathrm{i}T} - 1 - \mathrm{i}T)T^{-\alpha}}{-\alpha} + \int_0^T \mathrm{i}(\mathrm{e}^{\mathrm{i}t} - 1)\frac{t^{-\alpha}}{\alpha}\,\mathrm{d}t\right) \\
&\quad = -\mathrm{i}\,\mathrm{e}^{-\mathrm{i}(\alpha-1)\pi/2}\frac{\Gamma(2-\alpha)}{\alpha(\alpha-1)} = \mathrm{e}^{-\mathrm{i}\alpha\pi/2}\frac{\Gamma(2-\alpha)}{\alpha(\alpha-1)}
\end{aligned}$$

(again, the factor $\alpha^{-1}(\alpha-1)^{-1}\Gamma(2-\alpha)$ could be written more elegantly as $\Gamma(-\alpha)$). The proof now is completely analogous to that of (ii), but this time using Proposition 9.18(b) (rather than (a)) with the extended truncation function $c_m : \mathbb{R} \to \mathbb{R}$, $t \mapsto t$, which is possible since $\int_{|t|>1} |t|\,|t|^{-1-\alpha}\,\mathrm{d}t < \infty$.

(iv) Finally, let us consider the case $\alpha = 1$. Since $\lim_{T\to\infty}\int_0^T t^{-1}\sin(t)\,\mathrm{d}t = \pi/2$ (e. g., Billingsley [16, Ex. 18.4]), partial integration gives

$$\int_0^\infty t^{-2}(1 - \cos t)\,\mathrm{d}t = \lim_{T\to\infty,\delta\downarrow 0}\left(\frac{\cos(T) - 1}{T} - \frac{\cos(\delta) - 1}{\delta} + \int_\delta^T \frac{\sin t}{t}\,\mathrm{d}t\right) = \frac{\pi}{2}.$$

Define

$$c := \int_0^1 t^{-2}(\sin(t) - t)\,\mathrm{d}t + \int_1^\infty t^{-2}\sin(t)\,\mathrm{d}t.$$

Then, for $z > 0$, we obtain

$$\begin{aligned}
&\int_0^\infty (\mathrm{e}^{\mathrm{i}zt} - 1 - \mathrm{i}zt\mathbf{1}_{(0,1]}(t))\,t^{-2}\,\mathrm{d}t \\
&\quad = \int_0^\infty (\cos(zt) - 1)t^{-2}\,\mathrm{d}t + \mathrm{i}\int_0^{1/z}(\sin(zt) - zt)t^{-2}\,\mathrm{d}t + \mathrm{i}\int_{1/z}^\infty \sin(zt)\,t^{-2}\,\mathrm{d}t - \mathrm{i}\int_{1/z}^1 zt\,t^{-2}\,\mathrm{d}t \\
&\quad = z\left(\int_0^\infty (\cos r - 1)r^{-2}\,\mathrm{d}r + \mathrm{i}\int_0^1(\sin(r) - r)r^{-2}\,\mathrm{d}r + \mathrm{i}\int_1^\infty \sin(r)r^{-2}\,\mathrm{d}r - \mathrm{i}\log z\right) \\
&\quad = -\frac{\pi z}{2} + \mathrm{i}cz - \mathrm{i}z\log z.
\end{aligned}$$

By taking complex conjugates, we deduce a similar formula for $z < 0$, and both formulas can then be summarized as

$$\int_0^\infty (e^{izt} - 1 - izt\mathbf{1}_{(0,1]}(t))\, t^{-2}\, dt = -\frac{\pi|z|}{2} + icz - i|z| \log|z| \operatorname{sign}(z) \quad \forall z \in \mathbb{R} \setminus \{0\}.$$

The substitution $v = -t$ leads for $z \in \mathbb{R} \setminus \{0\}$ to

$$\int_{-\infty}^0 (e^{izt} - 1 - izt\mathbf{1}_{[-1,0)}(t))\, t^{-2}\, dt = -\frac{\pi|z|}{2} - icz + i|z| \log|z| \operatorname{sign}(z).$$

As in (ii) and (iii), we now see that a non-degenerate distribution μ is 1-stable if and only if there exist $\gamma \in \mathbb{R}$, $p_1 \geq 0$, $p_{-1} \geq 0$ with $p_{-1} + p_1 > 0$ such that $\widehat{\mu}(z)$ can be expressed as

$$\exp\left\{i\gamma z + \int_{\mathbb{R}} (e^{izt} - 1 - izt\mathbf{1}_{[-1,1]}(t))t^{-2}(p_{-1}\mathbf{1}_{(-\infty,0)}(t) + p_1\mathbf{1}_{(0,\infty)}(t))\, dt\right\},$$

which can be rewritten as

$$\exp\left\{-(p_1 + p_{-1})\frac{\pi|z|}{2} - i(p_1 - p_{-1})|z| \log|z| \operatorname{sign}(z) + i(p_1 - p_{-1})cz + i\gamma z\right\}.$$

The rest of the proof is now completely analogous to that of the case $\alpha < 1$ with

$$\sigma := (p_1 + p_{-1})\pi/2, \quad \beta := \frac{p_1 - p_{-1}}{p_1 + p_{-1}} \in [-1, 1] \quad \text{and} \quad \tau := (p_1 - p_{-1})c + \gamma. \qquad \square$$

Definition 14.9 (Parameterization when $d = 1$). (a) Suppose that $\alpha \in (0, 2)$ and μ is a non-degenerate α-stable distribution on $\mathbb{R}$. Then the parameters $\alpha, \beta \in [-1, 1], \sigma > 0$ and $\tau \in \mathbb{R}$, appearing in Theorem 14.8, are called the *index of stability*, the *skewness parameter*, the *scale parameter* and the *location parameter* of μ, respectively, and we write $\mu \sim S_\alpha(\sigma, \beta, \tau)$ to indicate that the characteristic function of μ is given by (14.4) or (14.5), respectively.

(b) When $d = 1$ and $\alpha = 2$, the right-hand side of (14.4) reduces to $\exp(-\sigma^2 z^2 + i\tau z)$, hence one also writes $S_2(\sigma, \beta, \tau)$ for an $N(\tau, 2\sigma^2)$-distribution, although the parameter $\beta \in [-1, 1]$ is irrelevant when $\alpha = 2$.

(c) Allowing $\sigma = 0$, the right-hand sides of (14.4) and (14.5) reduce to the characteristic function of the Dirac measure δ_τ, which is why one also writes $S_\alpha(0, \beta, \tau) = \delta_\tau$ (the parameters α and β are irrelevant here).

It should be noted that there is more than one parameterization of stable distributions used in the literature and also in R-packages, and the reader should be careful which one is used. We follow the parameterization used by Samorodnitsky and Taqqu [76].

Denoting the spherical part of a non-degenerate 1-dimensional stable distribution by $p_1\delta_1 + p_{-1}\delta_{-1}$, it was shown in the proof of Theorem 14.8 that the parameter β is given by $\beta = \frac{p_1 - p_{-1}}{p_1 + p_{-1}}$, which explains the term *skewness parameter*, since it reflects the skewness of the distribution in terms of the relation between the Lévy measure on $(0, \infty)$ and the

Lévy measure on $(-\infty, 0)$. Also, it follows readily from Theorem 14.8 that μ is symmetric around τ if and only if $\beta = 0$, which gives another justification for the term *skewness parameter*.

As a consequence of Theorem 14.8, we obtain the existence of densities of non-degenerate α-stable distributions:

Corollary 14.10 (Existence of density of stable distribution). *Let μ be a non-degenerate stable distribution on $\mathbb{R}$ with characteristic function $\widehat{\mu}$. Then $\int_{\mathbb{R}} |\widehat{\mu}(z)|\,dz < \infty$ and μ has a continuous Lebesgue density, given by*

$$f(t) = \frac{1}{2\pi} \int_{\mathbb{R}} e^{-izt} \widehat{\mu}(z)\,dz, \quad t \in \mathbb{R}.$$

Proof. From Theorem 14.8, we see that $|\widehat{\mu}(z)| \le e^{-|\sigma z|^\alpha}$ for all $z \in \mathbb{R}$, where $\sigma > 0$. Hence $\widehat{\mu}$ is integrable and it follows from the general theory of characteristic functions (see, e. g., Billingsley [16, Eqns. (26.19), (26.20)]) that μ has a continuous Lebesgue density with the stated form. Compare also Theorem 2.63(b) for the form of the density, which however assumes a priori a piecewise continuous density. □

Actually, the proof of Corollary 14.10 shows that the characteristic function of a non-degenerate α-stable distribution satisfies $\int_{\mathbb{R}} |z|^n |\widehat{\mu}(z)|\,dz < \infty$ for every $n \in \mathbb{N}$, hence it is known that the corresponding Lebesgue density is infinitely often continuously differentiable, with all derivatives tending to 0 as $|z| \to \infty$, see, e. g., Sato [79, Prop. 28.1] or Kawata [56, Thm. 11.6.2].

Figure 14.1 shows the effect on the symmetric stable probability densities as α changes from 3/4 to 2, and Figure 14.2 shows the 1-stable probability densities with skewness parameter $\beta = -1, 0$ and 1.

We shall now give some explicit examples of stable distributions.

Example 14.11 (Cauchy distribution). Consider a 1-stable distribution μ on $\mathbb{R}$ with skewness $\beta = 0$, scale parameter $\sigma > 0$ and location parameter $\tau \in \mathbb{R}$. Then the characteristic function of μ is given by $\widehat{\mu}(z) = \exp(-\sigma|z| + i\tau z)$. By Corollary 14.10, μ has a continuous Lebesgue density f, which is given by

$$\begin{aligned} f(t) &= \frac{1}{2\pi} \int_{-\infty}^{0} e^{i(\tau-t)z+\sigma z}\,dz + \frac{1}{2\pi} \int_{0}^{\infty} e^{i(\tau-t)z-\sigma z}\,dz \\ &= \frac{1}{2\pi}\left(\frac{1}{i(\tau-t)+\sigma} - \frac{1}{i(\tau-t)-\sigma}\right) = \frac{1}{\pi}\frac{\sigma}{(\tau-t)^2+\sigma^2}, \quad t \in \mathbb{R}. \end{aligned}$$

It follows that μ is a Cauchy distribution with *location parameter* $\tau \in \mathbb{R}$ and *scale parameter* $\sigma > 0$.

Example 14.12 (Lévy distribution). Suppose that $d = 1$, $\sigma > 0$ and μ is an $S_{1/2}(\sigma^2, 1, 0)$-distribution. By (14.4), the characteristic function of μ is given by $z \mapsto \widehat{\mu}(z) =$

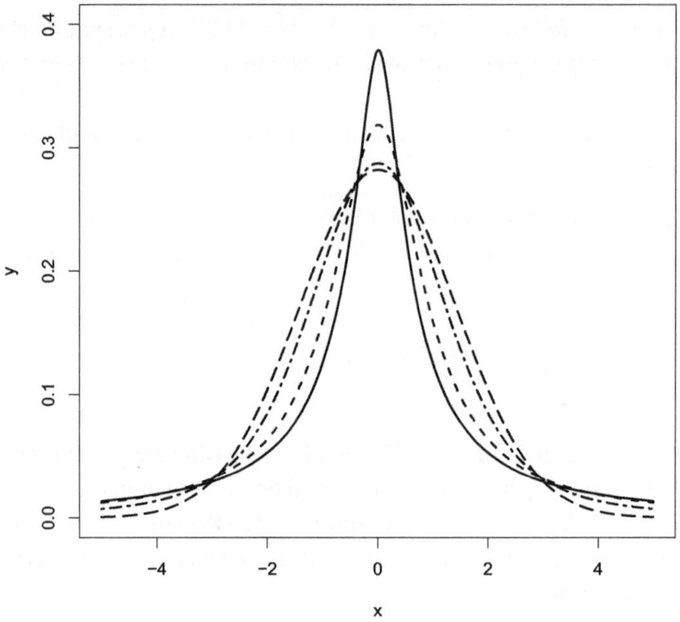

Figure 14.1: Symmetric stable probability densities of $S_{3/4}(1,0,0)$ (solid line), $S_1(1,0,0)$ (dashed), $S_{3/2}(1,0,0)$ (dot dashed) and $S_2(1,0,0) = N(0,2)$ (long dashed).

$\exp\{-\sigma|z|^{1/2}(1 - \mathrm{i}\ \mathrm{sign}(z))\}$, $z \in \mathbb{R}$. The distribution $S_{1/2}(\sigma^2, 1, 0)$ is called the *Lévy distribution* with *scale parameter* σ and *location parameter* 0. It can be shown (see Sato [79, Example 2.13] or Exercise 14.8) that an $S_{1/2}(\sigma^2,1,0)$-distribution has Lebesgue density $t \mapsto \frac{\sigma}{\sqrt{2\pi}} t^{-3/2} \mathrm{e}^{-\sigma^2/(2t)} \mathbf{1}_{(0,\infty)}(t)$. Consequently, the Lebesgue density of an $S_{1/2}(\sigma^2,1,\tau)$-distribution is given by $\frac{\sigma}{\sqrt{2\pi}}(t-\tau)^{-3/2}\mathrm{e}^{-\sigma^2/(2(t-\tau))}\mathbf{1}_{(\tau,\infty)}(t)$, $t \in \mathbb{R}$, and a distribution with this density is called a *Lévy distribution* with *scale parameter* σ and *location parameter* τ.

Apart from Lévy distributions, Cauchy distributions and normal distributions, no other univariate stable distributions whose density can be given in closed form are known to the authors.

Next, we shall define stable Lévy processes.

Definition 14.13 (Stable Lévy process). Let $L = (L_t)_{t\in I}$ be a one- or two-sided $\mathbb{R}^d$-valued Lévy process. Then L is called a *stable Lévy process*, if the distribution of L_1 is stable. When L_1 is α-stable with $\alpha \in (0,2]$, we also call L an *α-stable Lévy process*. If $d = 1$ and $\alpha \in (0,2]$, we write $L \sim S_\alpha(\sigma,\beta,\tau)$, if $\mathcal{L}(L_1) \sim S_\alpha(\sigma,\beta,\tau)$ with $\sigma \geq 0$, $\beta \in [-1,1]$ and $\tau \in \mathbb{R}$ (cf., Definition 14.9) and say that L is an *α-stable Lévy process with skewness parameter β, scale parameter σ and location parameter τ.*

The next proposition summarizes some properties of α-stable Lévy processes.

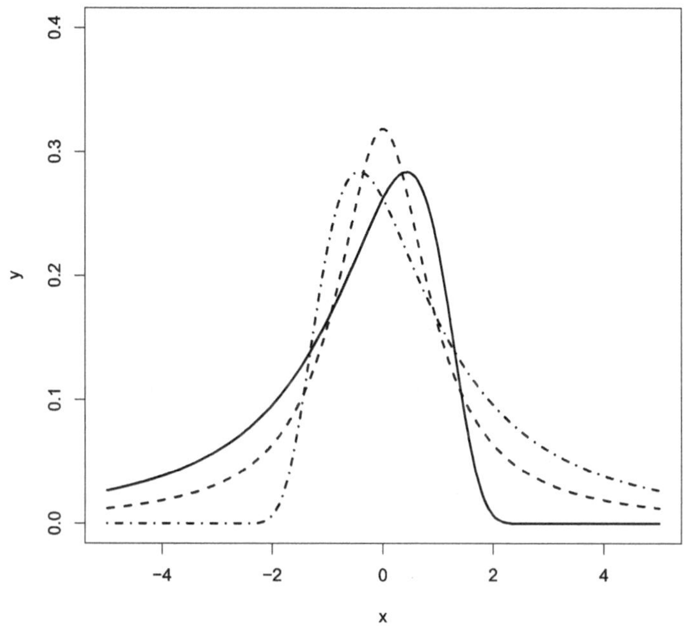

Figure 14.2: Probability densities of $S_1(1, -1, 0)$ (solid line), $S_1(1, 0, 0)$ (dashed) and $S_1(1, 1, 0)$ (dot dashed) illustrating the effect of the skewness parameter β.

Proposition 14.14 (Properties of α-stable Lévy processes). *Suppose that $\alpha \in (0, 2]$ and $L = (L_t)_{t\in I}$ is a one- or two-sided non-deterministic α-stable Lévy process in $\mathbb{R}^d$ with Lévy measure ν. Then the following statements are true.*

(a) *If $\alpha = 2$, then L is a Brownian motion.*
(b) *If $\alpha < 2$, then, almost surely, L has infinitely many jumps on compact non-degenerate intervals.*
(c) *If $\alpha < 2$ and $\beta > 0$, then $\mathbb{E}|L_t|^\beta < \infty$ for some $t \neq 0$ (equivalently for all $t \in I$), if and only if $\beta < \alpha$. In particular, L has finite expectation if and only if $\alpha > 1$.*
(d) *If $\beta > 0$ and $\alpha < 2$, then $\int_{|x|\leq 1} |x|^\beta\, \nu(\mathrm{d}x) < \infty$ if and only $\beta > \alpha$.*
(e) *L is of finite variation if and only if $\alpha < 1$.*
(f) *L is a subordinator (in the sense that all components are subordinators, cf. Remark 13.19(a)) if and only if $\alpha < 1$, the drift of L is in $[0, \infty)^d$ and the spherical part ρ of ν satisfies $\rho(S^{d-1} \setminus [0, \infty)^d) = 0$.*
(g) *If $d = 1$ and $L \sim S_\alpha(\sigma, \beta, \tau)$ with $\sigma > 0$, $\beta \in [-1, 1]$ and $\tau \in \mathbb{R}$, then L is a subordinator if and only if $\alpha < 1$, $\beta = 1$ and $\tau \geq 0$.*

Proof. (a) This is clear, since 2-stable distributions are normal distributions, and (b) follows immediately from Theorem 12.26(a), since the Lévy measure of a non-degenerate α-stable distribution (with $\alpha < 2$) is infinite by equation (14.1).

(c, d) These follow immediately from Lemma 14.7 and Corollary 11.14.

(e) Since Brownian motion is not of finite variation by Lemma 13.14, we can assume that $\alpha < 2$, in which case, the claim follows from (d) and Theorem 13.15.

(f) This follows immediately from Remark 13.19(a), part (e) and the fact that $\nu(\mathbb{R}^d \setminus [0,\infty)^d) = 0$ if and only if $\rho(S^{d-1} \setminus [0,\infty)^d) = 0$.

(g) This follows from (f), since, by (14.9), $\rho(\{-1\}) = 0$ if and only if $\beta = 1$, and since τ is the drift of L. □

We turn now to some specific examples of stable Lévy processes.

Example 14.15 (Cauchy process). Suppose that $d = 1$, $\sigma > 0$, $\tau \in \mathbb{R}$ and $L \sim S_1(\sigma, 0, \tau)$. By Example 14.11, L_1 is Cauchy distributed with location parameter τ and scale parameter σ, and the characteristic exponent of L (and L_1) is given by $z \mapsto -\sigma|z| + i\tau z$, $z \in \mathbb{R}$. Hence the characteristic exponent of L_t for $t > 0$ is given by $z \mapsto -\sigma t|z| + i\tau t z$, $z \in \mathbb{R}$, so that L_t is Cauchy distributed with location parameter τt and scale parameter σt. The Lebesgue density of L_t for $t > 0$ is then given by $s \mapsto f_t(s) = \frac{\sigma t}{\pi((\tau t - s)^2 + \sigma^2 t^2)}$, $s \in \mathbb{R}$. The process L is called a *Cauchy process (with location parameter τ and scale parameter σ)*. It is of infinite variation and its expectation is infinite. Figure 14.3 shows three sample paths of a Cauchy process L with location parameter 0 and scale parameter 1. In other words $L \sim S_1(1, 0, 0)$. Sometimes, general 1-stable processes $L \sim S_1(\sigma, \beta, \tau)$ with $\beta \in [-1, 1]$ and $\sigma > 0$ (i. e., not only with $\beta = 0$) are termed *Cauchy processes*, and those with $\beta = 0$ are termed *symmetric Cauchy processes*, as they are symmetric around τ.

Example 14.16 (Lévy subordinator). Suppose that $d = 1$, $\sigma > 0$ and that $L \sim S_{1/2}(\sigma^2, 1, 0)$. By Proposition 14.14(g), L is an 1/2-stable subordinator with drift 0. Its β-absolute moment is finite if and only if $\beta < 1/2$. Since, by (14.4), the characteristic exponent of L (and hence that of L_1) is given by $z \mapsto \Psi_1(z) = -\sigma|z|^{1/2}(1 - i\,\mathrm{sign}(z))$, $z \in \mathbb{R}$, the characteristic exponent of L_t for $t > 0$ is given by $z \mapsto \Psi_t(z) = -(t\sigma)|z|^{1/2}(1 - i\,\mathrm{sign}(z))$, $z \in \mathbb{R}$, which is the characteristic exponent of an $S_{1/2}(\sigma^2 t^2, 1, 0)$-distribution. Since the distribution $S_{1/2}(\sigma^2, 1, 0)$ is the Lévy distribution with scale parameter σ and location parameter 0, the subordinator $L \sim S_{1/2}(\sigma^2, 1, 0)$ is sometimes called a *Lévy subordinator*. Figure 14.4, on the left, shows three sample paths of the Lévy subordinator $S_{1/2}(1, 1, 0)$ on the time-interval $[0, 40]$ and, on the right, the same sample paths restricted to the time-interval $[0, 5]$. For longer time-intervals, the large jumps dwarf the smaller ones by even larger factors.

Remark 14.17. As for Gaussian processes, a general stochastic process (not necessarily a Lévy process) is called *α-stable* if all its finite dimensional distributions are α-stable. This is consistent with our definition of stable Lévy processes, as it can be shown (see Exercise 14.4) that if L is a Lévy process, then L_1 is α-stable if and only if all the finite dimensional distributions of L are α-stable, so that stable Lévy processes in the sense of Definition 14.13 are indeed stable.

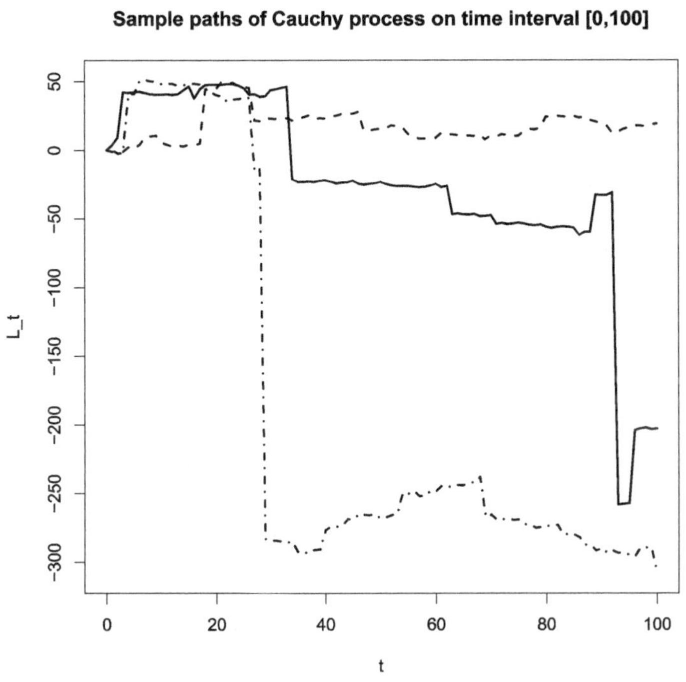

Figure 14.3: Three sample paths of the Cauchy process, $L \sim S_1(1, 0, 0)$, on the time-interval $[0, 100]$.

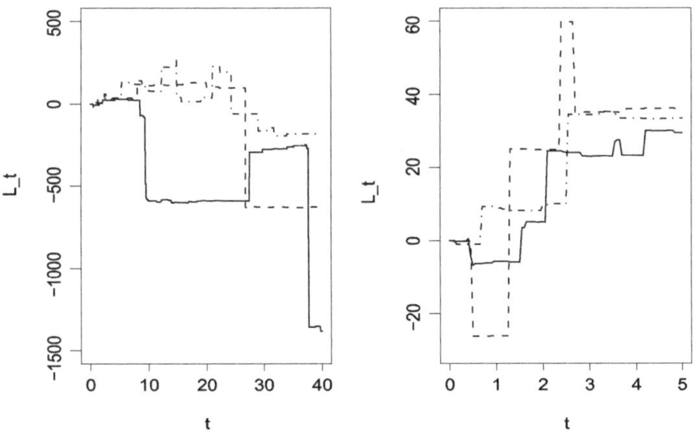

Figure 14.4: On the left are three sample paths of the Lévy subordinator $L \sim S_{1/2}(1, 1, 0)$ on the time-interval $[0, 40]$. The same three sample paths restricted to $[0, 5]$, shown on the right, give a clearer view of the fine-grained structure, as no really large jumps occur up to time 5.

14.2 The gamma process

The gamma process is an important subordinator. Suppose that $\alpha, \beta > 0$ and $L = (L_t)_{t\in I}$ is a one- or two-sided $\mathbb{R}$-valued Lévy process with standard characteristic triplet

$$(0, \alpha x^{-1} \exp(-\beta x)\mathbf{1}_{(0,\infty)}(x)\,\mathrm{d}x, \alpha\beta^{-1}(1-\mathrm{e}^{-\beta})).$$

Since the Lévy measure integrates x in a neighborhood of 0, the drift γ^0 of L exists and by (9.24) is given by

$$\gamma^0 = \frac{\alpha}{\beta}(1-\mathrm{e}^{-\beta}) - \int_0^1 x\,\alpha x^{-1}\mathrm{e}^{-\beta x}\,\mathrm{d}x = 0,$$

so that the characteristic triplet of L with respect to the extended truncation function $c_0(x) = 0$ is given by

$$(0, \alpha x^{-1} \exp(-\beta x)\mathbf{1}_{(0,\infty)}(x)\,\mathrm{d}x, 0)_0.$$

By Theorem 13.18, L is a subordinator. The characteristic triplet of L_t for $t > 0$ with respect to c_0 is then given by

$$(0, t\alpha x^{-1} \exp(-\beta x)\mathbf{1}_{(0,\infty)}(x)\,\mathrm{d}x, 0)_0.$$

By Example 13.23, L_t is gamma distributed with parameters αt and β for $t > 0$, which is why L is called a *gamma process (with parameters α and β)*. For each $t > 0$, Y_t therefore has probability density

$$f_t(x) = \frac{\beta^{\alpha t}}{\Gamma(\alpha t)} x^{\alpha t-1}\mathrm{e}^{-\beta x}\mathbf{1}_{(0,\infty)}(x).$$

By Example 13.23, the Laplace transform of L_t for $t > 0$ is given by

$$\mathrm{Lap}_{L_t}(u) = \left(\frac{\beta}{\beta+u}\right)^{\alpha t}, \quad u \geq 0,$$

and the Laplace exponent of L (i. e., of L_1) is given by

$$\Phi_L(u) = \alpha \log(1 + u/\beta), \quad u \geq 0.$$

By Example 13.28, the characteristic exponent Ψ_L of L is given by

$$\Psi_L(z) = -\alpha \log(1 - \mathrm{i}z/\beta) \quad \forall\, z \in \mathbb{R},$$

where log denotes the principal branch of the logarithm, and the characteristic function at time $t > 0$ is hence given by

$$\phi_{L_t}(z) = e^{t\Psi_L(z)} = (1 - iz/\beta)^{-\alpha t}, \quad \forall z \in \mathbb{R},$$

where $(\cdot)^{-\alpha t}$ denotes the principal branch of the $(-\alpha t)^{\text{th}}$-power.

The random variable $L_{1/\alpha}$ is gamma distributed with parameters 1 and β, i. e. exponentially distributed with parameter β. Consequently, $L_{1/\alpha}$ has expectation $1/\beta$ and variance $1/\beta^2$, so that

$$\mathbb{E}(L_t) = \frac{\alpha t}{\beta} \quad \text{and} \quad \operatorname{Var}(L_t) = \frac{\alpha|t|}{\beta^2} \quad \forall t \in I \tag{14.10}$$

by Theorem 11.15. Sometimes the definition of a gamma process restricts α to be 1, allowing only the parameter $\beta > 0$ to vary. Then L_1 has an exponential distribution with parameter $\beta > 0$. From the form of the Lévy measure ν (or from the form of its probability density) we see that L has finite absolute moments of all orders, and that the exponential moment $\mathbb{E}\, e^{bL_t}$ for $b, t > 0$ is finite if and only if $b < \beta$ (see Corollary 11.14). We also have $\int_{|x|\le 1} |x|^b\, \nu(dx) < \infty$ for all $b > 0$. Since the Lévy measure is infinite, L almost surely has infinitely many jumps on compact non-degenerate intervals. Figure 14.5 shows three sample paths of a gamma process L with parameters $\alpha = 0.2$ and $\beta = 0.3$.

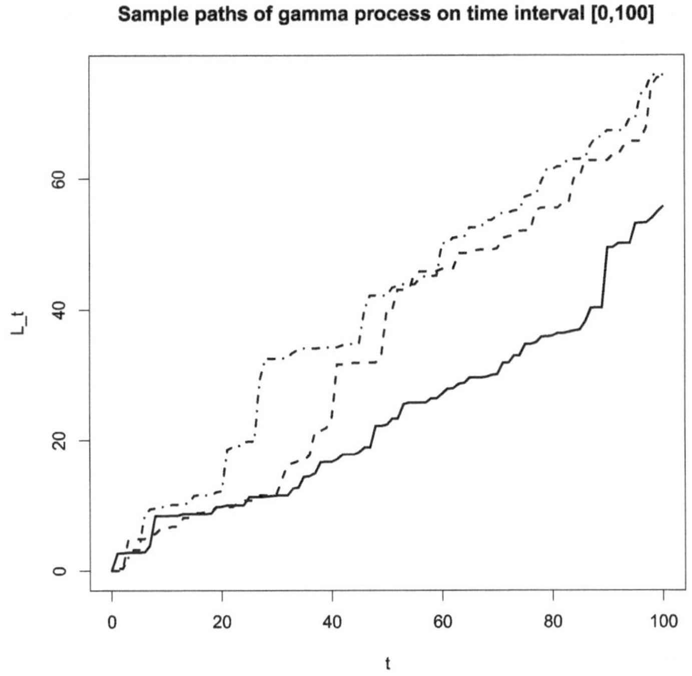

Figure 14.5: Three sample paths of the gamma process with $\alpha = .2$ and $\beta = .3$ on the time-interval $[0, 100]$. The first and second moments of L_1 are $\mathbb{E}(L_1) = \alpha/\beta = 2/3$ and $\operatorname{Var}(L_1) = \alpha/\beta^2 = 20/9$.

14.3 Processes obtained by subordination

An important way of generating new Lévy processes from existing ones is to subordinate a Lévy process by an independent subordinator. Subordination here means a time change described by the corresponding subordinator. The precise statement is as follows.

Theorem 14.18 (Subordination of Lévy processes). *Suppose that $I = [0,\infty)$ or $\mathbb{R}$, that $X = (X_t)_{t\in I}$ is an $\mathbb{R}^d$-valued Lévy process and that $L = (L_t)_{t\in I}$ is an $\mathbb{R}$-valued subordinator which is independent of X. Suppose also that X is strictly càdlàg, i. e., that all of its sample paths are càdlàg (this can always be achieved by choosing a suitable modification). Define the $\mathbb{R}^d$-valued stochastic process $Y = (Y_t)_{t\in I}$ by $Y_t = X_{L_t}$, i. e.,*

$$Y_t(\omega) := X_{L_t}(\omega) := X_{L_t(\omega)}(\omega)$$

for all ω in the sample space. Then Y is a Lévy process.

Proof. First, observe from Lemma 12.6 that $Y_t = X_{L_t}$ is indeed a random vector, so that $Y = (Y_t)_{t\in I}$ is a stochastic process. Secondly, since for almost all ω in the sample space the composition of the càdlàg path $X_.(\omega)$ with the non-decreasing càdlàg path $L_.(\omega)$ is again càdlàg, Y also has càdlàg paths. Clearly $Y_0 = 0$. To see that Y has independent and stationary increments, consider first the case $I = [0,\infty)$. Suppose that $t_0 \le t_1 \le t_2 \le \cdots \le t_n$, where $t_0 := 0$, and define $\vec{t} := (t_1,\dots,t_n)^T$ and

$$Y_{\vec{t}} := (Y_{t_1}^T - Y_{t_0}^T, Y_{t_2}^T - Y_{t_1}^T, \dots, Y_{t_n}^T - Y_{t_{n-1}}^T)^T.$$

Since both L and X are Lévy processes and since they are also mutually independent, we define

$$x := (x_1^T, x_2^T, \dots, x_n^T)^T \in (\mathbb{R}^d)^n$$

and can express the characteristic function of $Y_{\vec{t}}$ as

$$\begin{aligned}
&\mathbb{E}\, e^{i\langle x, Y_{\vec{t}}\rangle} \\
&= \mathbb{E}(\mathbb{E}(e^{i\langle x, Y_{\vec{t}}\rangle} | L_{t_1}, L_{t_2} - L_{t_1}, \dots, L_{t_n} - L_{t_{n-1}})) \\
&= \int_{[0,\infty)^n} \mathbb{E}(e^{i\langle x, Y_{\vec{t}}\rangle} | L_{t_1} = \tau_1, L_{t_2} - L_{t_1} = \tau_2, \dots, L_{t_n} - L_{t_{n-1}} = \tau_n) \\
&\qquad P_{(L_{t_1}, L_{t_2} - L_{t_1}, \dots, L_{t_n} - L_{t_{n-1}})}(d(\tau_1, \dots, \tau_n)) \\
&= \int_{[0,\infty)^n} \mathbb{E}\exp\{i\langle x, (X_{\tau_1}^T, X_{\tau_1+\tau_2}^T - X_{\tau_1}^T, \dots, X_{\tau_1+\dots+\tau_n}^T - X_{\tau_1+\dots+\tau_{n-1}}^T)^T\rangle\} \\
&\qquad (P_{L_{t_1}} \otimes P_{L_{t_2} - L_{t_1}} \otimes \cdots \otimes P_{L_{t_n} - L_{t_{n-1}}}) d(\tau_1, \dots, \tau_n)
\end{aligned}$$

$$
\begin{aligned}
&= \int_{[0,\infty)^n} \mathbb{E}\, e^{i\langle x_1, X_{\tau_1}\rangle}\, \mathbb{E}\, e^{i\langle x_2, X_{\tau_1+\tau_2}-X_{\tau_1}\rangle} \dots \mathbb{E}\, e^{i\langle x_n, X_{\tau_1+\dots+\tau_n}-X_{\tau_1+\dots+\tau_{n-1}}\rangle} \\
&\qquad P_{L_{t_1}}(d\tau_1)\, P_{L_{t_2}-L_{t_1}}(d\tau_2) \dots P_{L_{t_n}-L_{t_{n-1}}}(d\tau_n) \\
&= \int_{[0,\infty)^n} \mathbb{E}\, e^{i\langle x_1, X_{\tau_1}\rangle}\, \mathbb{E}\, e^{i\langle x_2, X_{\tau_2}\rangle} \dots \mathbb{E}\, e^{i\langle x_n, X_{\tau_n}\rangle} \\
&\qquad P_{L_{t_1}}(d\tau_1)\, P_{L_{t_2-t_1}}(d\tau_2) \dots P_{L_{t_n-t_{n-1}}}(d\tau_n) \\
&= \prod_{j=1}^n \int_{[0,\infty)} \mathbb{E}\, e^{i\langle x_j, X_{\tau_j}\rangle} P_{L_{t_j-t_{j-1}}}(d\tau_j) \\
&= \prod_{j=1}^n \int_{[0,\infty)} \mathbb{E}(\exp(i\langle x_j, X_{L_{t_j-t_{j-1}}}\rangle | L_{t_j-t_{j-1}} = \tau_j)) P_{L_{t_j-t_{j-1}}}(d\tau_j) \\
&= \prod_{j=1}^n \mathbb{E}\, e^{i\langle x_j, Y_{t_j-t_{j-1}}\rangle}.
\end{aligned}
\tag{14.11}
$$

Choosing $n = 2$ and defining $x_1 := 0$ in (14.11) gives

$$
\mathbb{E}\, e^{i\langle x_2, Y_{t_2}-Y_{t_1}\rangle} = \mathbb{E} \exp\left\{ i \left\langle \begin{pmatrix} 0 \\ x_2 \end{pmatrix}, \begin{pmatrix} Y_{t_1} \\ Y_{t_2} - Y_{t_1} \end{pmatrix} \right\rangle \right\} = \mathbb{E}\, e^{i\langle x_2, Y_{t_2-t_1}\rangle}
$$

for all $x_2 \in \mathbb{R}^d$, showing that $Y_{t_2} - Y_{t_1} \overset{d}{=} Y_{t_2-t_1}$, so that Y has stationary increments. Inserting this back into equation (14.11) then gives

$$
\mathbb{E}\, e^{i\langle x, Y_{\vec{t}}\rangle} = \prod_{j=1}^n \mathbb{E}\, e^{i\langle x_j, Y_{t_j}-Y_{t_{j-1}}\rangle} \quad \forall\, x = (x_1^T, \dots, x_n^T) \in (\mathbb{R}^d)^n,
$$

so that $(Y_t)_{t\geq 0}$ has independent increments and is therefore a Lévy process. The proof for $(Y_t)_{t\leq 0}$ when $I = \mathbb{R}$ is similar. Observing that $(Y_t)_{t\geq 0}$ and $(Y_t)_{t\leq 0}$ are independent and that $Y_{-1} \overset{d}{=} -Y_1$ (cf., Exercise 14.10), we conclude that Y is a Lévy process also when $I = \mathbb{R}$. □

Definition 14.19 (Subordinated process). Under the conditions of Theorem 14.18, the process Y is called the *process X subordinated by L* or, more specifically, the *process obtained by subordinating X with L*.

We can now determine the characteristic exponent and the distribution of the process Y obtained by subordinating X with L.

Theorem 14.20 (Charact. exponent and distr. of a subordinated process). *Suppose that X, L and Y are defined as in Theorem* 14.18.

(a) *Denote the characteristic exponents of X and Y by Ψ_X and Ψ_Y respectively. If γ_L^0 denotes the drift of L and ν_L its Lévy measure, as in (13.9), define an extension of the characteristic exponent Ψ_L of L to the closed upper half-plane of $\mathbb{C}$ by*

$$\Psi_L(w) := \mathrm{i}\gamma_L^0 w + \int_{(0,\infty)} (\mathrm{e}^{\mathrm{i}wt} - 1)\,\nu_L(\mathrm{d}t) \quad \forall\, w \in \mathbb{C} : \Im w \geq 0,$$

which, if $w \in \mathbb{R}$, is clearly the usual characteristic exponent of L. Then

$$\Im(-\mathrm{i}\,\Psi_X(z)) \geq 0 \quad \forall\, z \in \mathbb{R}^d$$

and

$$\Psi_Y(z) = \Psi_L(-\mathrm{i}\,\Psi_X(z)) \quad \forall\, z \in \mathbb{R}^d. \tag{14.12}$$

(b) *If $t > 0$ and $B \in \mathcal{B}_d$, then*

$$P(Y_t \in B) = \int_{[0,\infty)} P(X_s \in B)\, P_{L_t}(\mathrm{d}s). \tag{14.13}$$

Proof. (a) Since

$$\mathrm{e}^{\Re(\Psi_L(z))} = \left|\mathrm{e}^{\Psi_L(z)}\right| = \left|\mathbb{E}\,\mathrm{e}^{\mathrm{i}\langle z, L_1\rangle}\right| \leq 1 \quad \forall\, z \in \mathbb{R}^d$$

we have $\Re(\Psi_L(z)) \leq 0$ and hence $\Im(-\mathrm{i}\Psi_L(z)) \geq 0$ for all $z \in \mathbb{R}^d$. Now the same calculations that led to (14.11) show that

$$\mathbb{E}\,\mathrm{e}^{\mathrm{i}\langle z, Y_1\rangle} = \int_{[0,\infty)} \mathbb{E}\,\mathrm{e}^{\mathrm{i}\langle z, X_{\tau_1}\rangle}\, P_{L_1}(\mathrm{d}\tau_1) \quad \forall\, z \in \mathbb{R}^d,$$

so that

$$\begin{aligned}\mathrm{e}^{\Psi_Y(z)} = \mathbb{E}\,\mathrm{e}^{\mathrm{i}\langle z, Y_{t_1}\rangle} &= \int_{[0,\infty)} \mathrm{e}^{\tau_1 \Psi_X(z)}\, P_{L_1}(\mathrm{d}\tau_1) \\ &= \mathbb{E}\,\mathrm{e}^{L_1 \Psi_X(z)} = \mathbb{E}\,\mathrm{e}^{\mathrm{i}(-\mathrm{i}\Psi_X(z))L_1}.\end{aligned}$$

By (13.10), the latter is equal to $\mathrm{e}^{\Psi_L(-\mathrm{i}\Psi_X(z))}$, so that (14.12) follows by taking the distinguished logarithm (observe that $\Psi_L(-\mathrm{i}\Psi_X(z))$ is continuous in $z \in \mathbb{R}^d$, as seen in the proof of Proposition 13.26).

(b) Conditioning on $L_t = s \in [0,\infty)$, we obtain, for $t > 0$ and $B \in \mathcal{B}_d$,

$$
\begin{aligned}
P(Y_t \in B) &= \mathbb{E}\, P(X_{L_t} \in B | L_t) \\
&= \int_{[0,\infty)} P(X_{L_t} \in B | L_t = s)\, P_{L_t}(\mathrm{d}s) \\
&= \int_{[0,\infty)} P(X_s \in B)\, P_{L_t}(\mathrm{d}s),
\end{aligned}
$$

which is (14.13). □

It is also possible to calculate the characteristic triplet of the subordinated process, see Sato [79, Thm. 30.1].

When X is itself a subordinator, it is more convenient to work with Laplace exponents, i. e. with the negative of the logarithm of the Laplace transform, as defined in Definitions 13.21 and 13.25. This leads to the following corollary.

Corollary 14.21 (Laplace exponent of subordinated process). *Let $I = [0,\infty)$ or $\mathbb{R}$, and suppose that $X = (X_t)_{t\in I}$ and $L = (L_t)_{t\in I}$ are independent $\mathbb{R}$-valued subordinators. Suppose also that X is strictly càdlàg. Let $Y = (Y_t)_{t\in I}$ be the process X subordinated by L, i. e. $Y_t = X_{L_t}$ for $t \in I$. Then Y is a subordinator. Denote by Φ_L, Φ_X and Φ_Y the Laplace exponents of L, X and Y, respectively. Then*

$$
\Phi_Y(u) = \Phi_L(\Phi_X(u)) \quad \forall\, u \geq 0. \tag{14.14}
$$

Proof. By Theorem 14.18, Y is a Lévy process. Since X and L are both subordinators and since the composition of non-decreasing functions is again non-decreasing, also L has non-decreasing sample paths, i. e., is a subordinator. For $u \geq 0$, we find, with the aid of Corollary 13.24, that

$$
\begin{aligned}
\mathrm{Lap}_{Y_1}(u) = \mathbb{E}\, \mathrm{e}^{-uX_{L_1}} &= \int_{[0,\infty)} \mathbb{E}(\mathrm{e}^{-uX_\tau} | L_1 = \tau)\, P_{L_1}(\mathrm{d}\tau) \\
&= \int_{[0,\infty)} \mathrm{e}^{-\tau\Phi_X(u)} P_{L_1}(\mathrm{d}\tau).
\end{aligned}
$$

But the latter is equal to $\mathbb{E}\, \mathrm{e}^{-L_1\Phi_X(u)} = \mathrm{Lap}_{L_1}(\Phi_X(u))$. Taking the logarithm gives $-\Phi_Y(u) = -\Phi_L(\Phi_X(u))$, which is (14.14). □

Remark 14.22. It follows from equation (13.12) that if Ψ_L is the characteristic exponent of the subordinator L and Φ_L its Laplace exponent, then $\Phi_L(u) = -\Psi_L(\mathrm{i}u)$ for $u \geq 0$, the equality also holding if L is replaced by X or Y if X and Y are subordinators. Formally substituting $\mathrm{i}u$ for z (with $u \geq 0$) in (14.12) gives

$$
-\Phi_Y(u) = \Psi_Y(\mathrm{i}u) = \Psi_L(-\mathrm{i}\Psi_X(\mathrm{i}u)) = \Psi_L(-\mathrm{i}(-\Phi_X(u))) = -\Phi_L(\Phi_X(u)),
$$

which is (14.14). Using an argument involving holomorphic extensions, as in the argument used in Remark 13.27, the above formal substitution can be made rigorous to provide another proof of (14.14).

Of particular importance for us are processes subordinated by gamma processes, as they lead in particular to the variance gamma process.

Example 14.23 (Processes subordinated by a gamma process). Suppose that L is a gamma process with parameters $\alpha > 0$ and $\beta > 0$, as described in Section 14.2. We know that L has drift 0, Lévy measure $\nu_L(\mathrm{d}x) = \alpha x^{-1}\exp(-\beta x)\mathbf{1}_{(0,\infty)}(x)\,\mathrm{d}x$, characteristic exponent $\Psi_L(z) = -\alpha\log(1 - \mathrm{i}z/\beta)$ with log denoting the principal branch of the logarithm, where Ψ_L extends holomorphically to $\{z \in \mathbb{C} : \Im(z) > -\beta\}$, and that the Laplace exponent of L is given by $\Phi_L(u) = \alpha\log(1 + u/\beta)$ for $u \geq 0$. Moreover L_t has Lebesgue density $f_t(x) = \frac{\beta^{\alpha t}}{\Gamma(\alpha t)}x^{\alpha t-1}\mathrm{e}^{-\beta x}\mathbf{1}_{(0,\infty)}(x)$, $x \in \mathbb{R}$, for each $t > 0$. Now let X be an $\mathbb{R}^d$-valued Lévy process, independent of L, and let Y be the process X subordinated by L. An application of (14.13) then shows that

$$P(Y_t \in B) = \frac{\beta^{\alpha t}}{\Gamma(\alpha t)}\int_0^\infty P(X_s \in B)\, s^{\alpha t-1}\mathrm{e}^{-\beta s}\,\mathrm{d}s, \quad t > 0, B \in \mathcal{B}_d. \tag{14.15}$$

From (14.12), we see that the characteristic exponent Ψ_Y of Y can be expressed in terms of the characteristic exponent Ψ_X of X as

$$\Psi_Y(z) = -\alpha\log\left(1 - \frac{\Psi_X(z)}{\beta}\right) \quad \forall\, z \in \mathbb{R}^d, \tag{14.16}$$

and, when $d = 1$ and X is a subordinator with Laplace exponent Φ_X, Y has Laplace exponent given by

$$\Phi_Y(u) = \alpha\log\left(1 + \frac{\Phi_X(u)}{\beta}\right) \quad \forall\, u \geq 0.$$

To give a concrete example, suppose that $d = 1$ and that X is a symmetric and a-stable Lévy process of the form $X \sim S_a(\sigma, 0, 0)$ with $\sigma > 0$ and $a \in (0, 2]$. Then X has characteristic exponent $z \mapsto \Psi_X(z) = -\sigma^a|z|^a$, $z \in \mathbb{R}$, hence

$$\Psi_Y(z) = -\alpha\log(1 + \sigma^a|z|^a/\beta) \quad \forall\, z \in \mathbb{R}.$$

A distribution with such a characteristic function is called a *Linnik distribution* or a *geometric stable distribution.* When $a = 2$ it is a special case of the variance gamma distribution, which we now discuss in more detail.

Definition 14.24 (Variance gamma process and distribution). Suppose that $I = [0, \infty)$ or $\mathbb{R}$, $X = (X_t)_{t\in I}$ is an $\mathbb{R}$-valued Brownian motion with variance $\sigma^2 > 0$ and drift $\theta \in \mathbb{R}$, and that $L = (L_t)_{t\in I}$ is a gamma process with parameters $\alpha = \frac{1}{\kappa}$ and $\beta = \frac{1}{\kappa}$ where $\kappa > 0$. Then the process Y which is obtained by subordinating X with L is called a *variance gamma process with parameters* σ, κ, θ, written $Y \sim \mathrm{VG}(\sigma, \kappa, \theta)$. The distribution of Y_1 is called a *variance gamma distribution with parameters* σ, κ *and* θ and also written as $\mathrm{VG}(\sigma, \kappa, \theta)$.

By (14.10), the underlying gamma process L in the definition above satisfies $\mathbb{E}(L_1) = 1$ and $\mathrm{Var}(L_1) = \kappa$, which is the reason for the choice of the parameterization $\alpha = \beta = 1/\kappa$. When $\theta = 0$, the variance gamma distribution and variance gamma process were developed by Madan and Seneta [65] for the purpose of modelling stock returns and later generalized by Madan et al. [64] to general θ. In a formal sense, when $\theta = 0$, the variance gamma distribution arises by replacing the variance of a mean-zero normal distribution by a gamma-distributed random variable, which explains the name. Some important properties of the variance gamma process are given in the next theorem.

Theorem 14.25 (Properties of variance gamma process). *Suppose that* $I = [0, \infty)$ *or* $\mathbb{R}$ *and that* $Y = (Y_t)_{t\in I}$ *is a variance gamma process with parameters* $\sigma > 0$, $\kappa > 0$ *and* $\theta \in \mathbb{R}$. *Define*

$$A_+ := \frac{2}{\sqrt{\kappa^2\theta^2 + 2\sigma^2\kappa} + \kappa\theta} \quad \text{and} \quad A_- := \frac{2}{\sqrt{\kappa^2\theta^2 + 2\sigma^2\kappa} - \kappa\theta}.$$

(a) *Suppose that* $G = (G_t)_{t\in I}$ *and* $H = (H_t)_{t\in I}$ *are two independent gamma processes, that* G *has parameters* $1/\kappa$ *and* A_+, *and that* H *has parameters* $1/\kappa$ *and* A_-. *Then* Y *is equal in distribution to* $G - H$, *i. e.,* Y *can be represented as the difference of two independent gamma processes.*

(b) *The characteristic exponent of* Y *is given by*

$$\Psi_Y(z) = -\frac{1}{\kappa}\log(1 + \kappa\sigma^2 z^2/2 - \mathrm{i}\kappa\theta z) = -\frac{1}{\kappa}\log((1 - \mathrm{i}z/A_+)(1 + \mathrm{i}z/A_-))$$

for $z \in \mathbb{R}$, *and the characteristic function of* Y_t *at time* $t > 0$ *by*

$$\phi_{Y_t}(z) = (1 + \kappa\sigma^2 z^2/2 - \mathrm{i}\kappa\theta z)^{-t/\kappa} = (1 - \mathrm{i}z/A_+)^{-t/\kappa}(1 + \mathrm{i}z/A_-)^{-t/\kappa}$$

for $z \in \mathbb{R}$, *where* $(\cdot)^{-t/\kappa}$ *denotes the principal branch of the* $(-t/\kappa)^{\mathrm{th}}$*-power function. The Lévy measure of* Y *has a Lévy density given by*

$$x \mapsto \frac{1}{\kappa|x|}\left(\mathrm{e}^{-A_+x}\mathbf{1}_{(0,\infty)}(x) + \mathrm{e}^{-A_-|x|}\mathbf{1}_{(-\infty,0)}(x)\right), \quad x \in \mathbb{R},$$

and the Gaussian variance of Y *is zero. Furthermore,* Y *is of finite variation with drift* 0.

(c) Y *has moments of all orders, and the expectation and variance of* Y_t *are given by*

$$\mathbb{E}(Y_t) = \theta t \quad \text{and} \quad \mathrm{Var}\,(Y_t) = |t|(\sigma^2 + \kappa\theta^2) \quad \forall\, t \in I.$$

Furthermore, for $a \in \mathbb{R}$ *and* $t > 0$ *we have* $\mathbb{E}\,\mathrm{e}^{aY_t} < \infty$ *if and only if* $a \in (-A_-, A_+)$.

(d) *For each* $t \neq 0$, Y_t *is variance gamma distributed with parameters* $\sigma\sqrt{|t|}$, $\kappa/|t|$ *and* θt.
(e) *The distribution of* Y_1 *is symmetric if and only if* $\theta = 0$.

Proof. (a) Denote by Ψ_G, Ψ_{-H} and Ψ_H the characteristic exponents of G, $-H$ and H respectively. From Section 14.2, we know that

$$\Psi_G(z) = -\frac{1}{\kappa}\log(1 - \mathrm{i}z/A_+) \quad \text{and}$$
$$\Psi_{-H}(z) = \Psi_H(-z) = -\frac{1}{\kappa}\log(1 + \mathrm{i}z/A_-) \quad \forall\, z \in \mathbb{R}.$$

Hence $\Psi_{G-H}(z)$, the characteristic exponent of $G - H$ evaluated at $z \in \mathbb{R}$, is given by

$$\begin{aligned}
\Psi_{G-H}(z) &= -\frac{1}{\kappa}\log((1 - \mathrm{i}z/A_+)(1 + \mathrm{i}z/A_-)) \\
&= -\frac{1}{\kappa}\log\left(1 + \mathrm{i}z\left(\frac{1}{A_-} - \frac{1}{A_+}\right) + \frac{z^2}{A_-A_+}\right) \\
&= -\frac{1}{\kappa}\log\left(1 - \mathrm{i}\kappa\theta z + \frac{\sigma^2\kappa}{2}z^2\right).
\end{aligned}$$

But on the other hand, from equation (14.16) we see that this coincides with the characteristic exponent Ψ_Y of Y, so that $G - H$ and Y have the same characteristic exponents and hence are equal in distribution.

(b) The formula for the characteristic exponent as a consequence of (14.16) was already noted in the proof of (a), and the formula for the characteristic function is then clear. To obtain the form of the Lévy measure, we use the representation $Y \overset{d}{=} G - H$ from (a) and the fact that G and $-H$ have Lévy densities $x \mapsto \kappa^{-1}x^{-1}\exp(-A_+x)\mathbf{1}_{(0,\infty)}(x)$ and $x \mapsto \kappa^{-1}|x|^{-1}\mathrm{e}^{-A_-|x|}\mathbf{1}_{(-\infty,0)}(x)$ respectively (see Section 14.2 and Theorem 9.25). The Lévy density of Y is then the sum of the Lévy densities of G and $-H$. The fact that Y has finite variation with drift 0 follows also from the representation $Y \overset{d}{=} G - H$, since both G and H are subordinators with drift 0.

(c) That Y has moments of all orders follows from the representation $Y \overset{d}{=} G - H$ and the fact that G and H have moments of all orders. Furthermore, since

$$\mathbb{E}(G_t) = \frac{t}{\kappa A_+}, \quad \mathrm{Var}(L_t) = \frac{|t|}{\kappa A_+^2}, \quad \mathbb{E}(-H_t) = \frac{-t}{\kappa A_-}, \quad \mathrm{Var}(-H_t) = \frac{|t|}{\kappa A_-^2}$$

by (14.10), we readily find that the expectation and variance of Y_t are as stated in the theorem. The fact that if $t > 0$ the exponential moment $\mathbb{E}\,\mathrm{e}^{\alpha Y_t}$ is finite if and only if $\alpha \in (-A_-, A_+)$ follows readily from Corollary 11.14(d) and the specific form of the Lévy measure of Y.

(d) This follows immediately from the specific form of the characteristic exponent Ψ_Y and the identities $\Psi_{Y_t}(z) = t\Psi_Y(z)$ and $\Psi_{Y_{-t}}(z) = \Psi_{Y_t}(-z)$ for $t > 0$ and $z \in \mathbb{R}$.

(e) This is clear from the form of Ψ_Y and the fact that Y_1 is symmetric if and only if $\Psi_Y(-z) = \Psi_Y(z)$ for all $z \in \mathbb{R}$. □

The representation $Y = G - H$ in Theorem 14.25(a) provides a useful technique for simulation of variance gamma distributions and variance gamma processes. Three sample paths of the variance gamma process $\mathrm{VG}(\sqrt{20/9}, 1, 0)$ are shown in Figure 14.6.

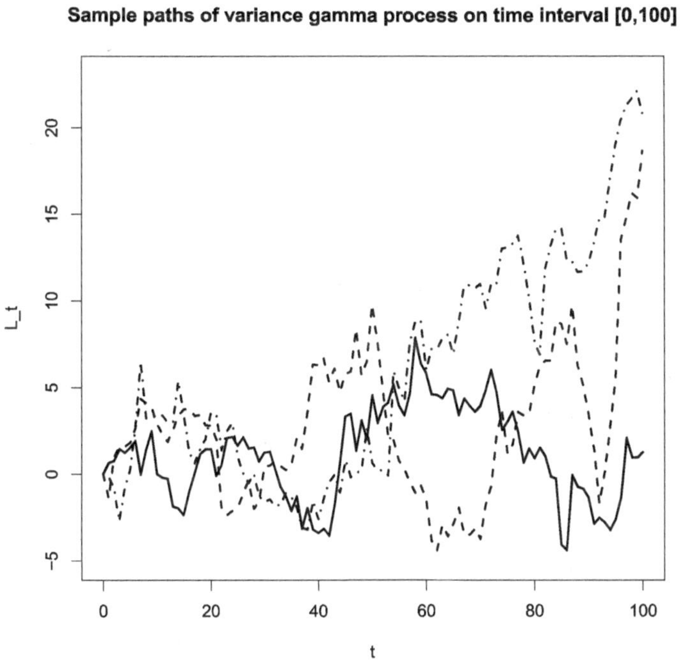

Figure 14.6: Three sample paths of the variance gamma process with parameters $\sigma = \sqrt{20/9}$, $\kappa = 1$ and $\theta = 0$ on the time-interval $[0, 100]$.

From equation (14.15), we see that a variance gamma distribution $Y_1 \sim \mathrm{VG}(\sigma, \kappa, \theta)$ has a probability density given by

$$f_{Y_1}(x) = \frac{1}{\kappa^{1/\kappa}\Gamma(1/\kappa)\sqrt{2\pi\sigma^2}} \int_0^\infty \mathrm{e}^{-(x-\theta s)^2/(2\sigma^2 s) - s/\kappa}\, s^{1/\kappa - 3/2}\, \mathrm{d}s.$$

This can be further simplified with the aid of modified Bessel functions, see Madan et al. [64, Thm. 1] for details.

Remark 14.26. Sometimes, a variance gamma process is parameterized by the parameters $\delta := 1/\kappa$ and $A_+ > 0$ and $A_- > 0$, as defined in Theorem 14.25, in which case $\delta > 0$

is called the *shape parameter* and $A_+, A_- > 0$ are called the *rate parameters*. The characteristic function of a gamma process at time $t > 0$ is then given by

$$\phi_{Y_t}(z) = (1 - \mathrm{i}z/A_+)^{-t\delta}(1 + \mathrm{i}z/A_-)^{-t\delta}, \quad z \in \mathbb{R}.$$

Remark 14.27. Let L^+ and L^- be two independent gamma processes with parameters $\alpha_+, \beta_+ > 0$ and $\alpha_-, \beta_- > 0$, respectively. Then the process $X = L^+ - L^-$ is called a *bilateral gamma process*. From the characteristic function of the gamma process we see that the characteristic function of X is given by

$$\phi_{X_t}(z) = (1 - \mathrm{i}z/\beta_+)^{-t\alpha_+}(1 + \mathrm{i}z/\beta_-)^{-t\alpha_-} \quad \forall\, t > 0\ \forall\, z \in \mathbb{R}.$$

By Theorem 14.25(a), a variance gamma process is a special case of a bilateral gamma process.

14.4 The inverse Gaussian and related Lévy processes

The jump rates of a gamma process and of a variance gamma process are infinite, but their jump-sizes are relatively small, since the Lévy densities behave like $c|x|^{-1}$ as $x \to 0$ for some constant $c > 0$. A popular subordinator which has infinite and moderate jump activity is the inverse Gaussian Lévy process, which we now introduce. Its Lévy density behaves like $x^{-3/2}$ as $x \downarrow 0$, so that it integrates x^β on $(0, 1]$ if and only if $\beta > 1/2$.

Definition 14.28 (Inverse Gaussian Lévy process and distribution). Suppose that $I = [0, \infty)$ or $\mathbb{R}$ and that $\delta > 0$ and $\gamma > 0$. Then the *inverse Gaussian distribution with parameters δ and γ* is an infinitely divisible probability distribution on $\mathbb{R}$ with Gaussian variance 0, drift 0 and Lévy measure having a Lévy density of the form

$$x \mapsto \frac{\delta}{\sqrt{2\pi}}\, \mathrm{e}^{-\gamma^2 x/2}\, x^{-3/2}\, \mathbf{1}_{(0,\infty)}(x), \quad x \in \mathbb{R}.$$

It is denoted by $\mathrm{IG}(\delta, \gamma)$. A Lévy process $L = (L_t)_{t\in I}$ is called an *inverse Gaussian Lévy process with parameters δ, γ*, written $L \sim \mathrm{IG}(\delta, \gamma)$, if L_1 is $\mathrm{IG}(\delta, \gamma)$-distributed.

Since the Lévy measure of an inverse Gaussian Lévy process L is supported on $[0, \infty)$ and integrates x over $(0, 1]$, and since L has drift zero, an inverse Gaussian Lévy process is a subordinator and an inverse Gaussian distribution is concentrated on $[0, \infty)$. Three sample paths of an inverse Gaussian Lévy process are shown in Figure 14.7.

The following theorem establishes the characteristic exponent and Laplace exponent of an inverse Gaussian distribution and an inverse Gaussian Lévy process.

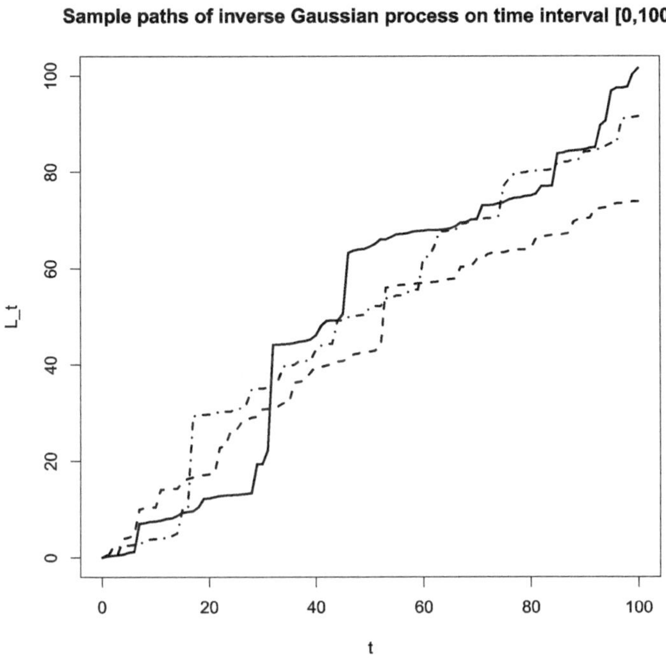

Figure 14.7: Three sample paths of the inverse Gaussian process with parameters $\gamma = \sqrt{0.3}$ and $\delta = 2\gamma/3$ on the time-interval $[0, 100]$. The mean and variance of L_1 are $2/3$ and $20/9$, respectively.

Theorem 14.29 (Characteristic and Laplace exponent of $\mathrm{IG}(\delta, \gamma)$). *Suppose that $\delta > 0$ and $\gamma > 0$. Then the Laplace exponent Φ and the characteristic exponent Ψ of the $\mathrm{IG}(\delta, \gamma)$-distribution and of the $\mathrm{IG}(\delta, \gamma)$-Lévy process are given by*

$$\Phi(u) = \delta\gamma\left(\sqrt{1 + 2u/\gamma^2} - 1\right) \quad \forall\, u \geq 0$$

and

$$\Psi(z) = \delta\gamma\left(1 - \sqrt{1 - 2\mathrm{i}z/\gamma^2}\right) \quad \forall\, z \in \mathbb{R},$$

respectively, where $\sqrt{\cdot}$ denotes the principal branch of the complex square root.

Proof. For $u \geq 0$, define

$$g(u) := \int_0^\infty (1 - \mathrm{e}^{-ux}) x^{-3/2}\, \mathrm{d}x.$$

Using the substitution $y = ux$ and Tonelli's theorem, we obtain, for $u > 0$,

$$
\begin{aligned}
g(u) &= \sqrt{u}\int_0^\infty (1-e^{-y})y^{-3/2}\,dy \\
&= \sqrt{u}\int_0^\infty\int_0^y e^{-v}\,dv\,y^{-3/2}\,dy = \sqrt{u}\int_0^\infty e^{-v}\int_v^\infty y^{-3/2}\,dy\,dv \\
&= 2\sqrt{u}\int_0^\infty e^{-v}v^{-1/2}\,dv = 2\sqrt{u}\,\Gamma(1/2) = 2\sqrt{\pi u},
\end{aligned}
$$

since $\Gamma(1/2) = \sqrt{\pi}$. Furthermore, $g(0) = 0$. Now suppose that $\delta > 0$ and $\gamma > 0$. The Laplace exponent of the $\mathrm{IG}(\delta,\gamma)$-distribution/process is then given by

$$
\begin{aligned}
\Phi(u) &= \frac{\delta}{\sqrt{2\pi}}\int_0^\infty (1-e^{-ux})\,e^{-\gamma^2 x/2}\,x^{-3/2}\,dx \\
&= \frac{\delta}{\sqrt{2\pi}}\left(\int_0^\infty (1-e^{-(u+\gamma^2/2)x})x^{-3/2}\,dx - \int_0^\infty (1-e^{-\gamma^2 x/2})x^{-3/2}\,dx\right) \\
&= \frac{\delta}{\sqrt{2\pi}}(g(u+\gamma^2/2) - g(\gamma^2/2)) \\
&= \delta\sqrt{2}(\sqrt{\gamma^2/2+u} - \sqrt{\gamma^2/2}) = \delta\gamma(\sqrt{1+2u/\gamma^2} - 1)
\end{aligned}
$$

for $u \geq 0$ as claimed.

Now define

$$
G(z) := \delta\gamma(1-\sqrt{1-2iz/\gamma^2}), \quad z \in \mathbb{C} : \Im(z) > -\gamma^2/2,
$$

where $\sqrt{\cdot}$ denotes the principal branch of the square root on $\mathbb{C}\setminus(-\infty,0]$ defined by $\sqrt{w} = \exp((\log w)/2)$, with log denoting the principal branch of the complex logarithm. Then G is holomorphic and $u \mapsto \exp(G(iu))$, $u \geq 0$, agrees with the Laplace transform of $\mathrm{IG}(\delta,\gamma)$, so that by Remark 13.27 the characteristic function of $\mathrm{IG}(\delta,\gamma)$ is given by $z \mapsto \exp(G(z))$, $z \in \mathbb{R}$. Since $G(0) = 0$, we conclude that G coincides with the characteristic exponent Ψ of $\mathrm{IG}(\delta,\gamma)$. □

Remark 14.30. (a) From Theorem 14.29, we see that the logarithm of the Laplace transform of $\mathrm{IG}(\delta,\gamma)$ at $u \geq 0$ is given by $-\Phi(u) = \delta\gamma(1-\sqrt{1+2u/\gamma^2})$. An easy calculation shows that $u \mapsto -\Phi(u)$, $u \in [0,\infty)$, is the inverse function of $v \mapsto g(v) := v^2/(2\delta^2) - \gamma v/\delta$, $v \in (-\infty,0]$. On the other hand, if X is a normally distributed random variable with mean γ/δ and variance $1/\delta^2$, then its Laplace transform $\mathbb{E}\,e^{-vX}$ at $v \leq 0$ is given by $\exp(-\gamma v/\delta + v^2/(2\delta^2)) = e^{g(v)}$. So the logarithm of the Laplace transform of $\mathrm{IG}(\delta,\gamma)$ is the inverse function of the logarithm of the Laplace transform of a normal distribution, which explains the name “inverse Gaussian”.

(b) Suppose now that $\delta > 0$, $\gamma > 0$ and $B = (B_t)_{t\geq 0}$ is Brownian motion with drift γ and variance 1. For $t > 0$, denote by T_t the first time when B reaches δt, i. e., $T_t := \inf\{s \geq 0 : B_s = \delta t\}$. Since B_s/s converges almost surely to $\gamma > 0$ as $s \to \infty$ by the strong law of large numbers (see Theorem 11.21), B_s converges almost surely to $+\infty$, and since B has continuous sample paths, the quantity T_t is almost surely finite, and it is easily seen that it is a random variable. It can be shown that T_t is $\mathrm{IG}(\delta t, \gamma)$-distributed (see, e. g., Bhattacharya and Waymire [15, Prop. 16.4] or Sato [79, Example 33.15]). Furthermore, it can be shown that the process $(T_t)_{t\geq 0}$ (with $T_0 := 0$) is an inverse Gaussian Lévy process with parameters δ and γ. See Sato [79, Ex. 33.15, Thm. 46.2, Ex. 46.6]; this relation to Brownian motion provides another justification for the name “inverse Gaussian”.

Let us collect some elementary properties of inverse Gaussian distributions and inverse Gaussian Lévy processes:

Proposition 14.31 (Properties of inverse Gaussian Lévy process). *Suppose that $I = [0, \infty)$ or $\mathbb{R}$ and let $L = (L_t)_{t\in I}$ be an inverse Gaussian Lévy process with parameters $\delta > 0$ and $\gamma > 0$. Then L_t is $\mathrm{IG}(t\gamma, \delta)$-distributed for every $t > 0$. Suppose that $a \in \mathbb{R}$. Then the exponential moment $\mathbb{E}\, e^{aL_t}$ for $t > 0$ is finite if and only if $a \leq \gamma^2/2$. Furthermore, L has moments of all orders and the mean and variance of L_t for $t \in I$ are given by*

$$\mathbb{E}(L_t) = \frac{\delta t}{\gamma} \quad \textit{and} \quad \mathrm{Var}(L_t) = \frac{\delta}{\gamma^3}|t|, \quad \textit{respectively.}$$

Proof. The fact that L_t is $\mathrm{IG}(t\delta, \gamma)$-distributed for $t > 0$ follows from the form of the characteristic exponent, since L_t has characteristic exponent given by $t\Psi_{L_1}(z)$. Suppose that $a \in \mathbb{R}$ and $t > 0$. From the form of the Lévy measure and Corollary 11.14(d) we see that the exponential moment $\mathbb{E}\, e^{aL_t}$ is finite if and only if $a \leq \gamma^2/2$ and in particular that L_t has moments of all orders. To calculate the expectation and variance when $t = 1$, an easy calculation shows that $\frac{d}{dz}\Psi_{L_1}(z)|_{z=0} = i\delta/\gamma$ and $\frac{d^2}{dz^2}\Psi_{L_1}(z)|_{z=0} = -\delta/\gamma^3$. Hence $\frac{d}{dz}e^{\Psi_{L_1}(z)}|_{z=0} = i\delta/\gamma$ and $\frac{d^2}{dz^2}e^{\Psi_{L_1}(z)}|_{z=0} = -\delta/\gamma^3 - \delta^2/\gamma^2$. The well-known relation between moments and derivatives of the characteristic function (e. g. Bauer [9, Thm. 8.3.3]) then shows that $\mathbb{E}(L_1) = \delta/\gamma$ and $\mathbb{E}L_1^2 = \delta/\gamma^3 + \delta^2/\gamma^2$, yielding $\mathrm{Var}\,(L_1) = \delta/\gamma^3$. Theorem 11.15 then yields the corresponding expressions for L_t with $t \in I$. □

Remark 14.32. (a) It can be shown that the inverse Gaussian distribution $\mathrm{IG}(\delta, \gamma)$ has a Lebesgue density given by

$$f(x) = \frac{\delta}{\sqrt{2\pi}}\, e^{\delta\gamma}\, x^{-3/2}\, e^{-(\gamma^2 x + \delta^2 x^{-1})/2} \mathbf{1}_{(0,\infty)}(x), \quad x \in \mathbb{R},$$

see Exercise 14.13.

(b) There is some ambiguity in the literature regarding the parameterization of the IG distribution. Ours is the same as that of Barndorff–Nielsen [6] and Applebaum [2,

Example 1.3.21], but there are others in use, see [52, Chapter 15]. The most common parametrization seems to be the one introducing the parameters $\mu := \delta/\gamma$ and $\lambda := \delta^2$, with which the density in (a) can be rewritten as

$$f(x) = \sqrt{\frac{\lambda}{2\pi}}\, x^{-3/2}\, \mathrm{e}^{-\lambda(x-\mu)^2/(2\mu^2 x)}\, \mathbf{1}_{(0,\infty)}(x), \quad x \in \mathbb{R}.$$

Note that some authors denote this distribution then by $\mathrm{IG}(\mu, \lambda)$.

The Lévy process obtained by subordinating a Brownian motion by an inverse Gaussian subordinator and adding a drift is called a *normal inverse Gaussian Lévy process*, defined as follows.

Definition 14.33 (Normal inverse Gaussian Lévy process and distr.). Suppose that $I = [0, \infty)$ or $\mathbb{R}$, that $\alpha > 0$, $\delta > 0$, $\beta \in (-\alpha, \alpha)$, $\mu \in \mathbb{R}$ and $W = (W_t)_{t \in I}$ is a Brownian motion with variance 1 and drift β. Suppose also that $L = (L_t)_{t\in I} \sim \mathrm{IG}(\delta, \sqrt{\alpha^2 - \beta^2})$ is an inverse Gaussian subordinator with parameters δ and $\sqrt{\alpha^2 - \beta^2}$. Let Y be the process W subordinated by L and define $X_t := Y_t + \mu t$ for $t \in I$, so that $X_t = W_{L_t} + \mu t$. Then X is called a *normal inverse Gaussian Lévy process with parameters α, β, μ, δ*, written $X \sim \mathrm{NIG}(\alpha, \beta, \mu, \delta)$. The distribution of X_1 is called a *normal inverse Gaussian distribution with parameters α, β, μ, δ*, written $X_1 \sim \mathrm{NIG}(\alpha, \beta, \mu, \delta)$.

The parameters α and β can be regarded as measures of tail-heaviness and skewness respectively. Using Theorems 14.20 and 14.29, it is easy to see that the $\mathrm{NIG}(\alpha, \beta, \mu, \delta)$-Lévy process $(X_t)_{t\in I}$ has characteristic exponent

$$\Psi(z) = \mathrm{i}\mu z + \delta\big(\sqrt{\alpha^2 - \beta^2} - \sqrt{\alpha^2 - (\beta + \mathrm{i}z)^2}\big), \quad z \in \mathbb{R},$$

and that X_t for $t > 0$ is $\mathrm{NIG}(\alpha, \beta, t\mu, t\delta)$-distributed; see Exercise 14.14. Using modified Bessel functions, it is possible to write down the density of a normal inverse Gaussian distribution explicitly. We refer to Barndorff–Nielsen [6] for further information and applications regarding normal inverse Gaussian Lévy processes.

Finally, in this section, let us briefly describe tempered stable processes and normal tempered stable processes. The Lévy density of an α-stable random variable with $\alpha \in (0, 2)$ is given by $\sigma_-|x|^{-1-\alpha}\mathbf{1}_{(-\infty,0)}(x) + \sigma_+ x^{-1-\alpha}\mathbf{1}_{(0,\infty)}(x)$ for $x \in \mathbb{R}$, where $\sigma_-, \sigma_+ \geq 0$. While behavior like $|x|^{-1-\alpha}$ as $x \to 0$ is often desirable as it allows infinite jump activity, as $|x| \to \infty$ this is often too restrictive as it implies that moments of order $\geq \alpha$ do not exist. A way to overcome this difficulty is to multiply the Lévy density by decreasing exponentials on both sides. Such a procedure is often called tempering and infinitely divisible distributions on $\mathbb{R}$ with Gaussian variance 0 and Lévy measures of the form

$$\nu(\mathrm{d}x) = \frac{\sigma_-}{|x|^{1+\alpha}}\mathrm{e}^{-\lambda_-|x|}\mathbf{1}_{(-\infty,0)}(x) + \frac{\sigma_+}{x^{1+\alpha}}\mathrm{e}^{-\lambda_+ x}\mathbf{1}_{(0,\infty)}(x)$$

are often called *tempered α-stable distributions with tempering parameters* $\lambda_-, \lambda_+ > 0$; here, $\sigma_-, \sigma_+ \geq 0$ and $\sigma_+ + \sigma_- > 0$. There are various generalizations of this tempering concept and we refer to Grabchak [48] for further information, but we shall restrict to the setting just described. The inverse Gaussian distribution $\mathrm{IG}(\delta, \gamma)$ is a tempered 1/2-stable distribution with tempering parameter $\lambda_+ = \gamma^2/2$ and $\sigma_- = 0$, $\sigma_+ = \delta/\sqrt{2\pi}$; the tempering parameter λ_- is irrelevant, since $\sigma_- = 0$.

When we have $\alpha \in (0,1)$, a non-negative drift and a tempered α-stable distribution with $\sigma_- = 0$, then the distribution is supported in $[0,\infty)$ and we speak of a *positive tempered α-stable distribution with tempering parameter* $\lambda_+ > 0$. Usually, we will assume that the drift is zero and denote the corresponding distribution and the corresponding subordinator then by $\mathrm{PTS}(\alpha, \sigma_+, \lambda_+)$. Mimicking the proof of Theorem 14.29, it is easy to show that a $\mathrm{PTS}(\alpha, \sigma_+, \lambda_+)$-distribution has Laplace exponent $\Phi(u) = \sigma_+\Gamma(-\alpha)(\lambda_+^\alpha - (u + \lambda_+)^\alpha)$ for $u \geq 0$ and characteristic exponent $\Psi(z) = -\sigma_+\Gamma(-\alpha)(\lambda_+^\alpha - (\lambda_+ - \mathrm{i}z)^\alpha)$ for $z \in \mathbb{R}$ (where $\Gamma(-\alpha) = \Gamma(1-\alpha)/(-\alpha)$), see Exercise 14.15.

A *normal tempered stable Lévy process* is then obtained by subordinating a Brownian motion by a positive tempered stable subordinator and adding a linear drift, just as the normal inverse Gaussian Lévy process was obtained by subordinating a Brownian motion by an inverse Gaussian subordinator and adding a drift. We omit the details.

14.5 Exercises

Exercise 14.1. Suppose that $\alpha \in (0,2)$ and ν is a Lévy measure on $\mathbb{R}^d$. Show that ν has a representation as in (14.1) with a finite measure ρ on S^{d-1} if and only if

$$\nu(B) = b^{-1}\,\nu(b^{-1/\alpha}B) \quad \forall\, b > 0\ \forall\, B \in \mathcal{B}_d.$$

Hint: To show that the above condition implies the representation (14.1), define ρ by $\rho(C) := \alpha\,\nu(\{x \in \mathbb{R}^d : |x| > 1, |x|^{-1}x \in C\})$ for $C \in \mathcal{B}(S^{d-1})$.

Exercise 14.2. Suppose that $\alpha \in (0,2]$, X is an $\mathbb{R}^d$-valued random vector with α-stable distribution, $C \in \mathbb{R}^{m\times d}$ and $\tau \in \mathbb{R}^m$. Show that the distribution of $CX + \tau$ is α-stable.
Hint: Use Exercise 14.1 and Theorem 9.25.

Exercise 14.3. Suppose that $\alpha \in (0,2]$ and $X_1, \ldots, X_n$ are independent random vectors in $\mathbb{R}^{d_1}, \ldots, \mathbb{R}^{d_n}$, respectively, such that each X_i has α-stable distribution. Show that $(X_1^T, \ldots, X_n^T)^T$ is α-stable distributed.

Exercise 14.4. Suppose that $\alpha \in (0,2]$ and L is an α-stable Lévy process. Show that all finite dimensional distributions of L are α-stable.
Hint: Use Exercises 14.2 and 14.3.

Exercise 14.5. Suppose that $d = 1$, $\alpha \in (0,2]$ and $\mu \sim S_\alpha(\sigma,\beta,\tau)$, where $\sigma > 0$ (so that μ is non-degenerate). Show that μ is symmetric (around the origin) if and only if $\tau = 0$ and $\beta = 0$.

Exercise 14.6. Let μ_1 and μ_2 be two non-degenerate distributions such that μ_1 is α_1-stable and μ_2 is α_2 stable. Show that $\mu_1 * \mu_2$ is stable if and only if $\alpha_1 = \alpha_2$, in which case, the index of stability of $\mu_1 * \mu_2$ is equal to $\alpha_1 = \alpha_2$.
Hint: For the "only if" direction, suppose that $\alpha_1 \neq \alpha_2$ and that $\mu_1 * \mu_2$ is α_3-stable. Then use (14.3), multiplied by n^{-1/α_i}, where α_i is the smallest of α_1, α_2 and α_3 and let $n \to \infty$.

Exercise 14.7. Suppose that $d = 1$, $\alpha \in (0,2]$ and $(\mu_n)_{n\in\mathbb{N}}$ is a sequence of α-stable distributions on $\mathbb{R}$ that converges weakly to some distribution μ on $\mathbb{R}^d$. Show that μ is also α-stable, i. e., the class of α-stable distributions on $\mathbb{R}$ is closed under weak convergence. (A similar statement is also true in higher dimensions).

Exercise 14.8. Suppose that $\sigma > 0$. Define the function $f : \mathbb{R} \to [0,\infty)$ by

$$f(t) := \frac{\sigma}{\sqrt{2\pi}} \mathrm{e}^{-\sigma^2/(2t)} t^{-3/2} \mathbf{1}_{(0,\infty)}(t).$$

(a) Show that f is the Lebesgue density of some probability measure concentrated on $[0,\infty)$. Denote this probability measure by μ.
(b) Show that the Laplace transform Lap_μ of μ satisfies the differential equation

$$\mathrm{Lap}'_\mu(u) = -\frac{\sigma}{\sqrt{2u}} \,\mathrm{Lap}_\mu(u) \quad \forall\, u > 0.$$

(c) Conclude that $\mathrm{Lap}_\mu(u) = \exp(-\sigma\sqrt{2u})$ for all $u \geq 0$.
(d) Deduce that μ is $S_{1/2}(\sigma^2, 1, 0)$-distributed, i. e., that μ has the Lévy distribution.

Hint: Use Remark 13.27 and an appropriate branch of the complex square root.

Exercise 14.9. Let $\alpha, \beta > 0$ and $L = (L_t)_{t\geq 0}$ be a gamma process with parameters α and β. Consider the function $g : \mathbb{R} \to \mathbb{R}$, $x \mapsto g(x) = (1 \wedge |x|^{-\alpha})\mathrm{e}^{\beta|x|}$. Show that $\mathbb{E}\, g(L_t) < \infty$ if and only if $t < 1$. In particular, conditions (i) and (ii) of Theorem 11.12 are not equivalent in this case. Deduce that g is not submultiplicative.

Exercise 14.10. Let $X = (X_t)_{t\in\mathbb{R}}$ be a two-sided $\mathbb{R}^d$-valued Lévy process and $L = (L_t)_{t\in\mathbb{R}}$ be an independent subordinator, and define $Y_t := X_{L_t}$ for $t \in \mathbb{R}$. Complete the proof of Theorem 14.18 by showing that $Y_{-1} \overset{d}{=} -Y_1$.

Exercise 14.11. Suppose that $I = [0,\infty)$ or $\mathbb{R}$, $Z = (Z_t)_{t\in I}$ is an $\mathbb{R}^d$-valued Lévy process and $L = (L_t)_{t\in I}$ and $X = (X_t)_{t\in I}$ are two $\mathbb{R}$-valued subordinators, such that X, Z and L are independent. Assume that both X and Z are strictly càdlàg. Denote the process X subordinated by L by X^L such that $X^L_t = X_{L_t}$ and similarly for other subordinated processes. Show that

$$(Z^X)^L = Z^{(X^L)},$$

i. e., subordinating Z first by X and then the obtained process again by L is the same as subordinating Z directly by the subordinated process X^L.

Exercise 14.12. Let X, Y and L be as in Theorem 14.18. Show that if X is of finite variation, then so is Y.

Exercise 14.13. Suppose that $\delta > 0$ and $\gamma > 0$. Define the function $f : \mathbb{R} \to [0, \infty)$ by

$$f(t) = \frac{\delta}{\sqrt{2\pi}} \mathrm{e}^{\delta\gamma} t^{-3/2} \mathrm{e}^{-(\gamma^2 t + \delta^2 t^{-1})/2} \mathbf{1}_{(0,\infty)}(t), \quad t \in \mathbb{R}.$$

Denote by $\mathrm{Lap}_f(u) := \int_0^\infty \mathrm{e}^{-ux} f(x)\,\mathrm{d}x$ the Laplace transform of the function f. Using Exercise 14.8(c), show that

$$\mathrm{Lap}_f(u) = \exp(-\delta\gamma(\sqrt{1 + 2u/\gamma^2} - 1)) \quad \forall\, u \geq 0.$$

Conclude that f is the Lebesgue density of the $\mathrm{IG}(\delta, \gamma)$-distribution.

Exercise 14.14. Suppose that $\alpha > 0$, $\delta > 0$, $\beta \in (-\alpha, \alpha)$, $\mu \in \mathbb{R}$ and consider the normal inverse Gaussian Lévy process $X = (X_t)_{t\geq 0}$ with parameters α, β, μ, δ. Show that X has characteristic exponent

$$\Psi(z) = \mathrm{i}\mu z + \delta(\sqrt{\alpha^2 - \beta^2} - \sqrt{\alpha^2 - (\beta + \mathrm{i}z)^2}), \quad z \in \mathbb{R},$$

and that X_t for $t > 0$ is $\mathrm{NIG}(\alpha, \beta, t\mu, t\delta)$-distributed.

Exercise 14.15. Suppose that $\alpha \in (0, 1)$, $\sigma > 0$ and $\lambda > 0$. Show that the positive tempered α-stable $\mathrm{PTS}(\alpha, \sigma, \lambda)$-distribution has Laplace exponent given by

$$\Phi(u) = \sigma\Gamma(-\alpha)(\lambda^\alpha - (u + \lambda)^\alpha) \quad \forall\, u \geq 0$$

and characteristic exponent

$$\Psi(z) = -\sigma\Gamma(-\alpha)(\lambda^\alpha - (\lambda - \mathrm{i}z)^\alpha) \quad \forall\, z \in \mathbb{R}.$$

Hint: Mimic the proof of Theorem 14.29.

15 Integration of deterministic functions with respect to Lévy processes

Abstract: In this chapter, we discuss integration of deterministic functions with respect to Lévy processes. Our definition of the integral is based on the fact that every Lévy process can be written as the sum of an orthogonal-increment process and an independent process of finite variation. We start with the definition of the Lebesgue–Stieltjes integral for Lévy processes of finite variation. We then move to the integral $\int_{\mathbb{R}} f(t)\,\mathrm{d}L_t$ when $f \in L^1(\mathbb{R};\mathbb{C}) \cap L^2(\mathbb{R};\mathbb{C})$ and L is a one-dimensional Lévy process and then extend this to the case of $f \in L^2(\mathbb{R};\mathbb{C}^{m\times d}) \cap L^1(\mathbb{R};\mathbb{C}^{m\times d})$ and $\mathbb{R}^d$-valued Lévy processes. While in this chapter we concentrate on the concept of integration and general properties such as Riemann–Stieltjes sums, partial integration or a characterization of integrability, in Chapter 16 we will derive the characteristic function of the integral and also discuss its distributional properties. There is a broader definition of the integral, due to Rajput and Rosiński [73, Thm. 2.7], which includes the one developed in this chapter (see Remark 16.5 in the next chapter). However, since the functions of interest to us, which are integrable in the sense of Rajput and Rosiński, are also integrable in the sense of Definition 15.10, we chose to use the approach taken in this chapter as it uses only the concepts of integration with respect to orthogonal increment and finite variation processes, the first of which was needed and developed earlier in the book.

15.1 Lebesgue–Stieltjes integral with respect to Lévy processes of finite variation

In this chapter, we define the integral of a deterministic Borel-measurable function f with respect to a two-sided Lévy process $L = (L_t)_{t\in\mathbb{R}}$. If $f \in L^2(\mathbb{R};\mathbb{C})$ and L is $\mathbb{R}$-valued with finite second moment and mean zero, then L is an OIP by Corollary 11.17, with distribution function $t \mapsto F(t) = t\,\mathrm{Var}(L_1)$ and we can define the integral $\int_{\mathbb{R}} f(t)\,\mathrm{d}L_t$ as in Section 4.3. While the square-integrability of f does not pose too big a restriction for our purposes, the requirement that L has finite second moment is not satisfied for all situations that are important to us. Hence we also need an integration concept for Lévy processes that do not have finite second moment. By Corollary 13.16, every $\mathbb{R}$-valued Lévy process L can be decomposed into an independent sum of two Lévy processes M and Z, where M is an OIP and Z is a finite variation process. The strategy now is to define a pathwise integral, namely the Lebesgue–Stieltjes integral, for Lévy processes of finite variation in this section, and then use the decomposition $L = M + Z$ for general Lévy processes in Section 15.2 to define the stochastic integral as a sum of an OIP integral and a Lebesgue–Stieltjes integral.

https://doi.org/10.1515/9783111325033-015

If $L = (L_t)_{t\in\mathbb{R}}$ is a Lévy process of finite variation, then by Theorems 13.15 and 13.1, a typical path of L is of the form

$$t \mapsto g(t) := \begin{cases} \gamma^0 t + \sum_{j\in J, 0<j\le t} c_j, & t \ge 0, \\ \gamma^0 t - \sum_{j\in J, t<j\le 0} c_j, & t < 0, \end{cases} \tag{15.1}$$

where $J \subset \mathbb{R}$ is countable and $(c_j)_{j\in J}$ is a sequence of real numbers such that $\sum_{j\in J\cap K} |c_j| < \infty$ for all compact sets $K \subset \mathbb{R}$. Namely, γ^0 is taken to be the drift of L, and for ω in the sample space Ω such that $t \mapsto L_t(\omega)$ is càdlàg and of finite variation with $L_0(\omega) = 0$, we can take $J = J(\omega) = \{j \in \mathbb{R} : \Delta L_j(\omega) \neq 0\}$ and $c_j = c_j(\omega) = \Delta L_j(\omega)$. Hence we only need to integrate with respect to functions g of the above form. To define the corresponding Lebesgue–Stieltjes integral, define the functions $g_1, g_2 : \mathbb{R} \to \mathbb{R}$ by

$$g_1(t) := \begin{cases} \sum_{j\in J, 0<j\le t, c_j>0} c_j, & t \ge 0, \\ -\sum_{j\in J, t<j\le 0, c_j>0} c_j, & t < 0, \end{cases}$$

and

$$g_2(t) := \begin{cases} \sum_{j\in J, 0<j\le t, c_j<0} |c_j|, & t \ge 0, \\ -\sum_{j\in J, t<j\le 0, c_j<0} |c_j|, & t < 0. \end{cases}$$

Then g_1 and g_2 are non-decreasing càdlàg functions. Further, define

$$g^+(t) := \begin{cases} g_1(t) + \gamma^0 t, & \gamma^0 \ge 0, \\ g_1(t), & \gamma^0 < 0, \end{cases} \quad \text{and} \quad g^-(t) := \begin{cases} g_2(t), & \gamma^0 \ge 0, \\ g_2(t) + |\gamma^0| t, & \gamma^0 < 0. \end{cases}$$

Then g^+ and g^- are non-decreasing càdlàg functions with

$$g = g^+ - g^-.$$

It is now well known that there exist unique measures μ^+ and μ^- on $(\mathbb{R}, \mathcal{B}_1)$ such that

$$\mu^+((a,b]) = g^+(b) - g^+(a) \quad \text{and} \quad \mu^-((a,b]) = g^-(b) - g^-(a) \quad \forall\, a \le b,$$

see, e. g., Billingsley [16, Thm. 12.4]. Denoting Lebesgue-measure on $\mathbb{R}$ by λ^1 and the Dirac measure at a point j by δ_j, these measures are given by

$$\mu^+ = \begin{cases} \gamma^0 \lambda^1 + \sum_{j\in J, c_j>0} c_j \delta_j, & \gamma^0 \ge 0, \\ \sum_{j\in J, c_j>0} c_j \delta_j, & \gamma^0 < 0, \end{cases}$$

and

$$\mu^- = \begin{cases} \sum_{j\in J, c_j<0} |c_j|\delta_j, & \gamma^0 \geq 0, \\ |\gamma^0|\lambda^1 + \sum_{j\in J, c_j<0} |c_j|\delta_j, & \gamma^0 < 0. \end{cases}$$

We now say that a Borel-measurable function $f : \mathbb{R} \to \mathbb{C}$ is *Lebesgue–Stieltjes-integrable with respect to* g^+, if it is integrable with respect to μ^+, in which case we define the *Lebesgue–Stieltjes integral* of f with respect to g^+ as the corresponding integral of f with respect to μ^+, i. e.,

$$\text{L. S.} \int_{\mathbb{R}} f(t)\, \mathrm{d}g^+(t) := \int_{\mathbb{R}} f(t)\, \mathrm{d}\mu^+(t),$$

and similarly, we define Lebesgue–Stieltjes integrability with respect to g^- and

$$\text{L. S.} \int_{\mathbb{R}} f(t)\, \mathrm{d}g^-(t) := \int_{\mathbb{R}} f(t)\, \mathrm{d}\mu^-(t).$$

Finally, we say that f is *Lebesgue–Stieltjes-integrable with respect to* g, if it is Legesgue–Stieltjes-integrable with respect to both g^+ and g^- (i. e., integrable with respect to both the measures μ^+ and μ^-), in which case, we define the *Lebesgue–Stieltjes integral* as

$$\begin{aligned} \text{L. S.} \int_{\mathbb{R}} f(t)\, \mathrm{d}g(t) &:= \text{L. S.} \int_{\mathbb{R}} f(t)\, \mathrm{d}g^+(t) - \int_{\mathbb{R}} f(t)\, \mathrm{d}g^-(t) \\ &= \int_{\mathbb{R}} f(t)\, \mathrm{d}\mu^+(t) - \int_{\mathbb{R}} f(t)\, \mathrm{d}\mu^-(t). \end{aligned}$$

Since the measures μ^+ and μ^- above have a very nice form, we see that f is Lebesgue–Stieltjes-integrable with respect to g^+ if and only if $\sum_{j\in J, c_j>0} c_j|f(t)| < \infty$ and (when $\gamma^0 > 0$) $f \in L^1(\mathbb{R}; \mathbb{C})$, in which case

$$\text{L. S.} \int_{\mathbb{R}} f(t)\, \mathrm{d}g^+(t) = \begin{cases} \gamma^0 \int_{\mathbb{R}} f(t)\, \mathrm{d}t + \sum_{j\in J, c_j>0} f(j)c_j, & \gamma^0 > 0, \\ \sum_{j\in J, c_j>0} f(j)c_j, & \gamma^0 \leq 0. \end{cases}$$

A similar characterization holds for the Lebesgue–Stieltjes integral with respect to g^-. Putting everything together, we have just arrived at the following result (see also Example 13.10 for the càdlàg and finite variation assertion below).

Theorem 15.1 (L. S.-integral with respect to jump functions with drift). *Suppose that* $\gamma^0 \in \mathbb{R}$, J *is a countable subset of* $\mathbb{R}$ *and* $(c_j)_{j\in J}$ *is a sequence of real numbers such that* $\sum_{j\in J\cap K} |c_j| < \infty$ *for all compact sets* $K \subset \mathbb{R}$. *Let the function* $g : \mathbb{R} \to \mathbb{R}$ *be given by* (15.1). *Then* g *is càdlàg and of finite variation, and a Borel-measurable function* $f : \mathbb{R} \to \mathbb{C}$ *is Lebesgue–Stieltjes-integrable with respect to* g *if and only* $\gamma^0 f \in L^1(\mathbb{R}; \mathbb{C})$ *and* $\sum_{j\in J} |f(t)c_j| < \infty$, *in which case the Lebesgue–Stieltjes integral is given by*

$$\mathrm{L.S.}\int_{\mathbb{R}} f(t)\,\mathrm{d}g(t) = \int_{\mathbb{R}} \gamma^0 f(t)\,\mathrm{d}t + \sum_{j\in J} f(j)c_j.$$

Using the Hahn–Jordan decomposition of finite signed measures (see, e. g., Cohn [33, Thm. 4.1.5])), the above machinery can be adapted to define Lebesgue–Stieltjes integrals with respect to more general càdlàg functions of finite variation, but we shall not go into detail as integration with respect to functions of the form (15.1) is sufficient for us.

Definition 15.2 (L. S.-integral w. r. t. Lévy process of finite variation). Let $L = (L_t)_{t\in\mathbb{R}}$ be an $\mathbb{R}$-valued Lévy process of finite variation with drift γ^0 defined on a probability space $(\Omega, \mathcal{F}, P)$ and $f : \mathbb{R} \to \mathbb{C}$ be a Borel-measurable function. By Theorem 13.15 and the simplified Lévy–Itô-decomposition (Theorem 13.1), there is some set $\Omega_0 \in \mathcal{F}$ with $P(\Omega_0) = 1$ such that the sample paths $t \mapsto L_t(\omega)$ are càdlàg for all $\omega \in \Omega_0$, such that $\sum_{-n<s\le n} |\Delta L_s(\omega)| < \infty$ for all $\omega \in \Omega_0$ and $n \in \mathbb{N}$ and such that

$$L_t(\omega) = \begin{cases} \gamma^0 t + \sum_{0<s\le t} \Delta L_s(\omega), & t \ge 0,\\ \gamma^0 t - \sum_{t<s\le 0} \Delta L_s(\omega), & t < 0. \end{cases}$$

for all $\omega \in \Omega_0$. Then we say that f is *Lebesgue–Stieltjes-integrable with respect to L*, if it is pathwise Lebesgue–Stieltjes-integrable with probability 1, i. e., if there is some set $\Omega_1 \in \mathcal{F}$, $\Omega_1 \subset \Omega_0$, with $P(\Omega_1) = 1$ such that for each $\omega \in \Omega_1$, the function f is Lebesgue–Stieltjes-integrable with respect to the path $t \mapsto L_t(\omega)$. In that case, by the *Lebesgue–Stieltjes integral of f with respect to L* we mean any random variable $Y : \Omega \to \mathbb{C}$, which is almost surely equal to the random variable

$$\begin{aligned}\omega &\mapsto \Big(\mathrm{L.S.}\int_{\mathbb{R}} f(t)\,\mathrm{d}L_t(\omega)\Big)\mathbf{1}_{\Omega_1}(\omega)\\ &= \Big(\int_{\mathbb{R}} \gamma^0 f(t)\,\mathrm{d}t + \sum_{t\in\mathbb{R},\Delta L_t(\omega)\neq 0} f(t)\Delta L_t(\omega)\Big)\mathbf{1}_{\Omega_1}(\omega)\end{aligned} \tag{15.2}$$

and each such random variable will be denoted by $\mathrm{L.S.}\int_{\mathbb{R}} f(t)\,\mathrm{d}L_t$.

The equality in equation (15.2) holds by Theorem 15.1. That the right-hand side of (15.2) indeed defines a random variable (i. e., is measurable) can be seen by using the modification L' of L, defined by $L'_t(\omega) := L_t(\omega)\mathbf{1}_{\Omega_1}(\omega)$, and applying Corollary 12.9 and Remark 12.11 to it.

Observe that we defined the Lebesgue–Stieltjes integral only almost surely, which is due to the fact that in general we have the paths of a Lévy process only almost surely.

We shall next characterize the Borel-measurable functions that are Lebesgue–Stieltjes-integrable with respect to a Lévy process of finite variation, and give some sufficient conditions for Lebesgue–Stieltjes integrability.

Theorem 15.3 (Characterization/sufficient cond. for L.-S. integrability). *Let $L = (L_t)_{t\in\mathbb{R}}$ be an $\mathbb{R}$-valued Lévy process of finite variation with Lévy measure ν and drift γ^0, and let $f : \mathbb{R} \to \mathbb{C}$ be a Borel-measurable function.*

(a) *The following statements are equivalent.*
  (i) *f is Lebesgue–Stieltjes-integrable with respect to L.*
  (ii) *$\sum_{t\in\mathbb{R};\Delta L_t\neq 0} |f(t)\Delta L_t| < \infty$ almost surely and $\gamma^0 f \in L^1(\mathbb{R};\mathbb{C})$.*
  (iii) *$\int_{\mathbb{R}}\int_{\mathbb{R}}(1\wedge|f(t)x|)\,\nu(\mathrm{d}x)\,\mathrm{d}t < \infty$ and $\gamma^0 f \in L^1(\mathbb{R};\mathbb{C})$, i. e.*

$$\int_{\mathbb{R}} |\gamma^0 f(t)|\,\mathrm{d}t + \int_{\mathbb{R}}\int_{\mathbb{R}}(1\wedge|f(t)x|)\,\nu(\mathrm{d}x)\,\mathrm{d}t < \infty.$$

If one and hence all of the equivalent conditions are satisfied, then

$$\left|\mathrm{L.\,S.}\int_{\mathbb{R}} f(t)\,\mathrm{d}L_t\right| \le \int_{\mathbb{R}} |\gamma^0 f(t)|\mathrm{d}t + \sum_{t\in\mathbb{R},\Delta L_t\neq 0} |f(t)\Delta L_t|.$$

(b) *If $f \in L^1(\mathbb{R};\mathbb{C})$ and L is of finite variation with $\mathbb{E}|L_1| < \infty$, then f is Lebesgue–Stieltjes-integrable with respect to L and $\mathbb{E}|\mathrm{L.\,S.}\int_{\mathbb{R}} f(t)\,\mathrm{d}L_t| < \infty$. Moreover,*

$$\mathbb{E}\left|\mathrm{L.\,S.}\int_{\mathbb{R}} f(t)\,\mathrm{d}L_t\right| \le \int_{\mathbb{R}} |f(t)|\,\mathrm{d}t\left(|\gamma^0| + \int_{\mathbb{R}} |x|\,\nu(\mathrm{d}x)\right) \tag{15.3}$$

and

$$\mathbb{E}\left(\mathrm{L.\,S.}\int_{\mathbb{R}} f(t)\,\mathrm{d}L_t\right) = \int_{\mathbb{R}} f(t)\,\mathrm{d}t\left(\gamma^0 + \int_{\mathbb{R}} x\,\nu(\mathrm{d}x)\right) = \int_{\mathbb{R}} f(t)\,\mathrm{d}t\;\mathbb{E}(L_1). \tag{15.4}$$

(c) *If $f \in L^1(\mathbb{R};\mathbb{C})$ with bounded support, then f is Lebesgue–Stieltjes-integrable with respect to L.*

Proof. (a) The equivalence of (i) and (ii) follows from Theorem 15.1 and that of (ii) and (iii) was shown in Theorem 12.36(a). The bound for the integral is clear from (15.2).

(b) Since $f \in L^1(\mathbb{R};\mathbb{C})$ and L is of finite variation with $\mathbb{E}|L_1| < \infty$, we have $\int_{\mathbb{R}} |x|\,\nu(\mathrm{d}x) < \infty$ and hence

$$\int_{\mathbb{R}}\int_{\mathbb{R}}(1\wedge|f(t)x|)\,\nu(\mathrm{d}x)\,\mathrm{d}t \le \int_{\mathbb{R}} |f(t)|\,\mathrm{d}t\int_{\mathbb{R}} |x|\,\nu(\mathrm{d}x) < \infty,$$

so that f is Lebesgue–Stieltjes-integrable with respect to L by (a). To see that the integral has finite expectation, define $g : \mathbb{R}\times\mathbb{R} \to [0,\infty)$ by $g(s,y) = |f(s)y|$ and $Z := \sum_{t\in\mathbb{R};\Delta L_t\neq 0} g(t,\Delta L_t)$. By Theorem 12.36(b), Z is infinitely divisible with drift 0, Gaussian variance 0 and Lévy measure ν_g given by $\nu_g(\mathrm{d}x) = g(\lambda^1\otimes\nu)(\mathrm{d}x)\mathbf{1}_{\mathbb{R}\setminus\{0\}}(x)$. From Proposition 9.18 and Theorem 11.15, we then conclude that

$$\mathbb{E}(Z) = \int_{\mathbb{R}} x\, \nu_g(\mathrm{d}x) = \int_{\mathbb{R}} |f(s)|\, \mathrm{d}s \int_{\mathbb{R}} |x|\, \nu(\mathrm{d}x) < \infty.$$

This establishes (15.3). When f is real-valued, equation (15.4) is obtained similarly by observing that $\int_{\mathbb{R}} f(s)\, \mathrm{d}L_s = \gamma^0 \int_{\mathbb{R}} f(t)\, \mathrm{d}t + \sum_{t\in\mathbb{R},\Delta L_t \neq 0} f(t)\Delta L_t$ and applying Theorem 12.36(b) with the function $h(s,y) = f(s)y$ rather than g. Finally, when f is complex valued, (15.4) is obtained by considering real and imaginary parts of f separately.

(c) When $f \in L^1(\mathbb{R};\mathbb{C})$ has bounded support, we conclude that

$$\int_{\mathbb{R}}\int_{\mathbb{R}} (1 \wedge |f(t)x|)\, \nu(\mathrm{d}x)\, \mathrm{d}t \le \int_{\mathbb{R}} |f(t)|\, \mathrm{d}t \int_{|x|\le 1} |x|\, \nu(\mathrm{d}x) + \int_{\operatorname{supp}(f)} \int_{|x|>1} \nu(\mathrm{d}x)\, \mathrm{d}t$$

which is finite, so that f is Lebesgue–Stieltjes-integrable with respect to L by (a). □

Remark 15.4 (Linearity of integral). The Lebesgue–Stieltes integral is linear with respect to both integrand and integrator. More precisely, suppose that $\lambda \in \mathbb{R}$, L and M are two Lévy processes of finite variation defined on the same probability space and that $L + \lambda M$ is a Lévy process (this is the case, e.g., if (L, M) is a bivariate Lévy process). Let $f, g : \mathbb{R} \to \mathbb{C}$ be Borel-measurable functions such that f is Lebesgue–Stieltjes-integrable with respect to both L and M and such that g is Lebesgue–Stieltjes-integrable with respect to L. Then $\lambda f + g$ is Lebesgue–Stieltjes-integrable with respect to L with

$$\mathrm{L.S.}\int_{\mathbb{R}} (\lambda f(t) + g(t))\, \mathrm{d}L_t = \lambda\, \mathrm{L.S.}\int_{\mathbb{R}} f(t)\, \mathrm{d}L_t + \mathrm{L.S.}\int_{\mathbb{R}} g(t)\, \mathrm{d}L_t$$

and f is Lebesgue–Stieltjes-integrable with respect to $L + \lambda M$ with

$$\mathrm{L.S.}\int_{\mathbb{R}} f(t)\, \mathrm{d}(L_t + \lambda M_t) = \mathrm{L.S.}\int_{\mathbb{R}} f(t)\, \mathrm{d}L_t + \lambda\, \mathrm{L.S.}\int_{\mathbb{R}} f(t)\, \mathrm{d}M_t.$$

This follows immediately from (15.2) and the corresponding integrability conditions involving absolute values stated in Theorem 15.3(a)(ii).

Example 15.5 (Integral of simple functions). Suppose that L is a Lévy process of finite variation and that $a < b$. Then $\mathbf{1}_{(a,b]}$ is Lebesgue–Stieltjes-integrable with respect to L by Theorem 15.3(c) and from equation (15.2) we obtain (with γ^0 denoting the drift of L)

$$\mathrm{L.S.}\int_{\mathbb{R}} \mathbf{1}_{(a,b]}(t)\, \mathrm{d}L_t = \gamma^0(b-a) + \sum_{a<t\le b} \Delta L_t = L_b - L_a.$$

Now if $f : \mathbb{R} \to \mathbb{C}$ is of the form $f(t) = \sum_{i=1}^n a_i \mathbf{1}_{(u_i,u_{i+1}]}(t)$ for some $u_1, \dots, u_{n+1} \in \mathbb{R}$ and $a_1, \dots, a_n \in \mathbb{C}$ with $-\infty < u_1 < u_2 < \cdots < u_{n+1} < \infty$ (i.e., $f \in \mathcal{D}$, the class of simple functions as defined in Step 1 of Section 4.3), then it follows from linearity of the integral that f is Lebesgue–Stieltjes-integrable with respect to f and that

$$\text{L. S.} \int_{\mathbb{R}} f(t)\, \mathrm{d}L_t = \sum_{i=1}^{n} \alpha_i (L_{u_{i+1}} - L_{u_i}).$$

The following is the dominated convergence theorem for Lebesgue–Stieltjes integrals with respect to a Lévy process of finite variation and will become very useful later.

Theorem 15.6 (Dominated convergence theorem for L.-S. integrals). *Let $L = (L_t)_{t\in\mathbb{R}}$ be an $\mathbb{R}$-valued Lévy process of finite variation. Let $f_n, f, g : \mathbb{R} \to \mathbb{C}$ ($n \in \mathbb{N}$) be Borel-measurable functions such that g is Lebesgue–Stieltjes-integrable with respect to L, f_n converges Lebesgue almost everywhere to f as $n \to \infty$ and $|f_n| \le |g|$ Lebesgue almost everywhere for each $n \in \mathbb{N}$. Then f and each of the f_n are Lebesgue–Stieltjes-integrable with respect to L, and* L. S. $\int_{\mathbb{R}} f_n(t)\, \mathrm{d}L_t$ *converges almost surely to* L. S. $\int_{\mathbb{R}} f(t)\, \mathrm{d}L_t$ *as $n \to \infty$.*

Proof. If the drift γ^0 of L is non-zero, then integrability of g with respect to L implies that $g \in L^1(\mathbb{R}; \mathbb{C})$ and hence that $f_n, f \in L^1(\mathbb{R}; \mathbb{C})$. The standard dominated convergence theorem for the Lebesgue integral then implies that $\int_{\mathbb{R}} \gamma^0 f_n(t)\, \mathrm{d}t$ converges to $\int_{\mathbb{R}} \gamma^0 f(t)\, \mathrm{d}t$ as $n \to \infty$. Hence the drift part presents no difficulty and we can assume without loss of generality that $\gamma^0 = 0$.

By modifying L if necessary, we can assume without loss of generality that $L_{0-} = L_0 = 0$ everywhere, so that in particular $\Delta L_0 = 0$ everywhere. Using the simplified Lévy–Itô decomposition (Theorem 13.1), we can then write

$$L_t = \begin{cases} \sum_{0<s\le t, \Delta L_s \ne 0} \Delta L_s, & t \ge 0, \\ -\sum_{s<t<0} \Delta L_s, & t < 0. \end{cases}$$

Define $B_1 := \mathbb{R} \setminus [-1, 1]$ and $B_m := \{x \in \mathbb{R} : 1/m < |x| \le 1/(m-1)\}$ for $m \in \{2, 3, \dots\}$. For $m \in \mathbb{N}$, define $L^m = (L_t^m)_{t\in\mathbb{R}}$ by

$$L_t^m := \begin{cases} \sum_{0<s\le t, \Delta L_s \in B_m} \Delta L_s, & t \ge 0, \\ -\sum_{t<s<0, \Delta L_s \in B_m} \Delta L_s, & t < 0. \end{cases}$$

Then L^m is a Lévy process by Theorem 12.13 (or also by Theorem 12.29(b)), and since its paths are obviously step functions, each L^m is either (almost surely equal to) the zero process or a compound Poisson process by Corollary 12.27. Denote by M the set of all $m \in \mathbb{N}$ such that L^m is a compound Poisson process. We can then write $L_t = \sum_{m\in M} L_t^m$, with the sum converging almost surely absolutely. For $m \in M$, denote the jump times of L^m by Γ_j^m, $j \in \mathbb{Z}$, where $\cdots < \Gamma_{-1}^m < \Gamma_0^m < 0 < \Gamma_1^m < \Gamma_2^m < \cdots$ as in Theorem 1.31, and the size of the jth jump of L^m by Y_j^m. Observe that each $|\Gamma_j^m|$ has a gamma distribution (as an independent sum of exponentially distributed random variables) that is, in particular, absolutely continuous with respect to Lebesgue measure.

Now let $C \in \mathcal{B}_1$ be a Borel set such that, as $n \to \infty$, $f_n(t)$ converges to $f(t)$ for every $t \in C$, where $|f_n(t)| \le |g(t)|$ for all $t \in C$ and $n \in \mathbb{N}$, where $|f(t)| \le |g(t)|$ for all $t \in \mathbb{N}$, and where $\mathbb{R} \setminus C$ has Lebesgue measure 0. Such a set C exists since f_n converges

Lebesgue almost everywhere to f and since f_n (and hence f) are Lebesgue-almost everywhere dominated by $|g|$. Since Γ_j^m is absolutely continuous with respect to Lebesgue measure ($m \in M, j \in \mathbb{Z}$), we have $P(\Gamma_j^m \in C) = 1$, hence $P(\Gamma_j^m \in C \ \forall \ (m,j) \in M \times \mathbb{Z}) = 1$. Hence, almost surely,

$$\begin{aligned}\sum_{t\in\mathbb{R},\Delta L_t\neq 0} |f_n(t)\Delta L_t| &= \sum_{m\in M}\sum_{j\in\mathbb{Z}} |f_n(\Gamma_j^m)Y_j^m| \\ &= \sum_{m\in M}\sum_{j\in\mathbb{Z}:\Gamma_j^m\in C} |f_n(\Gamma_j^m)Y_j^m|.\end{aligned}$$

The latter can be bounded everywhere (since $|f_n(t)| \leq |g(t)|$ for all $t \in C$) by $\sum_{m\in M}\sum_{j\in\mathbb{Z}:\Gamma_j^m\in C} |g(\Gamma_j^m)Y_j^m|$, which by the same reasoning is almost surely equal to $\sum_{t\in\mathbb{R};\Delta L_t\neq 0} |g(t)\Delta L_t|$, which is finite almost surely since g is Lebesgue–Stieltjes-integrable with respect to L. Hence f_n is Lebesgue–Stieltjes-integrable with respect to L, and by the same reasoning, so is f.

Now take a set Ω_1 of probability 1 such that

$$\sum_{m\in M}\sum_{j\in\mathbb{Z}:\Gamma_j^m(\omega)\in C} |g(\Gamma_j^m(\omega))Y_j^m(\omega)| < \infty \quad \forall\ \omega \in \Omega_1.$$

An application of Lebesgue's dominated convergence theorem for sums now implies that

$$\sum_{m\in M}\sum_{j\in\mathbb{Z}:\Gamma_j^m(\omega)\in C} f_n(\Gamma_j^m(\omega))Y_j^m(\omega) \to \sum_{m\in M}\sum_{j\in\mathbb{Z}:\Gamma_j^m(\omega)\in C} f(\Gamma_j^m(\omega))Y_j^m(\omega)$$

as $n \to \infty$ for all $\omega \in \Omega_1$, and since the left- and right-hand sides correspond to the corresponding Lebesgue–Stieltjes integrals, we obtain their desired almost sure convergence. □

As an immediate consequence we see that the Lebesgue–Stieltjes integral is unaffected by changes of the integrand on a Lebesgue null-set.

Corollary 15.7. *Let $L = (L_t)_{t\in\mathbb{R}}$ be an $\mathbb{R}$-valued Lévy process of finite variation and $f, h : \mathbb{R} \to \mathbb{C}$ be Borel-measurable functions such that f is Lebesgue almost everywhere equal to h. Assume further that h is Lebesgue–Stieltjes-integrable with respect to L. Then f is also Lebesgue–Stieltjes-integrable with respect to L, and*

$$\text{L. S.}\int_{\mathbb{R}} f(t)\,\mathrm{d}L_t = \text{L. S.}\int_{\mathbb{R}} h(t)\,\mathrm{d}L_t.$$

In particular, if f is Lebesgue almost everywhere equal to zero, then it is Lebesgue–Stieltjes-integrable with respect to L with $\text{L. S.}\int_{\mathbb{R}} f(t)\,\mathrm{d}L_t = 0$.

Proof. Lebesgue–Stieltjes integrability of f and the given formula follow from Theorem 15.6 on defining $f_n := h$ and $g := h$, since then f_n converges Lebesgue almost everywhere to f. The last assertion then follows by setting $h = 0$. □

Next, we show that (certain) Lebesgue–Stieltjes integrals can be approximated by Riemann(–Stieltjes) sums.

Theorem 15.8 (Riemann–Stieltjes sums). *Let $L = (L_t)_{t\in\mathbb{R}}$ be an $\mathbb{R}$-valued Lévy process of finite variation and $f : \mathbb{R} \to \mathbb{C}$ be a bounded Borel-measurable function with support contained in a compact interval $[a, b]$. Suppose that the set of discontinuities of f is a Lebesgue null-set. For each $n \in \mathbb{N}$ let $\pi_n = (t_{0,n}, \dots, t_{m(n),n})$ be a partition of $[a, b]$ as in Definition* 13.7 *and assume that*

$$\lim_{n\to\infty} \max\{|t_{j,n} - t_{j-1,n}| : j \in \{1, \dots, m(n)\}\} = 0.$$

For each $j \in \{1, \dots, m(n)\}$, suppose that $\xi_{j,n} \in [t_{j-1,n}, t_{j,n}]$. Then the Riemann–Stieltjes sums $\sum_{j=1}^{m(n)} f(\xi_{j,n})\,(L_{t_{j,n}} - L_{t_{j-1,n}})$ converge almost surely to the Lebesgue–Stieltjes integral $\mathrm{L.S.}\int_{\mathbb{R}} f(t)\,\mathrm{d}L_t$ as $n \to \infty$.

Proof. Define $f_n : \mathbb{R} \to \mathbb{C}$ by

$$f_n(s) := \sum_{j=1}^{m(n)} f(\xi_{j,n}) \mathbf{1}_{(t_{j-1,n}, t_{j,n}]}(s) \quad \text{for } n \in \mathbb{N},\ s \in \mathbb{R}.$$

Then

$$\mathrm{L.S.}\int_{\mathbb{R}} f_n(t)\,\mathrm{d}t = \sum_{j=1}^{m(n)} f(\xi_{j,n})\,(L_{t_{j,n}} - L_{t_{j-1,n}})$$

by Example 15.5. Further, the functions f_n are dominated by the function $g := \mathbf{1}_{[a,b]} \sup_{s\in[a,b]} |f(s)|$, which is Lebesgue–Stieltjes-integrable with respect to L by Theorem 15.3(c). Finally, if s_0 is a continuity point of f, then $f_n(s_0)$ converges to $f(s_0)$ as $n \to \infty$, so that f_n converges Lebesgue almost everywhere to f as $n \to \infty$. The claim then follows from Theorem 15.6. □

15.2 Integration with respect to one-dimensional Lévy processes

We now turn to the general integration concept. For that, let us first recall that if $M = (M_t)_{t\in\mathbb{R}}$ is an $\mathbb{R}$-valued Lévy process with $\mathbb{E}L_1^2 < \infty$ and $\mathbb{E}L_1 = 0$, then M is an orthogonal increment process by Corollary 11.17 with associated distribution function given by $t \mapsto t\ \mathrm{Var}(L_1)$. Hence, a Borel-measurable function $f : \mathbb{R} \to \mathbb{C}$ is integrable with respect to the orthogonal increment process M, as defined in Section 4.3, if $f \in L^2(\mathbb{R};\mathbb{C})$. (Observe that if $\operatorname{Var}(L_1) > 0$, this is actually an "if and only if" condition.) Let us denote the corresponding integral by

$$\text{O.I.P.} \int_{\mathbb{R}} f(t)\, \mathrm{d}L_t,$$

which in Section 4.3 was denoted by $I(f)$. We shall speak of it as the *OIP integral*. Observe that the OIP integral was defined as a limit in L^2, hence is defined only almost surely.

The idea now is to decompose a general one-dimensional Lévy process $L = (L_t)_{t\in\mathbb{R}}$ into two independent Lévy processes S and M, where S is of finite variation and M is an orthogonal increment process (such a decomposition exists by Corollary 13.16), and then to define the integral of f with respect to L (provided it exists) as the Lebesgue–Stieltjes integral of f with respect to S plus the orthogonal increment process integral of f with respect to M. To ensure that this procedure is well defined, we must first show that the corresponding value does not depend on the specific partition, which we shall do in the proof of the following theorem.

Theorem 15.9. *Let $L = (L_t)_{t\in\mathbb{R}}$ be an $\mathbb{R}$-valued Lévy process and $f \in L^2(\mathbb{R};\mathbb{C})$. Suppose that*

$$L = S^1 + M^1 \quad \text{and} \quad L = S^2 + M^2,$$

where S^1 and M^1 are independent Lévy processes, S^2 and M^2 are independent Lévy processes, and that S^1 and S^2 are of finite variation and M^1 and M^2 are orthogonal increment processes. Suppose that f is Lebesgue–Stieltjes-integrable with respect to both S^1 and S^2. Then

$$\begin{aligned} &\text{L.S.} \int_{\mathbb{R}} f(t)\, \mathrm{d}S^1(t) + \text{O.I.P.} \int_{\mathbb{R}} f(t)\, \mathrm{d}M^1(t) \\ &= \text{L.S.} \int_{\mathbb{R}} f(t)\, \mathrm{d}S^2(t) + \text{O.I.P.} \int_{\mathbb{R}} f(t)\, \mathrm{d}M^2(t). \end{aligned} \tag{15.5}$$

Proof. First consider the case when $f \in L^2(\mathbb{R};\mathbb{C})$ is bounded and has compact support contained in some interval $[a,b] \subset \mathbb{R}$. Any such f is Lebesgue–Stieltjes-integrable with respect to S^1 and S^2. Recall the space $\mathcal{D}$ of all simple functions of the form (4.10), which is dense in $L^2(\mathbb{R};\mathbb{C})$ (cf., Step 3 in Section 4.3). Hence there is a sequence $(f_n)_{n\in\mathbb{N}}$ of simple functions that converges in $L^2(\mathbb{R};\mathbb{C})$ to f as $n \to \infty$. By choosing a subsequence if necessary, we can assume that f_n converges also Lebesgue almost everywhere to f. Now let f be bounded by K and define $f_n' : \mathbb{R} \to \mathbb{C}$ by

$$f_n'(t) := f_n(t)\mathbf{1}_{(a,b]}(t)\mathbf{1}_{[0,K+1]}(|f_n(t)|),$$

which agrees with f_n when $t \in (a, b]$ and $|f_n(t)| \leq K + 1$ and is zero otherwise. Then f_n' is also a simple function for each $n \in \mathbb{N}$ and f_n' clearly also converges to f as $n \to \infty$, both in $L^2(\mathbb{R}; \mathbb{C})$ and Lebesgue almost everywhere. Then L. S. $\int_{\mathbb{R}} f_n'(t)\, \mathrm{d}S^1(t)$ converges almost surely to L. S. $\int_{\mathbb{R}} f(t)\, \mathrm{d}S^1(t)$ by Theorem 15.6 and O. I. P. $\int_{\mathbb{R}} f_n'(t)\, \mathrm{d}M^1(t)$ converges in $\mathcal{S}$ to O. I. P. $\int_{\mathbb{R}} f(t)\, \mathrm{d}M^1(t)$ by equation (4.14). Now if f_n' has a representation of the form $f_n' = \sum_{i=1}^{m(n)} a_{i,n} \mathbf{1}_{(u_{i,n}, u_{i+1,n}]}$ with $a_{i,n} \in \mathbb{C}$ and $a \leq u_{1,n} < \cdots < u_{m(n),n} \leq b$, then

$$\begin{aligned}
&\text{L. S.} \int_{\mathbb{R}} f_n'(t)\, \mathrm{d}S_t^1 + \text{O. I. P.} \int_{\mathbb{R}} f_n'(t)\, \mathrm{d}M_t^1 \\
&= \sum_{i=1}^{n} a_{i,n}(S^1_{u_{i+1,n}} - S^1_{u_{i},n}) + \sum_{i=1}^{n} a_{i,n}(M^1_{u_{i+1,n}} - M^1_{u_{i},n}) \\
&= \sum_{i=1}^{n} a_{i,n}(L_{u_{i+1,n}} - L_{u_{i},n})
\end{aligned}$$

by Example 15.5 and (4.11). It follows that L. S. $\int_{\mathbb{R}} f(t)\, \mathrm{d}S^1(t)$ + O. I. P. $\int_{\mathbb{R}} f(t)\, \mathrm{d}M^1(t)$ is the probability limit of $\sum_{i=1}^{n} a_{i,n}(L_{u_{i+1,n}} - L_{u_i,n})$ as $n \to \infty$. Similar reasoning applies to S^2 and M^2, and since the approximating sequence f_n' does not depend on the specific decomposition of L into $S^1 + M^1$ or $S^2 + M^2$, we obtain (15.5) when f is bounded with compact support.

Now let $f \in L^2(\mathbb{R}; \mathbb{C})$ be a general function that is Lebesgue–Stieltjes-integrable with respect to both S^1 and S^2. Define the sequence $(f_n)_{n \in \mathbb{N}}$ of complex-valued functions by

$$f_n(t) := f(t) \mathbf{1}_{[-n,n]}(t) \mathbf{1}_{[0,n]}(|f(t)|), \quad t \in \mathbb{R}.$$

Then each f_n is bounded and has compact support, and f_n converges both Lebesgue almost everywhere as well as in $L^2(\mathbb{R}; \mathbb{C})$ to f as $n \to \infty$. Since each f_n is dominated by f which is Lebesgue–Stieltjes-integrable with respect to both S^1 and S^2, Theorem 15.6 and equation (4.14) imply that L. S. $\int_{\mathbb{R}} f_n(t)\, \mathrm{d}S_t^1$ + O. I. P. $\int_{\mathbb{R}} f_n(t)\, \mathrm{d}M_t^1$ converges in probability to L. S. $\int_{\mathbb{R}} f(t)\, \mathrm{d}S_t^1$ + O. I. P. $\int_{\mathbb{R}} f(t)\, \mathrm{d}M_t^1$ as $n \to \infty$. A similar statement is true for the integrals with respect to S^2 and M^2. But since (15.5) is true for each f_n, it follows that it is also true for f as the corresponding probability limit when $n \to \infty$. □

We can now define the integral of deterministic functions f with respect to a general Lévy process. We will always assume that $f \in L^1(\mathbb{R}; \mathbb{C}) \cap L^2(\mathbb{R}; \mathbb{C})$, which is quite sufficient for our purposes and makes the integration theory much easier.

Definition 15.10 (Integral with respect to a Lévy process). Suppose that $L = (L_t)_{t \in \mathbb{R}}$ is a two-sided one-dimensional Lévy process and $f \in L^1(\mathbb{R}; \mathbb{C}) \cap L^2(\mathbb{R}; \mathbb{C})$. Then f is called *integrable with respect to* L, if there exists a decomposition $L = S + M$ of L into two independent Lévy processes S and M, where S is of finite variation and M has finite second moment and mean 0 (equivalent, since M is a Lévy process, to the statement that M is an OIP), such that f is Lebesgue–Stieltjes-integrable with respect to S. If f is integrable with respect to L, we define the *integral of f with respect to L* as

$$\int_{\mathbb{R}} f(t)\,\mathrm{d}L_t := \mathrm{L.\,S.} \int_{\mathbb{R}} f(t)\,\mathrm{d}S_t + \mathrm{O.\,I.\,P.} \int_{\mathbb{R}} f(t)\,\mathrm{d}M_t, \tag{15.6}$$

which by Theorem 15.9 does not depend on the specific choice of S and M.

Observe that the integral is defined only almost surely, since both the Lebesgue–Stieltjes integral and the orthogonal increment integral were defined only almost surely. However, the integral is a random variable.

As a corollary of Theorem 15.9 we obtain the following result.

Corollary 15.11 (Compatibility with previous integral definitions). *Suppose that $L = (L_t)_{t\in\mathbb{R}}$ is an $\mathbb{R}$-valued Lévy process and that $f \in L^1(\mathbb{R};\mathbb{C}) \cap L^2(\mathbb{R};\mathbb{C})$.*

(a) *If L is of finite variation and f is Lebesgue–Stieltjes-integrable with respect to L, then*

$$\int_{\mathbb{R}} f(t)\,\mathrm{d}L_t = \mathrm{L.\,S.} \int_{\mathbb{R}} f(t)\,\mathrm{d}L_t.$$

(b) *If $\mathbb{E}L_1^2 < \infty$ and $\mathbb{E}L_1 = 0$, then*

$$\int_{\mathbb{R}} f(t)\,\mathrm{d}L_t = \mathrm{O.\,I.\,P.} \int_{\mathbb{R}} f(t)\,\mathrm{d}L_t.$$

(c) *If L is of finite variation with $\mathbb{E}L_1^2 < \infty$ and $\mathbb{E}L_1 = 0$, then f is Lebesgue–Stieltjes-integrable with respect to L and*

$$\int_{\mathbb{R}} f(t)\,\mathrm{d}L_t = \mathrm{L.\,S.} \int_{\mathbb{R}} f(t)\,\mathrm{d}L_t = \mathrm{O.\,I.\,P.} \int_{\mathbb{R}} f(t)\,\mathrm{d}L_t.$$

Proof. Part (a) follows by taking the decomposition $S = L$ and $M = 0$ in (15.6), and (b) by taking the decomposition $S = 0$ and $M = L$. Part (c) follows from (a), (b) and Theorem 15.3(b) (for the fact that f is Lebesgue–Stieltjes-integrable with respect to L). □

Thanks to Corollary 15.11 there is no longer any need to distinguish between Lebesgue–Stieltes integrals, OIP integrals and general integrals with respect to Lévy processes, so that from now on we will omit the prefixes “L. S.” and “O. I. P.” for Lebesgue–Stieltjes integrals and orthogonal increment integrals and simply write $\int_{\mathbb{R}} f(t)\,\mathrm{d}L_t$.

Example 15.12 (Mean and variance of integral). Suppose that L has finite second moment and that $f \in L^1(\mathbb{R};\mathbb{C}) \cap L^2(\mathbb{R};\mathbb{C})$. Define $M_t := L_t - t\,\mathbb{E}(L_1)$ and $S_t := t\,\mathbb{E}(L_1)$. Then M is an OIP and f is Lebesgue–Stieltjes-integrable with respect to S by Theorem 15.3, since $f \in L^1(\mathbb{R};\mathbb{C})$ and is therefore integrable with respect to L. Since

$$\int_{\mathbb{R}} f(t)\,\mathrm{d}L_t = \mathbb{E}(L_1)\int_{\mathbb{R}} f(t)\,\mathrm{d}t + \int_{\mathbb{R}} f(t)\,\mathrm{d}M_t,$$

we see that $\int_{\mathbb{R}} f(t)\,\mathrm{d}L_t$ has finite second moment. Its mean is given by $\int_{\mathbb{R}} f(t)\,\mathrm{d}t$ and its variance by $\operatorname{Var}(L_1)\int_{\mathbb{R}} |f(t)|^2\,\mathrm{d}t < \infty$, the latter by (4.15), since the distribution function associated with the orthogonal-increment process M is $t \mapsto F_M(t) = t\operatorname{Var}(M_1)$ by Corollary 11.17.

Example 15.13 (Integral of simple functions). Let L be a Lévy process and $f : \mathbb{R} \to \mathbb{C}$ be a simple function of the form $f(t) = \sum_{i=1}^n a_i \mathbf{1}_{(u_i,u_{i+1}]}(t)$ for some $u_1,\dots,u_{n+1} \in \mathbb{R}$ and $a_1,\dots,a_n \in \mathbb{C}$ with $-\infty < u_1 < u_2 < \cdots < u_{n+1} < \infty$. Then f is integrable with respect to L and

$$\int_{\mathbb{R}} f(t)\,\mathrm{d}L_t = \sum_{i=1}^n a_i(L_{u_{i+1}} - L_{u_i}).$$

This follows for the Lebesgue–Stieltes part in the integral from Example 15.5 and for the OIP part in the integral from (4.11).

Remark 15.14. Since a measurable complex-valued function f is Lebesgue–Stieltjes-integrable with respect to a process S if and only if $|f|$ is Lebesgue–Stieltjes-integrable with respect to S, it is clear that $f \in L^1(\mathbb{R};\mathbb{C}) \cap L^2(\mathbb{R};\mathbb{C})$ is integrable with respect to a Lévy process L if and only if $|f|$ is integrable with respect to L. Further, $f \in L^1(\mathbb{R};\mathbb{C}) \cap L^2(\mathbb{R};\mathbb{C})$ is integrable with respect to L if and only if both $\Re(f)$ and $\Im(f)$ are integrable with respect to L. Finally, if $g : \mathbb{R} \to \mathbb{C}$ is Borel-measurable with $|g| \le |f|$ Lebesgue almost everywhere, then it is clear from Corollary 15.7 and Theorem 15.3(a) that g is also integrable with respect to L.

Suppose that $f \in L^1(\mathbb{R};\mathbb{C}) \cap L^2(\mathbb{R};\mathbb{C})$ is integrable with respect to L, i. e., there is a decomposition $L = S + M$ as specified in Definition 15.10 such that f is Lebesgue–Stieltjes-integrable with respect to S. Suppose now that another decomposition $L = S' + M'$ is given, where S' and M' are two independent Lévy processes such that S' is of finite variation and M' is an OIP. Does it follow that f is also Lebesgue–Stieltjes-integrable with respect to S', i. e., does the existence of some decomposition for which we have integrability imply that all decompositions give integrability? The answer to this question is affirmative and is included in the following characterization.

Theorem 15.15 (Characterization of integrability). *Suppose that $L = (L_t)_{t\in\mathbb{R}}$ is a two-sided $\mathbb{R}$-valued Lévy process and $f \in L^1(\mathbb{R};\mathbb{C}) \cap L^2(\mathbb{R};\mathbb{C})$. Then the following statements are equivalent.*

(i) *f is integrable with respect to L, i. e., there exists a decomposition $L = S + M$ of L into two independent Lévy processes S and M, where S is of finite variation and M is an OIP, such that f is Lebesgue–Stieltjes-integrable with respect to S.*

(ii) *For all decompositions $L = S + M$ of L into two independent Lévy processes S and M, where S is of finite variation and M is an OIP, the function f is Lebesgue–Stieltjes-integrable with respect to S.*

(iii) *The Lévy measure ν of L satisfies*

$$\int_{\mathbb{R}} \int_{|x|>1} (1 \wedge |f(t)x|)\, \nu(dx)\, dt < \infty. \tag{15.7}$$

Proof. That (ii) implies (i) is clear. Let us show that (i) implies (iii). Let $L = S + M$ as in (i). Denote the Lévy measures of S and M by ν_S and ν_M, respectively. Then $\nu = \nu_S + \nu_M$. Hence

$$\begin{aligned}&\int_{\mathbb{R}} \int_{|x|>1} (1 \wedge |f(t)x|)\, \nu(dx)\, dt \\ &\quad \le \int_{\mathbb{R}} \int_{|x|>1} (1 \wedge |f(t)x|)\, \nu_S(dx)\, dt + \int_{\mathbb{R}} \int_{|x|>1} |f(t)|\, |x|\, \nu_M(dx)\, dt.\end{aligned}$$

The integral involving ν_S is finite by Theorem 15.3(a), since f is Lebesgue–Stieltjes-integrable with respect to S, and the integral involving ν_M is finite since $f \in L^1(\mathbb{R};\mathbb{C})$ and M has finite expectation so that Corollary 11.14(a) applies. This establishes (15.7).

To see that (iii) implies (ii), let $L = S + M$ be an arbitrary decomposition of L as in (ii). Denote the Lévy measure of S by ν_S. Then $\nu_S \le \nu$ and since $f \in L^1(\mathbb{R};\mathbb{C})$ and $\int_{|x|\le 1} |x|\, \nu_S(dx) < \infty$ by Theorem 13.15 we conclude, using (15.7), that

$$\begin{aligned}&\int_{\mathbb{R}} \int_{\mathbb{R}} (1 \wedge |f(t)x|)\, \nu_S(dx)\, dt \\ &\quad \le \int_{\mathbb{R}} \int_{|x|\le 1} |f(t)|\, |x|\, \nu_S(dx)\, dt + \int_{\mathbb{R}} \int_{|x|>1} (1 \wedge |f(t)x|)\, \nu_S(dx)\, dt < \infty.\end{aligned}$$

Hence f is Lebesgue–Stieltjes-integrable with respect to S by Theorem 15.3. □

The following corollary is an immediate consequence.

Corollary 15.16. *Suppose that $L = (L_t)_{t\in\mathbb{R}}$ is a two-sided $\mathbb{R}$-valued Lévy process of finite variation and that $f \in L^1(\mathbb{R};\mathbb{C}) \cap L^2(\mathbb{R};\mathbb{C})$. Then f is integrable with respect to L if and only if it is Lebesgue–Stieltjes-integrable with respect to L.*

Remark 15.17. Theorem 15.15 rests heavily on the assumption that a priori we only allowed functions $f \in L^1(\mathbb{R};\mathbb{C}) \cap L^2(\mathbb{R};\mathbb{C})$. If this assumption is relaxed and one says that a general Borel-measurable function $f : \mathbb{R} \to \mathbb{C}$ is integrable with respect to L if there exists a decomposition $L = S+M$ of L into two independent Lévy processes, where S is of finite variation and M an OIP, such that f is Lebesgue–Stieltjes-integrable with respect to S and OIP-integrable with respect to the OIP M, then it is no longer true that for every other decomposition $L = S' + M'$ of L into two independent Lévy processes, where S' is of finite variation and M' is an OIP, that f will be Lebesgue–Stieltjes-integrable with respect to S' and OIP-integrable with respect to M'. See Exercise 15.1 for details.

From Theorem 15.15, we can deduce some useful conditions under which a function is integrable with respect to a given Lévy process:

Corollary 15.18 (Sufficient conditions for integrability). *Suppose that $L = (L_t)_{t\in\mathbb{R}}$ is a two-sided $\mathbb{R}$-valued Lévy process and $f \in L^1(\mathbb{R};\mathbb{C}) \cap L^2(\mathbb{R};\mathbb{C})$. Then each of the following conditions is sufficient for f to be integrable with respect to L.*
(i) $\mathbb{E}|L_1| < \infty$.
(ii) *f has bounded support.*
(iii) *There is $\varepsilon > 0$ such that* $\limsup_{|t|\to\infty} |f(t)|e^{\varepsilon|t|} < \infty$ *and* $\mathbb{E}\log^+|L_1| < \infty$.

In case (i), *the integral has finite expectation and*

$$\mathbb{E}\Big(\int_{\mathbb{R}} f(t)\,dL_t\Big) = \int_{\mathbb{R}} f(t)\,dt\;\mathbb{E}(L_1).$$

Proof. By Theorem 15.15, we have to show that (15.7) is satisfied under each of the conditions (i)–(iii). Under condition (i) this follows from the inequalities,

$$\int_{\mathbb{R}}\int_{|x|>1} (1\wedge|f(t)x|)\,\nu(dx)\,dt \le \int_{\mathbb{R}}\int_{|x|>1} |f(t)|\,|x|\,\nu(dx)\,dt < \infty$$

and it is clearly satisfied under condition (ii). The form of the expectation under condition (i) follows from Theorem 15.3(b). Now suppose that condition (iii) holds. Then there exists $T \ge 0$ such that $|f(t)| \le Ke^{-\varepsilon|t|}$ for all $|t| \ge T$ and, by choosing ε slightly smaller, we can assume without loss of generality that $K = 1$. An easy calculation shows that $\int_{\mathbb{R}}(1\wedge e^{-\varepsilon|t|}|x|)\,dt = (2+2\log|x|)/\varepsilon$ for every $|x| > 1$. Hence

$$\int_{\mathbb{R}\setminus[-T,T]}\int_{|x|>1} (1\wedge|f(t)x|)\,\nu(dx)dt \le \int_{|x|>1} \frac{2+2\log|x|}{\varepsilon}\,\nu_L(dx) < \infty$$

by Corollary 11.14(b). Since clearly $\int_{[-T,T]}\int_{|x|>1}(1\wedge|f(t)x|)\,\nu(dx)\,dt < \infty$, equation (15.7) follows. □

Theorem 15.15 can be used to show that the set of all integrable functions (that are in $L^1(\mathbb{R};\mathbb{C}) \cap L^2(\mathbb{R};\mathbb{C})$) with respect to L is a vector space.

Lemma 15.19 (Linearity of the integral). *Suppose that $L = (L_t)_{t\in\mathbb{R}}$ is a two-sided $\mathbb{R}$-valued Lévy process, $f, g \in L^1(\mathbb{R};\mathbb{C}) \cap L^2(\mathbb{R};\mathbb{C})$ and that $Q \in \mathbb{C}$.*
(a) *Suppose that f and g are integrable with respect to L. Then $Qf + g$ is integrable with respect to L and*

$$\int_{\mathbb{R}} (Qf(t) + g(t))\, \mathrm{d}L_t = Q \int_{\mathbb{R}} f(t)\, \mathrm{d}L_t + \int_{\mathbb{R}} g(t)\, \mathrm{d}L_t.$$

In particular, the set of functions in $L^1(\mathbb{R};\mathbb{C}) \cap L^2(\mathbb{R};\mathbb{C})$ that are integrable with respect to L form a vector space.

(b) *Suppose that f is integrable with respect to L and also integrable with respect to L', where L' is a Lévy process, which is assumed to be independent of L. Then f is also integrable with respect to the Lévy process $L + L'$, and*

$$\int_{\mathbb{R}} f(t)\, \mathrm{d}(L_t + L'_t) = \int_{\mathbb{R}} f(t)\, \mathrm{d}L_t + \int_{\mathbb{R}} f(t)\, \mathrm{d}L'_t.$$

Proof. (a) From Theorem 15.15(ii), we conclude that there is a decomposition $L = S + M$ such that f and g are both Lebesgue–Stieltjes-integrable with respect to S. The formula for $\int_{\mathbb{R}} (Qf(t) + g(t))\, \mathrm{d}L_t$ then follows from the corresponding results for Lebesgue–Stieltjes integrals (cf., Remark 15.4) and OIP integrals.

(b) If $L = S + M$ and $L' = S' + M'$ are partitions such that f is Lebesgue–Stieltjes-integrable with respect to S and S', then $L + L' = (S + S') + (M + M')$ provides a partition of $L + L'$ with the desired properties such that f is Lebesgue–Stieltjes-integrable with respect to $S + S'$. The formula then follows from the corresponding properties of the Lebesgue–Stieltjes (cf., Remark 15.4) and OIP integrals (it clearly holds when f is a simple function, and for general OIP integrals it follows by taking limits). □

Again, the a priori assumption $f \in L^1(\mathbb{R};\mathbb{C}) \cap L^2(\mathbb{R};\mathbb{C})$ is crucial for the vector space property of the integrable functions, see Exercise 15.2.

The stochastic integral is unaffected if the integrand is changed on a Lebesgue null set, a result stated more precisely in the following theorem.

Proposition 15.20 (Equality of integral for a. e. equal integrands). *Suppose that $L = (L_t)_{t\in\mathbb{R}}$ is an $\mathbb{R}$-valued Lévy process and let $f, h : \mathbb{R} \to \mathbb{C}$ be Borel-measurable such that f is Lebesgue almost everywhere equal to h. Assume that h is integrable with respect to L. Then also f is integrable with respect to L, and $\int_{\mathbb{R}} f(t)\, \mathrm{d}L_t = \int_{\mathbb{R}} h(t)\, \mathrm{d}L_t$. In particular, if f is Lebesgue, almost everywhere equal to zero, then it is integrable with respect to L with $\int_{\mathbb{R}} f(t)\, \mathrm{d}L_t = 0$.*

Proof. This follows immediately from Corollary 15.7 for the Lebesgue–Stieltes part in the integral and is clear for the OIP part in the integral from the construction of the OIP integral. □

The dominated convergence theorem for Lebesgue–Stieltjes integrals (Theorem 15.6) can be easily extended to integrals for general Lévy processes:

Theorem 15.21 (Dominated convergence thm. for integrals w. r. t. LPs). *Let $L = (L_t)_{t\in\mathbb{R}}$ be an $\mathbb{R}$-valued Lévy process. Suppose that $f_n, f, g \in L^1(\mathbb{R};\mathbb{C}) \cap L^2(\mathbb{R};\mathbb{C})$, where g is integrable with respect to L and f_n converges Lebesgue almost everywhere to f as $n \to \infty$ and $|f_n| \leq$*

$|g|$ *Lebesgue almost everywhere for each* $n \in \mathbb{N}$. *Then* f *and each of the* f_n *are integrable with respect to* L *and* $\int_{\mathbb{R}} f_n(t)\,\mathrm{d}L_t$ *converges in probability to* $\int_{\mathbb{R}} f(t)\,\mathrm{d}L_t$ *as* $n \to \infty$.

Proof. Since g is integrable with respect to L, there is a decomposition $L = S + M$ as in Definition 15.10 such that g is Lebesgue–Stieltjes-integrable with respect to S. Theorem 15.6 then implies that f_n and f are Lebesgue–Stieltjes-integrable with respect to S and hence integrable with respect to L and that $\int_{\mathbb{R}} f_n(t)\,\mathrm{d}S_t$ converges almost surely and hence in probability to $\int_{\mathbb{R}} f(t)\,\mathrm{d}S_t$. On the other hand, since $g \in L^2(\mathbb{R};\mathbb{C})$, the standard dominated convergence theorem implies that f_n converges in $L^2(\mathbb{R};\mathbb{C})$ to f and hence $\int_{\mathbb{R}} f_n(t)\,\mathrm{d}L_t$ converges in $\mathcal{S}$ (by equation (4.14)) and hence in probability, to $\int_{\mathbb{F}} f(t)\,\mathrm{d}L_t$, finishing the proof. □

The integral $\int_B f(t)\,\mathrm{d}L_t$ over a Borel set B is defined in the obvious way.

Definition 15.22 (Integral over sets). Suppose that $L = (L_t)_{t\in\mathbb{R}}$ is a two-sided $\mathbb{R}$-valued Lévy process and $f \in L^1(\mathbb{R};\mathbb{C}) \cap L^2(\mathbb{R};\mathbb{C})$. Suppose that $B \in \mathcal{B}_1$. Then f is said to be *integrable over* B *with respect to* L, if $\mathbf{1}_B f$ is integrable with respect to L. In this case, we set

$$\int_B f(t)\,\mathrm{d}L_t = \int_{\mathbb{R}} \mathbf{1}_B(t) f(t)\,\mathrm{d}L_t.$$

When $a, b \in [-\infty, \infty]$ with $a \le b$, we also define

$$\int_a^b f(t)\,\mathrm{d}L_t := \int_{(a,b)} f(t)\,\mathrm{d}L_t \quad \text{and}$$

$$\int_b^a f(t)\,\mathrm{d}L_t := -\int_a^b f(t)\,\mathrm{d}L_t = -\int_{(a,b)} f(t)\,\mathrm{d}L_t.$$

Corollary 15.18(ii) implies that $f \in L^1(\mathbb{R};\mathbb{C}) \cap L^2(\mathbb{R};\mathbb{C})$ is integrable over every bounded $B \in \mathcal{B}_1$, which we express by saying that every $f \in L^1(\mathbb{R};\mathbb{C}) \cap L^2(\mathbb{R};\mathbb{C})$ is *locally integrable with respect to* L. Since $\int_{\{a\}} f(t)\,\mathrm{d}L_t = 0$ (almost surely) for every $a \in \mathbb{R}$ by Proposition 15.20, we also have (almost surely)

$$\int_a^b f(t)\,\mathrm{d}L_t = \int_{[a,b]} f(t)\,\mathrm{d}L_t = \int_{(a,b]} f(t)\,\mathrm{d}L_t = \int_{[a,b)} f(t)\,\mathrm{d}L_t$$

for $a, b \in \mathbb{R}$ with $a \le b$.

From Theorem 15.15 it is clear that if $f \in L^1(\mathbb{R};\mathbb{C}) \cap L^2(\mathbb{R};\mathbb{C})$ and $B_1, B_2 \in \mathcal{B}_1$ are disjoint, then f is integrable over $B_1 \cup B_2$ if and only if it is integrable over both B_1 and B_2, in which case

$$\int_{B_1 \cup B_2} f(t)\,\mathrm{d}L_t = \int_{B_1} f(t)\,\mathrm{d}L_t + \int_{B_2} f(t)\,\mathrm{d}L_t$$

by Lemma 15.19.

As in Theorem 15.8 for Lebesgue–Stieltjes integrals, we can also approximate integrals with respect to Lévy processes over compact intervals by Riemann–Stieltjes sums, this time however with convergence in probability only.

Theorem 15.23 (Riemann–Stieltjes sums). *Suppose that $L = (L_t)_{t\in\mathbb{R}}$ is an $\mathbb{R}$-valued Lévy process and that a and b are in $\mathbb{R}$ with $a \le b$. Suppose also that $f : \mathbb{R} \to \mathbb{C}$ is a Borel-measurable function which is bounded on $[a, b]$ and such that the set of discontinuities of f in $[a, b]$ is a Lebesgue null set. For each $n \in \mathbb{N}$ let $\pi_n = (t_{0,n}, \dots, t_{m(n),n})$ be a partition of $[a, b]$ as in Definition 13.7 and assume that*

$$\lim_{n\to\infty} \max\{|t_{j,n} - t_{j-1,n}| : j \in \{1, \dots, m(n)\}\} = 0. \tag{15.8}$$

For each $j \in \{1, \dots, m(n)\}$, suppose that $\xi_{j,n} \in [t_{j-1,n}, t_{j,n}]$. Then f is integrable over $[a, b]$ with respect to L and the Riemann–Stieltjes sums $\sum_{j=1}^{m(n)} f(\xi_{j,n})\,(L_{t_{j,n}} - L_{t_{j-1,n}})$ converge in probability to $\int_{\mathbb{R}} f(t)\,\mathrm{d}L_t$ as $n \to \infty$.

Proof. The proof is completely analogous to the proof of Theorem 15.8, this time however using Theorem 15.21 instead of Theorem 15.6. □

As a consequence of Theorem 15.23 we can now establish a partial integration rule.

Theorem 15.24 (Partial integration). *Suppose that $L = (L_t)_{t\in\mathbb{R}}$ is an $\mathbb{R}$-valued Lévy process and that a and b are in $\mathbb{R}$ with $a \le b$. Suppose also that $f : \mathbb{R} \to \mathbb{C}$ is a Borel-measurable function such that the restriction $f_{|[a,b]}$ of f to $[a, b]$ is continuously differentiable. Denote its derivative by f'. Then*

$$\int_a^b f(t)\,\mathrm{d}L_t = f(b)L_b - f(a)L_a - \int_a^b f'(t)L_t\,\mathrm{d}t, \tag{15.9}$$

where the integral on the right-hand side of (15.9) *is a pathwise Lebesgue–integral.*

Proof. For each $n \in \mathbb{N}$, let $\pi_n = (t_{0,n}, \dots, t_{m(n),n})$ be a partition of $[a, b]$ as in Definition 13.7 such that (15.8) holds. Then, since $t_{0,n} = a$ and $t_{m(n),n} = b$, it is easily seen that

$$\sum_{j=1}^{m(n)} f(t_{j-1,n})\,(L_{t_{j,n}} - L_{t_{j-1,n}}) + \sum_{j=1}^{m(n)} (f(t_{j,n}) - f(t_{j-1,n}))L_{t_j} = f(b)L_b - f(a)L_a.$$

By Theorem 15.23, the first of these sums converges in probability to $\int_a^b f(t)\,\mathrm{d}L_t$. Hence (15.9) follows, if we can show that

$$X_n := \sum_{j=1}^{m(n)} (f(t_{j,n}) - f(t_{j-1,n}))L_{t_j}$$

converges in probability to $\int_a^b f'(t)L_t \, dt$ as $n \to \infty$. To see this, choose for each $j \in \{1, \ldots, m(n)\}$ a $\xi_{j,n} \in [t_{j-1,n}, t_{j,n}]$ such that $f(t_{j,n}) - f(t_{j-1,n}) = f'(\xi_{j,n})(t_j - t_{j-1})$, which is possible by the mean-value theorem. Denote the underlying probability space by $(\Omega, \mathcal{F}, P)$ and let $\Omega_1 \in \mathcal{F}$ be a set of probability 1 such that each of the paths $u \mapsto L_u(\omega)$ is càdlàg for $\omega \in \Omega_1$. Define the random function g_n by

$$g_n(\omega, u) := \sum_{j=1}^{m(n)} f'(\xi_{j,n}) \, L_{t_{j,n}}(\omega) \, \mathbf{1}_{(t_{j-1,n}, t_{j,n}]}(u), \quad u \in [a,b], \omega \in \Omega.$$

Then $X_n(\omega) = \int_a^b g_n(\omega, u) \, du$ for each $\omega \in \Omega_1$. But $\lim_{n\to\infty} g_n(\omega, u) = f'(u)L_u(\omega)$ for each fixed $\omega \in \Omega_1$ and $u \in [a,b]$ by continuity of f' and right-continuity of L. Since for each fixed $\omega \in \Omega_1$ the sequence of functions $(g_n(\omega, \cdot))_{n\in\mathbb{N}}$ is uniformly bounded, an application of the standard Lebesgue dominated convergence theorem shows that $\lim_{n\to\infty} X_n(\omega) = \int_a^b f'(u)L_u(\omega) \, du$ for each fixed $\omega \in \Omega_1$. This shows that X_n converges almost surely, and hence in probability, to $\int_a^b f'(u) \, L_u \, du$, as desired. □

The proof of Theorem 15.24 has also shown that there is a version of $\int_a^b f'(u)L_u \, du$ that is indeed a random variable (i. e., measurable), since it was shown to be an almost sure limit of random variables.

As pointed out in the comment following Theorem 15.1, it is possible to introduce general Lebesgue–Stieltjes integrals with respect to càdlàg functions of finite variation and not only with respect to functions of the form given in (15.1). A continuously differentiable function f is of finite variation, as seen in Example 13.9(b), and it can be shown that the integral $\int_a^b f'(t)L_t \, dt$ can be rewritten as a pathwise Lebesgue–Stieltjes integral $\int_a^b L_t \, df(t)$. With that notation, the formula (15.9) takes the form

$$\int_a^b f(t) \, dL_t = f(b)L_b - f(a)L_a - \int_a^b L_t \, df(t),$$

so that an integral of f with respect to L is transformed into an integral of L with respect to f plus some boundary terms. This explains the name *partial integration.*

15.3 The integral for matrix-valued integrands and $\mathbb{R}^d$-valued Lévy processes

Next we shall extend our definition to integrals of the form $\int_{\mathbb{R}} f(t) \, dL_t$, where $f = (f_{ij})_{i=1,\ldots,m; j=1,\ldots,d}$ is $\mathbb{C}^{m\times d}$-valued and L is an $\mathbb{R}^d$-valued Lévy process. This is done componentwise. Again, we restrict to functions that are integrable and square-integrable, i. e., which are in $L^1(\mathbb{R};\mathbb{C}^{m\times d})\cap L^2(\mathbb{R};\mathbb{C}^{m\times d})$.

Definition 15.25 (Integral w. r. t. a multivariate Lévy process). Let $L = ((L_t^1,\dots,L_t^d)^T)_{t\in\mathbb{R}}$ be a two-sided $\mathbb{R}^d$-valued Lévy process and the function $f = (f_{ij})_{i=1,\dots,m;j=1,\dots,d} : \mathbb{R}\to\mathbb{C}^{m\times d}$ be in $L^1(\mathbb{R};\mathbb{C}^{m\times d})\cap L^2(\mathbb{R};\mathbb{C}^{m\times d})$. Then f is said to be *integrable with respect to* L if, for all $i\in\{1,\dots,m\}$ and $j\in\{1,\dots,d\}$, f_{ij} is integrable with respect to L^j. In this case, the *integral of f with respect to L* is defined to be the $\mathbb{C}^m$-valued random vector given by

$$\int_{\mathbb{R}} f(t)\,\mathrm{d}L_t := \begin{bmatrix} \sum_{j=1}^d \int_{\mathbb{R}} f_{1j}(t)\,\mathrm{d}L_t^j \\ \vdots \\ \sum_{j=1}^d \int_{\mathbb{R}} f_{mj}(t)\,\mathrm{d}L_t^j \end{bmatrix}.$$

Further, the function f is said to be *integrable over a set $B\in\mathcal{B}_1$ with respect to L*, if $f\mathbf{1}_B$ is integrable with respect to L, and the integral is denoted by $\int_B f(t)\,\mathrm{d}L_t$. Integrals such as $\int_a^b f(t)\,\mathrm{d}L_t$ are defined as in the one-dimensional case.

Observe from the comment following Definition 15.22 that every function $f\in L^1(\mathbb{R};\mathbb{C}^{m\times d})\cap L^2(\mathbb{R};\mathbb{C}^{m\times d})$ is integrable over bounded sets $B\in\mathcal{B}_1$, which again we express by saying that it is *locally integrable* with respect to L. In particular, integrals such as $\int_a^b f(t)\,\mathrm{d}L_t$ and alike always exist. It is also clear that if $f\in L^1(\mathbb{R};\mathbb{C}^{m\times d})\cap L^2(\mathbb{R};\mathbb{C}^{m\times d})$ is such that $\|f\|$ is integrable with respect to each component L^j of L, $j=1,\dots,d$, then f is integrable with respect to L. Here, we denote by $\|f(t)\|$ the natural matrix norm of $f(t)$ with respect to the Euclidian norm defined by $\|f(t)\| := \sup_{w\in\mathbb{R}^d:|w|\le 1}|f(t)w|$, but by the equivalence of norms, it would be irrelevant which matrix norm is used in stating that $\|f\|$ is integrable with respect to the components L^j.

In the following theorem, we collect some sufficient conditions that ensure integrability.

Theorem 15.26 (Sufficient conditions for integrability). *Suppose that* $L = ((L_t^1,\dots,L_t^d)^T)_{t\in\mathbb{R}}$ *is a two-sided $\mathbb{R}^d$-valued Lévy process and let* $f\in L^1(\mathbb{R};\mathbb{C}^{m\times d})\cap L^2(\mathbb{R};\mathbb{C}^{m\times d})$. *Then each of the following conditions is sufficient for $\|f\|$ to be integrable with respect to each component L^j and hence for f to be integrable with respect to L:*

(i) $\mathbb{E}|L_1| < \infty$.

(ii) *f has bounded support.*

(iii) *There is $\varepsilon > 0$ such that* $\limsup_{|t|\to\infty}\|f(t)\|\,\mathrm{e}^{\varepsilon|t|} < \infty$ *and* $\mathbb{E}\log^+|L_1| < \infty$.

(iv) $\int_{\mathbb{R}}\int_{|x|>1}(1\wedge\|f(t)\|\,|x|)\,\nu(\mathrm{d}x)\,\mathrm{d}t < \infty$, *where ν denotes the Lévy measure of L.*

In case (i), *the integral has finite expectation and*

$$\mathbb{E}\Big(\int_{\mathbb{R}} f(t)\,\mathrm{d}L_t\Big) = \int_{\mathbb{R}} f(t)\,\mathrm{d}t\,\mathbb{E}(L_1).$$

Proof. (i), (ii), (iii) and the formula for the expectation in (i) follow immediately from the corresponding 1-dimensional results in Corollary 15.18 and the definition of the integral. For the proof of (iv), denote by ν_j the Lévy measure of L^j, where L^j is the jth component of L. Using the transformation theorem for image measures, we then obtain

$$\begin{aligned}&\int_{\mathbb{R}}\int_{z\in\mathbb{R}\setminus[-1,1]}(1\wedge\|f(t)\|\,|z|)\,\nu_j(\mathrm{d}z)\,\mathrm{d}t\\ &\quad\le\int_{\mathbb{R}}\int_{x\in\mathbb{R}^d:|x|>1}(1\wedge\|f(t)\|\,|x|)\,\nu(\mathrm{d}x)\,\mathrm{d}t,\end{aligned}$$

which is finite by assumption. Hence $\|f\|$ is integrable with respect to each L^j by Theorem 15.15, giving integrability of f with respect to L. □

We can also give a characterization of integrability:

Theorem 15.27 (Characterization of integrability; necessary condition). *Suppose that $L = (L_t)_{t\in\mathbb{R}}$ is a two-sided $\mathbb{R}^d$-valued Lévy process and the function $f = (f_{ij})_{i=1,\dots,m;j=1,\dots,d}$ is in $L^1(\mathbb{R};\mathbb{C}^{m\times d})\cap L^2(\mathbb{R};\mathbb{C}^{m\times d})$. Denote the Lévy measure of L by ν.*

(a) *Then f is integrable with respect to L if and only if*

$$\int_{\mathbb{R}}\int_{|x|>1}(1\wedge|f_{ij}(t)x_j|)\,\nu(\mathrm{d}x)\,\mathrm{d}t<\infty\quad\forall\,i\in\{1,\dots,m\},j\in\{1,\dots,d\},\tag{15.10}$$

equivalently if

$$\int_{\mathbb{R}}\int_{|x|>1}\left(1\wedge\sum_{j=1}^{d}|f_{ij}(t)x_j|\right)\nu(\mathrm{d}x)\,\mathrm{d}t<\infty\quad\forall\,i\in\{1,\dots,m\}.\tag{15.11}$$

(b) *If f is integrable with respect to L, then*

$$\int_{\mathbb{R}}\int_{|x|>1}(1\wedge|f(t)x|)\,\nu(\mathrm{d}x)<\infty.\tag{15.12}$$

Proof. (a) Denote the components of L by $L^1,\dots,L^d$, and the Lévy measure of L^j by ν_j. From Theorem 15.15 we see that f is integrable with respect to L if and only if $\int_{\mathbb{R}}\int_{\mathbb{R}\setminus[-1,1]}(1\wedge|f_{ij}(t)x_j|\,\nu_j(\mathrm{d}x_j)\,\mathrm{d}t<\infty$ for all $i\in\{1,\dots,m\}$ and $j\in\{1,\dots,d\}$. Since ν_j (restricted to $\mathbb{R}\setminus\{0\}$) is the image measure of ν under the projection $\pi_j:\mathbb{R}^d\to\mathbb{R}, x\mapsto x_j$, the integrals above can be rewritten as $\int_{\mathbb{R}}\int_{x\in\mathbb{R}^d:|x_j|>1}(1\wedge|f_{ij}(t)x_j|)\,\nu(\mathrm{d}x)\,\mathrm{d}t$. Since $\nu(\{x\in\mathbb{R}^d:|x|>1,|x_j|\le1\})<\infty$ and $f_{ij}\in L^1(\mathbb{R};\mathbb{C})$ we have $\int_{\mathbb{R}}\int_{|x|>1,|x_j|\le1}|f_{ij}(t)x_j|\,\nu(\mathrm{d}x)\,\mathrm{d}t<\infty$ for all i and j, hence f is integrable with respect to L if and only if (15.10) is satisfied. The latter condition is easily seen to be equivalent to (15.11).

(b) This follows immediately from (15.11), since

$$|f(t)x| = \left(\sum_{i=1}^m \left|\sum_{j=1}^d f_{ij}(t)x_j\right|^2\right)^{1/2} \leq \sum_{i=1}^m \left|\sum_{j=1}^d f_{ij}(t)x_j\right| \leq \sum_{i=1}^m \sum_{j=1}^d |f_{ij}(t)x_j|$$

and hence

$$1 \wedge |f(t)x| \leq \sum_{i=1}^m \left(1 \wedge \sum_{j=1}^d |f_{ij}(t)x_j|\right).$$ □

While equation (15.12) constitutes a necessary and sufficient condition for integrability when $m = d = 1$, for $d \geq 2$ it is only necessary but not sufficient, see Exercise 15.3 for an example.

The linearity assertions of Lemma 15.19 carry over without difficulty to the multivariate setting.

Lemma 15.28 (Linearity of the integral). *Let L and L' be two independent two-sided $\mathbb{R}^d$-valued Lévy processes, let $f, g \in L^1(\mathbb{R}; \mathbb{C}^{m\times d}) \cap L^2(\mathbb{R}; \mathbb{C}^{m\times d})$ and let $Q \in \mathbb{C}^{k\times m}$ and $C \in \mathbb{R}^{d\times d}$ be fixed matrices. We then have the following linearity properties.*

(a) *If f and g are integrable with respect to L, then $f+g$ and Qf are integrable with respect to L and*

$$\int_{\mathbb{R}} (f(t) + g(t))\,\mathrm{d}L_t = \int_{\mathbb{R}} f(t)\,\mathrm{d}L_t + \int_{\mathbb{R}} g(t)\,\mathrm{d}L_t,$$

$$\int_{\mathbb{R}} Qf(t)\,\mathrm{d}L_t = Q \int_{\mathbb{R}} f(t)\,\mathrm{d}L_t.$$

(b) *Suppose that $\|f\|$ is integrable with respect to each component L^j of L, $j = 1, \dots, d$. Then $\|f\|$ is integrable with respect to each component of the Lévy process CL, hence f is integrable with respect to CL and*

$$\int_{\mathbb{R}} f(t)\,\mathrm{d}(CL_t) = \int_{\mathbb{R}} f(t)C\,\mathrm{d}L_t.$$

(c) *Suppose that f is integrable with respect to both L and L'. Then f is integrable with respect to $L + L'$ and*

$$\int_{\mathbb{R}} f(t)\,\mathrm{d}(L_t + L_t') = \int_{\mathbb{R}} f(t)\,\mathrm{d}L_t + \int_{\mathbb{R}} f(t)\,\mathrm{d}L_t'.$$

Proof. This is an immediate consequence of Lemma 15.19 and the definition of the multivariate integral. □

As an immediate consequence of Lemma 15.28 and equation (15.10), we see that if $f \in L^1(\mathbb{R}; \mathbb{C}^{m\times d}) \cap L^2(\mathbb{R}; \mathbb{C}^{m\times d})$ is integrable over $B_1 \in \mathcal{B}_1$ and over $B_2 \in \mathcal{B}_1$ with respect to L, then it is also integrable over $B_1 \cup B_2$, and if B_1 and B_2 are additionally disjoint, then

$$\int_{B_1 \cup B_2} f(t)\,\mathrm{d}L_t = \int_{B_1} f(t)\,\mathrm{d}L_t + \int_{B_2} f(t)\,\mathrm{d}L_t.$$

Also, if f is integrable over some set $B \in \mathcal{B}_1$, then it is also integrable over any Borel-subset of B. Finally, we see that Proposition 15.20 carries over to the multidimensional setting without difficulty.

15.4 Exercises

Exercise 15.1. Let $N = (N_t)_{t\in\mathbb{R}}$ be a Poisson process with parameter 1 and define $L_t := N_t - t$. Then L is an OIP that is additionally of finite variation. Let $f_1 \in L^1(\mathbb{R}; \mathbb{C}) \setminus L^2(\mathbb{R}; \mathbb{C})$ with support $[0, 1]$ and define $f_2 : \mathbb{R} \to \mathbb{R}$ by $f_2(x) = x^{-1}\mathbf{1}_{(1,\infty)}(x)$. Show that f_1 is Lebesgue–Stieltjes-integrable with respect to L, but not OIP-integrable with respect to L, and that f_2 is OIP-integrable with respect to L, but not Lebesgue–Stieltjes-integrable with respect to L. Now prove Remark 15.17 by considering the decompositions $L = 0 + L$ and $L = L + 0$, with 0 being the zero-Lévy process.

Exercise 15.2. Let L, f_1 and f_2 be as in Exercise 15.1. Show that there is no decomposition $L = S + M$ of L into an independent sum of two Lévy processes, where S is of finite variation and M is an OIP, such that $f_1 + f_2$ is Lebesgue–Stieltjes-integrable with respect to S and OIP-integrable with respect to M.

Exercise 15.3. Let L^1 be a one-dimensional compound Poisson process with Lévy measure ν_1 such that ν_1 is supported in $[1, \infty)$ and such that $\int_{[1,\infty)} \sqrt{x}\,\nu_1(\mathrm{d}x) = \infty$. Consider the function $f_1 : \mathbb{R} \to \mathbb{R}$ defined by $f_1(t) = t^{-2}\mathbf{1}_{[1,\infty)}(t)$.
(a) Show that f_1 is not integrable with respect to L^1.
(b) Consider the function $f : \mathbb{R} \to \mathbb{R}^{1\times 2}$ defined by $f(t) = (t^{-2}, -t^{-2})$ and the two-dimensional compound Poisson process $L = (L^1, L^1)^T$. Denote by ν the Lévy measure of L. Show that f and ν satisfy (15.12), although f is not integrable with respect to L.

16 The distribution of the integral and consequences

Abstract: In this chapter, we continue the study of integration of deterministic functions with respect to Lévy processes. We show that $\int f(t)\,\mathrm{d}L_t$ is infinitely divisible for $\mathbb{R}^{m\times d}$-valued integrands and determine its characteristic exponent and characteristic triplet. We use these findings to study distributional properties of the integral, like independence of the integrals obtained over disjoint sets or shift-invariance in distribution. We also address Lévy driven continuous-time moving average processes and improper integrals.

16.1 The characteristic function of the integral

Let L be a two-sided Lévy process in $\mathbb{R}^d$. In this section, we show that $\int_{\mathbb{R}} f(t)\,\mathrm{d}L_t$ is infinitely divisible for $\mathbb{R}^{m\times d}$-valued integrable functions f and derive the characteristic triplet and characteristic exponent of the integral. From that result the distribution for general $\mathbb{C}^{m\times d}$-valued integrands can easily be obtained.

We begin with integrands, which are simple functions as in equation (4.10), generalized slightly to allow for matrix-valued integrands. A function $f : \mathbb{R} \to \mathbb{R}^{m\times d}$ is called an $\mathbb{R}^{m\times d}$-*valued simple function*, if it can be written in the form

$$f(t) = \sum_{i=1}^{n} \alpha_i \mathbf{1}_{(u_i,u_{i+1}]}(t), \quad t \in \mathbb{R}, \tag{16.1}$$

for some $u_1,\dots,u_{n+1} \in \mathbb{R}$ and $\alpha_1,\dots,\alpha_n \in \mathbb{R}^{m\times d}$ with $-\infty < u_1 < u_2 < \cdots < u_{n+1} < \infty$. We denote by $\mathcal{D}^{m,d}_{\text{real}}$ the class of all $\mathbb{R}^{m\times d}$-valued simple functions.

Example 16.1. Suppose that $f \in \mathcal{D}^{m,d}_{\text{real}}$ with representation as in (16.1) and that L is a two-sided $\mathbb{R}^d$-valued Lévy process. By Theorem 15.26(ii), f is integrable with respect to L, and from Example 15.13 and the definition of the multivariate integral, we see that

$$Y := \int_{\mathbb{R}} f(t)\,\mathrm{d}L_t = \sum_{i=1}^{n} \alpha_i (L_{u_{i+1}} - L_{u_i}).$$

From this representation we see that Y is $\sigma(L_t - L_s : s,t \in [u_1,u_{n+1}])$-measurable. It is infinitely divisible, since the random variables $\alpha_i(L_{u_{i+1}} - L_{u_i})$ are infinitely divisible (cf., Theorem 9.25) and independent. Denote by Ψ_L the characteristic exponent of L and by $(A_L,\nu_L,\gamma_L)_{c_1}$ the characteristic triplet of L with respect to a truncation function $c_1 : \mathbb{R}^d \to \mathbb{R}^d$. Since each $L_{u_{i+1}} - L_{u_i}$ has characteristic triplet $(u_{i+1}-u_i)(A_L,\nu_L,\gamma_L)_{c_1}$ by equation (10.3) and since the $L_{u_{i+1}} - L_{u_i}$ are independent, it follows from Theorem 9.25 and Example 9.15 that the characteristic exponent of Y is given by

https://doi.org/10.1515/9783111325033-016

$$\Psi_Y(z) = \sum_{i=1}^{n}(u_{i+1} - u_i)\Psi_L(a_i^T z), \quad z \in \mathbb{R}^m,$$

and the characteristic triplet $(A_Y, \nu_Y, \gamma_Y)_{c_2}$ by

$$\begin{aligned} A_Y &= \sum_{i=1}^{n}(u_{i+1} - u_i)a_i A_L a_i^T, \\ \nu_Y(C) &= \sum_{i=1}^{n}(u_{i+1} - u_i)\,\nu_L(\{x \in \mathbb{R}^d : a_i x \in C \setminus \{0\}\}), \quad C \in \mathcal{B}_m, \\ \gamma_Y &= \sum_{i=1}^{n}(u_{i+1} - u_i)\Big(a_i\gamma_L + \int_{\mathbb{R}^d}(c_2(a_i x) - a_i c_1(x))\,\nu_L(\mathrm{d}x)\Big) \end{aligned}$$

with respect to a fixed truncation function $c_2 : \mathbb{R}^m \to \mathbb{R}^m$. With later generalizations in mind, we rewrite these quantities as

$$\Psi_Y(z) = \int_{\mathbb{R}} \Psi_L(f(t)^T z)\,\mathrm{d}t, \quad z \in \mathbb{R}^m, \tag{16.2}$$

$$A_Y = \int_{\mathbb{R}} f(t)A_L f(t)^T\,\mathrm{d}t, \tag{16.3}$$

$$\nu_Y(C) = \int_{\mathbb{R}}\int_{\mathbb{R}^d} \mathbf{1}_{C\setminus\{0\}}(f(t)x)\,\nu_L(\mathrm{d}x)\,\mathrm{d}t, \quad C \in \mathcal{B}_m, \tag{16.4}$$

$$\gamma_Y = \int_{\mathbb{R}}\Big(f(t)\gamma_L + \int_{\mathbb{R}^d}(c_2(f(t)x) - f(t)c_1(x))\,\nu_L(\mathrm{d}x)\Big)\mathrm{d}t. \tag{16.5}$$

Our goal is to generalize the previous example to integrable functions $f \in L^1(\mathbb{R}; \mathbb{R}^{m\times d}) \cap L^2(\mathbb{R}; \mathbb{R}^{m\times d})$, but before doing that, we shall first show how to integrate with respect to the Lévy measure ν' defined in the following proposition. We shall later show that ν' is in fact the Lévy measure of $\int_{\mathbb{R}} f(t)\,\mathrm{d}L_t$.

Proposition 16.2. *Let ν be a measure on $\mathbb{R}^d$ and let $f : \mathbb{R}^d \to \mathbb{R}^{m\times d}$ be a Borel-measurable function. Define $\nu' : \mathcal{B}_m \to [0, \infty]$ by*

$$\nu'(C) := \int_{\mathbb{R}}\int_{\mathbb{R}^d} \mathbf{1}_{C\setminus\{0\}}(f(t)x)\,\nu(\mathrm{d}x)\,\mathrm{d}t, \quad C \in \mathcal{B}_m. \tag{16.6}$$

(a) *Then ν' is a measure on $\mathbb{R}^m$ with $\nu'(\{0\}) = 0$.*
(b) *For every Borel measurable function $g : \mathbb{R}^m \to [0, \infty]$ with $g(0) = 0$,*

$$\int_{\mathbb{R}^m} g(y)\,\nu'(\mathrm{d}y) = \int_{\mathbb{R}}\int_{\mathbb{R}^d} g(f(t)x)\,\nu(\mathrm{d}x)\,\mathrm{d}t. \tag{16.7}$$

Furthermore, a Borel measurable function $g : \mathbb{R}^m \to \mathbb{C}^k$ with $g(0) = 0$ is integrable with respect to ν' if and only $\int_{\mathbb{R}} \int_{\mathbb{R}^d} |g(f(t)x)|\, \nu(\mathrm{d}x)\, \mathrm{d}t < \infty$, and in that case (16.7) *holds.*

(c) *If ν is a Lévy measure and $f \in L^2(\mathbb{R}; \mathbb{R}^{m\times d})$ with bounded support, then ν' is also a Lévy measure.*

Proof. (a) It is obvious from the definition that $\nu'(\emptyset) = 0$, that $\nu'(\{0\}) = 0$ and that ν' is σ-additive and is hence a $[0, \infty]$-valued measure.

(b) If $g : \mathbb{R}^m \to [0, \infty]$ is of the form $g = \mathbf{1}_C$ with $0 \notin C$, then (16.7) holds since it is just a restatement of (16.6). It also holds for elementary functions of the form $g = \sum_{i=1}^n a_i \mathbf{1}_{C_i}$ with $a_i \in [0, \infty)$ and $C_i \in \mathcal{B}_m$ such that $0 \notin C_i$ by linearity of the integrals on both sides of (16.7). Finally, if $g : \mathbb{R}^m \to [0, \infty]$ with $g(0) = 0$ is Borel-measurable, we can choose a sequence $(g_n)_{n\in\mathbb{N}}$ of elementary functions of the above form such that $0 \le \cdots \le g_n \le g_{n+1} \le \cdots$ and (g_n) converges pointwise to g. Equation (16.7) then follows by monotone convergence. Decomposition of g into real and imaginary and then positive and negative parts and consideration of the separate components establishes the stated integrability condition for $g : \mathbb{R}^m \to \mathbb{C}^k$ with $g(0) = 0$, and the validity of equation (16.7), if g satisfies this condition.

(c) If $f \in L^2(\mathbb{R}; \mathbb{R}^{m\times d})$ and the support of f is contained in the compact interval $[a, b]$, it follows from (16.7) that

$$\begin{aligned}
&\int_{\mathbb{R}^m} (1 \wedge |y|^2)\, \nu'(\mathrm{d}y) \\
&= \int_a^b \int_{\mathbb{R}^d} (1 \wedge |f(t)x|^2)\, \nu(\mathrm{d}x)\, \mathrm{d}t \\
&\le \int_a^b \int_{\{x\in\mathbb{R}^d:|x|>1\}} 1\, \nu(\mathrm{d}x)\, \mathrm{d}t + \int_a^b \int_{\{x\in\mathbb{R}^d:|x|\le 1\}} \|f(t)\|^2 |x|^2\, \nu(\mathrm{d}x)\, \mathrm{d}t < \infty.
\end{aligned}$$

Hence ν' is a Lévy measure. □

The compact support condition in Assertion (c) of Proposition 16.2 can be relaxed at the cost of assuming certain integrability conditions on f and ν, see Exercise 16.1. More generally, in Theorem 16.3 below we will show that if $f \in L^1(\mathbb{R}; \mathbb{R}^{m\times d}) \cap L^2(\mathbb{R}; \mathbb{R}^{m\times d})$ is integrable with respect to L, then ν' will be a Lévy measure. However, it is generally not true that $f \in L^1(\mathbb{R}; \mathbb{R}^{m\times d}) \cap L^2(\mathbb{R}; \mathbb{R}^{m\times d})$ implies that ν' is a Lévy measure. An example is provided in Exercise 16.1(d).

In establishing the following important theorem, we show that $\int_{\mathbb{R}} f(s)\, \mathrm{d}L_s$ is infinitely divisible and specify its characteristic triplet. The function f is assumed to be $\mathbb{R}^{m\times d}$-valued rather than $\mathbb{C}^{m\times d}$-valued, since we defined infinite divisibility for $\mathbb{R}^m$-valued random vectors only. Again, we denote by $\|f(t)\|$ the natural matrix norm (with respect to the Euclidian norm) of $f(t)$, although by the equivalence of norms it is irrelevant which matrix norm is used.

Theorem 16.3 (Infinite divisibility and characteristic triplet of integral). *Let $L = (L_t)_{t\in\mathbb{R}}$ be a two-sided $\mathbb{R}^d$-valued Lévy process with characteristic exponent Ψ_L and characteristic triplet $(A_L, \nu_L, \gamma_L)_{c_1}$ with respect to a truncation function $c_1 : \mathbb{R}^d \to \mathbb{R}^d$. Let $f \in L^1(\mathbb{R}; \mathbb{R}^{m\times d}) \cap L^2(\mathbb{R}; \mathbb{R}^{m\times d})$ be integrable with respect to L and define*

$$Y := \int_{\mathbb{R}} f(t)\,\mathrm{d}L_t.$$

(a) *Then Y is infinitely divisible and $\sigma(L_t : t \in \mathbb{R})$-measurable (more precisely, a version of Y is $\sigma(L_t : t \in \mathbb{R})$-measurable). If, in addition, the support of f is contained in the compact interval $[a, b]$, then (a version of) Y is measurable with respect to $\sigma(L_t - L_s : s, t \in [a, b])$.*

(b) *Let $c_2 : \mathbb{R}^m \to \mathbb{R}^m$ be a fixed truncation function. Then*

$$\int_{\mathbb{R}} \|f(t)A_L f(t)^T\|\,\mathrm{d}t < \infty, \tag{16.8}$$

$$\int_{\mathbb{R}}\int_{\mathbb{R}^d} (1 \wedge |f(t)x|^2)\,\nu_L(\mathrm{d}x)\,\mathrm{d}t < \infty, \tag{16.9}$$

$$\int_{\mathbb{R}^d} |c_2(f(t)x) - f(t)c_1(x)|\,\nu_L(\mathrm{d}x) < \infty \quad \forall\, t \in \mathbb{R}, \tag{16.10}$$

$$\int_{\mathbb{R}} \left| f(t)\gamma_L + \int_{\mathbb{R}^d} (c_2(f(t)x) - f(t)c_1(x))\,\nu_L(\mathrm{d}x)\right| \mathrm{d}t < \infty \quad \text{and} \tag{16.11}$$

$$\int_{\mathbb{R}} |\Psi_L(f(t)^T z)|\,\mathrm{d}t < \infty \quad \forall\, z \in \mathbb{R}^m. \tag{16.12}$$

The characteristic triplet $(A_Y, \nu_Y, \gamma_Y)_{c_2}$ of Y with respect to c_2 is given by

$$A_Y = \int_{\mathbb{R}} f(t)A_L f(t)^T\,\mathrm{d}t, \tag{16.13}$$

$$\nu_Y(C) = \int_{\mathbb{R}}\int_{\mathbb{R}^d} \mathbf{1}_{C\setminus\{0\}}(f(t)x)\,\nu_L(\mathrm{d}x)\,\mathrm{d}t, \quad C \in \mathcal{B}_m, \tag{16.14}$$

$$\gamma_Y = \int_{\mathbb{R}} \left(f(t)\gamma_L + \int_{\mathbb{R}^d} (c_2(f(t)x) - f(t)c_1(x))\,\nu_L(\mathrm{d}x) \right)\mathrm{d}t, \tag{16.15}$$

and the characteristic exponent Ψ_Y of Y by

$$\Psi_Y(z) = \int_{\mathbb{R}} \Psi_L(f(t)^T z)\,\mathrm{d}t, \quad z \in \mathbb{R}^m. \tag{16.16}$$

Proof. We proceed in three steps. In the first two steps we prove all the assertions except for (16.10), (16.11) and (16.15), which refer to specific truncation functions c_1 and c_2. This is done in Step 1 under the additional assumptions that f is bounded with compact support, and in Step 2 for general integrable f. The remaining assertions are established in Step 3.

Step 1: *Proof of the assertions which do not refer to the specified truncation functions c_1 and c_2 when f is bounded with compact support.*

Since the assertions concerned do not depend on specific truncation functions, we are free in this step to use truncation functions that are continuous and satisfy (9.35). Assume that the support of f is contained in the compact interval $[a, b]$ and that f is bounded. Denote

$$K := 1 + \sup_{t\in[a,b]} \|f(t)\| < \infty.$$

An easy extension of Step 3 in Section 4.3 shows that $\mathcal{D}^{m,d}_{\text{real}}$ is dense in $L^2(\mathbb{R}; \mathbb{R}^{m\times d})$ with respect to $L^2(\mathbb{R}; \mathbb{R}^{m\times d})$-convergence. In particular, there exists a sequence $(f_n')_{n\in\mathbb{N}}$ in $\mathcal{D}^{m,d}_{\text{real}}$ which converges in $L^2(\mathbb{R}; \mathbb{R}^{m\times d})$ to f as $n \to \infty$. By taking a subsequence if necessary, we can assume that f_n' also converges Lebesgue almost everywhere to f. Now define the functions f_n by

$$f_n(t) := f_n'(t)\mathbf{1}_{(a,b]}(t)\mathbf{1}_{[0,K]}(\|f_n'(t)\|), \quad t \in \mathbb{R}, n \in \mathbb{N}.$$

Then (f_n) is again a sequence of simple functions that converges Lebesgue almost everywhere and in L^2 to f. Further, the support of each of the f_n is contained in $[a, b]$ and the f_n are uniformly bounded by K, hence so are all components of the f_n.

Write $f = (f^{ij})_{i=1,\dots,m;j=1,\dots,d}$, $f_n = (f_n^{ij})_{i=1,\dots,m;j=1,\dots,d}$ and $L = (L^1, \dots, L^d)^T$. By Theorem 15.21, $\int_{\mathbb{R}} f_n^{ij}(t)\,\mathrm{d}L_t^j$ converges in probability as $n \to \infty$ to $\int_{\mathbb{R}} f^{ij}(t)\,\mathrm{d}L^j(t)$ for all $i \in \{1, \dots, m\}$ and $j \in \{1, \dots, d\}$, hence $\int_{\mathbb{R}} f_n(t)\,\mathrm{d}L_t$ converges in probability to $\int_{\mathbb{R}} f(t)\,\mathrm{d}L_t$. Since the f_n are simple functions, the integrals $\int_{\mathbb{R}} f_n(t)\,\mathrm{d}L_t$ are measurable with respect to $\sigma(L_t - L_s : s, t \in [a, b])$ and infinitely divisible by Example 16.1, hence so is the probability limit $Y = \int_{\mathbb{R}} f(t)\,\mathrm{d}L_t$. This shows (a) when f is bounded with compact support.

Denote by $\nu'(C)$ for $C \in \mathcal{B}_m$ the right-hand side of (16.14). Further, denote by Ψ_n and $(A_n, \nu_n, \gamma_n)_{c_2}$ the characteristic exponent and characteristic triplet of $\int_a^b f_n(t)\,\mathrm{d}L_t$, respectively, which are given by (16.2)–(16.5). To see the specific form of Ψ_Y, observe that Ψ_L is continuous and hence bounded on compact intervals. Since f_n converges almost everywhere to f as $n \to \infty$ and, since the functions f_n are uniformly bounded by K, for each fixed $z \in \mathbb{R}^m$ the sequence of functions $(t \mapsto \Psi_L(f_n(t)^T z))_{n\in\mathbb{N}}$ is uniformly bounded and converges Lebesgue almost everywhere to the function $t \mapsto \Psi_L(f(t)^T z)$. Lebesgue's dominated convergence theorem therefore gives

$$\Psi_n(z) = \int_a^b \Psi_L(f_n(t)^T z)\,\mathrm{d}t \to \int_a^b \Psi_L(f(t)^T z)\,\mathrm{d}t \quad \text{as } n \to \infty,$$

for each fixed $z \in \mathbb{R}^m$, together with (16.12). On the other hand, Ψ_n converges locally uniformly to Ψ_Y by Corollary 8.23, so that $\Psi_Y(z) = \int_a^b \Psi_L(f(t)^T z)\,dt$, which gives (16.16).

Equations (16.8) and (16.9) are clear, since f is bounded with compact support and, by Proposition 16.2(c), ν' is a Lévy measure.

To see the specific form of the Gaussian variance and the Lévy measure, recall that we have chosen truncation functions c_1 and c_2 that are continuous and also satisfy (9.35). Then, by Theorem 9.31, as $n \to \infty$, $A_n + \int_{\mathbb{R}^m} c_2(y)c_2(y)^T\,\nu_n(dy) \to A_Y + \int_{\mathbb{R}^m} c_2(y)c_2(y)^T\,\nu_Y(dy)$ and $\int_{\mathbb{R}^m} g(y)\,\nu_n(dx) \to g(y)\,\nu_Y(dx)$ for all $g \in C_\sharp(\mathbb{R}^m;\mathbb{R})$ (recall that $C_\sharp(\mathbb{R}^m;\mathbb{R})$ is the set of all bounded continuous functions $f : \mathbb{R}^m \to \mathbb{R}$ such that $f(x) = 0$ for all x in a neighborhood of 0).

Now let $g : \mathbb{R}^m \to \mathbb{R}$ be a continuous function such that $|g(y)| \le M(1 \wedge |y|^2)$ for all $y \in \mathbb{R}^m$ and some constant M. This condition is satisfied in particular if $g \in C_\sharp(\mathbb{R}^m;\mathbb{R})$. Since $\|f_n(t)\| \le K$ for all n and t, we have $|f_n(t)x| \le K|x| \le 1$ whenever $|x| \le 1/K$. Consequently $|g(f_n(t)x)| \le M|f_n(t)x|^2 \le MK^2|x|^2$ if $|x| \le 1/K$. On the other hand, $|g(f_n(t)x)| \le M$ if $|x| > 1/K$, so that $|g(f_n(t)x)| \le MK^2|x|^2\mathbf{1}_{[0,1/K]}(|x|) + M\mathbf{1}_{(1/K,\infty)}(|x|)$. However, the latter is an integrable majorant of the functions $g(f_n(t)x)$ with respect to $\nu_L(dx) \otimes (dt)$ over $\mathbb{R}^d \times [a,b]$, and since $g(f_n(t)x) \to g(f(t)x)$ for $\nu(dx) \otimes (dt)$-almost all (x,t) in $\mathbb{R}^d \times [a,b]$ as $n \to \infty$, Lebesgue's dominated convergence theorem gives

$$\int_a^b \int_{\mathbb{R}^d} g(f_n(t)x)\,\nu_L(dx)\,dt \to \int_a^b \int_{\mathbb{R}^d} g(f(t)x)\,\nu_L(dx)\,dt \quad \text{as } n \to \infty.$$

Using (16.7) and (16.4), this can be rewritten as

$$\int_{\mathbb{R}^m} g(y)\,\nu_n(dy) \to \int_{\mathbb{R}^m} g(y)\,\nu'(dy) \quad \text{as } n \to \infty,$$

for all continuous g with $|g(y)| \le M(1 \wedge |y|^2)$. In particular, it holds for all $g \in C_\sharp(\mathbb{R}^m;\mathbb{R})$. Since also $\int_{\mathbb{R}^m} g(y)\,\nu_n(dy) \to \int_{\mathbb{R}^m} g(y)\,\nu_Y(dy)$ for $g \in C_\sharp(\mathbb{R}^m;\mathbb{R})$, we conclude that $\nu_Y = \nu'$, i. e., ν_Y has the form stated in (16.14).

To establish the stated form of A_Y, observe that c_2 is bounded, continuous and satisfies $|c_2(y)| \le M'(1 \wedge |y|)$ for all $y \in \mathbb{R}^m$ and some $M' > 0$ by (9.4). Hence all components of $y \mapsto c_2(y)c_2(y)^T$ are bounded by $M(1 \wedge |y|^2)$ for some M, and it follows that $\int_{\mathbb{R}^m} c_2(y)c_2(y)^T\,\nu_n(dy) \to \int_{\mathbb{R}^m} c_2(y)\,c_2(y)^T\,\nu'(dy) = \int_{\mathbb{R}^m} c_2(y)c_2(y)^T\,\nu_Y(dy)$ as $n \to \infty$. Since $A_n = \int_a^b f_n(t)A_L f_n(t)^T\,dt$ converges to $\int_a^b f(t)A_L f(t)^T\,dt$ as $n \to \infty$ by the L^2-convergence of f_n to f, we conclude that $A_n + \int_{\mathbb{R}^m} c_2(x)c_2(x)^T\,\nu_n(dx) \to \int_a^b f(t)A_L f(t)^T\,dt + \int_{\mathbb{R}^m} c_2(x)c_2(x)^T\,\nu_Y(dx)$. On the other hand, since it converges to $A_Y + \int_{\mathbb{R}^m} c_2(x)c_2(x)^T\,\nu_Y(dx)$, we conclude that A_Y is given by $\int_a^b f(t)A_L f(t)^T\,dt$, which is (16.13).

Step 2: Proof, for general integrable $f \in L^1(\mathbb{R};\mathbb{R}^{m\times d}) \cap L^2(\mathbb{R};\mathbb{R}^{m\times d})$, of the assertions which do not explicitly refer to the truncation functions c_1 and c_2.

As in Step 1, we are again free to use truncation functions that are continuous and satisfy (9.35). Let f be integrable with respect to L. We argue as in Step 1, but this time approximate with the functions f_n defined by

$$f_n(t) := f(t)\mathbf{1}_{[-n,n]}(t)\mathbf{1}_{[0,n]}(\|f(t)\|), \quad t \in \mathbb{R}, n \in \mathbb{N}.$$

They are bounded and have compact support and hence the assertions proved in Step 1 hold for them. Writing $f = (f^{ij})_{i=1,\dots,m;j=1,\dots,d}$, $f_n = (f_n^{ij})_{i=1,\dots,m;j=1,\dots,d}$ and $L = (L^1, \dots, L^d)^T$ as in Step 1 and observing that f_n^{ij} converges pointwise to f^{ij} and that $|f_n^{ij}| \le |f^{ij}|$ which is integrable with respect to L^j, an application of Theorem 15.21 shows that $\int_{\mathbb{R}} f_n^{ij}(t)\, \mathrm{d}L_t^j$ converges in probability as $n \to \infty$ to $\int_{\mathbb{R}} f^{ij}(t)\, \mathrm{d}L^j(t)$ for all $i \in \{1, \dots, m\}$ and $j \in \{1, \dots, d\}$. Hence, $\int_{\mathbb{R}} f_n(t)\, \mathrm{d}L_t$ converges in probability to $Y := \int_{\mathbb{R}} f(t)\, \mathrm{d}L_t$ as $n \to \infty$. Since each $\int_{\mathbb{R}} f_n(t)\, \mathrm{d}L_t$ is infinitely divisible, $\sigma(L_t : t \in \mathbb{R})$-measurable, and measurable with respect to $\sigma(L_t - L_s : s, t \in [a, b])$ if the support of f is contained in $[a, b]$, the same is true for Y and (a) is proved.

To show (b), denote by ν' the measure defined by the right-hand side of (16.14), by ν_Y the Lévy measure of Y and by ν_n the Lévy measure of $\int_{\mathbb{R}} f_n(t)\, \mathrm{d}L_t$. Define the finite measures ρ_n by $\rho_n(\mathrm{d}y) := (|y|^2 \wedge 1)\, \nu_n(\mathrm{d}y)$. Since $\mathcal{L}(\int_{\mathbb{R}} f_n(t)\, \mathrm{d}L_t)$ converges weakly, it follows from Lemma 9.12 that $(\rho_n)_{n \in \mathbb{N}}$ is uniformly bounded. Hence

$$\sup_{n \in \mathbb{N}} \int_{\mathbb{R}^m} (1 \wedge |y|^2) \nu_n(\mathrm{d}y) = \sup_{n \in \mathbb{N}} \rho_n(\mathbb{R}^m) < \infty.$$

Since $1 \wedge |f_n(t)x|^2$ converges $\nu_L(\mathrm{d}x) \otimes (\mathrm{d}t)$-almost everywhere to $1 \wedge |f(t)x|^2$ as $n \to \infty$ for $(x, t) \in \mathbb{R}^d \times \mathbb{R}$, an application of Fatou's lemma and (16.7) give

$$\begin{aligned}
\int_{\mathbb{R}^m} (1 \wedge |y|^2)\, \nu'(\mathrm{d}y) &= \int_{\mathbb{R}} \int_{\mathbb{R}^d} (1 \wedge |f(t)x|^2\, \nu_L(\mathrm{d}x)\, \mathrm{d}t \\
&= \int_{\mathbb{R}} \int_{\mathbb{R}^d} \liminf_{n \to \infty} (1 \wedge |f_n(t)x|^2)\, \nu_L(\mathrm{d}x)\, \mathrm{d}t \\
&\le \liminf_{n \to \infty} \int_{\mathbb{R}} \int_{\mathbb{R}^d} (1 \wedge |f_n(t)x|^2)\, \nu_L(\mathrm{d}x)\, \mathrm{d}t \\
&= \liminf_{n \to \infty} \int_{\mathbb{R}^m} (1 \wedge |y|^2)\, \nu_n(\mathrm{d}y) < \infty.
\end{aligned}$$

Hence (16.9) is satisfied and ν' is a Lévy measure.

Now let $g : \mathbb{R}^m \to \mathbb{R}$ be continuous with $|g(x)| \le M(1 \wedge |x|^2)$ for some M. Then

$$|g(f_n(t)x)| \le M(1 \wedge |f_n(t)x|^2) \le M(1 \wedge |f(t)x|^2) \quad \forall\, t \in \mathbb{R}, x \in \mathbb{R}^d, n \in \mathbb{N},$$

and the right-hand side is $\nu_L(dx)\otimes(dt)$-integrable, as we have just shown. Since $g(f_n(t)x)$ converges $\nu_L(dx)\otimes(dt)$-almost everywhere to $g(f(t)x)$, and since we have an integrable majorant, the same argument that was used in Step 1 can now be used to show that $\nu' = \nu_Y$, finishing the proof of (16.14). The same argument also applies to show that $\int_{\mathbb{R}^m} c_2(y)c_2(y)^T \nu_n(dy) \to \int_{\mathbb{R}^m} c_2(y)c_2(y)^T \nu_Y(dy)$ as $n \to \infty$.

Equation (16.8) is clear, since $f \in L^2(\mathbb{R}; \mathbb{R}^{m\times d})$. By Step 1, the Gaussian variance A_n of $\int_{\mathbb{R}} f_n(t)\, dL_t$ is given by $\int_{\mathbb{R}} f_n(t)A_L f_n(t)^T\, dt$. Using the dominated convergence theorem, we conclude that A_n converges to $\int_{\mathbb{R}} f(t)A_L f(t)^T\, dt$ so that $A_n + \int_{\mathbb{R}^m} c_2(y)c_2(y)^T \nu_n(dy) \to \int_{\mathbb{R}} f(t)A_L f(t)^T\, dt + \int_{\mathbb{R}^m} c_2(y)c_2(y)^T \nu_Y(dy)$ which, with (9.36), gives (16.13).

Next, we shall establish (16.12) and (16.16). Denote by Ψ_n the characteristic exponent of $\int_{\mathbb{R}} f_n(t)\, dL_t$. Since $\int_{\mathbb{R}} f_n(t)\, dL_t$ converges in probability to $\int_{\mathbb{R}} f(t)\, dL_t$, we find from Corollary 8.23 that

$$\Psi_Y(z) = \lim_{n\to\infty} \Psi_n(z) = \lim_{n\to\infty} \int_{\mathbb{R}} \Psi_L(\mathbf{1}_{[0,n]}(\|f(t)\|)f(t)^T z)\mathbf{1}_{[-n,n]}(t)\, dt$$

for all $z \in \mathbb{R}^m$. In particular, the limit on the right-hand side exists for all $z \in \mathbb{R}^m$, whenever $f \in L^1(\mathbb{R}; \mathbb{R}^{m\times d}) \cap L^2(\mathbb{R}; \mathbb{R}^{m\times d})$ is integrable with respect to L. However, in that case, $\mathbf{1}_B f$ is also integrable with respect to L for every $B \in \mathcal{B}_1$ by Theorem 15.27(a). Hence, for each $z \in \mathbb{R}^m$ and $B \in \mathcal{B}_1$, it is also true that $\int_{\mathbb{R}} \Psi_L(\mathbf{1}_{[0,n]}(\|f(t)\|)f(t)^T z)\mathbf{1}_{[-n,n]\cap B}(t)\, dt$ converges in $\mathbb{C}$ as $n \to \infty$. For given $z \in \mathbb{R}^m$ define $B_z := \{t \in \mathbb{R} : \Re(\Psi_L(f(t)^T z)) \geq 0\}$. Since $\Psi_L(0) = 0$ and Ψ_L is continuous, we see that

$$t \mapsto \Re(\Psi_L(\mathbf{1}_{[0,n]}(\|f(t)\|)f(t)^T z))\mathbf{1}_{[-n,n]}(t)\mathbf{1}_{B_z}(t), \quad n \in \mathbb{N},$$

is a non-decreasing sequence of non-negative functions that as $n \to \infty$ converges to $t \mapsto \Re(\Psi_L(f(t)^T z))\mathbf{1}_{B_z}(t)$. Hence, by monotone convergence,

$$\begin{aligned}\int_{B_z} |\Re(\Psi_L(f(t)^T z))|\, dt &= \int_{B_z} \Re(\Psi_L(f(t)^T z))\, dt \\ &= \lim_{n\to\infty} \int_{\mathbb{R}} \Re(\Psi_L(\mathbf{1}_{\|f(t)\|\leq n} f(t)^T z))\mathbf{1}_{[-n,n]\cap B_z}(t)\, dt.\end{aligned}$$

However, the right-hand side of this equation is finite, as seen above. A similar argument applied to $C_z := \{t \in \mathbb{R} : \Re(\Psi_L(f(t)^T z)) < 0\} = \mathbb{R} \setminus B_z$ yields the finiteness of $\int_{C_z} |\Re(\Psi_L(f(t)^T z))|\, dt < \infty$ and expresses it as a limit. Thus

$$\begin{aligned}\int_{\mathbb{R}} \Re(\Psi_L(f(t)^T z))\, dt &= \lim_{n\to\infty} \int_{\mathbb{R}} \Re(\Psi_L(\mathbf{1}_{\|f(t)\|\leq n} f(t)^T z))\mathbf{1}_{[-n,n]}(t)\, dt \\ &= \lim_{n\to\infty} \Re(\Psi_n(z)) = \Re(\Psi_Y(z)).\end{aligned}$$

Repeating the same argument for the imaginary part gives (16.12) and (16.16).

Step 3: Proof, for general integrable $f \in L^1(\mathbb{R};\mathbb{R}^{m\times d}) \cap L^2(\mathbb{R};\mathbb{R}^{m\times d})$, of the assertions referring to the specific truncation functions c_1 and c_2, i. e., of (16.10), (16.11) *and* (16.15).

Suppose that $f \in L^1(\mathbb{R};\mathbb{R}^{m\times d}) \cap L^2(\mathbb{R};\mathbb{R}^{m\times d})$ is integrable with respect to L. Equation (16.10) follows immediately from the boundedness of c_1 and c_2, equation (9.4) and the upper bound,

$$\begin{aligned}&|c_2(\zeta x) - \zeta c_1(x)|\\ &\quad \le |c_2(\zeta x) - \zeta x \mathbf{1}_{|\zeta x|\le 1}| + |\zeta x \mathbf{1}_{|\zeta x|\le 1} - \zeta x \mathbf{1}_{|x|\le 1}| + |\zeta x \mathbf{1}_{|x|\le 1} - \zeta c_1(x)|,\end{aligned}$$

for $x \in \mathbb{R}^d$ and $\zeta \in \mathbb{R}^{m\times d}$.

It remains to show (16.11) and (16.15). So let γ_L be the location parameter of L_1 with respect to c_1 and γ_Y the location parameter of Y with respect to c_2. From the Lévy–Khintchine representation, we can write, for each $z \in \mathbb{R}^m$,

$$\begin{aligned}z^T f(t)\gamma_L &= \langle \gamma_L, f(t)^T z\rangle\\ &= -\mathrm{i}\,\Psi_L(f(t)^T z) - \frac{\mathrm{i}}{2}\langle f(t)^T z, A_L f(t)^T z\rangle\\ &\quad + \mathrm{i}\int_{\mathbb{R}^d} \big(e^{\mathrm{i}\langle f(t)^T z, x\rangle} - 1 - \mathrm{i}\langle f(t)^T z, c_1(x)\rangle\big)\,\nu_L(\mathrm{d}x).\end{aligned}$$

Hence,

$$\begin{aligned}&z^T f(t)\gamma_L + z^T\int_{\mathbb{R}^d}\big(c_2(f(t)x) - f(t)c_1(x)\big)\,\nu_L(\mathrm{d}x)\\ &\quad= -\mathrm{i}\Psi_L(f(t)^T z) - \frac{\mathrm{i}}{2} z^T f(t) A_L f(t)^T z\\ &\qquad + \mathrm{i}\int_{\mathbb{R}^d}\big(e^{\mathrm{i}\langle z, f(t)x\rangle} - 1 - \mathrm{i}\langle z, c_2(f(t)x)\rangle\big)\,\nu_L(\mathrm{d}x).\end{aligned} \tag{16.17}$$

But

$$\begin{aligned}&\int_{\mathbb{R}}\int_{\mathbb{R}^d}\big|e^{\mathrm{i}\langle z, f(t)x\rangle} - 1 - \mathrm{i}\langle z, c_2(f(t)x)\rangle\big|\,\nu_L(\mathrm{d}x)\,\mathrm{d}t\\ &\quad = \int_{\mathbb{R}^m}\big|e^{\mathrm{i}\langle z, y\rangle} - 1 - \mathrm{i}\langle z, c_2(y)\rangle\big|\,\nu_Y(\mathrm{d}y)\end{aligned}$$

by Proposition 16.2(b) and (16.14), and the latter is finite since ν_Y is a Lévy measure. This, together with (16.17), (16.12) and (16.8), then implies (16.11) by choosing for z the canonical basis vectors of $\mathbb{R}^m$. Similarly, we obtain, for each $z \in \mathbb{R}^m$,

$$\int_{\mathbb{R}}\Big(z^T f(t)\gamma_L + z^T \int_{\mathbb{R}^d} (c_2(f(t)x) - f(t)c_1(x))\,\nu_L(\mathrm{d}x)\Big)\,\mathrm{d}t$$

$$= -\mathrm{i}\int_{\mathbb{R}} \Psi_L(f(t)^T z) - \frac{\mathrm{i}}{2} z^T \int_{\mathbb{R}} f(t) A_L f(t)^T \,\mathrm{d}t\, z$$

$$+\,\mathrm{i}\int_{\mathbb{R}}\int_{\mathbb{R}^d} (\mathrm{e}^{\mathrm{i}\langle z, f(t)x\rangle} - 1 - \mathrm{i}\langle z, c_2(f(t)x)\rangle)\,\nu_L(\mathrm{d}x)\,\mathrm{d}t$$

$$= -\mathrm{i}\Psi_Y(z) - \frac{\mathrm{i}}{2} z^T A_Y z + \mathrm{i}\int_{\mathbb{R}^m} (\mathrm{e}^{\mathrm{i}\langle z, y\rangle} - 1 - \langle z, c_2(y)\rangle)\,\nu_Y(\mathrm{d}y)$$

by (16.13), (16.14), (16.16) and Proposition 16.2(b). By the Lévy–Khintchine formula, the latter is equal to $z^T\gamma_Y$, thus proving (16.15). □

With the help of Theorem 16.3, we can also express the drift of $\int_{\mathbb{R}} f(t)\,\mathrm{d}L_t$ in terms of that of L whenever it exists.

Corollary 16.4 (Drift of integral). *Let $L = (L_t)_{t\in\mathbb{R}}$ be a two-sided $\mathbb{R}^d$-valued Lévy process with Lévy measure ν_L satisfying $\int_{|x|\le 1} |x|\,\nu_L(\mathrm{d}x) < \infty$ and denote the drift of L by γ_L^0. Let $f \in L^1(\mathbb{R};\mathbb{R}^{m\times d}) \cap L^2(\mathbb{R};\mathbb{R}^{m\times d})$ be integrable with respect to L and define $Y := \int_{\mathbb{R}} f(t)\,\mathrm{d}L_t$. Then the Lévy measure ν_Y of Y satisfies $\int_{|y|\le 1} |y|\,\nu_Y(\mathrm{d}y) < \infty$ if and only if $\int_{\mathbb{R}}\int_{\mathbb{R}^d}(1 \wedge |f(t)x|)\,\nu_L(\mathrm{d}x)\,\mathrm{d}t < \infty$. If this condition is satisfied, then the drift γ_Y^0 of Y is given by $\int_{\mathbb{R}} f(t)\,\mathrm{d}t\,\gamma_L^0$.*

Proof. The condition $\int_{|y|\le 1} |y|\,\nu_Y(\mathrm{d}y) < \infty$ is equivalent to finiteness of $\int_{\mathbb{R}^m}(1\wedge|y|)\,\nu_Y(\mathrm{d}y)$, which, by Proposition 16.2(b), is equivalent to $\int_{\mathbb{R}}\int_{\mathbb{R}^d}(1 \wedge |f(t)x|)\,\nu_L(\mathrm{d}x)\,\mathrm{d}t < \infty$. Denote by γ_L and γ_Y the location parameters of L and Y with respect to the standard truncation functions in $\mathbb{R}^d$ and $\mathbb{R}^m$, respectively. Then $\gamma_Y^0 = \gamma_Y - \int_{|y|\le 1} y\,\nu_Y(\mathrm{d}y)$ and $\gamma_L = \gamma_L^0 + \int_{|x|\le 1} x\,\nu_L(\mathrm{d}x)$ by (9.24). Inserting this in (16.15) then shows that $\gamma_Y^0 = \int_{\mathbb{R}} f(t)\gamma_L^0\,\mathrm{d}t$. □

Remark 16.5. Let $d = 1$ and L be an $\mathbb{R}$-valued Lévy process and let $f : \mathbb{R} \to \mathbb{R}$ be a Borel-measurable function. Then it is possible to show that (16.12) is satisfied if and only if there exists a sequence $(f_n)_{n\in\mathbb{N}}$ of functions $f_n : \mathbb{R} \to \mathbb{R}$ of the form $f_n = \sum_{i=1}^{k_n} \alpha_{i_n} \mathbf{1}_{B_i,n}$ with $\alpha_{i,n} \in \mathbb{R}$ and $B_{i,n}$ bounded Borel sets such that f_n converges Lebesgue-almost everywhere to f and such that $\int_B f_n(t)\,\mathrm{d}L_t$ converges in probability as $n \to \infty$ for *every* Borel set $B \in \mathcal{B}_1$ (cf. Rajput and Rosiński [73, Thm. 2.7] and Sato [78, Prop. 2.3]). The existence of such a sequence is used by Rajput and Rosiński [73, p. 460] as the *necessary condition* for the existence of the integral of *a general measurable function f* with respect to a Lévy process L. If it is satisfied, then $\int_B f(t)\,\mathrm{d}L_t$ is defined to be the probability limit of $\int_B f_n(t)\,\mathrm{d}L_t$. It can be shown that this limit is independent of the approximating sequence (f_n) and is frequently referred to as the *integral in the sense of Rajput and Rosiński* or as the *integral with respect to the homogeneous independently scattered random measure associated with L*. With the same methods that were used to prove Theorem 16.3, it is easy to show that every function $f \in L^1(\mathbb{R};\mathbb{R}) \cap L^2(\mathbb{R};\mathbb{R})$ that is integrable in the sense of Definition 15.10 is integrable in the sense of Rajput and Rosiński (with the same value of the integral). However, there are many more functions that are integrable in the sense of Rajput and Rosiński, in particular, functions that do not have to belong to $L^1(\mathbb{R};\mathbb{R}) \cap L^2(\mathbb{R};\mathbb{R})$. The integral in the sense of Rajput and Rosiński is the most commonly used for the integration of deterministic functions with respect to Lévy processes. However, since the functions of interest to us, which are integrable in the sense of Rajput and Rosiński, are also integrable in the sense of Definition 15.10, we chose to use the approach taken in the last chapter, as it uses only the concepts of integration with respect to orthogonal increment and finite variation processes, because the orthogonal increment integration concept was already needed and developed earlier in the book.

Remark 16.6 (Complex-valued integrands). Let $L = ((L_t^1, \dots, L_t^d)^T)_{t\in\mathbb{R}}$ be a two-sided $\mathbb{R}^d$-valued Lévy process and suppose that $f \in L^1(\mathbb{R};\mathbb{C}^{m\times d}) \cap L^2(\mathbb{R};\mathbb{C}^{m\times d})$ is integrable with respect to L. By definition, this means that f_{ij} is integrable with respect to L^j for every pair $(i,j) \in \{1\dots,m\} \times \{1,\dots,d\}$, equivalently that $\Re(f_{ij})$ and $\Im(f_{ij})$ are integrable with respect to L^j for all (i,j). The integral is then given by

$$\int_{\mathbb{R}} f(t)\,\mathrm{d}L_t = \int_{\mathbb{R}} \Re(f(t))\,\mathrm{d}L_t + \mathrm{i}\int_{\mathbb{R}} \Im(f(t))\,\mathrm{d}L_t.$$

Identifying $(z_1,\dots,z_m)^T \in \mathbb{C}^m$ with $(\Re z_1,\dots,\Re z_m, \Im z_1,\dots,\Im z_m)^T \in \mathbb{R}^{2m}$, the integral $\int_{\mathbb{R}} f(t)\,\mathrm{d}L_t$ can also be identified with the integral

$$\int_{\mathbb{R}} \begin{bmatrix} \Re(f(t)) & \mathbf{0}_{m\times d} \\ \mathbf{0}_{m\times d} & \Im(f(t)) \end{bmatrix} \mathrm{d}\begin{bmatrix} L_t \\ L_t \end{bmatrix},$$

i. e., with the integral of the $\mathbb{R}^{2m\times 2d}$-valued function $\left[\begin{smallmatrix} \Re f & \mathbf{0}_{m\times d} \\ \mathbf{0}_{m\times d} & \Im f \end{smallmatrix}\right]$ with respect to the $\mathbb{R}^{2d}$-valued Lévy process $\left[\begin{smallmatrix} L \\ L \end{smallmatrix}\right]$. Here, $\mathbf{0}_{m\times d}$ denotes the zero-matrix in $\mathbb{R}^{m\times d}$. It is obvious from this identification and Theorem 16.3, that if $f : \mathbb{R} \to \mathbb{C}^{m\times d}$ is integrable with respect to L, then $\int_{\mathbb{R}} f(t)\,\mathrm{d}L_t$ is measurable with respect to $\sigma(L_t : t \in \mathbb{R})$ and if, in addition, f has support contained in the compact interval $[a,b]$, then it is also measurable with respect to $\sigma(L_t - L_s : t, s \in [a,b])$. With this identification, Theorem 16.3 also provides us with the distribution of $\int_{\mathbb{R}} f(t)\,\mathrm{d}L_t$ via its characteristic function.

16.2 Distributional properties of the integral

In this section, we shall derive some important consequences of Theorem 16.3. Our first result is as follows.

Theorem 16.7 (Independence over disjoint sets). *If $L = (L_t)_{t\in\mathbb{R}}$ is an $\mathbb{R}^d$-valued Lévy process, $f \in L^1(\mathbb{R};\mathbb{C}^{m\times d}) \cap L^2(\mathbb{R};\mathbb{C}^{m\times d})$ is integrable with respect to L, and $B_1,\dots,B_n \in \mathcal{B}_1$ are pairwise disjoint, then the random vectors $\int_{B_1} f(t)\,\mathrm{d}L_t, \dots, \int_{B_n} f(t)\,\mathrm{d}L_t$ are independent.*

Proof. Thanks to Remark 16.6 we can immediately assume that f is $\mathbb{R}^{m\times d}$-valued. Let $B_1,\dots,B_n \in \mathcal{B}_1$ be pairwise disjoint. With the block matrix

$$F(t) := \operatorname{diag}(\mathbf{1}_{B_1}(t)f(t),\dots,\mathbf{1}_{B_n}(t)f(t)) \in \mathbb{R}^{mn\times dn},$$

we can write, by the definition of the integral, that

$$\begin{bmatrix} \int_{B_1} f(t)\,\mathrm{d}L_t \\ \vdots \\ \int_{B_n} f(t)\,\mathrm{d}L_t \end{bmatrix} = \int_{\mathbb{R}} \underbrace{\begin{bmatrix} \mathbf{1}_{B_1}(t)f(t) & & 0 \\ & \ddots & \\ 0 & & \mathbf{1}_{B_n}(t)f(t) \end{bmatrix}}_{=F(t)\in\mathbb{R}^{md\times nd}} \mathrm{d} \underbrace{\begin{bmatrix} L_t \\ \vdots \\ L_t \end{bmatrix}}_{\in\mathbb{R}^{nd}}.$$

Defining $Y_j := \int_{B_j} f(t)\,\mathrm{d}L_t$, $Y := (Y_1^T,\dots,Y_n^T)^T$ and $X_t := (L_t^T,\dots,L_t^T)^T \in \mathbb{R}^{nd}$, we see that $Y = \int_{\mathbb{R}} F(t)\,\mathrm{d}X_t$. Since

$$\mathbb{E}\,\mathrm{e}^{\mathrm{i}(x_1^T,\dots,x_n^T)X_1} = \mathbb{E}\,\mathrm{e}^{\mathrm{i}(x_1^T+\dots+x_n^T)L_1}, \quad x_1,\dots,x_n \in \mathbb{R}^d,$$

it follows that

$$\Psi_X((x_1^T,\dots,x_n^T)^T) = \Psi_L(x_1+\dots+x_n).$$

Hence (16.16) gives, for $z_1,\dots,z_n \in \mathbb{R}^m$,

$$\begin{aligned}
\mathbb{E}\,\mathrm{e}^{\mathrm{i}(z_1^T,\dots,z_n^T)Y} &= \exp\left\{\int_{\mathbb{R}} \Psi_X\left(F(t)^T \begin{bmatrix} z_1 \\ \vdots \\ z_n \end{bmatrix}\right)\mathrm{d}t\right\} \\
&= \exp\left\{\int_{\mathbb{R}} \Psi_X\left(\begin{bmatrix} \mathbf{1}_{B_1}(t)f(t)^T z_1 \\ \vdots \\ \mathbf{1}_{B_n}(t)f(t)^T z_n \end{bmatrix}\right)\mathrm{d}t\right\} \\
&= \exp\left\{\int_{\mathbb{R}} \Psi_L\left(\sum_{j=1}^n \mathbf{1}_{B_j}(t)f(t)^T z_j\right)\mathrm{d}t\right\} \\
&= \prod_{j=1}^n \exp\left\{\int_{B_j} \Psi_L(f(t)^T z_j)\,\mathrm{d}t\right\} \\
&= \prod_{j=1}^n \mathbb{E}\,\mathrm{e}^{\mathrm{i}z_j^T Y_j},
\end{aligned}$$

showing that $Y_1,\dots,Y_n$ are independent. □

A slight generalization of the previous result is the following:

Corollary 16.8. *If $L = (L_t)_{t\in\mathbb{R}}$ is an $\mathbb{R}^d$-valued Lévy process, $B_1, \dots, B_n \in \mathcal{B}_1$ are pairwise disjoint and $f_1, \dots, f_n \in L^1(\mathbb{R}; \mathbb{C}^{m\times d}) \cap L^2(\mathbb{R}; \mathbb{C}^{m\times d})$ are integrable with respect to L, then the integrals $\int_{B_1} f_1(t)\,\mathrm{d}L_t, \dots, \int_{B_n} f_n(t)\,\mathrm{d}L_t$ are independent.*

Proof. Define $g(t) := \sum_{j=1}^n \mathbf{1}_{B_j}(t) f_j(t)$. Then g is integrable with respect to L. Since $\int_{B_j} f_j(t)\,\mathrm{d}L_t = \int_{B_j} g(t)\,\mathrm{d}L_t$, the result then follows from Theorem 16.7. □

The next consequence of Theorem 16.3 is that the integral with respect to a Lévy process is σ-additive in a certain sense. In order to obtain the desired almost sure convergence, we need Theorem 16.9 below. We state it without proof, as it is well known and its proof can be found in many textbooks on probability. The equivalence of (i), (ii) and (iii) can be found, for example, in Chow and Teicher [32, Thm. 8.3.4], Feller [44, IX.9, Thms. 2 and 3], Kallenberg [55, Thm. 4.18], or Doob [37, Cor. III.2.2]. In these references it is only stated for $\mathbb{R}$-valued random variables, but considering each component separately, it can be established also for $\mathbb{R}^d$-valued random vectors and, by identifying $\mathbb{C}^d$ with $\mathbb{R}^{2d}$, also for $\mathbb{C}^d$-valued random vectors. The equivalence of (i) and (iv) follows from Kolmogorov's 0-1-law, since the set of convergence is in the tail-σ-algebra of the independent series and hence has probability 0 or 1.

Theorem 16.9 (Convergence of series with independent increments). *Let $\xi_1, \xi_2, \dots$ be independent $\mathbb{C}^d$-valued random vectors. Then the following statements are equivalent.*
(i) *The sum $\sum_{k=1}^n \xi_k$ converges almost surely to a $\mathbb{C}^d$-valued random vector as $n \to \infty$.*
(ii) *The sum $\sum_{k=1}^n \xi_k$ converges in probability to a $\mathbb{C}^d$-valued random vector as $n \to \infty$.*
(iii) *The sum $\sum_{k=1}^n \xi_k$ converges in distribution to a $\mathbb{C}^d$-valued random vector as $n \to \infty$.*
(iv) *The sum $\sum_{k=1}^n \xi_k$ converges on a set of positive probability to a $\mathbb{C}^d$-valued random vector as $n \to \infty$.*

The σ-additivity property of the integral can now be stated and proved.

Theorem 16.10 (σ-additivity of the integral). *If $L = (L_t)_{t\in\mathbb{R}}$ is a two-sided $\mathbb{R}^d$-valued Lévy process, $f \in L^1(\mathbb{R}; \mathbb{C}^{m\times d}) \cap L^2(\mathbb{R}; \mathbb{C}^{m\times d})$ is integrable with respect to L, and $B_1, B_2, \dots$ is a sequence of pairwise disjoint Borel sets in $\mathbb{R}$, then*

$$\int_{\bigcup_{k=1}^\infty B_k} f(t)\,\mathrm{d}L_t = \sum_{k=1}^{\infty} \int_{B_k} f(t)\,\mathrm{d}L_t,$$

with the sum converging almost surely.

Proof. By linearity of the integral, $\int_{\bigcup_{k=1}^n B_k} f(t)\,\mathrm{d}L_t = \sum_{k=1}^n \int_{B_k} f(t)\,\mathrm{d}L_t$ for $n \in \mathbb{N}$. Denote the components of f by f_{ij}. Since $f_{ij}\mathbf{1}_{\bigcup_{k=1}^n B_k}$ converges pointwise to $f_{ij}\mathbf{1}_{\bigcup_{k=1}^\infty B_k}$ as $n \to \infty$ and is dominated by $|f_{ij}|$, which is integrable with respect to L^j, for all i, j, Theorem 15.21 shows that $\int_{\bigcup_{k=1}^n B_k} f(t)\,\mathrm{d}L_t$ converges in probability to $\int_{\bigcup_{k=1}^\infty B_k} f(t)\,\mathrm{d}L_t$ as $n \to \infty$, i. e., $\sum_{k=1}^\infty \int_{B_k} f(t)\,\mathrm{d}L_t = \int_{\bigcup_{k=1}^\infty B_k} f(t)\,\mathrm{d}L_t$, with the sum converging in probability. However, as the integrals $\int_{B_k} f(t)\,\mathrm{d}L_t$ are independent by Theorem 16.7, Theorem 16.9 gives the desired almost sure convergence. □

Corollary 16.11. *If $L = (L_t)_{t\in\mathbb{R}}$ is an $\mathbb{R}^d$-valued Lévy process and $f \in L^1(\mathbb{R};\mathbb{C}^{m\times d}) \cap L^2(\mathbb{R};\mathbb{C}^{m\times d})$ is integrable with respect to L, then $\int_{(-n,n]} f(t)\,\mathrm{d}L_t$ (where $n \in \mathbb{N}$) converges almost surely to $\int_{\mathbb{R}} f(t)\,\mathrm{d}L_t$ as $n \to \infty$.*

Proof. Observe that $\int_{(-n,n]} f(t)\,\mathrm{d}L_t = \sum_{k=-n}^{n-1} \int_{(k,k+1]} f(t)\,\mathrm{d}L_t$, where the integrals $\int_{(k,k+1]} f(t)\,\mathrm{d}L_t$ form an independent sequence by Theorem 16.7. The claim then follows from Theorem 16.10. □

We can now show that integrability of a function f with respect to a Lévy process L depends only on f and on the distribution of L and that, if f is integrable with respect to L, then the distribution of $\int_{\mathbb{R}} f(t)\,\mathrm{d}L_t$ also depends only on f and the distribution of L.

Theorem 16.12. *If $L = (L_t)_{t\in\mathbb{R}}$ and $L' = (L'_t)_{t\in\mathbb{R}}$ are $\mathbb{R}^d$-valued Lévy processes with the same distribution and $f \in L^1(\mathbb{R};\mathbb{C}^{m\times d}) \cap L^2(\mathbb{R};\mathbb{C}^{m\times d})$, then f is integrable with respect to L if and only if it is integrable with respect to L', in which case the corresponding integrals have the same distribution, i. e.,*

$$\mathcal{L}\Big(\int_{\mathbb{R}} f(t)\,\mathrm{d}L_t\Big) = \mathcal{L}\Big(\int_{\mathbb{R}} f(t)\,\mathrm{d}L'_t\Big).$$

Proof. That integrability of f with respect to L is equivalent to integrability of f with respect to L' is clear from Theorem 15.27(a), since L and L' have the same Lévy measure. Since they also have the same characteristic exponent $\Psi_L = \Psi_{L'}$, equation (16.16) shows that $\int_{\mathbb{R}} f(t)\,\mathrm{d}L_t$ has the same distribution as $\int_{\mathbb{R}} f(t)\,\mathrm{d}L'_t$ when f is $\mathbb{R}^{m\times d}$-valued. The $\mathbb{C}^{m\times d}$-valued case then follows from Remark 16.6. □

The following presents a substitution rule for integrals with respect to Lévy processes. In the following, when $\alpha < 0$, the expression $L_{((t-)-\beta)/\alpha}$ means $\lim_{u\uparrow t} L_{(u-\beta)/\alpha}$.

Theorem 16.13 (Substitution rule/Time changed integral). *Suppose that $L = (L_t)_{t\in\mathbb{R}}$ is an $\mathbb{R}^d$-valued Lévy process, $f \in L^1(\mathbb{R};\mathbb{C}^{m\times d}) \cap L^2(\mathbb{R};\mathbb{C}^{m\times d})$ and $\alpha, \beta \in \mathbb{R}$ with $\alpha \neq 0$. Define $L' = (L'_t)_{t\in\mathbb{R}}$ by*

$$L'_t := \begin{cases} L_{(t-\beta)/\alpha} - L_{-\beta/\alpha}, & \alpha > 0, \\ L_{((t-)-\beta)/\alpha} - L_{((0-)-\beta)/\alpha}, & \alpha < 0. \end{cases}$$

Then L' is an $\mathbb{R}^d$-valued Lévy process. The function f is integrable with respect to L' if and only if the function $t \mapsto f(\beta + \alpha t)$ is integrable with respect to L, in which case

$$\int_{\mathbb{R}} f(\alpha t + \beta)\,\mathrm{d}L_t = \operatorname{sign}(\alpha) \int_{\mathbb{R}} f(t)\,\mathrm{d}L'_t. \tag{16.18}$$

In particular,

$$\int_a^b f(\alpha t+\beta)\,\mathrm{d}L_t = \int_{\alpha a+\beta}^{\alpha b+\beta} f(s)\,\mathrm{d}L'_s \quad \forall\, a,b\in\mathbb{R}.$$

Formal substitution of $s = \alpha t + \beta$ in the integrals on the left would give

$$\int_a^b f(\alpha t+\beta)\,\mathrm{d}L_t = \int_{\alpha a+\beta}^{\alpha b+\beta} f(s)\,\mathrm{d}L_{(s-\beta)/\alpha},$$

which, by the theorem, must be adjusted by subtraction of the constant random variable $L_{-\beta/\alpha}$ from L to ensure that $L'_0 = 0$ and the use of $(t-)$ and $(0-)$ in the definition of L'_t when $\alpha < 0$ to make L' càdlàg. Observe that for $\alpha = -1$, L' is the at time β time-reversed Lévy process L, which by Theorem 11.4 has the same distribution as $-L$. If $\alpha = 1$, then L' has the same distribution as L.

Proof. We only treat the case $\alpha < 0$, the case $\alpha > 0$ being similar but simpler. That L' is a Lévy process follows easily from the defining properties of a Lévy process (see also Theorem 11.4). Observe that $L'_1 \overset{d}{=} L_{(1-)/\alpha} \overset{d}{=} L_{1/\alpha} \overset{d}{=} -L_{1/|\alpha|}$. Hence, the Lévy measures ν and ν' of L and L' are related by $\nu'(B) = |\alpha|^{-1}\nu(-B)$ for $B\in\mathcal{B}_d$, see Theorem 9.25 and Lemma 10.23. Writing $f = (f_{ij})$, an easy substitution shows that

$$\int_{\mathbb{R}}\int_{|x|>1} (1\wedge |f_{ij}(t)x_j|)\,\nu'(\mathrm{d}x)\,\mathrm{d}t = \int_{\mathbb{R}}\int_{|x|>1} (1\wedge |f_{ij}(\alpha t+\beta)x_j|)\,\nu(\mathrm{d}x)\,\mathrm{d}t$$

for all $i\in\{1,\dots,m\}$ and $j\in\{1,\dots,d\}$. Theorem 15.27(a) then shows that f is integrable with respect to L' if and only if $t\mapsto f(\alpha t+\beta)$ is integrable with respect to L. To establish the substitution formula, consider first the case when $f = Q\mathbf{1}_{(u_1,u_2]}$ with $Q\in\mathbb{C}^{m\times d}$ and $u_1\le u_2$. Then

$$\begin{aligned}\int_{\mathbb{R}} f(\alpha t+\beta)\,\mathrm{d}L_t &= Q\int_{\mathbb{R}} \mathbf{1}_{[(u_2-\beta)/\alpha,(u_1-\beta)/\alpha)}(t)\,\mathrm{d}L_t\\ &= Q(L_{(u_1-\beta)/\alpha} - L_{(u_2-\beta)/\alpha})\end{aligned}$$

by Proposition 15.20 and Example 16.1. Since L is continuous in probability, the latter is (almost surely) equal to

$$Q(L_{((u_1-)-\beta)/\alpha} - L_{((u_2-)-\beta)/\alpha}) = Q(L'_{u_1} - L'_{u_2}) = -\int_{\mathbb{R}} f(s)\,\mathrm{d}L'_s.$$

From linearity it follows that $\int_{\mathbb{R}} f(\alpha t+\beta)\,\mathrm{d}L_t = -\int_{\mathbb{R}} f(s)\,\mathrm{d}L'_s$ for every simple function. Now let f be bounded with support contained in the compact interval $[a,b]$. Since the set of simple functions is dense in $L^2(\mathbb{R}; \mathbb{C}^{m\times d})$, as at the beginning of the proof of Theorem 15.9 we can find a uniformly bounded sequence (f_n) of simple functions with support contained in $[a, b]$ such that f_n converges in $L^2(\mathbb{R}; \mathbb{C}^{m\times d})$ and Lebesgue almost everywhere to f as $n \to \infty$. Applying Theorem 15.21 to each component separately then shows that equation (16.18) holds when f is bounded and has compact support. Finally, approximating a general integrable f with the functions $f^n = (f^n_{ij})$ defined by $f^n_{ij} = f_{ij}\mathbf{1}_{[-n,n]}\mathbf{1}_{[0,n]}(|f_{ij}|)$, Theorem 15.21 again shows that the formula (16.18) carries over to general integrable f. The assertion regarding integrals over (a, b) (or over (b, a) if $a > b$) is an immediate consequence of (16.18). □

We can now show the invariance in distribution of the integral under shifts of f.

Theorem 16.14 (Shift invariance in distribution). *Let $L = (L_t)_{t\in\mathbb{R}}$ be an $\mathbb{R}^d$-valued Lévy process and $f \in L^1(\mathbb{R}; \mathbb{C}^{m\times d}) \cap L^2(\mathbb{R}; \mathbb{C}^{m\times d})$ be integrable with respect to L. If $\beta \in \mathbb{R}$, then the function $t \mapsto f(t + \beta)$, $t \in \mathbb{R}$, is integrable with respect to L, and*

$$\mathcal{L}\Big(\int_{\mathbb{R}} f(t+\beta)\,\mathrm{d}L_t\Big) = \mathcal{L}\Big(\int_{\mathbb{R}} f(t)\,\mathrm{d}L_t\Big).$$

Proof. Define the Lévy process L' by $L'_t := L_{t-\beta} - L_{-\beta}$. Then L' has the same distribution as L, hence f is integrable with respect to L' and $\mathcal{L}(\int_{\mathbb{R}} f(t)\,\mathrm{d}L_t) = \mathcal{L}(\int_{\mathbb{R}} f(t)\,\mathrm{d}L'_t)$ by Theorem 16.12. By Theorem 16.13, $t \mapsto f(t + \beta)$ is integrable with respect to L and $\int_{\mathbb{R}} f(t + \beta)\,\mathrm{d}L_t = \int_{\mathbb{R}} f(t)\,\mathrm{d}L'_t$, giving the result. □

We end this section with expressions for the covariance of integrals of Lévy processes. We first treat the two-dimensional case.

Theorem 16.15 (Covariance of integrals). *If $((X_t, Z_t)^T)_{t\in\mathbb{R}}$ is an $\mathbb{R}^2$-valued Lévy process with finite second moment and $f, g \in L^1(\mathbb{R}; \mathbb{C}) \cap L^2(\mathbb{R}; \mathbb{C})$, then*

$$\operatorname{Cov}\Big(\int_{\mathbb{R}} f(t)\,\mathrm{d}X_t, \int_{\mathbb{R}} g(t)\,\mathrm{d}Z_t\Big) = \operatorname{Cov}(X_1, Z_1)\int_{\mathbb{R}} f(t)\overline{g}(t)\,\mathrm{d}t. \tag{16.19}$$

Proof. Observe that $\int_{\mathbb{R}} f(t)\,\mathrm{d}X_t$ and $\int_{\mathbb{R}} g(t)\,\mathrm{d}Z_t$ have finite second moments by Example 15.12.

(i) Consider first the case when $\mathbb{E}X_1 = \mathbb{E}Z_1 = 0$. Then X and Z are orthogonal increment processes. Suppose that $f, g \in \mathcal{D}$, the space of simple $\mathbb{C}$-valued functions as defined in Section 4.3. Choose a joint representation such that

$$(f(t), g(t)) = \sum_{i=1}^{n} (\alpha_i, \beta_i)\mathbf{1}_{(u_i,u_{i+1}]}, \quad t \in \mathbb{R},$$

where $-\infty < u_1 < \cdots < u_{n+1} < \infty$ and $\alpha_i, \beta_i \in \mathbb{C}$. Then $\int_{\mathbb{R}} f(t)\,\mathrm{d}X_t = \sum_{i=1}^n \alpha_i(X_{u_{i+1}} - X_{u_i})$ and $\int_{\mathbb{R}} g(t)\,\mathrm{d}Z_t = \sum_{j=1}^n \beta_j(Z_{u_{j+1}} - Z_{u_j})$. Since the intervals $(u_i, u_{i+1}]$ and $(u_j, u_{j+1}]$ are disjoint for

$i \neq j$, we conclude from the independent increments property and from Theorem 11.15 that

$$\begin{aligned}\operatorname{Cov}\Big(\int_{\mathbb{R}} f(t)\,\mathrm{d}X_t, \int_{\mathbb{R}} g(t)\,\mathrm{d}Z_t\Big) &= \sum_{i=1}^{n} \alpha_i \overline{\beta_i} \operatorname{Cov}(X_{u_{i+1}} - X_{u_i}, Z_{u_{i+1}} - Z_{u_i}) \\ &= \operatorname{Cov}(X_1, Z_1) \sum_{i=1}^{n} \alpha_i \overline{\beta_i}(u_{i+1} - u_i) \\ &= \operatorname{Cov}(X_1, Z_1) \int_{\mathbb{R}} f(t)\overline{g(t)}\,\mathrm{d}t.\end{aligned}$$

For general $f, g \in L^1(\mathbb{R}; \mathbb{C}) \cap L^2(\mathbb{R}; \mathbb{C})$ choose sequences $(f_n)_{n\in\mathbb{N}}$ and $(g_n)_{n\in\mathbb{N}}$ in $\mathcal{D}$ that converge in $L^2(\mathbb{R}; \mathbb{C})$ as $n \to \infty$ to f and g, respectively. By equation (4.14), $\int_{\mathbb{R}} f_n(t)\,\mathrm{d}X_t$ and $\int_{\mathbb{R}} g_n(t)\,\mathrm{d}Z_t$ converge in $\mathcal{S}$ to $\int_{\mathbb{R}} f(t)\,\mathrm{d}X_t$ and $\int_{\mathbb{R}} g(t)\,\mathrm{d}Z_t$, respectively. We have already shown that (16.19) holds when f and g are replaced by f_n and g_n, respectively. Making this replacement and letting $n \to \infty$ gives the required result.

(ii) In the general case, when it is not necessarily true that $\mathbb{E}X_1 = \mathbb{E}Z_1 = 0$, observe that $\mathbb{E} \int_{\mathbb{R}} f(t)\,\mathrm{d}X_t = \int_{\mathbb{R}} f(t)\,\mathrm{d}t\, \mathbb{E}(X_1)$ by Corollary 15.18, and hence that

$$\int_{\mathbb{R}} f(t)\,\mathrm{d}X_t - \mathbb{E}\int_{\mathbb{R}} f(t)\,\mathrm{d}X_t = \int_{\mathbb{R}} f(t)\,\mathrm{d}(X_t - t\mathbb{E}X_1)$$

by Lemma 15.19. The result then follows from (i), since $((X_t - t\mathbb{E}X_1, Y_t - t\mathbb{E}Y_1)^T)_{t\in\mathbb{R}}$ is an $\mathbb{R}^2$-valued Lévy process with finite second moment, mean 0 and the same covariance as $(X, Z)^T$. □

Recall that for a $\mathbb{C}^d$-valued random vector $U = (U^1, \ldots, U^d)^T$ and a $\mathbb{C}^k$-valued random vector $V = (V^1, \ldots, V^k)^T$ with finite second moment, the *cross-covariance matrix* $\operatorname{Cov}(U, V)$ *of* U *and* V is defined as

$$\begin{aligned}\operatorname{Cov}(U, V) &:= \mathbb{E}[(U - \mathbb{E}(U))\overline{(V - \mathbb{E}(V))^T}] = \mathbb{E}(U\overline{V}^T) - \mathbb{E}(U)\overline{\mathbb{E}(V)}^T \\ &= (\operatorname{Cov}(U^i, V^j))_{i=1,\ldots,d;j=1,\ldots,k} \in \mathbb{C}^{d\times k}.\end{aligned}$$

When $U = V$, we get the covariance matrix

$$\operatorname{Cov}(U) = \operatorname{Cov}(U, V) \in \mathbb{C}^{d\times d}$$

of U. Generalizing the previous result to higher dimensions, we get the cross-covariance matrix between $\int_{\mathbb{R}} f(t)\,\mathrm{d}L_t$ and $\int_{\mathbb{R}} g(t)\,\mathrm{d}L_t$.

Corollary 16.16 (Cross-covariance matrix of integrals). *Suppose that $L = (L_t)_{t\in\mathbb{R}}$ is an $\mathbb{R}^d$-valued Lévy process with finite second moment, that $f \in L^1(\mathbb{R}; \mathbb{C}^{m\times d}) \cap L^2(\mathbb{R}; \mathbb{C}^{m\times d})$ and* *that* $g \in L^1(\mathbb{R}; \mathbb{C}^{k\times d}) \cap L^2(\mathbb{R}; \mathbb{C}^{k\times d})$. *Then the cross-covariance matrix of* $\int_{\mathbb{R}} f(t)\,\mathrm{d}L_t$ *and* $\int_{\mathbb{R}} g(t)\,\mathrm{d}L_t$ *is given by*

$$\operatorname{Cov}\Big(\int_{\mathbb{R}} f(t)\,\mathrm{d}L_t, \int_{\mathbb{R}} g(t)\,\mathrm{d}L_t\Big) = \int_{\mathbb{R}} f(t)\operatorname{Cov}(L_1)\overline{g(t)^T}\,\mathrm{d}t \in \mathbb{C}^{m\times k}, \tag{16.20}$$

where $\operatorname{Cov}(L_1)$ *denotes the covariance matrix of* L_1. *In particular, the covariance matrix of* $\int_{\mathbb{R}} f(t)\,\mathrm{d}L_t$ *is given by*

$$\operatorname{Cov}\Big(\int_{\mathbb{R}} f(t)\,\mathrm{d}L_t\Big) = \int_{\mathbb{R}} f(t)\operatorname{Cov}(L_1)\overline{f(t)^T}\,\mathrm{d}t \in \mathbb{C}^{m\times m}.$$

Proof. By Example 15.12, $U := \int_{\mathbb{R}} f(t)\,\mathrm{d}L_t$ and $V := \int_{\mathbb{R}} g(t)\,\mathrm{d}L_t$ have finite second moments. Writing $U = (U^1,\dots,U^m)^T$, $V = (V^1,\dots,V^k)^T$ and $L = (L^1,\dots,L^d)^T$, Theorem 16.15 implies that the (i,j)-element of $\operatorname{Cov}(U,V)$ is given by

$$\begin{aligned}\operatorname{Cov}(U^i, V^j) &= \operatorname{Cov}\Big(\sum_{l=1}^{d} f_{il}(t)\,\mathrm{d}L_t^l, \sum_{n=1}^{d} g_{jn}(t)\,\mathrm{d}L_t^n\Big)\\ &= \sum_{l,n=1}^{d}\int_{\mathbb{R}} f_{il}(t)\operatorname{Cov}(L_1^l, L_1^n)\overline{g_{jn}(t)}\,\mathrm{d}t.\end{aligned}$$

However, the latter is equal to the (i,j)-element of $\int_{\mathbb{R}} f(t)\operatorname{Cov}(L)\overline{g(t)^T}\,\mathrm{d}t$, establishing (16.20). □

16.3 Continuous-time moving average processes

Of key importance to us are *Lévy driven continuous-time moving average processes*, in short *Lévy driven CMA*. These are processes of the form $\int_{\mathbb{R}} f(t-s)\,\mathrm{d}L_s$, where L is a two-sided $\mathbb{R}^d$-valued Lévy process and $f \in L^1(\mathbb{C}; \mathbb{C}^{m\times d}) \cap L^2(\mathbb{R}; \mathbb{C}^{m\times d})$ is integrable with respect to L. We begin by showing that they are strictly stationary. Here, Definition 1.25 generalizes in a straight-forward way to the multivariate setting, i. e., a $\mathbb{C}^m$-*valued* stochastic process $X = (X_t)_{t\in\mathbb{R}}$ is called *strictly stationary*, if its finite-dimensional distributions are shift-invariant, i. e., if the joint distributions of $(X_{t_1}^T,\dots,X_{t_k}^T)^T$ and $(X_{t_1+h}^T,\dots,X_{t_k+h}^T)^T$ are the same for all positive integers k and for all $t_1,\dots,t_k,h \in \mathbb{R}$.

Theorem 16.17 (Strict stationarity of Lévy driven CMA). *If* $L = (L_t)_{t\in\mathbb{R}}$ *is an* $\mathbb{R}^d$*-valued Lévy process and* $f \in L^1(\mathbb{R}; \mathbb{C}^{m\times d}) \cap L^1(\mathbb{R}; \mathbb{C}^{m\times d})$ *is integrable with respect to L, then* $s \mapsto f(t-s)$ *is integrable with respect to L for each* $t \in \mathbb{R}$ *and the process* $Y = (Y_t)_{t\in\mathbb{R}}$*, defined by* $Y_t = \int_{\mathbb{R}} f(t-s)\,\mathrm{d}L_s$*, is strictly stationary.*

Proof. Let L' be defined as in Theorem 16.13 with $\alpha = -1$ and $\beta = 0$. Since f is integrable with respect to L, $s \mapsto f(-s)$ is integrable with respect to L' by Theorem 16.13. By Theorem 11.4, L' has the same distribution as $-L$, and from Theorem 16.12 it follows that $s \mapsto f(-s)$ is integrable with respect to L. Hence $s \mapsto f(t-s)$ is integrable with respect to L by Theorem 16.14.

To see the strict stationarity, define, for fixed $t_1 < \cdots < t_n$, the block-matrix-valued function $F : \mathbb{R} \to \mathbb{C}^{mn\times dn}$ by $F(s) = \operatorname{diag}(f(t_1 - s), \ldots, f(t_n - s))$. Then, as in the proof of Theorem 16.7, we can write

$$\begin{bmatrix} Y_{t_1} \\ \vdots \\ Y_{t_n} \end{bmatrix} := \begin{bmatrix} \int_{\mathbb{R}} f(t_1 - s)\,\mathrm{d}L_s \\ \vdots \\ \int_{\mathbb{R}} f(t_n - s)\,\mathrm{d}L_s \end{bmatrix} = \int_{\mathbb{R}} \underbrace{\begin{bmatrix} f(t_1 - s) & & 0 \\ & \ddots & \\ 0 & & f(t_n - s) \end{bmatrix}}_{=F(s)\in\mathbb{C}^{md\times nd}} \mathrm{d} \underbrace{\begin{bmatrix} L_s \\ \vdots \\ L_s \end{bmatrix}}_{\in\mathbb{R}^{nd}}.$$

Similarly, for $\beta \in \mathbb{R}$ we obtain

$$\begin{bmatrix} Y_{t_1+\beta} \\ \vdots \\ Y_{t_n+\beta} \end{bmatrix} = \begin{bmatrix} \int_{\mathbb{R}} f(t_1 - (s-\beta))\,\mathrm{d}L_s \\ \vdots \\ \int_{\mathbb{R}} f(t_n - (s-\beta))\,\mathrm{d}L_s \end{bmatrix} = \int_{\mathbb{R}} F(s-\beta)\,\mathrm{d} \begin{bmatrix} L_s \\ \vdots \\ L_s \end{bmatrix}.$$

An application of Theorem 16.14 then shows that $(Y_{t_1+\beta}^T, \ldots, Y_{t_n+\beta}^T)^T$ has the same distribution as $(Y_{t_1}^T, \ldots, Y_{t_n}^T)^T$, giving the strict stationarity of Y. □

The following corollary is an important consequence.

Corollary 16.18. *If $L = (L_t)_{t\in\mathbb{R}}$ is an $\mathbb{R}^d$-valued Lévy process, $f : \mathbb{R} \to \mathbb{C}^{m\times d}$ is locally in $L^2(\mathbb{R}; \mathbb{C}^{m\times d})$, and $h > 0$, then $(\int_{(n-1)h}^{nh} f(nh - t)\,\mathrm{d}L_t)_{n\in\mathbb{Z}}$ is an i. i. d. sequence.*

Proof. Define $Y_n := \int_{((n-1)h,nh)} f(nh - t)\,\mathrm{d}L_t$. Then the Y_n are independent by Corollary 16.8. On the other hand, $\mathcal{L}(Y_n) = \mathcal{L}(Y_0)$ by Theorem 16.17, so that $(Y_n)_{n\in\mathbb{Z}}$ is i. i. d. □

As for strict stationarity, Definition 1.18 for weakly stationary processes generalizes in a straightforward way to the multivariate setting, namely, a $\mathbb{C}^m$-valued stochastic process $Y = (Y_t)_{t\in\mathbb{R}}$ is *weakly stationary* (or simply *stationary*), if $\mathbb{E}|Y_t|^2 < \infty$ for all $t \in \mathbb{R}$, if $\mathbb{E}(Y_t)$ does not depend on t and if the cross-covariance matrix $\operatorname{Cov}(Y_{t+h}, Y_t) \in \mathbb{C}^{m\times m}$ does not depend on $t \in \mathbb{R}$ but only on $h \in \mathbb{R}$. In that case, the *autocovariance function (acvf)* γ_Y of Y is defined for $h \in \mathbb{R}$ by

$$\gamma_Y(h) = \mathbb{E}\big[(Y_{t+h} - \mathbb{E}Y_{t+h})\overline{(Y_t - \mathbb{E}(Y_t))}^T\big] = \operatorname{Cov}(Y_{t+h}, Y_t) \in \mathbb{C}^{m\times m}.$$

Obviously, a strictly stationary process with finite second moments is also weakly stationary. For Lévy driven CMA processes, we now obtain:

Corollary 16.19 (Weak stationarity and the acvf of Lévy driven CMA). *Suppose that* $L = (L_t)_{t\in\mathbb{R}}$ *is an* $\mathbb{R}^d$*-valued Lévy process with* $\mathbb{E}|L_1|^2 < \infty$ *and that* $f \in L^1(\mathbb{R};\mathbb{C}^{m\times d}) \cap L^2(\mathbb{R};\mathbb{C}^{m\times d})$. *Then the continuous-time moving average process* $Y = (Y_t)_{t\in\mathbb{R}}$, *defined by* $Y_t := \int_{\mathbb{R}} f(t-s)\,\mathrm{d}L_s$, *is weakly and strictly stationary. The autocovariance function of* Y *is given by*

$$\gamma_Y(h) = \int_{\mathbb{R}} f(h+u)\,\mathrm{Cov}(L_1)\overline{f^T(u)}\,\mathrm{d}u \quad \forall\, h \in \mathbb{R},$$

where $\mathrm{Cov}(L_1) \in \mathbb{R}^{d\times d}$ *denotes the covariance matrix of* L_1. *If* $d = m = 1$, *then the spectral density of* Y *is given by* $\mathrm{Var}(L_1)|\mathcal{F}_2 f(u)|^2$, $u \in \mathbb{R}$, *where* $\mathcal{F}_2$ *denotes the* L^2*-Fourier transform, as defined in Theorem* 2.66(c).

Proof. By Example 15.12, f is integrable with respect to L under the stated conditions. Theorem 16.17 then implies that $s \mapsto f(t-s)$ is integrable with respect to L for each $t \in \mathbb{R}$ and that the process $Y = (Y_t)_{t\in\mathbb{R}}$ is strictly stationary. From Example 15.12 we see that $\mathbb{E}|Y_t|^2 < \infty$, so that Y is also weakly stationary. The form of the autocovariance function follows immediately from (16.20) using the substitution $u = t - s$ (this was already obtained in (5.5) when $m = d = 1$ and L was an OIP). The form of the spectral density when $m = d = 1$ then follows from Example 5.3. □

16.4 Improper integrals

In this section, we briefly introduce improper integrals with respect to Lévy processes, restricting attention to improper integrals obtained when the lower and upper bounds of integration tend to $-\infty$ and $+\infty$ respectively. Recall that a $\mathbb{C}^m$-valued stochastic process $X = (X_t)_{t\geq 0}$ *converges in probability to a random vector* C *as* $t \to \infty$ if $\lim_{t\to\infty} P(|C - X_t| > \varepsilon) = 0$ for all $\varepsilon > 0$. Since this is equivalent to $\lim_{t\to\infty} P(|C - X_{t_n}| > \varepsilon) = 0$ for all $\varepsilon > 0$ and all increasing sequences $(t_n)_{n\in\mathbb{N}}$ in $\mathbb{R}$ with $t_n \to \infty$, we see that $(X_t)_{t\in\mathbb{R}}$ converges in probability to C as $t \to \infty$ if and only if X_{t_n} converges in probability to C as $n \to \infty$ for every increasing sequence $(t_n)_{n\in\mathbb{N}}$ in $\mathbb{R}$ with $t_n \to \infty$.

Definition 16.20 (Improper integrals). Let $L = (L_t)_{t\in\mathbb{R}}$ be an $\mathbb{R}^d$-valued Lévy process and $f : \mathbb{R} \to \mathbb{C}^{m\times d}$ be locally in $L^2(\mathbb{R};\mathbb{C}^{m\times d})$ (so that, in particular, f is locally integrable with respect to L). We say that the *improper integral of* f *with respect to* L *exists* if both $\int_0^b f(t)\,\mathrm{d}L_t$ and $\int_a^0 f(t)\,\mathrm{d}L_t$ converge in probability to finite random vectors as $b \to \infty$ and $a \to -\infty$, respectively. We then denote the limits in probability by $\int_0^\infty f(t)\,\mathrm{d}L_t$ and $\int_{-\infty}^0 f(t)\,\mathrm{d}L_t$ and write

$$\int_{-\infty}^{\infty} f(t)\,\mathrm{d}L_t := \int_{-\infty}^{0} f(t)\,\mathrm{d}L_t + \int_{0}^{\infty} f(t)\,\mathrm{d}L_t,$$

which we call the *improper integral of f with respect to L*. Similarly, if $B \in \mathcal{B}_1$, we say that the *improper integral of f with respect to L over B exists*, if $\int_0^b f(t)\mathbf{1}_B(t)\,\mathrm{d}L_t$ and $\int_a^0 f(t)\mathbf{1}_B(t)\,\mathrm{d}L_t$ converge in probability to a finite random vector as $b \to \infty$ and $a \to -\infty$, respectively.

So far, for integrable functions f we used the notion $\int_{-\infty}^{\infty} f(t)\,\mathrm{d}L_t$ synonymously for $\int_{\mathbb{R}} f(t)\,\mathrm{d}L_t$, and similarly $\int_{-\infty}^{0}$ for the integral $\int_{(-\infty,0)}$. Below we shall see that in the case of integrability also the improper integral exists with the same value, so that $\int_{-\infty}^{\infty} f(t)\,\mathrm{d}L_t$ from Definition 16.20 is consistent with the previous usage.

Proposition 16.21. *Let $L = (L_t)_{t\in\mathbb{R}}$ be an $\mathbb{R}^d$-valued Lévy process and $f \in L^1(\mathbb{R};\mathbb{C}^{m\times d}) \cap L^2(\mathbb{R};\mathbb{C}^{m\times d})$ be integrable with respect to L. Then the improper integral of f with respect to L also exists, and in this case the integral $\int_{\mathbb{R}} f(t)\,\mathrm{d}L_t$ and the improper integral $\int_{-\infty}^{\infty} f(t)\,\mathrm{d}L_t$ coincide.*

Proof. Let $0 = t_0 < t_1 < t_2 < \cdots$ such that $t_n \to \infty$ as $n \to \infty$. Then $\int_0^{t_n} f(s)\,\mathrm{d}L_s = \sum_{j=1}^n \int_{t_{j-1}}^{t_j} f(s)\,\mathrm{d}L_s$, and it follows from Theorem 16.10 that the sum converges almost surely and hence in probability to $\int_{(0,\infty)} f(s)\,\mathrm{d}L_s$ as $n \to \infty$. Hence $\int_0^t f(s)\,\mathrm{d}L_s$ converges in probability to $\int_{(0,\infty)} f(s)\,\mathrm{d}L_s$ as $t \to \infty$. The proof for $\int_t^0 f(s)\,\mathrm{d}L_s$ when $t \to -\infty$ is similar. □

Remark 16.22. (a) It is important to be aware, when using improper integrals, that existence of the improper integral of f with respect to L over $\mathbb{R}$ does not imply the existence of the improper integral of f with respect to L over all Borel sets B. A simple example is given by the deterministic one-dimensional Lévy process $L_t = t$ and the function $f : \mathbb{R} \to \mathbb{R}$ defined by $f(t) := t^{-1}\sin(t)$ for $t \neq 0$ and $f(0) := 1$. It is well known that the improper integral of f with respect to L exists and that $\int_{-\infty}^{\infty} f(t)\,\mathrm{d}t = \pi$, see, e.g., Billingsley [16, Example 18.4]. However, if $B_0 := \bigcup_{n\in\mathbb{N}_0}(2\pi n + \pi/4, 2\pi n + 3\pi/4)$ then $\sin t \geq \sqrt{2}/2$ for $t \in B_0$ and the divergence of the harmonic series implies that $\lim_{b\to\infty} \int_{0+}^b \mathbf{1}_{B_0}(t)f(t)\,\mathrm{d}t = \infty$, so that the improper integral $\int_{-\infty}^{\infty} \mathbf{1}_{B_0}(t)f(t)\,\mathrm{d}t$ does not exist.

(b) While integrability of f with respect to L implies the existence of the improper integral $\int_{-\infty}^{\infty} f(t)\,\mathrm{d}L_t$, the converse is not true. This can be seen from the example given in (a) above, where the function f is not integrable with respect to $L_t = t$ (it is not even in $L^1(\mathbb{R};\mathbb{C})$).

(c) Sometimes it is helpful to regard integrals such as $\int_{\mathbb{R}} f(s)\,\mathrm{d}L_s$ as improper in order to evaluate them. For example, if f is càdlàg and of finite variation, one can use partial integration for $\int_a^b f(s-)\,\mathrm{d}L_s$ as in Theorem 15.24 and then consider the limits as $a \to -\infty$ and $b \to \infty$.

16.5 Exercises

Exercise 16.1. Let ν be a Lévy measure on $\mathbb{R}^d$, $f : \mathbb{R}^d \to \mathbb{R}^{m\times d}$ be a Borel-measurable function and define the measure ν' by (16.6).

(a) Show that if $\int_{|x|\le 1} |x|\,\nu(dx) < \infty$ and $f \in L^1(\mathbb{R};\mathbb{R}^{m\times d})$ with bounded support, then ν' is also a Lévy measure with $\int_{|y|\le 1} |y|\,\nu'(dy) < \infty$.

(b) Show that if $\int_{\mathbb{R}^d} |x|^2\,\nu(dx) < \infty$ and $f \in L^2(\mathbb{R};\mathbb{R}^{m\times d})$, then ν' is a Lévy measure with $\int_{\mathbb{R}^m} |y|^2\,\nu'(dy) < \infty$.

(c) Show that if $\int_{\mathbb{R}^d} |x|\,\nu(dx) < \infty$ and $f \in L^1(\mathbb{R};\mathbb{R}^{m\times d})$, then ν' is a Lévy measure with $\int_{\mathbb{R}^m} |y|\,\nu'(dy) < \infty$.

(d) Suppose that $d = m = 1$, define f by $f(t) = t^{-3/2}\mathbf{1}_{(1,\infty)}(t)$ and consider the Lévy measure ν defined by $\nu(dx) = x^{-\beta}\mathbf{1}_{(1,\infty)}(x)\,dx$, where $\beta \in (1, 5/3)$. Show that ν' is not a Lévy measure, although $f \in L^1(\mathbb{R};\mathbb{R}) \cap L^2(\mathbb{R};\mathbb{R})$.

Exercise 16.2. Let L be an $\mathbb{R}^d$-valued two-sided Lévy process and let $f : \mathbb{R} \to \mathbb{R}^{m\times d}$ be Borel-measurable and locally bounded. Then by (16.16), $\Psi_{\int_a^b f(t)\,dL_t}(z) = \int_a^b \Psi_L(f(t)^T z)\,dt$ for $a \le b$. Show that this formula does not extend to the case when $a > b$, i. e., that $\Psi_{\int_a^b f(t)\,dL_t}(z)$ is in general different from $\int_a^b \Psi_L(f(t)^T z)\,dt$ when $a > b$.

Exercise 16.3. Suppose that L is an $\mathbb{R}^d$-valued Lévy process, L' is an $\mathbb{R}^{d'}$-valued Lévy process, $f : \mathbb{R} \to \mathbb{R}^{m\times d}$ is integrable with respect to L and $f' : \mathbb{R} \to \mathbb{R}^{m'\times d'}$ is integrable with respect to L'. Show that if L and L' are independent, then so are $\int_B f(t)\,dL_t$ and $\int_{B'} f'(t)\,dL'_t$ for $B, B' \in \mathcal{B}_1$.

Exercise 16.4. Let $L = (L_t)_{t\in\mathbb{R}}$ be an $\mathbb{R}^d$-valued Lévy process and $f \in L^1(\mathbb{R};\mathbb{R}^{m\times d}) \cap L^2(\mathbb{R};\mathbb{R}^{m\times d})$ be an integrable function with respect to L and define $Y := \int_{\mathbb{R}} f(t)\,dL_t$.

(a) *[Moment conditions]* Suppose that $p > 0$, that $f \in L^p(\mathbb{R};\mathbb{R}^{m\times d})$, $\mathbb{E}|L_1|^p < \infty$ and, when $p < 2$, that f is also bounded. Show that $\mathbb{E}|Y|^p < \infty$.

(b) *[Log moment conditions]* Suppose that $\mathbb{E}\log^+|L_1| < \infty$ and that f is bounded with bounded support, or that $\mathbb{E}(\log^+|L_1|)^2 < \infty$ and that there are constants $K, \epsilon > 0$ such that $\|f(t)\| \le K e^{-\epsilon|t|}$ for all $t \in \mathbb{R}$. Show that $\mathbb{E}\log^+|Y| < \infty$.

(c) Suppose that $p \in (0, 2)$ and denote the Lévy measures of L and Y by ν_L and ν_Y, respectively. If $f \in L^p(\mathbb{R};\mathbb{R}^{m\times d})$, $\int_{|x|\le 1} |x|^p\,\nu_L(dx) < \infty$, and either $\mathbb{E}|L_1|^p < \infty$ or f has bounded support, show that $\int_{|y|\le 1} |y|^p\,\nu_Y(dy) < \infty$.

17 Ornstein–Uhlenbeck processes

Abstract: The multivariate Ornstein–Uhlenbeck process satisfying the equation $\mathrm{d}X_t = AX_t\,\mathrm{d}t + \mathrm{d}L_t$, where L is a two-sided $\mathbb{R}^d$-valued Lévy process, is studied in detail. Sufficient conditions for the existence of weakly and of strictly stationary solutions are derived, both in the one-dimensional and multi-dimensional cases. We also study causal solutions and Markov stability of multivariate Ornstein–Uhlenbeck processes.

17.1 Solution of the Ornstein–Uhlenbeck equation

In this chapter, we study in detail both univariate (also referred to as one-dimensional) and multivariate Ornstein–Uhlenbeck processes on $\mathbb{R}$ driven by a two-sided Lévy process L. Suppose that $\lambda < 0$, $x \in \mathbb{R}$ and that $L = (L_t)_{t\in\mathbb{R}}$ is an $\mathbb{R}$-valued Lévy process. Then the equation

$$X_t = x + \int_0^t \lambda X_s\,\mathrm{d}s + L_t, \quad t \in \mathbb{R}, \tag{17.1}$$

is called a (univariate) *Ornstein–Uhlenbeck* or *Langevin equation* and its solution a (Lévy driven) *Ornstein–Uhlenbeck process*. A short-hand notation for (17.1) is the differential form

$$\mathrm{d}X_t = \lambda X_t\,\mathrm{d}t + \mathrm{d}L_t, \quad t \in \mathbb{R}, \quad X_0 = x.$$

The Ornstein–Uhlenbeck equation, with L as Brownian motion, was proposed by Uhlenbeck and Ornstein [86] to model the velocity X_t of a small free particle moving in a fluid (a gas or a liquid) as the result of frequent random collisions with the moving molecules of the fluid, and whose motion is subject to a frictional force with strength determined by the value of $|\lambda|$. Brownian motion itself had been used as a model for the position of such a particle with no friction by Einstein [40]. The motivation behind Ornstein and Uhlenbeck's extension is that a moving particle is hit by more molecules moving in the opposite direction than by those moving in the same direction. Nowadays, Ornstein–Uhlenbeck processes have many applications, and Ornstein–Uhlenbeck processes driven by a Lévy process have become particularly popular in finance thanks to Barndorff-Nielsen and Shephard's seminal paper [7]. In this section, we solve the Ornstein–Uhlenbeck equation pathwise, jumping immediately to a multivariate version, in which λ is replaced by a complex square matrix A and L is multivariate.

Here is a formal definition of a Lévy driven Ornstein–Uhlenbeck process.

Definition 17.1 (OU-equation, OU-process). Suppose that $A \in \mathbb{C}^{d\times d}$ and $L = (L_t)_{t\in\mathbb{R}}$ is a two-sided $\mathbb{R}^d$-valued Lévy process defined on a probability space $(\Omega, \mathcal{F}, P)$. Then any $\mathbb{C}^d$-

https://doi.org/10.1515/9783111325033-017

valued càdlàg process $X = (X_t)_{t\in\mathbb{R}}$ on $(\Omega, \mathcal{F}, P)$ such that, for each $t \in \mathbb{R}$, X_t almost surely satisfies the equation

$$X_t = X_0 + \int_0^t AX_s\,\mathrm{d}s + L_t, \quad t \in \mathbb{R}, \tag{17.2}$$

is called a *(multivariate) Ornstein–Uhlenbeck process*, and we speak of X as *being driven by L* with *parameter A*. The integral equation (17.2) is called a *(multivariate) Ornstein–Uhlenbeck equation* or *(multivariate) Langevin equation*, abbreviated to *OU-equation*. The process X is correspondingly referred to as an *OU-process*.

The Ornstein–Uhlenbeck equation is often written in differential form as

$$\mathrm{d}X_t = AX_t\,\mathrm{d}t + \mathrm{d}L_t, \quad t \in \mathbb{R}, \tag{17.3}$$

which is short-hand notation for the integral equation (17.2).

If X satisfies (17.2), then for $s, t \in \mathbb{R}$,

$$X_t - X_s = \int_0^t AX_u\,\mathrm{d}u + L_t - \int_0^s AX_u\,\mathrm{d}u - L_s = \int_s^t AX_u\,\mathrm{d}u + (L_t - L_s).$$

Hence, the Ornstein–Uhlenbeck equation (17.2) can be written in the equivalent form,

$$X_t - X_s = \int_s^t AX_u\,\mathrm{d}u + (L_t - L_s) \quad \forall\, s, t \in \mathbb{R}.$$

(Observe that this reduces to (17.2) for $s = 0$ since $L_0 = 0$ a. s.)

We first ask whether, for any specified random variable x, there is a solution X of the equations (17.2) with $X_0 = x$ and, if such a solution exists, whether it is unique. As we shall see, the answer to both questions is yes. To specify the solution, we recall the definition of the exponential of a complex matrix $A \in \mathbb{C}^{d\times d}$ from (7.14) as $\mathrm{e}^A = \exp(A) := \sum_{k=0}^{\infty} \frac{1}{k!}A^k$, where the sum converges absolutely in each component. Also recall the properties of the matrix exponential stated in Section 7.2. With these preparations at hand, we can now solve the Ornstein–Uhlenbeck equation.

Theorem 17.2 (Solution of the multivariate OU-equation). *Suppose that $A \in \mathbb{C}^{d\times d}$, that $L = (L_t)_{t\in\mathbb{R}}$ is an $\mathbb{R}^d$-valued Lévy process, and that x_0 is a given random variable, defined on the same probability space as L. Then the multivariate Ornstein–Uhlenbeck equation* (17.2) *has an (almost surely) unique ($\mathbb{C}^d$-valued) càdlàg solution $X = (X_t)_{t\in\mathbb{R}}$, satisfying the condition $X_0 = x_0$. This solution is given by*

$$X_t = \mathrm{e}^{At}\left(x_0 + \int_0^t \mathrm{e}^{-As}AL_s\,\mathrm{d}s\right) + L_t, \quad t \in \mathbb{R}. \tag{17.4}$$

Furthermore, for each $t \in \mathbb{R}$, the solution X in (17.4) *can be rewritten as*

$$X_t = e^{At}\left(x_0 + \int_0^t e^{-As}\, dL_s\right), \quad t \in \mathbb{R}. \tag{17.5}$$

The integral in (17.4) is a pathwise Lebesgue–integral, while the integral in (17.5) is an integral with respect to the Lévy process L, and is therefore defined only almost surely. For this reason, we cannot deduce the càdlàg property of X from (17.5). We can however deduce it from (17.4). A more precise statement therefore is that the process defined by (17.5) has a càdlàg modification, which is an Ornstein–Uhlenbeck process.

Proof. (i) We first show that the process defined by the right-hand side of (17.4) is a solution of (17.2). If $X_t = e^{At}(x_0 + \int_0^t e^{-As} AL_s\, ds) + L_t$ then X is obviously càdlàg and $X_0 = x_0$. Furthermore, for each $t \in \mathbb{R}$, an application of (7.16) and Fubini's theorem yields

$$\begin{aligned} X_0 &+ \int_0^t AX_s\, ds + L_t \\ &= x_0 + \int_0^t Ae^{As} x_0\, ds + \int_0^t Ae^{As} \int_0^s e^{-Au} AL_u\, du\, ds + \int_0^t AL_s\, ds + L_t \\ &= e^{At} x_0 + \int_0^t \left(\int_u^t Ae^{As}\, ds \right) e^{-Au} AL_u\, du + \int_0^t AL_s\, ds + L_t \\ &= e^{At} x_0 + e^{At} \int_0^t e^{-Au} AL_u\, du + L_t = X_t, \end{aligned}$$

so that X defined by the right-hand side of (17.4) is indeed a solution of (17.2).

(ii) To show that the solution is unique, let $X = (X_t)_{t \in \mathbb{R}}$ and $Z = (Z_t)_{t \in \mathbb{R}}$ be two càdlàg solutions of (17.2) with $X_0 = Z_0 = x_0$. Define $Y = X - Z$. Then $Y_0 = 0$ and $Y_t = \int_0^t AY_s\, ds$ by (17.2). Since Y is càdlàg, the right-hand side of this equation is continuous, hence Y (as the left-hand side) is continuous. However, this implies that the right-hand side of the equation is differentiable, and hence so is Y. So the paths of Y satisfy the differential equation $\frac{d}{dt} Y_t = AY_t$. Defining $W_t := e^{-At} Y_t$ we obtain from (7.15)

$$\frac{d}{dt} W_t = e^{-At} \frac{d}{dt} Y_t - e^{-At} AY_t = e^{-At}\left(\frac{d}{dt} Y_t - AY_t\right) = 0.$$

Since $W_0 = 0$, we conclude $W_t = 0$ and hence $Y_t = 0$ for all $t \in \mathbb{R}$, implying that $X = Z$.

(iii) It remains only to show that the unique solution X can be rewritten as (17.5). If we denote the jth-component of any vector V by V^j and the (i,j)-element of any matrix B by B_{ij}, then for $u, t \in \mathbb{R}$ with $u \le t$ and $i \in \{1, \ldots, d\}$ we find, using partial integration (Theorem 15.24), that

$$
\begin{aligned}
&\left(\int_u^t \mathrm{e}^{-As}\,\mathrm{d}L_s\right)^i \\
&= \sum_{j=1}^n \int_u^t (\mathrm{e}^{-As})_{ij}\,\mathrm{d}L_s^j \\
&= \sum_{j=1}^n \left((\mathrm{e}^{-At})_{ij} L_t^j - (\mathrm{e}^{-Au})_{ij} L_u^j - \int_u^t \left(\frac{\mathrm{d}}{\mathrm{d}s}\mathrm{e}^{-As}\right)_{ij} L_s^j\,\mathrm{d}s \right) \\
&= (\mathrm{e}^{-At}L_t)^i - (\mathrm{e}^{-Au}L_u)^i + \sum_{j=1}^n \int_u^t (\mathrm{e}^{-As}A)_{ij} L_s^j\,\mathrm{d}s,
\end{aligned}
$$

so that

$$
\int_u^t \mathrm{e}^{-As}\,\mathrm{d}L_s = \mathrm{e}^{-At}L_t - \mathrm{e}^{-Au}L_u + \int_u^t \mathrm{e}^{-As}AL_s\,\mathrm{d}s \quad \forall\, u \le t \tag{17.6}
$$

and hence also for all $u > t$. Specializing this to $u = 0$ and inserting it into the right-hand side of (17.4) shows that (17.5) is true. □

We have seen in Theorem 17.2 that for any specified random variable X_0, Equation (17.1) has an almost surely unique càdlàg solution X, given by

$$
\begin{aligned}
X_t &= \mathrm{e}^{At}\left(X_0 + \int_0^t \mathrm{e}^{-As}AL_s\,\mathrm{d}s \right) + L_t \\
&= \mathrm{e}^{At}\left(X_0 + \int_0^t \mathrm{e}^{-As}\,\mathrm{d}L_s \right), \qquad t \in \mathbb{R}.
\end{aligned} \tag{17.7}
$$

This shows, in particular, that the multivariate Ornstein–Uhlenbeck process is uniquely determined by A, L and X_0. We hence also speak of X as *being driven by L* with *parameter A* and *specified value X_0 at time* 0.

When sampled at equidistant times, the Ornstein–Uhlenbeck process is a vector AR(1) process with i. i. d. noise, specified precisely in the following proposition.

Proposition 17.3 (OU-process sampled equidistantly). *Let X be the multivariate Ornstein–Uhlenbeck process driven by the $\mathbb{R}^d$-valued Lévy process L with parameter $A \in \mathbb{C}^{d\times d}$ and specified X_0. Then X satisfies the equation*

$$
X_t = \mathrm{e}^{A(t-s)}\left(X_s + \int_s^t \mathrm{e}^{-A(u-s)}\,\mathrm{d}L_u \right), \qquad t \ge s. \tag{17.8}
$$

In particular, for fixed $h > 0$, defining

$$Z_n^{(h)} := \int_{(n-1)h}^{nh} e^{A(nh-u)}\, dL_u, \quad n \in \mathbb{Z},$$

the sequence $(Z_n^{(h)})_{n\in\mathbb{Z}}$ *is i. i. d. and* $(X_{nh})_{n\in\mathbb{Z}}$ *satisfies the vector* AR(1) *equation*

$$X_{nh} = e^{Ah} X_{(n-1)h} + Z_n^{(h)}, \quad n \in \mathbb{Z},$$

with i. i. d. noise $(Z_n^{(h)})_{n\in\mathbb{Z}}$.

Proof. Equation (17.8) follows immediately from (17.7), since

$$X_t = e^{A(t-s)} e^{As} \left(X_0 + \int_0^s e^{-Au}\, dL_u + \int_s^t e^{-Au}\, dL_u \right),$$

which also gives the equation for X_{nh}. That $(Z_n^{(h)})_{n\in\mathbb{Z}}$ is i. i. d. follows from Corollary 16.18. □

Proposition 17.3 shows that Ornstein–Uhlenbeck processes constitute a natural continuous time family analogous to vector AR(1)-processes with i. i. d. noise since, when sampled on an arbitrary regular grid, they satisfy an AR(1)-equation with i. i. d. noise.

17.2 Stationary Ornstein–Uhlenbeck processes

In this section, we are interested in sufficient conditions for A and L under which it is possible to specify the random value X_0 in such a way that the resulting multivariate Ornstein–Uhlenbeck process X is strictly stationary or weakly stationary, and to determine under these conditions whether or not the choice of X_0 is unique.

We start with the case when either all eigenvalues of A have strictly negative real part or all eigenvalues of A have strictly positive real part.

Theorem 17.4 (Stationarity of OU-process for specific eigenvalues of A). *Suppose that* $A \in \mathbb{C}^{d\times d}$ *and* $L = (L_t)_{t\in\mathbb{R}}$ *is an* $\mathbb{R}^d$*-valued Lévy process on a probability space* $(\Omega, \mathcal{F}, P)$. *Consider the multivariate Ornstein–Uhlenbeck equation* (17.2) *with solution* $X = (X_t)_{t\in\mathbb{R}}$.

(a) *If all the eigenvalues of* A *have strictly negative real part and* $\mathbb{E} \log^+ |L_1| < \infty$, *then a random vector* X_0 *on* $(\Omega, \mathcal{F}, P)$ *can be chosen in such a way that* $(X_t)_{t\in\mathbb{R}}$ *is strictly stationary. Under this condition, the strictly stationary solution (including* X_0*) is almost surely unique and given by*

$$X_t = \int_{-\infty}^{t} e^{A(t-s)}\, dL_s, \quad t \in \mathbb{R}, \tag{17.9}$$

where the integrand is integrable with respect to L. A càdlàg version of this solution is given by

$$X_t = L_t + \int_{-\infty}^{t} A e^{A(t-s)} L_s \, ds, \quad t \in \mathbb{R}, \tag{17.10}$$

with the integral existing almost surely as a Lebesgue integral. Similarly, if all eigenvalues of A have strictly negative real part and $\mathbb{E}|L_1|^2 < \infty$, *then a random vector* X_0 *can be chosen to make X weakly stationary. Under this condition the weakly stationary solution is almost surely unique and given by* (17.9). *A càdlàg version is given by* (17.10).

(b) *If all the eigenvalues of A have strictly positive real part and* $\mathbb{E}\log^+|L_1| < \infty$, *then a random vector* X_0 *can be chosen in such a way that* $(X_t)_{t\in\mathbb{R}}$ *is strictly stationary. Under this condition, the stationary solution (including* X_0*) is almost surely unique and given by*

$$X_t = -\int_{t}^{\infty} e^{A(t-s)} \, dL_s, \quad t \in \mathbb{R}, \tag{17.11}$$

where the integrand is integrable with respect to L. A càdlàg version of this solution is given by

$$X_t = L_t - \int_{t}^{\infty} A e^{A(t-s)} L_s \, ds, \quad t \in \mathbb{R}, \tag{17.12}$$

with the integral existing almost surely as a Lebesgue integral. Similarly, if all eigenvalues of A have strictly negative real part and $\mathbb{E}|L_1|^2 < \infty$, *then a random vector* X_0 *can be chosen to make X weakly stationary. Under this condition, the weakly stationary solution is almost surely unique and given by* (17.11). *A càdlàg version is given by* (17.12).

Proof. (a) Denote the distinct eigenvalues of A by $\lambda_1, \ldots, \lambda_r$ and suppose that $\Re(\lambda_1), \ldots, \Re(\lambda_r) < 0$. Define

$$\varepsilon := -\max\{\Re(\lambda_1), \ldots, \Re(\lambda_r)\}/2 > 0.$$

(i) We shall first show that the solution given by (17.9) is indeed a strictly/weakly stationary solution when $\mathbb{E}\log^+|L_1| < \infty / \mathbb{E}(|L_1|^2) < \infty$. Since by (7.21) there exists, for each $t \in \mathbb{R}$, $K_{\varepsilon,t} > 0$ such that $\|e^{-As}\| \le K_{\varepsilon,t} e^{\varepsilon s}$ for all $s \le t$, the mapping $s \mapsto e^{-As}\mathbf{1}_{(-\infty,t]}(s)$ is integrable with respect to L by Theorem 15.26(iii). Define $X_t := e^{At}\int_{(-\infty,t]} e^{-As}\, dL_s$. Then $X_t = \int_{\mathbb{R}} e^{A(t-s)}\mathbf{1}_{[0,\infty)}(t-s)\, dL_s$ by linearity of the integral and it follows from Theorem 16.17 that X is strictly stationary, and by Corollary 16.19 weakly stationary if additionally $\mathbb{E}|L_1|^2 < \infty$. To see that X is a solution of (17.3), it suffices to observe that

$$e^{At}\left(X_0 + \int_0^t e^{-As}\, dL_s\right) = e^{At}\left(\int_{-\infty}^0 e^{-As}\, dL_s + \int_0^t e^{-As}\, dL_s\right) = X_t$$

for $t \in \mathbb{R}$, which is (17.5). In particular, it has a càdlàg modification by Theorem 17.2.

(ii) To see that (17.10) is a modification of (17.9), define X_t again by (17.9). By Proposition 16.21, X_t is the probability limit of $e^{At}\int_u^t e^{-As}\, dL_s$ as $u \to -\infty$, which by equation (17.6) can be rewritten as the probability limit of

$$L_t + e^{At}\left(-e^{-Au}L_u + \int_u^t e^{-As}AL_s\, ds\right) \tag{17.13}$$

as $u \to -\infty$. Since $\lim_{u\to-\infty} e^{\varepsilon u/2}|L_u| = 0$ a. s. by Proposition 11.22 and the finite log-moment of L, we can find for every $t \in \mathbb{R}$ an almost surely finite random constant $M_{\varepsilon,t}$ such that $|L_u| \le M_\varepsilon e^{-\varepsilon u/2}$ for all $u \le t$ so that

$$\|e^{-Au}L_u\| \le K_{\varepsilon,t}M_{\varepsilon,t}e^{\varepsilon u/2} \quad \text{and} \quad \|e^{-Au}AL_u\| \le K_{\varepsilon,t}M_{\varepsilon,t}\|A\|e^{\varepsilon u/2}$$

for all $u \le t$. We conclude that $e^{-Au}L_u$ converges almost surely to 0 as $u \to -\infty$, that the Lebesgue-integral $\int_{-\infty}^t e^{-As}AL_s\, ds$ exists almost surely and hence that $\int_u^t e^{-As}AL_s\, ds$ converges almost surely to $\int_{-\infty}^t e^{-As}AL_s\, ds$ as $u \to -\infty$. This shows that the probability limit in (17.13) is equal to (17.10), so that (17.10) is a modification of (17.9), and it obviously has càdlàg paths.

(iii) To see that the choice of X_0 implying strict/weak stationarity of X is almost surely unique, let $(X_t)_{t\in\mathbb{R}}$ be any strictly/weakly stationary solution. Then

$$X_0 = e^{-At}X_t - \int_0^t e^{-As}\, dL_s \quad \forall\, t \in \mathbb{R}$$

by (17.7). If X is strictly stationary, then X_t has the same distribution for every t, and since $e^{-At} \to 0$ as $t \to -\infty$ by (7.21), we see that $e^{-At}X_t$ converges in distribution and hence in probability to 0 as $t \to \infty$. Similarly, when X is weakly stationary, we observe that $e^{-At}X_t$ converges in $\mathcal{S}$ and hence in probability to 0 as $t \to -\infty$. Consequently, X_0 is the probability limit of $-\int_0^t e^{-As}\, dL_s$ as $t \to -\infty$ and in particular it is uniquely determined by L.

(b) The proof when all the eigenvalues of A have strictly positive parts is completely analogous to that of (a), with (7.20) replacing (7.21) and corresponding limits taken in the opposite direction from those taken in (a). □

We next calculate the autocovariance function of the weakly stationary solution under the conditions of Theorem 17.4:

Corollary 17.5 (Autocovariance function of stationary OU-process). *Suppose that $A \in \mathbb{C}^{d\times d}$, with eigenvalues all of which have either strictly negative real part or strictly positive real part. Let $L = (L_t)_{t\in\mathbb{R}}$ be an $\mathbb{R}^d$-valued Lévy process with $\mathbb{E}|L_1|^2 < \infty$ and denote by* $\mathrm{Cov}(L_1)$ *the covariance matrix of L_1. Let $X = (X_t)_{t\in\mathbb{R}}$ be the unique weakly stationary solution of the OU-equation* (17.3) *specified in Theorem* 17.4(a, b). *When all the eigenvalues of A have strictly negative real part, then the autocovariance function γ_X of X is given by*

$$\gamma_X(h) = \mathrm{Cov}(X_{t+h}, X_t) = \begin{cases} \mathrm{e}^{Ah} \int_0^\infty \mathrm{e}^{Au}\, \mathrm{Cov}(L_1)\, \mathrm{e}^{\overline{A}^T u}\, \mathrm{d}u, & h \geq 0, \\ \overline{\gamma_X(-h)}, & h < 0. \end{cases}$$

When all the eigenvalues of A have strictly positive real part, then γ_X is given by

$$\gamma_X(h) = \mathrm{Cov}(X_{t+h}, X_t) = \begin{cases} \mathrm{e}^{Ah} \int_{-\infty}^0 \mathrm{e}^{Au}\, \mathrm{Cov}(L_1)\, \mathrm{e}^{\overline{A}^T u}\, \mathrm{d}u, & h \leq 0, \\ \overline{\gamma_X(-h)}, & h > 0. \end{cases}$$

Proof. Suppose that all the eigenvalues of A have strictly negative real parts. Define $g(t) := \mathrm{e}^{At}\mathbf{1}_{[0,\infty)}(t)$ and observe that $X_t = \int_{\mathbb{R}} g(t-s)\, \mathrm{d}L_s$ by Theorem 17.4(a). It then follows from Corollary 16.19 that for $h \geq 0$,

$$\begin{aligned}\gamma_X(h) &= \int_{\mathbb{R}} \mathrm{e}^{A(h+u)}\mathbf{1}_{[0,\infty)}(h+u)\; \mathrm{Cov}(L_1)\, \mathrm{e}^{\overline{A}^T u}\mathbf{1}_{[0,\infty)}(u)\, \mathrm{d}u \\ &= \mathrm{e}^{Ah} \int_0^\infty \mathrm{e}^{Au}\; \mathrm{Cov}(L_1)\, \mathrm{e}^{\overline{A}^T u}\, \mathrm{d}u.\end{aligned}$$

The formula for $h < 0$ is clear. The proof when all the eigenvalues have strictly positive real parts is analogous, with g redefined as the corresponding kernel function, namely $g(t) := -\mathrm{e}^{At}\mathbf{1}_{(-\infty,0)}(t)$. □

In one dimension, the condition that all eigenvalues of A have either strictly negative real parts or strictly positive real parts reduces to $\Re(A) < 0$ or $\Re(A) > 0$, so that there is always a unique strictly stationary solution provided $\Re(A) \neq 0$ and $\mathbb{E}\log^+|L_1| < \infty$, and a unique weakly stationary solution provided $\Re(A) \neq 0$ and $\mathbb{E}|L_1|^2 < \infty$. Due to their importance, we collect these results in the following corollary and, for the weakly stationary solution, include also the spectral density.

Corollary 17.6 (Stationary 1-dimensional OU-processes). *Let X be a one-dimensional Ornstein–Uhlenbeck process with parameter $A \in \mathbb{C}$, driven by an $\mathbb{R}$-valued Lévy process L, defined on the probability space $(\Omega, \mathcal{F}, P)$.*

(a) *Suppose that $\mathbb{E}\log^+|L_0| < \infty$ and $\Re(A) \neq 0$. Then a random variable, $X_0 : \Omega \to \mathbb{C}$, can be chosen in such a way as to ensure that $(X_t)_{t\in\mathbb{R}}$ is strictly stationary. Under these conditions, the strictly stationary solution X (including X_0) is almost surely unique and given by*

$$X_t = \begin{cases} e^{At} \int_{-\infty}^{t} e^{-As} \, dL_s, & \Re(A) < 0, \\ -e^{At} \int_{t}^{\infty} e^{-As} \, dL_s, & \Re(A) > 0, \end{cases} \tag{17.14}$$

for $t \in \mathbb{R}$, where the integrand is integrable with respect to L in each case. A càdlàg modification of this solution is given by

$$X_t = \begin{cases} L_t + \int_{-\infty}^{t} A\, e^{A(t-s)} L_s \, ds, & \Re(A) < 0, \\ L_t - \int_{t}^{\infty} A\, e^{A(t-s)} L_s \, ds, & \Re(A) > 0, \end{cases} \tag{17.15}$$

for $t \in \mathbb{R}$, where the integrals exist almost surely as Lebesgue integrals.

(b) *Suppose that $\mathbb{E}|L_0|^2 < \infty$ and $\Re(A) \neq 0$. Then a random variable, $X_0 : \Omega \to \mathbb{C}$, can be chosen in such a way as to ensure that $(X_t)_{t\in\mathbb{R}}$ is weakly stationary. Under these conditions, the weakly stationary solution is almost surely unique and given by* (17.14), *or equivalently by* (17.15). *The autocovariance function of X is*

$$\gamma_X(h) = \frac{\operatorname{Var}(L_1)}{2|\Re(A)|} e^{-|\Re(A)h|} e^{i\Im(A)h}, \quad h \in \mathbb{R}, \tag{17.16}$$

and the spectral distribution function of X is absolutely continuous with spectral density f_X given by

$$f_X(\mu) = \frac{\operatorname{Var}(L_1)}{2\pi((\mu - \Im(A))^2 + (\Re(A))^2)}, \quad \mu \in \mathbb{R}. \tag{17.17}$$

Proof. All assertions follow from Theorem 17.4 apart from (17.16) and (17.17). For (17.16), using Corollary 17.5, we obtain for $\Re(A) < 0$ and $h \geq 0$

$$\begin{aligned} \gamma_X(h) &= \operatorname{Var}(L_1) e^{Ah} \int_0^\infty e^{2\Re(A)u} \, du \\ &= \frac{\operatorname{Var}(L_1)}{2|\Re(A)|} e^{Ah} = \frac{\operatorname{Var}(L_1)}{2|\Re(A)|} e^{-|\Re(A)h|} e^{i\Im(A)h}. \end{aligned}$$

Together with the relation $\gamma_X(-h) = \overline{\gamma_X(h)}$ this gives (17.16) when $\Re(A) < 0$. For the calculation of the spectral density, observe that $X_t = \int_{\mathbb{R}} g(t-s)\, dL_s$ with $g(t) = e^{At}\mathbf{1}_{[0,\infty)}(t)$ when $\Re(A) < 0$. Denoting the Fourier transform of g by $\mathcal{F}_2 g(\mu) = \frac{1}{\sqrt{2\pi}} \int_{\mathbb{R}} e^{-it\mu} g(t)\, dt$, it follows from Example 5.3 (or from Corollary 16.19) that the spectral density f_X of X satisfies $f_X(\mu) = \operatorname{Var}(L_1)|\mathcal{F}_2 g(\mu)|^2$ for $\mu \in \mathbb{R}$. But $\mathcal{F}_2 g(\mu) = (\sqrt{2\pi}(i\mu - A))^{-1}$, hence

$$f_X(\mu) = \frac{\operatorname{Var}(L_1)}{2\pi|i\mu - A|^2} = \frac{\operatorname{Var}(L_1)}{2\pi((\mu - \Im(A))^2 + (\Re(A))^2)}, \quad \mu \in \mathbb{R},$$

which is (17.17). The calculations when $\Re(A) > 0$ are similar. □

Remark 17.7. Consider the one-dimensional case, $d = 1$. If L is non-deterministic, then it can be shown that the conditions $\Re(A) \neq 0$ and $\mathbb{E} \log^+ |L_1| < \infty$ are not only sufficient, but also necessary for the existence of a strictly stationary solution of the Ornstein–Uhlenbeck equation, see, e. g., Brockwell and Lindner [26, Prop. 2.5] for a proof of this fact. Similarly, it can be shown that if L is non-deterministic, then $\mathbb{R}(A) \neq 0$ and $\mathbb{E}|L_1|^2 < \infty$ is not only sufficient, but also necessary for the existence of a weakly stationary solution to exist.

Returning to the multivariate case, by transforming A to block form, it is possible to characterize the existence of strictly and weakly stationary solutions of the multivariate Ornstein–Uhlenbeck equation for general $A \in \mathbb{C}^{d\times d}$. We shall not consider necessary conditions, but content ourselves with conditions that are sufficient for our subsequent applications.

Theorem 17.8 (Stationarity of multivariate OU-processes). *Suppose that $L = (L_t)_{t\in\mathbb{R}}$ is an $\mathbb{R}^d$-valued Lévy process on a probability space $(\Omega, \mathcal{F}, P)$ and let $A \in \mathbb{C}^{d\times d}$ be such that all the eigenvalues of A have non-zero real part. Consider the multivariate Ornstein–Uhlenbeck equation* (17.2) *with solution $X = (X_t)_{t\in\mathbb{R}}$.*

(a) *If $\mathbb{E}\log^+|L_1| < \infty$, then a random vector X_0 (on $(\Omega, \mathcal{F}, P)$) can be chosen in such a way that $(X_t)_{t\in\mathbb{R}}$ is strictly stationary, and the strictly stationary solution (including X_0) is almost surely unique.*

(b) *If $\mathbb{E}|L_1|^2 < \infty$, then a random vector X_0 (on $(\Omega, \mathcal{F}, P)$) can be chosen in such a way that $(X_t)_{t\in\mathbb{R}}$ is weakly stationary, and the weakly stationary solution (including X_0) is almost surely unique.*

(c) *To determine the unique strictly/weakly stationary solution, let $S \in \mathbb{C}^{d\times d}$ be an invertible matrix such that*

$$SAS^{-1} = B := \begin{bmatrix} B_1 & 0 \\ 0 & B_2 \end{bmatrix},$$

where all eigenvalues of $B_1 \in \mathbb{C}^{d_1\times d_1}$ have strictly negative real part and those of $B_2 \in \mathbb{C}^{d_2\times d_2}$ have strictly positive real part. The dimensions $d_i \in \{0, 1, \dots, d\}$ satisfy $d_1 + d_2 = d$. For example, B could be a Jordan matrix (with suitably ordered eigenvalues). Denote by $I_1 \in \mathbb{C}^{d_1\times d}$ the projection matrix that maps $(x_1, \dots, x_d)^T \in \mathbb{C}^d$ to $(x_1, \dots, x_{d_1})^T \in \mathbb{C}^{d_1}$ and by $I_2 \in \mathbb{C}^{d_2\times d}$ the projection matrix that maps $(x_1, \dots, x_d)^T \in \mathbb{C}^d$ to $(x_{d_1+1}, \dots, x_{d_1+d_2})^T \in \mathbb{C}^{d_2}$. Then, under the conditions of (a)/(b), *the almost surely unique strictly/weakly stationary solution is given by*

$$X_t = S^{-1} \begin{bmatrix} \int_{-\infty}^t e^{B_1(t-s)} I_1 S \, dL_s \\ -\int_t^\infty e^{B_2(t-s)} I_2 S \, dL_s \end{bmatrix}, \quad t \in \mathbb{R}, \tag{17.18}$$

a càdlàg modification of which is

$$X_t = L_t + S^{-1} \begin{bmatrix} \int_{-\infty}^t e^{B_1(t-s)} B_1 I_1 S L_s \, ds \\ -\int_t^\infty e^{B_2(t-s)} B_2 I_2 S L_s \, ds \end{bmatrix}, \quad t \in \mathbb{R}. \tag{17.19}$$

Here, the integrands in (17.18) *are integrable with respect to L and the integrals in* (17.19) *exist as Lebesgue integrals.*

Proof. Suppose first that $\mathbb{E}\log^+|L_1| < \infty$.

(i) To show that the right-hand side of (17.18) constitutes a strictly stationary solution of (17.2), define X_t by it. The integrals in (17.18) exist by the same reasoning that was used in the proof of Theorem 17.4. Now define

$$g(s) := S^{-1}\begin{bmatrix} \mathbf{1}_{[0,\infty)}(s)\mathrm{e}^{B_1 s}I_1 S \\ -\mathbf{1}_{(-\infty,0)}(s)\mathrm{e}^{B_2 s}I_2 S\end{bmatrix}$$
$$= S^{-1}\begin{bmatrix} \mathbf{1}_{[0,\infty)}(s)\mathrm{e}^{B_1 s} & 0 \\ 0 & -\mathbf{1}_{(0,\infty)}(s)\mathrm{e}^{B_2 s}\end{bmatrix} S \in \mathbb{C}^{d\times d}$$

for $s \in \mathbb{R}$. Then g is integrable with respect to L and $X_t = \int_{\mathbb{R}} g(t-s)\,\mathrm{d}L_s$ so that $(X_t)_{t\in\mathbb{R}}$ is strictly stationary by Theorem 16.17. To see that $(X_t)_{t\in\mathbb{R}}$ really satisfies (17.2), observe that X_t can be written as

$$S^{-1}\begin{bmatrix} \mathrm{e}^{B_1 t} & 0 \\ 0 & \mathrm{e}^{B_2 t}\end{bmatrix}\left(\begin{bmatrix} \int_{-\infty}^{0} \mathrm{e}^{-B_1 s}I_1 S\,\mathrm{d}L_s \\ -\int_0^{\infty} \mathrm{e}^{-B_2 s}I_2 S\,\mathrm{d}L_s\end{bmatrix} + \begin{bmatrix} \int_0^t \mathrm{e}^{-B_1 s}I_1 S\,\mathrm{d}L_s \\ -\int_t^0 \mathrm{e}^{-B_2 s}I_2 S\,\mathrm{d}L_s\end{bmatrix}\right)$$
$$= S^{-1}\mathrm{e}^{Bt}\left(SX_0 + \int_0^t \begin{bmatrix} \mathrm{e}^{-B_1 s} & 0 \\ 0 & \mathrm{e}^{-B_2 s}\end{bmatrix} S\,\mathrm{d}L_s\right)$$
$$= S^{-1}\mathrm{e}^{Bt}S\left(X_0 + \int_0^t S^{-1}\mathrm{e}^{-Bs}S\,\mathrm{d}L_s\right)$$
$$= \mathrm{e}^{At}\left(X_0 + \int_0^t \mathrm{e}^{-As}\,\mathrm{d}L_s\right).$$

This shows that (17.7) is satisfied, so that $(X_t)_{t\in\mathbb{R}}$ is indeed a solution of (17.2), and it admits a càdlàg modification by Theorem 17.2.

(ii) To show that (17.19) constitutes a càdlàg modification of (17.18), we mimic the proof of equation (17.6) and obtain

$$\left(\int_u^t \mathrm{e}^{-B_1 s}I_1 S\,\mathrm{d}L_s\right)^i = \sum_{j=1}^n \int_u^t (\mathrm{e}^{-B_1 s}I_1 S)_{ij}\,\mathrm{d}L_s^j$$
$$= \sum_{j=1}^n \left((\mathrm{e}^{-B_1 t}I_1 S)_{ij}L_t^j - (\mathrm{e}^{-B_1 u}I_1 S)_{ij}L_u^j - \int_u^t \left(\frac{\mathrm{d}}{\mathrm{d}s}\mathrm{e}^{-B_1 s}I_1 S\right)_{ij} L_s^j\,\mathrm{d}s\right)$$
$$= (\mathrm{e}^{-B_1 t}I_1 S L_t)^i - (\mathrm{e}^{-B_1 u}I_1 S L_u)^i + \sum_{j=1}^n \int_u^t (\mathrm{e}^{-B_1 s}B_1 I_1 S)_{ij}L_s^j\,\mathrm{d}s,$$

so that

$$\int_u^t e^{-B_1 s} I_1 S \, dL_s = e^{-B_1 t} I_1 S L_t - e^{-B_1 u} I_1 S L_u + \int_u^t e^{-B_1 s} B_1 I_1 S L_s \, ds$$

for all $u \le t$. As in part (ii) of the proof of Theorem 17.4(a), letting $u \to -\infty$ and multiplying from the left with $e^{B_1 t}$ results in

$$\int_{-\infty}^t e^{B_1(t-s)} I_1 S \, dL_s = I_1 S L_t + \int_{-\infty}^t e^{B_1(t-s)} B_1 I_1 S L_s \, ds,$$

with the integral on the right existing as a Lebesgue integral. Similarly, using limits in the opposite direction gives

$$-\int_t^\infty e^{B_2(t-s)} I_2 S \, dL_s = I_2 S L_t - \int_t^\infty e^{B_2(t-s)} B_2 I_2 S L_s \, ds,$$

and putting both equations together shows that (17.19) constitutes a modification of (17.18). That the solution specified by (17.19) is càdlàg is clear.

(iii) Now assume additionally that $\mathbb{E}|L_1|^2 < \infty$. In (i) we have seen that the solution X specified by (17.18) is a strictly stationary solution, which is a moving average process, and from Corollary 16.19 we derive that it is also weakly stationary.

(iv) It remains to show that X_0 and hence the strictly/weakly stationary solution are (almost surely) unique. For that let $X = (X_t)_{t \in \mathbb{R}}$ be any strictly/weakly stationary solution. Then

$$X_0 = e^{-At} X_t - \int_0^t e^{-As} \, dL_s = S^{-1} e^{-Bt} S X_t - \int_0^t S^{-1} e^{-Bs} S \, dL_s \quad \forall\, t \in \mathbb{R}$$

by (17.7). Multiplying this by $I_1 S$ from the left and observing that $I_1 e^{-Bs} = e^{-B_1 s} I_1$ results in

$$I_1 S X_0 = e^{-B_1 t} I_1 S X_t - \int_0^t e^{-B_1 s} I_1 S \, dL_s \quad \forall\, t \in \mathbb{R}.$$

If X is strictly stationary, then X_t has the same distribution for every t and, since $e^{-B_1 t} \to 0$ as $t \to -\infty$, we see that $e^{-B_1 t} I_1 S X_t$ converges in distribution and hence in probability to 0 as $t \to \infty$. Similarly, when X is weakly stationary, we observe that $e^{-B_1 t} I_1 S X_t$ converges in $\mathcal{S}$ and hence in probability to 0 as $t \to -\infty$. Consequently, $I_1 S X_0$ is the probability limit of $-\int_0^t e^{-B_1 s} I_1 S \, dL_s$ as $t \to -\infty$ and in particular it is uniquely determined by A and L. Similarly, $I_2 S X_0$ is the probability limit of $-\int_0^t e^{-B_2 s} I_2 S \, dL_s$ as $t \to +\infty$ and hence uniquely determined by A and L. This shows that SX_0 is uniquely determined by A and L and, since S is invertible, so is X_0. Since the Ornstein–Uhlenbeck process is determined by X_0, A and L, uniqueness of the strictly/weakly stationary solution follows. □

The stationary solution of Theorem 17.8 is a continuous-time moving average process (CMA) as the following corollary shows.

Corollary 17.9 (CMA representation). *Under the conditions and with the notation of Theorem 17.8, let X be the unique strictly/weakly stationary Ornstein–Uhlenbeck process with parameter A, driven by the Lévy process L. Define the kernel function g by*

$$g(t) := S^{-1}\begin{bmatrix} \mathrm{e}^{B_1 t} I_1 S \mathbf{1}_{[0,\infty)}(t) \\ -\mathrm{e}^{B_2 t} I_2 S \mathbf{1}_{(-\infty,0)}(t) \end{bmatrix}, \qquad t \in \mathbb{R}.$$

Then

$$X_t = \int_{\mathbb{R}} g(t-s)\,\mathrm{d}L_s,$$

so that X is a continuous-time moving average process with kernel function g. The function g can also be represented as

$$g(t) = S^{-1}\begin{bmatrix} \mathbf{1}_{[0,\infty)}(t)\, I_1 \\ -\mathbf{1}_{(0,\infty)}(t)\, I_2 \end{bmatrix} \mathrm{e}^{Bt} S = S^{-1}\begin{bmatrix} \mathbf{1}_{[0,\infty)}(t)\, I_1 \\ -\mathbf{1}_{(0,\infty)}(t)\, I_2 \end{bmatrix} S\mathrm{e}^{At}.$$

Proof. The fact that $X_t = \int_{\mathbb{R}} g(t-s)\,\mathrm{d}L_s$ follows directly from (17.18). The second representation of g follows from the fact that $I_1 \mathrm{e}^{Bt} = \mathrm{e}^{B_1 t} I_1$ and $I_2 \mathrm{e}^{Bt} = \mathrm{e}^{B_2 t} I_2$, and the third representation follows from the second, since $\mathrm{e}^{At} = S^{-1}\mathrm{e}^{Bt}S$. □

It is frequently desirable to have solutions X with the property that for each t, X_t depends only on the increments of L up to time t, and not on future increments. Such solutions are said to be *causal*, a concept made precise in the following definition.

Definition 17.10 (Causal OU-processes). Suppose that $X = (X_t)_{t\in\mathbb{R}}$ is a strictly or weakly stationary $\mathbb{C}^d$-valued Ornstein–Uhlenbeck process, driven by the $\mathbb{R}^d$-valued Lévy process L and with parameter $A \in \mathbb{C}^{d\times d}$. We say that X is *causal*, if for every $t \in \mathbb{R}$, X_t is measurable with respect to $\sigma(L_s - L_u : s, u \le t)$ (more precisely, if each X_t is almost surely equal to a $\sigma(L_s - L_u : s, u \le t)$-measurable random vector).

Observe that causality is a not a property of X itself, but of the pair (X, L). More precisely it is a property of the functional relationship between X and the increments of L. From Theorem 17.8 we obtain the following corollary.

Corollary 17.11 (Sufficient condition for causality of OU-process). *Let X be a strictly/weakly stationary $\mathbb{C}^d$-valued Ornstein–Uhlenbeck process driven by the $\mathbb{R}^d$-valued Lévy* *process L with parameter $A \in \mathbb{C}^{d\times d}$, where all eigenvalues of A have strictly negative real part and $\mathbb{E}\log^+ |L_1| < \infty / \mathbb{E}|L_1|^2 < \infty$. Then, X is causal.*

Proof. By Theorem 17.8 (or directly by Theorem 17.4), we have $X_t = \int_{-\infty}^t e^{A(t-s)}\, dL_s$. Hence X_t, as the probability limit of $e^{At} \int_v^t e^{-As}\, dL_s$ as $v \to -\infty$ (cf., Proposition 16.21), is $\sigma(L_s - L_u : s, u \le t)$-measurable by Theorem 16.3(a), showing causality. □

In the one-dimensional case, i. e., when $d = 1$, there is a converse result, stated below as a further corollary.

Corollary 17.12 (Necessary condition for causality of OU-process). *Let $X = (X_t)_{t\in\mathbb{R}}$ be a strictly/weakly stationary causal one-dimensional Ornstein–Uhlenbeck process with parameter $A \in \mathbb{C}$ such that $\Re(A) \neq 0$, driven by the $\mathbb{R}$-valued Lévy process L with $\mathbb{E}\log^+ |L_1| < \infty / \mathbb{E}|L_1|^2 < \infty$, respectively. Then, $\Re(A) < 0$ or L is deterministic.*

Proof. We shall assume in this proof that L is non-deterministic. and show that if $\Re(A) > 0$ we are led to a contradiction. (For deterministic L see Remark 17.13 below). Thus if $\Re(A) > 0$ then $X_t = -e^{At} \int_t^\infty e^{-As}\, dL_s$ by (17.14), so that X_t, as the probability limit of $-e^{At} \int_t^v e^{-As}\, dL_s$ as $v \to \infty$ (cf. Proposition 16.21), is $\sigma(L_s - L_u : s, u \ge t)$-measurable by Theorem 16.3(a). Since L is a Lévy process, $\sigma(L_s - L_u : s, u \ge t)$ and $\sigma(L_s - L_u : s, u \le t)$ are independent, hence X_t is independent of $\sigma(L_s - L_u : s, u \le t)$. On the other hand, by causality, X_t is $\sigma(L_s - L_u : s, u \le t)$-measurable, hence independent of itself, hence constant almost surely. In particular, $-X_0 = \int_0^\infty e^{-As}\, dL_s$ is constant almost surely. Denote the characteristic triplet of L by (σ^2, ν, γ). Since L is non-deterministic, we can find $\varepsilon > 0$ such that $\nu([-\varepsilon, \varepsilon]) + \sigma^2 > 0$. Let M and N be two independent Lévy processes such that M has characteristic triplet $(\sigma^2, \nu_{|[-\varepsilon,\varepsilon]}, \gamma)$ and N has characteristic triplet $(0, \nu_{|\mathbb{R}\setminus[-\varepsilon,\varepsilon]}, 0)$. Then $L \stackrel{d}{=} M + N$, M is non-deterministic and has finite variance, and $\int_0^\infty e^{-As}\, dL_s$ has the same distribution as $\int_0^\infty e^{-As}\, dM_s + \int_0^\infty e^{-As}\, dN_s$ (by (16.16), since $\Psi_L = \Psi_M + \Psi_N$). Since the two summands are independent (being measurable with respect to $\sigma(M_s - M_u : u, s \ge 0)$ and $\sigma(N_s - N_u : s, u \ge 0)$, respectively), they must also be constant. In particular, $\int_0^\infty e^{-As}\, dM_s$ is constant almost surely and hence has variance 0. On the other hand, by Example 15.12, the variance of $\int_0^\infty e^{-As}\, dM_s$ is given by $\int_0^\infty |e^{-As}|^2\, ds\, \mathrm{Var}(M_1)$. Since M_1 is non-deterministic, this is impossible, contradicting $\Re(A) > 0$. □

Remark 17.13. (a) Under the conditions of Corollary 17.12, it can happen that L is deterministic and $\Re(A) > 0$. Consider, for example, the case when $A = 1$ and $L_t = t$. Then $X_t = -e^t \int_t^\infty e^{-s}\, ds = -1$ defines a strictly and weakly stationary Ornstein–Uhlenbeck process by (17.14). Since X_t is constant, it is measurable with respect to $\sigma(L_s - L_u : s \le t) = \{\emptyset, \Omega\}$ for all $t \in \mathbb{R}$, and is therefore causal.

(b) Let X be a $\mathbb{C}^d$-valued strictly or weakly stationary Ornstein–Uhlenbeck process, driven by the $\mathbb{R}^d$-valued Lévy process L, with parameter $A \in \mathbb{C}^{d\times d}$, where all eigenvalues of A have non-zero real part and $\mathbb{E}\log^+ |L_1| < \infty$ (or $\mathbb{E}|L_1|^2 < \infty$). By Theorem 17.8, the unique stationary solution is given by (17.18). From this it is easily seen that X will be causal if and only if either all eigenvalues of A have strictly negative real part, or if (in the notation of Theorem 17.8), $d_2 \geq 1$ and $\int_0^\infty e^{-B_2 s} I_2 S \, dL_s$ is constant almost surely. Using methods similar to those that were used in the proof of Corollary 17.12, it can be shown that the latter is the case if and only if $I_2(\Re S)L$ and $I_2(\Im S)L$ are deterministic Lévy processes, i.e, if $I_2 S L_t$ is a deterministic process. As a concrete example, consider $A = \begin{pmatrix} -1 & 0 \\ 0 & 1 \end{pmatrix} \in \mathbb{R}^{2\times 2}$ and $L_t = (L_t^1, ct)^T \in \mathbb{R}^2$, where L^1 is a one-dimensional Lévy process with finite second moment and $c \in \mathbb{R}$. Then in Theorem 17.8(c) we can choose $d_1 = d_2 = 1$, $B = A$ and S the identity matrix to obtain $X_t = \begin{bmatrix} \int_{-\infty}^t e^{s-t} \, dL_s^1 \\ -c \end{bmatrix}$, which is strictly and weakly stationary and causal. In a similar fashion, more complicated examples can be constructed.

17.3 Markov-stable Ornstein–Uhlenbeck processes

A concept closely related to causality is the concept of stability in the sense of Markov processes, which we now explain in more detail. A *one-sided* (multivariate) Ornstein–Uhlenbeck is obtained by restricting a (two-sided) Ornstein–Uhlenbeck process $(X_t)_{t\in\mathbb{R}}$ to $t \in [0, \infty)$. By (17.8), a one-sided multivariate Ornstein–Uhlenbeck process $X = (X_t)_{t\geq 0}$ (with parameter A, driven by L), defined on a probability space $(\Omega, \mathcal{F}, P)$ satisfies

$$X_t = e^{A(t-s)}\left(X_s + \int_s^t e^{-A(u-s)} \, dL_u \right), \quad t \geq s \geq 0.$$

For $h \geq 0$, $y \in \mathbb{C}^d$ and $C \in \mathcal{B}(\mathbb{C}^d)$, define

$$p_h(y, C) := P\left(e^{Ah} y + e^{Ah} \int_0^h e^{-Au} \, dL_u \in C \right).$$

Then p_h is a stochastic kernel, and if X_0 is independent of $(L_s)_{s\geq 0}$, then for all $t \geq 0$, $(X_s)_{s\in[0,t]}$ is independent of $(L_s - L_t)_{s\geq t}$, and

$$P(X_t \in C | \sigma(X_s : s \leq t)) = P(X_t \in C | X_s) = p_{t-s}(X_s, C)$$

for all $C \in \mathcal{B}(\mathbb{C}^d)$ and $s \leq t$. This means that X has the *(elementary) Markov property*, and that X is a *homogeneous Markov process with transition probabilities* p_h (it is easy to show that the transition probabilities satisfy the Chapman–Kolmogorov equations). The independence of X_0 and $(L_s)_{s\geq 0}$ is satisfied in particular when X_0 is almost surely constant, i. e., when $X_0 = x$ a. s. for $x \in \mathbb{C}^d$, and sometimes one speaks of the family $(X_t^x)_{t\geq 0}$ of one-sided multivariate Ornstein–Uhlenbeck processes with $X_0^x = x$ a. s., indexed by $x \in \mathbb{C}^d$, as a Markov process or universal Markov process. We now define Markov-stability and its connection with stationarity in the context of a one-sided multivariate Ornstein–Uhlenbeck process.

Definition 17.14 (Markov-stable OU equation, limit distribution). Suppose that $A \in \mathbb{C}^{d\times d}$ and $L = (L_t)_{t\geq 0}$ is a one-sided $\mathbb{R}^d$-valued Lévy process. Then we say that the one-sided multivariate Ornstein–Uhlenbeck equation, i. e., (17.3) with t restricted to $[0,\infty)$, is *Markov-stable*, if there is a finite $\mathbb{C}^d$-valued random vector Z, such that, for every $x \in \mathbb{C}^d$, the solution $(X_t^x)_{t\geq 0}$ with initial value $X_0^x := x$ converges in distribution to Z as $t \to \infty$. The distribution of Z is obviously unique and is called the *limit distribution* of the Ornstein–Uhlenbeck equation (or of the universal Markov process $(X^x)_{x\in\mathbb{C}^d}$).

We use the expression *Markov-stable* to avoid confusion with stable Lévy processes which, as defined earlier, are Lévy processes whose marginal distribution at time 1 belongs to the stable family.

As we shall now show, if π is the limit distribution of a Markov-stable multivariate Ornstein–Uhlenbeck equation and if X_0 has the distribution π and is independent of $(L_t)_{t\geq 0}$, then the solution $(X_t)_{t\geq 0}$ is a strictly stationary Ornstein–Uhlenbeck process with marginal distributions π (here, strict stationarity of a multivariate one-sided process is defined in the obvious way by requiring shift-invariance of the finite-dimensional distributions under shifts of size $h \geq 0$ only). This is a particular case of a more general result which applies to arbitrary Markov processes (see Meyn and Tweedie [70, Section 10.1.1]).

Proposition 17.15 (Markov-stability and stationarity). *Suppose that the one-sided Ornstein–Uhlenbeck equation, i. e., (17.3) with t restricted to $[0,\infty)$, is Markov-stable with limit distribution π. If $X = (X_t)_{t\geq 0}$ is any solution with the property that X_0 has distribution μ and is independent of L (this can always be arranged by enlarging the probability space if necessary to support such a random vector) we shall denote X by X^μ. The following statements are then true.*

(a) *X_t^μ converges in distribution to π as $t \to \infty$ for every starting distribution μ (which is independent of L).*
(b) *X^π is strictly stationary.*

Proof. (a) Suppose that $C \in \mathcal{B}(\mathbb{C}^d)$ and $\pi(\partial C) = 0$, where ∂C denotes the topological boundary of C. Conditioning on $X_0 = x$, we obtain

$$\begin{aligned} P(X_t^\mu \in C) &= \int_{\mathbb{C}^d} P(X_t^\mu \in C|X_0 = x)\,\mu(\mathrm{d}x) = \int_{\mathbb{C}^d} P(X_t^x \in C)\,\mu(\mathrm{d}x) \\ &\to \int_{\mathbb{C}^d} \pi(C)\mu(\mathrm{d}x) = \pi(C) \quad \text{as} \quad t \to \infty \end{aligned}$$

by dominated convergence, showing that X_t^μ converges in distribution to π.

(b) Define $W_{t,s} := \int_t^s e^{A(s-u)}\,\mathrm{d}L_u$ for $s \geq t$. Suppose that $h \geq 0$ and $s_1,\dots,s_n \geq 0$. Observe that X_t^π and $(W_{t,t+h}, W_{t,t+h+s_1},\dots,W_{t,t+h+s_n})$ are independent for each $t \geq 0$. Since $X_t^\pi \xrightarrow{d} X_0^\pi$ as $t \to \infty$ by (a), and since $(W_{t,t+h}, W_{t,t+h+s_1},\dots,W_{t,t+h+s_n})$ has the same distribution as $(W_{0,h}, W_{0,h+s_1},\dots,W_{0,h+s_n})$ by Theorem 16.14, we see that

$$(X_t^\pi, W_{t,t+h}, W_{t,t+h+s_1}, \dots, W_{t,t+h+s_n}) \xrightarrow{d} (X_0^\pi, W_{0,h}, W_{0,h+s_1}, \dots, W_{0,h+s_n})$$

as $t \to \infty$ and hence (with "dlim" denoting limits in distribution)

$$\begin{aligned}
\begin{bmatrix} X_h^\pi \\ X_{h+s_1}^\pi \\ \vdots \\ X_{h+s_n}^\pi \end{bmatrix} &= \begin{bmatrix} e^{Ah} \\ e^{A(h+s_1)} \\ \vdots \\ e^{A(h+s_n)} \end{bmatrix} X_0^\pi + \begin{bmatrix} W_{0,h} \\ W_{0,h+s_1} \\ \vdots \\ W_{0,h+s_n} \end{bmatrix} \\
&= \operatorname*{dlim}_{t\to\infty} \left(\begin{bmatrix} e^{Ah} \\ e^{A(h+s_1)} \\ \vdots \\ e^{A(h+s_n)} \end{bmatrix} X_t^\pi + \begin{bmatrix} W_{t,t+h} \\ W_{t,t+h+s_1} \\ \vdots \\ W_{t,t+h+s_n} \end{bmatrix} \right) \\
&= \operatorname*{dlim}_{t\to\infty} \begin{bmatrix} X_{t+h}^\pi \\ X_{t+h+s_1}^\pi \\ \vdots \\ X_{t+h+s_n}^\pi \end{bmatrix}.
\end{aligned}$$

Substituting $v = t + h$, we see that the latter expression does not depend on h, showing that X^π is strictly stationary. □

We next define and characterize Markov-stability for two-sided multivariate Ornstein–Uhlenbeck equations.

Definition 17.16 (Two-sided Markov-stable OU equ., limit distribution). Suppose that $A \in \mathbb{C}^{d\times d}$ and $L = (L_t)_{t\in\mathbb{R}}$ is a two-sided $\mathbb{R}^d$-valued Lévy process. Then we say that the two-sided multivariate Ornstein–Uhlenbeck equation (17.3) is *Markov-stable*, if there is a finite $\mathbb{C}^d$-valued random vector Z, such that, for every $x \in \mathbb{C}^d$, the solution $(X_t^x)_{t\in\mathbb{R}}$ with $X_0^x := x$ converges in distribution to Z as $t \to \infty$. The distribution of Z is obviously unique and is called the *limit distribution* of the Ornstein–Uhlenbeck equation (17.3).

The following result relates two-sided Markov-stability with our conditions for strictly stationary causal Ornstein–Uhlenbeck processes found in Corollaries 17.11 and 17.12.

Theorem 17.17 (Two-sided Markov-stability and causality). *Suppose that $L = (L_t)_{t\in\mathbb{R}}$ is an $\mathbb{R}^d$-valued Lévy process, that $\mathbb{E}\log^+|L_1| < \infty$ and that $A \in \mathbb{C}^{d\times d}$. Then the multivariate Ornstein–Uhlenbeck equation* (17.3) *is Markov-stable if and only if all the eigenvalues of A have strictly negative real part. In that case, the limit distribution of the solution $(X_t)_{t\in\mathbb{R}}$ as $t \to \infty$ is the same as the marginal distribution of the strictly stationary Ornstein–Uhlenbeck process, i. e., $\mathcal{L}(\int_{-\infty}^0 e^{-As}\, dL_s)$.*

Proof. Suppose that all the eigenvalues of A have strictly negative real part. Then, in the notation of Proposition 17.3, $\int_0^t e^{A(t-u)}\,dL_u = Z_1^{(t)} \stackrel{d}{=} Z_0^{(t)} = \int_{-t}^0 e^{-Au}\,dL_u$. By Proposition 16.21, $\int_{-t}^0 e^{-Au}\,dL_u$ converges in probability to $\int_{-\infty}^0 e^{-Au}\,dL_u$ as $t \to \infty$. Since also $e^{At}x \to 0$ as $t \to \infty$ for all $x \in \mathbb{C}^d$, we obtain

$$\operatorname*{dlim}_{t\to\infty} X_t^x = \operatorname*{dlim}_{t\to\infty}\left(e^{At}x + \int_0^t e^{A(t-s)}\,dL_s \right)$$

$$= \operatorname*{dlim}_{t\to\infty} \int_{-t}^0 e^{-As}\,dL_s = \mathcal{L}\left(\int_{-\infty}^0 e^{-As}\,dL_s \right),$$

showing that $(X_t)_{t\in\mathbb{R}}$ is Markov-stable with limit distribution given by $\mathcal{L}(\int_{-\infty}^0 e^{-As}\,dL_s)$.

Conversely, suppose that the model (17.3) is Markov-stable, and again write $Z_1^{(t)} = \int_0^t e^{A(t-s)}\,dL_s$ and $Z_0^{(t)} = \int_{-t}^0 e^{-Au}\,dL_u$. Then there is a finite random vector Z such that $e^{At}x + Z_1^{(t)} \stackrel{d}{\to} Z$ as $t \to \infty$ for all $x \in \mathbb{C}^d$, and in particular for $x = 0$. Hence $Z_1^{(t)} \stackrel{d}{\to} Z$ as $t \to \infty$, implying that $(Z_1^{(n)})_{n\in\mathbb{N}}$ is tight. Now if $(e^{An}x)_{n\in\mathbb{N}}$ were unbounded for some $x \in \mathbb{C}^d$, then $(e^{An}x + Z_1^{(n)})_{n\in\mathbb{N}}$ could not be tight and so could not converge in distribution, a contradiction. Hence $(e^{An}x)_{n\in\mathbb{N}}$ is bounded for each $x \in \mathbb{C}^d$. Let v be an eigenvector of A corresponding to the eigenvalue $\lambda \in \mathbb{C}$. Then the equality $e^{An}v = e^{\lambda n}v$ and boundedness imply that $\Re(\lambda) \le 0$. Now if $\Re(\lambda) = 0$, we can choose a subsequence such that $e^{An_k}v = e^{\lambda n_k}v$ converges to some vector $w \in \mathbb{C}^d$ with $|w| = |v| \neq 0$ as $k \to \infty$. Since $e^{An_k}v + Z_1^{(n_k)}$ also converges in distribution to Z by assumption, this implies that $w = 0$, a contradiction. Hence all the eigenvalues of A have strictly negative real part. □

It is possible to show that the condition $\mathbb{E}\log^+|L_1| < \infty$ is also *necessary* for the multivariate Ornstein–Uhlenbeck equation to be Markov-stable, but we content ourselves with the slightly weaker statement of Theorem 17.17.

17.4 Exercises

Exercise 17.1. Let $a > 0$ and let $N = (N_t)_{t\in\mathbb{R}}$ be a Poisson process with jump rate $\lambda > 0$. Consider the stationary Poisson driven Ornstein–Uhlenbeck process $Y = (Y_t)_{t\in\mathbb{R}}$, given by $Y_t = \int_{-\infty}^t e^{-a(t-u)}\,dN_u$ of Example 1.24. Using Theorem 16.3, show that the characteristic exponent Ψ_Y of Y_t satisfies

$$\Psi_Y(z) = \frac{\lambda}{a}\sum_{j=1}^{\infty} \frac{1}{j}\frac{(iz)^j}{j!} \qquad \forall\, z \in \mathbb{R}.$$

Exercise 17.2. Prove the assertions made in Remark 17.13(b).

Exercise 17.3. Let $\alpha \in (0,2]$. Show that a probability distribution μ on $\mathbb{R}$ is α-stable if and only if there exists an α-stable $\mathbb{R}$-valued Lévy process $L = (L_t)_{t\in\mathbb{R}}$ such that μ is the marginal distribution of the strictly stationary Ornstein–Uhlenbeck process satisfying the OU-equation $\mathrm{d}X_t = -X_t\,\mathrm{d}t + \mathrm{d}L_t$.

Exercise 17.4. Let $L = (L_t)_{t\in\mathbb{R}}$ be an $\mathbb{R}^d$-valued Lévy process with finite log-moment. Let μ be the marginal distribution of the strictly stationary solution of the Ornstein–Uhlenbeck equation $\mathrm{d}X_t = -X_t\,\mathrm{d}t + \mathrm{d}L_t$. Show that for each $c \in (0,1)$ there exist independent random variables Z and Z_c such that

$$\mu = \mathcal{L}(Z) = \mathcal{L}(cZ + Z_c). \tag{17.20}$$

(Probability distributions on $\mathbb{R}^d$ satisfying (17.20) for all $c \in (0,1)$ are called *self-decomposable* distributions. It can be shown that a distribution is self-decomposable if and only if it is the marginal distribution of the strictly stationary solution of the Ornstein–Uhlenbeck equation $\mathrm{d}X_t = -X_t\,\mathrm{d}t + \mathrm{d}L_t$ for some Lévy process L with finite log-moment, see, e. g., Sato [79, Theorem 17.5].)

Exercise 17.5. Let $L = (L_t)_{t\in\mathbb{R}}$ and $M = (M_t)_{t\in\mathbb{R}}$ be two $\mathbb{R}^d$-valued non-deterministic Lévy processes and let $a > 0$. Suppose that the strictly stationary solutions of the two Ornstein–Uhlenbeck equations $\mathrm{d}X_t = -aX_t\,\mathrm{d}t + \mathrm{d}L_t$ and $\mathrm{d}Y_t = -aY_t\,\mathrm{d}t + \mathrm{d}M_t$ have the same marginal distribution μ. Show that L and M have the same distribution.
Hint: Express the characteristic exponent of μ in terms of the characteristic exponent Ψ_L of L, and similarly for M. Use this to express, for each $h > 0$, the distinguished logarithm of the characteristic function of the i. i. d. sequence $(Z_n^{(h)})_{n\in\mathbb{Z}}$ appearing in Proposition 17.3 in terms of Ψ_L, and similarly for M. Then take the derivative of this expression with respect to h at $h = 0$.

18 Lévy-driven CARMA processes: definition, existence, uniqueness and properties

Abstract: Just as ARMA processes play a central role in the representation of a stationary time series, $(X_n)_{n\in\mathbb{Z}}$, with discrete time parameter, CARMA processes play an analogous role in the representation of a stationary time series, $(Y_t)_{t\in\mathbb{R}}$, with continuous-time parameter. Lévy-driven CARMA processes are the continuous-time analogues of discrete-time ARMA processes driven by independent, identically distributed noise sequences. They include heavy-tailed and asymmetric time series and can be used to model processes with a wide variety of distributional and sample path properties. If the driving Lévy process $(L_t)_{t\in\mathbb{R}}$ has finite second moments, i. e., if $\mathbb{E}(L_1^2) < \infty$, then the CARMA process has rational spectral density and the results of Chapter 7 apply. However, since $\mathbb{E}(L_1^2)$ is not required to be finite, the class of Lévy-driven CARMA processes is much larger than that of those with $\mathbb{E}(L_1^2) < \infty$.

18.1 Introduction

Continuous-time autoregressive (CAR) processes have been of interest to physicists and engineers for many years (see, e. g., Fowler [46]). Early papers dealing with the properties and statistical analysis of such processes, and of the more general continuous-time autoregressive moving average (CARMA) processes, include those of Doob [36], Bartlett [8], Phillips [71] and Durbin [38]. In recent years, there has been a resurgence of interest in these processes and in continuous-time processes more generally, partly as a result of the very successful application of stochastic differential equation models to problems in finance, particularly to the pricing of options. The proliferation of high-frequency data, especially in fields such as finance and turbulence, has stimulated interest also in the connections between CARMA processes and the discrete-time processes obtained by sampling them at high frequencies, and the possible use of continuous-time models to suggest inferential methods for such data. CARMA models have also been utilized very successfully for the modelling of irregularly-spaced data (Jones [53, 54]). Recent applications include the CARMA interest rate model (Andresen et al. [1]), the application of stable CARMA processes to futures pricing in electricity markets (Benth et al. [11]) and applications to signal extraction (McElroy [68]). An asset price model with CAR(1) spot volatility was introduced in Barndorf–Nielsen and Shephard [7] and CARMA spot volatility models have been studied in Brockwell and Lindner [28] and Todorov and Tauchen [85]. The potential for further applications, with the proliferation of high-frequency data in so many fields, and further theoretical developments, particularly with respect to multivariate models (Marquardt and Stelzer [66]), nonlinear models, non-causal modelling, sampling and embedding remains broad and challenging.

https://doi.org/10.1515/9783111325033-018

In the next section, we shall define a Lévy-driven CARMA process (Definition 18.2) as a strictly stationary solution Y of the CARMA equations (18.4) and (18.5). Throughout this chapter, we make the following two assumptions:

Assumption A. The driving Lévy process L satisfies $\mathbb{E}\log^+|L_1| < \infty$.

Assumption B. The real parts of the zeros of $a(z)$ in (18.3) are all non-zero.

Under these assumptions, we already know from Theorem 17.8 (together with Remark 18.1 below) that there is a unique random vector $\mathbf{X}_0$ such that the corresponding unique solution $(\mathbf{X}_t)_{t\in\mathbb{R}}$ of the state equation (18.5) is strictly stationary. In Theorem 18.9, we derive explicit expressions for this solution $(\mathbf{X}_t)_{t\in\mathbb{R}}$ and its corresponding CARMA process Y defined by $Y = \mathbf{b}^T\mathbf{X}$. In Remark 18.11 we derive an equivalent expression for this solution Y. Using properties of the Δ-sampled sequence Y^Δ, defined for $\Delta > 0$, by $Y_n^\Delta = Y_{n\Delta}$, $n \in \mathbb{Z}$, we show, in Section 18.3, that Y is in fact the *unique* strictly stationary solution of the CARMA equations. This in turn leads to the alternative representation of $\mathbf{X}$ in Corollary 18.16.

Sections 18.4 and 18.7 discuss the properties of the CARMA process Y and the Δ-sampled sequence Y^Δ when $\mathbb{E}(|L_1|^2) < \infty$. In Section 18.5 we discuss causal CARMA processes and in Section 18.6 we introduce the canonical state vector and show that under mild conditions, a CARMA process can be expressed as a linear combination of dependent Ornstein–Uhlenbeck processes.

18.2 Definition and existence

A strictly stationary discrete-time autoregressive moving average (ARMA) process $(X_n)_{n\in\mathbb{Z}}$ is defined (see, e. g., Brockwell and Davis [22, Sections 3.1 and 13.3]) as a strictly stationary solution of difference equations of the form

$$\phi(B)X_n = \theta(B)W_n, \quad n \in \mathbb{Z}, \tag{18.1}$$

where $(W_n)_{n\in\mathbb{Z}}$ is an i. i. d. sequence of random variables and $\phi(B)$ and $\theta(B)$ are polynomials in the backward shift operator B, defined by $BX_n = X_{n-1}$. The polynomial ϕ is called the *autoregressive polynomial* and θ is the *moving average polynomial*. Equation (18.1) is a short-hand notation for (6.15). Necessary and sufficient conditions for the existence and uniqueness of a strictly stationary solution of (18.1) are given in Brockwell and Lindner [27].

To introduce a continuous-time analog of such a discrete-time sequence, we replace the i. i. d. sequence $(Z_n)_{n\in\mathbb{Z}}$ by the increments of a non-deterministic two-sided $\mathbb{R}$-valued Lévy process $L = (L_t)_{t\in\mathbb{R}}$. The difference equation (18.1) is then replaced by the *formal* differential equation,

$$a(D)Y_t = b(D)DL_t, \quad t \in \mathbb{R}, \tag{18.2}$$

where D denotes differentiation with respect to t, and $a(z)$ and $b(z)$ are polynomials of the form,

$$\begin{aligned} a(z) &= z^p + a_1 z^{p-1} + \cdots + a_p, \\ b(z) &= b_0 + b_1 z + \cdots + b_q z^q, \end{aligned} \tag{18.3}$$

with complex coefficients, $b_q = 1$ and $q < p$ (the same polynomials that appeared already in Definition 7.1). We shall refer to $a(z)$ as the *autoregressive polynomial* and to $b(z)$ as the *moving average polynomial.* Since the derivatives on the right-hand side of (18.2) do not exist in the usual sense, we give the equation a meaningful interpretation by rewriting it in the *state-space form* (or *state-space representation*),

$$Y_t = \mathbf{b}^T \mathbf{X}_t, \quad t \in \mathbb{R}, \tag{18.4}$$

where $\mathbf{X} = (\mathbf{X}_t)_{t\in\mathbb{R}}$ is a process with values in $\mathbb{C}^p$, satisfying the stochastic differential equation,

$$\mathrm{d}\mathbf{X}_t = A\mathbf{X}_t \,\mathrm{d}t + \mathbf{e}_p \,\mathrm{d}L_t, \quad t \in \mathbb{R}, \tag{18.5}$$

or equivalently,

$$\mathbf{X}_t - \mathbf{X}_s = \int_s^t A\mathbf{X}_u \,\mathrm{d}u + \mathbf{e}_p(L_t - L_s), \quad \forall s \le t \in \mathbb{R}, \tag{18.6}$$

with $A \in \mathbb{C}^{p\times p}$ and $\mathbf{e}_p, \mathbf{b} \in \mathbb{C}^p$ being defined as in equation (7.32), i. e., by

$$A = \begin{bmatrix} 0 & 1 & 0 & \cdots & 0 \\ 0 & 0 & 1 & \cdots & 0 \\ \vdots & \vdots & \vdots & \ddots & \vdots \\ 0 & 0 & 0 & \cdots & 1 \\ -a_p & -a_{p-1} & -a_{p-2} & \cdots & -a_1 \end{bmatrix}, \quad \mathbf{e}_p = \begin{bmatrix} 0 \\ 0 \\ \vdots \\ 0 \\ 1 \end{bmatrix}, \quad \mathbf{b} = \begin{bmatrix} b_0 \\ b_1 \\ \vdots \\ b_{p-2} \\ b_{p-1} \end{bmatrix}, \tag{18.7}$$

where $a_1, \dots, a_p, b_0, \dots, b_{p-1}$ are complex-valued coefficients satisfying $b_q = 1$ and $b_j = 0$ for $j > q$. For $p = 1$, the matrix A is to be understood as $A = (-a_1)$.

Remark 18.1. Since $\det(zI_p - A) = a(z)$ by Proposition 7.15(a), the eigenvalues of the matrix A are the same as the zeros of the autoregressive polynomial $a(z)$. We shall denote the distinct zeros by $\lambda_1, \dots, \lambda_r$ and their multiplicities by $m_1, \dots, m_r$, respectively. Thus $\sum_{i=1}^r m_i = p$.

The equations (18.4) and (18.5) will be referred to as the *observation* and *state equation*, respectively, and collectively as the *CARMA equations.*

Definition 18.2 (Lévy-driven CARMA process). Let $p \in \mathbb{N}$ and $q \in \mathbb{N}_0$ such that $q < p$. *A CARMA(p, q) process driven by the $\mathbb{R}$-valued Lévy process $(L_t)_{t\in\mathbb{R}}$ with autoregressive polynomial $a(z)$ and moving average polynomial $b(z)$* is a strictly stationary càdlàg solution $Y = (Y_t)_{t\in\mathbb{R}}$ of the CARMA equations (18.4) and (18.5). When we omit the orders p and q we also simply speak of a (Lévy-driven) *CARMA process*. When $q = 0$ we also speak of a CAR(p)*-process*.

Sometimes we will specify CARMA processes as strictly stationary solutions of the formal differential equation (18.2), by which we mean a strictly stationary solution of the corresponding CARMA equations (18.4) and (18.5). When the polynomials $a(z)$ and $b(z)$ factorize as $a(z) = \prod_{i=1}^{r}(z - \lambda_i)^{m_i}$ and $b(z) = \prod_{i=1}^{s}(z - \mu_i)^{n_i}$ with distinct zeros $\lambda_1, \ldots, \lambda_r$ and multiplicities m_i (so that $\sum_{i=1}^{r} m_i = p$) and similarly for $b(z)$, we also write

$$\prod_{i=1}^{r}(D - \lambda_i)^{m_i} Y_t = \left(\prod_{i=1}^{s}(D - \mu_i)^{n_i}\right) DL_t, \quad t \in \mathbb{R},$$

for (18.2). Also observe that CAR(1)-processes are nothing else than strictly stationary Ornstein-Uhlenbeck processes satisfying the differential equation $\mathrm{d}Y_t = -a_1 Y_t\,\mathrm{d}t + \mathrm{d}L_t$, since then $q = 0$ and $b_0 = 1$.

From Theorem 17.2 we immediately obtain the following result, which we restate here as a theorem.

Theorem 18.3 (Solution of state equation for given $\mathbf{X}_0$). *For any given random vector $\mathbf{X}_0 \in \mathbb{C}^p$, the state equation* (18.5) *has an (almost surely) unique solution, given by*

$$\mathbf{X}_t = \mathrm{e}^{At}\mathbf{X}_0 + \int_0^t \mathrm{e}^{A(t-u)}\mathbf{e}_p\,\mathrm{d}L_u, \quad t \in \mathbb{R}, \tag{18.8}$$

where $\int_0^t$ is interpreted as $-\int_t^0$ if $t < 0$. Moreover, the solution has a càdlàg-modification given by

$$\mathbf{X}_t = \mathrm{e}^{At}\left(\mathbf{X}_0 + \int_0^t \mathrm{e}^{-As}A\mathbf{e}_p L_s\,\mathrm{d}s\right) + L_t\,\mathbf{e}_p.$$

Remark 18.4. From (18.8) it also follows that

$$\mathbf{X}_t = \mathrm{e}^{A(t-s)}\mathbf{X}_s + \int_s^t \mathrm{e}^{A(t-u)}\mathbf{e}_p\,\mathrm{d}L_u, \quad \forall s, t \in \mathbb{R}, \tag{18.9}$$

as seen in equation (17.8).

Remark 18.5. If $(\mathbf{X}_t)_{t\in\mathbb{R}}$ is any $\mathbb{C}^p$-valued càdlàg process satisfying the state equation (18.5) and we denote the jth component of $\mathbf{X}_t$ by $X_t^{j-1}, j = 1, \ldots, p$, then the component equations of (18.6) can be written as

$$X_t^j - X_s^j = \int_s^t X_u^{j+1}\,\mathrm{d}u, \quad j = 0, \ldots, p-2,$$

and

$$X_t^{p-1} - X_s^{p-1} = -\int_s^t (a_1 X_u^{p-1} + a_2 X_u^{p-2} + \cdots + a_p X_u^0)\,\mathrm{d}u + L_t - L_s.$$

The last equation shows that X^{p-1} is càdlàg with the same jumps as the Lévy process L and the first $p-1$ equations identify X^j as the jth derivative of $X^0, j = 0, \ldots, p-2$, and X^{p-1} as the right-derivative of X^{p-2}. The component process X^0 is a solution of the CARMA$(p, 0)$ equations (18.4) and (18.5) with $b_0 = 1$ and $b_j = 0, j \geq 1$, and $Y = \mathbf{b}^T\mathbf{X}$ satisfies the CARMA(p, q) equations with autoregressive polynomial a, moving average polynomial b and driving Lévy process L. From the properties of $\mathbf{X}$ it follows that $Y = \mathbf{b}^T\mathbf{X}$ has continuous sample paths, unless $q = p-1$, in which case Y has the same jumps as both X^{p-1} and the driving Lévy process L.

From Theorem 17.8 we immediately obtain the following result, which we also restate as a theorem.

Theorem 18.6 (Strictly stationary solution of the state equation). *Suppose that* $\mathbb{E}\log^+|L_1| < \infty$ *and that all eigenvalues of A have non-zero real part (i. e., that Assumptions* A *and* B *hold). Then the random variable* $\mathbf{X}_0$ *can be chosen in such a way that the unique càdlàg solution* (18.8) *of the state equation* (18.5) *is strictly stationary. The choice of such* $\mathbf{X}_0$ *is (almost surely) unique.*

Our next goal is to express the strictly stationary solution $\mathbf{X}$ of the state equation (including $\mathbf{X}_0$) in terms of A and L when the conditions of Theorem 18.6 are satisfied. This was already achieved in Theorem 17.8(c) and Corollary 17.9, but for a companion matrix A we can derive the Jordan canonical decomposition needed for the computation of e^{At} more explicitly (recall the definition of the Jordan canonical decomposition from (7.17) and (7.18)). Clearly if $\mathbf{X}$ is the strictly stationary solution, then $Y := \mathbf{b}^T\mathbf{X}$ will be a corresponding strictly stationary solution of the CARMA equations.

Proposition 18.7 (Jordan canonical decomposition of A). *Let* $\lambda_i, i = 1, \ldots, r$, *with multiplicities* $m_i, i = 1, \ldots, r$, *respectively, denote the distinct zeros of the polynomial a (i. e., the eigenvalues of A). Then the Jordan canonical decomposition of the matrix A defined in* (18.7) *is*

$$A = RJT,$$

where

$$R = [R_1, R_2, \ldots, R_r] \in \mathbb{C}^{p\times p}, \quad T = R^{-1}$$

and the $p \times m_i$ matrix R_i is defined, for $i = 1, \ldots r$, by

$$R_i = \left[\frac{\boldsymbol{\phi}(\lambda_i)}{0!}, \frac{\boldsymbol{\phi}^{(1)}(\lambda_i)}{1!}, \ldots, \frac{\boldsymbol{\phi}^{(m_i-1)}(\lambda_i)}{(m_i-1)!}\right],$$

where

$$\boldsymbol{\phi}(z) := \left[1, z, \ldots, z^{p-1}\right]^T$$

and $\boldsymbol{\phi}^{(j)}$ denotes the jth derivative of $\boldsymbol{\phi}$. The matrix J is the block diagonal Jordan matrix

$$J = \begin{bmatrix} J_1 & 0 & \cdots & 0 \\ 0 & J_2 & \cdots & 0 \\ \vdots & \ddots & \ddots & \vdots \\ 0 & 0 & \cdots & J_r \end{bmatrix},$$

where J_i is the $m_i \times m_i$ Jordan matrix with each diagonal component equal to λ_i, each superdiagonal component equal to one and all other components equal to zero (as in equation (7.18)).

Proof. If λ is an eigenvalue of A with multiplicity $m \le p$, then $a(\lambda) = a^{(1)}(\lambda) = \cdots = a^{(m-1)}(\lambda) = 0$, giving the identities,

$$\binom{p}{j}\lambda^{p-j} = -a_1\binom{p-1}{j}\lambda^{p-1-j} - a_2\binom{p-2}{j}\lambda^{p-2-j} - \cdots - a_p\lambda^{-j}\binom{0}{j}$$

for $j = 0, 1, \ldots, m-1$, where $\binom{n}{j} := 0$ if $j > n$. Now define

$$\boldsymbol{\rho}_j := \boldsymbol{\phi}^{(j)}(\lambda)/j!, \quad j = 0, \ldots, m-1,$$

a non-zero $p \times 1$ vector, whose components are

$$(\boldsymbol{\rho}_j)_i = \binom{i}{j}\lambda^{i-j}, \quad i = 0, \ldots, p-1, j = 0, \ldots, m-1. \tag{18.10}$$

Using the above identities for $\binom{p}{j}\lambda^{p-j}$, $j = 0, \ldots, m-1$, it is easily shown that the ith component of the vector $A\boldsymbol{\rho}_j$ is

$$(A\boldsymbol{\rho}_j)_i = \binom{i+1}{j}\lambda^{i+1-j}, \quad i = 0, \ldots, p-1, j = 0, \ldots, m-1. \tag{18.11}$$

From (18.10) and (18.11) it follows at once that

$$(A - \lambda I_p)\boldsymbol{\rho}_j = \boldsymbol{\rho}_{j-1}, \quad j = 0, 1, \ldots, m-1,$$

where $\boldsymbol{\rho}_{-1} := \mathbf{0}$, and hence that $\boldsymbol{\rho}_0$ is a right eigenvector of A corresponding to the eigenvalue λ and that $\{\boldsymbol{\rho}_j, j = 0, \ldots, m-1\}$, is the corresponding chain of generalized eigenvectors.

It follows from these calculations that the columns of the matrix R are right eigenvectors and generalized eigenvectors of A corresponding to the eigenvalues $\lambda_1, \ldots, \lambda_r$. The proposition then follows from the standard representation of a matrix in Jordan canonical form (see, e. g., Lang [62, Sect. XI.6]). □

As a consequence, we obtain:

Corollary 18.8 (Explicit expression for e^{At}). *With the notations of Proposition* 18.7 *we can write*

$$\mathrm{e}^{At} = R\,\mathrm{e}^{tJ}\,T.$$

Writing $T = R^{-1}$ in block form as

$$T = \begin{bmatrix} T_1 \\ T_2 \\ \vdots \\ T_r \end{bmatrix},$$

where T_i is an $m_i \times p$ matrix, we have

$$\mathrm{e}^{At} = \sum_{i=1}^{r} R_i\,\mathrm{e}^{tJ_i}\,T_i \quad \text{and} \quad T_i R_j = \delta_{i,j} I_{m_i}, \tag{18.12}$$

where $\delta_{i,j} = 1$ if $i = j$ and zero otherwise, and I_{m_i} is the $m_i \times m_i$ identity matrix.

Proof. The first expression is clear from $At = R(Jt)T$ and $T = R^{-1}$, which then implies (18.12). □

The following theorem expresses the unique strictly stationary solution $\mathbf{X}$ of the state equation (18.5) in terms of L and A, and yields a corresponding strictly stationary solution $Y = \mathbf{b}^T\mathbf{X}$ of the CARMA equations. In Section 18.3 we shall prove that, under the conditions of the theorem, $Y = \mathbf{b}^T\mathbf{X}$ is in fact the *unique* strictly stationary solution of the CARMA equations (Theorem 18.15).

Theorem 18.9 (Strictly stat. state process and related CARMA process). *Suppose that the distinct zeros λ_i, $i = 1, \ldots, r$ (with corresponding multiplicities m_i), of the polynomial $a(z)$ all have non-zero real part and that $\mathbb{E}\log^+|L_1| < \infty$. Then, with the notations of Proposition* 18.7 *and Corollary* 18.8, *the process*

$$\mathbf{X}_t = \int_{\mathbb{R}} \mathbf{g}(t-u)\,\mathrm{d}L_u, \quad t \in \mathbb{R}, \tag{18.13}$$

where

$$\mathbf{g}(t) = \mathbf{g}_1(t)\mathbf{1}_{[0,\infty)}(t) - \mathbf{g}_2(t)\mathbf{1}_{(-\infty,0)}(t), \tag{18.14}$$

with

$$\mathbf{g}_1(t) = \sum_{i:\Re(\lambda_i)<0} R_i\, \mathrm{e}^{tJ_i} T_i \mathbf{e}_p \quad \textit{and} \quad \mathbf{g}_2(t) = \sum_{i:\Re(\lambda_i)>0} R_i\, \mathrm{e}^{tJ_i} T_i \mathbf{e}_p,$$

is the unique strictly stationary solution of (18.5) *and*

$$Y_t = \int_{\mathbb{R}} \mathbf{b}^T \mathbf{g}(t-u)\,\mathrm{d}L_u, \quad t \in \mathbb{R}, \tag{18.15}$$

is a corresponding strictly stationary solution of the CARMA equations. If, additionally, all zeros λ_i of $a(z)$ lie in the open left half-plane, then $\mathbf{g}_2(t) = 0$ and $\mathbf{g}_1(t) = \mathrm{e}^{At}\mathbf{e}_p$, so that

$$\mathbf{X}_t = \int_{-\infty}^{t} \mathrm{e}^{A(t-u)}\,\mathbf{e}_p\,\mathrm{d}L_u \quad \textit{and} \quad Y_t = \int_{-\infty}^{t} \mathbf{b}^T \mathrm{e}^{A(t-u)}\,\mathbf{e}_p\,\mathrm{d}L_u, \quad t \in \mathbb{R}.$$

Proof. This can be proved by identifying the kernel derived in Corollary 17.9 with $\mathbf{g}$ defined in (18.14), see Exercise 18.1, but in the following we present a more direct proof.

The existence of the integrals $\int_{(-\infty,t]}$ and $\int_{(t,\infty)}$ in the definition (18.13) of $\mathbf{X}_t$ follows from Theorem 15.26(iii) along with (7.19). The strict stationarity of $\mathbf{X}$ follows from Theorem 16.17.

We next show that $\mathbf{X}$ is the unique solution of (18.5) with $\mathbf{X}_0 = \int_{\mathbb{R}} \mathbf{g}(-u)\,\mathrm{d}L_u$. To do so we use the identities,

$$\mathrm{e}^{At}\mathbf{e}_p = \mathbf{g}_1(t) + \mathbf{g}_2(t), \quad t \in \mathbb{R}, \tag{18.16}$$

and

$$\mathrm{e}^{A(t-s)}\mathbf{g}_i(s-u) = \mathbf{g}_i(t-u), \quad s \le t \in \mathbb{R}, i = 1, 2, \tag{18.17}$$

both of which follow at once from (18.12) and the definitions of $\mathbf{g}_1$ and $\mathbf{g}_2$. Substituting the expression for $\mathbf{X}_0$ in the right-hand side of (18.8) and using the identities (18.16) and (18.17) shows that $\mathbf{X}$ as specified in the statement of the theorem is the unique solution of (18.8) with the specified value at time 0.

We have identified $\mathbf{X}$ as a strictly stationary solution of (18.8). The uniqueness of the strictly stationary solution $\mathbf{X}$ follows from Theorem 18.6 (and we can also find a càdlàg modification).

The corresponding strictly stationary CARMA process Y is obtained immediately from $\mathbf{X}$ using (18.4).

The assertions when all zeros of $a(z)$ lie in the open left half-plane are clear from (18.12). □

Definition 18.10. The function $g(t) := \mathbf{b}^T\mathbf{g}(t)$, with $\mathbf{g}$ as defined in (18.14), is called the *kernel* of the CARMA(p,q) process Y.

With the kernel g we have $Y_t = \int_{\mathbb{R}} g(t-s)\,\mathrm{d}L_s$. Clearly, when all zeros of $a(z)$ lie in the open left half-plane, then $g(t) = \mathbf{b}^T e^{At}\mathbf{e}_p\mathbf{1}_{[0,\infty)}(t)$.

Remark 18.11 (Alternative repr. of strictly stationary CARMA and kernel). Like the matrices R_i, the $m_i \times m_i$ matrices e^{tJ_i}, $i = 1,\dots,r$, in the definition of the kernel $\mathbf{g}$ have a simple explicit representation, as seen in (7.19). This shows that $\mathbf{b}^T e^{At}\mathbf{e}_p$ can be written in the form

$$\mathbf{b}^T e^{At}\mathbf{e}_p = \sum_{i:\Re(\lambda_i)<0} e^{\lambda_i t}\sum_{j=0}^{m_i-1} c_{\lambda_i,j}t^j + \sum_{i:\Re(\lambda_i)>0} e^{\lambda_i t}\sum_{j=0}^{m_i-1} c_{\lambda_i,j}t^j,$$

where the first and second sums on the right are $\mathbf{b}^T\mathbf{g}_1(t)$ and $\mathbf{b}^T\mathbf{g}_2(t)$, respectively. However, by equation (7.27), $\mathbf{b}^T e^{At}\mathbf{e}_p$ can also be expressed as

$$\mathbf{b}^T e^{At}\mathbf{e}_p = \sum_{i:\Re(\lambda_i)<0} \mathrm{Res}_{z=\lambda_i}\left[e^{zt}\frac{b(z)}{a(z)}\right] + \sum_{i:\Re(\lambda_i)>0} \mathrm{Res}_{z=\lambda_i}\left[e^{zt}\frac{b(z)}{a(z)}\right].$$

Calculating the residues, the linear independence of the functions $e^{\lambda_i t}t^j$, $t \in \mathbb{R}$, allows us to identify $\mathbf{b}^T\mathbf{g}_1(t)$ and $\mathbf{b}^T\mathbf{g}_2(t)$ with the first and second sums in the last equation and hence to write the stationary solution Y in (18.15) as

$$Y_t = \int_{\mathbb{R}} g(t-u)\,\mathrm{d}L_u, \tag{18.18}$$

where the kernel is represented as

$$\begin{aligned} g(t) = {} & \sum_{i:\Re(\lambda_i)<0} \mathrm{Res}_{z=\lambda_i}\left[e^{zt}\frac{b(z)}{a(z)}\right]\mathbf{1}_{[0,\infty)}(t) \\ & - \sum_{i:\Re(\lambda_i)>0} \mathrm{Res}_{z=\lambda_i}\left[e^{zt}\frac{b(z)}{a(z)}\right]\mathbf{1}_{(-\infty,0)}(t). \end{aligned} \tag{18.19}$$

If the zeros $\lambda_1,\dots,\lambda_r$ all have multiplicity one, i. e., if $r = p$, then g simplifies to

$$g(t) = \sum_{i:\Re(\lambda_i)<0} \frac{b(\lambda_i)}{a'(\lambda_i)}e^{\lambda_i t}\mathbf{1}_{[0,\infty)}(t) - \sum_{i:\Re(\lambda_i)>0} \frac{b(\lambda_i)}{a'(\lambda_i)}e^{\lambda_i t}\mathbf{1}_{(-\infty,0)}(t). \tag{18.20}$$

We have seen that under Assumptions A and B there exists a unique strictly stationary solution $\mathbf{X}$ of the state-equation (18.5), and that then $Y = (Y_t)_{t\in\mathbb{R}}$ defined by $Y_t = \mathbf{b}^T\mathbf{X}_t$ is a strictly stationary solution of the CARMA equations. However, it does not yet follow from this that the strictly stationary process Y is also unique. In theory, there could be a non-stationary state vector process $\widetilde{\mathbf{X}}$ solving (18.5) such that $\widetilde{Y} := \mathbf{b}^T\widetilde{\mathbf{X}}$ is strictly stationary and hence a CARMA process, and $\widetilde{Y}$ might then be different from Y (see also Remark 18.17). To prove that also Y is unique is the contents of the next section.

18.3 Uniqueness and the Δ-sampled sequence

Under Assumptions A and B we established, in Theorem 18.9, the existence of the CARMA process defined by (18.14) and (18.15) (or equivalently by (18.18) and (18.19)). To show that, under the same conditions on L and A, this is the unique strictly stationary process satisfying the CARMA equations (18.4) and (18.5), we need the definition of the Δ-sampled sequence and establish two preliminary lemmas.

Definition 18.12 (The Δ-sampled sequence). If $Y = (Y_t)_{t\in\mathbb{R}}$ is any (possibly vector-valued) random process and $\Delta > 0$, the Δ-*sampled sequence* Y^Δ of Y is defined as

$$Y_n^\Delta := Y_{n\Delta}, \quad n \in \mathbb{Z}.$$

If Y is any solution of the CARMA equations (18.4) and (18.5), then it follows at once from (18.9) that

$$Y_n^\Delta = \mathbf{b}^T\mathbf{X}_n^\Delta,$$

where

$$\mathbf{X}_n^\Delta = \mathrm{e}^{\Delta A}\mathbf{X}_{n-1}^\Delta + \mathbf{R}_n, \tag{18.21}$$

and

$$\mathbf{R}_n := \int_{(n-1)\Delta}^{n\Delta} \mathrm{e}^{(n\Delta-u)A}\mathbf{e}_p \,\mathrm{d}L_u.$$

Writing the autoregressive polynomial in (18.3) as $a(z) = \prod_{i=1}^r (z-\lambda_i)^{m_i}$ (with λ_i the distinct zeros and m_i their multiplicities), we introduce a new polynomial,

$$\Phi(z) := \prod_{i=1}^r \left(1-\mathrm{e}^{\lambda_i\Delta}z\right)^{m_i} = 1-\phi_1 z - \cdots - \phi_p z^p. \tag{18.22}$$

We can now derive difference equations satisfied by the sequences $(\mathbf{X}_n^\Delta)_{n\in\mathbb{Z}}$ and $(Y_n^\Delta)_{n\in\mathbb{Z}}$, as specified in the following lemma. Recall that we denote by B the backward shift operator defined by $BW_n = W_{n-1}$ for a stochastic process $(W_n)_{n\in\mathbb{Z}}$, and $\Phi(B)W_n$ with Φ as above is then obtained by substituting B formally for z, so that

$$\Phi(B)W_n = W_n - \phi_1 W_{n-1} - \cdots - \phi_p W_{n-p}.$$

Lemma 18.13 (Difference equations for Δ-sampled sequences). *If $Y_t = \mathbf{b}^T\mathbf{X}_t$, $t \in \mathbb{R}$, where $\mathbf{X} = (\mathbf{X}_t)_{t\in\mathbb{R}}$ is any solution of (18.5), B is the backward shift operator, and $\Phi(z)$ is defined as above, then*

$$\Phi(B)\mathbf{X}_n^\Delta = \sum_{k=0}^{p-1}\left(\mathrm{e}^{k\Delta A} - \sum_{j=1}^{k}\phi_j\,\mathrm{e}^{(k-j)\Delta A}\right)\mathbf{R}_{n-k} \tag{18.23}$$

and

$$\Phi(B)Y_n^\Delta = \mathbf{b}^T\sum_{k=0}^{p-1}\left(\mathrm{e}^{k\Delta A} - \sum_{j=1}^{k}\phi_j\,\mathrm{e}^{(k-j)\Delta A}\right)\mathbf{R}_{n-k}.$$

The latter can be written as

$$\Phi(B)Y_n^\Delta = Y_n^\Delta - \phi_1 Y_{n-1}^\Delta - \cdots - \phi_p Y_{n-p}^\Delta = Z_n^1 + Z_{n-1}^2 + \cdots + Z_{n-p+1}^p, \tag{18.24}$$

where

$$Z_n^k := \int_{(n-1)\Delta}^{n\Delta}\mathbf{b}^T\left(\mathrm{e}^{(k-1)\Delta A} - \sum_{j=1}^{k-1}\phi_j\mathrm{e}^{(k-1-j)\Delta A}\right)\mathrm{e}^{(n\Delta-u)A}\mathbf{e}_p\,\mathrm{d}L_u \tag{18.25}$$

for $k = 1,\dots,p$.

Proof. It suffices to prove (18.23), from which the remaining assertions follow. For that, we shall first show that for any $s \in \mathbb{N}_0$ and for any complex numbers $c_1,\dots,c_s$, we can write

$$\begin{aligned}\mathbf{X}_n^\Delta = {} & \sum_{k=1}^{s} c_k\mathbf{X}_{n-k}^\Delta + \left(\mathrm{e}^{s\Delta A} - \sum_{k=1}^{s} c_k\mathrm{e}^{(s-k)\Delta A}\right)\mathbf{X}_{n-s}^\Delta \\ & + \sum_{k=0}^{s-1}\left(\mathrm{e}^{k\Delta A} - \sum_{j=1}^{k} c_j\mathrm{e}^{(k-j)\Delta A}\right)\mathbf{R}_{n-k}.\end{aligned} \tag{18.26}$$

For $s = 0$ this is clear. To show that validity of the statement for any particular s implies validity for $s+1$, let c_{s+1} be an arbitrary complex number. We can then write, by (18.21),

$$\mathbf{X}_n^\Delta = \sum_{k=1}^{s} c_k \mathbf{X}_{n-k}^\Delta + \left(e^{s\Delta A} - \sum_{k=1}^{s} c_k e^{(s-k)\Delta A} \right) (e^{\Delta A}\mathbf{X}_{n-(s+1)}^\Delta + \mathbf{R}_{n-s})$$
$$+ \sum_{k=0}^{s-1} \left(e^{k\Delta A} - \sum_{j=1}^{k} c_j e^{(k-j)\Delta A} \right) \mathbf{R}_{n-k}$$
$$= \sum_{k=1}^{s+1} c_k \mathbf{X}_{n-k}^\Delta + \left(e^{(s+1)\Delta A} - \sum_{k=1}^{s} c_k e^{(s+1-k)\Delta A} - c_{s+1} \right) \mathbf{X}_{n-(s+1)}^\Delta$$
$$+ \sum_{k=0}^{s} \left(e^{k\Delta A} - \sum_{j=1}^{k} c_j e^{(k-j)\Delta A} \right) \mathbf{R}_{n-k},$$

completing the induction step.

Next, observe that from the Jordan canonical decomposition of $A = RJR^{-1}$ as in Proposition 18.7 and the specific form of $e^{-\Delta J}$, the characteristic polynomial of the matrix $e^{-\Delta A}$ is given by

$$\det(zI_p - e^{-\Delta A}) = \det(zI_p - e^{-\Delta J}) = \prod_{i=1}^{r} (z - e^{-\Delta\lambda_i})^{m_i}$$
$$= \Phi(z)(-1)^p \prod_{i=1}^{r} e^{-\Delta\lambda_i m_i}$$

so that the characteristic polynomial of $e^{-\Delta A}$ differs from Φ differ only by a constant factor. The Cayley–Hamilton theorem then implies $\Phi(e^{-\Delta A}) = 0$, so that

$$e^{-0\Delta A} - \phi_1 e^{-\Delta A} - \cdots - \phi_p e^{-p\Delta A} = 0.$$

Multiplying this by $e^{p\Delta A}$ gives

$$e^{p\Delta A} - \phi_1 e^{(p-1)\Delta A} - \cdots - \phi_p e^{0\Delta A} = 0,$$

and inserting this in (18.26) with $s = p$ and $c_k = \phi_k$ gives (18.23). □

The next lemma shows that strictly stationary solutions of autoregressive equations are determined by the coefficients in the difference equation and the noise, regardless if the noise is itself strictly stationary or not.

Lemma 18.14 (Uniqueness of strictly stat. sol. of autoregressive eqns.). *Let $\psi_1, \dots, \psi_p \in \mathbb{C}$ and $(V_n)_{n\in\mathbb{Z}}$ be a strictly stationary $\mathbb{C}$-valued process satisfying the autoregressive equations,*

$$V_n - \psi_1 V_{n-1} - \cdots - \psi_p V_{n-p} = Z_n, \quad n \in \mathbb{Z}, \tag{18.27}$$

where $(Z_n)_{n\in\mathbb{Z}}$ is some sequence of $\mathbb{C}$-valued random variables (not necessarily strictly stationary). Define the autoregressive polynomial by $\Psi(z) = 1 - \psi_1 z - \cdots - \psi_p z^p$ and suppose that $\Psi(\cdot)$ has no zeros on the unit circle. Denote by

$$1/\Psi(z) = \sum_{m\in\mathbb{Z}} c_m z^m$$

the Laurent expansion of $1/\Psi$ around zero, valid on $\{z \in \mathbb{C} : 1 - \varepsilon \le |z| \le 1 + \varepsilon\}$ for some $\varepsilon \in (0,1)$. The random variables V_n are then uniquely determined by the coefficients $\psi_1, \ldots, \psi_p$ and the sequence $(Z_{n-j})_{j\in\mathbb{Z}}$ as the limit in probability,

$$V_n = \operatorname*{plim}_{N\to\infty} \sum_{|j|\le N} c_j Z_{n-j}.$$

Proof. For each $N \in \mathbb{N}$ define the function f_N on $\{z \in \mathbb{C} : 1 - \varepsilon \le |z| \le 1 + \varepsilon\}$ by

$$f_N(z) := \sum_{|j|\le N} c_j z^j (1 - \psi_1 z - \cdots \psi_p z^p) =: \sum_{j=-N}^{N+p} b_{j,N} z^j.$$

By the uniform convergence of the Laurent series on this annulus, f_N converges uniformly to 1 there as $N \to \infty$, and from the representation of the Laurent coefficients as contour integrals (e. g., Conway [34, Theorem V.1.11]) the Laurent coefficients of f_N converge to those of the function 1, i. e.,

$$\lim_{N\to\infty} b_{j,N} = \begin{cases} 0, & j \ne 0, \\ 1, & j = 0. \end{cases}$$

Further, observe that

$$b_{j,N} = b_{j,N'} \quad \forall\, N' \ge N > p, j = -N + p, \ldots, N,$$

i. e., for fixed j, $b_{j,N}$ is constant for sufficiently large N. From the limit result, we hence see that

$$f_N(z) = 1 + \sum_{j=-N}^{-N+p-1} b_{j,N} z^j + \sum_{j=N+1}^{N+p} b_{j,N} z^j,$$

and that

$$\lim_{N\to\infty} \sup_{j\in\{-N,\ldots,-N+p-1\}\cup\{N+1,\ldots,N+p\}} |b_{j,N}| = 0$$

(the latter from the definition of the $b_{j,N}$ and the fact that $\lim_{|j|\to\infty} c_j = 0$). Since $(V_n)_{n\in\mathbb{Z}}$ is strictly stationary, it follows from Slutsky's theorem that for all indices $j(N) \in \{-N, \ldots, -N + p - 1\} \cup \{N + 1, \ldots, N + p\}$, the sequence $b_{j(N),N} V_{n-j(N)}$ converges in distribution and hence in probability to 0 as $N \to \infty$. Hence (with B the backward shift operator)

$$\begin{aligned} V_n &= \operatorname*{plim}_{N\to\infty}\left(V_n + \sum_{j=-N}^{-N+p-1} b_{j,N} V_{n-j} + \sum_{j=N+1}^{N+p} b_{j,N} V_{n+j} \right) \\ &= \operatorname*{plim}_{N\to\infty} f_N(B) V_n \\ &= \operatorname*{plim}_{N\to\infty} \sum_{|j|\le N} c_j B^j (1 - \psi_1 B - \cdots - \psi_p B^p) V_n. \end{aligned}$$

The claim then follows from the fact that $\Psi(B)V_n = Z_n$. □

We are now in a position to establish the uniqueness of the strictly stationary solution Y of the CARMA equations specified by (18.14) and (18.15), or equivalently by (18.18) and (18.19).

Theorem 18.15 (Uniqueness of CARMA process). *Suppose that the distinct zeros λ_i, $i = 1, \dots, r$ (with corresponding multiplicities m_i) of the polynomial $a(z)$ all have non-zero real part and that $\mathbb{E}\log^+|L_1| < \infty$. Then the process Y defined by* (18.14) *and* (18.15), *or equivalently by* (18.18) *and* (18.19), *is in fact the unique strictly stationary solution of the CARMA equations,* (18.4) *and* (18.5).

Proof. Let $Y = (Y_t)_{t\in\mathbb{R}}$ be some (strictly stationary) CARMA process satisfying the CARMA equations (18.4) and (18.5). By (18.24), the sampled sequence Y^Δ satisfies an equation of the form (18.27) with $\Psi(z)$ replaced by $\Phi(z)$ and Z_n replaced by $Z_n^1 + Z_{n-1}^2 + \cdots + Z_{n-p+1}^p$. Moreover, since $a(z)$ is assumed to be non-zero on the imaginary axis, $\Phi(z)$ satisfies the conditions imposed on $\Psi(z)$ in Lemma 18.14. Consequently, for every $\Delta > 0$ the strictly stationary sequence Y^Δ is uniquely determined by the coefficients in the CARMA equations and L, in particular, so are $Y_{-\Delta}$, Y_0 and Y_Δ. Since this is true for every $\Delta > 0$, each Y_t, $t \in \mathbb{R}$, is uniquely determined almost surely, and since the sample paths of Y are càdlàg, also Y itself is almost surely uniquely determined. □

As a consequence of Theorem 18.15, also the kernel of a CARMA process must be unique if the driving Lévy process is non-deterministic, i. e., if $h \in L^1(\mathbb{R};\mathbb{C}) \cap L^2(\mathbb{R};\mathbb{C})$ is another integrable function with respect to the non-deterministic Lévy process L, such that $\int_{\mathbb{R}} h(t-s)\,\mathrm{d}L_s = Y_t = \int_{\mathbb{R}} g(t-s)\,\mathrm{d}L_s$ with g the kernel of Definition 18.10, then $g = h$ Lebesgue almost everywhere, see Exercise 18.2.

With the aid of Theorem 18.15 and Remark 18.11 we can also derive an alternative representation of the strictly stationary state vector process.

Corollary 18.16 (Alternative representation of strictly stat. state vector). *If the distinct zeros λ_i, $i = 1, \dots, r$ (with corresponding multiplicities m_i) of the polynomial $a(z)$ all have non-zero real part and if $\mathbb{E}\log^+|L_1| < \infty$, then $(\mathbf{X}_t)_{t\in\mathbb{R}}$ as given by* (18.13) *and* (18.14) *has the alternative representation,*

$$\mathbf{X}_t = \int_{\mathbb{R}} \mathbf{g}(t-u)\,\mathrm{d}L_u,$$

where

$$\mathbf{g}(t) = \sum_{i:\Re(\lambda_i)<0} \mathrm{Res}_{z=\lambda_i}\left[e^{zt}\frac{(1,z,\dots,z^{p-1})^T}{a(z)}\right]\mathbf{1}_{[0,\infty)}(t) - \sum_{i:\Re(\lambda_i)>0} \mathrm{Res}_{z=\lambda_i}\left[e^{zt}\frac{(1,z,\dots,z^{p-1})^T}{a(z)}\right]\mathbf{1}_{(-\infty,0)}(t).$$

Proof. This follows from (18.19) and the fact that the jth component, $X^{j-1}, j = 1,\dots,p$, of $\mathbf{X}$ is the unique strictly stationary càdlàg solution of the CARMA equations with $b(z) = z^{j-1}$. □

Remark 18.17. We have shown that under the Assumptions A and B there exists a unique strictly stationary solution Y of the CARMA equations, in which case also a corresponding (unique) strictly stationary state vector $\mathbf{X}$ exists. When omitting Assumptions A and B, it has been shown in Brockwell and Lindner [26, Theorem 4.2] that if L is non-deterministic and if $\mathbf{b} \neq 0$, then the CARMA equations (18.4) and (18.5) have a strictly stationary solution if and only if $\mathbb{E}\log^+|L_1| < \infty$ and all singularities of the meromorphic function $z \mapsto b(z)/a(z)$ on the imaginary axis are removable. In particular, Assumption A (i. e., $\mathbb{E}\log^+|L_1| < \infty$) is necessary for a strictly stationary solution to exist, and if $a(z)$ and $b(z)$ do not have common zeros on the imaginary axis and L is non-deterministic, then Assumption B (real parts of the zeros of $a(z)$ all non-zero) is also necessary for a strictly stationary solution to exist. However, in general it is possible that zeros of $a(z)$ on the imaginary axis are factored out by corresponding zeros of $b(z)$. In [26, Theorem 4.2] it is shown that even in this case, the strictly stationary solution of the corresponding CARMA equations will still be unique, unless L is deterministic. It also follows from these results that Assumptions A and B are indeed necessary for a strictly stationary state vector to exist if L is non-deterministic; for if $\mathbf{X}$ is a strictly stationary solution of (18.5), then $\mathbf{b}^T X$ is a strictly stationary CARMA process for all $\mathbf{b} \in \mathbb{R}^p$, in particular also for some $\mathbf{b}$ such that $b(z)$ and $a(z)$ do not have common zeros. It follows that when $a(z)$ has zeros on the imaginary axis that are factored out by those of $b(z)$, and if the non-deterministic L has finite log-moment, then strictly stationary CARMA processes exist without a corresponding strictly stationary state vector process. More information on strictly stationary CARMA processes, where also the deterministic case is covered, is treated in [26].

Remark 18.18 (Structure of Y^Δ). Let $Y = (Y_t)_{t\in\mathbb{R}}$ be a strictly stationary CARMA process. Then Y^Δ is strictly stationary and the sequences $(Z_n^k)_{n\in\mathbb{Z}}$ from equation (18.25) are i. i. d. for each $k \in \{1,\dots,p\}$ by Corollary 16.18. From the definition of the Z_n^k it follows that the sequence $(W_n := Z_n^1 + Z_{n-1}^2 + \dots + Z_{n-p+1}^p)_{n\in\mathbb{Z}}$ is $(p-1)$-dependent, i. e., $(W_k)_{k\le n}$ and $(W_k)_{k\ge n+p}$ are independent for each $n \in \mathbb{Z}$, and $(W_n)_{n\in\mathbb{Z}}$ is obviously strictly stationary.

Hence, equations (18.24) and (18.25) identify Y^Δ as a discrete-time autoregression driven by the $(p-1)$-dependent strictly stationary sequence $(W_n)_{n\in\mathbb{Z}}$. Now if $\mathbb{E}|L_1|^2 < \infty$, then $\mathbb{E}|W_n|^2 < \infty$ and $\mathbb{E}|Y_n^\Delta|^2 < \infty$ by Example 15.12, so that Y^Δ is also weakly stationary, and by Proposition 3.2.1 of Brockwell and Davis [22], $(W_n - \mathbb{E}W_n)_{n\in\mathbb{Z}}$ is a moving average of order $s < p$ of uncorrelated (but not necessarily independent) white noise. (Actually, Proposition 3.2.1 in [22] is formulated only for real-valued $(p-1)$-dependent sequences, but it is easily seen from the proof that the result also holds for complex-valued $(p-1)$-dependent sequences.) This means that there is some white noise sequence $(Z_n)_{n\in\mathbb{Z}} \sim WN(0,\sigma^2)$ and coefficients $\theta_1,\dots,\theta_s \in \mathbb{C}$ such that $W_n - \mathbb{E}(W_n) = \Theta(B)Z_n$, where $\Theta(z) = 1 + \theta_1 z + \cdots + \theta_s z^s$. From (18.24) we then have, with Φ as defined in (18.22), that

$$\Phi(B)Y_n^\Delta = W_n = (W_n - \mathbb{E}W_n) + \mathbb{E}W_n = \Theta(B)Z_n + \mathbb{E}W_n.$$

Hence

$$\Phi(B)(Y_n^\Delta - \mathbb{E}Y_n^\Delta) = \Theta(B)Z_n + \mathbb{E}W_n - \Phi(1)\mathbb{E}Y_n^\Delta,$$

and taking expectations and both sides, we see $\mathbb{E}W_n = \Phi(1)\mathbb{E}Y_n^\Delta$, so that

$$\Phi(B)(Y_n^\Delta - \mathbb{E}Y_n^\Delta) = \Theta(B)Z_n. \tag{18.28}$$

This means that $Y^\Delta - \mathbb{E}Y_0^\Delta$ is an ARMA process of order (p,s). We shall refer to the problem of determining the moving average coefficients in this representation as *the sampling problem*, the solution of which is given below in Theorem 18.33.

18.4 Second-order properties when $\mathbb{E}|L_1|^2 < \infty$

Let $L = (L_t)_{t\in\mathbb{R}}$ be a non-zero $\mathbb{R}$-valued Lévy process with finite second moment, mean zero, and assume for the moment that all zeros of the autoregressive polynomial $a(z)$ have strictly negative real part and all zeros of the moving average process have non-positive real part, an assumption we used throughout Chapter 7 (see Remark 7.2). Let $Y = (Y_t)_{t\in\mathbb{R}}$ be the (unique strictly stationary) CARMA process satisfying (18.4) and (18.5). By Theorem 18.9, this solution is given by $Y_t = \int_{(-\infty,t]} \mathbf{b}^T e^{A(t-s)} \mathbf{e}_p \, dL_s$, $t \in \mathbb{R}$. From Corollary 7.16 and Definition 7.12 we recognize Y as a second-order CARMA process as defined in Definition 7.1. Hence, when all zeros of $a(z)$ have strictly negative real part and all zeros of $b(z)$ non-positive real part, the CARMA(p,q) process is a second-order CARMA process in the sense of Chapter 7, the results of which therefore apply also to Y.

The aim of this section is to show that when $\mathbb{E}|L_1|^2 < \infty$ and all zeros of $a(z)$ have *non-zero real part* (i. e., the real parts need not be strictly negative, but only non-zero), the Lévy-driven CARMA process Y as defined by (18.14) and (18.15) (equivalently, by (18.18) and (18.19)) still has rational spectral density as in (7.1), and is therefore also a second-order CARMA process in the sense of Definition 7.1 (but without Remark 7.2 being in force). Observe that while the second-order CARMA processes of Chapter 7 have representations as integrals with respect to orthogonal-increment processes, they do not in general have representations as integrals with respect to *independent-increment* processes like the CARMA process Y. And of course the only restriction on L_1 in (18.14) and (18.15) is that $\mathbb{E}(\log^+ |L_1|) < \infty$.

Let us first determine the mean of the CARMA process:

Lemma 18.19 (Mean of CARMA(p,q) process). *Let $L = (L_t)_{t\in\mathbb{R}}$ be an $\mathbb{R}$-valued Lévy process with $\mathbb{E}|L_1| < \infty$. Assume that all zeros of the autoregressive polynomial $a(z)$ have non-zero real part (this implies in particular that $a_p \neq 0$). Then the strictly stationary* CARMA(p,q) *process $Y = (Y_t)_{t\in\mathbb{R}}$ satisfying* (18.4) *and* (18.5) *and the corresponding strictly stationary state vector process $\mathbf{X} = (\mathbf{X}_t)_{t\in\mathbb{R}}$ both have finite mean given by*

$$\mathbb{E}(Y_t) = \frac{b_0}{a_p}\mathbb{E}(L_1) \quad \textit{and} \quad \mathbb{E}(\mathbf{X}_t) = \frac{\mathbb{E}(L_1)}{a_p}[1,0,\dots,0]^T,$$

respectively.

Proof. With the notations of Theorem 18.9, we have (18.13) with $\mathbf{g}$ given by (18.14). From Theorem 15.26(i), $\mathbf{X}_t$ has finite expectation given by

$$\mathbb{E}(\mathbf{X}_t) = \mathbb{E}(L_1)\int_{\mathbb{R}} \mathbf{g}(t-s)\,\mathrm{d}s = \mathbb{E}(L_1)\int_{\mathbb{R}} \mathbf{g}(s)\,\mathrm{d}s.$$

The last integral can be calculated as

$$\begin{aligned}\int_{\mathbb{R}} \mathbf{g}(s)\,\mathrm{d}s &= \sum_{i:\Re(\lambda_i)<0} R_i \int_0^\infty \mathrm{e}^{sJ_i}\,\mathrm{d}s\, T_i\mathbf{e}_p - \sum_{i:\Re(\lambda_i)>0} R_i \int_{-\infty}^0 \mathrm{e}^{sJ_i}\,\mathrm{d}s\, T_i\mathbf{e}_p \\ &= -\sum_{i:\Re(\lambda_i)\neq 0} R_i J_i^{-1} T_i \mathbf{e}_p \\ &= -RJ^{-1}T\mathbf{e}_p = -A^{-1}\mathbf{e}_p.\end{aligned}$$

Since $(-A)\frac{1}{a_p}[1,0,\dots,0]^T = \mathbf{e}_p$ the latter is equal to $\frac{1}{a_p}[1,0,\dots,0]^T$, giving the required form of $\mathbb{E}(\mathbf{X}_t)$ and hence of $\mathbb{E}(Y_t) = \mathbb{E}(\mathbf{b}^T\mathbf{X}_t)$. □

Next, let us calculate the Fourier transform of the kernel g.

Lemma 18.20 (Fourier transform of kernel). *Suppose that all zeros of the autoregressive polynomial $a(z)$ have non-zero real part and denote by g the kernel of a Lévy driven* CARMA(p,q) *process, given by* (18.19), *equivalently by $\mathbf{b}^T\mathbf{g}$ with $\mathbf{g}$ from* (18.14). *Then $g \in L^1(\mathbb{R};\mathbb{C}) \cap L^2(\mathbb{R};\mathbb{C})$ and the Fourier transform, $\mathcal{F}g(\omega) := \frac{1}{\sqrt{2\pi}}\int_{\mathbb{R}} \mathrm{e}^{-\mathrm{i}\omega t} g(t)\,\mathrm{d}t$, $\omega \in \mathbb{R}$, of g is*

$$\mathcal{F}g(\omega) = \frac{1}{\sqrt{2\pi}}\frac{b(\mathrm{i}\omega)}{a(\mathrm{i}\omega)}, \qquad \omega \in \mathbb{R}.$$

Proof. That $g \in L^1(\mathbb{R};\mathbb{C}) \cap L^2(\mathbb{R};\mathbb{C})$ is clear from (18.14). To calculate the Fourier transform, we use the representation (18.19) of the kernel. Denote

$$g_1(t) := \sum_{i:\mathbb{R}(\lambda_i)<0} \operatorname{Res}_{z=\lambda_i}\left[e^{zt}\frac{b(z)}{a(z)}\right]\mathbf{1}_{[0,\infty)}(t) \tag{18.29}$$

and

$$g_2(t) := \sum_{i:\mathbb{R}(\lambda_i)>0} \operatorname{Res}_{z=\lambda_i}\left[e^{zt}\frac{b(z)}{a(z)}\right]\mathbf{1}_{(-\infty,0)}(t), \tag{18.30}$$

so that $g(t) = g_1(t) - g_2(t)$ by (18.19). Denote by ρ_1 a simple closed curve in the open left half-plane that encircles all zeros of $a(z)$ in the left half-plane, and by ρ_2 a simple closed curve in the open right half-plane, encircling all zeros of $a(z)$ there, both in the counterclockwise sense. By the residue theorem,

$$g_1(t) = \frac{1}{2\pi i}\int_{\rho_1}\frac{b(z)}{a(z)}e^{zt}\,dz\,\mathbf{1}_{[0,\infty)}(t),\quad g_2(t) = \frac{1}{2\pi i}\int_{\rho_2}\frac{b(z)}{a(z)}e^{zt}\,dz\,\mathbf{1}_{(-\infty,0)}(t).$$

Observe that ρ_1 is strictly bounded away from the imaginary axis, so that e^{zt} decays exponentially as $t \to \infty$ uniformly for z on the trace of f. Hence, an application of Fubini's theorem yields for the Fourier transform $\mathcal{F}g_1$ of g_1

$$\mathcal{F}g_1(\omega) = \frac{1}{(2\pi)^{3/2}i}\int_0^\infty e^{-i\omega t}\int_{\rho_1}\frac{b(z)}{a(z)}e^{zt}\,dz\,dt = \frac{1}{(2\pi)^{3/2}i}\int_{\rho_1}\frac{b(z)}{a(z)(i\omega - z)}\,dt$$

for $\omega \in \mathbb{R}$. Applying the residue theorem again, we conclude

$$\mathcal{F}g_1(\omega) = -\frac{1}{\sqrt{2\pi}}\sum_{i:\mathbb{R}(\lambda_i)<0}\operatorname{Res}_{z=\lambda_i}\left[\frac{b(z)}{a(z)(z-i\omega)}\right],\quad \omega \in \mathbb{R}.$$

Similar reasoning applied to ρ_2 yields

$$\mathcal{F}g_2(\omega) = \frac{1}{\sqrt{2\pi}}\sum_{i:\mathbb{R}(\lambda_i)>0}\operatorname{Res}_{z=\lambda_i}\left[\frac{b(z)}{a(z)(z-i\omega)}\right]$$

so that

$$\mathcal{F}g(\omega) = -\frac{1}{\sqrt{2\pi}}\sum_{i:\mathbb{R}(\lambda_i)\neq 0}\operatorname{Res}_{z=\lambda_i}\left[\frac{b(z)}{a(z)(z-i\omega)}\right],\quad \omega \in \mathbb{R}.$$

Now let $\omega \in \mathbb{R}$ be fixed and $\rho_{3,s} : [0,2\pi] \to \mathbb{C}$ be defined by $\rho_{3,s}(u) := s\,e^{iu}$, where $s > 0$ is so large such that $s > |\omega|$ and such that $\rho_{3,s}$ encircles all zeros of $a(z)$. A final application of the residue theorem then shows that

$$\mathcal{F}g(\omega) = -\frac{1}{(2\pi)^{3/2}\mathrm{i}} \int\limits_{\rho_{3,s}} \frac{b(z)}{a(z)(z-\mathrm{i}\omega)}\,\mathrm{d}z + \frac{1}{\sqrt{2\pi}} \operatorname{Res}_{z=\mathrm{i}\omega}\left[\frac{b(z)}{a(z)(z-\mathrm{i}\omega)}\right].$$

However, the residue at $\mathrm{i}\omega$ is equal to $b(\mathrm{i}\omega)/a(\mathrm{i}\omega)$, and the integral over $\rho_{3,s}$ converges to 0 as $s \to \infty$, since $q \le p-1$, which gives the result. □

Remark 18.21. Similar to the calculations of Lemma 18.19, the Fourier transform of the kernel g can also be calculated using the representation (18.15) and the Jordan canonical decomposition of A. This leads to $\mathcal{F}g(\omega) = -\frac{1}{\sqrt{2\pi}}\mathbf{b}^T(A-\mathrm{i}\omega I_p)^{-1}\mathbf{e}_p$, which can be shown to be equal to the Fourier transform obtained in Lemma 18.20. See Exercise 18.4 for details.

We can now show that CARMA processes are second-order CARMA processes, even when the zeros of $a(z)$ are not restricted to lie in the open left half-plane.

Theorem 18.22 (Spectral density and ACVF of CARMA). *Let $(L_t)_{t\in\mathbb{R}}$ be a non-deterministic $\mathbb{R}$-valued Lévy process with finite variance* $\operatorname{Var}(L_1)$. *Assume that all zeros of the autoregressive polynomial $a(z)$ have non-zero real part. Then the strictly stationary* CARMA(p,q) *process $Y = (Y_t)_{t\in\mathbb{R}}$, satisfying* (18.4) *and* (18.5), *is also weakly stationary, and its spectral distribution function is absolutely continuous with spectral density given by*

$$f(\omega) = \frac{\operatorname{Var}(L_1)}{2\pi}\left|\frac{b(\mathrm{i}\omega)}{a(\mathrm{i}\omega)}\right|^2, \quad \omega \in \mathbb{R},$$

so that Y is a second-order CARMA(p,q) *process in the sense of Definition* 7.1. *The autocovariance function γ of Y is given by*

$$\gamma(h) = \frac{\operatorname{Var}(L_1)}{2\pi}\int\limits_{\mathbb{R}} \mathrm{e}^{\mathrm{i}h\omega}\left|\frac{b(\mathrm{i}\omega)}{a(\mathrm{i}\omega)}\right|^2 \mathrm{d}\omega, \quad h \in \mathbb{R}.$$

Proof. That Y is weakly stationary with the specified spectral density follows immediately from Corollary 16.19 and Lemma 18.20 (that Y is also mean-square continuous was established already in Example 4.11). Hence, Y is a second-order CARMA(p,q) process in the sense of Definition 7.1. The form of the autocovariance function is clear from Bochner's theorem (Theorem 5.1). □

Remark 18.23. We have established that when $\mathbb{E}|L_1|^2 < \infty$, the Lévy-driven CARMA process has the rational spectral density given by (7.1) and is therefore a CARMA process in the sense of Chapter 7, where further second-order properties and minimum mean-square error linear predictors of Y can be found. In particular, when all zeros of $a(z)$ lie in the open left half-plane, then the formulas derived there for the autocovariance function as in Propositions 7.6 or 7.26 can be applied immediately. If $a(z)$ has zeros in the right half-plane, say $a(z) = \prod_{i:\Re(\lambda_i)<0}(z-\lambda_i)^{m_i}\prod_{i:\Re(\lambda_i)>0}(z-\lambda_i)^{m_i}$, denote $\tilde{\lambda}_i := -\Re(\lambda_i) + \mathrm{i}\Im(\lambda_i) = \mathrm{i}\overline{(-\mathrm{i})\lambda_i}$ and

$$\widetilde{a}(z) := \prod_{i:\Re(\lambda_i)<0} (z-\lambda_i)^{m_i} \prod_{i:\Re(\lambda_i)>0} (z-\widetilde{\lambda}_i)^{m_i}.$$

Then all zeros of $\widetilde{a}(z)$ lie in the open left half-plane, and $|\widetilde{a}(\mathrm{i}\omega)| = |a(\mathrm{i}\omega)|$ for $\omega \in \mathbb{R}$. A similar reasoning can be applied to shift all zeros of $b(z)$ into the closed left half-plane, yielding a polynomial $\widetilde{b}(z)$. The Lévy driven CARMA process with polynomials $\widetilde{a}(z)$ and $\widetilde{b}(z)$ then has the same spectral density as the original one. Hence, all second-order properties derived in Chapter 7 can be applied by working with $\widetilde{a}(z)$ and $\widetilde{b}(z)$.

Remark 18.24 (CARMA autocorrelation function). By Theorem 18.22, the autocorrelation function of the Lévy-driven CARMA(p,q) process,

$$\frac{\gamma(h)}{\gamma(0)} = \frac{\int_{\mathbb{R}} e^{\mathrm{i}h\omega} |\frac{b(\mathrm{i}\omega)}{a(\mathrm{i}\omega)}|^2 \,\mathrm{d}\omega}{\int_{\mathbb{R}} |\frac{b(\mathrm{i}\omega)}{a(\mathrm{i}\omega)}|^2 \,\mathrm{d}\omega}, \qquad h \in \mathbb{R},$$

depends only on the CARMA parameters and not on the driving Lévy process. In particular, this means that it can be defined by the right-hand side of this equation, even when $\mathbb{E}|L_1|^2 = \infty$, although then the autocovariance function of the corresponding CARMA process is not defined. If the sample autocorrelation function of data from an infinite-variance CARMA process is compared with the model autocorrelation function, $\gamma/\gamma(0)$, the resemblance is frequently stronger than is observed when the corresponding comparison is made for a finite variance CARMA process.

18.5 Causality

In Definition 17.10, we called a strictly stationary Ornstein–Uhlenbeck process *causal* if at each time t it is measurable with respect to the increments of the Lévy process prior to t. We now give the corresponding definition for CARMA processes.

Definition 18.25 (Causal CARMA process). Let $L = (L_t)_{t\in\mathbb{R}}$ be an $\mathbb{R}$-valued Lévy process with $\mathbb{E}\log^+|L_1| < \infty$. Suppose that all zeros of the polynomial $a(z)$ have non-zero real part. Then the strictly stationary CARMA(p,q) process $Y = (Y_t)_{t\in\mathbb{R}}$, satisfying the CARMA equations (18.4) and (18.5), is said to be *causal*, if for every $t \in \mathbb{R}$, Y_t is measurable with respect to $\sigma(L_s - L_u : s,u \le t)$ (more precisely, if each Y_t is almost surely equal to a $\sigma(L_s - L_u : s,u \le t)$-measurable random variable).

As for the Ornstein–Uhlenbeck process, causality is not a property of Y itself, but of the pair (Y,L). Causality of CARMA processes can now be characterized as follows:

Theorem 18.26 (Characterization of causality and sufficient condition). *Suppose that $L = (L_t)_{t\in\mathbb{R}}$ is a non-deterministic $\mathbb{R}$-valued Lévy process with $\mathbb{E}\log^+|L_1| < \infty$, that all zeros of the polynomial $a(z)$ have non-zero real part and let $Y = (Y_t)_{t\in\mathbb{R}}$ be the unique strictly stationary* CARMA(p,q) *process satisfying the CARMA equations* (18.4) *and* (18.5). *Denote* *by g the corresponding kernel as given in* (18.19), *so that $Y_t = \int_{\mathbb{R}} g(t-u)\,\mathrm{d}L_u$. Then Y is causal if and only if $g(t) = 0$ for all $t < 0$, equivalently if*

$$\sum_{i:\Re(\lambda_i)>0} \operatorname{Res}_{z=\lambda_i}\left[e^{zt} \frac{b(z)}{a(z)} \right] = 0 \quad \forall\, t < 0, \tag{18.31}$$

where $\lambda_1, \ldots, \lambda_r$ denote the distinct zeros of $a(z)$. A sufficient condition for (18.31) *is that all zeros of $a(z)$ have strictly negative real part, in which case*

$$\mathbf{X}_t = \int_{-\infty}^{t} e^{A(t-u)} \mathbf{e}_p \,\mathrm{d}L_u$$

for the strictly stationary state vector and

$$Y_t = \int_{-\infty}^{t} \mathbf{b}^T e^{A(t-u)} \mathbf{e}_p \,\mathrm{d}L_u, \quad t \in \mathbb{R}. \tag{18.32}$$

Proof. If condition (18.31) holds, then the kernel g is supported in $[0,\infty)$ by (18.19), and (18.18) shows that $Y_t = \int_{-\infty}^{t} g(t-u)\,\mathrm{d}L_u$. Hence Y_t, as the probability limit of $\int_v^t g(t-u)\,\mathrm{d}L_u$ as $v \to -\infty$ (cf. Proposition 16.21) is $\sigma(L_s - L_u : s, u \leq t)$-measurable by Theorem 16.3(a). That (18.31) will be satisfied if all zeros of $a(z)$ have strictly negative real part is clear, and the representations for $\mathbf{X}_t$ and Y_t then follow from Theorem 18.9.

Conversely, assume that Y is a causal CARMA process. Define g_1 and g_2 as in (18.29) and (18.30), so that $g(t) = g_1(t) - g_2(t)$. Then

$$Y_0 = \int_{-\infty}^{0} g_1(-u)\,\mathrm{d}L_u - \int_{0}^{\infty} g_2(-u)\,\mathrm{d}L_u$$

by (18.18), the first summand is measurable with respect to $\sigma(L_s - L_u : u, s \leq 0)$ and the second with respect to $\sigma(L_s - L_u : u, s \geq 0)$. Since also Y_0 is measurable with respect to $\sigma(L_s - L_u : u, s \leq 0)$, we conclude that $\int_0^\infty g_2(-u)\,\mathrm{d}L_u$ is measurable with respect to $\sigma(L_s - L_u : u, s \leq 0)$, and since it is also measurable with respect to $\sigma(L_s - L_u : u, s \geq 0)$, which is independent of $\sigma(L_s - L_u : u, s \leq 0)$, Y_0 must be independent of itself, hence must be constant. Decomposing the non-deterministic L into two independent Lévy processes M and N, where M has finite variance and is non-deterministic, it follows in complete analogy to the proof of Corollary 17.12 that $\int_0^\infty |g_2(-u)|^2\,\mathrm{d}u = 0$, so that $g_2 = 0$ Lebesgue-almost everywhere. Since $t \mapsto g_2(t)$ is obviously continuous on $\mathbb{R} \setminus \{0\}$, we conclude that $g_2(t) = 0$ for all $t < 0$, which is (18.31). □

Remark 18.27. Observe that the condition that all zeros of $a(z)$ have strictly negative real part is *not* necessary for a CARMA process to be causal. This is easiest seen from (18.20) when all zeros of $a(z)$ have multiplicity 1 and those in the right half-plane are also zeros of $b(z)$. Then (18.31) is still satisfied. If all zeros of $a(z)$ have multiplicity 1 and $a(z)$ and $b(z)$ are assumed to have no common zeros, then it is seen from (18.20) that (18.31) is satisfied if and only if all zeros of $a(z)$ have strictly negative real part. The latter is also true when multiple zeros of $a(z)$ are allowed, but $b(z)$ and $a(z)$ are still assumed to have no common zeros, see Exercise 18.5. In view of this, in the literature a CARMA process is sometimes defined to be causal if all zeros of $a(z)$ have strictly negative real part. When the latter condition is satisfied, then also the strictly stationary state-vector process will be causal, as seen in Theorem 18.26.

Remark 18.28 (One-sided CARMA equations). The one-sided CARMA equations are (18.4) and (18.5) with $t \in \mathbb{R}$ replaced by $t \geq 0$ and $(L_t)_{t\in\mathbb{R}}$ replaced by the one-sided Lévy process $(L_t)_{t\geq 0}$.

If $\mathbb{E}\log^+|L_1| < \infty$ and the distinct zeros, $\lambda_1, \dots, \lambda_r$, of $a(z)$ all have strictly negative real part, so that the integral $\int_{[0,\infty)} e^{As}\mathbf{e}_p \, dL_s$ exists, it is clear from (18.8) and (18.9) that for *any* random vector $\mathbf{X}_0$ which is independent of the increments of $(L_u)_{u\geq 0}$, the solution $(\mathbf{X}_t)_{t\geq 0}$, of the state equation (18.5) on $[0,\infty)$ is Markovian and converges in distribution as $t \to \infty$ to $\int_{[0,\infty)} e^{As}\mathbf{e}_p \, dL_s$ (see Theorem 16.13 and the comment following it). The distribution of this integral is therefore the unique limit distribution of the Markov-stable process $(\mathbf{X}_t)_{t\geq 0}$ and if $\mathbf{X}_0$ is *any* random vector with this distribution and independent of $(L_u)_{u\geq 0}$, the corresponding unique solution of (18.5) on $[0,\infty)$ will be strictly stationary (see Proposition 17.15). Under these conditions, the corresponding CARMA process $Y_t = \mathbf{b}^T\mathbf{X}_t$, $t \geq 0$, will also be strictly stationary. Notice, however, that the strictly stationary solution, $(Y_t)_{t\geq 0}$, is unique only in *distribution* since $\mathbf{X}_0$ is required only to have the distribution of $\int_{[0,\infty)} e^{As}\mathbf{e}_p \, dL_s$ and to be independent of $(L_u)_{u\geq 0}$. This contrasts with the strictly stationary solution of the two-sided CARMA equations which is uniquely determined pathwise by $(L_t)_{t\in\mathbb{R}}$ and A.

If $a(z)$ has zeros in both the left and right halves of the complex plane, a corresponding strictly stationary solution of the one-sided CARMA equations is not uniquely determined pathwise by $(L_t)_{t\geq 0}$ and $a(z)$ and $b(z)$, and one can no longer require $\mathbf{X}_0$ to be independent of $(L_u)_{u\geq 0}$. However, a strictly stationary process on $[0,\infty)$ in this case can be obtained by extending $(L_t)_{t\geq 0}$ to $(L_t)_{t\in\mathbb{R}}$, where $L_t := -L^*_{(-t)-}$ for $t < 0$ and L^* is an independent copy of $(L_t)_{t\geq 0}$, see Theorem 10.17. Each extended L will determine a unique pathwise two-sided CARMA process $(Y_t)_{t\in\mathbb{R}}$ whose restriction to $[0,\infty)$ is a one-sided CARMA process. However, the sample paths of such processes will clearly vary in accordance with those of L^*.

On the other hand, if the zeros of $a(z)$ all lie in the right half of the complex plane, then the solution of the one-sided CARMA equations is uniquely determined by $(L_t)_{t\geq 0}$ and $a(z)$ and $b(z)$ and is given by $Y_t = \mathbf{b}^T\mathbf{X}_t$, where

$$\mathbf{X}_t = -\int_t^\infty e^{A(t-u)}\mathbf{e}_p \, dL_u, \quad t \geq 0,$$

see Theorem 17.4 for the form of the solution and the uniqueness of $\mathbf{X}$; the uniqueness of Y follows as in the proof of Theorem 18.15 from Lemma 18.14 by observing that the coefficients c_m in the Laurent expansion $\sum_{m\in\mathbb{Z}} c_m z^m$ of $1/\Phi(z)$ satisfy $c_m = 0$ for $m \in \mathbb{N}_0$ in this case.

In the modeling of continuous-parameter time series, causality of Y (more precisely, the *causality condition* that all zeros of $a(z)$ lie in the open left half-plane) is usually assumed, as it implies that the behavior of Y up to time t is not dependent on future increments of the driving process L. Moreover, if L has finite variance, then every L-driven non-causal CARMA process has the autocovariance function of a corresponding causal one (see Remark 18.23), so the causality condition is necessary for parameter identifiability based on (empirical) autocovariance functions. Further, the additional assumption that the zeros of $b(z)$ have strictly negative real parts is helpful for parameter identifiability (see Remark 18.23), which we call an *invertibility condition* or *minimum phase condition*. If a CARMA(p,q) process Y satisfies the causality and invertibility condition, and if the driving Lévy process has finite variance, then it has been shown in Theorem 7.19 and Proposition 7.20 that the strictly stationary state vector $\mathbf{X}_t$ is uniquely determined by $(Y_u)_{u\le t}$ and measurable with respect to $\sigma(Y_u : u \le t)$, and since $\mathbf{e}_p(L_t - L_s) = \mathbf{X}_t - \mathbf{X}_s - \int_s^t A\,\mathbf{X}_u\,du$ for $s \le t$ by (18.6), so are the increments $L_t - L_s$. This explains the name *invertibility condition*, since then increments of L up to time t can be recovered from the past of Y up to time t. Similar results hold for general strictly stationary Lévy driven CARMA processes under the invertibility condition, but without the finite variance condition, see Brockwell and Lindner [29, Theorem 2.2].

18.6 The canonical state vector

From time to time we shall need to make one or both of the following assumptions. In this section we shall make both.

Assumption C (Causality condition). The distinct zeros, $\lambda_1,\dots,\lambda_r$ of the autoregressive polynomial $a(z)$ satisfy $\Re\lambda_i < 0$, $i = 1,\dots,r$.

Assumption D (Distinct autoregressive zeros). The multiplicity of each zero of $a(z)$ is 1, i. e., $r = p$ and $\lambda_1,\dots,\lambda_p$ are the zeros of the autoregressive polynomial $a(z)$ that satisfy $\lambda_i \neq \lambda_j$ for $i \neq j$.

(Assumption D is invariably satisfied for models estimated from real data. Models with a zero λ of multiplicity $m(\lambda) > 1$ can be approximated by replacing the multiple zero by $m(\lambda)$ distinct zeros close to λ and allowing their distance from λ to converge to zero.)

Under these assumptions and Assumption A (i. e., $\mathbb{E}\log^+|L_1| < \infty$), it follows at once from (18.20) that the CARMA(p,q) process defined by (18.2) with autoregressive zeros

$\lambda_1, \dots, \lambda_p$ can be expressed as a linear combination of dependent and possibly complex-valued CAR(1) processes (i. e., strictly stationary OU-processes). Specifically,

$$Y_t = \sum_{k=1}^{p} Y_t^{(k)},$$

where

$$Y_t^{(k)} = \alpha_k \int_{-\infty}^{t} \mathrm{e}^{\lambda_k(t-u)}\, \mathrm{d}L_u, \quad k = 1, \dots, p, \tag{18.33}$$

and

$$\alpha_k = \frac{b(\lambda_k)}{a'(\lambda_k)}, \quad k = 1, \dots, p,$$

where a' is the derivative of the function a.

Definition 18.29 (The canonical state vector). Under the previous assumptions, define $R := (\lambda_j^{k-1})_{k,j=1,\dots,p}$, the matrix whose j^{th} column is the right eigenvector corresponding to the eigenvalue λ_j of A, $j = 1, \dots, p$ (see Proposition 18.7 and its proof), and let Λ be the $p \times p$ diagonal matrix whose k^{th} diagonal element is λ_k. Then the *canonical state vector* of the CARMA process $Y = (Y_t)_{t\in\mathbb{R}}$ is defined to be

$$\mathbf{Y}_t := b(\Lambda)R^{-1}\mathbf{X}_t, \quad t \in \mathbb{R},$$

where $\mathbf{X} = (\mathbf{X}_t)_{t\in\mathbb{R}}$ is the strictly stationary state vector process given by

$$\mathbf{X}_t = \int_{-\infty}^{t} \mathrm{e}^{A(t-u)}\, \mathbf{e}_p\, \mathrm{d}L_u.$$

Proposition 18.30 (Representation of canoncial state vector). *Under the previous assumptions, the canonical state vector is equal*

$$\mathbf{Y}_t = [Y_t^{(1)}, Y_t^{(2)}, \dots, Y_t^{(p)}]^T, \quad t \in \mathbb{R}.$$

Proof. By definition, $\mathbf{Y}_t = b(\Lambda)R^{-1} \int_{-\infty}^{t} \mathrm{e}^{A(t-u)}\mathbf{e}_p\, \mathrm{d}L_u$. Since $R^{-1}\mathrm{e}^{A(t-u)} = \mathrm{e}^{\Lambda(t-u)}R^{-1}$ by Proposition 18.7, we can write

$$\mathbf{Y}_t = b(\Lambda) \int_{-\infty}^{t} \mathrm{e}^{\Lambda(t-u)}R^{-1}\mathbf{e}_p\, \mathrm{d}L_u. \tag{18.34}$$

It remains to evaluate $R^{-1}\mathbf{e}_p$. For that, let $k \in \{1, \dots, p\}$ and $\rho_s : [0, 2\pi] \to \mathbb{C}$ be the circle with radius s given by $\rho_s(t) = s\,\mathrm{e}^{\mathrm{i}t}$, where s is large enough such that ρ_s encircles all zeros of $a(z)$. From the residue theorem, we then obtain

$$\sum_{j=1}^{p} \frac{\lambda_j^{k-1}}{a'(\lambda_j)} = \sum_{j=1}^{p} \mathrm{Res}_{z=\lambda_j}\left[\frac{z^{k-1}}{a(z)}\right] = \frac{1}{2\pi\mathrm{i}} \int_{\rho_s} \frac{z^{k-1}}{a(z)}\,\mathrm{d}z = \frac{1}{2\pi} \int_0^{2\pi} \frac{(s\,\mathrm{e}^{\mathrm{i}t})^k}{a(s\,\mathrm{e}^{\mathrm{i}t})}\,\mathrm{d}t.$$

Letting $s \to \infty$ we obtain from dominated convergence

$$\sum_{j=1}^{p} \frac{\lambda_j^{k-1}}{a'(\lambda_j)} = \begin{cases} 0, & k = 1, \dots, p-1, \\ 1, & k = p. \end{cases}$$

The definition of R then implies

$$R^{-1}\mathbf{e}_p = \left[\frac{1}{a'(\lambda_1)}\ \frac{1}{a'(\lambda_2)}\ \cdots\ \frac{1}{a'(\lambda_p)}\right]^T,$$

as can be seen by multiplying each side of this equation by R and using the above identities. Substituting this into (18.34) gives the required result. □

Example 18.31 (CARMA(2, 1) process). If $Y = (Y_t)_{t\in\mathbb{R}}$ is the Lévy-driven CARMA(2, 1) process satisfying

$$(D^2 + a_1 D + a_2)Y_t = (b_0 + D)DL_t,$$

where $a(z) = z^2 + a_1 z + a_2 = (z - \lambda_1)(z - \lambda_2)$ and Assumptions A, C and D are satisfied, then the kernel of Y is

$$g(t) = (\alpha_1 \mathrm{e}^{\lambda_1 t} + \alpha_2 \mathrm{e}^{\lambda_2 t})\mathbf{1}_{[0,\infty)}(t),$$

where $\alpha_k = (b_0 + \lambda_k)/(\lambda_k - \lambda_{3-k})$, $k = 1, 2$. Consequently, Y has the representation,

$$Y_t = \alpha_1 \int_{-\infty}^{t} \mathrm{e}^{\lambda_1(t-u)}\,\mathrm{d}L_u + \alpha_2 \int_{-\infty}^{t} \mathrm{e}^{\lambda_2(t-u)}\,\mathrm{d}L_u,$$

as a linear combination of the CAR(1) processes, $\int_{-\infty}^{t} \mathrm{e}^{\lambda_k(t-u)} dL_u$, $t \in \mathbb{R}$, $k = 1, 2$, or equivalently as the sum $Y_t^{(1)} + Y_t^{(2)}$ with $Y_t^{(k)}$ defined as in (18.33). From Proposition 18.30 and Definition 18.29 the canonical state vector is

$$\mathbf{Y}_t = \begin{bmatrix} Y_t^{(1)} \\ Y_t^{(2)} \end{bmatrix} = b(\Lambda)R^{-1}\mathbf{X}_t = \frac{1}{\lambda_2 - \lambda_1} \begin{bmatrix} \lambda_2(b_0 + \lambda_1) & -(b_0 + \lambda_1) \\ -\lambda_1(b_0 + \lambda_2) & b_0 + \lambda_2 \end{bmatrix} \mathbf{X}_t,$$

where $\mathbf{X}_t = \int_{-\infty}^{t} \mathrm{e}^{A(t-u)}\mathbf{e}_2\,\mathrm{d}L_u$.

18.7 The Δ-sampled sequence Y^Δ, when $\mathbb{E}|L_1|^2 < \infty$

In many applications, the values of a CARMA process Y are recorded at uniformly spaced times, say $t_k = k\Delta$, $k = 1, \dots, n$. We already know from (18.24), (18.25) and Remark 18.18 that the sampled sequence $Y_n^\Delta := Y_{n\Delta}$, $n \in \mathbb{Z}$, satisfies an autoregressive equation of order p, driven by a $(p-1)$-dependent sequence. If the driving Lévy process L satisfies $\mathbb{E}|L_1|^2 < \infty$, as we shall assume in this section, then by (18.28), $(Y_n^\Delta - \mathbb{E}(Y_n^\Delta))_{n\in\mathbb{Z}}$ is an ARMA(p, s) sequence for some $s < p$. The purpose of this section is to determine, for any specified second-order CARMA process Y with real coefficients, under Assumptions C and D, the parameters of the discrete-time ARMA sequence Y^Δ, where we will follow the exposition in [30, Section 2]. We already know, from (18.24) (alternatively, from (18.28)), the autoregressive coefficients of Y^Δ and, from Lemma 18.19, that $\mathbb{E}Y_t^\Delta = \mathbb{E}(L_1)b_0/a_p$, so the problem is to determine the moving average coefficients and the white noise variance of the ARMA sequence $(Y_n^\Delta)_{n\in\mathbb{Z}}$.

Let us start with the following general result, which is interesting in its own right and expresses the spectral density of the Δ-sampled sequence in terms of the spectral density of the underlying second-order continuous-time process, provided it exists.

Proposition 18.32 (Spectral density of Δ-sampled sequence). *Let $W = (W_t)_{t\in\mathbb{R}}$ be a $\mathbb{C}$-valued mean-square continuous weakly stationary process and assume that it has a spectral density $f : \mathbb{R} \to [0, \infty]$. Let $\Delta > 0$. Then the Δ-sampled sequence $W^\Delta = (W_n^\Delta)_{n\in\mathbb{Z}}$ also has a spectral density, f_Δ, which is given by*

$$f_\Delta(\omega) := \frac{1}{\Delta} \sum_{k=-\infty}^{\infty} f\left(\frac{\omega + 2\pi k}{\Delta}\right), \quad \omega \in (-\pi, \pi]. \tag{18.35}$$

Proof. Define $f_\Delta : (-\pi, \pi] \to [0, \infty]$ by the right-hand side of (18.35). Since $f \in L^1(\mathbb{R}; \mathbb{C})$, the function f_Δ is in $L^1((-\pi, \pi]; \mathbb{C})$ and an application of Lebesgue's dominated convergence theorem gives, for $n \in \mathbb{Z}$,

$$\begin{aligned}\int_{-\pi}^{\pi} \mathrm{e}^{in\omega} f_\Delta(\omega)\,\mathrm{d}\omega &= \frac{1}{\Delta} \sum_{k=-\infty}^{\infty} \int_{-\pi}^{\pi} \mathrm{e}^{in\omega} f\left(\frac{\omega + 2\pi k}{\Delta}\right) \mathrm{d}\omega \\ &= \sum_{k=-\infty}^{\infty} \int_{(2k-1)\pi/\Delta}^{(2k+1)\pi/\Delta} \mathrm{e}^{in(\Delta u - 2\pi k)} f(u)\,du \\ &= \int_{\mathbb{R}} \mathrm{e}^{in\Delta u} f(u)\,du.\end{aligned}$$

Since f is the spectral density of W, the latter is equal to the autocovariance function of W at lag Δn, hence to the autocovariance function of W^Δ at lag n. We have shown that $\mathrm{Cov}(Y_n^\Delta, Y_0^\Delta) = \int_{-\pi}^{\pi} \mathrm{e}^{in\omega} f_\Delta(\omega)\,\mathrm{d}\omega$ for all $n \in \mathbb{Z}$, meaning that f_Δ is the spectral density of Y^Δ. □

With these preparations, we can now determine the moving average coefficients and the white noise variance in (18.28) of the sampled CARMA process. Just as for CARMA processes with the differentiation operator D, we shall write $\prod_{j=1}^{p-1}(1-\eta_j B)Z_n$ to denote $\Theta(B)Z_n$, when $\prod_{j=1}^{p-1}(1-\eta_j z) = 1+\theta_1 z+\cdots+\theta_{p-1}z^{p-1}$ and alike (with B the backward shift operator).

Theorem 18.33 (Δ-sampled second-order CARMA. Sampling Theorem). *Suppose that L is an $\mathbb{R}$-valued Lévy process with $\mathbb{E}(L_1) = m$ and $\operatorname{Var}(L_1) = \nu^2 \in (0,\infty)$ and that Y is the strictly stationary* CARMA(p,q) *process defined by the formal equation,*

$$\prod_{j=1}^{p}(D-\lambda_j)Y_t = \prod_{j=1}^{q}(D-\mu_j)DL_t.$$

In addition to Assumptions C *and* D, *assume that the autoregressive and moving average polynomials, $a(z) = \prod_{j=1}^{p}(z-\lambda_j)$ and $b(z) = \prod_{j=1}^{q}(z-\mu_j)$, have real-valued coefficients. Define*

$$K_j := \frac{b(\lambda_j)b(-\lambda_j)}{a'(\lambda_j)a(-\lambda_j)}, \quad j = 1,\dots,p,$$

and the polynomial of degree $s \in \{0,\dots,p-1\}$,

$$k(z) := \sum_{j=1}^{p} K_j \sinh(\lambda_j\Delta) \prod_{i\in\{1,\dots,p\}\setminus\{j\}} (z-\cosh(\lambda_i\Delta)), \quad z \in \mathbb{C}.$$

Then $k(z)$ is not the zero-polynomial. Denoting by $k_1,\dots,k_s$ (when $s \neq 0$) the (not necessarily distinct) zeros of $k(z)$, exactly one of the values $k_j \pm \sqrt{k_j^2-1}$ has absolute value less than 1 and the other has absolute value greater than 1, for each $j = 1,\dots,s$. Define η_j for $j \in \{1,\dots,s\}$ to be the value of $k_j \pm \sqrt{k_j^2-1}$ with absolute value less than 1 and η_j, for $j \in \{s+1,\dots,p-1\}$, to be zero. Then $Y^\Delta = (Y_n^\Delta = Y_{n\Delta})_{n\in\mathbb{Z}}$ satisfies the causal invertible ARMA(p,s) *equations,*

$$\prod_{j=1}^{p}(1-e^{\lambda_j\Delta}B)(Y_n^\Delta - c) = \prod_{j=1}^{p-1}(1-\eta_j B)Z_n, \quad n \in \mathbb{Z}, \tag{18.36}$$

where, from Lemma 18.19, $c = \mathbb{E}(Y_n^\Delta) = \mathbb{E}(Y_t) = mb_0/a_p$, $(Z_n)_{n\in\mathbb{Z}}$ *is a real-valued zero-mean white noise (uncorrelated, but not necessarily independent) with variance*

$$\sigma^2 = \frac{\nu^2\prod_{j=1}^{p} e^{\lambda_j\Delta}}{\prod_{j=1}^{s}\eta_j}(-2)^{p-s}c_s > 0, \tag{18.37}$$

and $c_s \neq 0$ is the coefficient of z^s in $k(z)$. (If $\sum_{j=1}^{p} K_j \sinh(\lambda_j\Delta) \neq 0$, then $s = p-1$ and $c_s = \sum_{j=1}^{p} K_j \sinh(\lambda_j\Delta)$.) Further, the polynomials

$$\Phi(z) = \prod_{j=1}^{p}(1 - e^{\lambda_i \Delta} z) \quad \textit{and} \quad \Theta(z) := \prod_{j=1}^{s}(1 - \eta_j z), \quad z \in \mathbb{C},$$

have real-valued coefficients.

Proof. We first have a closer look at the ARMA representation (18.28). Since we assumed $a(z)$ to have real-valued coefficients, the zeros λ_j of $a(z)$ are either real or appear in pairs of complex conjugate numbers, so that from (18.22) we infer that the coefficients $\phi_1, \dots, \phi_p$ of the autoregressive polynomial Φ of $Y^\Delta - c$ are also real-valued. Since also the coefficients of $b(z)$ are assumed to be real-valued, we conclude that also the Z_n^k from (18.25) are real-valued, so that also $W_n = Z_n^1 + \cdots + Z_{n-p-1}^p$, $n \in \mathbb{Z}$, is real-valued and $(p-1)$-dependent. Proposition 3.2.1 in Brockwell and Davis [22] then shows that there are $s' \in \{0, \dots, p-1\}$, *real-valued coefficients* $\theta_1, \dots, \theta_{s'}$ and a real-valued white noise sequence $(Z_n)_{n \in \mathbb{Z}}$ with variance σ^2 such that $W_n - \mathbb{E}(W_n) = Z_n + \theta_1 Z_{n-1} + \cdots + \theta_{s'} Z_{n-s'}$ (we use s' for the moment and will later show that $s' = s$ with s as defined in the statement of the theorem). Observe that $\sigma^2 > 0$, for otherwise Z_n and hence W_n would be deterministic, hence so would be $Y_\Delta - c$ by Lemma 18.14, contradicting $\operatorname{Var}(Y_\Delta) = (2\pi)^{-1} \nu^2 \int_{\mathbb{R}} |b(\mathrm{i}\omega)/a(\mathrm{i}\omega)|^2 \,\mathrm{d}\omega > 0$ (cf., Theorem 18.22). The process $(Z_n)_{n \in \mathbb{Z}}$ is given by $Z_n = (W_n - \mathbb{E}(W_n)) - P_{\mathcal{M}_{n-1}}(W_n - \mathbb{E}(W_n))$, where $P_{\mathcal{M}_{n-1}}$ is the projection onto the closed linear span of $\{W_k - \mathbb{E}(W_k) : k \in \mathbb{Z}, k \le n-1\}$, see the proof of [22, Prop. 3.2.1]. Denoting

$$\Theta(z) = 1 + \theta_1 z + \cdots + \theta_{s'} z^{s'}, \quad z \in \mathbb{C},$$

we hence have $\theta_1, \dots, \theta_{s'} \in \mathbb{R}$ and $\Theta(B) Z_n = W_n - \mathbb{E}(W_n)$. From the definition of Z_n we see that the decomposition $W_n - \mathbb{E}(W_n) = (\sigma\Theta)(B)(\sigma^{-1} Z_n)$ satisfies the properties (i)–(vi) of Theorem 6.5, hence is the Wold decomposition of $W_n - \mathbb{E}(W_n)$. From (6.8) (alternatively, from (6.11)) we see that $\Theta(z) \neq 0$ for all $z \in \mathbb{C}$ with $|z| < 1$. Since $Y^\Delta - c$ is an ARMA(p, s') process with $\Phi(B)(Y_n^\Delta - c) = \Theta(B) Z_n$, by (6.16) the spectral density of $Y^\Delta - c$ is given by

$$f_\Delta(\omega) = \frac{\sigma^2}{2\pi} |\Theta(e^{-\mathrm{i}\omega})|^2 / |\Phi(e^{-\mathrm{i}\omega})|^2, \quad \omega \in (-\pi, \pi]. \tag{18.38}$$

On the other hand,

$$f_\Delta(\omega) = \frac{\nu^2}{2\pi\Delta} \sum_{k=-\infty}^{\infty} \frac{|b(\mathrm{i}(\omega + 2\pi k)/\Delta)|^2}{|a(\mathrm{i}(\omega + 2\pi k)/\Delta)|^2}$$

by Proposition 18.32. Since $a(z)$ and $b(z)$ have only finitely many zeros and are continuous, it is clear from this representation that $\inf_{\omega \in (-\pi,\pi]} f_\Delta(\omega) > 0$ and we conclude that $\Theta(e^{-\mathrm{i}\omega}) \neq 0$, so that Θ has no zeros on the unit circle (the argument just used is due to [20]). Hence, we can write

$$\Theta(z) = \prod_{j=1}^{s'}(1 - \eta_j' z), \quad z \in \mathbb{C},$$

with $\eta'_1, \dots, \eta'_{s'}$ (the reciprocals of the zeros of Θ, counted with multiplicity) all having absolute value less than 1 (we shall show later that $\eta'_j = \eta_j$ for $j \in \{1, \dots, s' = s\}$, but denote these zeros for the moment by η'_j). Since also $|e^{\lambda_j \Delta}| < 1$ for all $j \in \{1, \dots, p\}$, the ARMA(p, s') representation $\Phi(B)(Y_n^\Delta - c) = \Theta(B)Z_n$ is causal and invertible.

Our next goal is two find two representations of f_Δ and then equate them. For the first representation, observe that if $\eta'_j \in \mathbb{R}$, then

$$|1 - \eta'_j \mathrm{e}^{-\mathrm{i}\omega}|^2 = 1 + (\eta'_j)^2 - \eta'_j(\mathrm{e}^{\mathrm{i}\omega} + \mathrm{e}^{-\mathrm{i}\omega}) = 1 + (\eta'_j)^2 - 2\eta'_j \cos\omega.$$

If $\eta'_j \notin \mathbb{R}$, then also the complex conjugate $\overline{\eta'_j}$ of η'_j is the reciprocal of a zero of $\Theta(z)$ (since $\Theta(z)$ has real-valued coefficients), and we conclude that

$$|(1 - \eta'_j \mathrm{e}^{-\mathrm{i}\omega})(1 - \overline{\eta'_j} \mathrm{e}^{-\mathrm{i}\omega})|^2 = (1 + (\eta'_j)^2 - 2\eta'_j \cos\omega)(1 + \overline{\eta'_j}^2 - 2\overline{\eta'_j} \cos\omega).$$

Combining both cases, it follows that

$$|\Theta(\mathrm{e}^{-\mathrm{i}\omega})|^2 = \prod_{j=1}^{s'} (1 + (\eta'_j)^2 - 2\eta'_j \cos\omega).$$

Similar calculations applied to Φ give

$$|\Phi(\mathrm{e}^{-\mathrm{i}\omega})|^2 = \prod_{j=1}^{p} (1 + \mathrm{e}^{2\lambda_j \Delta} - 2\mathrm{e}^{\lambda_j \Delta} \cos\omega)$$

and inserting this in (18.38) shows

$$f_\Delta(\omega) = \frac{\sigma^2 \prod_{j=1}^{s'} (1 + (\eta'_j)^2 - 2\eta'_j \cos\omega)}{2\pi \prod_{j=1}^{p} (-2\mathrm{e}^{\lambda_j \Delta}) \prod_{j=1}^{p} (\cos\omega - \cosh(\lambda_j \Delta))}. \tag{18.39}$$

For the other representation of f_Δ, observe from (7.4) that the autocovariance function of the CARMA process Y is

$$\gamma_Y(h) = \nu^2 \sum_{j=1}^{p} K_j \mathrm{e}^{\lambda_j |h|}, \quad h \in \mathbb{R}.$$

The autocovariance function γ_Δ of the sampled sequence Y^Δ coincides at lag n, $n \in \mathbb{Z}$, with the autocovariance function of Y at lag $n\Delta$. Hence

$$\gamma_\Delta(n) = \nu^2 \sum_{j=1}^{p} K_j \mathrm{e}^{\lambda_j \Delta |n|}, \quad n \in \mathbb{Z}. \tag{18.40}$$

Since γ_Δ is the absolutely summable autocovariance function of Y^Δ, its spectral density can also be expressed as

$$f_\Delta(\omega) = \frac{1}{2\pi} \sum_{n=-\infty}^{\infty} \gamma_\Delta(n)\, \mathrm{e}^{-in\omega}, \qquad -\pi < \omega \le \pi,$$

see, e. g., Brockwell and Davis [22, Corollary 4.3.2]. Substituting (18.40) into this, we find that

$$\begin{aligned} f_\Delta(\omega) &= \frac{\nu^2}{2\pi} \sum_{j=1}^{p} K_j \Big(\frac{1}{1-\mathrm{e}^{\lambda_j\Delta - i\omega}} + \frac{1}{1-\mathrm{e}^{\lambda_j\Delta + i\omega}} - 1 \Big) \\ &= \frac{\nu^2}{2\pi} \sum_{j=1}^{p} \frac{K_j \sinh(\lambda_j\Delta)}{\cos\omega - \cosh(\lambda_j\Delta)} \\ &= \frac{\nu^2}{2\pi} \frac{k(\cos\omega)}{\prod_{j=1}^{p}(\cos\omega - \cosh(\lambda_j\Delta))}, \end{aligned}$$

with $k(z)$ as defined in the statement of the theorem (and the calculations also show that the denominators appearing above are different from zero for all $\omega \in (-\pi, \pi]$). Since $f_\Delta(\omega) \neq 0$, we conclude, in particular, that $k(z)$ cannot be the zero-polynomial. Comparing further with (18.39), we obtain

$$k(\cos\omega) = \frac{\sigma^2}{\nu^2}(-2)^{-p} \prod_{j=1}^{p} \mathrm{e}^{-\lambda_j\Delta} \prod_{j=1}^{s'} (1 + (\eta_j')^2 - 2\eta_j' \cos\omega), \qquad \omega \in (-\pi, \pi].$$

From this we deduce

$$c_s \prod_{j=1}^{s} (z - k_j) = k(z) = \frac{\sigma^2}{\nu^2} \frac{\prod_{j=1}^{s'}(-2\eta_j')}{\prod_{j=1}^{p}(-2\mathrm{e}^{\lambda_j\Delta})} \prod_{j=1}^{s'} \Big(z - \frac{1+(\eta_j')^2}{2\eta_j'} \Big), \qquad z \in \mathbb{C}.$$

This implies $s' = s$, $\sigma^2 = \nu^2 c_s (-2)^{p-s} (\prod_{j=1}^{p} \mathrm{e}^{\lambda_j\Delta}) / \prod_{j=1}^{s} \eta_j'$ and $1 + (\eta_j')^2 = 2\eta_j' k_j$ for $j \in \{1, \dots, s\}$, for a suitable enumeration of the zeros of $k(z)$. The equation $x^2 - 2xk_j + 1 = 0$ has the solutions $k_j \pm \sqrt{k_j^2 - 1}$, and we have seen that η_j' is one of the solutions and has absolute value less than 1. Since $(k_j + \sqrt{k_j^2 - 1})(k_j - \sqrt{k_j^2 - 1}) = 1$, the product of the two solutions is 1, hence the other solution must have absolute value greater than 1. This gives the assertions regarding the absolute values of $k_j \pm \sqrt{k_j^2 - 1}$ and shows that $\eta_j' = \eta_j$, finishing the proof. □

Remark 18.34. It may happen that the ARMA polynomials $\Phi(z)$ and $\Theta(z)$, appearing in Theorem 18.33, have common zeros and that hence $Y^\Delta - c$ also satisfies an ARMA equation of lower orders, in particular with autoregressive order less than p.

(a) The simplest example to illustrate this is when $a(z)$ and $b(z)$ have a common zero, λ_1 say. If $\lambda_1 \notin \mathbb{R}$, then also $\overline{\lambda_1}$ is a zero of $a(z)$ and $b(z)$, and factoring out $(z - \lambda_1)$ (or $(z - \lambda_1)(z - \overline{\lambda_1})$, respectively) from $a(z)$ and $b(z)$ shows that Y is also a CARMA process of lower orders (cf., Exercise 18.3), with autoregressive and moving average polynomial having real coefficients. By Theorem 18.33, Y^Δ then has a causal invertible ARMA representation with autoregressive order less than p, and it is known (e. g., [59, Satz 8.10]) that this can only happen if Φ and Θ have a common zero.

(b) Now suppose that $a(z)$ and $b(z)$ have no common zeros. Even though all zeros of $a(z)$ are distinct, it may happen that $e^{\lambda_i\Delta} = e^{\lambda_j\Delta}$ for some $i \neq j$, as is the case if λ_i and λ_j differ by an integer multiple of $2\pi i/\Delta$. It can be shown that again this implies that $\Phi(z)$ and $\Theta(z)$ have common zeros (in that case, Φ has a multiple zero at $e^{-\lambda_i\Delta}$, say with multiplicity $m_i \geq 2$, and if, additionally, $e^{-\lambda_i\Delta}$ were not also a zero of Θ, then the autocovariance function $\gamma_\Delta(n)$ would contain a non-vanishing summand of the form $|n|^{m_i-1}e^{\lambda_i\Delta|n|}$ for large n (as a consequence of [21, Equation (3.12)]), contradicting (18.40).) To give a concrete example, let Y be a CARMA(2, 1) process with $\lambda_1 = -1 + i\pi$ and $\lambda_2 = -1 - i\pi$ and $b(z) = z + 1$ and let $\Delta = 1$. Then a simple calculation shows that $K_1 = (2 + \pi^2 + i\pi)/(8(1 + \pi^2))$, $K_2 = (2 + \pi^2 - i\pi)/(8(1 + \pi^2))$, $k(z) = (2 + \pi^2)/(4(1 + \pi^2))\sinh(1)(z + \cosh(1))$, $\eta_1 = -e^{-1}$, $\Theta(z) = (1 + z/e)$ and $\Phi(z) = (1 + z/e)^2$, so that $-e$ is a common zero of Φ and Θ. Hence the ARMA(2, 1) equation $\Phi(B)(Y_n^\Delta - c) = \Theta(B)Z_n$ reduces to the AR(1) equation $(1+e^{-1}B)(Y_n^\Delta - c) = Z_n$. That Y_n^Δ here is an AR(1) process can also be seen from (18.40), according to which its autocovariance function is given by $\nu^2(2 + \pi^2)(4 + 4\pi^2)^{-1}(-e^{-1})^{|h|}$ for $h \in \mathbb{Z}$.

(c) Now assume, in addition to the assumptions of Theorem 18.33, that $a(z)$ and $b(z)$ do not have common zeros (i. e., $\lambda_i \neq \mu_j$ for all i, j), that $e^{\Delta\lambda_i} \neq e^{\Delta\lambda_j}$ whenever $i \neq j$, and that $\Re\mu_j < 0$ for all j (invertibility condition). Then $b(\lambda_j) \neq 0 \neq b(-\lambda_j)$ so that $K_j \neq 0$. Since the autocovariance function of a causal ARMA(p', s') process with $p' \leq p - 1$, for lags $n \geq \max(p', s'+1) - p'$, can be written as $\sum_{i=1}^{r'} \sum_{j=0}^{m(\xi_i)-1} \beta_{ij} n^j \xi_i^{-j}$, where $\xi_1, \ldots, \xi_{r'}$ are the distinct zeros of the autoregressive polynomial, with multiplicities $m(\xi_i)$ (cf. [22, Equation (3.3.10)]), we see from (18.40) and a linear independence argument that $Y^\Delta - c$ cannot have a representation as an ARMA(p', s') process with $p' \leq p - 1$. In particular, p is the minimal autoregressive order of the ARMA representation, and $\Phi(z)$ and $\Theta(z)$ cannot have common zeros. For identifiability reasons of the CARMA parameters in terms of Y^Δ, we shall hence often make the additional assumptions stated in this part (c).

Example 18.35 (CARMA(2, 1)). Consider the causal invertible CARMA(2, 1) process with $a(z) = (z - \lambda_1)(z - \lambda_2)$ and $b(z) = z - \mu$, where $\Re(\lambda_1), \Re(\lambda_2) < 0$, $\mu \in (-\infty, 0)$, $\lambda_1 \neq \lambda_2$, $\mu \neq \lambda_1$, $\mu \neq \lambda_2$, and $\lambda_1, \lambda_2 \in \mathbb{R}$ or $\lambda_2 = \overline{\lambda_1}$. We have $K_1 = [\mu^2 - \lambda_1^2]/[2\lambda_1(\lambda_1^2 - \lambda_2^2)]$ and $K_2 = [\mu^2 - \lambda_2^2]/[2\lambda_2(\lambda_2^2 - \lambda_1^2)]$. Provided $K_1 \sinh(\lambda_1\Delta) + K_2 \sinh(\lambda_2\Delta) \neq 0$, the parameter η_1 in the invertible representation (18.36) is therefore the root with absolute value less than 1 of the equation,

$$z^2 - 2k_1 z + 1 = 0,$$

where

$$k_1 = \frac{K_1 \sinh(\lambda_1\Delta)\cosh(\lambda_2\Delta) + K_2 \sinh(\lambda_2\Delta)\cosh(\lambda_1\Delta)}{K_1 \sinh(\lambda_1\Delta) + K_2 \sinh(\lambda_2\Delta)}.$$

From η_1 the white-noise variance σ^2 is found directly from (18.37) and the mean is $c = \mathbb{E}Y_n^\Delta = \frac{-\mu m}{\lambda_1\lambda_2}$. As $\Delta \to 0$, we easily find that $\eta_1 = 1+\mu\Delta+o(\Delta)$ and $\sigma^2 \sim \nu^2\Delta$ (and $K_1\sinh(\lambda_1\Delta)+K_2\sinh(\lambda_2\Delta) \neq 0$ for Δ small enough).

Remark 18.36 (High-frequency approximations). The factorization of $k(z)$, required in the application of Theorem 18.33, cannot be carried out in an algebraically explicit manner for CARMA(p, q) processes with $p > 5$, since then the polynomial $k(z)$ is of order greater than 4 (at least if $\sum_{j=1}^p K_j \sinh(\lambda_j\Delta) \neq 0$). However, general asymptotic expressions (as $\Delta \to 0$) for the moving average coefficients and white-noise variance in the representation (18.36) were found in [24] and [25] and applied to inference for the kernel of the underlying CARMA process when observations are made at times $\Delta, 2\Delta, \dots, n\Delta$, with Δ small and n large. Applying Theorem 1 of [25] to Example 18.35 we find that $\eta_1 = 1 + \mu\Delta + o(\Delta)$ and $\sigma^2 \sim \nu^2\Delta$ as $\Delta \to 0$ as in the example above.

18.8 Exercises

Exercise 18.1. Give an alternative proof of Theorem 18.9 by using Corollaries 17.9 and 18.8.

Exercise 18.2. Let Y be a CARMA(p, q) process with kernel g and being driven by a non-deterministic Lévy process L such that Assumptions A and B are satisfied. Let $h \in L^1(\mathbb{R};\mathbb{C}) \cap L^2(\mathbb{R};\mathbb{C})$ be integrable with respect to L such that $\int_{\mathbb{R}} h(t-s)\,\mathrm{d}L_s = Y_t$ $(= \int_{\mathbb{R}} g(t-s)\,\mathrm{d}L_s)$. Show that $h = g$ Lebesgue almost everywhere, so that the kernel of a CARMA process is unique.

Exercise 18.3. Let L be an $\mathbb{R}$-valued Lévy process with finite log-moment and suppose that $a(z)$ has no zeros on the imaginary axis. Consider the CARMA(p, q) process Y satisfying the CARMA(p, q) equations (18.4) and (18.5). Show that if $a(z)$ and $b(z)$ have a common zero λ, then Y is also a CARMA$(p-1, q-1)$-process driven by L with autoregressive polynomial $\tilde{a}(z) = a(z)/(z-\lambda)$ and moving average polynomial $\tilde{b}(z) = b(z)/(z-\lambda)$.

Exercise 18.4. Suppose that all zeros of the autoregressive polynomial $a(z)$ have non-zero real part. Denote by g the kernel of the Lévy driven CARMA(p, q) process.

(a) Applying the Jordan canoncial decomposition of A, as given in Proposition 18.7, show that the Fourier transform $\mathcal{F}g$ of g is given by

$$\mathcal{F}g(\omega) = -\frac{1}{\sqrt{2\pi}}\mathbf{b}^T(A - \mathrm{i}\omega I_p)^{-1}\mathbf{e}_p.$$

(b) Denote by $\mathbf{x} = [x_0, \dots, x_{p-1}]^T$ the unique solution of $-(A - \mathrm{i}\omega I_p)\mathbf{x} = \mathbf{e}_p$, where $\omega \in \mathbb{R}$. Show that $x_j = (\mathrm{i}\omega)^j x_0$ for $j \in \{1, \dots, p-1\}$ and that $a(\mathrm{i}\omega)x_0 = 1$. Deduce that $-\mathbf{b}^T(A - \mathrm{i}\omega I_p)^{-1}\mathbf{e}_p = b(\mathrm{i}\omega)/a(\mathrm{i}\omega)$.

Exercise 18.5. Let L be an $\mathbb{R}$-valued Lévy process with finite log-moment and suppose that $a(z)$ has no zeros on the imaginary axis. Consider the CARMA(p,q) process Y satisfying the CARMA equations (18.4) and (18.5).

(a) Suppose that $a(z)$ and $b(z)$ have no common zeros. Show that Y is causal if and only if all zeros of $a(z)$ have strictly negative real part.
Hint: If a meromorphic function f has a pole of order $n \in \mathbb{N}$ at some $\lambda \in \mathbb{C}$, then the residue of f at λ is given by $\frac{1}{(n-1)!}(\frac{\mathrm{d}}{\mathrm{d}z})^{n-1}h(\lambda)$, where $h(z) := (z-\lambda)^n f(z)$.

(b) Now do not necessarily assume that $a(z)$ and $b(z)$ have no common zeros. Expanding on the argument of (a), show that Y is causal if and only if all singularities of the meromorphic function $b(z)/a(z)$ in the open right half-plane are removable, meaning that every zero λ_i of $a(z)$ in the open right half-plane is also a zero of $b(z)$ of multiplicity as least as large as the multiplicity of λ_i as a zero of $a(z)$.

Exercise 18.6. Let $L = (L_t)_{t\in\mathbb{R}}$ be an $\mathbb{R}$-valued Lévy process with mean zero and finite non-zero variance and let Y be the CARMA(p,q) process satisfying the CARMA equations (18.4) and (18.5), where both the causality and the invertibility condition are assumed. The process Y is then given by $Y_t = \int_{-\infty}^t \mathbf{b}^T e^{A(t-u)}\mathbf{e}_p\,\mathrm{d}L_u$. Show that, apart from the constant factor $\sqrt{\operatorname{Var}(L_1)}$, this is a Wold–Karhunen representation of Y, i. e., that

$$Y_t = \int_{-\infty}^{t} \sqrt{\operatorname{Var}(L_1)}\,\mathbf{b}^T e^{A(t-u)}\mathbf{e}_p\,\mathrm{d}(L_u/\sqrt{\operatorname{Var}(L_1)})$$

is a Wold–Karhunen representation of Y.

Exercise 18.7. Let g be the kernel of a CARMA(p,q) process, as defined in (18.19), and assume that all of the zeros of $a(z)$ lie in the open left half-plane (Assumption C). Show that $g(0) = 1$ if $q = p-1$ and $g(0) = 0$ if $q < p-1$.

Exercise 18.8. Let $W = (W_t)_{t\in\mathbb{R}}$ be a $\mathbb{C}$-valued mean-square continuous weakly stationary process with spectral distribution function F. Let $\Delta > 0$. Modifying the proof of Proposition 18.32, show that the Δ-sampled sequence $W^\Delta = (W_n^\Delta)_{n\in\mathbb{Z}}$ has spectral distribution function F_Δ satisfying

$$F_\Delta(\omega_1) - F_\Delta(\omega_2) = \sum_{k=-\infty}^{\infty}\left(F\left(\frac{\omega_2 + 2\pi k}{\Delta}\right) - F\left(\frac{\omega_1 + 2\pi k}{\Delta}\right)\right)$$

for $-\pi < \omega_1 \le \omega_2 \le \pi$.

19 QML estimation for CARMA processes

Abstract: This chapter is devoted to quasi maximum likelihood estimation (QMLE) of the CARMA parameters and estimation of the Lévy increments of the driving process. Inference based on irregularly sampled data is dealt with in Section 19.1, but subsequent sections are concerned with inference based on observations made at regular intervals. An important aspect of this is the relation between the parameters of the CARMA process and the ARMA sequence obtained by sampling it at regular intervals. The sampling problem was dealt with by Theorem 18.33. In this chapter, we need a corresponding result for *embedding* an ARMA sequence into a CARMA process with matching autocorrelation structure at discrete lags. This is dealt with by Theorem 19.5. An alternative approach to estimation of CARMA parameters is via Whittle estimation, an excellent account of which is given by Fasen–Hartmann and Mayer [43]. However, we shall restrict attention here to QMLE (or its large-sample equivalent, least squares estimation). Illustrative examples are given with R programs, which can be copied and pasted into R, after first loading the package yuima (Brouste et al. [31]), to reproduce the results. Additional R programs to generate a variety of Lévy increments are included in Appendix A at the end of the book.

19.1 QML estimation for CARMA processes based on observations at fixed times $t_1, \ldots, t_n$

Parameter estimation for discrete-time ARMA sequences is usually based on maximization of the *Gaussian likelihood*, i. e., the likelihood of the observations under the assumption that the ARMA sequence is driven by a Gaussian white noise sequence. This estimation procedure is frequently referred to as *quasi maximum likelihood estimation* (*QML estimation*, or *QMLE*) and can be shown to have certain optimality properties (see Brockwell and Davis [22]), even when the underlying ARMA sequence is not Gaussian. We shall adopt the same procedure for CARMA processes.

When doing maximum likelihood estimation, we shall assume that all zeros of both the autoregressive polynomial $a(z)$ and the moving average polynomial $b(z)$ of the CARMA(p, q) process Y lie in the open left half-plane (causality condition and invertibility condition). We shall further assume then that the autoregressive polynomial $a(z)$ and the moving average polynomial $b(z)$ have real-valued coefficients, and that the driving Lévy process L has mean m and finite non-zero variance ν^2.

To compute the Gaussian likelihood we shall assume, whether it is true or not, that the driving Lévy process L is Gaussian with mean $\mathbb{E}(L_1) = m$ and $\mathrm{Var}(L_1) = \nu^2 \in (0, \infty)$ and determine the likelihood as a function of the coefficients of a causal CARMA(p, q) model and the parameters m and ν^2. The goal of QML estimation is then to maximize this function with respect to its $p + q + 2$ arguments.

https://doi.org/10.1515/9783111325033-019

With the aid of the state-space representation of Y and the discrete-time Kalman recursions, it is possible to develop an algorithm (see Jones [54], who applied this algorithm to the problem of modeling time series observed at irregularly-spaced times) for the recursive calculation of the Gaussian likelihood. The remainder of this section is devoted to the algorithm.

Denoting $B_t := (L_t - mt)/v$, the process $B = (B_t)$ is a Lévy process with mean 0 and variance 1, and since we assume it to be Gaussian, it is a standard Brownian motion. Denote by $W = (W_t)_{t\in\mathbb{R}}$ the CARMA(p,q) process driven by vB but otherwise the same parameters as Y, and denote the corresponding strictly stationary state vector process by $(\mathbf{U}_t)$. Let $(\mathbf{X}_t)$ be the strictly stationary state vector process of Y. Then, with A, $\mathbf{e}_p$ and $\mathbf{b}$ as defined in (18.7),

$$\mathbf{U}_t = v\int_{-\infty}^{t} e^{A(t-u)}\mathbf{e}_p\, dB_t,$$

$$\mathbf{X}_t = \int_{-\infty}^{t} e^{A(t-u)}\mathbf{e}_p\, d(vB_t + mt) = \mathbf{U}_t + m[1,0,\dots,0]^T/a_p,$$

$$W_t = \mathbf{b}^T\mathbf{U}_t \quad \text{and}$$

$$Y_t = \mathbf{b}^T\mathbf{X}_t = W_t + mb_0/a_p.$$

Now let $t_1 < t_2 < \cdots < t_n$ and observations $Y_{t_1} = y_1,\dots,Y_{t_n} = y_n$ be given. Denote the corresponding observations of W_{t_i} by $w_i = y_i - mb_0/a_p$, and the unobserved realizations of the corresponding state vectors by $\mathbf{x}_i$ and $\mathbf{u}_i$, respectively. From (18.4) and (18.9) we immediately obtain the following discrete state-space representation of $W_{t_1},\dots,W_{t_n}$,

$$W_{t_i} = \mathbf{b}^T\mathbf{U}_{t_i}, \quad i = 1,\dots,n, \tag{19.1}$$

and

$$\mathbf{U}_{t_{i+1}} = e^{A(t_{i+1}-t_i)}\mathbf{U}_{t_i} + v\int_{t_i}^{t_{i+1}} e^{A(t_{i+1}-u)}\mathbf{e}_p\, dB_u, \quad i = 1,2,\dots,n-1, \tag{19.2}$$

where each $\mathbf{U}_{t_i}$ has the multivariate normal distribution with mean zero and covariance matrix given by $v^2\int_0^\infty e^{At}\mathbf{e}_p\mathbf{e}_p^T e^{A^Tt}\,dt$, the latter from Corollary 17.5; the multivariate normal distribution follows from the construction of the OIP integral as a limit of integrals of simple functions with respect to Brownian motion. Observe that in the special case when $p = 1$, $\mathbf{b}$ and A take the scalar values 1 and $-a_1$, respectively.

The *observation equation* (19.1) and *state equation* (19.2) are in precisely the form required for application of the discrete-time Kalman recursions (see, e. g., Brockwell and Davis [22, Sections 12.1 and 12.2]) in order to compute numerically the best one-step linear predictors of $W_{t_1},\dots,W_{t_n}$, their mean-squared errors, and hence the Gaussian likelihood of the observations $w_1, \ldots, w_n$ in terms of ν^2 and the coefficients $\{a_j, 1 \le j \le p;\ b_j, 0 \le j < q\}$; from this we obtain the Gaussian likelihood of the observations $y_1, \ldots, y_n$ in terms of m, ν^2 and $\{a_j, 1 \le j \le p;\ b_j, 0 \le j < q\}$. Jones [54] used this representation, together with numerical maximization of the calculated Gaussian likelihood, to compute maximum Gaussian likelihood estimates of the parameters for time series with irregularly spaced data. A similar approach was used in a more general setting by Bergstrom [12].

To apply the Kalman recursions to (19.1) and (19.2), we rewrite them in the notation of Brockwell and Davis [22, Proposition 12.2.2], as

$$W_{t_i} = \mathbf{b}^T \mathbf{U}_{t_i}, \quad i = 1, 2, \ldots,$$

and

$$\mathbf{U}_{t_{i+1}} = F_i \mathbf{U}_{t_i} + \mathbf{V}_i, \quad i = 1, 2, \ldots,$$

where the matrices F_i and vectors $\mathbf{V}_i$ are given by

$$F_i = \mathrm{e}^{A(t_{i+1} - t_i)}$$

and

$$\mathbf{V}_i = \nu \int_{t_i}^{t_{i+1}} \mathrm{e}^{A(t_{i+1} - u)} \mathbf{e}_p \, \mathrm{d}B_u,$$

respectively. The vector $\mathbf{V}_i$ is multivariate Gaussian with mean zero and covariance matrix

$$Q_i = \nu^2 \int_0^{t_{i+1} - t_i} \mathrm{e}^{At} \mathbf{e}_p \mathbf{e}_p^T \mathrm{e}^{A^T t} \, \mathrm{d}t$$

and the vectors $\mathbf{V}_1, \mathbf{V}_2, \ldots,$ are independent (cf., Corollary 16.16 and Theorem 16.7).

Now let $\hat{\mathbf{X}}_{t_i}$, $\hat{\mathbf{U}}_{t_i}$, $\hat{Y}_{t_i}$ and $\hat{W}_{t_i}$ denote the componentwise orthogonal projections in $\mathcal{S}$ (cf., Section 2.6) of $\mathbf{X}_{t_i}$, $\mathbf{U}_{t_i}$, Y_{t_i} and W_{t_i} onto the closed linear span of $\{1, Y_{t_j}, j < i\}$. Since Y_{t_j} and W_{t_j} differ only by a constant, the latter is equal to the closed linear span of $\{1, W_{t_j}, j < i\}$. Given the realizations $y_1, \ldots, y_n$ (equivalently, $w_1, \ldots, w_n$), denote the corresponding realizations of the predictors by $\hat{\mathbf{x}}_i$, $\hat{\mathbf{u}}_i$, $\hat{y}_i$ and $\hat{w}_i$ (i. e., if $\hat{\mathbf{X}}_{t_i} = C_i[1, Y_{t_1}, \ldots, Y_{t_{i-1}}]^T$ for a matrix $C_i \in \mathbb{R}^{p \times i}$, then $\hat{\mathbf{x}}_i = C_i[1, y_1, \ldots, y_{i-1}]^T$ and similarly for the other quantities). Then by the Kalman recursions (see Brockwell and Davis [22, Prop. 12.2.2]), the predictors $\hat{\mathbf{U}}_{t_i}$ and their error covariance matrices $\Omega_i = \mathbb{E}[(\mathbf{U}_{t_i} - \hat{\mathbf{U}}_{t_i})(\mathbf{U}_{t_i} - \hat{\mathbf{U}}_{t_i})^T]$ are uniquely determined by the initial conditions (recalling that $\mathbb{E}\mathbf{U}_{t_1} = 0$),

$$\hat{\mathbf{U}}_{t_1} = 0, \quad \Pi_1 = \nu^2 \int_0^\infty \mathrm{e}^{At} \mathbf{e}_p \mathbf{e}_p^T \mathrm{e}^{A^T t} \, \mathrm{d}t, \quad \Psi_1 = 0_{p \times p}, \quad \Omega_1 = \Pi_1 - \Psi_1,$$

the recursions,

$$\begin{cases} \Delta_i = \mathbf{b}^T \Omega_i \mathbf{b} \\ \Theta_i = F_i \Omega_i \mathbf{b} \\ \Pi_{i+1} = F_i \Pi_i F_i^T + Q_i \\ \Psi_{i+1} = F_i \Psi_i F_i^T + \Delta_i^{-1} \Theta_i \Theta_i^T \\ \Omega_{i+1} = \Pi_{i+1} - \Psi_{i+1} \end{cases}$$

and

$$\hat{\mathbf{U}}_{t_{i+1}} = F_i \hat{\mathbf{U}}_{t_i} + \Theta_i \Delta_i^{-1} (W_{t_i} - \mathbf{b}^T \hat{\mathbf{U}}_{t_i}).$$

The best linear predictor of W_{t_i} in terms of $\{1, W_{t_k}, k < i\}$ is

$$\hat{W}_{t_i} = \mathbf{b}^T \hat{\mathbf{U}}_{t_i}$$

and its mean-squared error is

$$\mathbb{E}[(W_{t_i} - \hat{W}_{t_i})^2] = \Delta_i = \mathbf{b}^T \Omega_i \mathbf{b}$$

(from which it is easily seen that $\Delta_i > 0$). Notice that the mean-squared errors of $\hat{W}_{t_{i+1}}$ can be expressed as

$$\mathbb{E}[(W_{t_{i+1}} - \hat{W}_{t_{i+1}})^2] = r_i v^2, \quad i = 0, 1, 2, \dots, \tag{19.3}$$

where the coefficients r_i are equal to the corresponding mean-squared errors when v^2 is replaced in the Kalman recursions by 1. The matrices $D_{i+1} \in \mathbb{R}^{1\times(i+1)}$ (i. e., row vectors) such that $\hat{W}_{t_{i+1}} = D_{i+1}[1, W_{t_1}, \dots, W_{t_i}]^T$ are not affected by the value of v from the derived form of the predictors (observe also that the first entry of D_{i+1} is zero as a consequence of $\mathbb{E}W_{t_1} = \dots = \mathbb{E}W_{t_{i+1}} = 0$ and Exercise 2.12(a)). Hence we can use the Kalman recursions with v set equal to 1 to determine both the predictors and the constants r_i in (19.3). The realizations of the predictors, based on realizations $W_{t_1} = w_1, \dots, W_{t_n} = w_n$, can now be calculated as

$$\hat{\mathbf{u}}_{i+1} = F_i \hat{\mathbf{u}}_i + \Theta_i \Delta_i^{-1} (w_i - \mathbf{b}^T \hat{\mathbf{u}}_i) \quad \text{and} \quad \hat{w}_i = \mathbf{b}^T \hat{\mathbf{u}}_i.$$

The Gaussian likelihood of the observations $w_1, \dots, w_n$ under the Gaussian CARMA model W with parameters $\mathbf{a} := (a_1, \dots, a_p)^T$, $\mathbf{b} := (b_0, \dots, b_{q-1})^T$ and v^2, can then be written as (cf., Brockwell and Davis [22, equation (8.7.4) and Section 12.3]),

$$\begin{aligned} &L_W(\mathbf{a}, \mathbf{b}, v^2 | w_1, \dots, w_n) \\ &= \frac{1}{\sqrt{(2\pi v^2)^n r_0 \cdots r_{n-1}}} \exp\left\{ -\frac{1}{2v^2} \sum_{j=1}^{n} \frac{(w_j - \hat{w}_j)^2}{r_{j-1}} \right\}. \end{aligned}$$

Since $Y_{t_i} = W_{t_i} + mb_0/a_p, y_i = w_i + mb_0/a_p$ and $\hat{y}_i = \hat{w}_i + mb_0/a_p$, the Gaussian likelihood $L = L_Y$ of the observations $y_1, \ldots, y_n$ under the Gaussian CARMA model Y with parameters $\mathbf{a} := (a_1, \ldots, a_p)^T, \mathbf{b} := (b_0, \ldots, b_{q-1})^T$, m and ν^2, can then be written as

$$\begin{aligned}
&L(\mathbf{a}, \mathbf{b}, m, \nu^2 | y_1, \ldots, y_n) \\
&\quad = L_W(\mathbf{a}, \mathbf{b}, \nu^2 | y_1 - mb_0/a_p, \ldots, y_n - mb_0/a_p) \\
&\quad = \frac{1}{\sqrt{(2\pi\nu^2)^n r_0 \cdots r_{n-1}}} \exp\left\{ -\frac{1}{2\nu^2} \sum_{j=1}^{n} \frac{(y_j - \hat{y}_j)^2}{r_{j-1}} \right\}.
\end{aligned}$$

Differentiating $\log L(\mathbf{a}, \mathbf{b}, m, \nu^2 | y_1, \ldots, y_n)$ partially with respect to ν^2 and noting that $y_j - \hat{y}_j = w_j - \hat{w}_j = (1 - D_j[1, w_1, \ldots, w_{j-1}]^T) w_j$ and r_j do not depend on ν^2, we find that the maximum likelihood estimators $\hat{\mathbf{a}}, \hat{\mathbf{b}}, \hat{m}$ and $\hat{\nu}^2$ satisfy

$$\hat{\nu}^2 = n^{-1} S(\hat{\mathbf{a}}, \hat{\mathbf{b}}, \hat{m} | y_1, \ldots, y_n), \tag{19.4}$$

where

$$S(\mathbf{a}, \mathbf{b}, m | y_1, \ldots, y_n) = \sum_{j=1}^{n} (y_j - \hat{y}_j)^2 / r_{j-1},$$

and $\hat{\mathbf{a}}, \hat{\mathbf{b}}$ and $\hat{m}$ are the values of $\mathbf{a}, \mathbf{b}$ and m that minimize the *reduced likelihood*,

$$\ell(\mathbf{a}, \mathbf{b}, m | y_1, \ldots, y_n) = \log(n^{-1} S(\mathbf{a}, \mathbf{b}, m | y_1, \ldots, y_n)) + n^{-1} \sum_{j=1}^{n} \log r_{j-1}. \tag{19.5}$$

(Setting the derivative of $\log L$ with respect to ν^2 equal to zero, we find that maximum of $\log L$ must lie on the manifold described by $\nu^2 = n^{-1} S(\mathbf{a}, \mathbf{b}, m)$. Inserting this expression for ν^2 leads to the minimization problem of the reduced likelihood.) For any given parameter vectors $\mathbf{a}, \mathbf{b}$ and m, the reduced likelihood, $\ell(\mathbf{a}, \mathbf{b}, m | y_1, \ldots, y_n)$, can be computed from the data and the Kalman recursions as described above. A numerical search algorithm is then used to determine values of $\mathbf{a}, \mathbf{b}$ and m that minimize $\ell(\mathbf{a}, \mathbf{b}, m | y_1, \ldots, y_n)$. These are the QML estimators of $\mathbf{a}$ and $\mathbf{b}$ and m respectively. The corresponding QML estimator of ν^2 is obtained by substituting these values in (19.4). Notice that the reduced likelihood of the data under the model (18.2) depends only on $\mathbf{a}$, $\mathbf{b}$, m and the observed data, so that the QML estimation of $\mathbf{a}, \mathbf{b}$ and m can be carried out regardless of the nature of the driving Lévy process L (provided it has finite mean m).

Remark 19.1. We have assumed that the causality and invertibility conditions are satisfied. The causality condition was used for the derivation of the predictors in the Kalman recursions (giving independence of innovations and observed past). For each Brownian motion driven CARMA model satisfying the causality and invertibility condition, by Remark 18.23 there are non-causal and non-invertible Brownian motion driven models with the same mean and same spectral density, hence the same autocovariance function, and hence with the same Gaussian likelihood. This is one reason why we constrain to causal and invertible models and therefore also the maximization to models in which the zeros of both $a(z)$ and $b(z)$ are in the open left half-plane of $\mathbb{C}$. The restrictions on $b(z)$ also ensure that the state vector $\mathbf{X}_t$ and the increments $\{L_u - L_s, s \le u \le t\}$ are recoverable from the past observations $\{Y_u, u \le t\}$, as we have seen in Theorem 7.19 and Proposition 7.20.

Remark 19.2. The terms *weighted least squares (WLS) estimates* is frequently used to denote values of $\mathbf{a}$, $\mathbf{b}$ and m which minimize $S(\mathbf{a}, \mathbf{b}, m)$ and *least squares (LS) estimates* to denote values of $\mathbf{a}$, $\mathbf{b}$ and m that minimize $\sum_{j=1}^{n}(y_j - \hat{y}_j)^2$.

19.2 QML estimation for regularly sampled CARMA processes

In Section 19.1, we showed how the reduced Gaussian likelihood $\ell(\mathbf{a}, \mathbf{b}, m)$ defined in (19.5) and based on the observed values $y_1, \dots, y_n$ of $Y_{t_1}, \dots, Y_{t_n}$ can be computed, independently of the driving Lévy process, using the Kalman recursions. The QML estimators of the parameters $\mathbf{a}$, $\mathbf{b}$ and m are values that minimize $\ell(\mathbf{a}, \mathbf{b}, m)$. They are not unique, since to any causal invertible set of minimizing parameters there is a number of non-causal and/or non-invertible models with the same reduced likelihood. For this reason, we constrain the minimization of reduced likelihood to causal invertible models.

If the observations of Y are made at uniformly spaced times, say $t_k = k\Delta, k = 1, \dots, n$, the method of Section 19.1 clearly applies with the simplifications that $F_i = \mathrm{e}^{A\Delta}$ and $Q_i = \nu^2 \int_0^\Delta \mathrm{e}^{At}\mathbf{e}_p\mathbf{e}_p^T \mathrm{e}^{A^T t}\, dt$. In this case, there is yet another source of non-uniqueness of QML estimates due to the fact that integer multiples of $2\pi\mathrm{i}/\Delta$ can be added to the zeros of $a(z)$ without changing the Gaussian likelihood, as discussed in Remark 18.34(b). This phenomenon is called *aliasing* and we deal with it by constraining the minimization of the reduced likelihood to models such that the imaginary part of each zero λ_j of $a(z)$ lies in $(-\pi/\Delta, \pi/\Delta)$. This condition is always satisfied for Δ sufficiently small.

Using the fact (cf., Section 18.7) that the sequence Y^Δ observed when a CARMA(p,q) process Y is sampled at fixed intervals of length Δ is an ARMA(p,r) sequence (with $r < p$ and $r = p-1$ in all but exceptional cases), the idea was suggested by Phillips [71] and further developed by Durbin [38], Phillips [72], Robinson [74] and others, that to fit a CARMA model of autoregressive order p to regularly sampled observations, an ARMA$(p, p-1)$ model could be fitted to the uniformly spaced observations and a CARMA process determined with the same mean and autocovariance function restricted to the observation times as those of the fitted ARMA$(p, p-1)$. We shall refer to the determination of the corresponding CARMA model as the *embedding problem*, essentially the inverse of the *sampling problem* considered in Section 18.7. However, the embedding problem is complicated by the fact that there exist discrete-time ARMA processes which do not have the second-order structure of *any* regularly sampled CARMA process (see, e. g., Brockwell and Brockwell [20]), so that a QML ARMA($p, p-1$) model may not be embeddable in a CARMA model of autoregressive order p. However, if this is the case, the resulting CARMA model will be a QML CARMA model of autoregressive order p.

The estimation procedure thus consists of fitting, by maximizing the discrete-time Gaussian likelihood, a causal and invertible discrete-time ARMA($p, p-1$) model to the sampled data and then attempting to find a corresponding CARMA model of autoregressive order p whose Δ-sampled process has the same mean and autocovariance function as the ARMA model. If such a CARMA process Y exists, we shall say that U is *Δ-embeddable in Y*.

Theorem 19.5 of the next section gives conditions under which an ARMA(p, q) model can be embedded in a CARMA process of autoregressive order p and specifies the parameters of the CARMA process. This is the key to the implementation of the estimation procedure described in the preceding paragraph.

19.3 The embedding problem

Following the exposition in [30], the embedding theorem to be established in this section depends on the following characterization of CARMA autocovariance functions. Let Y be a causal CARMA(p, q) process with finite variance and real-valued coefficients and assume that the zeros $\lambda_1, \ldots, \lambda_p$ of the autoregressive polynomial $a(z)$ all have multiplicity 1. From Proposition 7.6(a), we know that the autocovariance function of Y has the form $\gamma(h) := \sum_{j=1}^{p} c_j \, e^{\lambda_j |h|}$, $h \in \mathbb{R}$, with complex coefficients $c_1, \ldots, c_p$, and is clearly real-valued. Further, if $a(z)$ does not have common zeros with $b(z)b(-z)$, then all of the c_j are different from zero. The following theorem establishes conditions under which such a linear combination of exponentials *is* in fact the autocovariance function of a CARMA process and specifies the parameters of a CARMA process which has γ as its autocovariance function.

Theorem 19.3 (Characterization of CARMA autocovariance functions). *Suppose that γ is a real-valued function of the form,*

$$\gamma(h) := \sum_{j=1}^{p} c_j \, e^{\lambda_j |h|}, \quad h \in \mathbb{R},$$

where $c_j \in \mathbb{C} \setminus \{0\}$, $\Re(\lambda_j) < 0$, $j = 1, \ldots, p$, and $\lambda_i \neq \lambda_j$ if $i \neq j$. Define the polynomial $g(z)$ by

$$g(z) := \sum_{j=1}^{p} c_j \lambda_j \prod_{m \neq j} (z + \lambda_m^2).$$

(a) *Then the following statements are equivalent:*
   (i) *γ is the autocovariance function of a mean-square continuous weakly stationary process.*

(ii) *γ is non-negative definite.*
(iii) *$g(z) \leq 0$ for all $z \geq 0$.*
(iv) *γ is the autocovariance function of a* CARMA(p', q) *process with finite variance for some $0 \leq q < p' \leq p$, where the autoregressive polynomial has no zeros on the imaginary axis.*
(v) *γ is the autocovariance function of a* CARMA(p', q) *process with finite variance for some $0 \leq q < p' \leq p$ such that all coefficients are real-valued, such that all zeros of the autoregressive polynomial lie in the open left half-plane and all zeros of the moving average polynomial lie in the closed left half-plane.*

(b) *If one and hence all of the equivalent conditions* (i)–(v) *above are satisfied, then the orders p' and q, the autoregressive polynomial $a(z)$, the moving average polynomial $b(z)$ as well as the variance $\nu^2 \in (0, \infty)$ of the driving Lévy process L of the CARMA process appearing in* (v) *are unique, and $a(z)$ and $b(z)$ do not have common zeros. We have $p' = p$, $a(z) = \prod_{j=1}^{p}(z - \lambda_j)$, q is the order of the polynomial $g(z)$ (which is not the zero-polynomial) and $\nu^2 = -2S > 0$, where S is the leading coefficient of $g(z)$. The moving average polynomial $b(z)$ is given by*

$$b(z) = \prod_{j=1}^{q}(1 + r_j e^{i\theta_j}),$$

where the numbers $r_j e^{i\theta_j}$, $j = 1, \ldots, q$, are found by writing $g(z) = S\prod_{j=1}^{q}(z + r_j^2 e^{2i\theta_j})$ with $r_j \geq 0$ and $-\pi/2 \leq \theta_j \leq \pi/2$. The numbers $r_j^2 e^{2i\theta_j}$ are real or appear in pairs of complex conjugates, hence a similar statement is true for the $r_j e^{i\theta_j}$ if $r_j^2 e^{2i\theta_j} \in \mathbb{C} \setminus (-\infty, 0)$; if $r_j^2 e^{2i\theta_j} \in (-\infty, 0)$ for some j (i. e. $r_j > 0$ and $\theta_j \in \{-\pi/2, \pi/2\}$), then the number of indices $k \in \{1, \ldots, q\}$ such that $r_k^2 e^{2i\theta_k} = r_j^2 e^{2i\theta_j}$ is even, and the corresponding $\theta_k \in \{-\pi/2, \pi/2\}$ have to be chosen in such a way that the $r_k e^{i\theta_k}$ also appear in pairs of complex conjugates. The CARMA process $(Y_t)_{t\in\mathbb{R}}$ appearing in (v) *is hence the* CARMA(p, q) *process satisfying the formal differential equation*

$$\prod_{j=1}^{p}(D - \lambda_j)Y_t = \prod_{j=1}^{q}(D + r_j e^{i\theta_j})DL_t. \tag{19.6}$$

(c) *The* CARMA(p, q) *process Y of* (v) *(i. e., by uniqueness, the one specified in* (19.6)*) satisfies the invertibility condition (i. e. all zeros of the moving average polynomial lie in the* open *left half-plane) if and only if $g(z) < 0$ for all $z \geq 0$.*

Proof. First observe that since γ is real-valued,

$$0 = \gamma(h) - \overline{\gamma(h)} = \sum_{j=1}^{p} c_j e^{\lambda_j |h|} - \sum_{j=1}^{p} \overline{c_j} e^{\overline{\lambda_j}|h|} \quad \forall\, h \in \mathbb{R}.$$

Using $c_j \neq 0$ and the linear independence of complex exponentials, we deduce that $c_j \in \mathbb{R}$ if $\lambda_j \in \mathbb{R}$, and that for j such that $\lambda_j \in \mathbb{C} \setminus \mathbb{R}$, there must exist an index $i \neq j$ such that $\lambda_j = \overline{\lambda_i}$ and $c_j = \overline{c_i}$. In particular, the constants $\lambda_1, \dots, \lambda_p$ are real-valued or appear in pairs of complex conjugates.

(a) That (v) implies (iv) is clear, and that (iv) implies (i) follows from the fact that CARMA processes are strictly stationary and those with finite variance also weakly stationary and mean-square continuous by Theorem 18.22. The equivalence of (i) and (ii) is an immediate consequence of Corollary 3.2, Remark 3.3 and Proposition 3.6 since γ is continuous and even.

If (i) holds, then, since γ is continuous and in $L^1(\mathbb{R}; \mathbb{C})$, Theorem 5.15 shows that Y has a spectral density given by

$$f(\omega) = \frac{1}{2\pi} \int_{\mathbb{R}} \gamma(h) e^{-i\omega h} \, dh, \quad \omega \in \mathbb{R}.$$

Direct evaluation of the integral defining f gives

$$f(\omega) = -\frac{\sum_{j=1}^p c_j \lambda_j \prod_{m \neq j} (\lambda_m^2 + \omega^2)}{\pi \prod_{j=1}^p (\lambda_j^2 + \omega^2)} = -\frac{g(\omega^2)}{\pi \prod_{j=1}^p (\lambda_j^2 + \omega^2)}, \quad \omega \in \mathbb{R},$$

and non-negativity of f and the fact that the λ_j are real or appear in pairs of complex conjugates implies condition (iii).

For the implication that (iii) implies (i), define $f(\omega) := -\frac{g(\omega^2)}{\pi \prod_{j=1}^p (\lambda_j^2 + \omega^2)}$ for $\omega \in \mathbb{R}$. Then $f \geq 0$ by assumption, and it is clearly integrable, hence by Bochner's theorem (Theorem 5.1), it is the spectral density of some autocovariance function $\tilde{\gamma} : \mathbb{R} \to \mathbb{C}$ of a MSC stationary process, i. e., $\tilde{\gamma}(h) = \int_{\mathbb{R}} e^{ih\omega} f(\omega) \, d\omega$. On the other hand, as seen above, $\sqrt{2\pi} f$ is the Fourier transform of γ, and the Fourier inversion theorem (Theorem 2.63) shows $\gamma = \tilde{\gamma}$, giving (i). Also observe that g cannot be the zero-polynomial, for otherwise f and hence γ would be the zero-functions. We have seen that (i)–(iii) are equivalent and that (v) implies (iv), which in turn implies (i).

Now let the equivalent conditions (i)–(iii) be satisfied. Let us show that these imply (v) and that a possible choice for the CARMA process is given by (19.6). Since $g(z)$ is not the zero-polynomial, we can factorize it as

$$g(z) = S \prod_{j=1}^q (z + r_j^2 e^{2i\theta_j}),$$

where $r_j \geq 0$, $-\pi/2 \leq \theta_j \leq \pi/2$, $q \in \{0, \dots, p-1\}$ is the order of the polynomial $g(z)$ and S is the coefficient of z^q, which is necessarily strictly negative. (If $\sum_{j=1}^p c_j \lambda_j \neq 0$ then $q = p-1$ and $S = \sum_{j=1}^p c_j \lambda_j$.) Since $g(z)$ is real-valued on $[0, \infty)$, it has real coefficients, hence the numbers $r_j^2 e^{2i\theta_j}$ are real or appear in pairs of complex conjugates. Hence so do the numbers $r_j e^{i\theta_j}$ if they lie in $\mathbb{C} \setminus (-\infty, 0)$. If $r_j^2 e^{2i\theta_j} \in (-\infty, 0)$ for some j, i. e., if $r_j > 0$ and $\theta_j \in \{-\pi/2, \pi/2\}$, then g has a zero at $-r_j^2 e^{2i\theta_j} \in (0, \infty)$. Since $g(z) \leq 0$ for all $z \geq 0$, this can only happen if the multiplicity of this zero is even. Hence we can choose the $\theta_j \in \{-\pi/2, \pi/2\}$ also in this case in such a way that all the $r_j e^{i\theta_j}$ are real or appear in pairs of complex conjugates. In particular, the polynomial $b(z) := \prod_{j=1}^q (z + r_j e^{i\theta_j})$ has real-valued coefficients, as does the polynomial $a(z) := \prod_{j=1}^p (z - \lambda_j)$. Now define again

$$f(\omega) := -\frac{g(\omega^2)}{\pi \prod_{j=1}^p (\lambda_j^2 + \omega^2)} = \frac{-2S}{2\pi} \frac{\prod_{j=1}^q (\omega^2 + r_j^2 e^{2i\theta_j})}{\prod_{j=1}^p (\lambda_j^2 + \omega^2)} = \frac{-2S}{2\pi} \frac{|b(i\omega)|^2}{|a(i\omega)|^2}.$$

By Theorem 18.22, f is then the spectral density of the CARMA(p, q) process $(Y_t)_{t\in\mathbb{R}}$ satisfying the formal differential equation $a(D)Y_t = b(D)DL_t$, with $\operatorname{Var}(L_1) = -2S$. Since $\sqrt{2\pi} f$ is the Fourier transform of γ as seen above, we see that γ is the autocovariance function of Y. That all zeros of $a(z)$ lie in the open left half-plane and all zeros of $b(z)$ in the closed left half-plane is clear.

(b) We have already seen that the CARMA(p, q) process defined by (19.6) satisfies (v), so we only have to show uniqueness and that the autoregressive and moving average polynomial do not have common zeros. So let Y be a CARMA(p', q) process satisfying $a(D)Y_t = b(D)DL_t$, where $a(z)$ and $b(z)$ have real-valued coefficients and degrees $p' \leq p$ and $q < p'$ with zeros lying in the open and closed left half-plane, respectively, and L has variance $\nu^2 > 0$, then we deduce from (7.2) and the facts that $c_1, \ldots, c_p \neq 0$ and $\lambda_i \neq \lambda_j$ for $i \neq j$, that $p' = p$ and that the zeros of $a(z)$ all have multiplicity 1. Hence by (7.3) they coincide with $\lambda_1, \ldots, \lambda_p$. This shows $a(z) = \prod_{j=1}^p (z - \lambda_j)$. From equation (7.4) we deduce $c_j = \nu^2 \frac{b(\lambda_j)b(-\lambda_j)}{a'(\lambda_j)a(-\lambda_j)}$ for $j = 1, \ldots, p$. Since $c_j \neq 0$, $a(z)$ and $b(z)$ do not have common zeros and we see that $\nu^2 b(\lambda_j)b(-\lambda_j)$, $j = 1, \ldots, p$, are uniquely determined by γ. Now define the polynomial $B(z) := \nu^2 b(z)b(-z)$. This has real-valued coefficients, degree $2q$ and leading coefficient ν^2. The values of $B(z)$ at $\lambda_1, \ldots, \lambda_p$ are known, and since $B(z)$ is symmetric, also its values at $-\lambda_1, \ldots, -\lambda_p$ are known. Hence the values of $B(z)$ at $2p$ distinct points are known, and since $B(z)$ has degree less than or equal to $2(p - 1)$, we conclude that $B(z)$ is uniquely determined by γ, in particular, so are its degree $2q$ and its leading coefficient ν^2. Since all zeros of $b(z)$ lie in the closed left half-plane, we can factorize $b(z) = \prod_{j=1}^q (z - \mu_j)$ with $\Re(\mu_j) < 0$ for $j = 1, \ldots, r$, where $r \leq q$, and $\Re(\mu_j) = 0$ for $j = r + 1, \ldots, q$. Since the coefficients of $b(z)$ are real, the zeros μ_j are real or appear in pairs of complex conjugates. In particular, if $\mu_j \in i\mathbb{R} \setminus \{0\}$ is a zero of $b(z)$, then so is $-\mu_j$. Hence the factorization of $B(z)$ can be written as

$$\begin{aligned} B(z) &= \nu^2 (-1)^q \prod_{j=1}^r (z - \mu_j) \prod_{j=1}^r (z + \mu_j) \prod_{j=r+1}^q (z - \mu_j)(z + \mu_j) \\ &= \nu^2 (-1)^q \prod_{j=1}^r (z - \mu_j) \prod_{j=1}^r (z + \mu_j) \prod_{j=r+1}^q (z - \mu_j)^2. \end{aligned}$$

Since $B(z)$ is uniquely determined by γ, so are its zeros, and its zeros on the imaginary axis have twice the multiplicity of the corresponding zeros of $b(z)$. It follows that also $\mu_1,\dots,\mu_q$ and hence $b(z)$ are uniquely determined by γ. Since the process specified in (19.6) has the required properties, we conclude $b(z) = \prod_{j=1}^{q}(z + r_j e^{i\theta_j})$.

(c) By (b), the zeros of $b(z)$ are given by $-r_j e^{i\theta_j}$. They clearly lie in the *closed* left half-plane, since $\theta_j \in [-\pi/2, \pi/2]$. The invertibility condition is hence satisfied if none of the zeros lies on the imaginary axis, i. e., if $b(i\omega) \neq 0$ for all $\omega \in \mathbb{R}$. But since $g(\omega^2) = S|b(i\omega)|^2$, as seen in the proof of (a), this is clearly equivalent to $g(z) < 0$ for all $z \geq 0$. □

We shall need the following result for ARMA sequences.

Proposition 19.4 (Autocovariance function of ARMA processes). *Suppose that $(U_n)_{n\in\mathbb{Z}}$ is the* ARMA(p,s) *sequence with mean c satisfying the equations, with $p > s$,*

$$\phi(B)(U_n - c) = \theta(B)Z_n, \quad n \in \mathbb{Z}, \tag{19.7}$$

where $(Z_n)_{n\in\mathbb{Z}}$ is zero-mean white noise with variance $\sigma^2 > 0$, B is the backward shift operator, $\phi(z) := \prod_{j=1}^{p}(1 - \xi_j z)$ and $\theta(z) := \prod_{j=1}^{s}(1 - \eta_j z)$. Suppose also that $\xi_i \neq \xi_j$ for $i \neq j$, that $\xi_i \neq \eta_j$ for all i and j, that $0 < |\xi_i| < 1$ and $0 < |\eta_j| < 1$ for all i and j, and that the coefficients of the polynomials $\phi(z)$ and $\theta(z)$ are real-valued. Define

$$G_j = -\xi_j \frac{\theta(\xi_j)\theta(\xi_j^{-1})}{\phi(\xi_j)\phi'(\xi_j^{-1})}, \quad j = 1,\dots,p. \tag{19.8}$$

Then the autocovariance function γ of $(U_n)_{n\in\mathbb{Z}}$ can be expressed as

$$\gamma(h) = \sigma^2 \sum_{j=1}^{p} G_j\, \xi_j^{|h|}, \quad h \in \mathbb{Z}.$$

Proof. See Brockwell [21, Equation (3.14)], or Exercise 19.1. □

Observe that as a consequence of the real-valued coefficients of $\phi(z)$ and $\theta(z)$, the ξ_j are real or appear in pairs of complex conjugates, and consequently the same is true for the G_j from (19.8).

If the ARMA sequence in (19.7) is Δ-embeddable in a CARMA(p,q) process for some $q < p$ there exist parameters $m = \mathbb{E}(L_1)$, $v^2 = \operatorname{Var}(L_1)$ and $\boldsymbol{a} = (a_1,\dots,a_p,b_0,\dots,b_{q-1})^T$, or equivalently $\boldsymbol{\theta} = (m, v^2, \lambda_1,\dots,\lambda_p,\mu_1,\dots,\mu_q)^T$, such that the autocovariance at lag $n\Delta$ and the mean of a CARMA process with these parameters coincides with those of U at lag n. The following theorem gives conditions under which this is the case and specifies the parameters of a CARMA process into which the ARMA process is Δ-embeddable.

Theorem 19.5 (Embedding an ARMA sequence. Embedding Theorem). *Let $\Delta > 0$ and suppose that $(U_n)_{n\in\mathbb{Z}}$ is the* ARMA(p,s) *sequence specified in* (19.7), *with the same assumptions* *and notations valid as in Proposition 19.4. Additionally assume that $\xi_j \notin (-\infty, 0)$ for all j. With G_j as in (19.8), define the polynomial of degree $q \leq p-1$,*

$$g(z) := \sigma^2 \Delta^{-1} \sum_{j=1}^{p} G_j \log(\xi_j) \prod_{m \neq j} (z + \Delta^{-2} (\log \xi_m)^2),$$

where log *denotes the principal branch of the complex logarithm. (Observe that $q = p-1$ when $\sum_{j=1}^{p} G_j \log(\xi_j) \neq 0$.) Then the polynomial g has real-valued coefficients, is not the zero-polynomial, and the following statements* (i) *and* (ii) *are equivalent:*

(i) *The process $(U_n)_{n\in\mathbb{Z}}$ is Δ-embeddable into a* CARMA(p', q') *process Y with finite variance and real-valued coefficients such that $q' \leq p' \leq p$, such that all zeros of the autoregressive CARMA polynomial lie in the half-strip $\{z \in \mathbb{C} : \Re(z) < 0, |\Im(z)| < \pi/\Delta\}$ and all zeros of the moving average polynomial in the closed left half-plane.*

(ii) *$g(z) \leq 0$ for all $z \geq 0$.*

If one and hence all of the equivalent conditions (i) *and* (ii) *are satisfied, and if $(U_n)_{n\in\mathbb{Z}}$ is Δ-embeddable into Y subject to the conditions of* (i)*, then the mean and variance of the Lévy process driving Y are uniquely determined, as are the orders p' and q' and the autoregressive polynomial $a(z)$ and the moving average polynomial $b(z)$. We have $p' = p$, $q' = q$ and $a(z) = \prod_{j=1}^{p}(z - \Delta^{-1} \log \xi_j)$. The moving average polynomial $b(z)$ is found by writing $g(z) = S \prod_{j=1}^{q}(z + r_j^2 e^{2i\theta_j})$, with $r_j \geq 0$ and $-\pi/2 \leq \theta_j \leq \pi/2$, and setting $b(z) = \prod_{j=1}^{q}(z + r_j e^{i\theta_j})$ (with corresponding complex conjugate choices of the $r_j e^{i\theta_j}$, as in Theorem 19.3(b)). The unique mean m and variance v^2 of the driving Lévy process are given by*

$$m = c(-\Delta)^{-p} \frac{\prod_{j=1}^{p} \log \xi_j}{\prod_{j=1}^{r} r_j e^{i\theta_j}} \quad \text{and} \quad v^2 = -2S. \tag{19.9}$$

Hence $Y = (Y_t)_{t\in\mathbb{R}}$ satisfies the formal differential equation

$$\prod_{j=1}^{p} (D - \Delta^{-1} \log \xi_j) Y_t = \prod_{j=1}^{q} (D + r_j e^{i\theta_j}) D L_t. \tag{19.10}$$

The autoregressive polynomial $a(z)$ and the moving average polynomial $b(z)$ of Y do not have common zeros. Finally, the zeros of the moving average polynomial $b(z)$ all lie in the open *left half-plane (i. e., Y satisfies the invertibility condition) if and only if $g(z) < 0$ for all $z \geq 0$.*

Proof. By Proposition 19.4, the autocovariance function of the ARMA process defined by (19.7) is

$$\kappa(h) = \sigma^2 \sum_{j=1}^{p} G_j\, \xi_j^{|h|}, \quad h \in \mathbb{Z}.$$

Define the function $\gamma : \mathbb{R} \to \mathbb{C}$ by

$$\gamma(h) := \sigma^2 \sum_{j=1}^{p} G_j\, \mathrm{e}^{|h|\Delta^{-1}\log \xi_j}, \quad h \in \mathbb{R}. \tag{19.11}$$

The values $\gamma(h\Delta)$ clearly coincide with $\kappa(h)$ for $h \in \mathbb{Z}$. Since $\Delta^{-1}\log\xi_j$, $j = 1,\dots,p$ are pairwise disjoint and are real or appear in pairs of complex conjugates, and since $G_j = \overline{G_k}$ when $\xi_j = \overline{\xi_k}$, we conclude that γ is real-valued. Also observe that all of the G_j are different from zero by the assumptions of Proposition 19.4.

To show that (i) implies (ii), let $Y = (Y_t)_{t\in\mathbb{R}}$ be a CARMA(p',q') process as in (i), in which $(U_n)_{n\in\mathbb{Z}}$ is Δ-embeddable, and denote its autocovariance function by γ_Y. Then $\gamma_Y(\Delta h) = \kappa(h)$ for all $h \in \mathbb{Z}$. Denote the distinct zeros of the autoregressive polynomial of Y by $\lambda_1,\dots,\lambda_r$ and their multiplicities by n_j, so that $p' = \sum_{i=1}^{r} n_j$. By (7.2), $\gamma_Y(h) = \sum_{j=1}^{r}\sum_{k=0}^{n_j-1} c_{j,k} h^k \mathrm{e}^{\lambda_j h}$ for all $h \geq 0$ with constants $c_{j,k} \in \mathbb{C}$. Hence

$$\sigma^2 \sum_{j=1}^{p} G_j\, \xi_j^{h} = \kappa(h) = \gamma_Y(\Delta h) = \sum_{j=1}^{r}\sum_{k=0}^{n_j-1} c_{j,k}\Delta^k h^k (\mathrm{e}^{\Delta\lambda_j})^h \quad \forall\, h \in \mathbb{N}_0.$$

Since $\Im(\lambda_j) \in (-\pi/\Delta, \pi/\Delta)$, the quantities $\mathrm{e}^{\Delta\lambda_j}, j = 1,\dots,r$, are all distinct. Since sequences of the form $(h^{k_l}\alpha_l^h)_{h\in\mathbb{N}_0}$ with $k_l \in \mathbb{N}_0$ are linearly independent (e. g., Brockwell and Davis [22, Theorem 3.6.2]) we conclude that $r = p$, $n_j = 1$ for all $j = 1,\dots,p$, and (after a suitable enumeration) that $\xi_j = \mathrm{e}^{\Delta\lambda_j}$ and $\sigma^2 G_j = c_{j,0}$. However, this implies $\lambda_j = \Delta^{-1}\log\xi_j$ and hence $\gamma_Y(h) = \gamma(h)$ for all $h \in \mathbb{R}$. Hence γ is the autocovariance function of some CARMA(p',q') process and Theorem 19.3 implies that $g(z) \leq 0$ for all $z \geq 0$, which is (ii). Further, we have seen that any CARMA(p',q') subject to the conditions of (i) must have autocovariance function γ.

To show that (ii) implies (i), assume that $g(z) \leq 0$ for all $z \geq 0$. By Theorem 19.3, γ is the autocovariance function of the CARMA(p,q) process Y specified by (19.10) and (19.9). Since $\gamma(\Delta h) = \kappa(h)$ for all $h \in \mathbb{Z}$ and since $\mathbb{E}Y_t = c$ by Lemma 18.19, we conclude that $(U_n)_{n\in\mathbb{Z}}$ is Δ-embeddable into the process $(Y_t)_{t\in\mathbb{R}}$. The zeros of the autoregressive polynomial of Y clearly lie in the half-strip $\{z \in \mathbb{C} : \Re(z) < 0, |\Im(z)| < \pi/\Delta\}$ and the zeros of the moving average polynomial in the closed left half-plane. This shows (i). Also observe from Theorem 19.3 that the obtained autoregressive polynomial and moving average polynomial do not have common zeros, and that all zeros of the moving average polynomial lie in the *open* left half-plane if and only if $g(z) < 0$ for all $z \geq 0$. Note also that the obtained process Y is the unique CARMA(p',q') process subject to the conditions of Theorem 19.3(a)(v) having γ as autocovariance function, and since any CARMA(p',q')

process subject to the condition (i) into which $(U_n)_{n\in\mathbb{Z}}$ is Δ-embeddable must have autocovariance function γ, as seen above, we obtain the uniqueness assertion. □

Example 19.6 (The case $p = 2$). If $p = 2$, then, defining $\lambda_j := \Delta^{-1}\log\xi_j, j = 1, 2$, the condition $g(z) \leq 0$ for all $z \geq 0$ can be expressed as

$$G_1\lambda_1 + G_2\lambda_2 \leq 0 \quad \text{and} \quad G_1\lambda_2 + G_2\lambda_1 \leq 0,$$

where, when $s = p - 1 = 1$ and $\theta(z) = 1 - \eta z$,

$$G_1 = -\xi_1\frac{(1-\eta\xi_1)(1-\eta/\xi_1)}{(\xi_2-\xi_1)(1-\xi_1^2)(1-\xi_1\xi_2)}$$

and G_2 is the same with subscripts 1 and 2 interchanged (when $s = 0$ and $\theta(z) = 1$ the above formulas are still true when setting $\eta := 0$). If these conditions are satisfied, then the ARMA process is embeddable in the CARMA process satisfying (19.10). When $G_1\lambda_1 + G_2\lambda_2 < 0$, this process is the CARMA(2, 1) process, given by

$$(D - \lambda_1)(D - \lambda_2)Y_t = (D - \mu)DL_t,$$

where

$$\mu = -\left(\lambda_1\lambda_2\frac{G_1\lambda_2 + G_2\lambda_1}{G_1\lambda_1 + G_2\lambda_2}\right)^{1/2} \quad \text{and} \quad \mathrm{Var}(L_1) = -2\sigma^2(G_1\lambda_1 + G_2\lambda_2).$$

When $G_1\lambda_1 + G_2\lambda_2 = 0$, it is the CAR(2) process given by

$$(D - \lambda_1)(D - \lambda_2)Y_t = DL_t,$$

where

$$\mathrm{Var}(L_1) = -2\sigma^2\lambda_1\lambda_2(G_1\lambda_2 + G_2\lambda_1).$$

Remark 19.7. If $g(z) > 0$ for some $z \geq 0$, then the function γ in (19.11) is not non-negative definite and is therefore not the autocovariance function of any CARMA process. However, there may be other choices of the logarithms, equal modulo$(2\pi i)$ to the principial branch used in the theorem, for which the condition $g(z) \leq 0$ for all $z \geq 0$ is satisfied. In this case, the conclusions hold with the revised values of the logarithms, however, with the imaginary parts of the autoregressive zeros no longer being bounded by π/Δ.

Remark 19.8. Since $g(z)$ is a polynomial, the condition that $g(z) \leq 0$ for all $z \geq 0$ is easily checked by evaluating $g(0)$, $\lim_{z\to\infty} g(z)$ and the values of g at each of the positive values of z where $g'(z) = 0$. It is automatically satisfied if Y^Δ is known to be the Δ-sampled sequence of some CARMA process with autoregressive zeros in the strip $\{z \in \mathbb{C} : \Re(z) < 0, |\Im(z)| < \pi/\Delta\}$.

Remark 19.9 (The case of strictly negative ξ_i). In Theorem 19.5, we excluded the case that some of the ξ_i are strictly negative, since if ξ_i was strictly negative for some i, then necessarily λ_i satisfying $e^{\lambda_i \Delta} = \xi_i$, would not be real, and since $\xi_j \neq \xi_i$ for all $j \neq i$, there would be no complex conjugate of λ_i within the other λ_j, so that the derived autoregressive polynomial $a(z)$ would not have real-valued coefficients. For example, the AR(1) process $U_n + 0.5U_{n-1} = Z_n$ with $(Z_n)_{n\in\mathbb{Z}}$ being zero-mean white noise with variance $\sigma^2 > 0$ is clearly not 1-embeddable into a CAR(1) process with real-valued coefficients, since the autoregressive root would necessarily have to be of the form $\lambda = -\log 2 + (2l+1)\pi i$ with some $l \in \mathbb{Z}$. The autocovariance function of $(U_n)_{n\in\mathbb{Z}}$ is $\gamma(h) = \frac{4}{3}\sigma^2(-0.5)^{|h|}$ for $h \in \mathbb{Z}$, and by (7.4), at integer times this coincides with the autocovariance function of the CAR(2) process given by $(D + \log 2 + i\pi)(D + \log 2 - i\pi)Y_t = DL_t$, where L has mean zero and variance $\operatorname{Var}(L_1) = 16\sigma^2(\log 2)(\pi^2 + (\log 2)^2)/3$. Observe that the autoregressive polynomial $a(z) = (z + \log 2 + i\pi)(z + \log 2 - i\pi) = (z + \log 2)^2 + \pi^2$ has real-valued coefficients. Hence $(U_n)_{n\in\mathbb{Z}}$ is 1-embeddable into the CAR(2) process Y, although it is not 1-embeddable into a CAR(1) process with real-valued coefficients. We shall not go into further details of the embedding problem of ARMA(p,s) processes into CARMA processes of autoregressive order higher than p.

Example 19.10 (Sampling and recovering a CARMA(2, 1) process).

(a) Consider the Δ-sampled sequence Y^Δ, $\Delta = 0.1$, of the stationary CARMA(2, 1) process defined by the formal differential equation,

$$(D + 0.5)(D + 1)Y_t = (D + 0.25)DL_t,$$

where D as usual denotes differentiation with respect to t and the driving Lévy process L has moments $\mathbb{E}L_1 = 0$ and $\mathbb{E}(L_1^2) = 1$. Taking $\lambda_1 = -0.5$ and $\lambda_2 = -1$, we find immediately that the constants K_1 and K_2 in the Sampling Theorem 18.33 are given by

$$K_1 = -0.25 \quad \text{and} \quad K_2 = 0.625.$$

Since $K_1 \sinh(-0.05) + K_2 \sinh(-0.1) \neq 0$, we deduce from the theorem that the Δ-sampled sequence Y^Δ with $\Delta = 0.1$ satisfies the ARMA(2, 1) equation,

$$(1 - e^{-0.05}B)(1 - e^{-0.1}B)Y_n^\Delta = (1 - \eta B)Z_n, \quad \{Z_n\} \sim \mathrm{WN}(0, \sigma^2), \tag{19.12}$$

where η and σ^2 are to be determined. From Example 18.35, we find that $k_1 = 1.0003132479211$. Hence, by Theorem 18.33,

$$\eta = k_1 - \sqrt{k_1^2 - 1} = 0.9752813889$$

and

$$\sigma^2 = 2(K_1 \sinh(0.05) + K_2 \sinh(0.1))\, e^{-0.15}/\eta = 0.08842703.$$

(b) Suppose, on the other hand, that we were to start with the ARMA sequence for Y^Δ (where $\Delta = 0.1$) defined by (19.12) with η and σ^2 as specified, and wish to find a CARMA$(2, q)$ process of which it is the Δ-sampled sequence. To do so, we apply Theorem 19.5 or, more simply (since $p = 2$), the results of Example 19.6.
Thus $\lambda_1 = \Delta^{-1} \log \xi_1 = -0.5$, $\lambda_2 = \Delta^{-1} \log \xi_2 = -1$, $\eta = 0.9752813889$ and $\sigma^2 = 0.08842703$. From Example 19.6, we readily find that

$$G_1 = -2.8271899 \quad \text{and} \quad G_2 = 7.0679747,$$

and that

$$G_1\lambda_1 + G_2\lambda_2 < 0 \quad \text{and} \quad G_1\lambda_2 + G_2\lambda_1 < 0.$$

Hence the ARMA sequence is Δ-embeddable in the CARMA(2, 1) process defined by

$$(D - \lambda_1)(D - \lambda_2)Y_t = (D - \mu)DL_t,$$

where $\lambda_1 = -0.5$, $\lambda_2 = -1$, $\mu = -(\lambda_1\lambda_2 \frac{G_1\lambda_2+G_2\lambda_1}{G_1\lambda_1+G_2\lambda_2})^{1/2} = -0.25$ and $\mathrm{Var}(L_1) = -2\sigma^2(G_1\lambda_1 + G_2\lambda_2) = 1$, as, of course, we already knew from part (a).

19.4 Examples, stochastic volatility

All R programs and data sets presented in this and the next section can be downloaded from

https://doi.org/10.1515/9783111325033.

In this section, we illustrate the application of QML estimation as described in the previous sections by applying it, using the package R, to two series connected with stochastic volatility. The use of Lévy driven CARMA processes in this connection was initiated by the seminal paper of Barndorff-Nielsen and Shephard [7], who introduced a stationary Lévy driven Ornstein-Uhlenbeck (i. e., a CARMA(1, 0)) process to replace the constant spot volatility in the geometric Brownian motion model for the price process $(P_t)_{t\geq 0}$ of an asset such as a financial stock.

The geometric Brownian motion model for $(P_t)_{t\geq 0}$ is defined by the Itô stochastic differential equation,

$$\mathrm{d}P_t = P_t(\mu\,\mathrm{d}t + \sqrt{\mathrm{v}}\,\mathrm{d}B_t), \quad \text{with } P_0 > 0,$$

where $\mu \in \mathbb{R}$, $\mathrm{v} > 0$ and B is standard Brownian motion. In terms of $X_t := \log P_t$ (the *log asset price*), this can be written, using Itô's formula, as

$$\mathrm{d}X_t = (\mu - \mathrm{v}/2)\,\mathrm{d}t + \sqrt{\mathrm{v}}\,\mathrm{d}B_t. \tag{19.13}$$

The parameter v is known as the *volatility* and is a measure of variance per unit time.

In the model of Barndorff-Nielsen and Shephard [7], v is replaced by the second-order CARMA(1, 0) process (i. e., Ornstein–Uhlenbeck process), independent of B, and known as the *spot volatility process*,

$$V_t = \int_{(-\infty,t]} e^{\lambda(t-u)}\, dL_u, \quad \lambda < 0,$$

where the Lévy process L must be non-decreasing, i. e. a subordinator, in order to ensure the non-negativity of the spot volatility V. By (17.16), the autocorrelation function of the process V so defined is $e^{\lambda|h|}$, $h \in \mathbb{R}$. Unlike the geometric Brownian motion model, this model allows the sample paths of spot volatility both to vary randomly with time and to exhibit jumps.

In order to extend the class of possible autocorrelation functions for spot volatility beyond the limited two-sided exponential functions associated with CARMA(1, 0) processes, we can replace the latter by non-negative CARMA processes. Non-negativity of the spot-volatility V in general requires the driving Lévy process to be non-decreasing and the kernel of the CARMA process to be non-negative. The *generalized Barndorff-Nielsen and Shephard (GBS) model* for log asset prices then becomes

$$dX_t = (\mu - V_t/2)\, dt + \sqrt{V_t}\, dB_t, \tag{19.14}$$

where V is a non-negative second-order CARMA process.

If time is measured in days then the *integrated volatility for day n* is $I_n = \int_{n-1}^{n} V_t\, dt$.

The spot volatility V_t is not an observable quantity but the integrated volatility sequence, $I_n = \int_{n-1}^{n} V_t\, dt$, can be estimated from the increments of the log asset price, $X_t = \log P_t$, over small subintervals of the interval $(n-1, n]$. For the geometric Brownian motion model (19.13), measuring time in days and defining

$$X_{n,j} := X_{n-1+j/N}, \quad j = 0, \ldots, N,$$

and

$$\Delta_{n,j} = X_{n,j} - X_{n,j-1}, \quad j = 1, \ldots N,$$

for some large positive integer N, it is easily shown that $\sum_{j=1}^{N} \Delta_{n,j}^2$ converges in mean-square to the integrated volatility I_n for day n as $N \to \infty$ (Exercise 19.2). This sum of squared increments with N large, adjusted to allow for market irregularities, is known as the *daily realized volatility* R_n *for day n* and is used to estimate the integrated volatility I_n, not only for the geometric Brownian motion model, but also for the more general model (19.14).

Example 19.11 (The Deutsche Mark/US Dollar Exchange rate). Figure 19.1 below shows the daily realized volatilities, R_n, $n = 1, \ldots, 3045$, of the Deutsche Mark/US Dollar (DM/\$) exchange rate series covering the period from December 1, 1986 through June 30, 1999 (kindly provided to us by Viktor Todorov). Missing data, weekends, fixed holiday and similar calendar effects were removed, leaving the total of 3045 days. The following

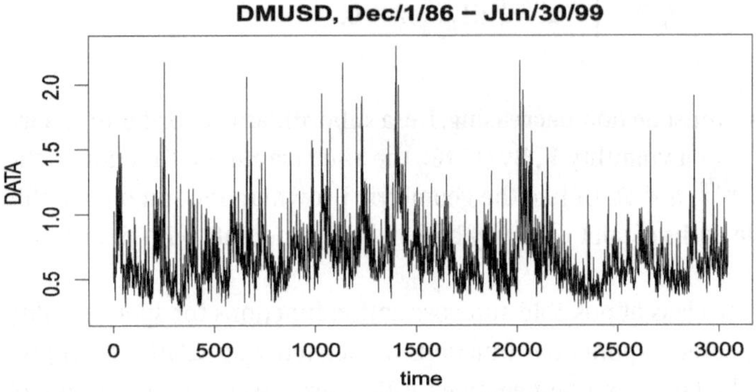

Figure 19.1: Daily realized volatilities for the DM/US\$ exchange rate data.

R programs were used to fit a QML CARMA model to the data. We measure time in days, so that we use $\Delta = 1$ for the embedding procedure. First, DMUSD.txt must be stored as a single-column text file in the R working directory. The file DMUSD.txt can be downloaded from

https://doi.org/10.1515/9783111325033.

STEP 1 plots the data and stores it as the R-vector DATA.

```
# STEP 1: READ DATA FROM A TEXT FILE IN THE WORKING DIRECTORY OF R.
# To find the working directory type getwd() .
DATA=scan('DMUSD.txt', what=numeric(), quiet=TRUE)
time=c(1:(length(DATA)))
plot(time,DATA,main="DMUSD,␣Dec/1/86␣-␣Jun/30/99",type="l")
```

STEP 2 prints the Akaike information criterion (AIC) and Bayesian information criterion (BIC) values for ARMA$(p, p-1)$ models for DATA with values of p up to 5. Here, AIC is defined as $-2(\log L_{\text{ARMA}}) + 2k$, where L_{ARMA} is the maximized likelihood and $k = p + (p-1) + 1 + 1 = 2p + 1$ is the number of parameters. The BIC value is defined as $-2(\log L_{\text{ARMA}}) + 2k \log(n)$, where n is the sample size.

```
# STEP 2: CHOOSE p< 6 BY MINIMIZING AIC or BIC OF FITTED ARMA(p,p-1).
n=length(DATA);
for(p in 1:5)
   {q=p-1;fitted=arima(DATA,c(p,0,q)); k=2*p+1;
    AIC<-fitted$aic;  BIC=AIC+2*k*(log(n)-1);
    cat("The␣AIC␣for␣p=", p, "is", AIC, "and␣the␣BIC␣is", BIC, "\n")}
```

In this example, both AIC and BIC are minimized, when $p = 2$, so that we choose an ARMA(2, 1) model.

STEPS 3 calculates the QML estimated ARMA(2, 1) model.

```
# STEP 3: p=2. QMLE OF ARMA(2,1) COEFFS AND WHITE NOISE VARIANCE incl. the mean.
# phi are autoregressive coefficients, theta the moving average coefficients
# WNV is the white noise variance, intercept the mean
p=2; q=p-1; fitted=arima(DATA,c(p,0,q))
phi=fitted$coef[1:p]; theta=fitted$coef[(p+1):(p+q)]
WNV=sqrt(fitted$sigma^2); Intercept=fitted$coef[(p+q+1)];
# Observe that fitted$sigma is the white noise variance, NOT the standard dev.
AIC<-fitted$aic; Intercept<-Re(polyroot(c(Intercept,-1)))
phi; theta
WNV
Intercept
```

It gives the QMLE discrete ARMA(2, 1) model for the daily data,

$$(1 - 1.2321B + 0.2564B^2)(Y_n - 0.6698) = (1 - 0.8243B)Z_n, \tag{19.15}$$

where $(Z_n) \sim \mathrm{WN}(0, 0.0341)$.

STEP 4 uses Example 19.6 to check for 1-embeddability of the ARMA sequence, i. e., that

$$G_1\lambda_1 + G_2\lambda_2 \leq 0 \quad \text{and} \quad G_1\lambda_2 + G_2\lambda_1 \leq 0,$$

and determines (if the ARMA model is embeddable and additionally $G_1\lambda_1 + G_2\lambda_2 < 0$) the parameters of the corresponding QML CARMA model for DATA using Theorem 19.5. Observe that the program does not check for the other assumptions of Theorem 19.5, such as the roots of the discrete autoregressive polynomial being distinct or non-negative. That the discrete moving average and autoregressive polynomials cannot have common zeros is clear, since before we have chosen the model with the smallest AIC and BIC, hence if there were common zeros, then a lower order model would have been chosen.

```
# STEP 4: p=2. QMLE OF CARMA ZEROS AND COEFFICIENTS, EL1 and Var(L1)
# delta is the parameter of delta-embedding
# here, delta=1
delta = 1
psi<- -rev(phi)
roots<- polyroot(c(psi,1)); lambda<- log(roots)/delta
xi<- roots; eta<- polyroot(c(theta,1))
G1<--xi[1]*(1-eta*xi[1])*(1-eta/xi[1])/((xi[2]-xi[1])*(1-xi[1]^2)*(1-xi[1]*xi[2]))
G2<--xi[2]*(1-eta*xi[2])*(1-eta/xi[2])/((xi[1]-xi[2])*(1-xi[2]^2)*(1-xi[2]*xi[1]))
H1<- G1*lambda[1]+G2*lambda[2];H2<- G1*lambda[2]+G2*lambda[1]
H1;H2
# Embeddable if and only if H1 and H2 are both less than or equal to zero.
# If H1 strictly neg., obtain CARMA(2,1) with previously specified coefficients.
mu<- -sqrt(lambda[1]*lambda[2]*H2/H1)
a<- c(0,0); a[1]<- -(lambda[1]+lambda[2]); a[2]<- lambda[1]*lambda[2]
b0<- -mu; b1<- 1; b<- c(b0,b1)
VarL1<- -2*WNV*H1
EL1<- Intercept  * a[2]/b[1]
```

```
cat("CARMA␣AR␣zeros,␣lambda[1],␣lambda[2]:",lambda,"\n")
cat("CARMA␣MA␣zeros,␣mu[1]:",mu,"\n")
cat("CARMA␣AR␣coefficients,␣a1,a2:",Re(a),"\n")
cat("CARMA␣MA␣coefficients,␣b0,b1:",Re(b),"\n")
cat("Variance␣of␣L(1):",Re(VarL1),"\n")
cat("Mean␣of␣L(1):",Re(EL1),"\n")
```

The resulting QML CARMA model is

$$(D^2 + 1.3612D + 0.04449)Y_t = (0.2058 + D)DL_t, \tag{19.16}$$

where $\mathbb{E}(L_1) = 0.1448$ and $\mathrm{Var}(L_1) = 0.08340$. We observe finally that this model is almost identical to the one fitted by Brockwell et al. [23] by least squares using the Kalman recursions, namely,

$$(D^2 + 1.3619D + 0.04429)Y_t = (0.2061 + D)DL_t. \tag{19.17}$$

Remark 19.12 (Integration of CARMA processes). In the last example, we fitted a CARMA(2, 1) model to the daily realized volatility sequence. In order to fit a CARMA(2, 1) model to the *spot* volatility based on the same data, Brockwell and Lindner [28] showed that the daily integrated (or daily realized) volatility sequence corresponding to a second-order CARMA(2, 1) spot volatility is a second-order ARMA(2, 2) sequence and, by fitting such a model to the daily realized volatility, arrived at a CARMA(2, 1) model for the spot volatility. We refer to [28] for details.

Example 19.13 (The VIX index). The Cboe Volatility Index, or VIX, is a real-time market index representing the market's expectations for volatility over the coming 30 days. Investors use the VIX to measure the level of risk, fear, or stress in the market when making investment decisions. Figure 19.2 shows the daily closing VIX index for 349 trading days (January 19, 2021 – June 6, 2022), originally found from

https://finance.yahoo.com/quote/%5EVIX/history?p=%5EVIX.

The data can be downloaded as vix.txt from

https://doi.org/10.1515/9783111325033.

To fit a QML CARMA model to the data, we carry out the same steps that were used in Example 19.11. Again, we measure time in days and set $\Delta = 1$ for the embedding procedure. To start with, the data set vix.txt must be stored as a single-column text file in the R working directory. STEP 1 plots the data and stores it as the R-vector DATA.

```
# STEP 1: READ DATA FROM A TEXT FILE IN THE WORKING DIRECTORY OF R.
# To find the working directory type getwd() .
DATA=scan('vix.txt', what=numeric(), quiet=TRUE)
time=c(1:(length(DATA)))
plot(time,DATA,main="VIX␣index,␣Jan␣/19/21␣--␣June/6/22",type="l")
```

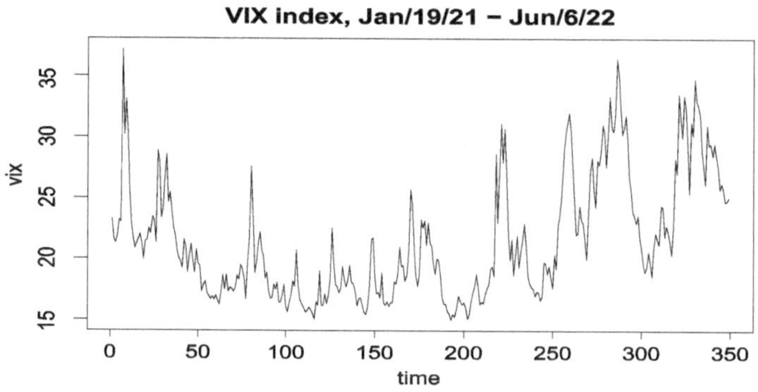

Figure 19.2: The daily VIX index, January 19th, 2021 – June 6th, 2022.

STEP 2 prints the AIC and BIC values for ARMA$(p, p-1)$ models for DATA with values of p up to 5. The program is completely the same as the one in Example 19.11. Running the program, both AIC and BIC are minimized when $p = 1$, so that we choose a CARMA$(1,0)$ = CAR(1) model.

STEP 3 calculates the QMLE estimated discrete time AR(1) model.

```
# STEP 3: p=1. QMLE OF AR(1) PARAMETERS
p=1; q=p-1; fitted=arima(DATA,c(p,0,q))
phi=fitted$coef[1:p];ph<-polyroot(c(phi[1],-1))
WNV=sqrt(fitted$sigma^2); Intercept=fitted$coef[(p+1)]
AIC<-fitted$aic; Intercept<-polyroot(c(Intercept,-1))
"QMLE␣AR(1)␣PARAMETERS"
cat("Estimated␣AR␣coefficient,␣phi",Re(phi),"\n")
cat("Estimated␣White-noise␣variance",Re(WNV),"\n")
cat("Estimated␣Intercept",Re(Intercept),"\n")
```

The obtained AR(1) model is

$$(1 - 0.90684B)(Y_n - 21.615) = Z_n,$$

where (Z_n) is zero-mean white noise with variance WNV = 4.3134.

STEP 4 determines (since $\phi > 0$) the corresponding QML CARMA$(1,0)$ = CAR(1) model for the data using Theorem 19.5.

```
# STEP 4: p=1. QMLE OF CAR(1) PARAMETERS (provided ph>0)
delta = 1
lambda<- polyroot(c(log(phi[1]),-delta)); a1<- -lambda
varL1<- -2*lambda*WNV/(1-ph^2); EL1<- -lambda*Intercept
"QMLE␣CAR(1)␣PARAMETERS"
cat("Estimated␣CAR␣coefficient,␣a1",Re(a1),"\n")
cat("Estimated␣variance␣of␣L(1)",Re(varL1),"\n")
cat("Estimated␣mean␣of␣L(1)",Re(EL1),"\n")
```

The resulting QML CAR(1) model is $(D - 0.09779)Y_t = DL_t$, where $\mathbb{E}(L_1) = 2.1137$ and $\mathrm{Var}(L_1) = 4.7489$.

In the next example, we shall apply Theorem 19.5 to QML estimation of a regularly sampled simulated causal CARMA(3, 2) process.

Example 19.14 (A simulated causal CARMA(3, 2) process). In the next section, we will describe how CARMA processes can be simulated in R. Right now, we use the data set CARMA32.txt, which was produced with the method of Section 19.5. It gives 3000 consecutive data points of a 1-sampled simulated sample path of the CARMA(3, 2) process

$$(D^3 + 0.5D^2 + 1.08D + 0.104)Y_t = (D^2 + 0.8D + 0.15)DL_t, \tag{19.18}$$

where L_1 has the gamma distribution with mean 2/3 and variance 20/9. The data set can be downloaded from

https://doi.org/10.1515/9783111325033.

Our goal in this example is to estimate the model parameters. Again, CARMA32.txt must be stored as a single-column text file in the R working directory. Then STEP 1 plots the data and stores is as the R-vector DATA.

```
# STEP 1: READ DATA AS A VECTOR INTO R FROM A TEXT FILE IN YOUR WORKING DIRECTORY.
# To find your working directory type getwd()) .
DATA<-scan('CARMA32.txt', what=numeric(), quiet=TRUE);delta<-1
CARMA32 <- DATA
time<- c(1:(length(DATA)))
plot(time,CARMA32,main="Sampled simulated gamma-driven CARMA32, delta=1",type="l")
```

The corresponding plot can be seen in Figure 19.3. Then in STEP 2, the AIC and BIC values

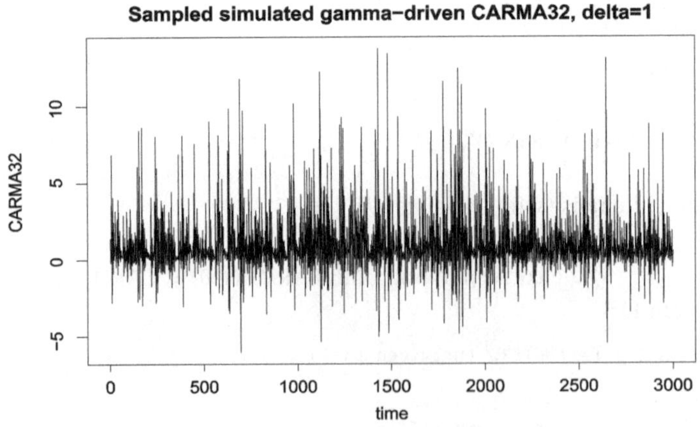

Figure 19.3: The CARMA(3, 2) simulated series sampled at time-intervals of length $\Delta = 1$.

for ARMA$(p, p-1)$ models for this data set with values of p up to 5 are calculated. Again, the program is completely the same as the one in Example 19.11. Running the program, both the AIC and BIC values are minimized for $p = 3$. STEP 3 then calculates the QMLE estimated discrete time ARMA(3, 2) model and also provides some parameters needed for the QML estimation of the CARMA(3, 2) model in STEP 4.

```
# STEP 3: p=3. QML ESTIMATORS OF ARMA PARAMETERS.
delta = 1
p=3;q=p-1;fitted=arima(DATA,c(p,0,q))
phi=fitted$coef[1:p];theta=fitted$coef[(p+1):(p+q)]
WNV=sqrt(fitted$sigma^2); AIC<-fitted$aic
Intercept=fitted$coef[(p+q+1)];Intercept=polyroot(c(Intercept,-1))
roots<-polyroot(c(-phi[3],-phi[2],-phi[1],1));lambda<-log(roots)/delta
xi1<-roots[1];xi2<-roots[2];xi3<-roots[3]
th1<-polyroot(c(-theta[1],1));th2<-polyroot(c(-theta[2],1))
ph1=polyroot(c(-phi[1],1));ph2=polyroot(c(-phi[2],1))
ph3=polyroot(c(-phi[3],1));xi=c(xi1,xi2,xi3)
cat("AR coeffs ,phi[j]:",phi,"\n");cat("MA coeffs, theta[j]:",theta,"\n");
cat("White noise variance:",WNV,"\n");cat("Intercept:",Re(Intercept),"\n")
```

The obtained QML ARMA(3, 2) model is

$$(1 - 1.7439B + 1.4062B^2 - 0.5904)(Y_n - 0.9374) = (1 - 1.2387B + 0.3432B^2)Z_n,$$

where (Z_n) is zero-mean white noise with variance WNV = 2.1459. The corresponding QML CARMA(3, 2) model for the data is now obtained in STEP 4.

```
# STEP 4: p=3.  QML ESTIMATORS OF CARMA PARAMETERS
G1<- -xi1*(1+th1*xi1+th2*xi1^2)*(1+th1/xi1+th2/(xi1^2))
G1<- G1/(1-ph1*xi1-ph2*xi1^2-ph3*xi1^3);G1<- G1/(-ph1-2*ph2/xi1-3*ph3/xi1^2)
G2<- -xi2*(1+th1*xi2+th2*xi2^2)*(1+th1/xi2+th2/(xi2^2))
G2<- G2/(1-ph1*xi2-ph2*xi2^2-ph3*xi2^3);G2<- G2/(-ph1-2*ph2/xi2-3*ph3/xi2^2)
G3<- -xi3*(1+th1*xi3+th2*xi3^2)*(1+th1/xi3+th2/(xi3^2))
G3<- G3/(1-ph1*xi3-ph2*xi3^2-ph3*xi3^3);G3<- G3/(-ph1-2*ph2/xi3-3*ph3/xi3^2)
co<-c(0,0);co[2]<- G1*lambda[1]+G2*lambda[2]+G3*lambda[3]
co[1]<-G1*lambda[1]*(lambda[2]^2+lambda[3]^2);
co[1]<-co[1]+G2*lambda[2]*(lambda[1]^2+lambda[3]^2);
co[1]<-co[1]+G3*lambda[3]*(lambda[1]^2+lambda[2]^2)
co0<- G1*lambda[1]*lambda[2]^2*lambda[3]^2+G2*lambda[2]*lambda[1]^2*lambda[3]^2;
co0 <- co0 + G3*lambda[3]*lambda[1]^2*lambda[2]^2
co[2]<-WNV*co[2];co[1]<-WNV*co[1];co0<-WNV*co0
# g(z)= co[2]*z^2+co[1]*z+co0 (embeddability theorem).
# To check embeddability check that g(z)<or=0 for z>or=0
# The following steps assume embeddability.
z<-polyroot(c(co0,co[1],co[2]));z<-sqrt(-z)
a1=-(lambda[1]+lambda[2]+lambda[3])
a2=lambda[1]*lambda[2]+lambda[1]*lambda[3]+lambda[2]*lambda[3]
a3=-lambda[1]*lambda[2]*lambda[3];a=c(a1,a2,a3);b=c(z[1]*z[2],z[1]+z[2],1)
cat("AR zeros,lambda[j]:",lambda,"\n");cat("MA zeros,mu[j]:",-z,"\n")
cat("AR coeffs,a1,a2,a3:",Re(a),"\n");cat("MA coeffs,b0,b1,b2:",Re(b),"\n")
cat("Var(L(1)):",Re(-2*co[2]),"\n");cat("EL(1):",Re(Intercept*a3/b[1]),"\n")
```

The obtained QML CARMA(3, 2) model is

$$(D^3 + 0.5269D^2 + 1.1429D + 0.1018)Y_t = (D^2 + 0.9011D + 0.1487)DL_t,$$

where L_1 has mean 0.6416 and variance 2.0417, as compared with the model (19.18) generating the data.

19.5 Simulation of causal CARMA processes

In this section, we give a sequence of R-programs which can be adapted to generate simulated causal CARMA(p, q) sample paths, with $p \leq 5$ and with a variety of different driving Lévy processes. The latter are obtained from the R-package yuima, which must first be loaded. Also observe that the R-package expm must first be loaded. All R-programs can be downloaded from

https://doi.org/10.1515/9783111325033.

In order to generate a simulated causal CARMA sample path on the time interval $(0, T]$, we use a discretization of the kernel representation (18.32) with kernel g, namely

$$\begin{aligned} Y_{nh} &= \sum_{i=0}^{\tau/h} g(ih)(L_{nh-ih} - L_{nh-ih-h}) \\ &= \sum_{j=n-\tau/h}^{n} \mathbf{b}^T e^{A(n-j)h} \mathbf{e}_p (L_{jh} - L_{(j-1)h}), \quad n = 1, \dots, T/h, \end{aligned}$$

to simulate the process at times nh up to time T, truncating the kernel at a suitably chosen τ. The discretization step size h is such that $1/h \in \mathbb{N}$. In this section we set $h = 0.01$, $T = 3000$ and $\tau = 100$. The linearly interpolated continuous-time function is then taken as our simulated sample path. The truncation time $\tau > 0$ of the discretized kernel is chosen to make its absolute value at τ a negligible fraction (say less than 10^{-5}) of its maximum absolute value on the interval $[0, \tau]$.

Programs 1–3 were already presented in Section 7.4. Program 1 specifies the total time T, the discretization step size h, the sampling interval delta, the truncation time tau for the kernel, the required orders, p and q (<p), and the required CARMA model, either by specifying the zeros of the autorgressive and moving average polynomials or by entering the coefficients directly. In the version shown, we use the CARMA(3, 2) model (19.18) of Example 19.14 and specify it by the zeros of the polynomials. The autoregressive zeros must all have strictly negative real parts to ensure the causality condition, and, for the invertibility condition, the moving average zeros must also have strictly negative real parts. There must also be no common autoregressive and moving average zeros and complex zeros must occur in conjugate pairs.

Program 2 generates the kernel of the chosen model up to time tau and prints the ratio of the value at tau to the maximum absolute value up to tau. This ratio can be made appropriately small (say, less than 10^{-5}) if necessary by re-running Programs 1 and 2 with larger tau.

Program 3 generates and plots the autocorrelation function of the CARMA model up to lag tau.

The Lévy increments are generated by R in Program 4 after loading the package yuima of Brouste et al. [31], for which more information is available in Iacus and Mercuri [49] or the book of Iacus and Yoshida [50].

```
# Program 4: Gamma: Increments of L
# BE SURE TO LOAD THE R-PACKAGE 'yuima' FROM NOW ON
# Parameters: del >0, shape
# gam > 0, rate
# EL(1) = m = del/gam
# Var(L1) = v = del/gam^2.
# This determines (tau+T)/h gamma increments, LI on h-intervals
# up to time tau+T. LIshort are the increments up to time 10.
del <- .2; gam<-.3; m<- del/gam; v<- del/gam^2
set.seed(879);nrep = (tau+T)/h;t=h
time<-c(1:(tau/h+T/h))*h
LI<-rgamma(nrep,del*t,gam)
plot(time,LI,main="Gamma␣increments␣on␣h-intervals␣to␣time␣tau+T",type="l")
Sys.sleep(5)
LIshort<-LI[1:(10/h)];time<-c(1:(10/h))*h
plot(time,LIshort,
    main="Gamma␣increments␣on␣h-intervals␣to␣time␣10",type="l")
```

In the version shown, the driving Lévy process is a gamma process with mean 2/3 and variance 20/9 as in Example 19.14, the increments of which are generated in Program 4. The driving process can be changed by replacing Program 4 with any of the alternative versions given in Appendix A at the end of the book.

Program 5 applies the CARMA kernel to the Lévy increments to generate the sample path of the CARMA process Y.

```
# Program 5: Applies the CARMA kernel to the Levy increments
# The filter is now applied to get the simulated CARMA z evaluated
# at h-increments from time tau+h to time tau+T. Re-labelling
# the time index gives the simulated CARMA Y from time h to time T
# in time-steps of h.
z<-filter(LI,g,sides=1); plot(z)
Y<-c(1:(T/h));
Y[1:(T/h)]<-z[((tau+h)/h):((tau/h)+T/h)]
time<-c(1:(T/h))*h
plot(time,Y,main="CARMA␣process␣on␣h-intervals␣to␣time␣T", type="l")
Sys.sleep(5)
Yshort<-Y[1:(10/h)];time<-c(1:(10/h))*h
plot(time,Yshort,main="CARMA␣process␣on␣h-intervals␣to␣time␣10",type="l")
write(Y, ncolumns=1, "Y32.txt")
# Y is now saved to the text file Y32.txt in the work directory.
```

The specific simulated process generated by Programs 1–5 is the process defined by (19.18) which, sampled with $\Delta = 1$, provided the data used in Example 19.14.

Remark 19.15. At this point, the simulation is complete and the sample path at integer multiples of h is stored in the working directory as the text file Y32.txt. However, Programs 7, 8 and 9 compute a fitted QMLE model based on the simulated data sampled at integer multiples of delta (in this case 1) in order to compare the fitted model with the original. In particular, if a different value of the sampling interval delta, e. g., 0.1, is chosen you will see (Exercise 19.4) that although the parameters of the sampled ARMA sequence are quite different, the estimated parameters of the fitted CARMA model are much the same. This highlights one of the virtues of CARMA as opposed to ARMA modelling as the CARMA model is essentially the same at different, sufficiently fine sampling frequencies, while the ARMA models, fitted to the sequences sampled at different frequencies, are quite different.

Program 6 compares the model ACF with the sample ACF of Y. It is simply a check that the sample ACF is consistent with that of the model generating the data and is not required for the subsequent estimation of parameters. The true time lag in the graph is the indicated Lag times h. Hence, with h = 0.01, the ACFs plotted at Lag 2000 are in reality the ACFs at lag 20.

```
# PROGRAM 6: Superpose CARMA model ACF on sample ACF of Y
ACF = acf (Y, lag = (tau/h)-1)
plot(ACF,main="Sample/Model␣(red)␣ACF␣of␣Y.␣Time␣lag=Lag*h", type="l")
ulim=tau/h; t<-c(1:(ulim-1));
lines(t,kappa[2:ulim],col="red",type="l")
```

Program 7 samples Y at time intervals specified by the chosen value of delta, which in this specific example is 1.0 and saves the sampled series as the text file CARMA32.txt. With the "set.seed(879)" command in Program 4 this reproduces exactly the data set that was used in Example 19.14.

```
# PROGRAM 7: Gives the delta-sampled sequence Ysamp
time<-c(1:(T/delta));index<-c(1:(T/delta))*delta/h
Ysamp<-Y[index];
plot(time,Ysamp,main="Ysamp,␣delta=␣1",type="l")
write(Ysamp, ncolumns=1, "CARMA32.txt")
# Ysamp is now saved to the text file CARMA32.txt in the work directory.
```

Program 8 fits a QML estimated ARMA($p, p-1$) sequence to the sampled data with p in this case set to 3, since we are interested in seeing how well we recover the generating CARMA(3, 2) model. Program 8 is nothing else than STEP 3 of Example 19.14, with "DATA" replaced by "Ysamp". (If a sampling interval delta different from 1 is used, then the value of the parameter delta = 1 has to be replaced.)

```
# Program 8: QML estimators of ARMA parameters
# Same as STEP 3 of Example 19.14 with "DATA" replaced by "Ysamp"
DATA <- YSamp
# Then use the code of STEP 3 of Example 19.14,
# with delta set to the sampling frequency specified in Program 1.
```

Program 9 uses Theorem 19.5 to embed the fitted ARMA model (if possible) in a CARMA model with autoregressive order p (here, $p = 3$), thereby determining a QML model with autoregressive order p for the sampled CARMA data. Again, Program 9 is nothing else than STEP 4 of Example 19.14 and is hence not reprinted here. Running Programs 1–9, we get the same estimated CARMA(3, 2) model as in Example 19.14.

It is most illuminating to run the programs one by one but, once the model, T, tau, delta and h have been specified, all nine can be run without a break.

19.6 Asymptotic distribution of QML estimators for a CARMA($p, p - 1$) process

CARMA processes of order $(p, p-1)$, $p \geq 1$, are of particular importance because they are the only ones capable of exhibiting sample paths with jumps. In fact, even CARMA(2, 1) processes have been found to be useful in financial econometrics (see, e. g., Todorov and Tauchen [85] and Todorov [84]).

In this section, we shall suppose that Y is a specified second-order CARMA(p, q) process, as is assumed in QML estimation, with real-valued coefficients

$$\boldsymbol{a} = (a_1, \dots, a_p, b_0, \dots, b_{p-2})^T$$

(recalling that $b_q = b_{p-1} = 1$). Denoting the zeros of the polynomials $a(z) = z^p + a_1 z^{p-1} + \cdots + a_p$ and $b(z) = b_0 + b_1 z + \cdots + b_{p-1} z^{p-1}$ by $\lambda_1, \dots, \lambda_p$ and $\mu_1, \dots, \mu_{p-1}$ respectively, we define

$$\boldsymbol{\beta} := (\lambda_1, \dots, \lambda_p, \mu_1, \dots \mu_{p-1})^T.$$

We shall assume that $\lambda_1, \dots, \lambda_p$ and $\mu_1, \dots, \mu_{p-1}$ have strictly negative real parts, $e^{\lambda_i} \neq e^{\lambda_j}$ for $i \neq j$ and $\lambda_i \neq \mu_j$ for all pairs i and j. We assume also that $\mathfrak{I}(\lambda_j) \in (-\pi/\Delta, \pi/\Delta)$ for each j, where Δ is the interval between successive sampled observations $Y_j^\Delta, j = 1, \dots, n$, on which inference is to be based. This last condition is satisfied if Δ is small enough.

Suppose now that Y^Δ is the ARMA($p, p-1$) sequence obtained by sampling Y at time intervals of length $\Delta > 0$ (and that the moving average order is indeed $p - 1$ and not smaller). Then the Sampling Theorem (Theorem 18.33) allows us to determine the reciprocal zeros $\xi_1, \dots, \xi_p$ and $\eta_1, \dots, \eta_{p-1}$ of the ARMA autoregressive and moving average polynomials, $\phi(z) = (1 - \phi_1 z - \cdots - \phi_p z^p)$ and $\theta(z) = (1 + \theta_1 z + \cdots + \theta_{p-1} z^{p-1})$, respectively. We shall denote by $\boldsymbol{\gamma}$ and $\boldsymbol{\delta}$ the parameter vectors,

$$\boldsymbol{\gamma} = (\xi_1, \dots, \xi_p, \eta_1, \dots, \eta_{p-1})^T$$

and

$$\boldsymbol{\delta} = (\phi_1, \dots, \phi_p, \theta_1, \dots, \theta_{p-1})^T.$$

The mappings from $\boldsymbol{a}$ to $\boldsymbol{\beta}$ and from $\boldsymbol{\gamma}$ to $\boldsymbol{\delta}$ are straightforward, but the mapping from $\boldsymbol{\beta}$ to $\boldsymbol{\gamma}$ requires the Sampling Theorem, the difficult part being the determination of $\eta_1, \dots, \eta_{p-1}$. However, the Sampling Theorem makes this feasible numerically so that the overall mapping from $\boldsymbol{a}$ to $\boldsymbol{\delta}$ can be determined. We shall denote this mapping by M. By our assumptions, and by the Sampling Theorem and the Embedding Theorem, the mapping M is a C^1-diffeomorphism in a neighborhood of the true coefficient vector $\boldsymbol{a}$.

Observe that if $\boldsymbol{a}$ is the true coefficient vector of Y then the QMLE $\hat{\boldsymbol{a}}$ of $\boldsymbol{a}$ based on observations of Y^Δ is $M^{-1}\hat{\boldsymbol{\delta}}$, where $\hat{\boldsymbol{\delta}}$ is the QMLE of $\boldsymbol{\delta}$ based on the same observations of Y^Δ.

Now suppose additionally that the driving Lévy process is a Brownian motion with mean m and variance $\nu^2 \in (0, \infty)$. Then the CARMA process Y and hence the Δ-sampled sequence Y^Δ are Gaussian. Since Y^Δ is an invertible and causal ARMA process with mean c, the white noise $(Z_n)_{n\in\mathbb{Z}}$ in the representation $\phi(B)(Y_n^\Delta - c) = \theta(B)Z_n$ can be written as $Z_n = \sum_{n=0}^{\infty} \pi_j(Y_{n-j}^\Delta - c)$, where $\sum_{j=0}^{\infty} \pi_j z^j = \phi(z)/\theta(z)$ for all $|z| \le 1$, see Brockwell and Davis [22, Theorem 3.1.2]. This implies that also $(Z_n)_{n\in\mathbb{Z}}$ is Gaussian, and since it is white noise, it must be independent, so that $(Z_n)_{n\in\mathbb{Z}}$ is i. i. d. $N(0, \sigma^2)$ for some $\sigma^2 > 0$. It then follows from equation (8.8.2) of Brockwell and Davis [22, p. 258], that

$$\sqrt{n}(\hat{\boldsymbol{\delta}}_n - \boldsymbol{\delta}) \xrightarrow{d} N(\mathbf{0}, W(\boldsymbol{\delta})), \quad \text{as } n \to \infty, \tag{19.19}$$

where

$$W(\boldsymbol{\delta}) = \begin{bmatrix} \mathbb{E}\mathbf{U}_n\mathbf{U}_n^T & \mathbb{E}\mathbf{U}_n\mathbf{V}_n^T \\ \mathbb{E}\mathbf{V}_n\mathbf{U}_n^T & \mathbb{E}\mathbf{V}_n\mathbf{V}_n^T \end{bmatrix}^{-1}, \tag{19.20}$$

$\mathbf{U}_n = (U_n, \dots, U_{n+1-p})^T$, $\mathbf{V}_n = (V_n, \dots, V_{n+2-p})^T$ and $(U_n)_{n\in\mathbb{Z}}$, $(V_n)_{n\in\mathbb{Z}}$ are the autoregressive processes,

$$\phi(B)U_n = \sigma^{-1}Z_n$$

and

$$\theta(B)V_n = \sigma^{-1}Z_n,$$

where $(\sigma^{-1}Z_n)_{n\in\mathbb{Z}}$ is i. i. d. $N(0,1)$. (Actually, the result in [22] is formulated when QML maximization is performed with *fixed* expectation c of Y^Δ, i. e., QMLE estimation is performed over the parameter vector $(\boldsymbol{\delta}, \sigma^2)$ and not over the parameter vector $(\boldsymbol{\delta}, c, \sigma^2)$ *including the mean*, as we describe it here. However, as shown in Bao [5], the asymptotic covariance matrix remains the same when the mean is included in the QMLE estimation, so that we can indeed apply the result in [22]. Also observe that the variance σ^2 plays no role for the QMLE of $\boldsymbol{\delta}$, since $\hat{\boldsymbol{\delta}}$ is obtained as a minimizer of the reduced likelihood, in which σ^2 does not enter, as described in Section 19.1.)

To translate (19.19) into the corresponding result for $\hat{\boldsymbol{a}}$, we need the Jacobian

$$J=\frac{\partial\boldsymbol{a}}{\partial\boldsymbol{\delta}}=\begin{bmatrix}\frac{\partial a_1}{\partial\phi_1} & \cdots & \frac{\partial a_1}{\partial\phi_p} & \frac{\partial a_1}{\partial\theta_1} & \cdots & \frac{\partial a_1}{\partial\theta_{p-1}}\\ \vdots & & \vdots & \vdots & & \vdots\\ \frac{\partial a_p}{\partial\phi_1} & \cdots & \frac{\partial a_p}{\partial\phi_p} & \frac{\partial a_p}{\partial\theta_1} & \cdots & \frac{\partial a_p}{\partial\theta_{p-1}}\\ \frac{\partial b_0}{\partial\phi_1} & \cdots & \frac{\partial b_0}{\partial\phi_p} & \frac{\partial b_0}{\partial\theta_1} & \cdots & \frac{\partial b_0}{\partial\theta_{p-1}}\\ \vdots & & \vdots & \vdots & & \vdots\\ \frac{\partial b_{p-2}}{\partial\phi_1} & \cdots & \frac{\partial b_{p-2}}{\partial\phi_p} & \frac{\partial b_{p-2}}{\partial\theta_1} & \cdots & \frac{\partial b_{p-2}}{\partial\theta_{p-1}}\end{bmatrix},\qquad(19.21)$$

which can also be calculated numerically with the aid of the Embedding Theorem 19.5, by incrementing, one by one, each of $\phi_1,\dots,\phi_p$ and $\theta_1,\dots,\theta_{p-1}$, and computing the coefficients $a_1,\dots,a_p$ and $b_0,\dots,b_{p-2}$ of the modified CARMA process obtained by embedding the modified ARMA process. Alternatively, the Sampling Theorem 18.33 can be used in an analogous way to compute J^{-1}. From the Jacobian matrix J and the matrix $W(\boldsymbol{\delta})$, the delta-method (e. g., Brockwell and Davis [22, Proposition 6.4.3]) implies the following asymptotic result for the QML estimator $\hat{\boldsymbol{a}}$,

$$\sqrt{n}(\hat{\boldsymbol{a}}_n-\boldsymbol{a})\xrightarrow{d}N(\boldsymbol{0},JW(\boldsymbol{\delta})J^T),\quad\text{as } n\to\infty.$$

In particular, $\hat{\boldsymbol{a}}$ is consistent, i. e.,

$$\hat{\boldsymbol{a}}_n\xrightarrow{P}\boldsymbol{a}\quad\text{as } n\to\infty.$$

Remark 19.16. To derive the asymptotics of the QMLE, we assumed the driving Lévy process to be a Brownian motion. This implied that the noise in the ARMA representation of Y^Δ was not only white noise, but actually i. i. d. white noise, so that we could apply the results for QMLE estimation of ARMA parameters with i. i. d. noise. A similar reasoning could have been carried out for general second-order CAR(1) processes, since also then the driving noise of the AR(1) process is i. i. d. For general CARMA processes and driving Lévy processes, however, the white noise in the ARMA representation of Y^Δ will not be i. i. d. Nevertheless, using results of Francq and Zakoïan [47] and of Marquardt and Stelzer [66], the asymptotic normality of the QMLE estimator can be established under light conditions on the driving Lévy process, see Brockwell et al. [23, Proposition 5], and Schlemm and Stelzer [80, Theorem 3.4] in the multivariate case. Observe however that the derived form of the asymptotic covariance matrix there is not as explicit as the one derived for the Gaussian case, which is why we usually use the asymptotic covariance matrix derived under the assumption of Gaussianity.

Example 19.17 (Application to the daily realized volatility model). Recall the fitted CARMA(2, 1) model (19.16) for the 3045 realized volatilities of the daily Deutsche Mark/US dollar exchange rates of Example 19.11, namely

$$(D^2 + 1.3612D + 0.04449)Y_t = (D + 0.20584)DL_t,$$

with $\mathbb{E}(L_1) = 0.1448$, $\mathrm{Var}(L_1) = 0.08340$ and t measured in days (since the sampling interval $\Delta = 1$ represents one day). Our aim in this example is to use the asymptotic theory presented above, as the number of observations goes to infinity with $\Delta = 1$, to estimate the (Gaussian) standard errors of the estimated coefficients $\hat{a}_1 = 1.3612$, $\hat{a}_2 = 0.04449$ and $\hat{b}_0 = 0.20584$. We use the following R programs. The first is an R-function that calculates the ARMA parameters of the sampled sequence in terms of the CARMA parameters, using the Sampling Theorem.

```
# FUNCTION CARMA(2,1) parameters to ARMA(2,1) parameters via Sampling Theorem,
# in particular via Example 18.35
# Arguments are a=c(a1,a2) autoregressive coefficients and
# mu=b0 moving average coefficient,  delta=sampling interval,
# output of function is vector with AR coefficients phi1, phi2 and moving average
# coefficient theta
carmaTOarma <- function(a,mu,delta){
   A=matrix(rep(0,len=4),nrow=2)
   lambda=c(0,0)
   A[1,1]=0;A[1,2]=1;A[2,1]=-a[2];A[2,2]=-a[1]
   v=eigen(A)$values
   lambda[1]=v[2]; lambda[2]=v[1]
   K1= (mu^2-lambda[1]^2)/(2*lambda[1]*(lambda[1]^2-lambda[2]^2))
   K2= (mu^2-lambda[2]^2)/(2*lambda[2]*(lambda[2]^2-lambda[1]^2))
   k1=K1*sinh(lambda[1]*delta)*cosh(lambda[2]*delta)
   k1=k1+ K2*sinh(lambda[2]*delta)*cosh(lambda[1]*delta)
   H1=K1*sinh(lambda[1]*delta)+K2*sinh(lambda[2]*delta)
   k1=k1/H1
   eta=k1-sqrt(k1^2-1)
   # for the specific data example this is the root with abs. value less than 1
   theta= -eta
   xi= exp(delta*lambda)
   ph1 <- xi[1]+xi[2]; ph2 <- - xi[1] * xi[2]
   return(c(ph1,ph2,theta))
   }
```

The next commands calculate the AR and MA coefficients of the sampled sequence in our specific example.

```
# SAMPLE
delta=1
a=c(1.3612,0.04449); mu=.20584; b=c(mu,1)
cat("CARMA AR coeffs, a[j]:",a,"\n");cat("CARMA MA coeffs, b[j]:",b,"\n");
out <- carmaTOarma(a,mu,delta)
cat("AR coefficients, phi1, phi2:", out[1], out[2], "\n")
cat("MA coefficient theta", out[3], "\n")
# Store original a and b,  mu, phi and theta.
a1=a[1];a2=a[2];b0=mu
phi1=out[1];phi2=out[2]; theta1 = out[3]; mu1=mu
```

This recovers the ARMA(2, 1) model of (19.15), as it should. For this process, the asymptotic covariance matrix of (19.20) is easily calculated as

$$W(\boldsymbol{\delta}) = \begin{bmatrix} g & \frac{\phi_1 g}{1-\phi_2} & \frac{1}{1+\phi_1\theta_1-\phi_2\theta_1^2} \\ \frac{\phi_1 g}{1-\phi_2} & g & \frac{-\theta_1}{1+\phi_1\theta_1-\phi_2\theta_1^2} \\ \frac{1}{1+\phi_1\theta_1-\phi_2\theta_1^2} & \frac{-\theta_1}{1+\phi_1\theta_1-\phi_2\theta_1^2} & \frac{1}{1-\theta_1^2} \end{bmatrix}^{-1}, \tag{19.22}$$

where

$$g = \mathrm{Var}(U_0) = \frac{1-\phi_2}{1-\phi_2-\phi_1^2-\phi_1^2\phi_2-\phi_2^2+\phi_2^3},$$

see Exercise 19.7. The following program calculates $W(\boldsymbol{\delta})$.

```
# ASYMPTOTIC COVARIANCE MATRIX W OF QMLE's of phi1, phi2, theta
V=matrix(rep(0,len=9),nrow=3)
R = 1/(1-phi2-phi1^2-phi1^2*phi2 - phi2^2 + phi2^3)
V[1,2]= phi1 * R; V[2,1] = V[1,2]
V[1,1]= (1-phi2) * R ; V[2,2] = V[1,1]
V[1,3] = 1/(1+ phi1 * theta1 - phi2 * theta1^2); V[3,1] = V[1,3]
V[2,3] = - theta1 * V[1,3]; V[3,2] = V[2,3]
V[3,3] = 1/(1-theta1^2)
W = solve(V); W
```

The Jacobi matrix J in (19.21) is now calculated numerically with the aid of the function carmaTOarma.

```
# JACOBIAN, J, OF MAPPING FROM phi1, phi2, theta TO a1,a2, b0
inc=.0001
I=matrix(rep(0,len=9),nrow=3)
out <- carmaTOarma(c(a1+inc,a2),mu1,delta)
I[1,1]=(out[1]-phi1)/inc; I[2,1]=(out[2]-phi2)/inc; I[3,1]=(out[3]-theta1)/inc
out <- carmaTOarma(c(a1,a2+inc),mu1,delta)
I[1,2]=(out[1]-phi1)/inc; I[2,2]=(out[2]-phi2)/inc; I[3,2]=(out[3]-theta1)/inc
out <- carmaTOarma(c(a1,a2),mu1+inc,delta)
I[1,3]=(out[1]-phi1)/inc; I[2,3]=(out[2]-phi2)/inc; I[3,3]=(out[3]-theta1)/inc
J=solve(I); J
```

The asymptotic covariance matrix and the standard deviations for 3450 observations are now easily obtained.

```
# ASYMPTOTIC COVARIANCE MATRIX Sigma OF QMLE's of a1, a2, b0
Sigma=J %*% W %*% t(J)
Sigma
# Standard errors for 3045 observations
sqrt(Sigma[1,1]/3045)
sqrt(Sigma[2,2]/3045)
sqrt(Sigma[3,3]/3045)
```

For the 3045 observations of the Deutsche Mark/US dollar data the estimated standard deviations of $\hat{a}_1$, $\hat{a}_2$ and $\hat{b}_0$ are $\sqrt{(30.563779/3045)} = 0.1002$, $\sqrt{(0.3722699/3045)} = 0.0111$ and $\sqrt{(2.3082275/3045)} = 0.0275$, respectively. For estimates based on 2000 successive observations of the same model, these estimated standard deviations become 0.1236, 0.0136 and 0.03397, respectively.

Remark 19.18 (Alternative approach to estimation error). Instead of using the asymptotic results above, one could simulate a large number, say N, of realizations of the fitted model on the interval $(0, T]$ with sampling interval Δ, compute QML estimates of the parameters for each realization, and compute the sample variance of the resulting sets of N parameter estimates. This approach was used by Brockwell et al. [23], who simulated 1000 independent realizations of the model (19.17) on the time interval $[0, 2000]$ with $\Delta = 1$ and $\Delta = 0.1$. Observe that (19.17) is very close to (19.16). The estimated standard deviations computed by this lengthy procedure are found to be closely compatible with the asymptotic estimates found above. In particular, for the gamma-driven model with $\Delta = 1$, the estimated standard deviations obtained in this way were 0.13262, 0.01603 and 0.03742 as compared with the values 0.1237, 0.0136 and 0.03399, calculated with the method of the previous example.

19.7 Estimating the Lévy increments

In order to suggest an appropriate parametric model for the driving Lévy process, it is necessary to recover the realization of L from a given realization of Y on $[0, T]$ for given or estimated values of $\{a_j, 1 \le j \le p;\ b_j, 0 \le j < q\}$ (we assume the causality and invertibility condition, that the multiplicities of the zeros of the autoregressive polynomial are all one, and that all coefficients are real-valued). In Theorem 7.19 and Proposition 7.20 we already recovered the state vector $\mathbf{X}$ of a second-order CARMA process from the infinite past using mean-square differentiation. Using differentiation, we now use a related argument for the determination of $\mathbf{X}_t$ (and then of L_t) in terms of the process Y observed continuously on $[0, T]$ and a known or plausible value for $\mathbf{X}_0$. The key to recovering the corresponding realization of $\mathbf{X}$ and then L on $[0, T]$ is the observation that equations (18.4) and (18.5) imply an equation for the state vector $\mathbf{X}_t$ in terms of Y, namely,

$$\mathrm{d}\mathbf{X}_t^{(q)} - B\mathbf{X}_t^{(q)}\,\mathrm{d}t = \mathbf{e}_q\, Y_t\,\mathrm{d}t, \tag{19.23}$$

where $\mathbf{X}_t^{(q)} = (X_t^0, \ldots, X_t^{q-1})^T$ is the upper q-truncation of $\mathbf{X}_t$,

$$B = \begin{bmatrix} 0 & 1 & 0 & \cdots & 0 \\ 0 & 0 & 1 & \cdots & 0 \\ \vdots & \vdots & \vdots & \ddots & \vdots \\ 0 & 0 & 0 & \cdots & 1 \\ -b_0 & -b_1 & -b_2 & \cdots & -b_{q-1} \end{bmatrix} \in \mathbb{R}^{q\times q}, \quad \mathbf{e}_q = \begin{bmatrix} 0 \\ 0 \\ \vdots \\ 0 \\ 1 \end{bmatrix} \in \mathbb{R}^q$$

and $B = -b_0$ if $q = 1$. By an argument similar to the one used in the proof of Theorem 17.2, the solution of (19.23) satisfies

$$\mathbf{X}_t^{(q)} = \mathrm{e}^{Bt}\mathbf{X}_0^{(q)} + \int_0^t \mathrm{e}^{B(t-u)}\mathbf{e}_q\, Y_u\,\mathrm{d}u, \tag{19.24}$$

and the remaining $(p-q)$ components of $\mathbf{X}_t$ are obtained from the recursions $X_t^j = DX_t^{j-1}$, $j = q, \ldots, p-1$, where X^{j-1} denotes the $(j-1)^{\text{st}}$ (right-)derivative with respect to t of the first component X^0 of $\mathbf{X}$. This completes the determination of $\mathbf{X}_t$ in terms of $\mathbf{X}_0$ and continuously observed values of Y on $[0, t]$.

Having determined the vector $\mathbf{X}_t$ on $[0, T]$, and knowing (or having estimated) the coefficients $\{a_1, \ldots, a_p; b_0, \ldots, b_{q-1}\}$, one way to recover L_t on $[0, T]$ is to use the last component of equation (18.5), according to which

$$\mathrm{d}X_t^{p-1} = [-a_p, \ldots, -a_1]\,[X_t^0, \ldots, X_t^{p-1}]^T\,\mathrm{d}t + \mathrm{d}L_t,$$

leading to

$$\begin{aligned} L_t &= X_t^{p-1} - X_0^{p-1} + \sum_{j=0}^{p-1} a_{p-j} \int_0^t X_s^j\,\mathrm{d}s \\ &= X_t^{p-1} - X_0^{p-1} + a_p \int_0^t X_s^0\,\mathrm{d}s + \sum_{j=1}^{p-1} a_{p-j}(X_t^{j-1} - X_0^{j-1}). \end{aligned}$$

Another way is to use the canonical state vector $\mathbf{Y}_t$ from Definition 18.29. By Proposition 18.30, the rth component of the canonical state vector is

$$Y_t^{(r)} = \alpha_r Z_t^{(r)},$$

where

$$\alpha_r = \frac{b(\lambda_r)}{a'(\lambda_r)} \quad \text{and} \quad Z_t^{(r)} = \int_{(-\infty,t]} \mathrm{e}^{\lambda_r(t-u)}\,\mathrm{d}L_u.$$

So $Z^{(r)}$ is the CAR(1) process defined by

$$\mathrm{d}Z_t^{(r)} = \lambda_r Z_t^{(r)}\mathrm{d}t + \mathrm{d}L_t.$$

Consequently

$$\alpha_r\,\mathrm{d}L_t = \alpha_r\,\mathrm{d}Z_t^{(r)} - \lambda_r \alpha_r Z_t^{(r)}\mathrm{d}t = \mathrm{d}Y_t^{(r)} - \lambda_r Y_t^{(r)}\,\mathrm{d}t,$$

i. e.,

$$\alpha_r L_t = Y_t^{(r)} - Y_0^{(r)} - \lambda_r \int_0^t Y_u^{(r)}\,\mathrm{d}u.$$

Using the definition of α_r, we obtain p equivalent representations of L_t in terms of the observations of the CARMA process on $[0, t]$, namely

$$L_t = \frac{a'(\lambda_r)}{b(\lambda_r)}\left[Y_t^{(r)} - Y_0^{(r)} - \lambda_r \int_0^t Y_u^{(r)}\, du \right], \quad r = 1, \dots, p. \tag{19.25}$$

Thus if Y were observed continuously on $[0, t]$ we could use any one of these p simple equations to recover L from the realization of Y, the value of $\mathbf{X}_0$ and the parameters of the CARMA process. For calculations, it is simplest to choose a value of r for which λ_r is real and small in absolute value so that the contribution of the integral term in (19.25) is small compared with $Y_t^{(r)} - Y_0^{(r)}$.

Example 19.19 (Application to a CARMA(2, 1) process). In this case, writing $\mathbf{X}_t = [X_t^0, X_t^1]^T$, we have $\mathbf{X}_t^{(q)} = X_t^0$ and equation (19.24) takes the form

$$X_t^0 = X_0^0\, e^{-b_0 t} + \int_0^t e^{-b_0(t-u)} Y_u\, du. \tag{19.26}$$

The remaining components of $\mathbf{X}_t$ consist only of X_t^1, which is given by

$$X_t^1 = DX_t^0 = -b_0 X_t^0 + Y_t. \tag{19.27}$$

Given $\mathbf{X}_0$ and $\{Y_t, 0 \le t \le T\}$, (19.26) determines $\{X_t^0, 0 \le t \le T\}$ and then (19.27) determines $\{X_t^1, 0 \le t \le T\}$. We replace the unobserved $\mathbf{X}_0$ by $\overline{Y}[1\ 0]^T/b_0$ where $\overline{Y}$ is the sample mean of the observed values of the CARMA process. The canonical state vector $\mathbf{Y}_t$ is given by Definition 18.29 and $\{L_t, 0 \le t \le T\}$ is determined by either of the two equations (19.25).

In practice, of course, we do not observe Y continuously and we must replace the integrals appearing in the exact solution with approximating sums.

Remark 19.20. Brockwell et al. [23] used the daily realized volatility series of Example 19.11 and the fitted model (19.17) to estimate the corresponding Lévy increments $L_n - L_{n-1}$. The corresponding estimates of the zeros of the autoregressive polynomial were $\hat{\lambda}_1 = -0.03334$ and $\hat{\lambda}_2 = -1.32856$. The estimates of the Lévy increments $L_n - L_{n-1}$ were obtained by constructing the realization of the component CAR(1) process, corresponding to the estimated autoregressive root $\hat{\lambda}_1 = -0.03334$ and proceeding as described in Example 19.19. Setting the negative estimates to zero, Brockwell et al. [23] obtained the empirical distribution function. This was compared with distribution functions corresponding to four different members of the generalized inverse Gaussian family, each chosen to have the same mean 0.1783 and variance 0.06746. Of the four processes considered, the best fitting was the gamma process, suggesting the appropriateness of the model (19.17) driven by the gamma process L with $\mathbb{E}(L_1) = 0.1783$ and $\operatorname{Var}(L_1) = 0.06746$.

19.8 Exercises

Exercise 19.1. Prove Proposition 19.4.
Hint: Use that $f(\omega) = (2\pi)^{-1}\sigma^2|\theta(e^{-i\omega})|^2/|\phi(e^{-i\omega})|^2$, $\omega \in (-\pi, \pi]$, is the spectral density of Y. Calculate $\gamma(h) = \int_{-\pi}^{\pi} e^{i\omega h} f(\omega)\, d\omega$ with the aid of the residue theorem, similar to the proof of (7.4).

Exercise 19.2. Show that $\sum_{j=1}^{N} \Delta_{n,j}^2$ converges in mean-square to the integrated volatility I_n for day n as $N \to \infty$, as asserted just before Example 19.11.

Exercise 19.3. Use the R-programs in Section 19.5 to determine the parameters of the sampled ARMA(3, 2) sequence and of the QML fitted CARMA model found in that section.

Exercise 19.4. Repeat Exercise 19.3, but with the sampling interval delta = 0.1 instead of 1.

Exercise 19.5. Repeat Exercise 19.3, again with sampling interval delta = 1, but replacing Program 4 by its counterpart in Appendix A, generating inverse Gaussian instead of gamma-distributed Lévy increments. Choose the parameters of the driving inverse Gaussian process to have the same values of $\mathbb{E}(L_1)$ and $\mathrm{Var}(L_1)$ as the gamma process in Exercise 19.3.

Exercise 19.6. Repeat Exercise 19.5, but with the sampling interval delta = 0.1 instead of 1.

Exercise 19.7. Prove equation (19.22).
Hint: Use Brockwell and Davis [22, equations (3.3.8) and (3.3.9)] for the calculation of g, or alternatively [22, Example 8.8.1].

A Appendix: R Programs for generation of Lévy increments

In the simulation programs of Section 19.5, Program 4 can be replaced by any of the various Lévy-increment generators shown on the following pages (again, the R-package Yuima has to be loaded first). For reference, we list their characteristic functions, $\phi_{L_t}(z) = \mathbb{E}\exp(izL_t)$, $z \in \mathbb{R}$ and $t > 0$, below.

Each of the programs can also be run independently, but it is then necessary first to specify T, tau and h, e. g., T = 1000; h = 0.01; tau = 100.

(a) *Alpha-stable* (cf., Definition 14.9)

$$\phi_{L_t}(z) = \exp(t[i\gamma z - |\sigma z|^\alpha(1 - i\beta\ \mathrm{sign}(z)s)]),$$
$$\alpha \in (0,2], \beta \in [-1,1], \sigma > 0, \gamma \in \mathbb{R},$$

where α = index of stability, β = skewness, σ = scale, γ = location and

$$s = \begin{cases} \tan\frac{\pi\alpha}{2}, & \alpha \neq 1, \\ -\frac{2}{\pi}\log|z|, & \alpha = 1. \end{cases}$$

```
# Program 4: Alpha-stable: Increments of L
# Stable parameters: alpha in (0,2], beta in [-1,1]
# sigma in (0,infinity), gam in (-infinity,infinity).
# alpha: stability index, beta: skewness, sigma: scale, gam: location.
# Mean, m=gam if alpha>1.
# For positive stable alpha in (0,1),beta=1, gam in [0,infty)
# This program determines (tau+T)/h alpha-stable increments,
# LI on h-intervals up to time tau+T.
# LIshort shows the increments up to time 10.
alpha <- 0.5; beta <- 1; sigma<-1; gam<-0;
if(alpha>1){m<-gam}
set.seed(879); nrep = (tau+T)/h;t=h
time<-c(1:(tau/h+T/h))*h
LI<-rstable(nrep,alpha,beta,sigma*t^(1/alpha),gam*t)
plot(time,LI,
    main="Alpha-stable␣increments␣on␣h-intervals␣to␣time␣tau+T",type="l")
Sys.sleep(5)
LIshort<-LI[1:(10/h)];time<-c(1:(10/h))*h
plot(time,LIshort,
    main="Alpha-stable␣increments␣on␣h-intervals␣to␣time␣10",type="l")
```

(b) *Cauchy* (cf., Example 14.11)
Alpha-stable with $\alpha = 1, \beta = 0, \sigma > 0$ and $\gamma \in \mathbb{R}$.

(c) *Positive stable* (cf., Proposition 14.14(g))
Alpha-stable with $\alpha \in (0,1), \beta = 1, \sigma > 0$ and $\gamma \geq 0$.

(d) *Gaussian* (Alpha-stable with $\alpha = 2$, i. e., BM(μ, σ^2), cf., equation (1.4))

$$\phi_{L_t}(z) = \exp\left[itz\mu - \frac{t\sigma^2 z^2}{2}\right], \quad \mu \in \mathbb{R}, \sigma > 0.$$

https://doi.org/10.1515/9783111325033-020

```
# Program 4: Gaussian process: Increments of L
# Parameters: mean in (-infty,infty)
# sd > 0,
# EL(1) = m = mean
# Var(L(1)) = v = sd^2.
# This determines (tau+T)/h Gaussian increments,
# LI on h-intervals up to time tau+T.
# LIshort shows the increments up to time 10.
mean <- 0; sd<- 1; m<- mean; v<- sd^2
set.seed(879); nrep = (tau+T)/h;t=h
time<-c(1:(tau/h+T/h))*h
LI<-rnorm(nrep,mean*t,sd*sqrt(t))
plot(time,LI,
    main="Gaussian␣increments␣on␣h-intervals␣to␣time␣tau+T",type="l")
Sys.sleep(5)
LIshort<-LI[1:(10/h)];time<-c(1:(10/h))*h
plot(time,LIshort,
    main="Gaussian␣increments␣on␣h-intervals␣to␣time␣10",type="l")
```

(e) *Positive tempered stable* (PTS$(\alpha, \sigma, \lambda)$, cf., end of Section 14.4)

$$\phi_{L_t}(z) = \exp\big[t\sigma\Gamma(-\alpha)\big((\lambda - \mathrm{i}z)^{\alpha} - \lambda^{\alpha}\big)\big], \quad \alpha \in (0,1), \sigma > 0, \lambda > 0.$$

```
# Program 4: Positive tempered stable: Increments of L
# Parameters: alpha in (0,1), sigma > 0, lambda > 0.
# alpha: exponent, sigma: scale, lambda: tempering parameter.
# E(L(1)) = m = sigma*Gamma(1-alpha)/lambda^(1-alpha).
# Var(L(1))= v = sigma*Gamma(2-alpha)/lambda^(2-alpha).
# This determines (tau+T)/h Positive tempered stable increments,
# LI on h-intervals up to time tau+T.
# LIshort shows the increments up to time 10.
alpha <- 0.5;sigma <- 0.2;lambda <- 2
m<- sigma*gamma(1-alpha)/lambda^(1-alpha)
v<- sigma*gamma(2-alpha)/lambda^(2-alpha)
set.seed(879); nrep = (tau+T)/h;t=h
time<-c(1:(tau/h+T/h))*h
LI<-rpts(nrep,alpha,sigma*t,lambda)
plot(time,LI,
    main="PTStable␣increments␣on␣h-intervals␣to␣time␣tau+T",type="l")
Sys.sleep(5)
LIshort<-LI[1:(10/h)];time<-c(1:(10/h))*h
plot(time,LIshort,
    main="PTStable␣increments␣on␣h-intervals␣to␣time␣10",type="l")
```

(f) *Normal tempered stable* ($L_t = \mu t + \beta S_t + W_{S_t}$, where $S_t \sim$ PTS$(\alpha, \sigma, \lambda)$ and W is standard Brownian motion independent of S, cf., end of Section 14.4 or Iacus and Yoshida [50, p. 179])

$$\phi_{L_t}(z) = \exp\{\mathrm{i}t\mu + t\sigma\Gamma(-\alpha)[(\lambda - \mathrm{i}\beta z + z^2/2)^{\alpha} - \lambda^{\alpha})]\}, \quad \mu \in \mathbb{R}, \beta \in \mathbb{R}.$$

```
# Program 4: Normal tempered stable: Increments of L
# Parameters: alpha in (0,1), sigma > 0, lambda > 0. (PTStable params)
# alpha: exponent, sigma: scale, lambda: tempering param.
# beta: drift of BM, mu: added drift, Lambda: dimension
# See Iacus and Yoshida p. 179. X_t = mu*t + beta*S_t + W_{S_t}
```

```
# This determines (tau+T)/h Normal tempered stable increments,
# LI on h-intervals up to time tau+T.
# LIshort shows the increments up to time 10.
alpha <- 0.5; sigma <- 0.2;lambda=.1;beta=1;mu=1
Lambda <- matrix(1,1,1)
set.seed(879); nrep = (tau+T)/h;t=h
time<-c(1:(tau/h+T/h))*h
Li=rnts(nrep,alpha,sigma*t,lambda,beta,mu*t,Lambda)
plot(time,LI,
    main="NTStable␣increments␣on␣h-intervals␣to␣time␣tau+T",type="l")
Sys.sleep(5)
LIshort<-LI[1:(10/h)];time<-c(1:(10/h))*h
plot(time,LIshort,
    main="NTStable␣increments␣on␣h-intervals␣to␣time␣10",type="l")
```

(g) *Gamma* (with parameters δ and γ, cf., Section 14.2)

$$\phi_{L_t}(z) = (1 - \mathrm{i}z/\gamma)^{-t\delta}, \quad \gamma > 0, \delta > 0.$$

```
# Program 4: Gamma: Increments of L
# Parameters: del >0, shape
# gam > 0, rate
# EL(1) = m = del/gam
# Var(L1) = v = del/gam^2.
# This determines (tau+T)/h gamma increments, LI on h-intervals
# up to time tau+T. LIshort are the increments up to time 10.
del <- .2; gam<-.3; m<- del/gam; v<- del/gam^2
set.seed(879);nrep = (tau+T)/h;t=h
time<-c(1:(tau/h+T/h))*h
LI<-rgamma(nrep,del*t,gam)
plot(time,LI,
    main="Gamma␣increments␣on␣h-intervals␣to␣time␣tau+T",type="l")
Sys.sleep(5)
LIshort<-LI[1:(10/h)];time<-c(1:(10/h))*h
plot(time,LIshort,
    main="Gamma␣increments␣on␣h-intervals␣to␣time␣10",type="l")
```

(h) *Bilateral gamma* ($L_t = X_t^+ - X_t^-$ where X^+ and X^- are independent gamma processes, cf., Remark 14.27)

$$\phi_{L_t}(z) = (1 - \mathrm{i}z/\gamma_+)^{-t\delta_+}(1 + \mathrm{i}z/\gamma_-)^{-t\delta_-}, \quad \gamma_\pm > 0, \delta_\pm > 0.$$

```
# Program 4: Bilateral gamma: Increments of L
# Parameters: gamma.plus>0, gamma.minus>0, rates
# delta.plus>0, delta.minus>0, shapes
# E(L(1)) = m = delta.plus/gamma.plus-delta.minus/gamma.minus
# Var(L(1))= v = delta.plus/(gamma.plus)^2 + delta.minus/(gamma.minus)^2
# This determines (tau+T)/h biliteral gamma increments,
# LI on h-intervals up to time tau+T.
# LIshort shows the increments up to time 10
delta.plus <- .2;gamma.plus<-.3;delta.minus<-2;gamma.minus<-.6
m<- delta.plus/gamma.plus-delta.minus/gamma.minus
v<- delta.plus/(gamma.plus)^2 + delta.minus/(gamma.minus)^2
set.seed(127); nrep = (tau+T)/h;t=h
time<-c(1:(tau/h+T/h))*h
LI<-rbgamma(nrep,delta.plus*t,gamma.plus,delta.minus*t,gamma.minus)
```

```
plot(time,LI,
    main="B-gamma␣increments␣on␣h-intervals␣to␣time␣tau+T",type="l")
Sys.sleep(5)
LIshort<-LI[1:(10/h)];time<-c(1:(10/h))*h
plot(time,LIshort,
    main="B-gamma␣increments␣on␣h-intervals␣to␣time␣10",type="l")
```

(i) *Variance gamma* (Bilateral gamma with $\delta_- = \delta_+ = \delta$, cf., Theorem 14.25)

$$\phi_{L_t}(z) = (1 - \mathrm{i}z/\gamma_+)^{-t\delta}(1 + \mathrm{i}z/\gamma_-)^{-t\delta}.$$

```
# Program 4: Variance gamma : Increments of L
# Parameters: gamma.plus>0, gamma.minus>0, rates
# del>0, shape
# E(L(1)) = m = del/gamma.plus-del/gamma.minus
# Var(L(1))= v = del/(gamma.plus)^2 + del/(gamma.minus)^2
# This determines (tau+T)/h variance gamma increments,
# LI on h-intervals up to time tau+T.
# LIshort shows the increments up to time 10
del <- 2;gamma.plus<-.3;gamma.minus<-.6
m<- del/gamma.plus-del/gamma.minus
v<- del/(gamma.plus)^2 + del/(gamma.minus)^2
set.seed(127); nrep = (tau+T)/h;t=h
time<-c(1:(tau/h+T/h))*h
LI<-rbgamma(nrep,del*t,gamma.plus,del*t,gamma.minus)
plot(time,LI,
    main="V-gamma␣increments␣on␣h-intervals␣to␣time␣tau+T",type="l")
Sys.sleep(5)
LIshort<-LI[1:(10/h)];time<-c(1:(10/h))*h
plot(time,LIshort,
    main="V-gamma␣increments␣on␣h-intervals␣to␣time␣10",type="l")
```

(j) *Inverse Gaussian* (L_t = time for BM$(\gamma, 1)$ to reach level $\delta t > 0$, cf., Theorem 14.29 and Remark 14.30(b))

$$\phi_{L_t}(z) = \exp[t\gamma\delta(1 - \sqrt{1 - 2\mathrm{i}z/\gamma^2})], \quad \gamma > 0, \delta > 0.$$

```
# Program 4: Inverse Gaussian: Increments of L
# Parameters: del >0, shape
# gam > 0, shape/mean
# EL(1) = m = del/gam
# Var(L1) = v = del/gam^3.
# This determines (tau+T)/h Inverse Gaussian increments,
# LI on h-intervals up to time tau+T.
# LIshort shows the increments up to time 10.
del<- 2.23657; gam<- 3.34016; m<- del/gam ; v<- del/gam^3
set.seed(879); nrep = (tau+T)/h;t=h
time<-c(1:(tau/h+T/h))*h
LI<-rIG(nrep,del*t,gam)
plot(time,LI,
    main="Inverse␣Gaussian␣increments␣on␣h-intervals␣to␣time␣tau+T",
    type="l")
Sys.sleep(5)
LIshort<-LI[1:(10/h)];time<-c(1:(10/h))*h
plot(time,LIshort,
    main="Inverse␣Gaussian␣increments␣on␣h-intervals␣to␣time␣10",
    type="l")
```

(k) *Normal inverse Gaussian* ($L_t = \mu t + \beta S_t + W_{S_t}$, where $S_t \sim \mathrm{IG}(\delta, \gamma)$ with $\delta > 0$, $\gamma = \sqrt{\alpha^2 - \beta^2}$, $\alpha > 0, \beta \in \mathbb{R}, |\beta| < \alpha$, and W is standard Brownian motion independent of S, cf., Definition 14.33)

$$\phi_{L_t}(z) = \exp\big[it\mu z + t\delta(\sqrt{\alpha^2 - \beta^2} - \sqrt{\alpha^2 - (\beta + iz)^2})\big].$$

```
# Program 4: Normal inverse Gaussian: Increments of L
# Parameters: alpha>0, tail heaviness; beta with |beta|<alpha, skewness
# del>0, mu real-valued.
# L(1) has the distribution of (mu+beta*S(1)+W(S(1))), where
# S is an inverse Gaussian process with parameters
# del and gam=sqrt(alpha^2-beta^2)),
# and W is standard Brownian motion independent of S.
# E(L(1)) = mu + del*beta/gam
# Var(L(1))= del*alpha^2/gam^3
# This determines (tau+T)/h NIG increments,
# LI on h-intervals up to time tau+T.
# LIshort shows the increments up to time 10.
del <- 2; alpha <- 1.5; beta <- -0.7; mu <- 3; gam<- sqrt(alpha^2-beta^2)
m<- mu+del*beta/gam; v<- del*alpha^2/gam^3
set.seed(879); nrep = (tau+T)/h;t=h
time<-c(1:(tau/h+T/h))*h
LI<-rNIG(nrep,alpha,beta,del*t,mu*t)
plot(time,LI,main="NIG␣increments␣on␣h-intervals␣to␣time␣tau+T",type="l")
Sys.sleep(5)
LIshort<-LI[1:(10/h)];time<-c(1:(10/h))*h
plot(time,LIshort,
    main="NIG␣increments␣on␣h-intervals␣to␣time␣10",type="l")
```

(l) *Poisson* (cf., Theorem 10.9)

$$\phi_{L_t}(z) = \exp\big[t\lambda(e^{iz} - 1)\big], \quad \lambda > 0.$$

```
# Program 4: Poisson process : Increments of L
# Parameters: lam > 0 is the jump rate
# E(L(1)) =m= lam
# Var(L(1))=v= lam
# This determines (tau+T)/h Poisson increments,
# LI on h-intervals up to time tau+T.
# LIshort shows the increments up to time 10.
lam<- 1; m<- lam; v<- lam
set.seed(879);nrep = (tau+T)/h;t=h
time<-c(1:(tau/h+T/h))*h
LI <- rpois(nrep,lam*t)
plot(time, LI,
    main="Poisson␣increments␣on␣h-intervals␣to␣time␣tau+T",type="l")
Sys.sleep(5)
LIshort<-LI[1:(10/h)];time<-c(1:(10/h))*h
plot(time,LIshort,
    main="Poisson␣increments␣on␣h-intervals␣to␣time␣10",type="l")
```

(m) *Compound Poisson with exponential jumps* (cf., Theorem 10.9)

$$\phi_{L_t}(z) = \exp\big[t\lambda(\phi(z) - 1)\big], \quad \lambda > 0,$$

where

$$\phi(z) = \gamma/(\gamma - \mathrm{i}z), \quad \gamma > 0.$$

```
# Program 4: Compound Poisson, Exponential jumps: Increments of L
# Parameters: lam > 0 is the jump rate
# mj = mean jump size = 1/(param. gamma of exp. jump size distribution)
# E(L(1)) = m = lam*mj
# Var(L(1))= v = lam*2*mj^2
# This determines (tau+T)/h compound Poisson increments,
# LI on h-intervals up to time tau+T.
# LIshort shows the increments up to time 10.
lam<- 4; mj<- 1
m<- lam*mj; v<- lam*2*mj^2
set.seed(879);nrep = (tau+T)/h;t=h
time<-c(1:(tau/h+T/h))*h
LI<-rep(0,nrep)
N <- rpois(nrep,lam*t)
for (i in 1 : nrep)
        {if(N[i]>0)
                {x <- mj*rexp(N[i]); LI[i] <- sum(x)}}
plot(time,LI,main="CP␣increments␣on␣h-intervals␣to␣time␣tau+T",type="l")
Sys.sleep(5)
LIshort<-LI[1:(10/h)];time<-c(1:(10/h))*h
plot(time,LIshort,
    main="CP␣increments␣on␣h-intervals␣to␣time␣10",type="l")
```

(n) *Compound Poisson with Gaussian jumps* (cf., Theorem 10.9)

$$\phi_{L_t}(z) = \exp[t\lambda(\phi(z) - 1)], \quad \lambda > 0,$$

where

$$\phi(z) = \exp\left[\mathrm{i}z\mu - \frac{\sigma^2 z^2}{2}\right], \quad \mu \in \mathbb{R}, \sigma > 0.$$

```
# Program 4: Compound Poisson, Gaussian jumps: increments of L
# Parameters: lam > 0, jump rate; mu in (-infty,infty), mean jump size;
# sd > 0 is the jump size standard deviation
# E(L(1)) =m= lam*mu; Var(L(1))=v= lam*(sd^2+mu^2)
# This determines (tau+T)/h compound Poisson increments,
# LI on h-intervals up to time tau+T.
# LIshort shows the increments up to time 10.
lam<-4; mu<- 0;sd<- .5; m<- lam*mu ; v<- lam*(sd^2+mu^2)
set.seed(879);nrep = (tau+T)/h;t=h
time<-c(1:(tau/h+T/h))*h
LI<-rep(0,nrep)
N <- rpois(nrep,lam*t)
for (i in 1 : nrep)
        {if(N[i]>0)
                {x <- mu + sd*rnorm(N[i]); LI[i] <- sum(x)}}
plot(time,LI,main="CP␣increments␣on␣h-intervals␣to␣time␣tau+T",type="l")
Sys.sleep(5)
LIshort<-LI[1:(10/h)];time<-c(1:(10/h))*h
plot(time,LIshort,
    main="CP␣increments␣on␣h-intervals␣to␣time␣10",type="l")
```

Bibliography

[1] Andresen, A., Benth, F. E., Koekebakker, S., Zakamulin, V. (2014) The CARMA interest rate model. *Int. J. Theor. Appl. Finance* **17**, 1450008, 27pp.

[2] Applebaum, D. (2009) *Lévy Processes and Stochastic Calculus*. 2nd ed., Cambridge University Press, Cambridge.

[3] Arató, M. (1982) *Linear Stochastic Systems with Constant Coefficients. A Statistical Approach*. Lecture Notes in Control and Information Sciences **45**. Springer, Berlin.

[4] Bachelier, L. (1900) Théorie de la spéculation. *Ann. Sci. Éc. Norm. Supér.* **17**, 21–86.

[5] Bao, Y. (2018) The asymptotic covariance matrix of the QMLE in ARMA models. *Econom. Rev.* **37**, 309–324.

[6] Barndorff-Nielsen, O. E. (1997) Normal inverse Gaussian distributions and stochastic volatility modelling. *Scand. J. Stat.* **24**, 1–13.

[7] Barndorff-Nielsen, O. E. and Shephard, N. (2001) Non-Gaussian Ornstein–Uhlenbeck-based models and some of their uses in financial economics. *J. R. Stat. Soc., Ser. B, Stat. Methodol.*. **63**, 167–241.

[8] Bartlett, M. S. (1946) On the theoretical specification and sampling properties of autocorrelated time series. *Suppl. J. R. Stat. Soc.* **8**, 27–41.

[9] Bauer, H. (1981) *Probability Theory and Elements of Measure Theory*. 2nd ed., Academic Press, London–New York.

[10] Bauer, H. (2001) *Measure and Integration Theory*. De Gruyter, Berlin.

[11] Benth, F. E., Klüppelberg C., Müller, G. and Vos, L. (2014) Futures pricing in electricity markets based on stable CARMA spot models. *Energy Econ.* **44**, 392–406.

[12] Bergstrom, A. R. (1985) The estimation of parameters in non-stationary higher-order continuous-time dynamic models. *Econom. Theory* **1**, 369–385.

[13] Bernstein, D. S. (2009) *Matrix Mathematics. Theory, Facts and Formulas*. 2nd ed., Princeton University Press, Princeton.

[14] Betser, A., Cohen, N. and Zeheb, E. (1995) On solving the Lyapunov and Stein equations for a companion matrix. *Syst. Control Lett.* **25**, 211–218.

[15] Bhattacharya, R. and Waymire, E. C. (2021) *Random Walk, Brownian Motion, and Martingales*. Springer, Cham.

[16] Billingsley, P. (1995) *Probability and Measure*. 3rd ed., Wiley-Interscience, New York.

[17] Billingsley, P. (1999) *Convergence of Probability Measures*. 2nd ed., Wiley-Interscience, New York.

[18] Black, F. and Scholes, M. (1973) The pricing of options and corporate liabilities. *J. Polit. Econ.* **81**, 637–654.

[19] Brémaud, P. (2014) *Fourier Analysis and Stochastic Processes*, Springer, Cham.

[20] Brockwell, A. E. and Brockwell, P. J. (1999) A class of non-embeddable ARMA processes. *J. Time Ser. Anal.* **20**, 483–486.

[21] Brockwell, P. J. (2001) Continuous-time ARMA processes. In: *Handbook of Statistics* **19**. *Stochastic Processes: Theory and Methods*, D. N. Shanbhag and C. R. Rao (eds.), pp. 249–276, Elsevier, Amsterdam.

[22] Brockwell, P. J. and Davis, R. A. (1991) *Time Series: Theory and Methods*. 2nd ed., Springer, New York.

[23] Brockwell, P. J., Davis, R. A. and Yang, Y. (2011) Estimation for non-negative Lévy-driven CARMA processes. *J. Bus. Econ. Stat.* **29**, 250–259.

[24] Brockwell, P. J., Ferrazzano, V. and Klüppelberg C. (2012) High-frequency sampling of a continuous-time ARMA process. *J. Time Ser. Anal.* **33**, 152–160.

[25] Brockwell, P. J., Ferrazzano, V. and Klüppelberg C. (2013) High-frequency sampling and kernel estimation for continuous-time moving average processes. *J. Time Ser. Anal.* **35**, 385–404.

[26] Brockwell, P. J. and Lindner, A. (2009) Existence and uniqueness of stationary Lévy-driven CARMA processes. *Stoch. Process. Appl.* **119**, 2660–2681.

[27] Brockwell, P. J. and Lindner, A. (2010) Strictly stationary solutions of autoregressive moving average equations. *Biometrika* **97**, 765–772.

https://doi.org/10.1515/9783111325033-021

[28] Brockwell, P. J. and Lindner, A. (2013) Integration of CARMA processes and spot volatility modelling. *J. Time Ser. Anal.* **34**, 156–167.
[29] Brockwell, P. J. and Lindner, A. (2015) Prediction of Lévy-driven CARMA processes. *J. Econom.* **189**, 263–271.
[30] Brockwell, P. J. and Lindner, A. (2019). Sampling, embedding and inference for CARMA processes. *J. Time Ser. Anal.* **40**, 163–181.
[31] Brouste, A., Fukasawa, M., Hino, H., Iacus, S. M., Kamatani, K., Koike, Y., Masuda, H., Nomura, R., Ogihara, T., Shimuzu, Y., Uchida, M. and Yoshida, N. (2014) The Yuima project: a computational framework for simulation and inference of stochastic differential equations. *J. Stat. Softw.* **57**, 1–51.
[32] Chow, Y. S. and Teicher, H. (1997). *Probability Theory. Independence, Interchangeability, Martingales*. 3rd ed., Springer, New York.
[33] Cohn, D. L. (2013) *Measure Theory*. 2nd ed., Birkhäuser, New York.
[34] Conway, J. B. (1978) *Functions of One Complex Variable I*. 2nd ed., Springer, New York.
[35] Doob, J. L. (1937) Stochastic processes depending on a continuous parameter. *Trans. Am. Math. Soc.* **42**, 107–140.
[36] Doob, J. L. (1944) The elementary Gaussian processes. *Ann. Math. Stat.* **15**, 229–282.
[37] Doob, J. L. (1953) *Stochastic Processes*. Wiley, New York.
[38] Durbin, J. (1961) Efficient fitting of linear models for continuous stationary time series from discrete data. *Bull. Inst. Int. Stat.* **38**, 273–281.
[39] Duren, P. L. (1970) *Theory of H^p Spaces*. Academic Press, New York–London.
[40] Einstein, A. (1905) Über die von der molekularkinetischen Theorie der Wärme geforderte Bewegung von in ruhenden Flüssigkeiten suspendierten Teilchen. *Ann. Phys.* **322**, 549–560.
[41] Eller, J. (1987) On functions of companion matrices. *Linear Algebra Appl.* **96**, 191–210.
[42] Elstrodt, J. (1996) *Maß- und Integrationstheorie*. Springer, Berlin.
[43] Fasen-Hartmann, V. and Mayer, C. (2022) Whittle estimation for continuous-time stationary state space models with finite second moments. *Ann. Inst. Stat. Math.* **74**, 233–270.
[44] Feller, W. (1971) *An Introduction to Probability Theory and its Applications. Vol. II*. 2nd ed., Wiley, New York–London–Sydney.
[45] Folland, G. B. (1992) *Fourier Analysis and its Applications*, Wadsworth and Brooks / Cole, Pacific Grove.
[46] Fowler, R. H. (1936) *Statistical Mechanics*. Cambridge University Press, Cambridge.
[47] Francq, C. and Zakoïan, J.-M. (1998) Estimating linear representations of nonlinear processes. *J. Stat. Plan. Inference* **68**, 145–165.
[48] Grabchak, M. (2016) *Tempered Stable Distributions. Stochastic Models for Multiscale Processes*, Springer, Cham.
[49] Iacus, S. M. and Mercuri, L. (2015) Implementation of Lévy CARMA model in Yuima package. *Comput. Stat.* **30**, 1111–1141.
[50] Iacus, S. M. and Yoshida, N. (2018) *Simulation and Inference for Stochastic Processes with YUIMA. A Comprehensive R Framework for SDEs and Other Stochastic Processes*. Springer, Cham.
[51] Itô, K. (2004) *Stochastic Processes. Lectures Given at Aarhus University* (O. E. Barndorff-Nielsen and K. Sato, eds.), Springer, Berlin.
[52] Johnson, N. L. and Kotz, S. (1970) *Distributions in Statistics. Continuous Univeriate Distributions, Vol. 1*. Houghton Mifflin Co., Boston.
[53] Jones, R. H. (1981) Fitting a continuous time autoregression to discrete data. In: *Applied Time Series Analysis II, Proceedings of the Second Applied Time Series Symposium Held in Tulsa, Oklahoma, March 3–5, 1980*, D. F. Findley (ed.), pp. 651–682, Academic Press, New York.
[54] Jones, R. H. (1985) Time series analysis with unequally spaced data. In: *Time Series in the Time Domain*, Handbook of Statistics **5**, E. J. Hannan, P. R. Krishnaiah and M. M. Rao (eds.), pp. 157–177, North Holland, Amsterdam.
[55] Kallenberg, O. (2002) *Foundations of Modern Probability*. 2nd ed., Springer, New York.
[56] Kawata, T. (1972) *Fourier Analysis in Probability Theory*. Academic Press, New York–London.

[57] Klenke, A. (2008) *Probability Theory. A Comprehensive Course*. Springer, London.
[58] Klüppelberg, K., Lindner, A. and Maller, R. (2004) A continuous-time GARCH process driven by a Lévy process: stationarity and second-order behaviour. *J. Appl. Probab.* **41**, 601–622.
[59] Kreiß, J.-P. and Neuhaus, G. (2006) *Einführung in die Zeitreihenanalyse. Statistik und ihre Anwendungen*. Springer, Berlin.
[60] Kuelbs, J. (1973) A representation theorem for symmetric stable processes and stable measures on *H*, *Z. Wahrscheinlichkeitstheor. Verw. Geb.* **26**, 259–271.
[61] Kyprianou, A. E. (2006) *Introductory Lectures on Fluctuations of Lévy Processes with Applications*, Springer, Berlin.
[62] Lang, S. (1987) *Linear Algebra*. 3rd ed., Springer, New York.
[63] Lindner, A. and Maller, R. (2005). Lévy inegrals and the stationarity of generalised Ornstein–Uhlenbeck processes. *Stoch. Process. Appl.* **115**, 1701–1722.
[64] Madan, D. B., Carr, P. P. and Chang, E. C. (1998) The variance gamma process and option pricing. *Eur. Finance Rev.* **2**, 79–105.
[65] Madan, D. B. and Seneta, E. (1990) The variance gamma (V. G.) model for share returns. *J. Bus.* **63**, 511–524.
[66] Marquardt, T. and Stelzer, R. (2007) Multivariate CARMA processes. *Stoch. Process. Appl.* **117**, 96–120.
[67] Mattner, L. (2001) Complex differentiation under the integral. *Nieuw Arch. Wiskd. (5)* **2**, 32–35.
[68] McElroy, T. (2013) Forecasting continuous-time processes with applications to signal extraction. *Ann. Inst. Stat. Math.* **65**, 439–456.
[69] Merton, R. (1973) Theory of rational option pricing. *Bell J. Econ. Manag. Sci.* **4**, 141–183.
[70] Meyn, S. and Tweedie, R. L. (2009) *Markov Chains and Stochastic Stability*. 2nd ed., Cambridge University Press, Cambridge.
[71] Phillips, A. W. (1959) The estimation of parameters in systems of stochastic differential equations. *Biometrika* **46**, 67–76.
[72] Phillips, P. C. B. (1973) The problem of identification in finite parameter continuous time models. *J. Econom.* **1**, 351–362.
[73] Rajput, B. S. and Rosiński, J. (1989) Spectral representations of infinitely divisible processes. *Probab. Theory Relat. Fields* **82**, 451–487.
[74] Robinson, P. M. (1977). The construction and estimation of continuous time models and discrete approximations in econometrics. *J. Econom.*, **6**, 173–197.
[75] Rudin, W. (1987) *Real and Complex Analysis*, 3rd ed., McGraw-Hill, New York.
[76] Samorodnitsky, G. and Taqqu, M. S. (1994) *Stable Non-Gaussian Random Processes. Stochastic Models with Infinite Variance*, Chapman & Hall, New York.
[77] Samuelson, P. A. (1965) Rational theory of warrant pricing. *Ind. Manage. Rev.* **6**, 13–32.
[78] Sato, K. (2006) Monotonicity and non-monotonicity of domains of stochastic integral operators. *Probab. Math. Stat.* **26**, 23–39.
[79] Sato, K. (2013) *Lévy Processes and Infinitely Divisible Distributions*. Revised edition, Cambridge University Press, Cambridge.
[80] Schlemm, E. and Stelzer, R. (2012) Quasi maximum likelihood estimation for strongly mixing state space models and multivariate Lévy-driven CARMA processes, *Electron. J. Stat.* **6**, 2185–2234.
[81] Shiryaev, A. N. (1996) *Probability*. 2nd ed., Springer, New York.
[82] Simmons, G. F. (1963) *Introduction to Topology and Modern Analysis*. McGraw-Hill, New York–San Francisco.
[83] Steutel, F. W. and van Harn, K. (2004) *Infinite Divisibility of Probability Distributions on the Real Line*, Marcel Dekker, New York.
[84] Todorov, V. (2011) Econometric analysis of jump-driven stochastic volatility models. *J. Econom.* **160**, 12–21.
[85] Todorov, V. and Tauchen, G. (2006) Simulation methods for Lévy-driven continuous-time autoregressive moving average (CARMA) stochastic volatility models. *J. Bus. Econ. Stat.* **24**, 455–469.
[86] Uhlenbeck, G. E. and Ornstein, L. S. (1930) On the theory of Brownian motion. *Phys. Rev.* **36**, 823–841.

Index

https://doi.org/10.1515/9783111325033-022

De Gruyter Studies in Mathematics

Volume 97
Ştefan Ovidiu I. Tohăneanu
Commutative Algebra Methods for Coding Theory, 2024
ISBN 978-3-11-121292-0, e-ISBN 978-3-11-121479-5, e-ISBN (ePUB) 978-3-11-121538-9

Volume 25
Karl H. Hofmann, Sidney A. Morris
The Structure of Compact Groups. A Primer for the Student – A Handbook for the Expert, 5th Edition, 2023
ISBN 978-3-11-117163-0, e-ISBN 978-3-11-117260-6, e-ISBN (ePUB) 978-3-11-117405-1

Volume 96
Francesco Aldo Costabile, Maria Italia Gualtieri, Anna Napoli
Polynomial Sequences. Basic Methods, Special Classes, and Computational Applications, 2023
ISBN 978-3-11-075723-1, e-ISBN 978-3-11-075724-8, e-ISBN (ePUB) 978-3-11-075732-3

Volume 54
Yasushi Ishikawa
Stochastic Calculus of Variations, For Jump Processes, 3rd Edition 2023
ISBN 978-3-11-067528-3, e-ISBN 978-3-11-067529-0, e-ISBN (ePUB) 978-3-11-067532-0

Volume 95
Marko Kostić
Metrical Almost Periodicity and Applications to Integro-Differential Equations, 2023
ISBN 978-3-11-123303-1, e-ISBN 978-3-11-123387-1, e-ISBN (ePUB) 978-3-11-123417-5

Volume 94
Nora Doll, Hermann Schulz-Baldes, Nils Waterstraat
Spectral Flow. A Functional Analytic and Index-Theoretic Approach, 2023
ISBN 978-3-11-116989-7, e-ISBN 978-3-11-117247-7, e-ISBN (ePUB) 978-3-11-117308-5

Volume 93
Brian Street
Maximal Subellipticity, 2023
ISBN 978-3-11-108517-3, e-ISBN 978-3-11-108564-7, e-ISBN (ePUB) 978-3-11-108594-4

Volume 92
Min Ru
Minimal Surfaces through Nevanlinna Theory, 2023
ISBN 978-3-11-099982-2, e-ISBN 978-3-11-098955-7, e-ISBN (ePUB) 978-3-11-098996-0

